HIV-1: MOLECULAR BIOLOGY AND PATHOGENESIS

VIRAL MECHANISMS

HIV-1: MOLECULAR BIOLOGY AND PATHOGENESIS

VIRAL MECHANISMS

The First of a Two-Volume Set

Edited by

Kuan-Teh Jeang

Laboratory of Molecular Biology
NIAID/NIH
Bethesda, Maryland

ADVANCES IN
PHARMACOLOGY

VOLUME 48

ACADEMIC PRESS

A Harcourt Science and Technology Company

San Diego San Francisco New York Boston London Sydney Tokyo

This book is printed on acid-free paper. 

Academic Press
A Harcourt Science and Technology Company
525 B Street, Suite 1900, San Diego, California 92101-4495, USA
http://www.academicpress.com

Academic Press
Harcourt Place, 32 Jamestown Road, London NW1 7BY, UK
http://www.hbuk.co.uk/ap/

International Standard Book Number: 0-12-032949-2

PRINTED IN THE UNITED STATES OF AMERICA
00 01 02 03 04 05 EB 9 8 7 6 5 4 3 2 1

Contents

Multiple Biological Roles Associated with the Repeat (R) Region of the HIV-1 RNA Genome

Ben Berkhout

HIV Accessory Proteins: Multifunctional Components of a Complex System

Stephan Bour and Klaus Strebel

Role of Chromatin in HIV-1 Transcriptional Regulation

Carine Van Lint

NF-κB and HIV: Linking Viral and Immune Activation

Arnold B. Rabson and Hsin-Ching Lin

Rev Protein and Its Cellular Partners

Jørgen Kjems and Peter Askjaer

HIV-1 Nef: A Critical Factor in Viral-Induced Pathogenesis

A. L. Greenway, G. Holloway, and D. A. McPhee

Nucleocapsid Protein of Human Immunodeficiency Virus as a Model Protein with Chaperoning Functions and as a Target for Antiviral Drugs

Jean-Luc Darlix, Gaël Cristofari, Michael Rau, Christine Péchoux, Lionel Berthoux, and Bernard Roques

Bioactive CD4 Ligands as Pre- and/or Postbinding Inhibitors of HIV-1

Laurence Briant and Christian Devaux

Coreceptors for Human Immunodeficiency Virus and Simian Immunodeficiency Virus

Keith W. C. Peden and Joshua M. Farber

Contributors

Numbers in parentheses indicate the pages on which the authors' contributions begin.

Adriana Albini (229) Istituto Nazionale per la Ricerca sul Cancro 16132 Genova, Italy

Peter Askjaer (251) Department of Molecular and Structural Biology, University of Aarhus, DK-8000 Aarhus C, Denmark

Ben Berkhout (29) Department of Human Retrovirology, Academic Medical Center, University of Amsterdam, 1105 AZ Amsterdam, The Netherlands

Lionel Berthoux (345) LaboRetro, Unité de Virologie Humaine INSERM 412, Ecole Normale Supérieure de Lyon, 69364 Lyon, France, Phamacochimie moléculaire et structurale INSERM 266, 75270 Paris, France

Stephan Bour (75) Laboratory of Molecular Microbiology, National Institute of Allergy and Infectious Diseases, National Institutes of Health, Bethesda, Maryland 20892-0460

Laurence Briant (373) Laboratoire Infections Rétrovirales et Signalisation Cellulaire, CNRS EP J0004, Institut de Biologie, 34060 Montpellier, France

Gaël Cristofari (345) LaboRetro, Unité de Virologie Humaine INSERM 412, Ecole Normale Supérieure de Lyon, 69364 Lyon, France, Phamacochimie moléculaire et structurale INSERM 266, 75270 Paris, France

Jean-Luc Darlix (345) LaboRetro, Unité de Virologie Humaine INSERM 412, Ecole Normale Supérieure de Lyon, 69364 Lyon, France, Phamacochimie moléculaire et structurale INSERM 266, 75270 Paris, France

Christian Devaux (373) Laboratoire Infections Rétrovirales et Signalisation Cellulaire, CNRS EP J0004, Institut de Biologie, 34060 Montpellier, France

Joshua M. Farber (409) Laboratory of Clinical Investigation, National Institute of Allergy and Infectious Diseases, National Institutes of Health, Bethesda, Maryland 20892

Anne Gatignol (209) Institut Cochin de Génétique Moléculaire, 75014 Paris, France, Molecular Oncology Group, McGill AIDS Centre, Lady Davis Institute for Medical Research, Montréal, QC, H3T IE2 Canada

A. L. Greenway (299) AIDS Cellular Biology Unit, Macfarlane Burnet Centre for Medical Research, Fairfield, Victoria, Australia

G. Holloway (299) AIDS Cellular Biology Unit, Macfarlane Burnet Centre for Medical Research, Fairfield, Victoria, Australia

Kuan-Teh Jeang (209) Molecular Virology Section, Laboratory of Molecular Microbiology, National Institute for Allergy and Infectious Diseases, National Institutes of Health, Bethesda, Maryland 20892-0460

Jørgen Kjems (251) Department of Molecular and Structural Biology, University of Aarhus, DK-8000 Aarhus C, Denmark

Andrew M. L. Lever (1) University of Cambridge, Department of Medicine, Addenbrooke's Hospital, Cambridge CB2 2QQ, United Kingdom

Hsin-Ching Lin (161) Department of Molecular Genetics and Microbiology, Center for Advanced Biotechnology and Medicine and Cancer Institute of New Jersey, University of Medicine and Dentistry of New Jersey, Robert Wood Johnson Medical School, Piscataway, New Jersey 08854

D. A. McPhee (299) AIDS Cellular Biology Unit, Macfarlane Burnet Centre for Medical Research, Fairfield, Victoria, Australia

Christine Péchoux (345) LaboRetro, Unité de Virologie Humaine INSERM 412, Ecole Normale Supérieure de Lyon, 69364 Lyon, France, Phamacochimie moléculaire et structurale INSERM 266, 75270 Paris, France

Keith W. C. Peden (409) Laboratory of Retrovirus Research, Center for Biologics Evaluation and Research, Food and Drug Administration

Arnold B. Rabson (161) Department of Molecular Genetics and Microbiology, Center for Advanced Biotechnology and Medicine and Cancer Institute of New Jersey, University of Medicine and Dentistry of New Jersey, Robert Wood Johnson Medical School, Piscataway, New Jersey 08854

Michael Rau (345) LaboRetro, Unité de Virologie Humaine INSERM 412, Ecole Normale Supérieure de Lyon, 69364 Lyon, France, Phamacochimie moléculaire et structurale INSERM 266, 75270 Paris, France

Bernard Roques (345) LaboRetro, Unité de Virologie Humaine INSERM 412, Ecole Normale Supérieure de Lyon, 69364 Lyon, France, Phamacochimie moléculaire et structurale INSERM 266, 75270 Paris, France

Klaus Strebel (75) Laboratory of Molecular Microbiology, National Institute of Allergy and Infectious Diseases, National Institutes of Health, Bethesda, Maryland 20892-0460

Carine Van Lint (121) Département de Biologie Moléculaire, Laboratoire de Chimie Biologique, Institut de Biologie et de Médecine Moléculaires, Université Libre de Bruxelles, 6041 Gosselies, Belgium

Preface

As we enter a new millennium, it is clear that the AIDS pandemic is not going away. United Nations AIDS statistics indicate that in 1999 alone 2.6 million individuals succumbed to this disease. Cumulatively, since 1980, 16 million have died from AIDS, and despite remarkable medical advances, HIV-1 infections remain on the increase. Today, more than 33 million persons globally are living with HIV-1; 5.6 million of these were newly infected in 1999. Regrettably, the disease burden is highest in nations that have the most limited medical resources. Hence, 25% of all adults in Botswana, Zimbabwe, Swaziland, and Namibia are AIDS carriers. Every minute, in sub-Saharan Africa, 10 new individuals become infected with HIV-1. Worldwide, 95% of HIV-1 infections are in developing countries.

In view of these rather daunting numbers, one might think that researchers have made little progress in the study of HIV. This, in fact, is far from the truth. Since its initial identification and isolation in 1983, HIV-1 has become one of the best-elucidated viruses. Arguably, today we understand more about the workings of HIV-1 than of any other virus. In parallel, chemotherapeutic advances in the treatment of AIDS have also been impressive. From the combined efforts of many investigators, we currently have a large armamentarium of specific anti-HIV-1 reverse transcriptase (RT) and protease inhibitors. These antivirals work. Mortality from AIDS in developed countries that use RT and protease inhibitors has been significantly reduced. However, it is equally evident that, as yet, no chemotherapeutic regimen is curative.

How then might one view the AIDS question in the coming years? Better chemical antivirals are unlikely to be the final answer. The cost of chemical antivirals (currently around US$ 20,000 per person per year) makes this route prohibitive for developing nations. Chemical antivirals also have serious side effects that render questionable the ability of patients to sustain lifelong therapy. Additionally, we do not yet fully understand the ultimate scope of multi-drug-resistant viruses that are emerging in RT- and protease-inhibitor-treated individuals. Considered thusly, it is understandable how one prevailing view is that a practical, globally applicable solution for AIDS rests with the development of effective mass vaccination. However, the possibility that there are other yet-thought-of means for resolving the AIDS pandemic cannot be excluded.

This two-volume set of *Advances in Pharmacology* brings together 26 teams of authors for the purpose of describing where we have been in HIV-1 research and to explore where we might want to go in the future. The goal was to combine expositions on fundamental mechanisms of viral expression and replication with findings on viral pathogenesis in animal models and applications of chemotherapeutics in human patients. The authors were asked to survey the structures and functions of all the open reading frames of the HIV-1 genome as well as the roles of several noncoding regulatory RNA sequences. More importantly, each was encouraged to propose new ways for therapeutic intervention against HIV-1. In this regard, many interesting and novel ideas on HIV vaccines, gene therapy, and small-molecule inhibitors for viral envelope–cell fusion, viral assembly, integration, and gene expression are presented.

It is no accident that most of the authors in this set are some of the younger (late-30s, mid-40s), albeit authoritative, researchers on HIV. Likely, AIDS will defy a quick solution. Considering this and the fact that a major goal of this set is to explore new ideas rather than simply review past progress, I particularly wanted to assemble colleagues who, after having proposed interesting solutions for HIV, will be around to test, refine, and execute those ideas even if such were to require 10, 20, or 30 years. With some measure of luck, I and most of my co-authors will be here when the AIDS pandemic is solved. I wait with anticipation and interest to see whose ideas raised in these two volumes will be the ones that withstand the test of time.

Putting together this set has been an interesting learning experience for me. I thank all the authors for their enthusiastic participation. At the outset, I had thought that it would be difficult to recruit a sufficient number of busy researchers to write chapters for this project. However, I was pleasantly surprised when 26 of the 30 invited colleagues promptly agreed to contribute. Along the way, another colleague remarked to me that he did not want to write a chapter because "nobody important reads HIV books anymore." That comment taken, it remains my hope that some "unimportant" readers of these two volumes might nevertheless be spurred by its content to do important work in the fight against AIDS.

In closing, I thank Tom August for inviting me to edit these volumes. I am grateful to the late George Khoury, who asked me 14 years ago to work on the HIV Tat protein, and to Malcolm Martin, with whom I have discussed and debated various aspects of HIV biology for the past 12 years. Tari Paschall, Judy Meyer, and Destiny Irons from Academic Press have been wonderfully helpful, and Michelle Van and Lan Lin have provided excellent secretarial assistance. Finally, I appreciate the endless patience and understanding of my wife (Diane) and children (David, Diana, and John), who have, year-upon-year, put up with the abnormal working hours of an HIV researcher.

Kuan-Teh Jeang
January 14, 2000

Andrew M. L. Lever

University of Cambridge
Department of Medicine
Addenbrooke's Hospital
Cambridge CB2 2QQ, United Kingdom

HIV RNA Packaging and Lentivirus-Based Vectors

I. Introduction

The study of the mechanism by which the viral RNA genome is incorporated into the virus particle in lentiviruses has been a relatively undersubscribed activity in the panoply of HIV molecular biological research, overshadowed by studies on HIV-1 envelope and the regulatory proteins, Tat and Rev. By comparison, the research output on RNA packaging and RNA packaging signals has been considerably less. Vectors based on HIV have been in use since only 1990. When initially produced, their low titer and the biohazards involved in handling them restricted their use to a small number of laboratories. A major surge in interest occurred when pseudotyping of HIV-1-based vectors with alternative envelopes, and in particular the vesicular stomatitis virus (VSV) G protein, was demonstrated to produce high-titer, stable, storable, and highly infectious lentiviral vectors (Akkina *et al.*, 1996; Naldini *et al.*, 1996) with the lentivirus specific property of

Advances in Pharmacology, Volume 48

1054-3589/00 $35.00

delivery and integration of genes into growth arrested cells *in vitro* and *in vivo* (Andresson *et al.*, 1993). The use of lentivirus vectors has proceeded despite very incomplete knowledge of many aspects of the encapsidation process of the viral (or vector) RNA. Little by little, these gaps are being filled in. We are now gaining an understanding of the mechanism of genomic RNA-protein recognition in lentiviruses in which, surprisingly, significant differences are being discovered in different stages of the process in quite closely related lentiviruses. This knowledge may have implications for the utility of different lentiviruses as gene vectors.

II. HIV RNA Packaging

A. HIV RNA Species

As detailed elsewhere in this volume, the integrated provirus of HIV-1 is transcribed by the RNA polymerase II of the infected cell to produce a full-length RNA transcript which, in the early stages of the infection cycle, is spliced to produce RNAs encoding the viral regulatory proteins including Tat and Rev (Cullen *et al.*, 1989). In the late stage of the viral lifecycle, through accumulation of Rev and its interaction with the Rev response element at the 3′ end of the envelope gene, RNAs containing this sequence are stabilized, splicing is avoided or limited to removal of the 5′ Gag/Pol exon alone, and unspliced and singly spliced RNAs are then exported to the cytoplasm. Here they are translated to produce the structural and regulatory proteins of the virus, the unspliced RNA coding for the Gag and Pol proteins. Thus, in the cytoplasm, there is a coincidental appearance of the major structural proteins of the capsid, whose coding RNA is itself the genomic RNA which will be packaged into the viral particle.

RNA packaging in all retroviruses is highly specific, particularly considering that, because of the nature of their replication cycle, retroviruses are dependent on, and cannot switch off, host cell processes such as transcription and translation, resulting in an abundance of host cell RNA and protein in the infected cell. It is estimated that the viral genomic RNA constitutes 1% or less of the total messenger RNA in the cytoplasm of the cell, yet it is packaged virtually specifically while cellular messenger RNAs and spliced viral RNAs are excluded from the virion.

Packaging in complex retroviruses such as lentiviruses is more complicated than in simple retroviruses. In the latter, only two major retroviral RNA species are present in the cell to compete for packaging, and one of these, the envelope RNA, contains a leader sequence. Once ribosomal capture and scanning of this latter RNA has begun, the appearance of the leader peptide will sequester away the singly spliced RNA attached to the ribosome to the rough endoplasmic reticulum and, thus, isolate this RNA away from the vicinity of the viral core (Gag) proteins which are being translated from unspliced RNA on free cytoplasmic ribosomes. Thus, in simple retroviruses, the only viral RNA encountered by the Gag protein may be the unspliced (genomic) RNA. The large number of additional spliced species of RNA produced in the replication cycle of complex retroviruses means that there are a large number of competing RNAs which might be packaged and many of these will be translated on cytoplasmic ribosomes along with Gag, so that physical separation (which still segregates away envelope RNA) cannot substantially contribute to the specificity of packaging.

In simple and complex retroviruses, specific capture of the genome implies that this RNA has recognition motifs—packaging signals—to distinguish it from other cellular and viral RNAs. These signals are referred to as "Ψ" or sometimes as "E."

B. Proteins Involved in RNA Capture

Transport of the RNA from the nucleus to the (as yet unidentified) region of the cell where capture by the viral proteins occurs is a poorly understood process. For the genomic RNA of lentiviruses, the Rev/RRE system is involved but it is quite likely that a number of other cellular and possibly viral chaperone proteins are involved in coating and protecting the RNA genome prior to its incorporation into the virus during encapsidation. RNA protein interaction is undoubtedly initiated to a major extent through interaction of the nucleocapsid (NC) region of Gag with the viral RNA packaging signal (Swanstrom *et al.*, 1997). A number of studies have shown that deletions and mutations in the NC zinc fingers will disrupt packaging (Aldovini *et al.*, 1990; Gorelick *et al.*, 1990; Dannull *et al.*, 1994; Dorfman *et al.*, 1993), implicating these motifs and flanking basic residues in RNA capture. *In vitro* studies of the interaction between HIV Gag proteins and the 5′ leader sequence have been performed by direct *in vitro* binding assays (Clever *et al.*, 1995; Berkowitz *et al.*, 1994), by selex (Berglund *et al.*, 1997), by footprinting and protection of the RNA target from nuclease attack (Damgaard *et al.*, 1998; Zeffman *et al.*, 2000), and also using the yeast 3-hybrid system (Bacharach *et al.*, 1998). These have confirmed the ability of NC and also of the Gag polyprotein to bind to the leader region with particular elements of the RNA apparently involved in high-affinity binding. RNA/NC binding has recently also been elegantly modeled by NMR (De Guzman *et al.*, 1998). Although NC protein is a cleavage product of the Gag polyprotein and is not usually generated until after budding and protease-mediated cleavage, a small amount of NC is detectable in the cytoplasm of infected cells and it is conceivable that NC alone is involved in RNA recognition. *In vitro* studies have also shown, however, that for HIV the uncleaved Gag polyprotein is capable of capturing the RNA and, in fact, captures and protects it better than a protease-containing Gag expressor, which gives rise to Gag particles which are subsequently processed (Kaye *et al.*, 1995). The most probable model involves full-length Gag polyprotein capturing the RNA, but the site within the cell at which this occurs is not defined as yet.

Recent work involving cross-packaging between HIV-1 and HIV-2 identified the P2 domain of the HIV-1 Gag protein (immediately N terminal to NC) as playing a critical role in encapsidation (Kaye *et al.*, 1998). HIV-2 Gag containing both P2 and NC from HIV-1 in place of the HIV-2 homologous region was capable of packaging HIV-1 and HIV-2 RNA, whereas a polyprotein consisting of the HIV-1 NC alone substituted into the HIV-2 Gag, packaged both much less efficiently. The P2 could credibly be involved in RNA capture or orientation of the zinc fingers to facilitate RNA capture. Alternatively, it may have a completely different function, for example, triggering Gag–Gag recognition and interaction before or after RNA capture.

C. RNA Packaging Signals in HIV-1

Following the demonstration in other retroviruses of encapsidation signals in the 5′ leader reviewed in Linial and Miller (Linial *et al.*, 1990), these were first sought in HIV-1 in the analogous region downstream of the major splice donor in an area which is unique to the unspliced RNA. Three groups (Aldovini *et al.*, 1990; Lever *et al.*, 1989; Clavel *et al.*, 1990) initially generated deletion mutants in this region, each of which demonstrated a packaging defect when assessed in a variety of cell types, including lymphocytes in which the virus was replicating. The packaging defect was associated with a proportional replication defect and, although the reduction in encapsidation of RNA was often profound, none of the deletions described abrogated packaging (or replication) entirely, suggesting that there is functional redundancy in the system and that other areas either contribute to packaging or can substitute for the deleted region.

In seeking to elucidate the mechanism of RNA protein recognition involved in packaging, deletion mutagenesis and secondary structural analysis of the HIV-1 leader has been pursued by a number of different groups (Harrison *et al.*, 1992, 1998; Pappalardo *et al.*, 1998; McBride *et al.*, 1997; Das *et al.*, 1997; Sakaguchi *et al.*, 1993; Clever *et al.*, 1999) and evidence has been presented for an extended series of RNA secondary structure motifs (Berkhout, 1996) beginning at the transcriptional start site with the TAR stem loop. This is followed by a long complex multibranch structure involving the primer binding site (PBS) followed by three stem loops (Fig. 1), the first of which (SL1) has a palindromic sequence in the terminal loop. The second (SL2) has the splice donor motif at the tip and the third (SL3) is the structure which has most consistently been identified as the major packaging signal motif (Ψ). A fourth stem loop involving the initiation codon of Gag has been modeled by some groups but not by others. There is some evidence

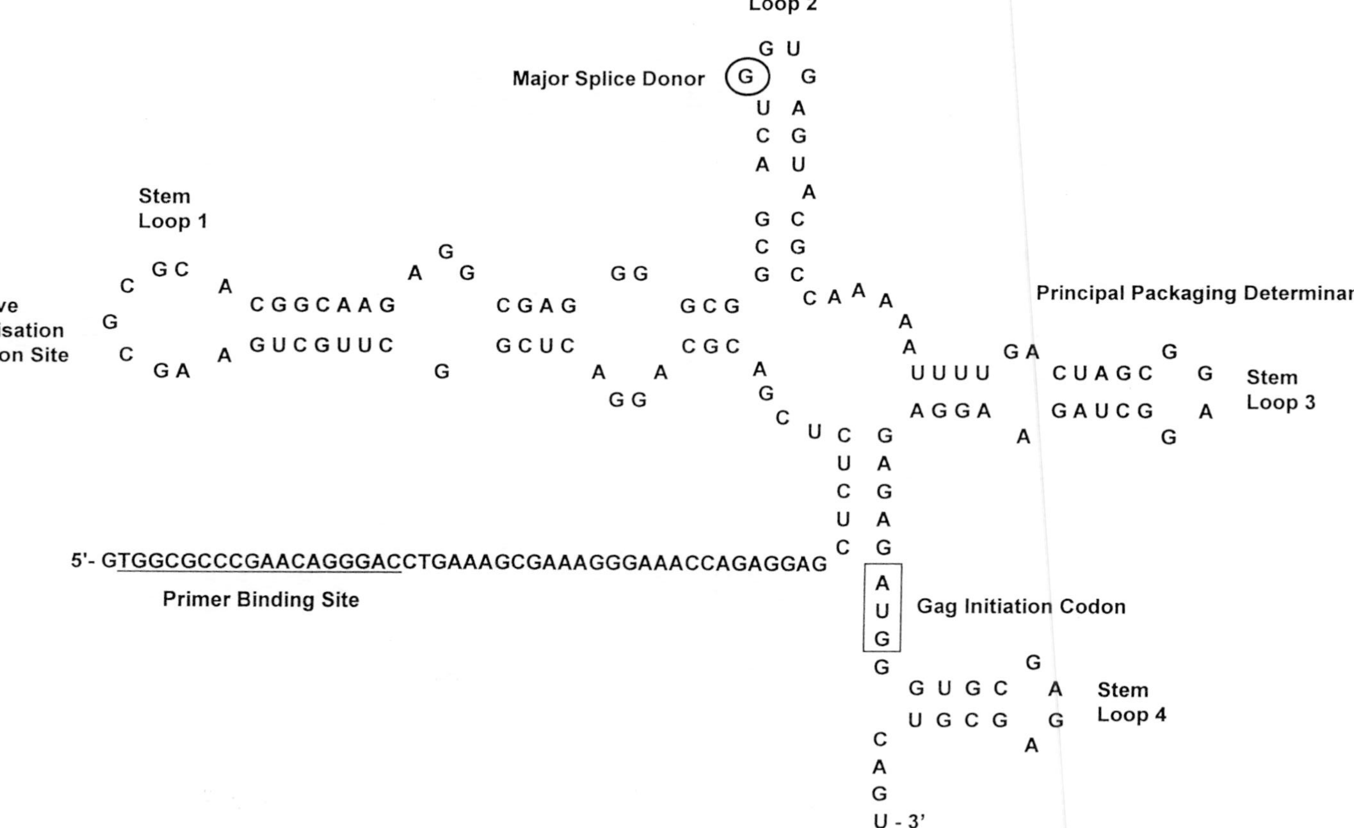

FIGURE 1 Secondary structure model of the major packaging signal region of HIV-1. This model is from an original structure generated by biochemical and computational analysis (Harrison, 1992) and has been modified by Sakaguchi (1993) and further refined by mutagenic studies (Harrison, 1998).

for it having a role in packaging (McBride *et al.*, 1997), although others have found it dispensable (Harrison *et al.*, 1998). Deletion mutants of many of these structures in the leader have been associated with a defect in RNA packaging. In some cases, studies have been performed measuring the end point of virion RNA concentration but without accurate comparison of the corresponding cytoplasmic genomic RNA levels. Thus, the role of other factors, such as RNA stability, which contribute to the eventual quantity of RNA available for packaging in the virus particle, may not have been fully considered. This becomes very obvious when one mathematically summates the percentage defect, in RNA packaging reported for all of the deletion mutants described. The result comes out at a figure much higher than 100%, implying that many of the defects which have been identified do not necessarily interfere directly with binding of Gag protein to the viral RNA but may affect packaging indirectly by altering factors such as subcellular localization of the RNA, RNA stability, or the ability of the genomic RNA to bind to other cellular chaperone proteins. Thus, the concept of a packaging "pathway" can be considered (Fig. 2) in which the route of genomic RNA during its transport from the proviral transcription site through to export in the budding virus particle is considered as a continuum along which many influences may act. *Some* of these will affect the RNA protein interaction directly but many may not, but perturbing any of them may lead to a phenotype of diminished genomic RNA content of virions.

In vitro binding assays of Gag protein and the HIV-1 viral RNA leader have largely confirmed the requirement for binding to the SL1 and SL3 stem loops, although binding can also be detected elsewhere in the leader (Damgaard *et al.*, 1998). In a number of studies, identification of single discrete packaging signal regions has not been possible and the conclusion has been drawn that either the packaging signal of HIV-1 is multipartite or, equally likely, that the binding to the individual RNA motifs is context dependent (Berkowitz *et al.*, 1995; McBride *et al.*, 1996). These hypotheses are clearly not mutually exclusive.

Outside of the major packaging signal (SL3), regions such as the TAR stem loop (Clever *et al.*, 1999) and the R-U5 region (Das *et al.*, 1997) have been shown to contribute to RNA encapsidation. The major secondary packaging signal is, however, almost certainly the stem loop involved in dimer linkage (SL1) (Laughrea *et al.*, 1994; Marquet *et al.*, 1994; Skripkin *et al.*, 1994; Awang *et al.*, 1993). This region has a terminal palindromic loop which has now been modeled by a number of groups as being able to undergo self–self interaction (Paillart *et al.*, 1997; Haddrick *et al.*, 1996; Mujeeb *et al.*, 1998) with the corresponding region in the packaged sister genomic strand. This region is termed the dimer initiation site (DIS). The process of melting together of the two strands is, as yet, unclear and it is not certain at which stage in the packaging pathway this occurs (Greatorex *et al.*, 1998). There is evidence from HIV (Feng *et al.*, 1996) and other

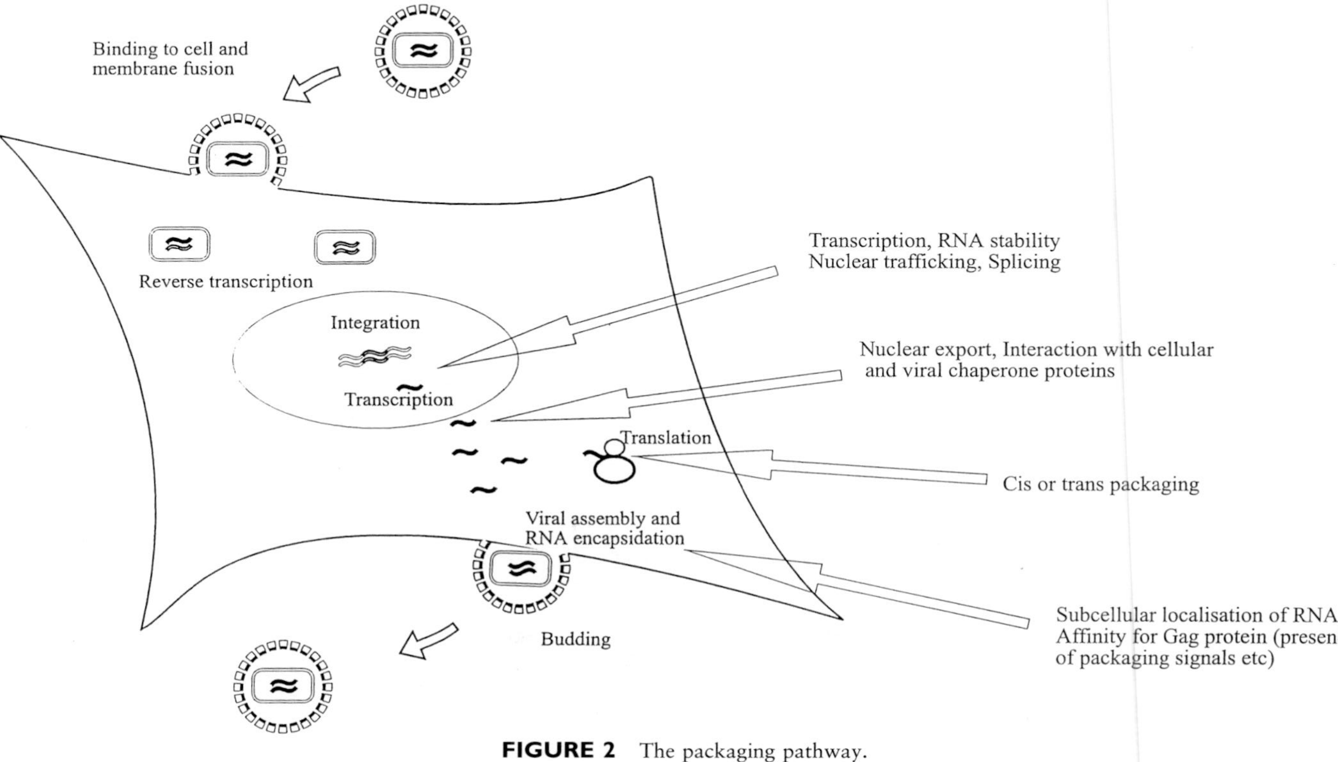

FIGURE 2 The packaging pathway.

retroviruses (Brahic *et al.*, 1975; Fu *et al.*, 1993) that relatively loose dimer formation occurs prior to virus budding and that this is consolidated following budding and proteolytic cleavage of the core proteins. Thus, a relatively weak initial interaction involving the dimer initiation structure may subsequently be cemented by further intermolecular links, perhaps facilitated by newly cleaved nucleocapsid (NC) protein (Fu *et al.*, 1994).

D. RNA Dimerization and Packaging

RNA dimerization is closely linked with encapsidation in HIV-1 and several groups have shown that disruptive mutations to the dimer linkage stem loop have a profoundly deleterious effect on packaging (Berkowitz *et al.*, 1995; Paillart *et al.*, 1996; Berkhout and van Wamel, 1996; Clever *et al.*, 1997). In particular, disruption of the base pairing in SL1 subtending the terminal palindrome disrupts packaging. However, restoration of the potential Watson–Crick base pairs, by exchanging the sequences on the two sides of the helical stem, restores replication and packaging to wild-type levels (Harrison *et al.*, 1998). Intriguingly, truncation of SL1 to leave a shorter palindrome-containing stem loop has relatively little effect on encapsidation (personal observations).

The splice donor stem loop (SL2) appears to contribute relatively little to encapsidation, although it is clearly critical for control of splicing.

Stem loop 3 has been modeled differently by different groups. Some groups have described a relatively short five-base-pair helix subtending a tetraloop. Our own modeling has consistently identified a second bulge below this and a further four-base-pair helix proximally (Fig. 2) (Zeffman *et al.*, 2000). The structure of this region has been solved by NMR in association with a bound nucleocapsid protein subunit (De Guzman *et al.*, 1998) and without protein (Zeffman *et al.*, 2000). The nature of the interaction between the RNA and the protein is intriguing in that the terminal four bases GGAG appear to bond with the basic residues at the constriction of one of the zinc fingers of NC using the identical hydrogen bonding normally involved in Watson–Crick pairing. It will be interesting to see whether this turns out to be a more widespread form of interaction between single-stranded RNA and RNA binding proteins other than NC. In the virus, NC very likely binds nonspecifically along the whole length of the genomic RNA and the interaction which has been modeled to date confirms that NC is capable of binding to unpaired RNA in this particular stem loop *in vitro*. The protein involved in RNA capture is most likely the uncleaved full-length Gag protein and recent *in vitro* evidence (Zeffman *et al.*, 2000) suggests that this binds to both the terminal tetraloop and to a G-rich bulge in SL3 and that subsequently cooperative binding occurs analogous to the binding of Rev to the RRE. However, unlike the latter, Gag binding appears to unwind the packaging signal helical regions.

Until recently, searches for sequence homologies in packaging signal regions had revealed only limited regions of similarity including certain tetraloop motifs found in primate and nonprimate retroviruses (Konings *et al.*, 1992; Harrison *et al.*, 1995). However, close study of the leader sequence of a number of primate lentiviruses reveals the frequent occurrence of a sequence GGNG(R), which appears multiple times in the RNA leader region of each virus, notably at the tip of SL3 (Harrison *et al.*, 1992, 1998). It is plausible that this motif in a variety of contexts can act as a packaging signal, although the most efficient context is that of the SL3 helix. The existence of a number of these motifs would explain the functional redundancy in the system which leads to only partial packaging defects when SL3 or other regions are deleted. This is quite different from murine leukemia viruses, in which discrete packaging signal deletions produce a virtual ablation of packageability of the RNA. The requirement for multiple GGNG(R) motifs might also provide support for the advantages of some degree of RNA dimerization prior to packaging in that it would double the number of motifs available for initiation of Gag binding. It would still be consistent with the fact that dimerization of HIV-1 and other lentiviruses such as Maedi–Visna may not be absolutely essential for packaging (Haddrick *et al.*, 1996; Brahic *et al.*, 1975).

E. Accessory Packaging Signals

Deletion mutations causing packaging defects have been described in various parts of the 5′ leader outside of the SL1 and SL3 stem loops and *in vitro* RNA protein binding studies have implicated other regions in encapsidation including the primer binding site. These may act in part by altering RNA transport or stability or access to the packaging pathway. The *gag* gene sequence has also been implicated in packaging. Published lentivirus vector studies suggest that there is an optimal length of *gag* gene which needs to be incorporated into a vector to maximize packaging efficiency (Parolin *et al.*, 1994). This has been assumed to be a compromise between incorporating additional cis-acting packaging signals while excluding portions of the instability sequence known to exist in the *gag* gene which render the RNA prone to splicing in the absence of Rev. Inclusion of more *gag* sequence may enhance packaging further (personal observations). Conceivably, this might, in part, be due to increasing the amount of Gag protein available in the cell to make virion particles, although *in vitro* evidence has suggested that the packaging capacity of an HIV-1 producing cell is not saturated, even in wild-type infection (Richardson *et al.*, 1993). This latter study also suggested that a region at the 3′ end of the envelope gene was essential for packaging HIV RNA. This region encompasses the Rev Responsive Element and Rev RRE interactions are obviously essential for export and stability of the full-length genomic RNA. Further analysis of this region demonstrated that the Rev Responsive Element was not the sole contributor (Kaye *et al.*, 1995) and that regions of the envelope gene incorporated into vectors could rescue otherwise unpackagable vectors. The mechanism by which this is accomplished is not clear and spacer sequence effects in some of the vectors could not be excluded, the inserted sequence possibly distancing a foreign gene from the 5′ packaging signal where it might otherwise disrupt the base pairing and the consequent secondary and tertiary structure of the RNA. In other vectors, however, this region has been shown to be dispensable.

F. HIV-2 Packaging Signals

There is conflicting evidence as to the site and size of packaging signals in HIV-2. One early study suggested that a packaging signal existed in a position analogous to that of HIV-1 between the splice donor and the Gag ATG and that, in addition, this region encompassed a translational inhibitory sequence (Garzino Demo *et al.,* 1995). Two studies, one using slot-blot analysis (McCann *et al.,* 1997) and a second using RNase protection (Kaye *et al.,* 1999), indicate that the major packaging signal of HIV-2 is in a region 5′ to the splice donor and that the region between the splice donor and the Gag ATG makes a definite but relatively minor contribution. Supporting evidence for this is the fact that HIV-1 can cross-package HIV-2 RNA but, in doing so, captures both spliced and unspliced RNA with equal efficiency (Kaye *et al.,* 1998). Capture of spliced HIV-2 RNA by HIV-1 suggests that a fully functional signal exists in the region 5′ to the splice donor. Two studies have described cross-packaging of RNA from other lentiviruses by HIV-1 (Kaye *et al.,* 1998; Rizvi *et al.,* 1993) and, in one of these (Rizvi), SIV RNA was packaged by HIV-1 as well. A very extensive deletion between splice donor and the Gag ATG does appear to have a deleterious effect on packaging (Poeschla *et al.,* 1998) and the situation may be analogous to HIV-1 in which context-dependent components of the packaging signal exist 5′ and 3′ to the splice donor. In a study demonstrating helper virus free packaging of HIV-2 vectors, a combined deletion mutation upstream and downstream appeared to have the most profound packaging defect (Arya *et al.,* 1998). Interestingly, this deletion caused a decrease in the viral particle production by around 50%, a phenomenon seen in the largest packaging deletions in HIV-1, suggesting that functions other than packaging are affected by extensive mutations such as these.

G. Packaging versus Translation

The full-length genomic RNA of retroviruses also codes for the Gag and Gag/Pol polyproteins. There is conflicting evidence of the effect of the Ψ region on translation. *In vitro* transcription translation studies suggested that the Ψ stem loop structure acts as a significant inhibitor of translation

(Miele *et al.*, 1996), whereas cellular studies did not confirm an important effect here (McBride *et al.*, 1997).

There has been considerable debate as to whether a single RNA species can subsume both translation and genomic functions, whether there are completely separate pools of full-length RNA for translation and for encapsidation, or whether the system is flexible and both mechanisms occur. This question was originally addressed by experiments in murine leukemia virus by Levin *et al.* (Levin *et al.*, 1974) in which actinomycin D treatment of virus-producing cells was used as a transcriptional inhibitor. Translation was not inhibited and they were able to show that, with time, viral particles continued to be produced but these lacked genomic RNA. The conclusion was that viral RNA which was involved in translation for production of structural proteins was then not accessible for subsequent use for packaging. While separate pools of RNA for translation and packaging may exist, it is possible that actinomycin D treatment of cells inhibits transcription of genes coding for accessory cellular proteins which are required to chaperone RNA into the packaging pathway such that it has access to the Gag proteins. There may be perfectly packagable RNA in the actinomycin D-treated cells which has been translated and would otherwise be encapsidated but for the lack of these accessory cellular factors.

This question has recently been highlighted in HIV in which comparisons of the predominant mode of packaging of HIV-1 and HIV-2 have shown substantial differences (Kaye *et al.*, 1999). The major packaging signal of HIV-1 is downstream of the splice donor, whereas that of HIV-2 is upstream of the splice donor and, thus, is found in all of the HIV-2 RNA species. HIV-1 Gag is capable of picking up RNAs in trans, whereas HIV-2 RNA seems to be packaged preferentially in cis, i.e., cotranslationally. This mechanism would compensate for the presence of the Ψ site on all HIV-2 RNA species and regain selectivity of packaging for the genomic RNA which encodes Gag. Recent work suggests that in lentiviruses genomic RNA can be translated and subsequently packaged unlike the situation seen in simple retroviruses (Dorman and Lever, personal observations).

H. Lentiviral Vectors

I. Retroviral Vector Systems

Murine and avian retroviruses (Miller, 1992; Vile *et al.*, 1995) have been engineered for use as gene vectors and successfully used for many years. The first evidence that HIV-1 could be used to transfer a heterologous gene was work by Terwilliger *et al.* (Terwilliger *et al.*, 1989) in which the chloramphenicol acetyl transferase (CAT) gene was substituted into a deleted Nef open-reading frame. This generated a replication-competent HIV-1 which expressed CAT in all the cells it infected.

The great advantage of lentivirus-based vectors over other retroviruses is their ability to transduce growth-arrested and terminally differentiated cells. The viral components conferring this property are not fully defined but probably include nuclear localization signals (NLS) on specific components of the preintegration complex. An NLS has been identified on the virus Matrix Protein (Bukrinsky *et al.,* 1993), on Integrase (Gallay *et al.,* 1997), and also on Vpr (Heinzinger *et al.,* 1994) and there is some evidence for involvement of the karyopherin pathway (Gallay *et al.,* 1996). There appears to be some functional redundancy in the system, as Vpr is dispensable in some systems.

As with other retroviral vector systems, lentiviral vectors require a number of specific components (Fig. 3). The vector construct expresses an RNA containing the gene which is required to be transduced together with the appropriate cis-acting signals to ensure packaging of the vector RNA into the viral particle and successful reverse transcription and integration into the target cell. Complementing this are the packaging constructs (usually more than one) which between them are capable of expressing all the structural, enzymatic, and regulatory viral proteins required for production of an infectious viral particle. Ideally, the RNAs encoding these latter proteins should not incorporate a packaging signal, thus excluding their RNAs from the viral particle. All or none of these constructs may be stably transfected into cells. If packaging constructs alone are stably expressed, this is termed a packaging cell line. If the vector is stably expressed with the packaging construct(s), this is a producer line.

The vast majority of work on lentiviral vectors has been performed using HIV-1. There are a small number of publications describing HIV-2-based vectors and, more recently, the nonprimate lentiviruses, FIV, CAEV and EIAV, have been investigated for their potential to generate lentiviral vector systems. These also appear capable of delivering genes to nondividing cells although, as yet, the efficiency is generally less than that of HIV-1. In all of these viruses, pseudotyping with the VSV-G envelope is now widely used to broaden the tropism and increase the utility of these vector systems.

2. Vector Design

The cis-acting sequences required in a lentiviral vector are two long terminal repeat sequences to permit integration, a primer binding site for initiation of reverse transcription, an intact polypurine tract for second-strand synthesis during reverse transcription, and a functional packaging signal. The size of RNA, which can be accommodated in a viral particle, has not been formally tested. Early experiments using CAT cassettes of varying sizes in a replication-competent HIV-1 demonstrated a fall-off in infectivity with larger cassettes notably when the genome was larger than the wild-type genome (Terwilliger *et al.,* 1989). It was suggested that this might relate to decreasing efficiency of packaging with genomic size. Since

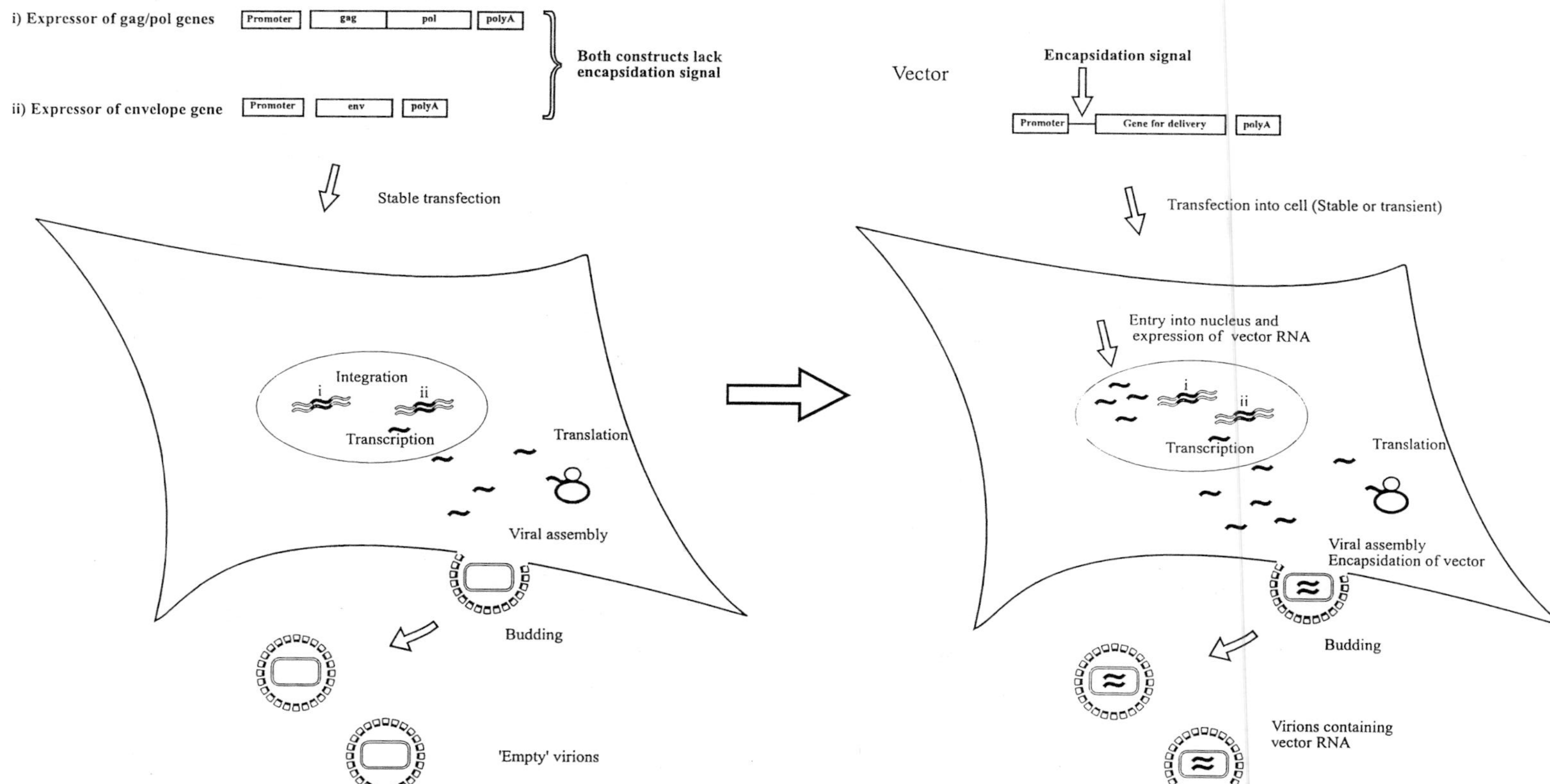

FIGURE 3 Conventional generation of stable packaging cell lines and introduction of packageable retroviral vectors.

that time, anecdotal reports have suggested that enlarging the genome beyond the size of the wild-type virus genome does not significantly impair packaging. There appears to be no lower limit on the size of vector which can be packaged other than that imposed by inclusion of the above mentioned cis-acting sequences.

Inserted genes may use the viral LTR as a promoter or, alternatively, heterologous promoters within the vector, although these risk problems with promoter competition. The 3′ LTR is generally used to polyadenylate genes inserted in sense orientation. Heterologous polyadenylation sequences are required for those expressed from internal promoters in reverse orientation to the viral LTR.

As in murine- and avian-based vectors (Olsen *et al.*, 1994; Julias *et al.*, 1995; Yu *et al.*, 1986), self-inactivating lentiviral vectors have been designed in which a mutation at the 3′ long terminal repeat has been introduced (Miyoshi *et al.*, 1998; Zufferey *et al.*, 1998). The 3′ LTR acts as the template for both long terminal repeats during reverse transcription and, hence, the promoter region of the 5′ LTR in the vector in the target cell becomes mutated. This gives additional flexibility to vector design, allowing independence from Tat transactivation and increased promoter choice for control of the level or tissue specificity of expression.

3. *Packaging Cell Lines*

A number of groups have sought to generate stable packaging cell lines based on lentiviruses (Haselhorst *et al.*, 1998; Carroll *et al.*, 1994; Yu *et al.*, 1996; Srinivasakumar *et al.*, 1997; Corbeau *et al.*, 1996; Kaul *et al.*, 1998). Both stable and inducible promoters have been used and the results are summarized in Table 1. In general, long-term expression of lentiviral

TABLE I A Summary of Lentiviral Packaging Cell Lines[a]

Author	*Particle-producing cells*	*Vector*	*Titre*	*Comments*
Richardson	Jurkat (± Tat)	Various	10^4	Stable
Carroll	Vero	T− R−	10^2	Stable
Srinavasakumar	Cos	T+	10^3	Stable
	(CMTS)	R ±		
		Nef ±		
Haselhorst	SW480	T+	10^2	Stable
	MDS	R+		
Corbeau	HeLa	T− R−	10^5	Stable
				Full-length helper ΔΨ
Yu	HeLa	T− R−	10^3	Inducible
Kaul	HeLa	T+ R+	10^4	Inducible
Kafri	293	T− R−	10^5–10^6	Inducible
				VSV pseudotyped

[a] Abbreviations: T, Tat; R, Rev; ΔΨ, packaging signal deletion.

proteins in cells *in vitro* has been problematic, although inducible expression long term has been documented (Kaul *et al.*, 1998). A number of lentiviral proteins are toxic to cells. The HIV envelope protein can lead to syncitia formation and cytopathicity in cells expressing CD4 (Kowalski *et al.*, 1987) but also in cells which apparently do not express detectable CD4 (Haselhorst *et al.*, 1998). The Vpr protein is capable of inducing cell-cycle arrest (Macreadie *et al.*, 1995; He *et al.*, 1995; Re *et al.*, 1995). The HIV-1 protease is toxic to some mammalian cell lines (Krausslich, 1992). All of these and the probable toxic effects of other lentiviral proteins have made production of stable lines difficult. Those which have been produced commonly lose viral protein expression at a rapid rate (Fig. 4) (RT counts may fall by up to 10% per week; personal observations) and, thus, their useful life span is limited. Envelopes such as VSV-G, which have been used to pseudotype lentiviral cores, are also cytopathic and cannot be expressed stably. Recently, a packaging cell line has been described in which the VSV-G protein expression and HIV *gag* and *pol* gene products are controlled by a tetracycline inducible promoter (Kafri *et al.*, 1999), thus giving a time window of 3–4 days following induction during which vector production is possible before the cytopathic effects of the rhabdovirus envelope and lentivirus core proteins take effect. Interestingly, tetracycline inducibility was soon lost but expression could be regenerated by sodium butyrate, an inhibitor of histone

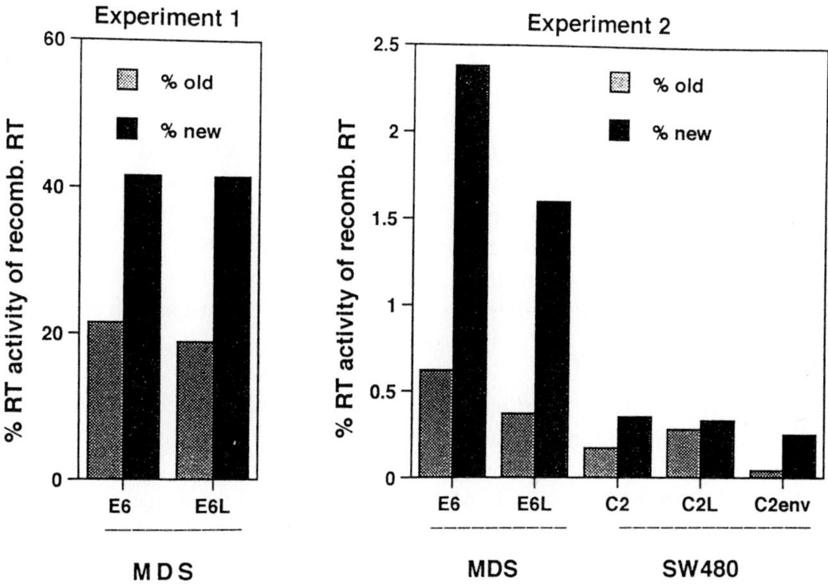

FIGURE 4 Decline in RT activity of stably expressing packaging cell lines; new, within 2 weeks of establishment; old, greater than 2 months following establishment. Data from Haselhorst, D, Ph.D. thesis, University of Cambridge, 1997.

deacetylation. Chromatinization of retroviral inserts may be an important problem for both producer cells and for the vector in the targeted cell line.

One stable packaging cell line has been based on a full-length provirus with deletions in the packaging signal region and envelope gene (Corbeau *et al.*, 1996). This cell line encapsidates vectors efficiently and produces particles at high titer. A concern must be the risk of a relatively limited series of recombination events occurring between vector and RNA from this full-length packaging construct generating a novel vector. Use of an HIV-1 envelope in this system would clearly require stringent monitoring for regeneration of replication-competent virus.

In contrast to vectors based on murine leukemia virus and RSV, the major successes in lentiviral vectors production and use have come through utilizing transient cotransfection of packaging constructs with vectors (Fig. 5).

4. Transient Cotransfection and Vector Production

With accumulating data regarding HIV-1 packaging signals, a number of groups have generated HIV-1-based vectors targeted using the virus' own envelope glycoproteins. An early study using transient cotransfection of vector and Gag/Pol and HIV-1 envelope expressing genes apparently achieved vector titers of 10^5 cfu/ml (Poznansky *et al.*, 1991), although others have not since achieved this level using HIV-1 proteins alone. Transient cotransfection of a 3-plasmid packaging combination using an HIV-1 Gag/Pol and HIV-1 envelope and an HIV-1-based vector confirmed the possibility of helper virus free gene transfer (Richardson *et al.*, 1995). This work also demonstrated that, using a full-length helper virus, a single deletion in its packaging signal as the only attenuating mutation was insufficient to prevent recombination with a packaging signal-containing vector and that replication competent recombinants were rapidly generated. An increasing number of studies have used heterologous envelope proteins and shown that high-titer vectors can be generated by transient cotransfection of a lentiviral gag/pol expressor, a VSV-G envelope, and a packagable HIV-1-based vector. In some cases the Rev Protein is expressed separately from a fourth construct.

5. Enhancing Vector Titer

Despite the ability of the cotransfection system to generate vectors in reproducibly high titer as assessed on transformed cell lines, most groups have found it necessary to incorporate additional steps to increase the titer or infectivity of their vector stock. Most commonly used is ultracentrifugation, following which the virion pellet is resuspended in a small volume (Akkina *et al.*, 1996; Naldini *et al.*, 1996). Some groups routinely use two sequential ultracentrifugations to concentrate vectors in this way. Alternatives to this include PEG precipitation which can lead to a 5- to 10-fold increase in concentration of the vector stock. Column chromatography (Matsuoka *et*

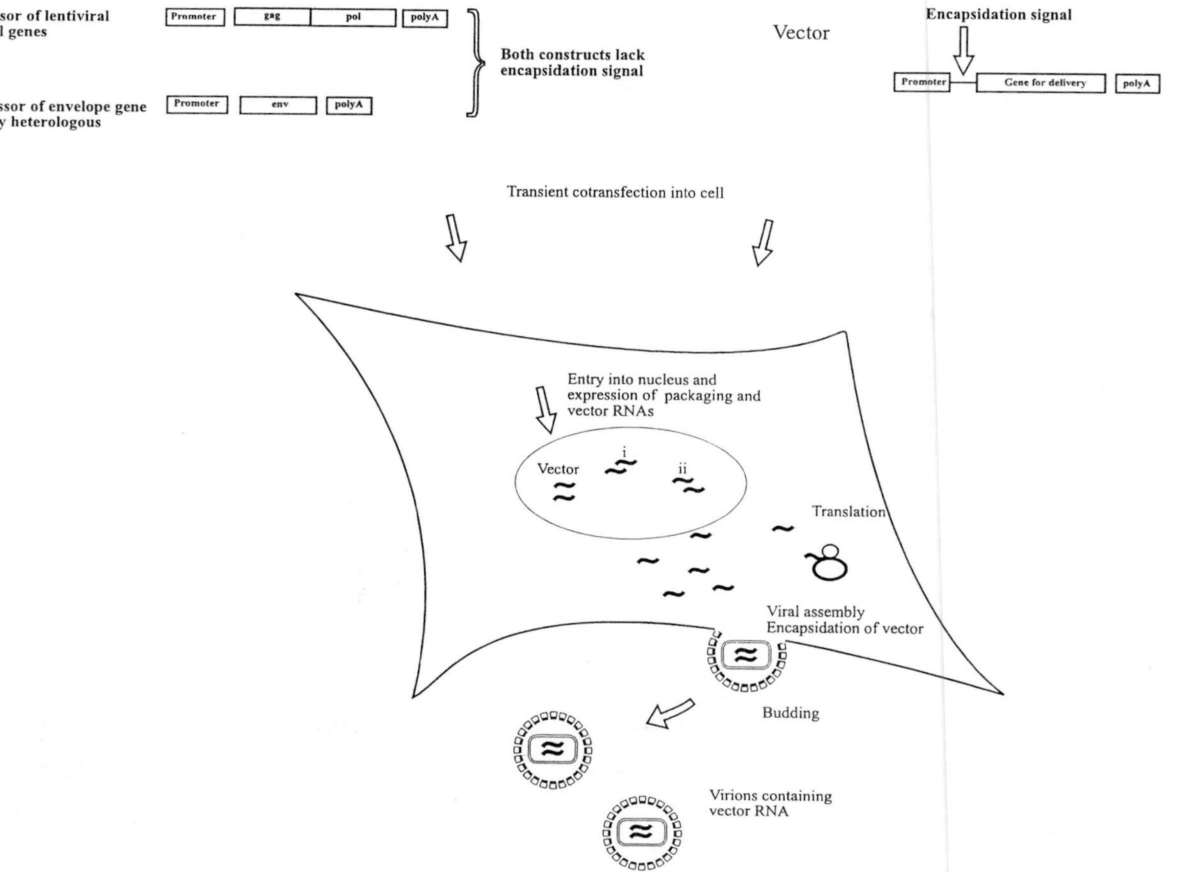

FIGURE 5 Generation of lentivirus vectors using transient cotransfection of vector and packaging constructs.

al., 1998) has also been used, achieving a 20-fold concentration, and simple ultrafiltration (Reiser *et al.,* 1996) has also been employed. The widespread usage of concentrating technologies indicates that, despite titers of 10^6 and above on unconcentrated vector stocks, even higher titers are required for efficient gene transfer particularly in *in vivo* models.

As with many viruses which require deoxynucleotides for synthesis of their genome, the availability of these may be a limiting factor particularly in cells which are not undergoing mitosis. Pretreatment of vector stocks with dNTP mixtures can help to progress the small amount of reverse transcription which commonly takes place in the virion particle and enhances complete provirus synthesis once the target cell is infected (Blomer *et al.,* 1997; Goldman *et al.,* 1997).

6. Role of Accessory Lentiviral Proteins

Like all complex retroviruses, HIV encodes a number of small open reading frames from which regulatory proteins are produced. A number of these have very important roles in the retroviral lifecycle. Nef, for example, appears to be critical for *in vivo* replication (Harris, 1996), Vpr may contribute to nuclear targeting (Heinzinger *et al.,* 1994), and Vpu may contribute to efficient viral assembly and export (Klimkait *et al.,* 1990). The roles of Tat and Rev are also well documented. Inclusion of regulatory proteins may have theoretical advantages in the construction of an HIV-based vector for enhancement of viral assembly, export, or infectivity and there may be better expression of the delivered gene in the target cell if Tat and Rev are present. Despite the clear importance of these genes for wild-type virus replication, there seems to be little firm evidence that proteins other than Tat and Rev make a significant contribution to vector efficiency. Despite one report to the contrary (Blomer *et al.,* 1997), most research suggests that Vif, Vpr, Vpu, and Nef (Kim *et al.,* 1998; Sutton *et al.,* 1998; Zufferey *et al.,* 1997; Reiser *et al.,* 1996) are dispensable for vector transfer in direct comparison experiments. One report even finds that Vif and Vpr may be deleterious to transduction of hemopoietic stem cells (Sutton *et al.,* 1998). The effect of accessory proteins may be tissue-dependent, one study finding that neuronal transduction did not require Vpr or Vif (Kafri *et al.,* 1997); whereas, in the absence of these two proteins, gene transfer into liver cells was significantly impaired.

VSV pseudotypes appear to enter target cells through an endocytotic route rather than the fusion route of wild-type HIV (Aiken, 1997) and gp120 enveloped vectors. This alternative route of entry appears to confer Nef independence, suggesting perhaps that Nef is only required for a process such as disassembly when entry is by fusion.

It has clearly been shown that the presence of Tat will enhance the expression of an LTR-driven gene in the target cell, and if any instability sequences remain in the vector, entry into the packaging pathway will require the presence of Rev in the producer line and expression may be dependent on the presence of Rev in the target cell line. A number of groups have now shown production and successful packaging and transfer of minimal HIV-based vectors which contain virtually no viral coding sequence (Kim *et al.*, 1998; Obaru *et al.*, 1996) in which the genes are under the control of heterologous promoters. Such "stripped out" vectors have the advantage of minimizing sequence homology between vector and packaging constructs and, thus, limiting the possibilities of recombination.

7. Genes Delivered by Lentiviral Vectors

Most studies to date have concentrated on validating gene transfer by the use of marker genes. The enhanced green fluorescent protein (EGFP) is a widely used and reliable marker and easily detectable to validate successful gene delivery. It also permits FACS sorting of gene-transduced cells. A handful of studies have tested transfection of a therapeutic gene. An intracellular antibody against the IL-2 receptor has been shown to be deliverable (Richardson *et al.*, 1998) and to sequester the IL-2 receptor in the endoplasmic reticulum. Anti-HIV ribozyme and decoy RNAs have been delivered (Dropulic *et al.*, 1996). A conditional lethal gene, thymidine kinase (Obaru *et al.*, 1996), has been delivered to CD4 cells with the eventual therapeutic aim of targeting HTLV-1-transformed adult T-cell leukemia lymphoma, a malignancy of CD4+ cells. The cystic fibrosis transmembrane regulatory has also been delivered using lentiviral vectors (Goldman *et al.*, 1997). In bronchial organ culture experiments lentiviruses were shown to be more efficient than murine-vector-mediated delivery. Unfortunately, delivery to differentiated bronchial epithelium appeared to be very inefficient using either vector system. Interleukin 4 has been delivered to pancreatic islet (Gallichan *et al.*, 1998) cells and has been shown to protect them against autoimmune destruction in the diabetes-prone mouse. The efficacy of the BCL-XL gene (Blomer *et al.*, 1998) in protecting cholinergic neurones has also been demonstrated after gene delivery using a lentiviral vector.

8. Cellular Targets of Lentiviral Vectors

Tropism clearly depends to a large extent on the envelope proteins employed. There has been little exploitation of tissue-specific promoters as yet. Using the native HIV-1 envelope, delivery is limited to cells expressing CD4 and the appropriate chemokine receptor. This may be useful if one wishes to deliberately target such cells as in the case of the delivery of anti-HIV genes to inhibit HIV viral replication or to target T cells for other purposes such as those transformed by HTLV-1. Other than this, it is a limitation, although one group has shown that it is possible to render cells transiently susceptible to transduction by HIV enveloped vectors using short-term expression of the CD4 molecule in the target cell (Miyake *et al.*, 1996), delivered by an adenovirus vector as a pretreatment. Successful pseudotyping of lentiviral vectors with murine leukemia virus (MuLV) envelope has shown that (Spector *et al.*, 1990), using the amphotropic envelope, it is possible to widen the tropism of HIV-1-based vectors. This may be of use since the MuLV envelope has been extensively modified to produce variants which affect tropism (Cosset *et al.*, 1995; Valsesia Wittmann *et al.*, 1996).

The majority of lentivirus-based vectors now are pseudotyped with the rhabdovirus envelope from VSV, giving a wide cellular tropism. The majority of transformed cell lines in the laboratory have been shown to be transducible by VSV-G pseudotyped lentiviruses with titers comparable to other retroviral and nonretroviral vectors. Delivery to primary cells has also been achieved, some of the most spectacular successes being the successful delivery and prolonged expression of genes in cells of the central nervous system. Stable gene expression was demonstrated for periods of over 6 months in cells which morphologically were demonstrated to be terminally differentiated neurones (Kafri *et al.*, 1997; Blomer *et al.*, 1997; Zufferey *et al.*, 1997).

Delivery to hepatocytes *in vivo* has also been demonstrated with expression again persisting for 5–6 months (Kafri *et al.*, 1997). On analysis, some 3–4% of hepatocytes appeared to be expressing the gene.

Other groups have documented T-cell transduction *in vitro* and of great importance was the demonstration that hematopoietic stem cells (HSC) appeared to be transduced with up to 50% efficiency with apparent enhancement of long-term expression if the accessory proteins Vif and Vpr were deleted (Sutton *et al.*, 1998). Transduction of HSC with maintenance of pluripotentiality has also been documented (Uchida *et al.*, 1998). Gene delivery has been demonstrated to the lung (Goldman *et al.*, 1997), skin (Reiser *et al.*, 1996), pancreatic islets (Gallichan *et al.*, 1998; Ju *et al.*, 1998), retina (Miyoshi *et al.*, 1997), and skeletal (Kafri *et al.*, 1997) and cardiac muscle (Rebolledo *et al.*, 1998).

As yet, the *in vivo* studies have been limited to experimental animals, although transduction of primary human cells *in vitro* has also been demonstrated and there is no reason to believe that the same efficiency will not be found *in vivo* in humans.

9. *HIV-2-Based Vectors*

A small number of publications from a few research groups have shown that HIV-2 can be used as a vector (Poeschla *et al.*, 1996, 1998; Corbeau *et al.*, 1998; Sadaie *et al.*, 1998; Arya *et al.*, 1998). Human macrophages were transduced using a chimeric system of an HIV-1 helper packaging an HIV-2 vector (Corbeau *et al.*, 1998). Such cross-packaging has been observed by others but intriguingly appears to be nonreciprocal, with HIV-1 capable of packaging HIV-2 vectors but the reverse not being possible (Kaye *et al.*, 1998). The pure HIV-2 system appears to transduce both dividing and nondividing cells. Transduction was shown in growth-arrested transformed cell lines and the postmitotic NTN2 neuronal cell line (Poeschla *et al.*,

1998). Transduction of hematopoietic stem cells was less certain, with some evidence of transduction and reverse transcription of the vector to DNA but without expression of the transduced gene.

10. Vectors Based on Other Lentiviruses

EIAV is being researched as a vector (Olsen, 1998; Mitrophanous *et al.*, 1999). However, there are, as yet, only two publications describing its use. In these, a conventional three-plasmid system is used with a Gag/Pol expressor driven by a heterologous promoter (CMV) and flanked to the 3′ end by a heterologous PolyA. It is cotransfected with a CMV-driven VSV-G envelope and the EIAV vector. Empirically, the R-U5 and leader regions are incorporated in the vector together with the first portion of the *gag* gene, the assumption that this must incorporate the packaging signal of EIAV (which has not been localized). A second downstream promoter drives the reporter gene expressed in the target cell. Successful gene transfer was demonstrated to both dividing and partially (Aphidicolin) growth-arrested cells, the latter being much more efficient than using a murine vector. Titers of greater than 10^5 cfu/ml are described. However, for successful gene transfer, 50-fold concentration by ultracentrifugation appears to be required.

A vector system based on FIV has also been published (Poeschla *et al.*, 1998) which is able to gene transduce growth-arrested cells.

CAEV has also been used as a lentivirus vector. CAEV-based vectors have been packaged by replication-competent CAEV and delivered to goat synovial membrane cells (Mselli Lakhal *et al.*, 1998). These have developed more conventional CMV-driven packaging constructs expressing the Gag/Pol, Vif, and Tat genes and pseudotyped these with the VSV-G envelope. A CAEV-based vector system has been demonstrated to be able to transduce growth-arrested cell lines (De Rocquigny and Heard, personal communication).

III. Summary

Since the mid-1990s, the number of publications on lentivirus-based vectors has expanded dramatically as people have realized the opportunity that they represent. High-titer helper-virus free transfer of genes to nondividing cells is a reality and it can only be a short time before clinical trials are initiated. The most efficient vector to date appears to be HIV-1 and it is no coincidence that this is the virus in which there is the greatest theoretical understanding of the encapsidation process and viral assembly. Basic studies in the other viruses are at an earlier stage and this is reflected to some extent in their relative inefficiency. Emphasis is placed in some publications on non-HIV-based vector systems having the additional safety feature of a viral vector not based on a human pathogen. As yet, this is largely a cosmetic advantage in that no system would be used which was capable of regenerating a full-length wild-type HIV and the vectors all have single round replication kinetics. More important will be elucidation of the mechanism of packaging in the different lentiviruses. Cis and trans packaging preferences may influence efficiency. Accurate delineation of packaging signals will be important. Most influential, however, will be a deeper understanding of all the viral and cellular factors involved in the packaging pathway.

References

Aiken, C. (1997). Pseudotyping human immunodeficiency virus type 1 (HIV-1) by the glycoprotein of vesicular stomatitis virus targets HIV-1 entry to an endocytic pathway and suppresses both the requirement for Nef and the sensitivity to Cyclosporin A. *J. Virol.* **71,** 5871–5877.

Akkina, R. K., Walton, R. M., Chen, M. L., Li, Q. X., Planelles, V. and Chen, I. S. Y. (1996). High efficiency gene transfer into CD34(+) cells with a human immunodeficiency virus type 1-based retroviral vector pseudotyped with vesicular stomatitis virus envelope glycoprotein G. *J. Virol.* **70,** 2581–2585.

Aldovini, A., and Young, R. A. (1990). Mutations of RNA and protein sequences involved in human immunodeficiency virus type 1 packaging result in production of noninfectious virus. *J. Virol.* **64,** 1920–1926.

Andresson, O. S., Elser, J. E., Tobin, G. J., Greenwood, J. D., Gonda, M. A., Georgsson, G., Andresdottir, V., Benediktsdottir, E., Carlsdottir, H. M., and Mantyla, E. O. (1993). Nucleotide sequence and biological properties of a pathogenic proviral molecular clone of neurovirulent visna virus. *Virology* **193,** 89–105.

Arya, S. K., Zamani, M., and Kundra, P. (1998). Human immunodeficiency virus type 2 lentivirus vectors for gene transfer: Expression and potential for helper virus-free packaging. *Hum. Gene Ther.* **9,** 1371–1380.

Awang, G., and Sen, D. (1993). Mode of dimerisation of HIV-1 genomic RNA. *Biochemistry* **32,** 11453–11457.

Bacharach, E., and Goff, S. P. (1998). Binding of the human immunodeficiency virus type 1 Gag protein to the viral RNA encapsidation signal in the yeast three-hybrid system. *J. Virol.* **72,** 6944–6949.

Berglund, J. A., Charpentier, B., and Rosbash, M. (1997). A high affinity binding site for the HIV-1 nucleocapsid protein. *Nucleic Acids Res.* **25,** 1042–1049.

Berkhout, B. (1996). Structure and function of the Human Immunodeficiency Virus leader RNA. *Prog. Nucl. Acid Res. Mol. Biol.* **54,** 1–34.

Berkhout, B., and van Wamel, J. L. B. (1996). Role of the DIS hairpin in replication of Human Immunodeficiency Virus Type 1. *J. Virol.* **70,** 6723–6732.

Berkowitz, R. D., and Goff, S. P. (1994). Analysis of binding elements in the human immunodeficiency virus type 1 genomic RNA and nucleocapsid protein. *Virology* **202,** 233–246.

Berkowitz, R. D., Hammarskjold, M. L., Helga Maria, C., Rekosh, D., and Goff, S. P. (1995). 5′ regions of HIV-1 RNAs are not sufficient for encapsidation: Implications for the HIV-1 packaging signal. *Virology* **212,** 718–723.

Blomer, U., Kafri, T., Randolph More, L., Verma, I. M., and Gage, F. H. (1998). Bcl-xL protects adult septal cholinergic neurons from axotomized cell death. *Proc. Natl. Acad. Sci. USA* **95,** 2603–2608.

Blomer, U., Naldini, L., Kafri, T., Trono, D., Verma, I. M., and Gage, F. H. (1997). Highly efficient and sustained gene transfer in adult neurons with a lentivirus vector. *J. Virol.* **71,** 6641–6649.

Brahic, M., and Vigne, R. (1975). Properties of visna virus particles harvested at short time intervals: RNA content, infectivity and ultrastructure. *J. Virol.* **15**, 1222–1230.

Bukrinsky, M. I., Haggerty, S., Dempsey, M. P., Sharova, N., Adzhubel, A., Spitz, L., Lewis, P., Goldfarb, D., Emerman, M., and Stevenson, M. (1993). A nuclear localisation signal within HIV-1 matrix protein that governs infection of non-dividing cells. *Nature* **365**, 666–669.

Carroll, R., Lin, J. T., Dacquel, E. J., Mosca, J. D., Burke, D. S., and St-Louis, D. C. (1994). A human immunodeficiency virus type 1 (HIV-1) based retroviral vector system utilising stable HIV-1 packaging cell lines. *J. Virol.* **68**, 6047–6051.

Clavel, F., and Orenstein, J. M. (1990). A mutant of human immunodeficiency virus with reduced RNA packaging and abnormal particle morphology. *J. Virol.* **64**, 5230–5234.

Clever, J. L., and Parslow, T. G. (1997). Mutant human immunodeficiency virus type 1 genomes with defects in RNA dimerization or encapsidation. *J. Virol.* **71**, 3407–3414.

Clever, J., Sassetti, C., and Parslow, T. G. (1995). RNA secondary structure and binding sites for gag gene products in the 5′ packaging signal of human immunodeficiency virus type 1. *J. Virol.* **69**, 2101–2109.

Clever, J. L., Eckstein, D. A., and Parslow, T. G. (1999). Genetic dissociation of the encapsidation and reverse transcription functions in the 5′ R region of human immunodeficiency virus type 1. *J. Virol.* **73**, 101–109.

Corbeau, P., Kraus, G., and Wong Staal, F. (1998). Transduction of human macrophages using a stable HIV-1/HIV-2-derived gene delivery system. *Gene Therapy* **5**, 99–104.

Corbeau, P., Kraus, G., and Wong-Staal, F. (1996). Efficient gene transfer by a human immunodeficiency virus type 1 (HIV-1)-derived vector utilising a stable HIV packaging cell line. *Proc. Natl. Acad. Sci. USA* **93**, 14070–14075.

Cosset, F. L., Morling, F. J., Takeuchi, Y., Weiss, R. A., Collins, M. K., and Russell, S. J. (1995). Retroviral retargeting by envelopes expressing an N-terminal binding domain. *J. Virol.* **69**, 6314–6322.

Cullen, B. R., and Greene, W. C. (1989). Regulatory pathways governing HIV-1 replication. *Cell* **58**, 423–426.

Damgaard, C. K., Dyhr Mikkelsen, H., and Kjems, J. (1998). Mapping the RNA binding sites for human immunodeficiency virus type 1 Gag and NC proteins within the complete HIV-1 and -2 untranslated leader regions. *Nucleic Acids Res.* **26**, 3667–3676.

Dannull, J., Surovoy, A., Jung, G., and Moelling, K. (1994). Specific binding of HIV-1 nucleocapsid protein to Psi-RNA *In-Vitro* requires N-terminal zinc-finger and flanking basic-amino-acid residues. *Embo J.* **13**, 1525–1533.

Das, A. T., Klaver, B., Klasens, B. I. F., van Wamel, J. L. B., and Berkout, B. (1997). A conserved hairpin motif in the R-U5 region of the human immunodeficiency virus type 1 RNA genome is essential for replication. *J. Virol.* **71**, 2346–2356.

De Guzman, R. N., Wu, Z. R., Stalling, C. C., Pappalardo, L., Borer, P. N., and Summers, M. F. (1998). Structure of the HIV-1 nucleocapsid protein bound to the SL3 psi-RNA recognition element. *Science* **279**, 384–388.

Dorfman, T., Luban, J., Goff, S. P., Haseltine, W. A., and Gottlinger, H. G. (1993). Mapping of functionally important residues of a cysteine-histidine box in the human immunodeficiency virus type 1 nucleocapsid protein. *J. Virol.* **67**, 6159–6169.

Dropulic, B., Hermankova, M., and Pitha, P. M. (1996). A conditionally replicating HIV-1 vector interferes with wild-type HIV-1 replication and spread. *Proc. Natl. Acad. Sci. USA* **93**, 11103–11108.

Feng, Y.-X., Copeland, T. D., Henderson, L. E., Gorelick, R. J., Bosche, W. J., Levin, J. G., and Rein, A. (1996). HIV-1 nucleocapsid protein induces "maturation" of dimeric retroviral RNA *in vitro*. *Proc. Natl. Acad. Sci. USA* **93**, 7577–7581.

Fu, W., and Rein, A. (1993). Maturation of dimeric viral RNA of Moloney Murine Leukaemia virus. *J. Virol.* **67**, 5443–5449.

Fu, W., Gorelick, R. J., and Rein, A. (1994). Characterization of Human Immunodeficiency Virus Type 1 dimeric RNA from wild-type and protease-defective virions. *J. Virol.* **68,** 5013–5018.

Gallay, P., Hope, T., Chin, D., and Trono, D. (1997). HIV-1 infection of nondividing cells through the recognition of integrase by the importin/karyopherin pathway. *Proc. Natl. Acad. Sci. USA* **94,** 9825–9830.

Gallay, P., Stitt, V., Mundy, C., Oettinger, M., and Trono, D. (1996). Role of the karyopherin pathway in human immunodeficiency virus type 1 nuclear import. *J. Virol.* **70,** 1027–1032.

Gallichan, W. S., Kafri, T., Krahl, T., Verma, I. M., and Sarvetnick, N. (1998). Lentivirus-mediated transduction of islet grafts with interleukin 4 results in sustained gene expression and protection from insulitis. *Hum. Gene Ther.* **9,** 2717–2726.

Garzino Demo, A., Gallo, R. C., and Arya, S. K. (1995). Human immunodeficiency virus type 2 (HIV-2): Packaging signal and associated negative regulatory element. *Hum. Gene Ther.* **6,** 177–184.

Goldman, M. J., Lee, P.-S., Yang, J.-S., and Wilson, J. M. (1997). Lentiviral vectors for gene therapy of cystic fibrosis. *Hum. Gene Ther.* **8,** 2261–2268.

Gorelick, R. J., Nigida, S. M., Jr., Bess, J. W., Jr., Arthur, L. O., Henderson, L. E., and Rein, A. (1990). Noninfectious human immunodeficiency virus type 1 mutants deficient in genomic RNA. *J. Virol.* **64,** 3207–3211.

Greatorex, J., and Lever, A. (1998). Retroviral RNA dimer linkage. *J. Gen. Virol.* **79,** 2877–2882.

Haddrick, M., Lear, A. L., Cann, A. J., and Heaphy, S. (1996). Evidence the a kissing loop structure facilitates genomic RNA dimerisation in HIV-1. *J. Mol. Biol.* **259,** 58–68.

Harris, M. (1996). From negative factor to a critical role in virus pathogenesis: The changing fortunes of Nef. *J. Gen. Virol.* **77,** 2379–2392.

Harrison, G. P., and Lever, A. M. L. (1992). The human immunodeficiency virus type 1 packaging signal and major splice donor region have a conserved stable secondary structure *J. Virol.* **66,** 4144–4153.

Harrison, G. P., Hunter, E., and Lever, A. M. L. (1995). Secondary structure model of the Mason-Pfizer Monkey Virus 5′ leader sequence: Identification of a structural motif common to a variety of retroviruses. *J. Virol.* **69,** 2175–2186.

Harrison, G. P., Miele, G., Hunter, E., and Lever, A. M. L. (1998). Functional analysis of the core packaging signal in a permissive cell line. *J. Virol.* **72,** 5886–5896.

Haselhorst, D., Kaye, J. F., and Lever, A. M. L. (1998). Development of cell lines stably expressing human immunodeficiency virus type 1 proteins for studies in encapsidation and gene transfer. *J. Gen. Virol.* **79,** 231–237.

He, J., Choe, S., Walker, R., Di-Marzio, P., Morgan, D. O., and Landau, N. R. (1995). Human immunodeficiency virus type 1 viral protein R (Vpr) arrests cells in the G2 phase of the cell cycle by inhibiting p34 cdc2 activity. *J. Virol.* **69,** 6705–6711.

Heinzinger, N. K., Bukinsky, M. I., Haggerty, S. A., Ragland, A. M., Kewalramani, V., Lee, M. A., Gendelman, H. E., Ratner, L., Stevenson, M., and Emerman, M. (1994). The Vpr protein of human immunodeficiency virus type 1 influences nuclear localisation of viral nucleic acids in nondividing host cells. *Proc. Natl. Acad. Sci. USA* **91,** 7311–7315.

Ju, Q., Edelstein, D., Brendel, M. D., Brandhorst, D., Brandhorst, H., Bretzel, R. G., and Brownlee, M. (1998). Transduction of non-dividing adult human pancreatic beta cells by an integrating lentiviral vector. *Diabetologia* **41,** 736–739.

Julias, J. G., Hash, D., and Pathak, V. K. (1995). E-vectors: Development of novel self-inactivating and self-activating retroviral vectors for safer gene therapy. *J. Virol.* **69,** 6839–6846.

Kafri, T., Blomer, U., Peterson, D. A., Gage, F. H., and Verma, I. M. (1997). Sustained expression of genes delivered directly into liver and muscle by lentiviral vectors. *Nature Gen.* **17,** 314–317.

Kafri, T., van Pragg, H., Ouyang, L., Gage, F. H., and Verma, I. M. (1999). A packaging cell line for lentivirus vectors. *J. Virol.* **73,** 576–584.

Kaul, M., Yu, H., Ron, Y., and Dougherty, J. P. (1998). Regulated lentiviral packaging cell line devoid of most viral *cis*-acting sequences. *Virology* **249,** 167–174.

Kaye, J. F., and Lever, A. M. L. (1998). Nonreciprocal packaging of human immunodeficiency virus type 1 and type 2 RNA: A possible role for the p2 domain of Gag in RNA encapsidation. *J. Virol.* **72,** 5877–5885.

Kaye, J. F., and Lever, A. M. L. (1999). Human immunodeficiency virus types 1 and 2 use different mechanisms for selection of unspliced RNA for encapsidation. *J. Virol.* **73,** 3023–3031.

Kaye, J. F., Richardson, J. H., and Lever, A. M. (1995). cis-acting sequences involved in human immunodeficiency virus type 1 RNA packaging. *J. Virol.* **69,** 6588–6592.

Kim, V. N., Mitrophaneous, K., Kingsman, S. M., and Kingsman, A. J. (1998). Minimal requirement for a lentivirus vector based on human immunodeficiency virus type 1. *J. Virol.* **72,** 811–816.

Klimkait, T., Strebel, K., Hoggan, M. D., Martin, M. A., and Orenstein, J. M. (1990). The human immunodeficiency virus type 1-specific protein Vpu is required for efficient virus maturation and release. *J. Virol.* **64,** 621–629.

Konings, D. A., Nash, M. A., Maizel, J. V., and Arlinghaus, R. B. (1992). Novel GACG-hairpin motif in the 5′ untranslated region of type C retroviruses related to murine leukaemia virus. *J. Virol.* **66,** 632–640.

Kowalski, M., Potz, J., Basiripour, L., Dorfman, T., Goh, W. C., Terwilliger, E., Dayton, A., Rosen, C., Haseltine, W., and Sodroski, J. (1987). Functional regions of the envelope glycoprotein of human immunodeficiency virus type 1. *Science* **237,** 1351–1355.

Krausslich, H. G. (1992). Specific inhibitor of human immunodeficiency virus proteinase prevents the cytotoxic effects of a single-chain proteinase dimer and restores particle formation. *J. Virol.* **66,** 567–572.

Laughrea, M., and Jette, L. (1994). A 19-nucleotide sequence upstream of the 5′ major splice donor is part of the dimerization domain of the Human Immunodeficiency Virus 1 genomic RNA. *Biochemistry* **33,** 13464–13474.

Lever, A., Gottlinger, H., Haseltine, W., and Sodroski, J. (1989). Identification of a sequence required for efficient packaging of human immunodeficiency virus type 1 RNA into virions. *J. Virol.* **63,** 4085–4087.

Levin, J. G., Grimley, P. M., Ramseur, J. M., and Berezesky, I. K. (1974). Deficiency of 60 to 70S RNA in murine leukemia virus particles assembled in cells treated with actinomycin D. *J. Virol.* **14,** 152–161.

Linial, M. L., and Miller, A. D. (1990). Retroviral RNA Packaging: Sequence Requirements and Implications. *Curr. Top. Microbiol. Immunol.* **157,** 125–152.

Macreadie, I. G., Castelli, L. A., Hewish, D. R., Kirkpatrick, A., Ward, A. C., and Azad, A. A. (1995). A domain of human immunodeficiency virus type 1 Vpr containing repeated H (S/F) RIG amino acid motifs causes cell growth arrest and structural defects. *Proc. Natl. Acad. Sci. USA* **92,** 2770–2774.

Marquet, R., Paillart, J. C., Skripkin, E., Ehresmann, C., and Ehresmann, B. (1994). Dimerization of human immunodeficiency virus type 1 RNA involves sequences located upstream of the splice donor site. *Nucleic Acids Res.* **22,** 145–151.

Matsuoka, H., Miyake, K., and Shimada, T. (1998). Improved methods of HIV vector mediated gene transfer. *Int. J. Haematol.* **67,** 267–273.

McBride, M. S., and Panganiban, A. T. (1996). The human immunodeficiency virus type 1 encapsidation site is a multipartite RNA element composed of functional hairpin structures. *J. Virol.* **70,** 2963–2973.

McBride, M. S., and Panganiban, A. T. (1997). Position dependence of functional hairpins important for human immunodeficiency virus type 1 RNA encapsidation *in vivo*. *J. Virol.* **71,** 2050–2058.

McCann, E. M., and Lever, A. M. (1997). Location of cis-acting signals important for RNA encapsidation in the leader sequence of human immunodeficiency virus type 2. *J. Virol.* **71**, 4133–4137.

Miele, G., Mouland, A., Harrison, G. P., Cohen, E., and Lever, A. M. (1996). The human immunodeficiency virus type 1 5′ packaging signal structure affects translation but does not function as an internal ribosome entry site structure. *J. Virol.* **70**, 944–951.

Miller, A. D. (1992). Retroviral vectors. *Curr. Top. Microbiol. Immunol.* **158**, 1–24.

Mitrophanous, K. A., Yoon, S., Rohll, J. B., Patil, D., Wilkes, F. J., Kim, V. N., Kingsman, S. M., Kingsman, A. J., and Mazarakis, N. D. (1999). Stable gene transfer to the nervous system using a non-primate lentiviral vector. *Gene Ther.* **6**, 1808–1818.

Miyake, K., Tohyama, T., and Shimada, T. (1996). Two-step gene transfer using an adenoviral vector carrying the CD4 gene and human immunodeficiency viral vectors. *Hum. Gene Ther.* **7**, 2281–2286.

Miyoshi, H., Blomer, U., Takahashi, M., Gage, F. H., and Verma, I. M. (1998). Development of a self-inactivating lentivirus vector. *J. Virol.* **72**, 8150–8157.

Miyoshi, H., Takahashi, M., Gage, F. H., and Verma, I. M. (1997). Stable and efficient gene transfer into the retina using an HIV-based lentiviral vector. *Proc. Natl. Acad. Sci. USA* **94**, 10319–10323.

Mselli Lakhal, L., Favier, C., Teixeira, M. F. D., Chettab, K., Legras, C., Ronfort, C., Verdier, G., Mornex, J. F., and Chebloune, Y. (1998). Defective RNA packaging is responsible for low transduction efficiency of CAEV-based vectors. *Arch. Virol.* **143**, 681–695.

Mujeeb, A., Clever, J. L., Billeci, T. M., James, T. L., and Parslow, T. G. (1998). Structure of the dimer initiation complex of HIV-1 genomic RNA. *Nat. Struct. Biol.* **5**, 432–436.

Naldini, L., Blomer, U., Gallay, P., Ory, D., Mulligan, P., Gage, F. H., Verma, I. M., and Trono, D. (1996). *In vivo* gene delivery and stable transduction of nondividing cells by a lentiviral vector. *Science* **272**, 263–267.

Obaru, K., Fujii, S., Matsushita, S., Shimada, T., and Takatsuki, K. (1996). Gene therapy for adult T cell leukemia using human immunodeficiency virus vector carrying the thymidine kinase gene of herpes simplex virus type 1. *Hum. Gene Ther.* **7**, 2203–2208.

Olsen, J. C. (1998). Gene transfer vectors derived from equine infectious anemia virus. *Gene Ther.* **5**, 1481–1487.

Olsen, P., Nelson, S., and Dornburg, R. (1994). Improved self-inactivating retroviral vectors derived from spleen necrosis virus. *J. Virol.* **68**, 7060–7066.

Paillart, J. C., Berthoux, L., Ottmann, M., Darlix, J. L., Marquet, R., Ehresmann, B., and Ehresmann, C. (1996). A dual role of the putative RNA dimerization initiation site of human immunodeficiency virus type 1 in genomic RNA packaging and proviral DNA synthesis. *J. Virol.* **70**, 8348–8354.

Paillart, J.-C., Westhof, E., Ehresmann, C., Ehresmann, B., and Marquet, R. (1997). Non-canonical interactions in a kissing loop complex: the dimerization initiation site of HIV-1 genomic RNA. *J. Mol.Biol.* **270**, 36–49.

Pappalardo, L., Kerwood, D. J., Pelczer, I., and Borer, P. N. (1998). Three-dimensional folding of an RNA hairpin required for packaging HIV-1. *J. Mol. Biol.* **282**, 801–818.

Parolin, C., Dorfman, T., Palu, G., Gottlinger, H., and Sodroski, J. (1994). Analysis in human immunodeficiency virus type 1 vectors of cis-acting sequences that affect gene transfer into human lymphocytes. *J. Virol.* **68**, 3888–3895.

Poeschla, E., Corbeau, P., and Wongstaal, F. (1996). Development Of HIV Vectors For Anti-HIV Gene-Therapy. *Proc. Natl. Acad. Sci. USA* **93**, 11395–11399.

Poeschla, E., Gilbert, J., Li, X., Huang, S., Ho, A., and Wong-Stall, F. (1998a). Identification of a Human Immunodeficiency Virus Type 2 (HIV-2) encapsidation determinant and transduction of nondividing human cells by HIV-2-based lentivirus vectors. *J. Virol.* **72**, 6527–6536.

Poeschla, E. M., Wong Staal, F., and Looney, D. J. (1998b). Efficient transduction of nondividing human cells by feline immuodeficiency virus lentiviral vectors. *Nature Med.* **4**, 354–357.

Poznansky, M., Lever, A., Bergeron, L., Haseltine, W., and Sodroski, J. (1991). Gene transfer into human lymphocytes by a defective human immunodeficiency virus type 1 vector. *J. Virol.* **65**, 532–536.

Re, F., Braaten, D., Franke, E. K., and Luban, J. (1995). Human immunodeficiency virus type 1 Vpr arrests the cell cycle in G2 by inhibiting the activation of p34 cdc2-cyclin B. *J. Virol.* **69**, 6859–6864.

Rebolledo, M. A., Drogstad, P., Chen, F. H., Shannon, K. M., and Klitzner, T. S. (1998). Infection of human fetal cardiac myocytes by a human immunodeficiency virus-1-derived vector. *Circ. Res.* **83**, 738–742.

Reiser, J., Harmison, G., Kluepfel-Stahl, S., Brady, R. O., Karlsson, S., and Schubert, M. (1996). Transduction of nondividing cells using pseudotyped defective high-titer HIV type 1 particles. *Proc. Natl. Acad. Sci. USA* **93**, 15266–15271.

Richardson, J. H., Child, L. A., and Lever, A. M. (1993). Packaging of human immunodeficiency virus type 1 RNA requires cis-acting sequences outside the 5′ leader region. *J. Virol.* **67**, 3997–4005.

Richardson, J. H., Hofmann, W., Sodroski, J. G., and Marasco, W. A. (1998). Intrabody-mediated knockout of the high-affinity IL-2 receptor in primary human T cells using a bicistronic lentivirus vector. *Gene Ther.* **5**, 635–644.

Richardson, J. H., Kaye, J. F., Child, L. A., and Lever, A. M. (1995). Helper virus-free transfer of human immunodeficiency virus type 1 vectors. *J. Gen. Virol.* **76**, 691–696.

Rizvi, T. A., and Panganiban, A. T. (1993). Simian immunodeficiency virus RNA is efficiently encapsidated by human immunodeficiency virus type 1 particles. *J. Virol.* **67**, 2681–2688.

Sadaie, M. R., Zamani, M., Whang, S., Sistron, N., and Arya, S. K. (1998). Towards developing HIV-2 lentivirus-based retroviral vectors for gene therapy: dual gene expression in the context of HIV-2 LTR and Tat. *J. Med. Virol.* **54**, 118–128.

Sakaguchi, K., Zambrano, N., Baldwin, E. T., Shapiro, B. A., Erickson, J. W., Omichinski, J. G., Clore, G. M., Gronenborn, A. M., and Appella, E. (1993). Identification of a binding site for the human immunodeficiency virus type 1 nucleocapsid protein. *Proc. Natl. Acad. Sci. USA* **90**, 5219–5223.

Skripkin, E., Paillart, J. C., Marquet, R., Ehresmann, B., and Ehresmann, C. (1994). Identification of the primary site of the human immunodeficiency virus type 1 RNA dimerization *in vitro*. *Proc. Natl. Acad. Sci. USA* **91**, 4945–4949.

Spector, D. H., Wade, E., Wright, D. A., Koval, V., Clarke, C., Jaquish, D., and Spector, S. A. (1990). Human immunodeficiency virus pseudotypes with expanded cellular and species tropism. *J. Virol.* **64**, 2298–2308.

Srinivasakumar, N., Chazal, N., Helga-Maria, C., Prasad, S., Hammarskjold, M. L., and Rekosh, D. (1997). The effect of viral regulatory protein expression on gene delivery by human immunodeficiency virus type 1 vectors produced in stable packaging cell lines. *J. Virol.* **71**, 5841–5848.

Sutton, R. E., Wu, H. T. M., Rigg, R., Bohnlein, E., and Brown, P. O. (1998). Human immunodeficiency virus type 1 vectors efficiently transduce human hematopoietic stem cells. *J. Virol.* **72**, 5781–5788.

Swanstrom, R., and Wills, J. W. (1997). Synthesis, assembly and processing of viral protein. *In* "Retroviruses," pp. 263–334. Cold Spring Harbor Press, Cold Spring Harbor, NY.

Terwilliger, E. F., Godin, B., Sodroski, J. G., and Haseltine, W. A. (1989). Construction and use of a replication-competent human immunodeficiency virus (HIV-1) that expresses the chloramphenicol acetyltransferase enzyme. *Proc. Natl. Acad. Sci. USA* **86**, 3857–3861.

Uchida, N., Sutton, R. E., Friera, A. M., He, D., Reitsma, M. J., Chang, W. C., Veres, G., Scollay, R., and Weissman, I. L. (1998). HIV, but not murine leukemia virus, vectors mediate high efficiency gene transfer into freshly isolated G0/G1 human hematopoietic stem cells. *Proc. Natl. Acad. Sci. USA* **95**, 11939–11944.

Valsesia Wittmann, S., Morling, F. J., Nilson, B. H., Takeuchi, Y., Russell, S. J., and Cosset, F. L. (1996). Improvement of retroviral retargeting by using amino acid spacers between an additional binding domain and the N terminus of Moloney murine leukemia virus SU. *J. Virol.* **70**, 2059–2064.

Vile, R. G., and Russell, S. J. (1995). Retroviruses as vectors. *In* "Gene Therapy" (A. M. L. Lever and P. Goodfellow, Ed.), Vol. 51, pp. 12–30. Churchill Livingstone, London.

Yu, H., Rabson, A. B., Kaul, M., Ron, Y., and Dougherty, J. P. (1996). Inducible human immunodeficiency virus type 1 packaging cell lines. *J. Virol.* **70**, 4530–4537.

Yu, S. F., von Ruden, T., Kantoff, P. W., Garber, C., Seiberg, M., Ruther, U., Anderson, W. F., Wagner, E. F., and Gilboa, E. (1986). Self-inactivating retroviral vectors designed for transfer of whole genes into mammalian cells. *Proc. Natl. Acad. Sci. USA* **83**, 3194–3198.

Zeffman, A., Massard, S., Varani, G., and Lever, A. M. L. (2000). The major HIV-1 packaging signal is an extended bulged stem loop whose structure is ALTGRG on interaction with the Gag polyprotein. *J. Mol. Biol.* **297**, 877–893.

Zufferey, R., Dull, T., Mandel, R. J., Bukovsky, A., Quiroz, D., Naldini, L., and Trono, D. (1998). Self-inactivating lentivirus vector for safe and efficient *in vivo* gene delivery. *J. Virol.* **72**, 9873–9880.

Zufferey, R., Nagy, D., Mandel, R. J., Naldini, L., and Trono, D. (1997). Multiply attenuated lentiviral vector achieves efficient gene delivery *in vivo*. *Nature Biotech.* **15**, 871–875.

Ben Berkhout

Department of Human Retrovirology
Academic Medical Center
University of Amsterdam
1105 AZ Amsterdam, The Netherlands

Multiple Biological Roles Associated with the Repeat (R) Region of the HIV-1 RNA Genome

I. Introduction

Reverse transcription of a retroviral RNA genome produces a double-stranded DNA copy that is longer than the RNA template at both the 5′ and 3′ ends. This additional genetic information is generated in an intricate, discontinuous reverse transcription mechanism that includes two specialized strand-transfer steps of the nascent cDNA onto redundant sequence elements. It is for this reason that retroviruses encode a repeat (R) region that constitutes the extreme 5′ and 3′ end of the viral RNA genome (Fig. 1). The length of this R region varies significantly among retroviruses; it can be as short as 16 nucleotides for the mouse mammary tumor virus (MMTV) and as long as 228 nucleotides for the human T-cell leukemia virus (HTLV-I). Besides this elementary function in reverse transcription, the R region of several retroviruses encodes RNA signals that regulate other steps of the viral replication cycle. The R region forms both the extreme 5′ and 3′ end

Advances in Pharmacology, Volume 48

1054-3589/00 $35.00

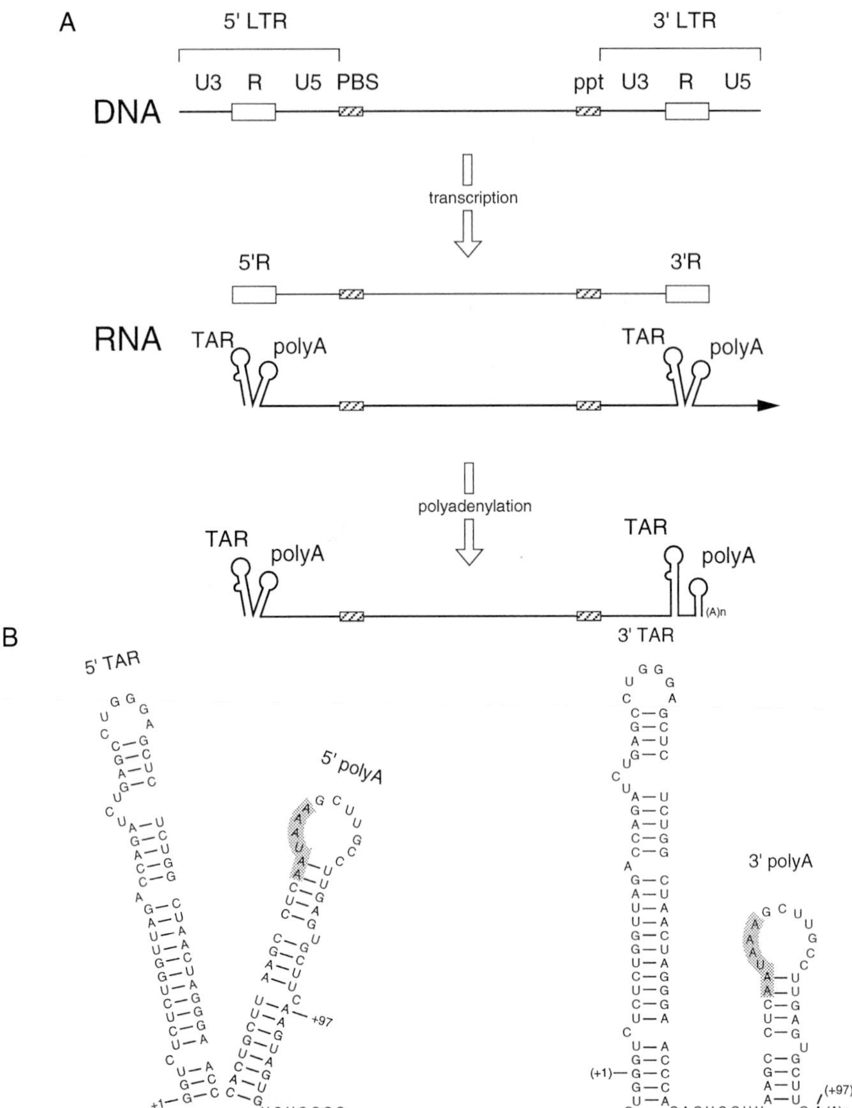

FIGURE 1 The R region of the HIV-1 RNA genome encodes a tandem hairpin motif. (A) The retroviral DNA provirus is shown with the two long terminal repeats (5′ and 3′ LTR). The primer-binding site (PBS) site flanks the 5′ LTR, and the polypurine tract (ppt) flanks the 3′ LTR. The LTRs are split in the U3, R, and U5 domains. Transcription starts at the U3–R border, and mRNA polyadenylation takes place at the R–U5 border. Thus, the extreme ends of the mature viral transcript are formed by the 5′R and 3′R elements, and both encode two adjacent stem–loop structures, the TAR and polyA hairpin. Polyadenylation within the 3′R will slightly rearrange the RNA structure (shortening of the polyA hairpin, extension of the TAR hairpin, with several single-stranded nucleotides in between the two stems, see B for details). (B) Secondary structure model of the tandem hairpin motif in 5′R and 3′R of the

of the retroviral RNA genome, and these two motifs could in theory have distinct replicative functions at either end of the genome. For instance, the 5′R region of the human immunodeficiency virus type 1 (HIV-1) encodes the trans-acting responsive (TAR) hairpin, which forms the binding site for the viral Tat protein and cellular cofactors that induce transcription of the viral promoter. Some retroviruses with an extended R region encode the polyadenylation signal within the R region, but this signal should be active exclusively in the 3′R. This necessitates differential regulation of polyadenylation to either repress the 5′R signal or to enhance the 3′R signal. Recent evidence that the RNA structure of the R region plays an important role in this regulatory mechanism is discussed. For HTLV-I, the 3′R element functions as a binding site for the viral Rex protein that regulates splicing of the viral transcript. Furthermore, this highly structured Rex responsive element (RexRE) is also involved in regulated polyadenylation. These "early" functions of RNA signals in the R region are used to control gene expression in virus-producing cells, but there is also evidence for "late" functions in virus-infected cells. The role in reverse transcription was mentioned above, but R region motifs have also been suggested to modulate the processes of RNA dimerization and packaging. In this chapter, which is not intended to be encyclopedic, the pleiotropy of functions that have been attributed to retroviral R-region elements are discussed. The particular focus is on the two R-region hairpins that border the HIV-1 genome.

The 97-nucleotide R region of the HIV-1 genome encodes the TAR and polyA hairpin (Fig. 1A). The detailed secondary structure model of this tandem hairpin motif is shown for the 5′R and 3′R context of the mature, polyadenylated HIV-1 transcript (Fig. 1B). Both the TAR and polyA hairpins are supported by computer predictions, biochemical probing experiments, comparative sequence analysis of phylogenetically distinct virus isolates, and replication studies with virus mutants. Experiments that support these RNA secondary structures are not reviewed; instead the focus is on the function(s) of molecular signals in the R region. The conformation of the two HIV-1 hairpins in 3′R is predicted to be slightly different as a result of cleavage and polyadenylation at a position 19 nucleotides downstream of the AAUAAA hexamer signal (Fig. 1, nucleotide 97 in 3′R). As a result, the polyA hairpin is shortened by eight base pairs, which allows the TAR hairpin to be extended by two base pairs. A multitude of replicative roles

---

mature, polyadenylated HIV-1 transcript. The transcription start site is marked +1, and numbers in both the 5′R and 3′R refer to this position. The AAUAAA polyadenylation signal is marked by a gray box. Polyadenylation occurs at position 97 in the 3′R, but repression of 5′ polyadenylation is not absolute, yielding a short polyadenylated transcript of 97 nucleotides with a polyA tail. In the 5′R, the two hairpins are connected without a single unpaired nucleotide, raising the possibility of coaxial stacking.

TABLE I Putative Functions of the Tandem RNA Hairpins Encoded by the HIV-1 R Region

		5′R motif		*3′R motif*	
	Replication step (section in text)	*TAR*	*PolyA*	*TAR*	*PolyA*
HIV-1 producing cell ("early")	LTR transcription (II)	X[a]			
	Polyadenylation (III)				
	Blockade 5′R signal		X		
	Activation 3′R signal			X	
	mRNA translation (IV)	X			
HIV-1 infected cell ("late")	RNA dimerization/packaging (V)	X		X	
	Reverse transcription (VI)				
	Initiation (VI,A)	X			
	Elongation (VI,B)	X	X		
	Strand transfer (VI,C)	X[b]	X[b]	X	X

[a] The TAR DNA sequence also encodes transcription motifs.
[b] The cDNA copy may also play a regulatory role.

have been proposed for these hairpins as part of either the 5′R or 3′R, and these functions are listed in Table I. It is generally thought that the transcription function of TAR is exclusive for the 5′R motif and that the AAUAAA signal is recognized solely in 3′R. However, the 3′ TAR structure may also have a function(s) in the viral life cycle, e.g., in the process of polyadenylation. Furthermore, there is recent evidence that the 5′ polyA signal is used at a low efficiency, which produces a unique viral transcript that is extremely short, but with a cap at the 5′ end and a polyA tail at the 3′ end. Many additional RNA signals that control virus replication are clustered in the untranslated leader region of the RNA genome, and several functional interactions between the 5′R signals and sequence elements that are located further downstream in the leader region are discussed. An RNA secondary structure model for the complete HIV-1 leader is shown in Fig. 2. There is significant evidence for most parts of this secondary structure

FIGURE 2 Secondary structure model of the complete HIV-1 leader RNA. Shown is the entire untranslated leader of the RNA genome of HIV-1 isolate LAI. Several, but not all of the RNA secondary structure elements of this model are supported by biochemical, phylogenetic, and/or virological experiments, but there remains a general lack of knowledge on the tertiary structure of this RNA (Berkhout, 1996). The 5′ end of the transcript (position +1) has a cap structure (m7G), and the leader extends to the AUG startcodon of the *gag* gene at position 336. The R region runs from position +1 to 97, which is the site of mRNA cleavage/polyadenylation in the 3′R context. Several important replicative signals are marked, e.g., the PBS (position 182 to 199), the hairpin constituting the dimerization initiation signal (DIS), the major splice donor (SD, marked by an arrow), and the RNA domain that forms the core packaging signal (Ψ).

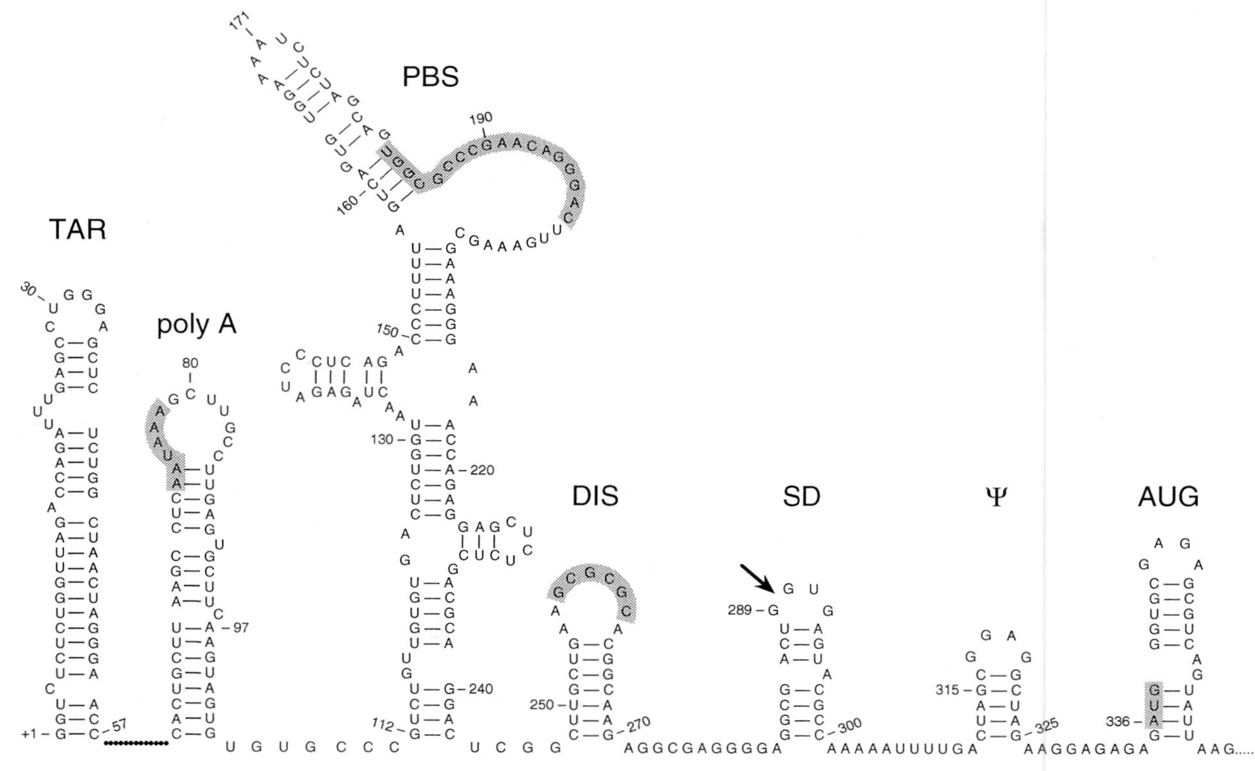

TAR
poly A
PBS
DIS
SD
Ψ
AUG

model, but there is little information on the tertiary interactions and the three-dimensional conformation of this molecule. One possibility is that the stems of the TAR and polyA hairpins stack coaxially. This idea is supported by phylogenetic evidence from viral isolates with diverse sequences that fold similar hairpins that are connected without a single nucleotide between the two stems. The possibility that the TAR and polyA hairpins interact structurally does imply that there may also be a functional interaction between these two RNA motifs, which obviously complicates the mutational dissection of these molecular signals. The putative functions of the HIV-1 R region structures in the early phase of the replication cycle are addressed, including transcription (Section II), polyadenylation (Section III), and translation (Section IV), and, subsequently, roles in the late phase, including RNA dimerization/packaging (Section V) and reverse transcription (Section VI), are discussed.

II. Transcription

The TAR RNA hairpin structure in the 5′R is important for optimal transcription from the viral promoter that is located within the 5′ long terminal repeat (LTR). However, it is also possible that the 3′ LTR of an integrated provirus is transcriptionally active in a TAR-dependent manner, thereby triggering the expression of downstream cellular genes (Klaver and Berkhout, 1994c; Raineri and Senn, 1992; Greger *et al.*, 1998). A detailed TAR phylogeny has been presented previously for all human and simian immunodeficiency virus groups (HIV and SIV) (Berkhout, 1992). For instance, viruses that belong to the HIV-2 group and several SIV groups contain a relatively complicated, branched stem–loop structure as the TAR element compared with the one-stem structure of HIV-1. An updated TAR phylogeny is presented in Fig. 3, which includes the HIV-1 subtypes of the major group M. The group M viruses that comprise the current global pandemic have diversified during their spread worldwide. These isolates have been grouped on the basis of the genomic sequences and can be divided into 10 distinct subtypes or clades termed A through J (Myers *et al.*, 1995). Isolates from different subtypes may differ by 30–40% in the amino acid sequence of the Env protein, whereas variations range from 5 to 20% within a subtype. The nucleotide changes in comparison with the TAR sequence of the prototype LAI strain of subtype B are marked by a black box in the TAR phylogeny. It is clear that the subtypes have distinctive mutations in TAR, in particular in the lower stem region. Most sequence changes represent base-pair variation (e.g., A-U to G-U) or base-pair covariation (e.g., A-U to G-C) and therefore do not disturb the TAR conformation. The TAR motifs of two other HIV-1 groups that are confined to a more restricted geographical area in Africa were included. That is the outlier group O (isolate ANT-

70) and the recently identified group N (isolate YBF-30) (Simon *et al.* 1998). The group O TAR motif shows seven mutated base pairs in the lower stem region, but this evolution can also be interpreted as deletion of three base pairs and a concomitant insertion of 3 base pairs (Berkhout, 1992). The TAR element of group N is genetically similar to that of group M viruses. In particular, group N shares several nucleotide changes with the A/G recombinant (Fig. 3). The related SIVcpz virus is included for comparison because it represents the SIV counterpart of HIV-1 (Gao *et al.*, 1999). These results demonstrate the conservation of both structural and sequence information in the TAR element.

The upper part of the TAR structure, constituting the stem with a U-rich bulge and a hexanucleotide loop, is critical for transcriptional activation. The TAR–Tat interaction is only sensitive to mutations in the bulge and surrounding base pairs, but *trans*-activation requires additional sequences in the TAR loop (Dingwall *et al.*, 1989; Berkhout and Jeang, 1989; Cullen, 1995). These results suggested the involvement of a loop-specific cellular cofactor. Insight into the role of TAR RNA in Tat binding and transcriptional activation has recently been advanced by the identification of cyclin T1 as the factor that interacts with the activation domain of Tat and that mediates loop-specific binding of Tat to TAR (Wei *et al.*, 1998; Garber *et al.*, 1998). Indeed, the Tat-associated kinase complex (TAK) that is present in nuclear extracts contains cyclin T1 as well as its partner kinase CDK9 (Yang *et al.*, 1996, 1997; Gold *et al.*, 1998; Mancebo *et al.*, 1997; Zhu *et al.*, 1997; Chun and Jeang, 1996). It has been proposed that TAR/Tat-mediated transcriptional activation occurs through phosphorylation of the carboxy-terminal domain (CTD) of RNA polymerase II by the CDK9 kinase (reviewed in Jeang, 1998). This does not exclude other mechanisms of transcriptional activation of the HIV-1 LTR promoter. For instance, there is recent evidence that Tat can recruit histone acetyltransferases to the integrated provirus to activate transcription (Benkirane *et al.*, 1998; Marzio *et al.*, 1998; Hottiger and Nabel, 1998). The mechanism of Tat/TAR-mediated LTR transcription is not discussed in further detail; instead, putative additional roles of TAR RNA, either as sequence or structured RNA motif, are the focus.

The idea that TAR may play an additional role in the viral life cycle was originally suggested in replication studies with mutant viruses (Klaver and Berkhout, 1994b; Rounseville *et al.*, 1996; Harrich *et al.*, 1995). In particular, truncated TAR motifs lacking the bottom stem region did not support efficient virus replication, but were shown to be transcriptionally competent in LTR-reporter gene assays. However, there has been some controversy concerning the transcriptional activity of truncated TAR motifs. Whereas transfection studies with LTR-CAT constructs performed in COS cells indicated that a truncated TAR is fully active in transcription, subsequent studies in a variety of other cell types, including the T-cell lines used for HIV-1 replication studies, demonstrated a significant reduction of viral

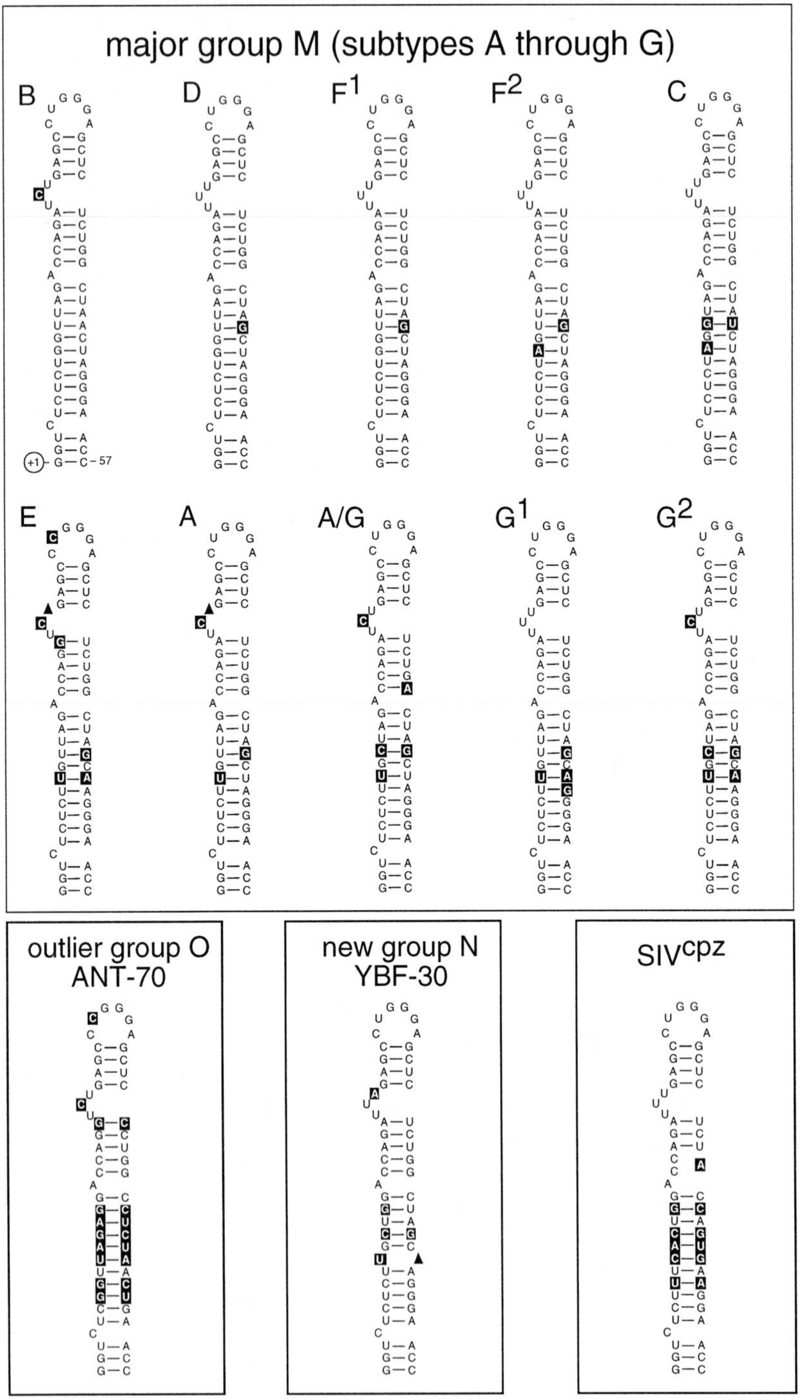

major group M (subtypes A through G)
B
D
F1
F2
C
+1
57
E
A
A/G
G1
G2
outlier group O
ANT-70
new group N
YBF-30
SIVcpz

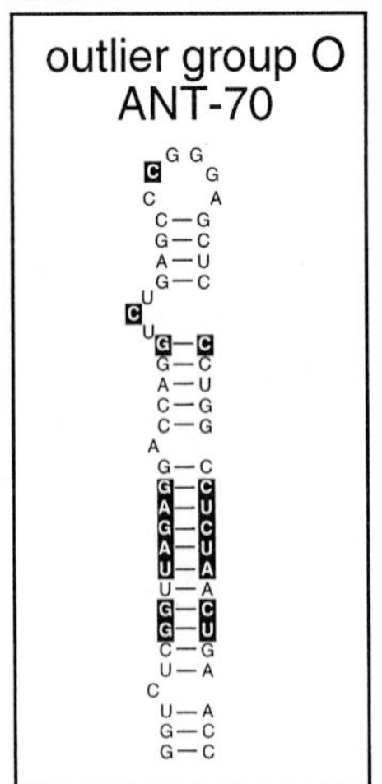

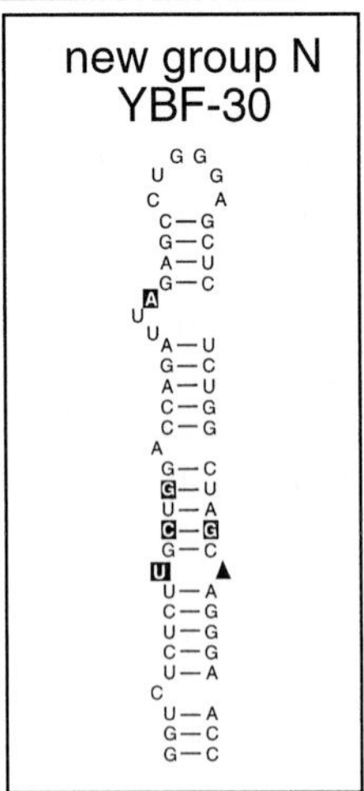

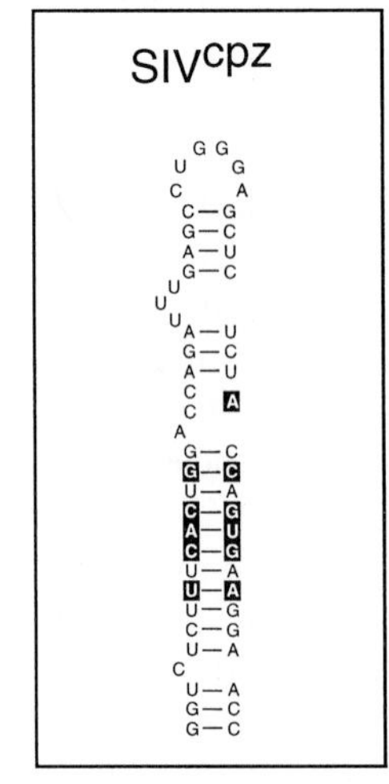

gene expression (Verhoef *et al.*, 1997b). Thus, the replication defect of viruses with a truncated TAR hairpin is caused, at least in part, by a transcriptional defect (Das *et al.*, 1998). Putative other functions of TAR are discussed in the subsequent sections, but it is essential to realize that any defect that is observed in later stages of the replication cycle may, at least partially, result from the reduced level of intracellular HIV-1 RNA.

III. Polyadenylation

At the 3′ end of the retroviral RNA genome, a polyadenylation (polyA) site is recognized by cellular enzymes to produce a polyadenylated transcript. The AAUAAA hexamer is the almost-invariant polyA signal that is positioned about 15 nucleotides upstream of the site of mRNA cleavage. A poorly conserved GU-rich element is usually positioned 20 to 30 nucleotides downstream of the hexamer (Bohnlein *et al.*, 1989; Hart *et al.*, 1985b; Gil and Proudfoot, 1987; Hart *et al.*, 1985a; Kessler *et al.*, 1986; McDevitt *et al.*, 1986; McLauchlan *et al.*, 1985). The AAUAAA hexamer binds the cleavage and polyadenylation specificity factor (CPSF), and the downstream element interacts with the cleavage stimulation factor (CstF) (Gilmartin and Nevins, 1991; Keller *et al.*, 1991; Takagaki *et al.*, 1992; MacDonald *et al.*, 1994; Takagaki *et al.*, 1997). Some retroviruses with an extended R region encode the polyA hexamer signal within the R region such that it is present at both the 5′ and 3′ end of the viral transcript. This necessitates differential regulation either to repress recognition of the 5′ polyA signal or to enhance usage of the 3′ signal.

HIV-1 encodes the polyA signal in the R region and has been reported to have both regulatory features. Usage of the 3′ polyA site is promoted by an upstream enhancer motif in the U3 region that is uniquely present at the 3′ end of viral transcripts (Valsamakis *et al.*, 1991, 1992; Brown *et al.*, 1991; DeZazzo *et al.*, 1991; Gilmartin *et al.*, 1992; Gilmartin *et al.*, 1995). This upstream sequence element (USE) appears to stabilize binding of CPSF to the AAUAAA hexamer motif (Gilmartin *et al.*, 1995). Repression of the 5′ polyA site is mediated by several mechanisms. The 5′ polyA site is acti-

FIGURE 3 Phylogeny of TAR structures in different HIV-1 subtypes. The nucleotide sequence of the LTR region of subtypes A through G of the HIV-1 major group M was used to fold the TAR hairpin structure. The TAR hairpin of subtype B isolate LAI was used as prototype, and the subtype-specific changes that are observed in the majority of sequences are indicated by a black box. Note that most subtype B viruses encode a UCU bulge, whereas the prototype LAI strain has a UUU bulge. Nucleotide deletion is marked by ▲. A/G is a recombinant virus. Two clusters within subtypes F and G were recognized. Representative TAR structures of two other HIV-1 groups (M and N) and the simian counterpart SIV^{cpz} are included in the lower part of the figure.

vated when moved further downstream in the transcript, indicating that this site is repressed because it is positioned too close to the transcription initiation site (Cherrington and Ganem, 1992; Weichs an der Glon *et al.*, 1991). A possible mechanistic explanation for this effect was recently provided by the observation that polyadenylation factors gain access to the nascent transcript through interaction with the RNA polymerase II complex (McCracken *et al.*, 1997). In this scenario, the polyadenylation factors will bind the elongating RNA polymerase II only after synthesis of the 5′R region is completed, thereby avoiding recognition of the 5′R signal. The 5′ polyA site is also negatively influenced by the major splice donor signal (SD) that is present in the downstream leader region (Fig. 2, position 289), as mutational inactivation of the SD triggered usage of the 5′ polyA site (Ashe *et al.*, 1995). This repression is mediated by binding of the U1 snRNP to the SD (Ashe *et al.*, 1997). Although it is currently unknown how the splicing machinery influences the process of polyadenylation, this example adds to the growing list of complex control circuits that affect splicing and polyadenylation (Colgan and Manley, 1997). Thus, a complex interplay of positive and negative signals control HIV-1 polyadenylation, but there is recent evidence for an additional level of complexity in that the polyA hairpin structure itself is critical for regulation.

There is compelling evidence for a role of the polyA hairpin in virus replication. Although the sequence of this part of the viral genome varies significantly among different HIV and SIV strains, all viruses can fold a similar hairpin of comparable thermodynamic stability (Berkhout *et al.*, 1995a). The phylogenetic conservation suggested a critical role for this structured RNA motif in virus replication, and this was confirmed in studies with mutant viruses. Opening of the hairpin structure in both the 5′R and 3′R did severely affect HIV-1 replication (Das *et al.*, 1997, 1999; Clever *et al.*, 1999). Through prolonged culturing of these mutants, revertant viruses with improved replication capacity were obtained. Analysis of such phenotypic revertants revealed that additional mutations had been introduced into the sequences encoding the polyA hairpin. Although different mutations were observed in individual revertants, all nucleotide changes had in common that they restored the hairpin conformation with an approximate wild-type stability (Berkhout *et al.*, 1997). These results demonstrate an absolute requirement for this structured RNA motif in virus replication. However, these results do not distinguish among a function of the 5′R or 3′R motif, and these results do not reveal the exact function of the hairpin motif.

A negative effect of stable RNA structure on the efficiency of polyadenylation was first demonstrated in transient transfection assays with reporter constructs containing the wild-type HIV-1 polyA hairpin and mutants thereof (Klasens *et al.*, 1998). Consistent with this result, sequences lacking stable secondary structure were obtained in *in vitro* evolution experiments that selected for functional variants of the HIV-1 polyA site (Graveley *et al.*, 1996a,b). Subsequently, *in vitro* protein binding studies were performed to test which step of the polyadenylation mechanism is affected (Klasens *et al.*, 1999b). Because the AAUAAA hexamer motif itself is partially occluded by base pairing (Fig. 1), a likely possibility is that binding of CPSF to this sequence motif is inhibited, thereby blocking the initial step of the polyadenylation reaction. An example of an electrophoretic mobility shift assay (EMSA) is shown in Fig. 4. Wild-type and mutant HIV-1 transcripts with polyA hairpins of various thermodynamic stabilities were tested for *in vitro* binding of polyadenylation factors, which were provided either as purified proteins or as nuclear extract (Klasens *et al.*, 1999b). The wild-type transcript and the mutant RNAs with a destabilized hairpin bound the polyadenylation factors efficiently, but mutant A with the stabilized polyA hairpin did not form the "polyA complex" (Fig. 4). Additional mutations observed in the revertant viruses (sample A2, A4, and A7) restored the stability of this mutant hairpin and thereby the binding capacity. These results suggest an inverse correlation between the stability of the polyA hairpin and its ability to interact with polyadenylation factors. These results were obtained in the presence of the USE-enhancer, but a different pattern was described without USE (Klasens *et al.*, 1999b). Whereas the wild-type HIV-1 transcript depends on this enhancer for optimal interaction with the polyadenylation factors, the destabilized polyA mutants remain fully active without enhancer. This result indicates that the wild-type RNA structure represses the polyA site and that the USE can overcome this repression in the 3′R.

The 5′ polyA hairpin was also destabilized in the context of the complete HIV-1 provirus (Das *et al.*, 1999). This mutation activated premature 5′ polyadenylation such that approximately 30–40% of the viral transcripts use the 5′ polyA site. Obviously, this effect coincided with a concomitant decline in production of full-length viral transcripts. In the 3′ context with the upstream USE-enhancer, the wild-type polyA hairpin did not interfere with efficient polyadenylation. However, 3′ polyadenylation can be inhibited by mutations that further stabilize the hairpin. These combined results suggest that the role of the polyA hairpin is to create a regulatable polyA site, which is repressed in the presence of inhibitory signals (5′R context) and activated in the presence of an enhancer (3′R context). It was proposed that the thermodynamic stability of the polyA hairpin in fine-tuned to allow this on–off switching of polyadenylation, and this modulating role is illustrated in Fig. 5. The idea that RNA structure plays a critical role in differential HIV-1 polyadenylation does not replace the existing models, but rather provides a mechanistic explanation for 5′ down-regulation and 3′ up-regulation. The HIV-1 polyA signal represents an inherently efficient poly-adenylation sequence in reporter constructs, and the role of the polyA hairpin RNA structure is to repress the efficiency of this otherwise constitutive polyA site. The thermodynamic stability of the polyA hairpin needs to be delicately

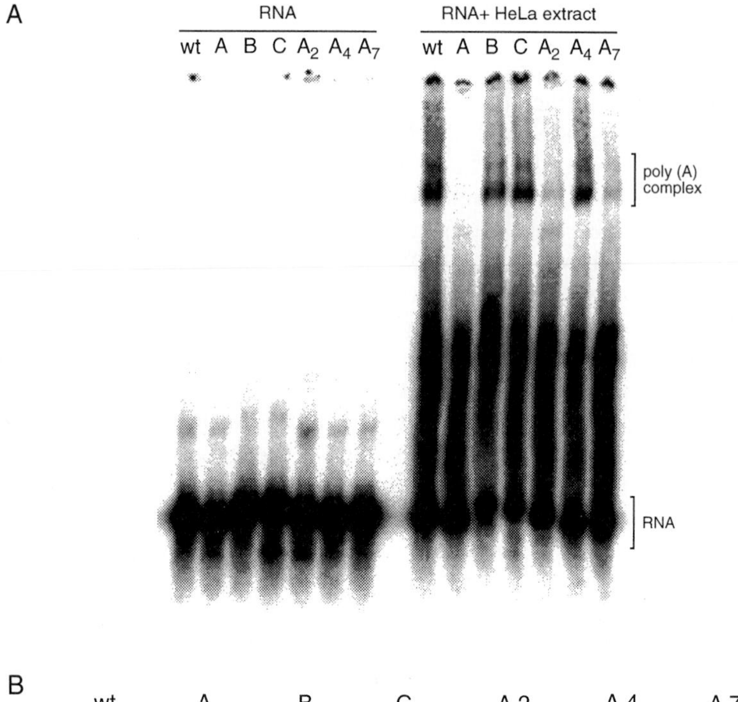

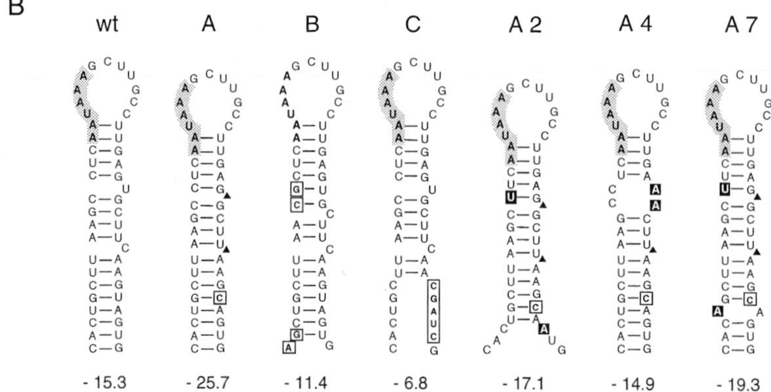

FIGURE 4 Local RNA structure occludes the AAUAAA polyadenylation signal. (A) EMSA was performed with wild-type and mutant HIV-1 transcripts and HeLa nuclear extract as source of polyadenylation factors. The HIV-1 transcripts (position −54 to +134) contain the upstream USE-enhancer, but differ in the stability of the polyA hairpin structure (see B). The mock-incubated RNA samples are included for comparison. The position of the naked RNA and the RNA–protein complex is indicated on the right. More details of this experimental system have been described recently (Klasens *et al.*, 1999b). (B) RNA structure of the wild-type polyA hairpin and several mutants/revertants. The polyA signal AAUAAA is marked by shading. The thermodynamic stability of the structures is presented (ΔG in kcal/mol). In mutant A, the hairpin is stabilized by deletion (▲) of two bulged nucleotides and one nucleotide substitution (boxed). In mutants B and C, destabilizing mutations were introduced into the left and right side of the stem, respectively. A2, A4, and A7 are revertants of mutant A, and the mutations that mediate the reversion phenotype are marked by black boxes.

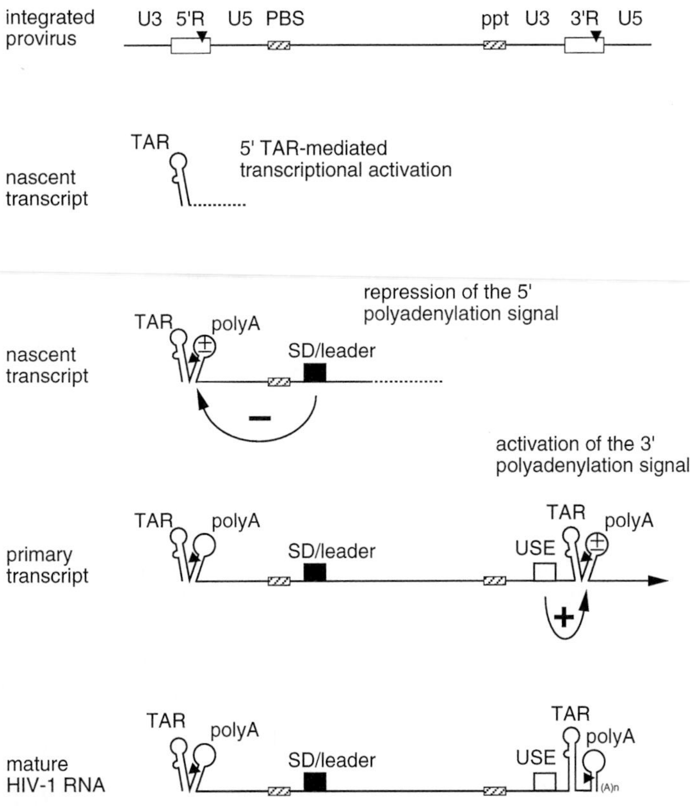

FIGURE 5 The TAR and polyA hairpins regulate viral gene expression as part of the nascent transcript. Shown are the proviral HIV-1 DNA and different phases of the nascent/growing RNA chain; the primary transcript; and the mature, 3′ polyadenylated HIV-1 RNA. The position of the AAUAAA polyadenylation signal in the R region of the DNA provirus and in the hairpin structure of the RNA transcript is indicated by a black triangle. The TAR and polyA hairpins are supposed to fold immediately after synthesis (see the text). 5′ TAR facilitates Tat-mediated transcriptional activation as part of the nascent transcript. The polyA hairpin is instrumental in repression of the 5′ polyA signal (indicated by the ± sign in the loop of the hairpin), but nearly complete repression (− sign) requires the presence of other repressive signals, including the downstream SD/leader sequences (black box). The polyA hairpin also puts the 3′ polyA site in an unfavorable context (± sign), but this deficiency is overcome (+ sign) by the USE-enhancer (open box). The USE is encoded in the U3 region at position −22 to −5, but an additional stimulatory element has been described for sequences −104 to −69 (Valsamakis *et al.*, 1991, 1992). Polyadenylation in 3′R produces the mature HIV-1 RNA, in which both the TAR and polyA hairpin are predicted to have a slightly rearranged base-pairing scheme (see Fig. 1 for details).

balanced to allow nearly complete repression of the 5′ polyA site, yet full activity of the 3′ polyA site. This may explain the apparent conservation of the stability of this structured RNA motif among virus isolates (Berkhout *et al.,* 1995a). Although a more stable hairpin in the 5′R is beneficial for virus production because 5′ polyadenylation is blocked more efficiently, this structure will be inherited after a single round of virus replication in the 3′R, where it will cause a significant replication defect by inhibing 3′ polyadenylation. These divergent regulatory pathways for the 5′R and 3′R is now discussed in further mechanistic detail.

A. Repression of the 5′ PolyA Site

Inhibition of 5′ polyadenylation requires the rapid folding of the polyA hairpin structure on the nascent viral transcript in order to occlude the AAUAAA signal and thereby delay binding of polyadenylation factors that are part of the elongating RNA polymerase II complex (Fig. 5). In fact, this enzyme complex should have advanced up to position 120 on the HIV-1 template before the nucleotides that form the base-paired stem of the polyA hairpin are extruded from the elongating enzyme, which encompasses approximately 16 nucleotides of the nascent transcript (Komissarova and Kashlev, 1998). Rapid folding seems possible because the approximate time scale for the formation of such secondary structure is in the 10^{-4}- to 10^{-5}-s range (Sclavi *et al.,* 1998; Batey and Doudna, 1998). Nevertheless, the hairpin structure will be in equilibrium with the open form ("breathing"), and the AAUAAA signal will eventually be exposed. It is likely that the delayed interaction with polyadenylation factors is sufficient for other repressive mechanisms to become operative. For instance, the growing RNA chain will fold a higher order structure that restricts the accessibility of the 5′ polyA site (Klasens *et al.,* 1999b), and binding of U1 snRNP to the major SD site in the HIV-1 leader RNA leads to suppression of the 5′ polyA site (Ashe *et al.,* 1997). Almost complete repression of the 5′ polyA site is achieved by the combination of these repressive effects, but folding of the polyA hairpin is critical in the initial phase because it buys the time that is required for the other repression mechanisms to become effective. Interestingly, the TAR hairpin is also thought to perform its transcriptional function as part of the nascent viral transcript (Berkhout *et al.,* 1989). This cotranscriptional action of the TAR and polyA hairpins in the 5′R is depicted in Fig. 5. These adjacent RNA structures may be composed of particular nucleotide sequences that allow a rapid and ordered base pairing of the two stem segments, perhaps through restricting the ability to fold a "kinetic trap" in the native folding pathway. As has been shown previously, the ribozyme sequence of hepatitis delta virus is designed to reduce the possibility of kinetic trapping of inactive RNA conformations (Perrotta *et al.,* 1999). The potential of the two hairpins to stack coaxially (Berkhout, 1996) may also influence the kinetics of folding of this RNA domain that is critical for enhancement of transcription (TAR hairpin) and repression of premature 5′ polyadenylation (polyA hairpin).

Despite this potent inhibition of the 5′ polyA site that is imposed by multiple repressive mechanisms, it was measured that approximately 5–10% of the transcripts are prematurely polyadenylated in cells transfected with the wild-type HIV-1 proviral construct (Das *et al.,* 1999). We assume that the short 5′ polyadenylated viral transcript will have a 5′ cap structure and this RNA may actually represent a stably expressed transcript that is exported to the cytoplasm. Although speculative, this short HIV-1 RNA may have a function in the infected cell. For instance, the RNA could base-pair with cellular nucleic acids or form the binding site for cellular proteins. The HIV-1 transcript resembles several cellular and viral transcript forms. First, the short HIV-1 transcript is similar to RNA polymerase III transcripts of the Y RNA family, which are 85 to 112 nucleotides in length (Farris *et al.,* 1999). Second, the HIV-1 transcript resembles the abortive, nonpolyadenylated HIV-1 transcripts of approximately 60 nucleotides that are observed in some HIV-1 LTR transcription assays without Tat protein (Kao *et al.,* 1987). These RNAs are likely to be artificial because they are expressed exclusively when LTR transcription is fired, despite the absence of Tat protein, from replicating LTR-reporter plasmids (Kao *et al.,* 1987; Jeang *et al.,* 1993). Third, the 5′ polyadenylated HIV-1 transcript does also resemble RNA decoy molecules that were previously designed to inhibit transcriptional activation of HIV-1 gene expression through binding of the Tat protein and/or cellular cofactors (Sullenger *et al.,* 1990; Sullenger *et al.,* 1991; Berkhout and van Wamel, 1995; Yamamoto *et al.,* 1997). The short HIV-1 transcript may also gain access to virion particles, in particular, because the TAR motif contributes to packaging, at least in the context of the full-length viral genome (Section V). In virions, the short R region derived transcript may play a regulatory role in reverse transcription because the R region is genuinely involved in this process. In particular, an additional 5′R molecule may inhibit the strand-transfer reaction by annealing to the (−)ssDNA product, thereby forming a "dead-end" RNA–DNA duplex (Section VI,C,2).

B. Activation of the 3′ PolyA Site

A mechanistic model for recognition of the 3′ HIV-1 polyA site is now presented that incorporates several experimental findings (Fig. 6). The presence of an inhibitory RNA conformation in the HIV-1 polyA site necessitates the presence of an enhancer near the 3′R. The USE-enhancer is positioned in the upstream U3 region and is necessary for efficient polyadenylation of HIV-1 transcripts (Valsamakis *et al.,* 1991, 1992; Brown *et al.,* 1991; DeZazzo *et al.,* 1991). Although the USE does not have a sequence that is similar to that of the AAUAAA signal, the enhancer does act through binding of CPSF (Gilmartin *et al.,* 1995). Thus, the wild-type HIV-1 RNA template may use the upstream USE to overcome the structure-imposed deficiency of

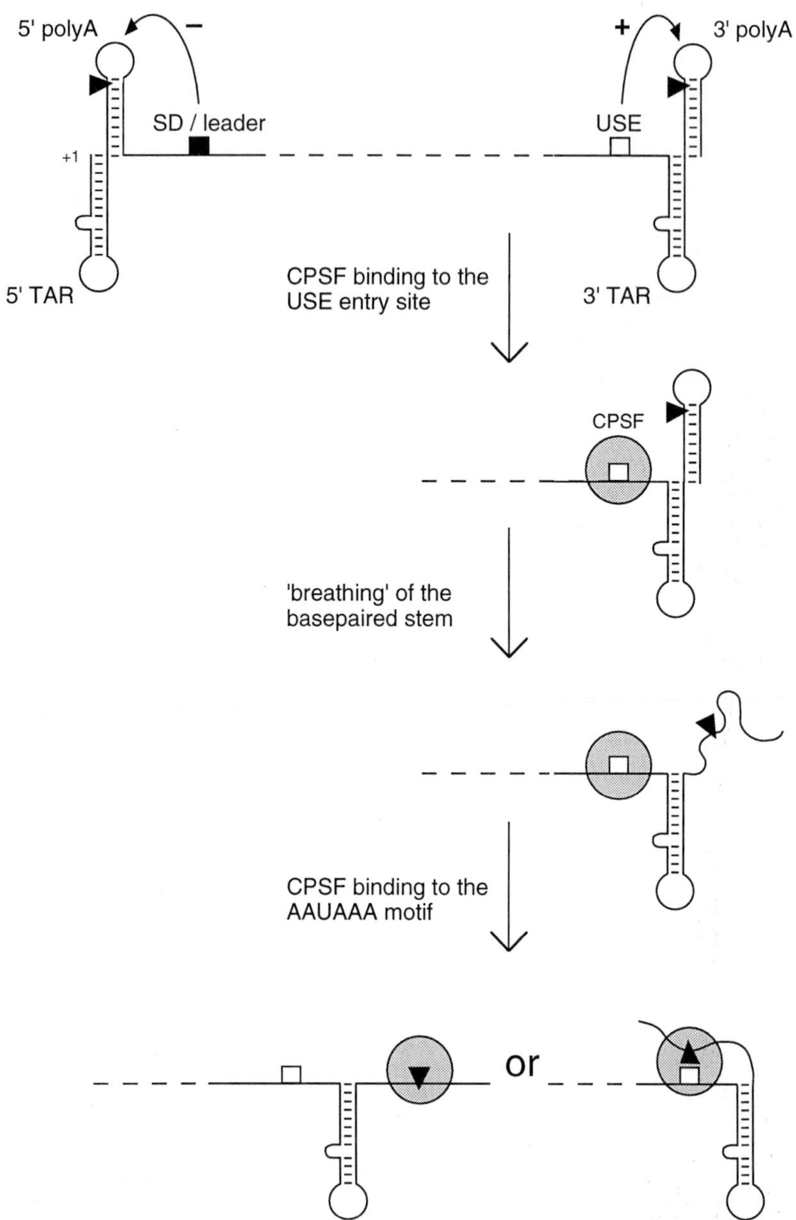

FIGURE 6 Model of activation of the 3′R polyadenylation site. The key regulatory motifs that control polyadenylation are indicated. The polyA hairpin structure that occludes the AAUAAA hexamer (black triangle) is drawn for both the 5′ and 3′ polyA site. In the 5′ setting, the repressive SD/leader elements (−sign) are marked by a black box. In the 3′ setting, the USE enhancer element (+) is shown as an open box. The USE acts as binding site or entry

CPSF-binding to the polyA signal. It has been demonstrated previously that the USE element is necessitated by the suboptimal sequence context of the HIV-1 polyA site (Gilmartin *et al.*, 1995), but these results are now interpreted in the structural context of the RNA template. For instance, RNA mutants with a destabilized hairpin do not require the USE enhancer, presumably because their AAUAAA motif is fully accessible (Klasens *et al.*, 1999b). It has been suggested that the presence of the USE may enable CPSF to identify the correct AAUAAA hexamer (Gilmartin *et al.*, 1995). The USE acts as the entry site for CPSF, which will subsequently bind the AAUAAA signal upon transient opening or "breathing" of the polyA hairpin. This mechanistic model can be modified further to include the TAR hairpin to spatially juxtapose the two CPSF-binding sites, that is, the USE and the polyA signal, as there is experimental evidence for such an accessory role of TAR (Gilmartin *et al.*, 1992; Tiley and Cullen, 1991). CPSF may exchange the USE for the AAUAAA sequence, and this substitution may be driven by a higher affinity for the latter sequence. Alternatively, CPSF may interact simultaneously with the two RNA motifs (both situations are depicted in Fig. 6). Note that an entry site is beneficial only if the polyadenylation steps subsequent to recognition of the USE are irreversible or very fast compared with the reverse reaction, which is a likely scenario (Keller, 1995).

C. RNA Structure in the PolyA Site of Other Retroviruses

A very similar regulatory mechanism is likely to be utilized by other HIV and SIV virus groups because they all encode a comparable polyA hairpin structure (Berkhout *et al.*, 1995a). However, restricted binding of CPSF to the AAUAAA signal is unlikely for HIV-2 because the signal is exposed in the single-stranded loop region (Fig. 7B). Instead, it is possible that HIV-2 polyadenylation is regulated by base pairing of the GU-rich downstream element. The combination of these two sequence elements defines a core polyA site (Fig. 7A). The polyadenylation factors CPSF and CstF bind specifically to the AAUAAA hexamer and the downstream element, respectively, but the presence of CPSF is required for efficient binding of CstF, and the interaction of CPSF with the RNA is stabilized by CstF. The stability of this RNA–CPSF–CstF ternary complex correlates with the

---

site for CPSF (gray circle). CPSF subsequently binds the AAUAAA hexamer upon transient opening or breathing of the polyA stem. The TAR hairpin is located immediately upstream of the polyA hairpin. The two hairpins may stack coaxially and perhaps TAR acts as spacer to juxtapose the USE and AAUAAA signals. Two situations are depicted; CPSF either exchanges the AAUAAA sequence for the USE sequence, or CPSF binds the two RNA sequences simultaneously.

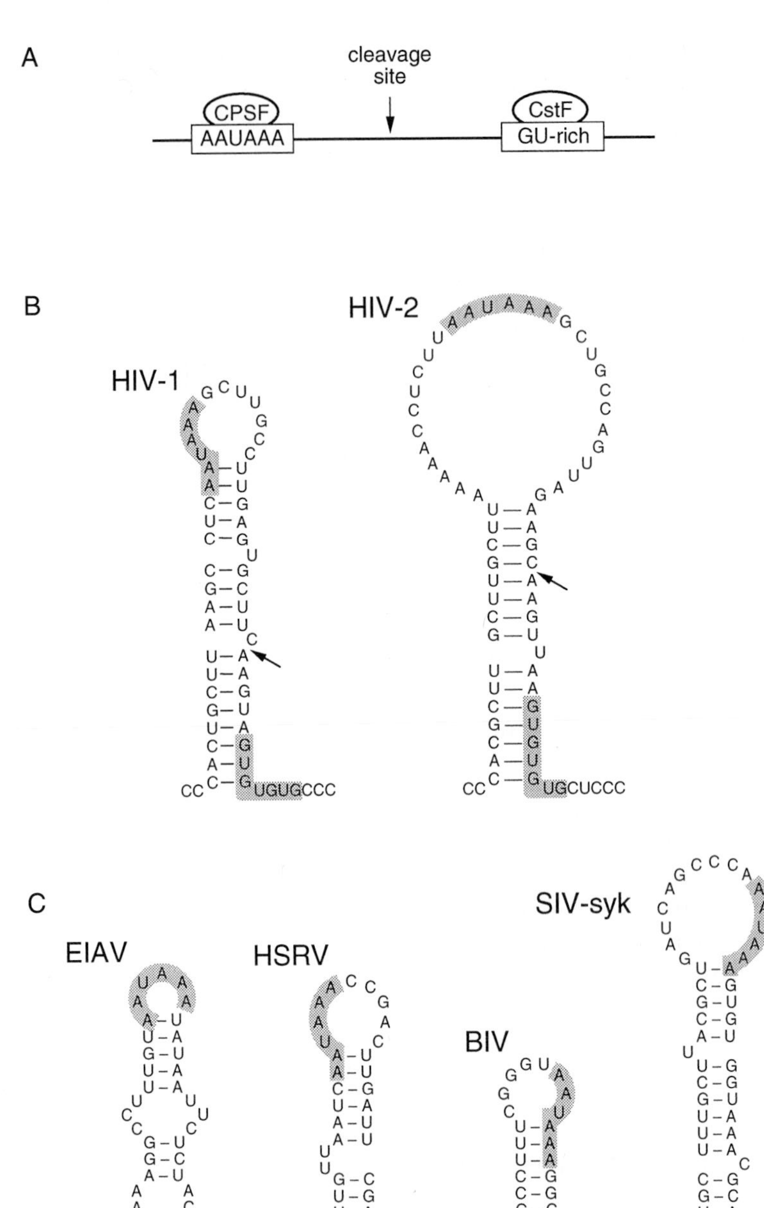
A
cleavage site
CPSF
AAUAAA
CstF
GU-rich
B
HIV-1
HIV-2
C
EIAV
HSRV
BIV
SIV-syk

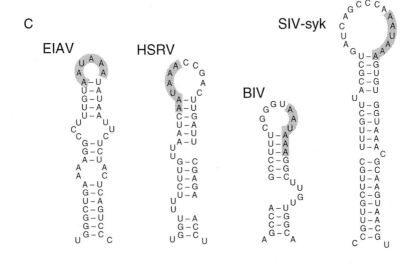

efficiency of mRNA processing, and this ensemble forms an mRNA processing unit upon association of cleavage factors and polyA polymerase (Weiss *et al.*, 1991). Thus, it is possible that polyadenylation is sensitive to base pairing of either the CPSF or CstF binding site, but the latter possibility remains to be tested with HIV-2 RNA templates.

Retroviruses that do not belong to the immunodeficiency virus group may use a similar mechanism of regulated gene expression. Likely candidates are retroviruses with a relatively extended R region that includes the AAUAAA signal. Retroviral genomes that meet these criteria were screened, and remarkably similar hairpin structures could be drawn for other retroviruses of the lentivirus and spumavirus groups (Das *et al.*, 1999). Figure 7C shows the structured polyA site of the human spumaretrovirus (HSRV) and lentiviruses such as bovine immunodeficiency virus (BIV) and the equine infectious anemia virus (EIAV). There is considerable variation in the thermodynamic stability of these retroviral RNA structures, but stability is merely one of the many parameters that may control the efficiency of these polyA sites. These variables include the AAUAAA and GU-rich signals and their accessibility, and the presence of enhancer or silencer elements. These results suggest that regulation of polyadenylation by RNA structure is widespread among the lentivirus and spumavirus groups and that this mechanism may represent a more common retroviral strategy. HTLV-I, a representative of the oncoretroviruses, also uses RNA secondary structure to regulate polyadenylation (Seiki *et al.*, 1983; Ahmed *et al.*, 1991; Bar-Shira *et al.*, 1991). In this case, a complex RNA structure is formed at the 3′ end of the viral genome that juxtaposes the AAUAAA hexamer and the actual cleavage/polyadenylation site, which are separated by 274 nucleotides in the linear sequence. These examples demonstrate the versatile use of RNA structure as a key component of regulatory circuits to control retroviral replication.

FIGURE 7 RNA structure and occlusion of the polyA signal in other lentiviruses and spumaviruses. (A) A polyA site consists of the AAUAAA signal and a downstream GU-rich element, which form the binding sites for CPSF and CstF, respectively (see text). The site of mRNA cleavage (arrow) is in between these two signals. (B) Both sequence elements are marked by a gray box in the HIV-1 and HIV-2 polyA hairpin structures. In HIV-1, both the AAUAAA signal and the GU-rich element are partially occluded by base pairing. In HIV-2, the AAUAAA signal is exposed in the single-stranded loop, but most of the GU-rich element is engaged in base pairing. The cleavage site is marked with an arrow. (C) The polyA hairpin of the SIV-syk isolate is shown, but similar stem-loop structures were predicted for all HIV and SIV variants (Berkhout *et al.*, 1995a). Other lentiviruses (EIAV and BIV) and spumaviruses (HSRV) also encode a relatively extended R region that includes the AAUAAA polyadenylation signal, and a hairpin structure that occludes part of the hexamer motif (marked by a gray box) can be proposed for these viruses (Das *et al.*, 1999).

IV. mRNA Translation

The 5′ TAR motif forms the extreme 5′ end of all HIV-1 mRNAs and has been suggested to affect translation by several means, including restriction of the 5′ cap accessibility to translation initiation factors. Because the polyA hairpin has been suggested to stack coaxially on the TAR stem, this structure is also likely to be in close approximation to the 5′ cap. In this section, experiments that have suggested a role for the 5′ TAR hairpin structure in the process of translation are briefly reviewed. Most of these results have been obtained in *in vitro* assay systems. It is well-established that *in vivo* studies with TAR-mutated constructs are complicated by the fact that any TAR mutation, even in the lower stem region, will have an effect on the transcription level (Section II). Likewise, disruption of the polyA hairpin will reduce the amount of intracellular HIV-1 RNA through activation of premature 5′ polyadenylation (Section III). Thus, it is critical that translation efficiencies are not based on the actual virus production levels (e.g., CA-p24 value in the culture supernatant) but rather are corrected for the amount of mRNA template that is available in the cytoplasm.

Although it is generally believed that the TAR element is an essential transcription motif that mediates the Tat response (Section II), there have been numerous reports of posttranscriptional effects exerted by this element. A translational component of Tat/TAR-mediated activation of HIV-1 gene expression has been reported initially (Cullen, 1986). The 5′TAR structure was also shown to interfere with mRNA translation in *Xenopus* oocytes (Braddock *et al.,* 1990, 1993) and in cell-free assays (Parkin *et al.,* 1988; SenGupta *et al.,* 1990; Viglianti *et al.,* 1992), and this repression could be overcome by the Tat protein. Two mechanistic explanations have been proposed for TAR-mediated repression of translation. First, the 5′ terminal TAR hairpin may inhibit translation in *cis* by interfering with the binding of translation initiation factors or ribosomes to the mRNA cap structure (Parkin *et al.,* 1988). Second, TAR may activate the double-stranded RNA-dependent kinase PKR (Edery *et al.,* 1989; SenGupta *et al.,* 1990; Roy *et al.,* 1991; McCormack and Samuel, 1995). The activated form of this kinase phosphorylates and thereby inactivates the translation initiation factor eIF-2, causing inhibition of translation in *trans.* The Tat protein has been reported to inhibit activation of PKR by TAR RNA (Roy *et al.,* 1990; Maitra *et al.,* 1994), and recent results obtained in experiments with a peptide Tat antagonist are consistent with the idea that Tat protein can also modulate the level of HIV-1 mRNA translation (Choudhury *et al.,* 1999). TAR RNA interacts with several cellular proteins that may regulate these mechanisms (Masuda and Harada, 1993; Rothblum *et al.,* 1995). For instance, the TAR RNA binding protein TRBP (Gatignol *et al.,* 1991) was demonstrated to interact with the PKR protein kinase (Benkirane *et al.,* 1997; Cosentino *et al.,* 1995). Another regulatory interaction is with the LA autoantigen, which alleviates translation repression imposed by the HIV-1 leader RNA (Svitkin *et al.*, 1994; Chang *et al.*, 1994).

A recent transfection study confirms that mutant HIV-1 mRNAs with a destabilized 5′ TAR structure are translated more efficiently than the wild-type transcript (Das *et al.*, 1998). Opening of the lower TAR stem reduced the intracellular RNA level, but protein production was not decreased compared with the wild-type control. A very similar pattern has been described for HIV-1 constructs with an opened 5′ polyA hairpin (Das *et al.*, 1999). Despite a reduction in the level of cellular HIV-1 RNA because of premature 5′ polyadenylation, no loss of viral protein production was measured. These combined results may indicate that both 5′R structures negatively affect the process of HIV-1 mRNA translation. Obviously, these results should be viewed with caution because it is unknown whether the mutations also affect other stages of RNA metabolism, e.g., nuclear-cytoplasmic transport, which may affect the mRNA pool. It is possible that translational repression represents a retroviral strategy to balance the RNA pools for translation and packaging. This scenario may provide the optimal amount of viral proteins and genomic RNA and therefore ultimately control the production of infectious virus. Previous findings with murine leukemia virus (MLV) suggest that full-length viral RNA is routed to either a pool for translation or a pool for packaging and that the translated RNA is not recycled in the form of genomic RNA that is packaged in virions (Levin and Rosenak, 1976). For the Rous sarcoma virus, it has been reported that this sorting mechanism is mediated by the viral Gag proteins (Sonstegard and Hackett, 1996).

V. RNA Dimerization/Packaging

These two processes are discussed jointly because the mechanisms may be intrinsically linked (Fu *et al.*, 1994), e.g., mutations in the primary RNA dimerization signal have been reported to affect the efficiency of RNA packaging (Berkhout and van Wamel, 1996; Greatorex and Lever, 1998). There is some direct evidence for an involvement of 5′R sequences in retroviral RNA dimerization. First, *in vitro* RNA dimerization studies indicated that the 5′R region with the TAR and polyA hairpins is involved in dimerization of HIV-2 transcripts (Berkhout *et al.*, 1993). Second, electron microscopic studies are consistent with the involvement of 5′R sequences in formation of the HIV-1 RNA dimer (Hoglund *et al.*, 1997). Furthermore, both the TAR and polyA hairpin have been suggested to be involved in RNA packaging because mutation of these elements reduced the virion RNA content (McBride *et al.*, 1997; McBride and Panganiban, 1996; Clever *et al.*, 1999; Das *et al.*, 1997, 1998). However, the level of intracellular HIV-1 RNA is also reduced by these mutations. Therefore, the ratio of virion RNA to intracellular HIV-1 RNA seems a better measure of the packaging efficiency than the ratio of virion RNA to virion protein. When this actual packaging efficiency was calculated, it was found that the 5′ TAR motif contributes moderately to packaging (Das *et al.*, 1998). This packaging function of TAR was shown to be independent of the Tat protein (McBride *et al.*, 1997). For mutants with a destabilized 5′ polyA hairpin, the reduced amount of virion RNA correlated perfectly with the reduction of intracellular HIV-1 RNA, which is caused by activation of the 5′ polyA site (Das *et al.*, 1999). Therefore, only the TAR hairpin remains a candidate accessory packaging signal.

Although the exact function of the R region hairpins in packaging of the viral genome remains to be determined, it is possible that these structures are part of the packaging signal that is recognized by the viral Gag protein during virion assembly (Berkowitz *et al.*, 1996; Damgaard *et al.*, 1998; De Guzman *et al.*, 1998; Clever *et al.*, 1995). Alternatively, the effect may be more indirect, as the complete leader RNA may be required for correct folding and presentation of the actual packaging signal that is supposed to be located in the downstream part of the untranslated leader RNA. With the results of a large set of mutational analyses in mind (Aldovini and Young, 1990; Clavel and Orenstein, 1990; Harrison and Lever, 1992; Clever *et al.*, 1995, 1999; Harrison and Lever, 1992; McBride and Panganiban, 1996; Richardson *et al.*, 1993; Berkhout and van Wamel, 1996; Berkowitz and Goff, 1994; McBride and Panganiban, 1997; McBride *et al.*, 1997; Clever and Parslow, 1997; Banks *et al.*, 1998), it may indeed be more appropriate to consider the entire untranslated leader region as the packaging signal, and it is likely that the complete leader region is required to fold a specific tertiary RNA structure that is recognized in the process of RNA encapsidation (Berkhout, 1996). The presence of multiple accessory packaging signals in parts of the genome that are present in subgenomic, spliced HIV-1 RNAs (e.g., 5′ TAR and 3′ TAR) can explain the relative abundance of subgenomic RNAs in viral particles, in particular in mutants with a defect in the core packaging signal (McBride and Panganiban, 1996; McBride *et al.*, 1997; Clever *et al.*, 1999). Finally, cross-species packaging has been demonstrated between HIV-1 and SIV strains, suggesting that a very similar mechanism is involved (Rizvi and Panganiban, 1993). Recent experiments indicate that RNA packaging is mechanistically different for HIV-2, in which the packaging signals are located in a more upstream part of the leader RNA (Kaye and Lever, 1998).

A modest contribution to packaging was also reported for the TAR element of the 3′R, but not for the 3′ polyA hairpin (Das *et al.*, 1998, 1999). The finding that both the 5′ and 3′ end of the viral genome contribute to packaging may suggest a functional interaction between the two ends of the RNA. The possibility of a physical interaction between the 5′ cap structure and the polyA tail cannot be excluded because such a phenomenon has been proposed for cellular transcripts to explain the effects of the polyA tail and

3′ untranslated region on translation initiation (Jackson and Standart, 1990; Beelman and Parker, 1995; Preiss and Hentze, 1998; Craig *et al.*, 1998). The intravirion structure of the RNA genome remains an enigma, but there is recent evidence for a highly ordered structure (Takasaki *et al.*, 1997). A close proximity of the 5′ and 3′ ends may be particularly advantageous for retroviral genomes to facilitate the intricate first strand-transfer step of reverse transcription (Section VI).

VI. Reverse Transcription

One obvious function of the retroviral R-region sequences is in the process of reverse transcription (Gilboa *et al.*, 1979; Telesnitsky and Goff, 1993, 1997). Reverse transcription is initiated near the 5′ end of the RNA genome at the primer-binding site (PBS, see Fig. 1). This reaction is primed by a $tRNA^{lys3}$ molecule that is bound to the PBS, and a cDNA of the 5′R region is synthesized. This intermediate is termed the minus-strand strong-stop DNA or (−)ssDNA. Through removal of the RNA template by the RNaseH domain of the elongating reverse transcriptase enzyme (RT), the nascent cDNA is released and therefore able to anneal to the 3′R region in the strand-transfer reaction. Reverse transcription can subsequently elongate over the RNA template to generate a full-length (−)strand cDNA that serves as a template for (+)strand DNA synthesis. Several steps of reverse transcription have been suggested to be either positively or negatively influenced by the structured RNA motifs in the HIV-1 R region. The 5′ TAR element has been proposed to stimulate the initiation step in which the $tRNA^{lys3}$ primer is annealed and extended (Section VI,A). Furthermore, stable structure in the template RNA can interfere with efficient elongation of the RT enzyme (Section VI,B). Most importantly, 5′R and 3′R play an essential role in reverse transcription as template (donor) for the synthesis of (−)ssDNA and as acceptor in the strand-transfer reaction (Section VI,C).

A. Initiation of Reverse Transcription

Detailed studies with mutant viruses suggested that TAR is involved in reverse transcription (Harrich *et al.*, 1996; Clever *et al.*, 1999). The exact role of TAR in reverse transcription is presently unknown, but TAR could affect this mechanism at several levels. TAR has been proposed to influence the initiation step of reverse transcription (Harrich *et al.*, 1996; Clever *et al.*, 1999), although this result could not be confirmed in another study (Das *et al.*, 1998). There is also some evidence from *in vitro* assays for the importance of HIV-1 5′R sequences for efficient initiation of reverse transcription (Arts *et al.*, 1994). Theoretically, TAR could influence binding of the $tRNA^{Lys3}$ primer onto the genomic RNA. Although this occurs at the

PBS that is located far downstream of TAR (Fig. 1), additional interactions between the tRNA-primer and upstream viral RNA sequences have been proposed for several retroviruses (Marquet *et al.,* 1995).

There are several factors that complicate a quantitative analysis with mutant virion particles to investigate the role of TAR in reverse transcription. It was mentioned in Sections II and V that TAR mutations affect the level of intracellular HIV-1 RNA and intravirion RNA. Obviously, the latter value should be measured accurately because it represents the actual template for reverse transcription. For instance, an RT-PCR protocol that does not discriminate between spliced and unspliced HIV-1 transcripts is not appropriate because spliced HIV-1 RNAs are encapsidated with high efficiency compared with nonviral RNA (Berkowitz *et al.,* 1995; Clever *et al.,* 1999). Furthermore, HIV-1 mutants that package less full-length RNA genome seem to compensate for this defect by increased packaging of spliced HIV-1 RNAs (Schwartz *et al.,* 1997; Clever *et al.,* 1999). It therefore cannot be excluded that the packaging defect of TAR-mutated transcripts is underestimated in one study (Harrich *et al.,* 1996). The same technical problem was apparent in studies with HIV-1 variants with a mutated 5′ polyA hairpin. These virions also produce less reverse transcription products, but recent evidence from several laboratories indicate that this effect is caused by the reduced level of RNA template packaged in the particles (Clever *et al.,* 1999; Das *et al.,* 1999).

Another complicating factor is that reverse transcription cannot be studied appropriately within virions that have an RNA genome with dissimilar 5′R and 3′R regions because this may profoundly affect the strand-transfer step (Berkhout *et al.,* 1995b). Such adverse effects may have hampered one of these studies (Clever *et al.,* 1999). HIV-1 mutants with identical TAR changes in both the 5′R and 3′R were tested in one study (Das *et al.,* 1998). This study measured fewer reverse-transcription products, both in infected cells and in *in vitro* assays with the RNA template that was extracted from virion particles. However, the reduced amount of cDNA products correlated accurately with the diminished amount of RNA template. Thus, no net effect of TAR on reverse transcription was apparent. It has also been suggested that the viral Tat protein plays a role in initiation of reverse transcription, and it was proposed that the function of TAR in this process is to tether this protein to the reverse transcription machinery (Huang *et al.,* 1994; Harrich *et al.,* 1997; Ulich *et al.,* 1999). Tat is able to shuttle between the nucleus and cytoplasm (Stauber and Pavlakis, 1998), suggesting that Tat has the potential to perform additional functions in the viral replication cycle. It was mentioned in Section IV that Tat enhances the translational efficiency of HIV-1 mRNAs. For a direct role in reverse transcription, the cytoplasmic Tat should also be incorporated into virions, but there is no proof for this. As there is also some evidence against a role for the Tat protein in replication steps other than activation of LTR transcription (Verhoef *et al.*, 1997a; Kim *et al.*, 1998), further experimentation is required to ascertain the proposed roles of Tat protein and TAR RNA in reverse transcription.

B. Elongation of Reverse Transcription

Reverse transcription of the retroviral genome has to proceed through some highly structured regions of the RNA template. Stable RNA structures have been reported to interfere with reverse transcription with a variety of structured RNA templates (Wu *et al.*, 1996; Pathak and Temin, 1992; Suo and Johnson, 1997a,b, 1998; Klasens *et al.*, 1999a). All these studies indicate that stable RNA structures interfere with efficient elongation of the RT enzyme, as judged by the appearance of pause cDNA products. A direct relation was apparent for the stability of template RNA structure and the extent of RT pausing, and structure-induced pausing was more pronounced at high Mg^{2+} concentrations, which is known to stabilize RNA secondary structures (Klasens *et al.*, 1999a). However, the rules of RT pausing are not likely to be simple, as particular template sequences are also known to influence elongation of reverse transcription (Klarmann *et al.*, 1993; Abbotts *et al.*, 1993; Ji *et al.*, 1994).

Not all sites of RT pausing are located precisely at the base of stem–loop structures, and stops were observed frequently approximately 6 nucleotides ahead of the RNA duplex (Harrison *et al.*, 1998; Klasens *et al.*, 1999a). Because the template/primer-bound RT enzyme covers 7 template nucleotides upstream and 22 nucleotides downstream of the cDNA extension point (Wöhrl *et al.*, 1995), this "early" stop may reflect the collision of the most frontal RT domain with the base-paired stem (Fig. 8B). These stops mimic those observed when the elongating RT enzyme is blocked by a bulky mRNA-bound ribosome in the "toeprinting" assay (Hartz *et al.*, 1988). Another study (Suo and Johnson, 1997a) reported that RNA structure-induced pausing occurs precisely at the base of the stem region, with a direct correlation of the degree of RT pausing and the free energy required for melting of the individual base pairs (Fig. 8A). Thus, some hairpins cause RT to stop ahead of the base-paired stem, whereas others apparently are able to enter the RT enzyme as an intact stem region up to the site of polymerization. The ability of RNA hairpins to penetrate into the RT enzyme may simply depend on the dimensions of the RNA structure, as illustrated in Fig. 8. For instance, early stops were observed on templates with a relatively lengthy polyA hairpin of 17 base pairs (Klasens *et al.*, 1999a).

RT pausing has also been described to occur following the copying of a template region of stable secondary structure (Harrison *et al.*, 1998). Because this part of the template RNA is expected to be melted or even degraded by RNaseH during reverse transcription, it has been suggested that this effect is mediated by the formation of a secondary structure in the nascent cDNA (Harrison *et al.*, 1998). Interestingly, there is recent evidence

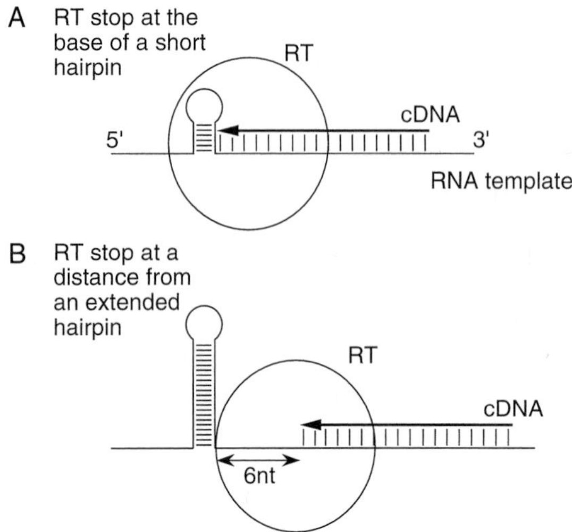

FIGURE 8 Extended RNA hairpins block the elongating RT enzyme at a distance. The elongating RT enzyme encounters either a short hairpin in the template RNA (A) or an extended hairpin (B). The situation depicted in A indicates that the small hairpin can penetrate the RT enzyme up to the catalytic site, causing termination of cDNA synthesis at the base of the stem region (Suo and Johnson, 1997a). An "early" RT stop at a six-nucleotide distance from the base-paired stem is shown in B. This situation may be specific for extended hairpin structures (Harrison *et al.,* 1998; Klasens *et al.,* 1999a).

that the natural HIV-1 (−)ssDNA molecule adopts a stable secondary structure (Jeeninga *et al.,* 1998). If cDNA structure can indeed influence the elongation properties of the RT enzyme, it is tempting to speculate on the putative role of the (−)ssDNA conformation in the strand-transfer reaction, in which the (−)ssDNA is transferred from 5′R to 3′R of the HIV-1 RNA genome (Section VI,C).

The *in vitro* studies convincingly demonstrate that the RT polymerase has problems penetrating regions of the template with stable RNA structure, but this does not necessarily mean that the same problem is encountered *in vivo.* For instance, addition of the viral nucleocapsid (NC) protein to the *in vitro* assay reduced the level of structure-induced pausing of the HIV-1 RT enzyme (Ji *et al.,* 1996; Drummond *et al.,* 1997; Klasens *et al.,* 1999a). Similar results have been reported with the MLV RT enzyme and templates with RNA structures of various stability (Wu *et al.,* 1996). The NC induces conformational changes in nucleic acids through altering energy barriers of duplex melting and annealing (Darlix *et al.,* 1995; Tsuchihashi and Brown, 1994; Herschlag, 1995). The observed resolution of pause sites by NC is consistent with the idea that NC causes RNA structures to unfold more readily and thus supports the proposed property of NC to lower the thermodynamic stability of RNA secondary structures. Alternatively, there may be a direct NC–RT interaction (Peliska *et al.*, 1994; Cameron *et al.*, 1997; Lener *et al.*, 1998), but the functional substitution of NC protein for an unrelated bacterial RNA chaperon StpA argues against this possibility (Negroni and Buc, 1999). Because of the abundance of NC in virion particles, it is unlikely that major problems are encountered during elongation of the HIV-1 R region in virus-infected cells (Klasens *et al.*, 1999a). Furthermore, there is genetic evidence that the entire 5′R is copied by RT before strand-transfer occurs (Klaver and Berkhout, 1994a; Kulpa *et al.*, 1997; Kim *et al.*, 1997). Thus, retroviruses can use relatively extended RNA structures as molecular signal in their genome without causing reverse transcription problems. Efficient reverse transcription through structured templates is facilitated in part by the NC protein and perhaps other virion cofactors, and the proposed helicase activity associated with RT may also play a role (Collett *et al.*, 1978). On the other hand, the requirement to perform efficient reverse transcription may set the upper limit of the stability of these genomic RNA signals.

C. Strand-Transfer during Reverse Transcription

Relatively little is known about the strand-transfer step of reverse transcription, but there is accumulating evidence that the base-pairing complementarity is not the only parameter that is important. Few experiments have been performed to study strand-transfer in the context of the virus-infected cells. Experiments with RNaseH-minus virus mutants indicated that incomplete removal of the 5′R template inhibits the ability of the (−)ssDNA to participate in strand-transfer (Blain and Goff, 1996; Tanese *et al.*, 1991). Excessively stable RNA structures in the 3′R have been shown to interfere with strand-transfer (Berkhout *et al.*, 1995b). Furthermore, the introduction of multiple 3′R motifs triggered transfer to all copies, indicating that the (−)ssDNA does not locate the 3′R by a directional mechanism. It is generally thought that the tertiary structure of the virion RNA genome facilitates reverse transcription and the strand-transfer step. Although there is little evidence for this idea, it has been shown recently that the dimeric state of the RNA genome is required for efficient strand-transfer (Berkhout *et al.*, 1998). For instance, the three-dimensional RNA conformation may actually juxtapose the 5′R and 3′R, which are 9 kb apart in the linear sequence. Because there is no bias for strand-transfer to occur intra- or intermolecularly (Jones *et al.*, 1994; Hu and Temin, 1990; van Wamel and Berkhout, 1998), this putative “5′–3′ communication” should not discriminate between the two template strands. Besides a putative role of the template RNA structure, there may also be protein cofactors that facilitate this critical reverse transcription step, and NC protein is a likely candidate (see below).

Most studies have addressed strand-transfer in well-controlled *in vitro* systems with a donor RNA (5′R) and an acceptor RNA (3′R) as illustrated in Fig. 9. Strand-transfer critically depends on the NC protein, except in reactions with short nucleic acid templates (Luo and Taylor, 1990; Allain *et al.,* 1994; Darlix *et al.,* 1993; Tsuchihashi and Brown, 1994; You and McHenry, 1994; Cameron *et al.,* 1997; Guo *et al.,* 1997; Peliska and Benkovic, 1992; Peliska *et al.,* 1994; Rodriquez-Rodriquez *et al.,* 1995; Negroni and Buc, 1999). Several strand-transfer mechanisms have been proposed. For instance, RT pausing has been reported to trigger recombination or strand-transfer of partially extended cDNA molecules (Pathak and Temin, 1992; DeStefano, 1994; Wu *et al.,* 1995; Kim *et al.,* 1997). It was suggested that RNaseH activity degrades the template while RT stalls at the pause site, thereby favoring the separation of the nascent cDNA strand from the template and subsequent annealing onto the acceptor RNA. Alternatively, it has been suggested that RT can interact with the acceptor strand prior to departure from the donor strand (Peliska and Benkovic, 1992). Consistent with the idea of a three-strand intermediate is the recent finding that the RT enzyme can bind a template strand in addition to the primer–template duplex (Canard *et al.,* 1997). Mechanistically different models have been proposed by others (DeStefano, 1994).

There is recent evidence that special features in the retroviral R region stimulates strand-transfer. For instance, hardly any strand-transfer was observed with templates with a randomized repeat region (Allain *et al.,* 1998). For the relatively short R region of Moloney murine leukemia virus, it was also shown that the integrity of the entire R region is essential for strand-transfer (Kulpa *et al.,* 1997). A base-pairing-independent mechanism has been proposed to facilitate the specific guidance of the (−)ssDNA to the 3′R (Topping *et al.,* 1998). These combined results strongly suggest that retroviral strand-transfer is not a simple base-pairing reaction. Given the presence of well-conserved hairpins in the R region of HIV-SIV viruses, but also in the relatively short R region of animal retroviruses such as MLV (Cupelli *et al.,* 1998), it is tempting to suggest a role for nucleic acid structure in strand-transfer. The putative roles of the HIV-1 R-region hairpins in strand-transfer (either as part of the donor RNA, the newly synthesized (−)ssDNA, or the acceptor RNA) are discussed, but the evidence that the HIV-1 (−)ssDNA adopts a specific conformation is presented first.

1. A Special Conformation of the HIV-1 (−)ssDNA

The HIV-1 (−)ssDNA migrates unusually fast on native agarose gels (Jeeninga *et al.,* 1998), a phenomenon that is generally attributed to compactness of the nucleic acid molecule. The idea that the (−)ssDNA is compactly folded is supported by the finding that the typical gel migration pattern is lost by heat- or formamide-denaturation prior to electrophoresis (Jeeninga *et al.,* 1998). Furthermore, the drug actinomycin D, which is a potent inhibitor of the strand-transfer reaction (Guo *et al.*, 1998; Davis *et al.*, 1998; Jeeninga *et al.*, 1998), shifts the (−)ssDNA from the folded to the unfolded conformation (Jeeninga *et al.*, 1998). This drug binds to the single-stranded DNA (Fig. 9) and thereby prevents both intramolecular base pairing (as in the fast-migrating conformer) and intermolecular base pairing (as in the strand-transfer reaction). A secondary structure model of the HIV-1 (−)ssDNA is presented in Fig. 10. This structure prediction was generated on the computer with a base-pairing algorithm for single-stranded DNA (SantaLucia, 1998; SantaLucia and Allawi, 1997). The HIV-1 (−)ssDNA folds two hairpin structures that are the approximate "mirror image" of the TAR and polyA hairpins encoded by the (+)strand HIV-1 RNA. These two (−)ssDNA hairpins are termed anti-TAR and anti-polyA. Differences between the RNA- and DNA-folding scheme are caused in part by the fact that GoU basepairs in the RNA (marked in the TAR and polyA hairpins) will produce C-A mismatches in the (−)ssDNA. Furthermore, the base of the two structures is altered in the (−)ssDNA because the complement of the G-triplet within TAR (marked in gray) is predicted to base-pair with the complement of the C-triplet downstream of the polyA hairpin (also marked in gray).

2. *Putative Roles of the HIV-1 (−)ssDNA in Strand-Transfer*

It is possible that the capacity of the HIV-1 (−)ssDNA to self-anneal is important for efficient release from the donor RNA template. Although one may expect that the newly synthesized (−)ssDNA will not be in a duplex with the donor RNA template because of removal of the latter molecule by the RNaseH activity of the RT enzyme, previous reports have indicated that the template RNA is not degraded completely during reverse transcription (Peliska and Benkovic, 1992; Gopalakrishnen *et al.*, 1992; DeStefano *et al.*, 1991; Fu and Taylor, 1992; Champoux, 1993; Telesnitsky and Goff, 1993; Topping *et al.*, 1998). RNaseH cleavage occurs infrequently, leaving RNA fragments of considerable length (e.g., 15 to 100 nucleotides, see Fig. 9). Many of these RNA fragments will not dissociate spontaneously from the newly synthesized (−)ssDNA, thus posing a problem for strand-transfer. Consistent with this idea, it has been suggested that removal of the donor RNA is the rate-limiting step of strand-transfer (Peliska and Benkovic, 1992). Release of the donor RNA from the (−)ssDNA may be facilitated by the proposed helicase activity associated with the RT enzyme (Collett *et al.*, 1978). Alternatively, the ability of the (−)ssDNA to self-anneal may initiate its release from the donor RNA fragments. As discussed in Section VI,B, a related mechanism may explain the stops observed following reverse transcription of a template region with stable secondary structure (Harrison *et al.*, 1998). A very similar strategy is used by some single-stranded RNA viruse to preclude the formation of stable double-stranded RNA duplexes during genome replication (van Duin, 1994; Beekwilder *et al.*, 1995). For

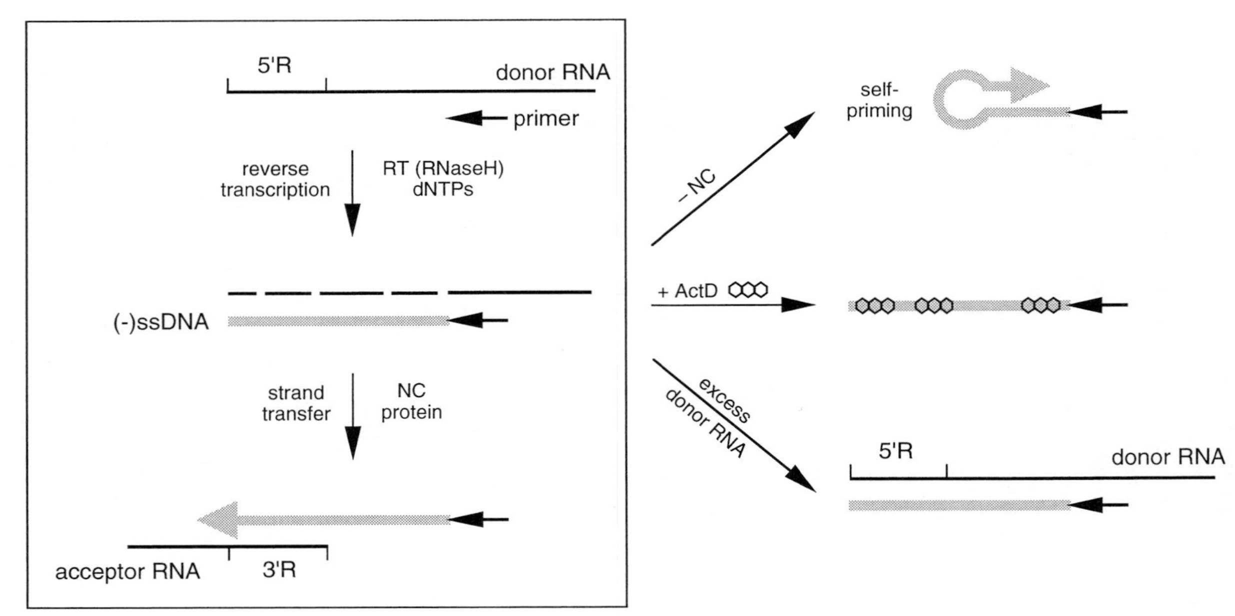
5'R
donor RNA
primer
reverse transcription
RT (RNaseH) dNTPs
(-)ssDNA
strand transfer
NC protein
acceptor RNA
3'R
– NC
self-priming
+ ActD
excess donor RNA
5'R
donor RNA

both RNA viruses and retroviruses, self-annealing of the newly synthesized strand may allow the separation of the mother and daughter strands.

Another possible function of (−)ssDNA structure is in the subsequent step of strand-transfer, in which the (−)ssDNA anneals to the acceptor RNA. It is tempting to speculate that the conformation of both the (−)ssDNA and acceptor RNA are involved in strand-transfer, and a theoretical model is presented in Fig. 10. Because the RNA and (−)ssDNA structures are nearly mirror images of each other, there is complete complementarity between the loop regions of the TAR RNA hairpin and the anti-TAR DNA structure as well as between the polyA RNA hairpin and the anti-polyA DNA structure. Thus, base-pairing interactions between the loops may represent the initial contact, a mechanism that is very similar to the "loop–loop kissing" interaction during HIV-1 RNA dimerization (Skripkin *et al.,* 1994). Subsequently, base pairs should be opened in both molecules to facilitate the formation of additional intermolecular base pairs, eventually resulting in a perfect cDNA–RNA duplex. This base-pair rearrangement is likely to be catalyzed by the nucleic-acid-chaperon activity of the NC protein (Tsuchihashi and Brown, 1994; Herschlag, 1995; Lapadat-Tapolsky *et al.,* 1995; Rein *et al.,* 1998; Darlix *et al.,* 1995). The NC is able to lower the energy barrier for breakage and reformation of base pairs, thereby catalyzing the formation of nucleic acid conformations, either intra- or inter-molecularly, with the maximal number of base pairs. In this scenario, NC protein is critical for the formation of the extended cDNA–RNA duplex, which is a critical intermediate for successful elongation of reverse transcription. Consistent with this idea, it has been shown that NC stimulates the annealing of (−)ssDNA onto the acceptor RNA (You and McHenry, 1994; Allain *et al.,* 1994; Guo *et al.,* 1997; DeStefano, 1996).

FIGURE 9 The *in vitro* strand-transfer reaction and aberrant reaction products. A schematic *in vitro* strand-transfer reaction is shown within the box. The assay is usually performed with two different RNA templates, the 5′R donor RNA (e.g., position +1/+292) and the 3′R acceptor RNA (e.g., position −54/+97, followed by a polyA tail). The R region (position +1/+97) is present in both templates. HIV-1 RT extends the primer with dNTPs up to the 5′ end of the donor RNA. The 5′R template is partially degraded by the RNaseH activity of RT, and the newly synthesized cDNA or (−)ssDNA (drawn as a thick gray line) will anneal to 3′R in a strand-transfer reaction that is catalyzed by the viral NC protein. Subsequent reverse transcription will generate an extended cDNA product. Although the dogma has been that strand-transfer is controlled solely by a base-pairing interaction between the newly synthesized (−)ssDNA and the acceptor RNA, there is accumulating evidence that other features of the nucleic acids involved play an important role (see the text for further details). Three situations that interfere with strand-transfer are shown on the right. Without NC protein, the (−)ssDNA will self-anneal and prime (+) strand cDNA synthesis. The drug actinomycin D (drawn as a three-ring structure) binds to the (−)ssDNA and thereby prevents base pairing, either with itself (as in self-priming) or with the acceptor RNA (as in strand-transfer). Excess donor RNA is particularly toxic for strand-transfer because it forms a preferential target for annealing of the (−)ssDNA, thus forming a "dead-end" duplex.

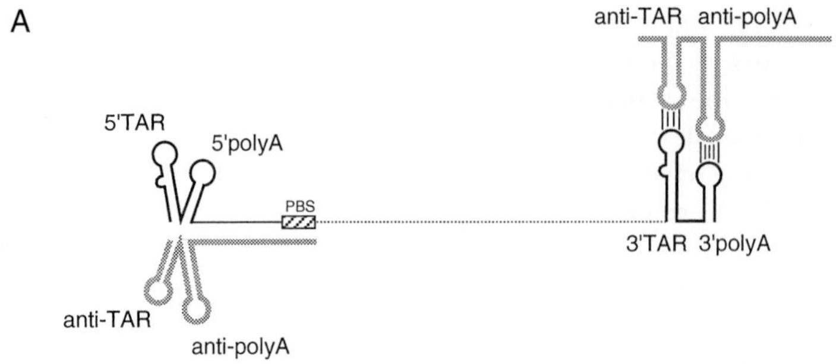

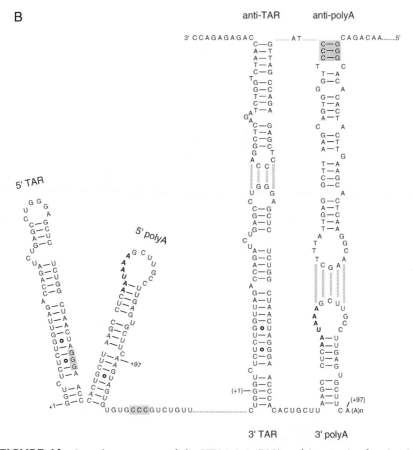

FIGURE 10 Secondary structure of the HIV-1 (−)ssDNA and its putative function in strand-transfer. (A) Shown is the HIV-1 RNA genome with the tandem hairpin in the 5′R and 3′R region. The (−)ssDNA is initiated from the PBS position and is shown as a thick gray line. This cDNA molecule is folded as the approximate mirror image of the TAR and polyA hairpins in the RNA. These DNA hairpins are termed anti-TAR and anti-polyA. The situation before and after strand-transfer is shown, with the (−)ssDNA base paired to 5′R and 3′R, respectively. It is speculated that the (−)ssDNA makes the initial contact with 3′R through

Some further insight was gained in strand-transfer reactions performed at unnatural conditions (shown in Fig. 9, right). For instance, the (−)ssDNA was shown to fold back onto itself in reactions without NC protein, resulting in self-priming and the addition of (+)strand sequences to the (−)ssDNA (Guo *et al.,* 1997). This aberrant reaction reinforces the idea that the HIV-1 (−)ssDNA is likely to self-anneal (Fig. 9). Strand-transfer can be inhibited effectively by the drug actinomycin D, which binds the (−)ssDNA and thereby precludes any base-pairing interaction (Fig. 9). Actinomycin D not only blocks the aberrant pathways of self-annealing and self-priming, but the drug also blocks the productive base pairing of the (−)ssDNA with the acceptor RNA during strand-transfer (Guo *et al.,* 1998; Davis *et al.,* 1998; Jeeninga *et al.,* 1998). Finally, it is important to appreciate that the efficiency of *in vitro* strand-transfer reactions is very low compared with that of the *in vivo* reaction in virus-infected cells, in which, for instance, no significant amount of (−)ssDNA intermediate is observed (Telesnitsky and Goff, 1997; Guo *et al.,* 1997; Cameron *et al.,* 1997; You and McHenry, 1994; Allain *et al.,* 1994; Peliska and Benkovic, 1992; Peliska *et al.,* 1994; Rodriquez-Rodriquez *et al.,* 1995; Telesnitsky and Goff, 1993; DeStefano, 1995). The inefficiency of some *in vitro* reactions is caused by the usage of an excess donor RNA, which is toxic because it is the perfect, fully complementary base-pairing partner for the (−)ssDNA (Fig. 9). Formation of such "dead-end" (−)ssDNA-donor RNA duplexes will compete effectively with the formation of productive (−)ssDNA-acceptor RNA duplexes.

VII. Perspective and Future Directions

This review deals with the pleiotropy of functions that have been attributed to overlapping RNA signals encoded by the R-region of the HIV-1 genome. Because of the superimposed demands of multiple essential viral replication steps, it is difficult to study these functions separately in experi-

---

base pairing of the exposed loop sequences. (B) Detailed model of the (−)ssDNA structure and the proposed cDNA–RNA interaction ("loop–loop kissing"). DNA structure prediction and free-energy minimization were performed with the SantaLucia parameters (SantaLucia, 1998; SantaLucia and Allawi, 1997). The non-Watson–Crick base pairs (G-U) are marked in the RNA because they will form a mismatch (C-A) in the complementary DNA. The G-triplet in TAR and the C-triplet downstream of the polyA hairpin are marked in gray. The complements of these sequence motifs are predicted to be base paired in the (−)ssDNA, thereby extending the stem region of the anti-polyA hairpin compared with that of the polyA RNA hairpin and truncating the stem of the anti-TAR hairpin compared with that of the TAR RNA hairpin. The thermodynamic stability of the RNA and DNA hairpin structures is as follows: TAR ($\Delta G = -24.8$ kcal/mol), polyA ($\Delta G = -15.3$ kcal/mol), anti-TAR ($\Delta G = -9.0$ kcal/mol), and anti-polyA ($\Delta G = -10.5$ kcal/mol).

mental infections with mutant viruses. Several examples were presented in which "late" replication defects are in fact an indirect consequence of an "early" defect. For instance, reduced LTR transcription (5′ TAR mutant) or increased 5′ polyadenylation (5′ polyA mutants) will reduce the level of intracellular HIV-1 RNA, and this has an influence on the amount of RNA that is available for translation and encapsidation into virions. Another way to experimentally dissect the overlapping RNA signals and their multitude of functions is in well-defined *in vitro* assays that mimic a single step of the replication cycle. However, some of these simplified systems do not accurately mimic the *in vivo* context in virus-infected cells.

Despite these experimental limitations, it seems safe to conclude that the R-region encodes multiple essential RNA signals. The tandem hairpin motif controls viral gene expression. In particular, the TAR hairpin is involved in 5′ LTR transcription, and the polyA hairpin inhibits premature polyadenylation in the 3′R. Furthermore, the 5′R and 3′R play an intrinsic role in strand-transfer during reverse transcription. R-region sequences and/or structures have been suggested to play other roles in virus replication, but these functions need to be verified by further experimentation. The proposed roles range from simple mechanisms, such as protection of the HIV-1 RNA against cellular exonucleases by the tandem hairpin motif, to elaborate mechanisms, such as the role of RNA and (−)ssDNA structures in the strand-transfer reaction (Fig. 10). I hope that speculations I have indulged in at some points will provoke experiments that are needed to test some of these fascinating possibilities.

Although there is fairly good evidence for the secondary structure of RNA signals within the R-region, it should be emphasized that very little is known about the actual three-dimensional folding of these signals. Because there is accumulating evidence for functional interactions between different replicative signals on the untranslated leader RNA, it is likely that there are such three-dimensional interactions. For instance, the functional and molecular interaction of splicing and polyadenylation signals was discussed (Section III), and the influence of several leader RNA regions on the process of initiation of reverse transcription was reviewed (Section VI,A). Such interactions are apparently not restricted to sequences that are encoded by the untranslated leader region. For instance, we speculated on a possible interaction between the 5′R and 3′R ends of the viral genome, which may facilitate the first strand-transfer during reverse transcription. All these possibilities remain to be worked out in a variety of experimental systems, but it seems of utmost importance that structural information on the tertiary structure of the leader RNA becomes available through biophysical studies. Eventually, one would like to comprehend the conformation of the condensed RNA genome within virion particles. This picture includes the NC protein that coats the viral RNA, and the putative structural effects that are imposed by the PBS-annealed tRNA primer and the dimeric conformation of the viral RNA genome. Elucidation of these molecular structures and how they are involved in regulated viral replication may be the object of future studies. This will not only present us with novel molecular mechanisms of viral gene expression, but may also pave the way for recognition of similar mechanisms in cellular biology. For instance, the recent developments in the field of nuclear export of RNA transcripts once again demonstrates how much basic knowledge can be derived from the study of viral model systems. The detailed molecular understanding of retroviral replication may also provide a basis for the development of novel antivirals.

Acknowledgments

The author is particularly grateful to people in the laboratory, past and present, whose work is summarized here: Nancy Beerens, Atze Das, Hendrik Huthoff, Rienk Jeeninga, Bianca Klasens, Bep Klaver, Maike Thiesen, Koen Verhoef, and Jeroen van Wamel. I thank Rienk Jeeninga and Bianca Klasens for reading the manuscript, Atze Das for thoughtful comments that helped to improve the manuscript, and Wim van Est for professional artwork. I am indebted to Anders Virtanen, Andrew Lever, Alexander Gultyaev, Jan van Duin, Kuan-Teh Jeang, Jean-Luc Darlix, Judith Levin, and Louis Henderson for promoting this work through encouragement, scientific discussions, and generous sharing of materials. The review covers information available up to January, 1999. The limited scope of the review and the recent widespread activity in this field do not allow us to cite all the references. We do apologize to those whose work we may have inadvertently failed to mention. Research in the Berkhout laboratory is supported by the Dutch Organization for Scientific Research (NWO) and the Dutch AIDS Fund (AIDS Fonds).

References

Abbotts, J., Bebenek, K., Kunkel, T. A., and Wilson, S. H. (1993). Mechanism of HIV-1 reverse transcriptase: Termination of processive synthesis on a natural DNA template is influenced by the sequence of the template-primer stem. *J. Biol. Chem.* **268,** 10312–10323.

Ahmed, Y. F., Gilmartin, G. M., Hanly, S. M., Nevins, J. R., and Greene, W. C. (1991). The HTLV-I Rex response element mediates a novel form of mRNA polyadenylation. *Cell* **64,** 727–737.

Aldovini, A., and Young, R. A. (1990). Mutations of RNA and protein sequences involved in human immunodeficiency virus type 1 packaging results in production of noninfectious virus. *J. Virol.* **64,** 1920–1926.

Allain, B., Lapadat-Tapolsky, M., Berlioz, C., and Darlix, J.-L. (1994). Trans-activation of the minus-strand DNA transfer by nucleocapsid protein during reverse transcription of the retroviral genome. *EMBO J.* **13,** 973–981.

Allain, B., Rascle, J.-B., De Rocquigny, H., Roques, B., and Darlix, J.-L. (1998). cis Elements and trans-acting factors required for minus-strand DNA transfer during reverse transcription of the genomic RNA of murine leukemia virus. *J. Virol.* **72,** 225–235.

Arts, E. J., Li, X., Gu, Z., Kleiman, L., Parniak, M. A., and Wainberg, M. A. (1994). Comparison of deoxyoligonucleotide and $tRNA^{Lys3}$ as primers in an endogenous human immunodeficiency virus-1 *in vitro* reverse transcription/template-switching reaction. *J. Biol. Chem.* **269,** 14672–14680.

Ashe, M. P., Griffin, P., James, W., and Proudfoot, N. J. (1995). Poly (A) site selection in the HIV-1 provirus: Inhibition of promoter-proximal polyadenylation by the downstream major splice donor site. *Genes Dev.* **9,** 3008–3025.

Ashe, M. P., Pearson, L. H., and Proudfoot, N. J. (1997). The HIV-1 5′ LTR poly (A) site is inactivated by U1 snRNP interaction with the downstream major splice donor site. *EMBO J.* **16,** 5752–5763.

Banks, J. D., Yeo, A., Green, K., Cepeda, F., and Linial, M. L. (1998). A minimal avian retroviral packaging sequence has a complex structure. *J. Virol.* **72,** 6190–6194.

Bar-Shira, A., Panet, A., and Honigman, A. (1991). An RNA secondary structure juxtaposes two remote genetic signals for human T-cell leukemia virus type 1 RNA 3′-end processing. *J. Virol.* **65,** 5165–5173.

Batey, R. T., and Doudna, J. A. (1998). The parallel universe of RNA folding. *Nat. Struct. Biol.* **5,** 337–340.

Beekwilder, M. J., Nieuwenhuizen, R., and van Duin, J. (1995). Secondary structure model for the last two domains of single-stranded RNA phage Qβ. *J. Mol. Biol.* **247,** 903–917.

Beelman, C. A., and Parker, R. (1995). Degradation of mRNA in eukaryotes. *Cell* **81,** 179–183.

Benkirane, M., Neuveut, C., Chun, R. F., Smith, S. M., Samuel, C. E., Gatignol, A., and Jeang, K.-T. (1997). Oncogenic potential of TAR RNA binding protein TRBP and its regulatory interaction with RNA-dependent protein kinase PKR. *EMBO J.* **16,** 611–624.

Benkirane, M., Chun, R. F., Xiao, H., Ogryzko, V. V., Howard, B., Nakatani, Y., and Jeang, K.-T.(1998). Activation of integrated provirus requires histone acetyltransferase. p300 and P/CAF are coactivators for HIV-1 Tat. *J. Biol. Chem.* **273,** 24898–24905.

Berkhout, B. (1992). Structural features in TAR RNA of human and simian immunodeficiency viruses: A phylogenetic analysis. *Nucleic Acids Res.* **20,** 27–31.

Berkhout, B. (1996). Structure and function of the Human Immunodeficiency Virus leader RNA. *Progr. Nucl. Acid. Res. Mol. Biol.* **54,** 1–34.

Berkhout, B., and Jeang, K. T. (1989). Trans activation of human immunodeficiency virus type 1 is sequence specific for both the single-stranded bulge and loop of the trans-acting-responsive hairpin: a quantitative analysis. *J. Virol.* **63,** 5501–5504.

Berkhout, B., and van Wamel, J. L. (1995). Inhibition of human immunodeficiency virus expression by sense transcripts encoding the retroviral leader RNA. *Antiviral Res.* **26,** 101–115.

Berkhout, B., and van Wamel, J. L. B. (1996). Role of the DIS hairpin in replication of human immunodeficiency virus type 1. *J. Virol.* **70,** 6723–6732.

Berkhout, B., Das, A. T., and van Wamel, J. L. B. (1998). The native structure of the HIV-1 RNA genome is required for the first strand-transfer of reverse transcription. *Virology* **249,** 211–218.

Berkhout, B., Klaver, B., and Das, A. T. (1995a). A conserved hairpin structure predicted for the poly (A) signal of human and simian immunodeficiency viruses. *Virology* **207,** 276–281.

Berkhout, B., van Wamel, J., and Klaver, B. (1995b). Requirements for DNA strand transfer during reverse transcription in mutant HIV-1 virions. *J. Mol. Biol.* **252,** 59–69.

Berkhout, B., Klaver, B., and Das, A. T. (1997). Forced evolution of a regulatory RNA helix in the HIV-1 genome. *Nucleic Acids Res.* **25,** 940–947.

Berkhout, B., Oude Essink, B. B., and Schoneveld, I. (1993). *In vitro* dimerization of HIV-2 leader RNA in the absence of PuGGAPuA motifs. *FASEB J.* **7,** 181–187.

Berkhout, B., Silverman, R. H., and Jeang, K. T. (1989). Tat trans-activates the human immunodeficiency virus through a nascent RNA target. *Cell* **59,** 273–282.

Berkowitz, R. D., and Goff, S. P. (1994). Analysis of binding elements in the human immunodeficiency virus type 1 genomic RNA and nucleocapsid protein. *Virology* **202,** 233–246.

Berkowitz, R., Fisher, J., and Goff, S. P. (1996). RNA packaging. *Curr. Top. Microbiol. Immunol.* **214,** 177–218.

Berkowitz, R. D., Ohagen, A., Hoglund, S., and Goff, S. P. (1995). Retroviral nucleocapsid domains mediate the specific recognition of genomic viral RNAs by chimeric Gag polyproteins during RNA packaging *in vivo*. *J. Virol.* **69,** 6445–6456.

Blain, S. W., and Goff, S. P. (1996). Effects on DNA synthesis and translocation caused by mutations in the RNase H domain of Moloney murine leukemia virus reverse transcriptase. *J. Virol.* **69**, 4440–4452.

Bohnlein, S., Hauber, J., and Cullen, B. R. (1989). Identification of a U5-specific sequence required for efficient polyadenylation within the human immunodeficiency virus long terminal repeat. *J. Virol.* **63**, 421–424.

Braddock, M., Thorburn, A. M., Chambers, A., Elliot, G. D., Anderson, G. J., Kingsman A. J., and Kingsman, S. M. (1990). A nuclear translational block imposed by the HIV-1 U3 region is relieved by the Tat–TAR interaction. *Cell* **62**, 1123–1133.

Braddock, M., Powell, R., Blanchard, A. D., Kingsman, A. J., and Kingsman, S. M. (1993). HIV-1 TAR RNA-binding proteins control TAT activation of translation in Xenopus oocytes. *FASEB J.* **7**, 214–222.

Brown, P. H., Tiley, L. S., and Cullen, B. R. (1991). Efficient polyadenylation within the human immunodeficiency virus type 1 long terminal repeat requires flanking U3-specific sequences. *J. Virol.* **65**, 3340–3343.

Cameron, C. E., Ghosh, M., LeGrice, S. F. J., and Benkovic, S. J. (1997). Mutations in HIV reverse transcriptase which alter RNaseH activity and decrease strand transfer efficiency are suppressed by HIV nucleocapsid protein. *Proc. Natl. Acad. Sci. USA* **94**, 6700–6705.

Canard, B., Sarfati, R., and Richardson, C. C. (1997). Binding of RNA template to a complex of HIV-1 reverse transcriptase/primer/template. *Proc. Natl. Acad. Sci. USA* **94**, 11279–11284.

Champoux, J. J. (1993). "Reverse Transcriptase" (A. M. Skalka and S. P. Goff, Eds.), pp. 103–117. Cold Spring Harbor Laboratory Press, Cold Spring Harbor, New York.

Chang, Y. N., Kenan, D. J., Keene, J. D., Gatignol, A., and Jeang, K.-T. (1994). Direct interactions between autoantigen La and human immunodeficiency virus leader RNA. *J. Virol.* **68**, 7008–7020.

Cherrington, J., and Genem, D. (1992). Regulation of polyadenylation in human immunodeficiency virus (HIV): Contributions of promoter proximity and upstream sequences. *EMBO J.* **11**, 1513–1524.

Choudhury, I., Wang, J., Stein, S., Rabson, A., and Leibowitz, M. J. (1999). Translational effects of peptide antagonists of Tat protein of human immunodeficiency virus type 1. *J. Gen. Virol.* **80**, 777–782.

Chun, R. F., and Jeang, K.-T. (1996). Requirements for RNA polymerase II carboxy-terminal domain for activated transcription of human retroviruses human T-cell lymphotropic virus I and HIV-1. *J. Biol. Chem.* **271**, 27888–27894.

Clavel, F., and Orenstein, J. M. (1990). A mutant of human immunodeficiency virus with reduced RNA packaging and abnormal particle morphology. *J. Virol.* **64**, 5230–5234.

Clever, J. L., and Parslow, T. G. (1997). Mutant human immunodeficiency virus type 1 genomes with defects in RNA dimerization or encapsidation. *J. Virol.* **71**, 3407–3414.

Clever, J. L., Eckstein, D. A., and Parslow, T. G. (1999). Genetic dissociation of the encapsidation and reverse transcription functions in the 5′R region of human immunodeficiency virus type 1. *J. Virol.* **73**, 101–109.

Clever, J. L., Sassetti, C., and Parslow, T. G. (1995). RNA secondary structure and binding sites for gag gene products in the 5′ packaging signal of human immunodeficiency virus type 1. *J. Virol.* **69**, 2101–2109.

Colgan, D. F., and Manley, J. L. (1997). Mechanism and regulation of mRNA polyadenylation. *Genes Dev.* **11**, 2755–2766.

Collett, M. S., Leis, J. P., Smith, M. S., and Faras, A. J. (1978). Unwinding-like activity associated with avaian retrovirus RNA-directed DNA polymerase. *J. Virol.* **26**, 498–509.

Cosentino, G. P., Venkatesan, S., Serluca, F. C., Green, S. R., Mathews, M. B., and Sonenberg, N. (1995). Double-stranded-RNA-dependent protein kinase and TAR RNA-binding protein form homo- and heterodimers *in vivo*. *Proc. Natl. Acad. Sci. USA* **92**, 9445–9449.

Craig, A. W. B., Haghighat, A., Yu, A. T. K., and Sonenberg, N. (1998). Interaction of polyadenylate-binding protein with the eIF4G homologue PAIP enhances translation. *Nature* **392,** 520–523.

Cullen, B. R. (1986). Trans-activation of human immunodeficiency virus occurs via a bimodal mechanism. *Cell* **46,** 973–982.

Cullen, B. R. (1995). Regulation of HIV gene expression. *AIDS* **9,** S19–S32.

Cupelli, L., Okenquist, S. A., Trubetskoy, A., and Lenz, J. (1998). The secondary structure of the R region of a murine leukemia virus is important for stimulation of long terminal repeat driven gene expression. *J. Virol.* **72,** 7807–7814.

Damgaard, C. K., Dyhr-Mikkelsen, H., and Kjems, J. (1998). Mapping the RNA binding sites for human immunodeficiency virus type-1 Gag and NC proteins within the complete HIV-1 and -2 untranslated leader regions. *Nucleic Acids Res.* **26,** 3667–3676.

Darlix, J.-L., Vincent, A., Gabus, C., De Rocquigny, H., and Roques, B. (1993). Trans-activation of the 5′ to 3′ viral DNA strand transfer by nucleocapsid protein during reverse transcription of HIV-1 RNA. *Genetics* **316,** 763–771.

Darlix, J.-L., Lapadat-Tapolsky, M., De Rocquigny, H., and Roques, B. P. (1995). First glimpses at structure–function relationships of the nucleocapsid protein of retroviruses. *J. Mol. Biol.* **254,** 523–537.

Das, A. T., Klaver, B., Klasens, B. I. F., van Wamel, J. L. B., and Berkhout, B. (1997). A conserved hairpin motif in the R-U5 region of the human immunodeficiency virus type 1 RNA genome is essential for replication. *J. Virol.* **71,** 2346–2356.

Das, A. T., Klaver, B., and Berkhout, B. (1998). The 5′ and 3′ TAR elements of the human immunodeficiency virus exert effects at several points in the virus life cycle. *J. Virol.* **72,** 9217–9223.

Das, A. T., Klaver, B., and Berkhout, B. (1999). A hairpin structure in the R region of the Human Immunodeficiency Virus type 1 RNA genome is instrumental in polyadenylation site selection. *J. Virol.* **73,** 81–91.

Davis, W. R., Gabbara, S., Hupe, D., and Peliska, J. A. (1998). Actinomycin D inhibition of DNA strand transfer reactions catalyzed by HIV-1 reverse transcriptase and nucleocapsid protein. *Biochemistry* **37,** 14213–14221.

De Guzman, R. N., Rong Wu, Z., Stalling, C. C., Pappalardo, L., Borer, P. N., and Summers, M. F. (1998). Structure of the HIV-1 nucleocapsid protein bound to the SL3 psi-RNA recognition element. *Science* **279,** 384–388.

DeStefano, J. J. (1994). Kinetic analysis of strand transfer from internal regions of heteropolymeric RNA templates by human immunodeficiency virus reverse transcriptase. *J. Mol. Biol.* **243,** 558–567.

DeStefano, J. J. (1995). Human immunodeficiency virus nucleocapsid protein stimulates strand transfer from internal regions of heteropolymeric RNA templates. *Arch. Virol.* **140,** 1775–1789.

DeStefano, J. J. (1996). Interaction of Human Immunodeficiency Virus Nucleocapsid protein with a structure mimicking a replication intermediate. *J. Biol. Chem.* **271,** 16350–16356.

DeStefano, J. J., Buiser, R. G., Mallaber, L. M., Myers, T. W., Bambara, R. A., and Fay, P. J. (1991). Polymerization and RNase H activities of the reverse transcriptases of avian myeloblastosis, human immunodeficiency, and Moloney murine leukemia viruses are functionally uncoupled. *J. Biol. Chem.* **266,** 7423–7431.

DeZazzo, J. D., Kilpatrick, J. E., and Imperiale, M. J. (1991). Involvement of long terminal repeat U3 sequences overlapping the transcription control region in human immunodeficiency virus type 1 mRNA 3′ end formation. *Mol. Cell Biol.* **11,** 1624–1630.

Dingwall, C., Ernberg, I., Gait, M. J., Green, S. M., Heaphy, S., Karn, J., Lowe, A. D., Singh, M., Skinner, M. A., and Valerio, R. (1989). Human Immunodeficiency Virus 1 tat protein binds trans-activating-responsive region (TAR) RNA *in vitro. Proc. Natl. Acad. Sci. USA* **86,** 6925–6929.

Drummond, J. E., Mounts, P., Gorelick, R. J., CasasFinet, J. R., Bosche, W. J., Henderson, L. E., Waters, D. J., and Arthur, L. O. (1997). Wild-type and mutant HIV type 1 nucleocapsid proteins increase the proportion of long cDNA transcripts by viral reverse transcriptase. *AIDS Res. Hum. Retroviruses* **13**, 533–543.

Edery, I. R., Petryshyn, R., and Sonenberg, N. (1989). Activation of double-stranded RNA dependent kinase (dsI) by the TAR region of HIV-1 mRNA: A novel translational control mechanism. *Cell* **56**, 303–312.

Farris, A. D., Koelsch, G., Pruijn, G. J. M., van Venrooij, W. J., and Harley, J. B. (1999). Conserved features of Y RNAs revealed by automated phylogenetic secondary structure analysis. *Nucleic Acids Res.* **27**, 1070–1078.

Fu, T. B., and Taylor, J. (1992). Wshen retroviral Reverse Transcriptase reach the end of their RNA templates. *J. Virol.* **66**, 4271–4278.

Fu, W., Gorelick, R. J., and Rein, A. (1994). Characterization of human immunodeficiency virus type 1 dimeric RNA from wild-type and protease-defective virions. *J. Virol.* **68**, 5013–5018.

Gao, F., Bailes, E., Robertson, D. L., Chen, Y., Rodenburg, C. M., Michael, S. F., Cummins, L. B., Arthur, L. O., Peeters, M., Shaw, G. M., Sharp, P. M., and Hahn, B. H. (1999). Origin of HIV-1 in the chimpanzee Pan troglodytes troglodytes. *Nature* **397**, 436–441.

Garber, M. E., Wei, P., KewalRamani, V. N., Mayall, T. P., Herrmann, C. H., Rice, A. P., Littman, D. R., and Jones, K. A. (1998). The interaction between HIV-1 Tat and human cyclin T1 requires zinc and a critical cysteine residue that is not conserved in the murine CycT1 protein. *Genes Dev.* **12**, 3512–3527.

Gatignol, A., Buckler-White, A., Berkhout, B., and Jeang, K. T. (1991). Characterization of a human TAR RNA-binding protein that activates the HIV-1 LTR. *Science* **251**, 1597–1600.

Gil, A., and Proudfoot, N. J. (1987). Position-dependent sequence elements downstream of AAUAAA are required for efficient rabbit β-Globin mRNA 3′ end formation. *Cell* **49**, 399–406.

Gilboa, E., Mitra, S. W., Goff, S., and Baltimore, D. (1979). A detailed model of reverse transcription and tests of crucial aspects. *Cell* **18**, 93–100.

Gilmartin, G. M., and Nevins, J. R. (1991). Molecular analysis of two poly(A) site-processing factors that determine the recognition and efficiency of cleavage of the pre-mRNA. *Mol. Cell Biol.* **11**, 2432–2438.

Gilmartin, G. M., Fleming, E. S., and Oetjen, J. (1992). Activation of HIV-1 pre-mRNA 3′ processing in vitro requires both an upstream element and TAR. *EMBO J.* **11**, 4419–4428.

Gilmartin, G. M., Fleming, E. S., Oetjen, J., and Graveley, B. R. (1995). CPSF recognition of an HIV-1 mRNA 3′-processing enhancer: multiple sequence contacts involved in poly(A) site definition. *Genes Dev.* **9**, 72–83.

Gold, M. O., Yang, X., Herrmann, C. H., and Rice, A. P. (1998). PITALRE, the catalytic subunit of TAK, is required for human immunodeficiency virus Tat transactivation *in vivo*. *J. Virol.* **72**, 4448–4453.

Gopalakrishnan, V., Peliska, J. A., and Benkovic, S. J. (1992). Human immunodeficiency virus type 1 reverse transcriptase: Spatial and temporal relationship between the polymerase and RNase H activities. *Proc. Natl. Acad. Sci. USA* **89**, 10763–10767.

Graveley, B. R., Fleming, E. S., and Gilmartin, G. M. (1996a). Restoration of both structure and function to a defective poly(A) site by *in vitro* selection. *J. Biol. Chem.* **271**, 33654–33663.

Graveley, B. R., Fleming, E. S., and Gilmartin, G. M. (1996b). RNA structure is a critical determinant of poly(A) site recognition by cleavage and polyadenylation specificity factor. *Mol. Cell Biol.* **16**, 4942–4951.

Greatorex, J. S., and Lever, A. M. L. (1998). Retroviral RNA dimer linkage. *J. Gen. Virol.* **79**, 2877–2882.

Greger, I. H., Demarchi, F., Giacca, M., and Proudfoot, N. J. (1998). Transcriptional interference perturbs the binding of Sp1 to the HIV-1 promoter. *Nucleic Acids Res.* **26**, 1294–1300.

Guo, J., Henderson, L. E., Bess, J., Kane, B., and Levin, J. G. (1997). Human immunodeficiency virus type 1 nucleocapsid protein promotes efficient strand transfer and specific viral DNA synthesis by inhibiting TAR-dependent self-priming from minus-strand strong-stop DNA. *J. Virol.* **71,** 5178–5188.

Guo, J., Wu, T., Bess, J., Henderson, L. E., and Levin, J. G. (1998). Actinomycin D inhibits Human Immunodeficiency Virus type 1 minus-strand transfer in *in vitro* and endogenous reverse transcription assays. *J. Virol.* **72,** 6716–6724.

Harrich, D., Mavankal, G., Mette-Snider, A., and Gaynor, R. B. (1995). Human immunodeficiency virus type 1 TAR element revertant viruses define RNA structures required for efficient viral gene expression and replication. *J. Virol.* **69,** 4906–4913.

Harrich, D., Ulich, C., and Gaynor, R. B. (1996). A critical role for the TAR element in promoting efficient human immunodeficiency virus type 1 reverse transcription. *J. Virol.* **70,** 4017–4027.

Harrich, D., Ulich, C., Garcia-Martinez, L. F., and Gaynor, R. B. (1997). Tat is required for efficient HIV-1 reverse transcription. *EMBO J.* **16,** 1224–1235.

Harrison, G. P., and Lever, A. M. L. (1992). The human immunodeficiency virus type 1 packaging signal and major splice donor region have a conserved stable secondary structure. *J. Virol.* **66,** 4144–4153.

Harrison, G. P., Mayo, M. S., Hunter, E., and Lever, A. M. L. (1998). Pausing of reverse transcriptase on retroviral RNA templates is influenced by secondary structures both 5′ and 3′ of the catalytic site. *Nucleic Acids Res.* **26,** 3433–3442.

Hart, R. P., McDevitt, M. A., Ali, H., and Nevins, J. R. (1985a). Definition of essential sequences and functional equivalence of elements downstream of the Adenovirus E2A and the early Simian Virus 40 polyadenylation sites. *Mol. Cell. Biol.* **5,** 2975–2983.

Hart, R. P., McDevitt, M. A., and Nevins, J. R. (1985b). Poly(A) site cleavage in a HeLa nuclear extract is dependent on downstream sequences. *Cell* **43,** 677–683.

Hartz, D., McPheeters, D. S., Traut, R., and Gold, L. (1988). Extension inhibition analysis of translation initiation complexes. *Methods Enzymol.* **164,** 419–425.

Herschlag, D. (1995). RNA chaperones and the RNA folding problem. *J. Biol. Chem.* **270,** 20871–20874.

Hoglund, S., Ohagen, A., Goncalves, J., Panganiban, A. T., and Gabuzda, D. (1997). Ultrastructure of HIV-1 genomic RNA. *Virology* **233,** 271–279.

Hottiger, M. O., and Nabel, G. J. (1998). Interaction of the human immunodeficiency virus type 1 Tat with the transcriptional coactivators p300 and CREB binding protein. *J. Virol.* **72,** 8252–8256.

Huang, L. M., Joshi, A., Willey, R., Orenstein, J., and Jeang, K. T. (1994). Human immunodeficiency viruses regulated by alternative trans-activators: Genetic evidence for a novel non-transcriptional function of Tat in virion infectivity. *EMBO J.* **13,** 2886–2896.

Hu, W.-S., and Temin, H. M. (1990). Retroviral recombination and reverse transcription. *Science* **250,** 1277–1233.

Jackson, R. J., and Standart, N. (1990). Do the poly(A) tail and 3′ untranslated region control mRNA translation? *Cell* **62,** 15–24.

Jeang, K.-T. (1998). Tat, Tat-associated kinase, and transcription. *J. Biomed. Sci.* **5,** 24–27.

Jeang, K. T., Berkhout, B., and Dropulic, B. (1993). Effects of integration and replication on transcription of the HIV-1 long terminal repeat. *J. Biol. Chem.* **268,** 24940–24949.

Jeeninga, R. E., Huthoff, H. T., Gultyaev, A. P., and Berkhout, B. (1998). The mechanism of Actinomycin D-mediated inhibition of HIV-1 reverse transcription. *Nucleic Acids Res.* **26,** 5472–5479.

Ji, J., Hoffmann, J. S., and Loeb, L. (1994). Mutagenicity and pausing of HIV reverse transcriptase during HIV plus-strand DNA synthesis. *Nucleic Acids Res.* **22,** 47–52.

Ji, X., Klarmann, G. J., and Preston, B. D. (1996). Effect of human immunodeficiency virus type 1 (HIV-1) nucleocapsid protein on HIV-1 reverse transcriptase activity *in vitro. Biochem.* **35,** 132–143.

Jones, J. S., Allan, R. W., and Temin, H. M. (1994). One retroviral RNA is sufficient for synthesis of viral DNA. *J. Virol.* **68,** 207–216.

Kao, S.-Y., Calman, A. F., Luciw, P. A., and Peterlin, B. M. (1987). Antitermination of transcription within the long terminal repeat of HIV-1 by tat gene product. *Nature* **330,** 489–493.

Kaye, J. F., and Lever, A. M. L. (1998). Nonreciprocal packaging of human immunodeficiency virus type 1 and type 2 RNA: a possible role for the p2 domain of Gag in RNA encapsidation. *J. Virol.* **72,** 5877–5855.

Keller, W., Bienroth, S., Lang, K. M., and Christofori, G. (1991). Cleavage and polyadenylation factor CPSF specifically interacts with the pre-mRNA 3′ processing signal AAUAAA. *EMBO J.* **10,** 4249

Keller, W. (1995). No end yet to messenger RNA 3′ processing! *Cell* **81,** 829–832.

Kessler, M. M., Beckendorf, R. C., Westhafer, M. A., and Nordstrom, J. L. (1986). Requirement of AAUAAA and adjacent downstream sequences for SV40 early polyadenylation. *Nucleic Acids Res.* **14,** 4939–4952.

Kim, J. K., Palaniappan, C., Wu, W., Fay, P., and Bambara, R. A. (1997). Evidence for a unique mechanism of strand transfer from the transactivation region of HIV-1. *J. Biol. Chem.* **272,** 16769–16777.

Kim, V. N., Mitrophanous, K., Kingsman, S. M., and Kingsman, A. J. (1998). Minimal requirements for a lentivirus vector based on human immunodeficiency virus type 1. *J. Virol.* **72,** 811–816.

Klarmann, G. J., Schauber, C. A., and Preston, B. D. (1993). Template-directed pausing of DNA synthesis by HIV-1 reverse transcriptase during polymerization of HIV-1 sequences *in vitro*. *J. Biol. Chem.* **268,** 9793–9802.

Klasens, B. I. F., Das, A. T., and Berkhout, B. (1998). Inhibition of polyadenylation by stable RNA secondary structure. *Nucleic Acids Res.* **26,** 1870–1876.

Klasens, B. I. F., Huthoff, H. T., Das, A. T., Jeeninga, R. E., and Berkhout, B. (1999a). The effect of template RNA structure on elongation by HIV-1 reverse transcriptase. *Biochim. Biophys. Acta,* **1444,** 355–370.

Klasens, B. I. F., Thiesen, M., Virtanen, A., and Berkhout, B. (1999b). The ability of the HIV-1 AAUAAA signal to bind polyadenylation factors is controlled by local RNA structure. *Nucleic Acids Res.* **27,** 446–454.

Klaver, B., and Berkhout, B. (1994a). Premature strand transfer by the HIV-1 reverse transcriptase during strong-stop DNA synthesis. *Nucleic Acids Res.* **22,** 137–144.

Klaver, B., and Berkhout, B. (1994b). Evolution of a disrupted TAR RNA hairpin structure in the HIV-1 virus. *EMBO J.* **13,** 2650–2659.

Klaver, B., and Berkhout, B. (1994c). Comparison of 5′ and 3′ long terminal repeat promoter function in human immunodeficiency virus. *J. Virol.* **68,** 3830–3840.

Komissarova, N., and Kashlev, M. (1998). Functional topography of nascent RNA in elongation intermediates of RNA polymerase. *Proc. Natl. Acad. Sci. USA* **95,** 14699–14704.

Kulpa, D., Topping, R., and Telesnitsky, A. (1997). Determination of the site of first strand transfer during Moloney murine leukemia virus reverse transcription and identification of strand transfer-associated reverse transcriptase errors. *EMBO J.* **16,** 856–865.

Lapadat-Tapolsky, M., Pernelle, C., Borie, C., and Darlix, J.-L. (1995). Analysis of the nucleic acid annealing activities of nucleocapsid protein from HIV-1. *Nucleic Acids Res.* **23,** 2434–2441.

Lener, D., Tanchou, V., Roques, B. P., Le Grice, S. F. J., and Darlix, J.-L. (1998). Involvement of HIV-1 nucleocapsid protein in the recruitment of reverse transcriptase into nucleoprotein complexes formed *in vitro*. *J. Biol. Chem.* **273,** 33781–33786.

Levin, J. G., and Rosenak, M. J. (1976). Synthesis of murine leukemia virus proteins associated with virions assembled in actinomycin-D-treated cells: Evidence for persistence of viral messenger RNA. *Proc. Natl. Acad. Sci. USA* **73,** 1154–1158.

Luo, G., and Taylor, J. (1990). Template switching by reverse transcriptase during DNA synthesis. *J. Virol.* **64,** 4321–4328.

MacDonald, C., Wilusz, J., and Shenk, T. (1994). The 64-kilodalton subunit of the CstF polyadenylation factor binds to premRNAs downstream of the cleavage site and influences cleavage site location. *Mol. Cell Biol.* **14,** 6647–6654.

Maitra, R. K., McMillan, N. A. J., Desai, S., McSwiggen, J., Hovanessian, A. G., Sen, G., Williams, B. R. G., and Silverman, R. H. (1994). HIV-1 TAR RNA has an intrinsic ability to activate interferon-inducible enzymes. *Virology* **204,** 823–827.

Mancebo, H. S. Y., Lee, G., Flygare, J., Tomassini, J., Luu, P., Zhu, Y., Peng, J., Blau, C., Hazuda, D., Price, D., and Flores, O. (1997). P-TEFb kinase is required for HIV Tat transcriptional activation *in vivo* and *in vitro*. *Genes Dev.* **11,** 2633–2644.

Marquet, R., Isel, C., Ehresmann, C., and Ehresmann, B. (1995). tRNAs as primer of reverse transcriptases. *Biochimie* **77,** 113–124.

Marzio, G., Tyagi, M., Gutierrez, M. I., and Giacca, M. (1998). HIV-1 Tat transactivator recruits p300 and CREB-binding protein histone acetyltransferases to the viral promoter. *Proc. Natl. Acad. Sci. USA* **95,** 13519–13524.

Masuda, T., and Harada, S. (1993). Modulation of host cell nuclear proteins that bind to HIV-1 trans-activation-responsive element RNA by phorbol ester. *Virology* **192,** 696–700.

McBride, M. S., and Panganiban, A. T. (1996). The human immunodeficiency virus type 1 encapsidation site is a multipartite RNA element composed of functional hairpin structures. *J. Virol.* **70,** 2963–2973.

McBride, M. S., and Panganiban, A. T. (1997). Position dependence of functional hairpins important for human immunodeficiency virus type 1 RNA encapsidation *in vivo*. *J. Virol.* **71,** 2050–2058.

McBride, M. S., Schwartz, M. D., and Panganiban, A. T. (1997). Efficient encapsidation of human immunodeficiency virus type 1 vectors and further characterization of cis elements required for encapsidation. *J. Virol.* **71,** 4544–4554.

McCormack, S. J., and Samuel, C. E. (1995). Mechanism of interferon action: RNA-binding of full-length and R-domain forms of the RNA-dependent protein kinase PKR-determination of Kd values for VAi and TAR RNAs. *Virology* **206,** 511–519.

McCracken, S., Fong, N., Yankulov, K., Ballantyne, S., Pan, G., Greenblatt, J., Patterson, S. D., Wickens, M., and Bentley, D. L. (1997). The C-terminal domain of RNA polymerase II couples mRNA processing to transcription. *Nature* **385,** 357–361.

McDevitt, M. A., Hart, R. P., Wong, W. W., and Nevins, J. R. (1986). Sequences capable of restoring poly(A) site function define two distinct downstream elements. *EMBO J.* **5,** 2907–2913.

McLauchlan, J., Gaffney, D., Whitton, J. L., and Clements, J. B. (1985). The consensus sequence YGTGTTYY located downstream from the AATAAA signal is required for efficient formation of mRNA 3′ termini. *Nucleic Acids Res.* **13,** 1347–1368.

Myers, G., Korber, B., Hahn, B. H., Jeang, K.-T., Mellors, J. H., McCutchan, F. E., Henderson, L. E., and Pavlakis, G. N. (1995). AnonymousTheoretical Biology and Biophysics Group, Los Alamos National Laboratory, Los Alamos, New Mexico.

Negroni, M., and Buc, H. (1999). Recombination during reverse transcription: An evaluation of the role of the nucleocapsid protein. *J. Mol. Biol.* **286,** 15–31.

Parkin, N. T., Cohen, E. A., Darveau, A., Rosen, C., Haseltine, W., and Sonenberg, N. (1988). Mutational analysis of the 5′ noncoding region of human immunodeficiency virus type 1: Effects of secondary structure. *EMBO J.* **7,** 2831–2837.

Pathak, V. K., and Temin, H. M. (1992). 5-Azacytidine and RNA secondary structure increase the retrovirus mutation rate. *J. Virol.* **66,** 3093–3100.

Peliska, J. A., and Benkovic, S. J. (1992). Mechanism of DNA strand transfer reactions catalyzed by HIV-1 reverse transcriptase. *Science* **258,** 1112–1118.

Peliska, J. A., Balasubramanian, S., Giedroc, D. P., and Benkovic, S. J. (1994). Recombinant HIV-1 nucleocapsid protein accelerates HIV-1 reverse transcriptase catalyzed DNA strand transfer reactions and modulates RNaseH activity. *Biochemistry* **33,** 13817–13823.

Perrotta, A. T., Nikiforova, O., and Been, M. D. (1999). A conserved bulged adenosine in a peripheral duplex of the antigenomic HDV self-cleaving RNA reduces kinetic trapping of inactive conformations. *Nucleic Acids Res.* **27,** 795–802.

Preiss, T., and Hentze, M. W. (1998). Dual function of the messenger RNA cap structure in poly(A)-tail-promoted translation in yeast. *Nature* **392,** 516–520.

Raineri, I., and Senn, H.-P. (1992). HIV-1 promoter insertion revealed by selective detection of chimeric provirus-host gene transcripts. *Nucleic Acids Res.* **20,** 6261–6266.

Rein, A., Henderson, L. E., and Levin, J. G. (1998). Nucleic-acid-chaperone activity of retroviral nucleocapsid proteins: Significance for viral replication. *Trends Biochem. Sci.* **23,** 297–301.

Richardson, J. H., Child, L. A., and Lever, A. M. L. (1993). Packaging of human immunodeficiency virus type 1 RNA requires cis-acting sequences outside the 5′leader region. *J. Virol.* **67,** 3997–4005.

Rizvi, T. A., and Panganiban, A. T. (1993). Simian immunodeficiency virus RNA is efficiently encapsidated by human immunodeficiency virus type 1 particles. *J. Virol.* **67,** 2681–2688.

Rodriquez-Rodriquez, L., Tsuchihashi, Z., Fuentes, G. M., Bambara, R. A., and Fay, P. J. (1995). Influence of Human Immunodeficiency Virus nucleocapsid protein on synthesis and strand transfer by the reverse transcriptase in vitro. *J. Biol. Chem.* **270,** 15005–15011.

Rothblum, C. J., Jackman, J., Mikovits, J., Shukla, R. R., and Kumar, A. (1995). Interaction of nuclear protein p140 with human immunodeficiency virus type 1 TAR RNA in mitogen-activated primary human T lymphocytes. *J. Virol.* **69,** 5156–5163.

Rounseville, M. P., Lin, H. C., Agbottah, E., Shukla, R. R., Rabson, A. B., and Kumar, A. (1996). Inhibition of HIV-1 replication in viral mutants with altered TAR RNA stem structures. *Virology* **216,** 411–417.

Roy, S., Agy, M., Hovanessian, A. G., Sonenberg, N., and Katze, M. G. (1991). The integrity of the stem structure of human immunodeficiency virus type 1 Tat-responsive sequence of RNA is required for interaction with the interferon-induced 68,000-Mr protein kinase. *J. Virol.* **65,** 632–640.

Roy, S., Katze, M. G., Parkin, N. T., Edery, I., Hovanessian, A. G., and Sonenberg, N. (1990). Control of the interferon-induced 68-kilodalton protein kinase by the HIV-1 tat gene product. *Science* **247,** 1216–1219.

SantaLucia, J. J. (1998). A unified view of polymer, dumbbell, and oligonucleotide DNA nearest-neighbor thermodynamics. *Proc. Natl. Acad. Sci. USA* **95,** 1460–1465.

SantaLucia, J. J., and Allawi, H. T. (1997). Thermodynamics and NMR of internal GT mismatches in DNA. *Biochemistry* **36,** 10581–10594.

Schwartz, M. D., Fiore, D., and Panganiban, A. T. (1997). Distinct functions and requirements for the Cys-His boxes of the human immunodeficiency virus type 1 nucleocapsid protein during RNA encapsidation and replication. *J. Virol.* **71,** 9295–9305.

Sclavi, B., Sullivan, M., Chance, M. R., Brenowitz, M., and Woodson, S. A. (1998). RNA folding at millisecond intervals by synchrotron hydroxyl radical footprinting. *Science* **279,** 1940–1943.

Seiki, M., Hattori, S., Hirayama, Y., and Yoshida, M. (1983). Human adult T-cell leukemia virus: Complete nucleotide sequence of the provirus genome integrated in leukemia cell DNA. *Proc. Natl. Acad. Sci. USA* **80,** 3618–3622.

SenGupta, D. N., Berkhout, B., Gatignol, A., Zhou, A. M., and Silverman, R. H. (1990). Direct evidence for translational regulation by leader RNA and Tat protein of human immunodeficiency virus type 1. *Proc. Natl. Acad. Sci. USA* **87,** 7492–7496.

Simon, F., Mauclere, P., Roques, P., Loussert-Ajaka, I., Muller-Trutwin, M. C., Saragosti, S., Georges-Courbot, M. C., Barre-Sinoussi, F., and Brun-Vezinet, F. (1998). Identification of a new human immunodeficiency virus type 1 distinct from group M and group O. *Nature Med.* **4,** 1032–1037.

Skripkin, E., Paillart, J. C., Marquet, R., Ehresmann, B., and Ehresmann, C. (1994). Identification of the primary site of the human immunodeficiency virus type 1 RNA dimerization in vitro. *Proc. Natl. Acad. Sci. USA* **91,** 4945–4949.

Stauber, R. H., and Pavlakis, G. N. (1998). Intracellular trafficking and interactions of the HIV-1 Tat protein. *Virology* **252,** 126–136.

Sonstegard, T. S., and Hackett, P. B. (1996). Autogenous regulation of RNA translation and packaging by Rous sarcoma virus Pr76Gag. *J. Virol.* **70,** 6642–6652.

Sullenger, B. A., Gallardo, H. F., Ungers, G. E., and Gilboa, E. (1990). Overexpression of TAR sequences renders cells resistant to human immunodeficiency virus replication. *Cell* **63,** 601–608.

Sullenger, B. A., Gallardo, H. F., Ungers, G. E., and Gilboa, E. (1991). Analysis of trans-acting response decoy RNA-mediated inhibition of human immunodeficiency virus type 1 transactivation. *J. Virol.* **65,** 6811–6816.

Suo, Z., and Johnson, K. A. (1997a). Effect of RNA secondary structure of RNA cleavage catalyzed by HIV-1 Reverse Transcriptase. *Biochemistry* **36,** 12468–12476.

Suo, Z., and Johnson, K. A. (1997b). Effect of RNA secondary structure on the kinetics of DNA synthesis catalyzed by HIV-1 Reverse Transcriptase. *Biochemistry* **36,** 12459–12467.

Suo, Z., and Johnson, K. A. (1998). RNA secondary structure switching during DNA synthesis catalyzed by HIV-1 Reverse Transcriptase. *Biochemistry* **36,** 14778–14785.

Svitkin, Y. V., Pause, A., and Sonenberg, N. (1994). LA autoantigen alleviates translational repression by the 5′ leader sequence of the human immunodeficiency virus type 1 RNA. *J. Virol.* **68,** 7001–7007.

Takagaki, Y., MacDonald, C. C., Shenk, T., and Manley, J. L. (1992). The human 64-kDa polyadenylation factor contains a ribonucleoprotein-type RNA binding domain and unusual auxiliary motifs. *Proc. Natl. Acad. Sci. USA* **89,** 1403–1407.

Takagaki, Y., Seipelt, R. L., Peterson, M. L., and Manley, J. L. (1997). The polyadenylation factor CstF-64 regulates alternative processing of IgM heavy chain pre-mRNA during B-cell differentiation. *Cell* **87,** 941–952.

Takasaki, T., Kurane, I., Aihara, H., Ohkawa, N., and Yamaguchi, J. (1997). Electron microscopic study of human immunodeficiency virus type 1 (HIV-1) core structure: Two RNA strands in the core of mature and budding particles. *Arch. Virol.* **142,** 375–382.

Tanese, N., Telesnitsky, A., and Goff, S. P. (1991). Abortive reverse transcription by mutants of Moloney murine leukemia virus deficient in the reverse transcriptase-associated RNase H function. *J. Virol.* **65,** 4387–4397.

Telesnitsky, A., and Goff, S. P. (1993). "Reverse Transcriptase" (A. M. Skalka and S. P. Goff, Eds.), 4th ed., pp. 49–83. Cold Spring Harbor Laboratory Press, Cold Spring Harbor, New York.

Telesnitsky, A., and Goff, S. P. (1997). "Retroviruses" (J. M. Coffin, S. H. Hughes, and H. E. Varmus, Eds.), pp. 121–160. Cold Spring Harbor Laboratory Press, Cold Spring Harbor, New York.

Tiley, L. S., and Cullen, B. R. (1991). Effect of RNA secondary structure on polyadenylation site selection. *Genes Dev.* **5,** 1277–1284.

Topping, R., Demoitie, M. A., Shin, N. H., and Telesnitsky, A. (1998). Cis-acting elements required for strong stop acceptor template selection during Moloney murine leukemia virus reverse transcription. *J. Mol. Biol.* **281,** 1–15.

Tsuchihashi, Z., and Brown, P. O. (1994). DNA strand exchange and selective DNA annealing promoted by the human immunodeficiency virus type 1 nucleocapsid protein. *J. Virol.* **68,** 5863–5870.

Ulich, C., Dunne, A., Parry, E., Hooker, C. W., Gaynor, R. B., and Harrich, D. (1999). Functional domains of Tat required for efficient human immunodeficiency type 1 reverse transcription. *J. Virol.* **73,** 2499–2508.

Valsamakis, A., Zeichner, S., Carswell, S., and Alwine, J. C. (1991). The human immunodeficiency virus type 1 polyadenylylation signal: A 3′ long terminal repeat element upstream of the AAUAAA necessary for efficient polyadenylylation. *Proc. Natl. Acad. Sci. USA* **88,** 2108–2112.

Valsamakis, A., Schek, N., and Alwine, J. C. (1992). Elements upstream of the AAUAAA within the human immunodeficiency virus polyadenylation signal are required for efficient polyadenylation *in vitro*. *Mol. Cell Biol.* **12,** 3699–3705.

van Duin, J. (1994). "Encyclopedia of Virology" (R. G. Webster and A. Granoff, Eds.), pp. 1334–1339. Academic Press, London.

van Wamel, J. L. B., and Berkhout, B. (1998). The first strand transfer during HIV-1 reverse transcription can occur either intramolecularly or intermolecularly. *Virology* **244,** 245–251.

Verhoef, K., Koper, M., and Berkhout, B. (1997a). Determination of the minimal amount of Tat activity required for human immunodeficiency virus type 1 replication. *Virology* **237,** 228–236.

Verhoef, K., Tijms, M., and Berkhout, B. (1997b). Optimal Tat-mediated activation of the HIV-1 LTR promoter requires a full-length TAR RNA hairpin. *Nucleic Acids Res.* **25,** 496–502.

Viglianti, G. A., Rubinstein, E. C., and Graves, K. L. (1992). Role of the TAR RNA splicing in translational regulation of simian immunodeficiency virus from rhesus macaques. *J. Virol.* **66,** 4824–4833.

Wei, P., Garber, M. E., Fang, S.-M., Fisher, W. H., and Jones, K. A. (1998). A novel CDK9-associated C-type cyclin interacts directly with HIV-1 Tat and mediates its high-affinity, loop-specific binding to TAR RNA. *Cell* **92,** 451–462.

Weichs an der Glon, C., Monks, J., and Proudfoot, N. J. (1991). Occlusion of the HIV poly(A) site. *Genes Dev.* **5,** 244–253.

Weiss, E. A., Gilmartin, G. M., and Nevins, J. R. (1991). Poly(A) site efficiency reflects the stability of complex formation involving the downstream element. *EMBO J.* **10,** 215–219.

Wöhrl, B. M., Georgiadis, M. M., Telesnitzky, A., Hendrickson, W. A., and Le Grice, S. F. J. (1995). Footprint analysis of replicating murine leukemia virus reverse transcriptase. *Science* **267,** 96–99.

Wu, W., Blumberg, B. M., Fay, P. J., and Bambara, R. A. (1995). Strand transfer mediated by human immunodeficiency virus reverse transcriptase *in vitro* is promoted by pausing and results in misincorporation. *J. Biol. Chem.* **270,** 325–332.

Wu, W., Henderson, L. E., Copeland, T. D., Gorelick, R. J., Bosche, W. J., Rein, A., and Levin, J. G. (1996). Human Immunodeficiency Virus type I nucleocapsid protein reduces reverse transciptase pausing at a secondary structure near the Murine Leukemia Virus polypurine tract. *J. Virol.* **70,** 7132–7142.

Yamamoto, R., Koseki, S., Ohkawa, J., Murakami, K., Nishikawa, S., Taira, K., and Kumar, P. K. R. (1997). Inhibition of transcription by the TAR RNA of HIV-1 in a nuclear extract of HeLa cells. *Nucleic Acids Res.* **25,** 3445–3450.

Yang, X., Herrmann, C. H., and Rice, A. P. (1996). The human immunodeficiency virus Tat proteins specifically associate with TAK *in vivo* and require the carboxyl-terminal domain of RNA polymerase II for function. *J. Virol.* **70,** 4576–4584.

Yang, X., Gold, M. O., Tang, D. N., Lewis, D. E., Aguilar-Cordova, E., Rice, A. P., and Herrmann, C. H. (1997). TAK, an HIV Tat-associated kinase, is a member of the cyclin-depedent family of protein kinases and is induced by activation of peripheral blood lymphocytes and differentiation of promonocytic cell lines. *Proc. Natl. Acad. Sci. USA* **94,** 12331–12336.

You, J. C., and McHenry, C. S. (1994). Human immunodeficiency virus nucleocapsid protein accelerates strand transfer of the terminally redundant sequences involved in reverse transcription. *J. Biol. Chem.* **269,** 31491–31495.

Zhu, Y., Pe'ery, T., Peng, J., Ramanathan, Y., Marshall, N., Marshall, T., Amendt, B., Mathews, M. B., and Price, D. H. (1997). Transcriptional elongation factor P-TEFb is required for HIV-1 Tat transactivation *in vitro*. *Genes Dev.* **11,** 2622–2632.

Stephan Bour and Klaus Strebel

Laboratory of Molecular Microbiology
National Institute of Allergy and Infectious Diseases
National Institutes of Health
Bethesda, Maryland 20892-0460

HIV Accessory Proteins: Multifunctional Components of a Complex System

I. Introduction

In addition to the prototypical retroviral *gag, pol,* and *env* genes, primate lentiviruses, including HIV, encode a number of accessory genes (Fig. 1) that perform functions not provided by the host cell. The *vif, vpr,* and *nef* genes are expressed in most HIV-1, HIV-2, and SIV isolates. In contrast, the *vpu* gene is found exclusively in HIV-1 isolates and in one SIV isolate, SIVcpz (Huet *et al.,* 1990). The *vpx* gene, on the other hand, is not found in HIV-1 isolates but is common to HIV-2 and most SIV isolates. Defects in accessory genes are frequently not correlated with a detectable impairment of virus replication in continuous cell lines, in contrast to primary cell types, which more closely reflect the *in vivo* situation. The molecular basis for this cell-type specific role of some of the accessory proteins remains largely unknown. However, it is increasingly clear that these proteins indeed exert important functions in their relevant target cells *in vivo,* and most of the

Advances in Pharmacology, Volume 48

1054-3589/00 $35.00

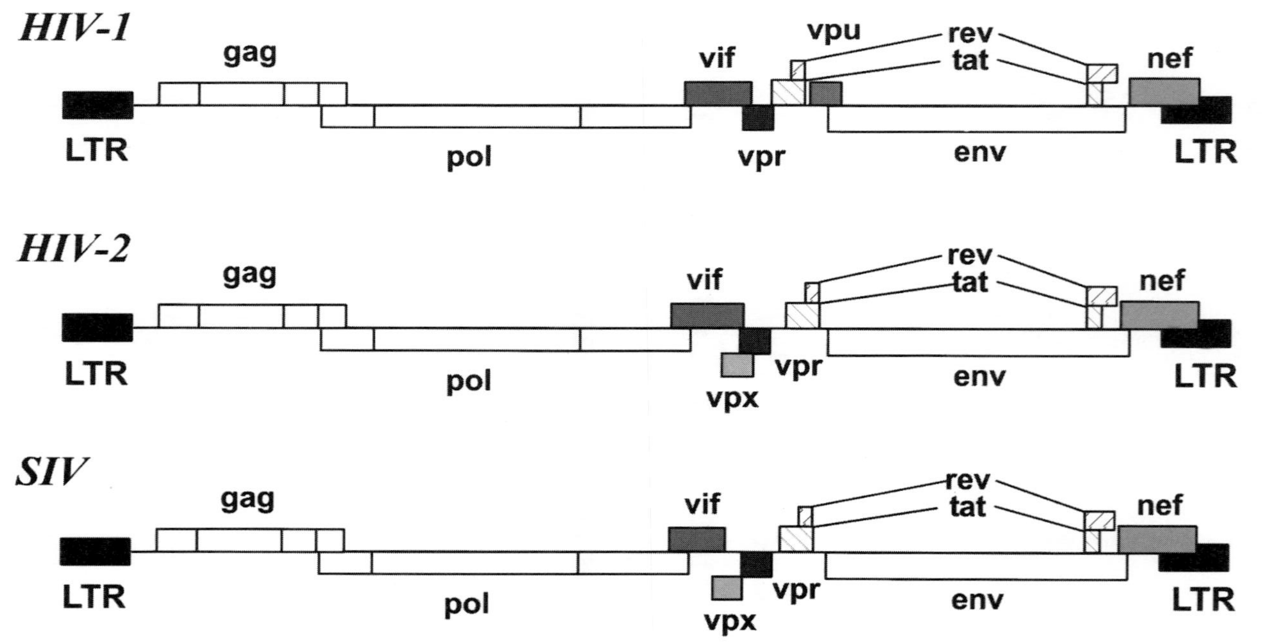

FIGURE 1 Genome organization of primate lentiviruses.

HIV accessory proteins in fact exert multiple independent functions. This chapter attempts to summarize our current knowledge of the function of HIV accessory proteins.

II. Vif: A Potent Regulator of Viral Infectivity

Vif is encoded by all lentiviruses except equine infectious anemia virus (Oberste and Gonda, 1992). Its gene product is a 23-kDa basic protein which is produced late in the infection cycle in a Rev-dependent manner (Garrett *et al.,* 1991; Schwartz *et al.,* 1991). Deletions in *vif* have been associated with a reduction or loss of viral infectivity (Fisher *et al.,* 1987; Strebel *et al.,* 1987; Kishi *et al.,* 1992), a phenomenon, which is largely host-cell dependent (Fan and Peden, 1992; Gabuzda *et al.,* 1992; Blanc *et al.,* 1993; Sakai *et al.,* 1993; von Schwedler *et al.,* 1993; Borman *et al.,* 1995) and can vary in its extent by several orders of magnitude (Fisher *et al.,* 1987; Strebel *et al.,* 1987; Kishi *et al.,* 1992). In permissive cell types, such as HeLa, COS, C8166, Jurkat, U937, or SupT1 (Fan and Peden, 1992; Gabuzda *et al.,* 1992; Sakai *et al.,* 1993), production of infectious particles does not require a functional *vif* gene product. In contrast, *vif*-deficient viruses produced from nonpermissive cells, such as H9, CEM, PBMC, or macrophages (Fan and Peden, 1992; Gabuzda *et al.,* 1992; Sova and Volsky, 1993; von Schwedler *et al.,* 1993; Gabuzda *et al.,* 1994; Borman *et al.,* 1995; Courcoul *et al.,* 1995), are noninfectious regardless of the permissiveness of the target cells (Gabuzda *et al.,* 1992; von Schwedler *et al.,* 1993; Borman *et al.,* 1995). These findings are consistent with two possible mechanisms. First, permissive cells may express a cellular factor that is functionally homologous to Vif, thus eliminating the requirement for a functional *vif* gene. Second, it is possible that permissive cell types lack an inhibitory factor that is present in restrictive cell types where it could interfere with the formation of infectious virions unless it is neutralized by Vif. In either case, infectivity of Vif-defective viruses is a function of the virus-producing cell rather than the target cell. Neither a positive nor a negative cellular factor(s) have thus far been identified; however, two recent studies seem to indicate the presence of an inhibitory factor in restrictive cell types (Simon *et al.,* 1998a; Madani and Kabat, 1998). Both studies are based on the analysis of heterokaryons formed between permissive and nonpermissive cell types. The basic assumption was that the presence of a putative Vif-like (positive) factor in permissive cells would confer a permissive phenotype to heterokaryons, while, conversely, the presence of an inhibitory factor in restrictive cell types would be expected to bestow a restrictive phenotype. Interestingly, both studies found the phenotype of such heterokaryons to be restrictive for Vif-defective viruses. These findings therefore are more consistent with the presence of an inhibitory factor in restrictive cell types, even though the rapid loss of a potential positive Vif-like factor in permissive cells following formation of heterokaryons cannot be entirely ruled out (Simon *et al.*, 1998a; Madani and Kabat, 1998).

Several lines of evidence suggest that Vif exerts its function through interactions with species-specific host cell factor(s) (Simon *et al.*, 1995). Accordingly, HIV-1 Vif was able to regulate infectivity of HIV-1, HIV-2, and SIV_{AGM} in human cells while SIV_{AGM} Vif was inactive in human cells—even on SIV substrates—but was active in African green monkey cells (Simon *et al.*, 1998c). Similarly, the identification of an HIV-2 isolate, $HIV2_{KR}$, whose *vif*-defective variants exhibit cell-type restrictions that are distinct from those observed for *vif*-deficient HIV-1, point to the involvement of cellular factor(s) for Vif function (Reddy *et al.*, 1995). Nevertheless, the observation that *vif* genes from HIV-1, HIV-2, and SIV are capable of functional complementation in appropriate cellular backgrounds (Reddy *et al.*, 1995; Simon *et al.*, 1995, 1998c) suggests a common mechanistic basis for Vif function.

As noted above, the defect in viruses synthesized by restrictive cells in the absence of Vif cannot be complemented by the presence of Vif in recipient cells (von Schwedler *et al.*, 1993; Borman *et al.*, 1995). It was proposed that Vif is required at the time of particle production in the host cell for regulating virus assembly or maturation (Gabuzda *et al.*, 1992; Blanc *et al.*, 1993; Sakai *et al.*, 1993; von Schwedler *et al.*, 1993; Borman *et al.*, 1995). It is conceivable that Vif is involved in the posttranslational modification of one or several virion components. This is supported by the observation of morphological aberrancies in Vif defective virions produced in restrictive cells, which were not observed in wild-type virions or in Vif-defective virions produced in permissive cells (Hoglund *et al.*, 1994; Borman *et al.*, 1995). In fact, one model proposes, as discussed in more detail below, that Vif regulates virus infectivity by delaying the activation of the HIV protease, thus preventing premature processing of the Gag and Gag/Pol polyprotein precursors (Kotler *et al.*, 1997).

III. Functional Domains in Vif

The biochemical function of Vif is still unclear. Using a vaccinia virus-derived expression system for the expression of Vif in HeLa cells and supported by *in vitro* kinase assays on recombinant Vif, Yang *et al.* identified multiple phosphorylation sites in Vif, including Ser_{144}, Thr_{155}, and Thr_{188} (Yang *et al.*, 1996) (Fig. 2). In addition, Vif was found to be a substrate for mitogen-activated protein kinase (MAPK), which was reported to regulate phosphorylation of Vif at two additional sites, Thr_{96} and Ser_{165} (Yang and Gabuzda, 1998) (Fig. 2). Mutation of two of the phosphorylation sites (Thr_{96} and Ser_{144}), which are highly conserved, abolished Vif function, suggesting that Vif phosphorylation may be important for its biological activity

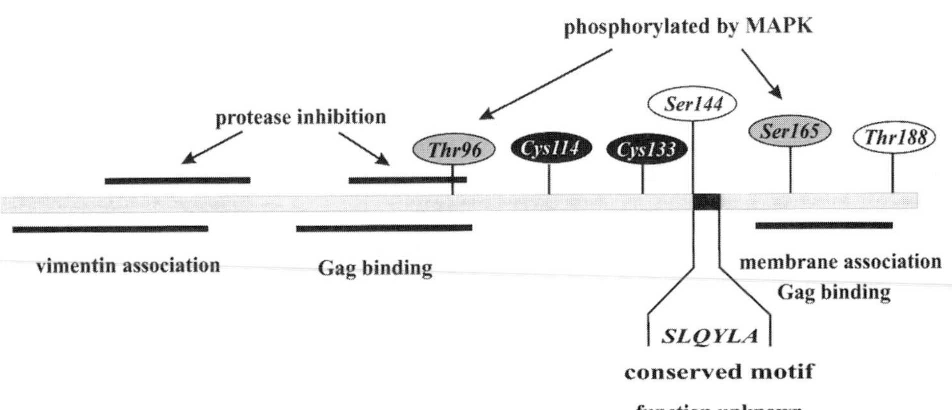

FIGURE 2 Functional domains in Vif.

(Yang *et al.,* 1996; Yang and Gabuzda, 1998). Other important sequences critical for Vif function include two conserved cysteine residues at positions 114 and 133 (Ma *et al.,* 1994; Sova *et al.,* 1997). However, experimental evidence suggests that neither intracellular nor virus-associated Vif utilize Cys_{114} and Cys_{133} for the formation of sulfhydryl bonds (Sova *et al.,* 1997). Nevertheless, the fact that mutation of either cysteine results in complete loss of Vif function (Ma *et al.,* 1994; Simon *et al.,* 1999) points to an important function of these residues. A stretch of basic amino acids located near the C-terminus of Vif was found to be critical for interactions with cellular membranes (Goncalves *et al.,* 1995) and the $Pr55^{Gag}$ precursor (see below) (Fig. 2). Membrane association of Vif is sensitive to trypsin treatment and is therefore presumably mediated through an interaction with a putative membrane-associated protein(s) (Goncalves *et al.,* 1995).

Vif regulation of viral infectivity may be related to its ability to interact with the $Pr55^{Gag}$ precursor (Bouyac *et al.,* 1997). Vif/Gag interactions were abolished in a Vif mutant lacking the C-terminal 22 amino acids (Bouyac *et al.,* 1997), suggesting an involvement of this C-terminal basic domain in Vif, previously identified as a membrane binding domain (Goncalves *et al.,* 1994, 1995). Using an insect cell system, Huvent *et al.* identified four discrete Gag-binding sites in Vif, which included residues T_{68}-L_{81} (site I) and W_{89}-P_{100} (site II) in the central domain and residues P_{162}-R_{173} (III) and P_{177}-M_{189} (IV) at the C terminus (Huvent *et al.,* 1998). Substitutions in site I and deletion of site IV were detrimental to Vif encapsidation, whereas substitution of basic residues for alanine in sites III and IV had a positive effect. The data suggest a direct intracellular Gag–Vif interaction and could point to a $Pr55^{Gag}$-mediated membrane-targeting pathway for Vif (Huvent *et al.,* 1998). The Vif-interacting domains in $Pr55^{Gag}$ were analyzed by screening of a phage-display library (Huvent *et al.,* 1998). The Vif-binding domain in $Pr55^{Gag}$ defined by this procedure spanned residues H_{421}-T_{470} and includes the C-terminal region of nucleocapsid (NC), including the second zinc finger, the intermediate spacer peptide sp2, and the N-terminal half of the p6 domain. Deletions in these Gag domains significantly decreased the Vif encapsidation efficiency, and complete deletion of NC abolished Vif encapsidation (Huvent *et al.*, 1998).

IV. Vif Associates with the Cytoskeleton

Aside from the proposed association of Vif with cellular membranes, the intracellular localization of Vif was found to be affected by the presence of the intermediate filament (IF) vimentin (Karczewski and strebel, 1996). Fractionation of acutely infected T-cells or transiently transfected HeLa cells revealed the existence of soluble, cytoskeletal, and detergent-extractable forms of Vif. Confocal microscopic analysis of Vif-expressing HeLa cells suggests that Vif is predominantly present in the cytoplasm and closely colocalizes with the intermediate filament vimentin. The close association of Vif with vimentin is evidenced by the fact that treatment of cells with drugs affecting the structure of vimentin filaments similarly affected the localization of Vif (Karczewski, 1996). Mutational analysis of Vif suggests that the N-terminal domain of Vif is important for vimentin association (Fig. 2; Strebel, unpublished). The association of Vif with vimentin severely alters the structure of the IF network and can lead to its complete collapse in a perinuclear region (Karczewski and Strebel, 1996). This effect of Vif on vimentin was found to be reversible and dependent on the microtubule network (Strebel, unpublished). Interestingly, the proper establishment of vimentin networks in normal fibroblasts was also found to require stable (detyrosinated) microtubules (Gurland and Gunderson, 1995). It is therefore possible that the observed effects of Vif on vimentin structure result from the disruption of such vimentin/microtubule interactions by Vif. In vimentin-negative cells, immunocytochemical analysis revealed significant staining of the nucleus and the nuclear membrane in addition to diffuse cytoplasmic staining, supporting the notion that the association of Vif with vimentin significantly affects its subcellular distribution (Karczewski and Strebel, 1996). Colocalization of Vif and vimentin was also observed in acutely infected H9 cells (Strebel, unpublished). However, this observation was restricted to single infected cells and little or no association of Vif with the cytoskeleton was observed in multinucleated giant cells (Simon *et al.*, 1997; Strebel, unpublished). It is possible that Vif does not directly interact with vimentin but does so through unknown vimentin-associated factor(s) whose subcellular distribution may be altered in multinucleated giant cells. Preliminary evidence suggests that association of Vif with vimentin is cell-cycle dependent (Strebel, unpublished). This could provide an explanation for the reported failure to detect vimentin association of Vif in syncytia of infected

H9 cells (Simon *et al.*, 1997) and could, furthermore, explain the reversible nature of the Vif-induced changes of the cytoskeleton. At any rate, it remains to be shown whether the observed Vif-induced alteration of the cytoskeleton in virus-producing cells is correlated with Vif's role in regulating viral infectivity or whether it reflects an independent activity of Vif.

V. Vif as a Possible Regulator of Gag/Pol Polyprotein Processing

Based on its similarity to a family of cysteine proteases, Guy and coworkers initially proposed that Vif might act as a protease, targeting the cytoplasmic domain of the Env glycoprotein (Guy *et al.*, 1991). However, even though Vif has been implicated in regulating incorporation of Env into virions (Sakai *et al.*, 1993; Borman *et al.*, 1995), a proteolytic activity of Vif has so far not been demonstrated and the processing of the C-terminal end of gp41 as suggested by Guy and coworkers (Guy *et al.*, 1991) could not be confirmed by others (Gabuzda *et al.*, 1992; von Schwedler *et al.*, 1993). In fact, a more recent model proposes that Vif acts as an inhibitor of the HIV protease and functions to prevent premature processing of the Gag/Pol polyprotein precursor (Potash *et al.*, 1998; Baraz *et al.*, 1998; Kotler *et al.*, 1997; Friedler *et al.*, 1999). This proposed function of Vif is based on the observation that expression of Vif or an N-terminal Vif-derived peptide was able to inhibit autoprocessing of truncated Gag/Pol polyproteins in *E. coli* and also inhibited processing of a synthetic model peptide *in vitro* (Kotler *et al.*, 1997). In addition, synthetic peptides corresponding to an N-terminal domain of Vif (residues 30–65 and 78–98) were found to inhibit HIV-1 replication in peripheral blood lymphocytes (Potash *et al.*, 1998; Friedler *et al.*, 1999) (Fig. 2). Despite the obvious effect of Vif (or Vif-derived peptides) on Gag/Pol processing *in vitro,* the *in vivo* function of Vif, both with respect to the regulation of Gag processing, as well as its site of action, remains controversial (Simm *et al.*, 1995; Fouchier *et al.*, 1996; Bouyac *et al.*, 1997; Ochsenbauer *et al.*, 1997; Strebel, unpublished).

VI. Virion-Associated Vif May Have a Crucial Role in Regulating Viral Infectivity

Aside from its proposed function in HIV-infected cells, Vif may also have a virion-associated activity. Like Vpr and Nef, Vif is packaged into virus particles (Liu *et al.*, 1995; Karczewski and Strebel, 1996; Camaur and Trono 1996; Borman *et al.*, 1995). Unlike Vpr, however, which is packaged in significant quantities through specific interaction with the p6 domain in Gag (Lu *et al.*, 1993, 1995; Paxton *et al.*, 1993; Kondo *et al.*, 1995; Lavallee *et al.*, 1994; Selig *et al.*, 1999), Vif incorporation into virions was reported to be nonselective. While there is general agreement on the fact that Vif is incorporated into virions, there is an ongoing discussion regarding the absolute amounts of Vif packaged. There is general consensus that the amounts of Vif packaged into virions are low and vary depending on the intracellular expression levels of Vif (Liu *et al.*, 1995; Karczewski and Strebel, 1996; Camaur and Trono, 1996; Simon *et al.*, 1998b; Dettenhofer and Yu, 1999). Nevertheless, virion-associated Vif is found in tight association with the viral core (Liu *et al.*, 1995; Karczewski and Strebel, 1996) and consistently copurifies with viral reverse transcriptase, integrase, and unprocessed $Pr55^{Gag}$ (Strebel, unpublished). It is thus likely that Vif is a component of the HIV nucleoprotein complex and as such could, despite its low abundance in virions, perform a crucial function during the early phase of viral infection.

One hypothetical model that invokes an active role of Vif during a preintegration step in virus replication involves the transport of the nucleoprotein or preintegration complex from the plasma membrane to the nucleus. Active transport of nucleocapsids from the cell surface to the nucleus has been reported for other viruses such as herpes simplex virus (Sodeik *et al.*, 1997) or human foamy virus (Saib, 1997). In the case of herpes simplex virus, viral nucleocapsids were found to associate with the minus-end directed microtubule motor dynein for active transport along microtubules to the perinuclear region (Sodeik *et al.*, 1997). Similarly, infection of cells by human foamy virus was found to result in a microtubule-dependent centrosomal accumulation of Gag proteins (Saib *et al.*, 1997).

Several lines of evidence suggest that virion-associated Vif could be similarly involved in active nuclear targeting of viral nucleoprotein complexes. First, it is well documented that *vif*-defective HIV-1 particles produced in nonpermissive cells can bind to and penetrate host cells but are impaired in viral DNA synthesis (Sova and Volsky, 1993; von Schwedler *et al.*, 1993; Borman *et al.*, 1995; Courcoul *et al.*, 1995; Nascimbeni *et al.*, 1998; Goncalves *et al.*, 1996), indicating that Vif regulates a postentry step in the early phase of the virus life cycle prior to integration. Second, Vif is packaged into HIV particles in close association with the viral core (Liu *et al.*, 1995; Karczewski and Strebel, 1996) and is thus most likely part of the nucleoprotein complex (Strebel, unpublished). Third, Vif has the ability to associate with intermediate filaments such as vimentin, which connect nuclear and plasma membranes (Karczewski and Strebel, 1996). Finally, in transient transfection studies, expression of Vif was found to induce reversible, microtubule-dependent alterations in the vimentin structure, which resulted in perinuclear accumulation of Vif and vimentin and presumably involved the activity of microtubule motor molecules (Strebel, unpublished). These data suggest that similar to the herpes simplex virus model, HIV-1Vif could potentially act as a linker between viral nucleoprotein complexes and cellular microtubule motor molecules for active transport to the nucleus.

As a consequence, active nuclear targeting of nucleoprotein complexes by Vif would significantly increase the efficiency of virus infection and thus could explain the observed effect of Vif on viral infectivity. While the parallels of such a proposed function of Vif to the herpes simplex or foamy virus systems are intriguing, it remains to be shown whether the effect of Vif on viral infectivity is indeed a direct consequence of its presence in virions or, instead, is the manifestation of an activity of Vif performed during particle assembly in the virus producing cell.

VII. VPR and Vpx

Vpr and Vpx are two small viral accessory proteins that accumulate in the nucleus of infected cells. Vpr and Vpx are remarkably similar both with respect to amino acid sequence and function (reviewed in Kappes, 1995). However, unlike Vpr, which is encoded by each lentiviral group, Vpx is restricted to HIV-2 and some SIV viruses, specifically SIV_{smm} and SIV_{mac} (for review see Kappes, 1995). The origin of *vpx* is still under discussion. It was originally proposed that *vpr* and *vpx* genes arose by gene duplication during evolution of the HIV-2/SIVsmm group (Tristem *et al.*, 1990). Alternatively, *vpx* could have been the result of a horizontal transfer of *vpr* from the SIVagm group to the HIV-2 group involving nonhomologous (Sharp *et al.*, 1996) or homologous (Tristem *et al.*, 1998) recombination events. In HIV-2/SIV_{sm} viruses, Vpr and Vpx exert distinct functions: Vpx was found to be necessary and sufficient for nuclear import of the preintegration complex in nondividing cells, while Vpr blocked cell cycle progression past the G_2 phase (Fletcher *et al.*, 1996). In contrast, in HIV-1 both functions are catalyzed by Vpr. In fact, most of our knowledge concerning the function of Vpr and Vpx comes from the analysis of HIV-1 Vpr, which is the main focus of this chapter.

Vpr and Vpx are both efficiently and specifically incorporated into virions through interactions with the C-terminal p6 domain of the $Pr55^{Gag}$ precursor (Lu *et al.*, 1993; Paxton *et al.*, 1993; Kondo *et al.*, 1995; Lu *et al.*, 1995; Lavallee *et al.*, 1994; Selig *et al.*, 1999). Packaging of HIV-1 Vpr involves a leucine triplet repeat motif $(LXX)_4$ in p6 (Lu *et al.*, 1995; Kondo and Gottlinger, 1996). HIV-1 Vpr has also been reported to directly interact with the zinc finger domain in NCp7 (Li *et al.*, 1996; de Rocquigny *et al.*, 1997). Packaging of HIV-2 Vpx, like HIV-1 Vpr, is dependent on sequences in the p6 domain of the HIV-2 Gag precursor (Pancio and Ratner, 1998; Wu *et al.*, 1994; Selig *et al.*, 1999). In contrast, the minimal sequence required for SIV_{sm} Vpr and Vpx binding to Gag was found upstream of the equivalent $(LXX)_3$ domain of the SIV_{sm} p6 domain (Selig *et al.*, 1999). The role of virus-associated Vpr is unclear. However, the observation that Vpr-induced arrest of cells in G_2 (thereby facilitating transcriptional activation as discussed below) does not require *de novo* synthesis of Vpr but can be induced, albeit less efficiently, by virus-associated Vpr (Poon *et al.*, 1998) and could point to an early function of Vpr.

Based on structural analyses as well as computer modeling (Schuler *et al.*, 1999; Subbramanian *et al.*, 1998b), Vpr is predicted to contain two helical domains. Helix 1 (Fig. 3) extends from residues 17 to 34 of HIV-1 Vpr and includes sequences critical for virion association and oligomerization (Mahalingam *et al.*, 1995b,c; Yao *et al.*, 1995; Selig *et al.*, 1999; Di Marzio *et al.*, 1995; Zhao 1994a,b). Helix-2 (residues 53–78) forms an amphipathic α-helix and is important for nuclear translocation of Vpr (Di Marzio *et al.*, 1995; Mahalingam *et al.*, 1995a; Subbramanian *et al.*, 1998b). Helix 2 also contains a leucine zipper-type motif important for Vpr dimerization, which is thought to be mediated by coiled-coil hydrophobic interactions of helices 2 (Schuler *et al.*, 1999). The structure of the C-terminus of Vpr (residues 79–96) is less well defined. It contains a relatively high number of basic amino acids and has been implicated in nucleic acid binding (Zhang *et al.*, 1998; Schuler *et al.*, 1999) and cell cycle arrest (Di Marzio *et al.*, 1995; Macreadie *et al.*, 1996; Forget *et al.*, 1998).

Like other HIV accessory proteins, Vpr appears to regulate multiple independent functions that involve different domains in the protein. Thus, Vpr has been implicated in transcriptional activation of the HIV LTR (Cohen *et al.*, 1990; Felzien *et al.*, 1998; Wang *et al.*, 1995; Subbramanian *et al.*, 1998a) and coactivation of the human glucocorticoid receptor (Kino *et al.*, 1999). Furthermore, Vpr has a well-documented role in cell-cycle regulation (Jowett *et al.*, 1995; He *et al.*, 1995; Planelles *et al.*, 1996; Re *et al.*,

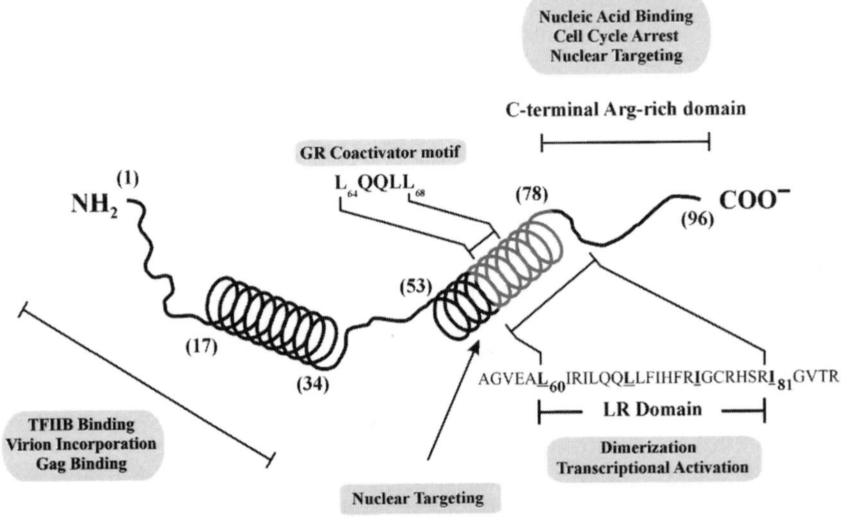

FIGURE 3 Structural domains in HIV-1 Vpr.

1995; Rogel *et al.*, 1995) and has been implicated in nuclear targeting of preintegration complexes in terminally differentiated macrophages (Heinzinger *et al.*, 1994; Connor *et al.*, 1995; Balotta *et al.*,1993). Finally, low-level expression of Vpr in constitutive cell lines was found to confer antiapoptotic properties to such cells (Conti *et al.*, 1998; Fukumori *et al.*, 1998). It was suggested that this function of Vpr is the result of a Vpr-induced upregulation of bcl-2 an oncogene with antiapoptotic activities (Conti *et al.*, 1998).

VIII. Vpr as a Transcriptional Activator

Vpr contains a leucine-rich domain (LR) with a leucine-zipperlike structure (Zhao *et al.*, 1994a,b; Wang *et al.*, 1996). The LR domain encompasses residues L_{60} to I_{81} and is located in helix2. The LR domain was found to facilitate protein–protein interaction and may be important for dimerization of Vpr (Wang *et al.*, 1996). In addition, mutations in the LR region (as well as the C-terminal basic domain of Vpr) were found to interfere with transcriptional activation by Vpr (Forget *et al.*, 1998), supporting a role for the LR domain in this process. The precise mechanism of Vpr-mediated transcriptional regulation remains unclear but is believed to be linked to its ability to induce G_2 arrest (Forget *et al.*, 1998; Goh *et al.*, 1998; Subbramanian *et al.*, 1998a,b; Felzien *et al.*, 1998). Vpr induces a transcriptional activation during the G_2 phase of the cell cycle that results in increased virus production (Goh *et al.*, 1998; Felzien *et al.*, 1998) and it was proposed that G_2 arrest in T cells may in fact be sufficient to account for the increased transcription from the HIV LTR (Goh *et al.*, 1998). Transcriptional activation by Vpr is presumably mediated through p300, a transcriptional coactivator that regulates NF-κB and the basal transcription machinery (Felzien *et al.*, 1998). P300 is a potent coactivator of the p65 (RelA) subunit of NFkB (Gerritsen *et al.*, 1997; Perkins *et al.*, 1997). Since p300 is expressed in limiting quantities relative to RelA (Hottiger *et al.*, 1998; Ravi *et al.*, 1998), RelA-regulated transcriptional activation is sensitive to factors competing for p300 binding. Interestingly, cyclin-B1/cdc2 complexes (whose activity was found to be affected by Vpr) were found to interact with the C-terminus of p300. This suggests that Vpr may induce transcriptional activation through regulation of the activity of cyclin-B1/cdc2 complexes rather than by directly interacting with p300 (Felzien *et al.*, 1998) (Fig. 4).

IX. VPR as a Coactivator of the Glucocorticoid Receptor

The effect of Vpr on nuclear receptors such as the glucocorticoid receptor (GR) (Refaeli *et al.*, 1995; Kino *et al.*, 1999) is distinct from LTR activation

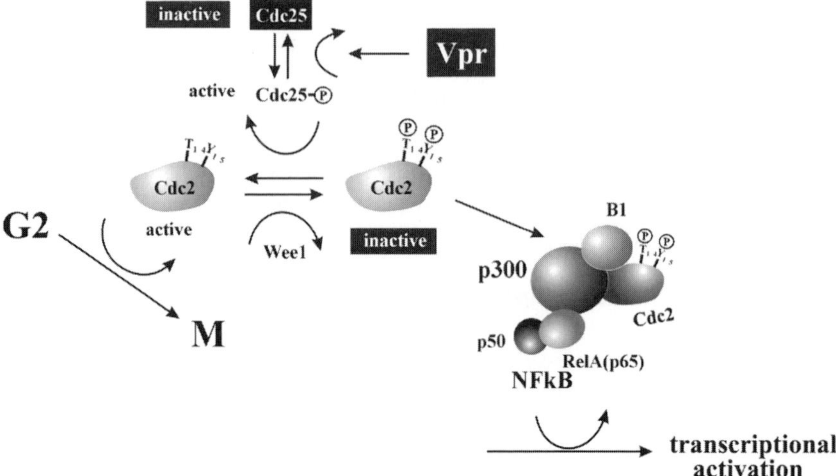

FIGURE 4 Vpr-mediated cell cycle arrest and transcriptional activation.

(Kino *et al.,* 1999). Unlike activation of the LTR, which is correlated with Vpr's ability to induce cell-cycle arrest (Forget *et al.,* 1998), glucocorticoid coreceptor activity of Vpr is distinct from its cell-cycle arrest function and involves a direct interaction with GR through a coactivator motif (LXXLL), located in helix-2 (residues 64–68) (Kino *et al.,* 1999). The LXXLL motif in Vpr is found in other nuclear receptor coregulators and constitutes a signature motif required for the interaction of proteins with nuclear hormone receptors (Heery *et al.,* 1997). Mutation of L_{64} in the LXXLL motif of Vpr (Fig. 3) was found to abolish coactivator function of Vpr while retaining its ability to induce cell-cycle arrest (Kino *et al.,* 1999). In contrast, an R_{80} mutant was unable to induce G_2 arrest but retained coactivator function (Kino *et al.,* 1999). Thus, the glucocorticoid coactivator activity of Vpr is functionally distinct from its cell-cycle inhibitory function. HIV-2 and SIVmac239 Vpr and Vpx proteins do not contain an LXXLL coactivator motif and are unable to act as GR coactivator, supporting the role of this motif (Kino, 1999). In addition to its interaction with GR through the LXXLL motif, Vpr was found to interact with the cellular transcription factor TFIIB (Agostini *et al.,* 1996). Deletion of residues 1–35 in Vpr abolished TFIIB binding, suggesting that binding of Vpr to TFIIB is mediated by sequences located near the N-terminus of Vpr (Kino *et al.,* 1999). Like other Vpr-interacting proteins, TFIIB contains a WXXF motif, which appears to be required for its interaction with Vpr (BouHamdan *et al.,* 1998).

X. Vpr-Induced Cell-Cycle Arrest

Probably the most conserved function of Vpr is its ability to arrest or delay the cell cycle in G_2/M (reviewed in Re and Luban, 1997). This activity presumably reflects a signal-transduction phenomenon and does not require the presence of Vpr in the nucleus (Subbramanian *et al.,* 1998b). Furthermore, G_2 arrest does not require *de novo* synthesis of Vpr but can be induced, albeit less efficiently, by virus-associated Vpr (Poon *et al.,* 1998). Interestingly, it appears that replication-deficient viruses are capable of inducing cell-cycle arrest through a mechanism that requires virus entry but not active virus replication (Poon *et al.,* 1998). Nonproductive infection of $CD4^+$ T cells by Vpr-containing particles was therefore proposed to be a mechanism by which HIV could suppress an effective immune response through inhibition of the clonal expansion of $CD4^+$ T cells (Poon *et al.,* 1998). How Vpr induces a cell-cycle arrest in G_2 is not completely understood. A model incorporating the current state of knowledge is shown in Fig. 4. One of the obvious effects of Vpr expression is the hyperphosphorylation and thus inactivation of Cdc2 (Jowett *et al.,* 1995; He *et al.,* 1995; Re *et al.,* 1995; Bartz *et al.,* 1996). While G_2 arrest is a direct consequence of the inhibition of the Cdc2 kinase activity, it is unlikely that this is achieved by direct interaction with Vpr. First, there is no experimental evidence for such an interaction (Re *et al.,* 1995; He *et al.,* 1995). Second, it was observed that an upstream regulator of Cdc2, the phosphatase Cdc25C, is itself in the inactive, unphosphorylated state (Re *et al.,* 1995) and is regulated by upstream kinase/phosphatase networks (Fig. 4; reviewed in Coleman, 1994). Last, the fact that treatment of cells with okadaic acid, an inhibitor of protein phosphatases 1 and 2A, was found to relieve a Vpr-induced cell-cycle arrest, suggests an involvement of upstream regulators of Cdc2 (Re *et al.,* 1995). This is also consistent with the observation that Vpr/NCp7 complexes can directly activate protein phosphatase-$2A_0$ (Tung *et al.,* 1997), which in a yeast model was found to be a negative regulator of the G_2–M transition.

XI. Nuclear Import

Unlike oncoretroviruses, which depend on the breakdown of the nuclear envelope at mitosis for efficient integration into the host genome (Humphries and Temin, 1974), HIV and other lentiviruses have the ability to infect nondividing cells (Bukrinsky *et al.,* 1993). In addition to the viral matrix (MA) and integrase (Int) products, Vpr has been implicated in facilitating nuclear translocation of preintegration complexes. One of the early steps of HIV-1 nuclear import is the recognition of nuclear localization signals (NLS) on components of the preintegration complex. Nuclear import of proteins containing typical monopartite or bipartite nuclear localization signals (NLS) is mediated by an heterodimeric protein complex, composed of importin α (Imp-α) and importin α (Imp-β), also referred to as karyopherin α and β. Impβ binds the NLS while Imp-β mediates translocation through the nuclear pore complex (reviewed in Pemberton *et al.,* 1998). Both MA and Int carry NLS sequences and have been shown to interact with members of the importin/karyopherin family (Gallay *et al.*, 1996, 1997; Bukrinsky and Haffar, 1997), suggesting a role for both MA and Int in the connection of viral preintegration complexes with the nuclear import machinery. Vpr was also proposed to be involved in nuclear transport of preintegration complexes by interacting with Imp-α in an NLS-independent manner (Heinzinger *et al.*, 1994; Popov *et al.*, 1998). In fact, the presence of Vpr appeared to stabilize Imp-α/MA complexes, rendering them resistant to competition with NLS-containing peptides (Popov *et al.*, 1998). Thus, it is possible that the role of Vpr during nuclear import is to stabilize the NLS-dependent interaction of MA and Int with nuclear import factors. Such a model of a cooperative action of MA, Int, and Vpr is consistent with the notion that mutations in MA and Vpr were found to display additive rather than synergistic effects during the infection of nondividing cells (Freed *et al.*, 1997). Also, the observed association of Vpr with nuclear pore complexes (Fouchier *et al.*, 1998; Vodicka *et al.*, 1998) as well as the recent observation that Vpr, at least *in vitro,* facilitates binding of preintegration complexes to nuclear pore complexes in an Imp-β-dependent manner further support such a model (Popov *et al.*, 1998). Vpr does not contain a canonical NLS yet still localizes to the nucleus of infected cells. Mutational analysis of Vpr suggests an involvement of helix 2 in the nuclear targeting of Vpr (Di Marzio *et al.*, 1995; Mahalingam *et al.*, 1995a, 1997; Yao, 1995; Subbramanian *et al.*, 1998b). More recently, Jenkins *et al.* suggested that the C-terminal domain of Vpr may have an additional and independent role for nuclear targeting and the authors found that the two nuclear targeting signals in Vpr are distinct and use different receptors for nuclear entry (Jenkins *et al.*, 1998).

XII. The Multifunctional Nef Protein

HIV-1 Nef is a 206-amino-acid protein synthesized early in the viral life cycle (Klotman *et al.*, 1991; Munis *et al.*, 1992; Guatelli *et al.*, 1990). Most isolates of HIV-2 and SIV also encode functional Nef proteins with an average length of 263 residues. Nef is localized to the intracellular side of cellular membranes by virtue of a posttranslationally added N-terminal myristic acid (Fig. 5). This membrane association was shown to be critical for Nef function (Greenway *et al.*, 1994; Harris and Neil, 1994; Aldrovandi *et al.*, 1998; Chowers *et al.*, 1994). According to NMR structural analysis, Nef is composed of an N-terminal unstructured domain followed by a dense core held together by a polyproline type II helix, three α-helices and five antiparallel β-sheets (Lee *et al.*, 1996; Grzesiek *et al.*, 1996). The core structure is interrupted between residues 148 and 180 by a solvent-exposed

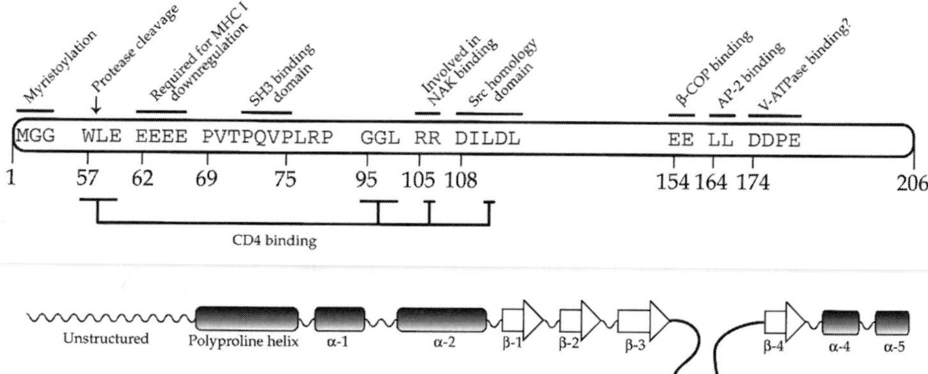

FIGURE 5 Structural domains and amino acid sequences involved in HIV-1 Nef action.

flexible loop that contains residues important for Nef interactions with cellular proteins (Fig. 5).

Numerous biological activities have been attributed to Nef. Some are still controversial in their modalities or relevance while others may be viewed as the indirect consequence of what has emerged as the three main functions of Nef: (1) internalization and lysosomal degradation of cell-surface CD4 receptor, (2) enhancement of viral particle infectivity, and (3) acceleration of disease progression in both humans and primate animal models.

XIII. Downregulation of Cell-Surface CD4

The downregulation of cell-surface expression of CD4 by Nef contributes to receptor interference (Benson *et al.*, 1993), a process by which the cellular receptor is masked to prevent superinfection (reviewed in Geleziunas *et al.*, 1994). Most retroviruses rely exclusively on the high-affinity interaction between their envelope glycoprotein (Env) and cellular receptor to achieve a state of superinfection immunity. The two main mechanisms involve either intracellular trapping of the receptor molecules following the formation of Env–receptor complexes in the endoplasmic reticulum (ER) or functional blocking of cell-surface receptors with endogenously produced Env proteins. HIV-1 is unique in the number and variety of mechanisms it employs to achieve complete removal of cell-surface CD4, its main cellular receptor (reviewed in Bour *et al.*, 1995). In addition to the well-documented inhibition of CD4 trafficking following the formation of complexes with the Env precursor gp160 (Crise *et al.*, 1990; Bour *et al.*, 1991; Jabbar and Nayak, 1990), two viral accessory proteins, Vpu and Nef, participate in

CD4 cell-surface depletion. Although both proteins ultimately target CD4 for degradation, Vpu acts on CD4 in the endoplasmic reticulum (Willey, 1992a), while Nef preferentially acts on cell-surface CD4 (Garcia and Miller, 1991). Neither Vpu nor Nef possess the catalytic activities necessary to directly degrade CD4. Instead, they function as adapter molecules, linking CD4 to key host factors involved in cellular degradation pathways.

Endocytosis of the CD4 receptor is a normal physiological response to T-cell activation by antigen-presenting cells (for review see Marsh and Pelchen-Matthews, 1996). Removal of cell-surface CD4 under these physiological conditions involves recruitment into clathrin-coated pits, endocytosis, and delivery to lysosomes for degradation. Several signals that operate through the cytoplasmic tail of CD4 play a critical role in regulating the rate of internalization. First, binding of the protein tyrosine kinase p56Lck to the CD4 cytoplasmic domain stabilizes CD4 at the cell surface by preventing its entry into coated pits (Pelchen-Matthews *et al.*, 1992). In contrast, phosphorylation by protein kinase C (PKC) of serine residues at positions 408 and 415 in the CD4 cytoplasmic domain following T-cell activation or treatment with phorbol esters induces the rapid internalization of CD4. Phosphorylation of the cytoplasmic serine residues induces the release of Lck from the CD4 cytoplasmic tail, which in turn leads to active recruitment of CD4 into coated pits (Pelchen-Matthews *et al.*, 1993). The molecular basis for this recruitment is not fully understood, although a plausible model suggests that the dissociation of Lck as well as the presence of phosphorylated serine residues might generate conformational changes in the CD4 cytoplasmic domain that allow a critical dileucine motif to interact with the AP-2 clathrin adapter complex (Marsh and Pelchen-Matthews, 1996). Following its internalization in early endosomes, CD4 has two possible fates depending on the nature of the endocytic signal. Most of the nonphosphorylated CD4 internalized by bulk-flow endocytosis is recycled to the plasma membrane while CD4 species bearing phosphorylated cytoplasmic serine residues are directed to late endosomes and lysosomes where they are degraded (Fig. 6). Nef has been shown to subvert this natural cellular pathway, bypassing the requirement for CD4 phosphorylation, and artificially enhancing the rate of cell-surface CD4 turnover. The major effect of Nef on CD4 is manifested by an enhanced rate of internalization followed by accumulation of the endocytosed receptor in an acidic compartment with characteristics of early endosomes (Schwartz *et al.*, 1995a,b) or lysosomes (Sanfridson *et al.*, 1994; Rhee and Marsh, 1994; Aiken *et al.*, 1994). Nef was also reported, albeit to a lesser extent, to prevent CD4 transport from the trans-Golgi network (TNG) to the cell surface (Brady *et al.*, 1993; Mangasarian *et al.*, 1997). The activity of Nef on CD4 is to a large extent cell-type and species independent and operates on CD4 molecules of human, primate, and murine origin (Garcia *et al.*, 1993). In addition, Nef proteins from both laboratory strains and patient isolates of HIV-1 as well as Nef from SIV and HIV-2 isolates

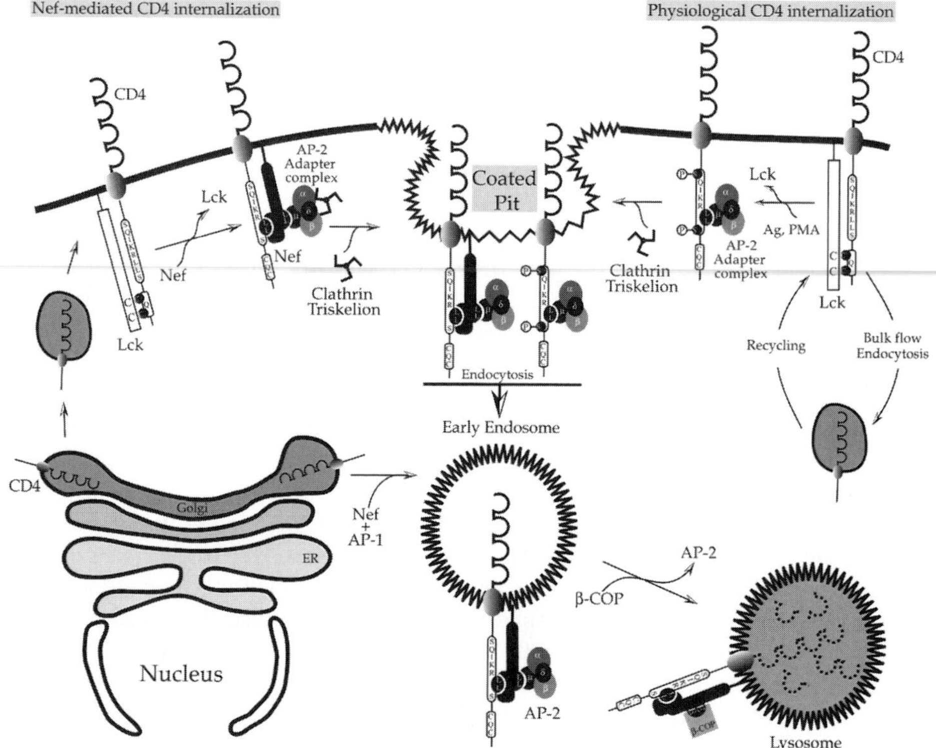

FIGURE 6 Mechanism of Nef-mediated CD4 down-regulation.

show the ability to decrease cell-surface expression of CD4 (Anderson *et al.*, 1993; Benson *et al.*, 1993; Mariani and Skowronski, 1993; Sanfridson *et al.*, 1994). Nef does not appear to generate a global perturbation of the cell surface that would lead to nonspecific internalization of cell-surface proteins. It rather targets a small number of membrane-associated cell-surface components through specific protein–protein interactions. In the case of CD4, the target specificity of Nef was explained by the requirement for specific sequences in the CD4 cytoplasmic domain. Substituting the CD8 cytoplasmic domain for that of CD4 conferred Nef susceptibility to the otherwise stable CD8 molecule (Garcia *et al.*, 1993). Deletion mutagenesis of such chimeric molecules further showed the requirement for the same dileucine motif at position 413–414 in the CD4 cytoplasmic tail also involved in PMA-induced internalization (Aiken *et al.*, 1994). In contrast, the cysteine motif required for interaction with Lck as well as the serine residues involved in PKC-mediated CD4 internalization were shown to be dispensable (Aiken *et al.*, 1994; Anderson *et al.*, 1994). The presence of specific amino acid sequences in CD4 responsible for susceptibility to Nef suggests that specific interactions between the two molecules may be necessary to trigger internalization. *In vitro* evidence points to such an interaction (Grzesiek *et al.*, 1996; Rossi *et al.*, 1996; Harris and Neil, 1994), although CD4 and Nef have not yet been successfully coimmunoprecipitated from lysates of cells of human or primate origin. While the notion of Nef directly or indirectly interacting with CD4 is generally accepted, the question as to whether Nef interacts with Lck is still debated. Dissociation of Lck from the CD4 cytoplasmic tail is an obligatory event preceding CD4 internalization. Accordingly, an increase in the pool of free Lck was observed in the presence of Nef and correlated with enhanced CD4 internalization (Rhee and Marsh, 1994; Anderson *et al.*, 1994). However, while it appears that Lck dissociation is an important step in the mode of action of Nef, it is not mechanistically required. Indeed, CD4 molecules lacking the CQC signal required for Lck binding are efficiently internalized by Nef (Anderson *et al.*, 1994). Furthermore, Nef is functional in nonlymphoid cells lacking Lck expression (Garcia *et al.*, 1993).

Nef-mediated internalization of CD4 is thus dependent on similar signals in CD4 than those involved in the physiological turnover of the receptor from the cell surface, i.e., a dileucine motif in the CD4 cytoplasmic tail and the dissociation of Lck. This suggests that the mechanism by which Nef generates the internalization of CD4 is similar to that observed after PMA treatment, with the major difference that it is independent of serine phosphorylation of CD4 (Aiken *et al.*, 1994; Garcia and Miller, 1991). It therefore appears that Nef promotes internalization of CD4 through its normal endocytosis route while removing the regulatory requirement for cytoplasmic serine phosphorylation. Recent experiments shed light on how Nef might perform this function. Indeed, the use of a chimeric CD4 molecule bearing Nef in place of its cytoplasmic domain demonstrated that Nef could directly connect to the cellular trafficking machinery (Mangasarian *et al.*, 1997). Despite the absence of the CD4-specific internalization signals, such chimeric molecules showed cell-surface behavior similar to that of wild-type CD4 in the presence of Nef. CD4–Nef chimeras were also able to induce the endocytosis of wild-type but not mutated CD4 molecules lacking the cytoplasmic dileucine motif. These data further clarify the requirement for the CD4 dileucine motif for Nef susceptibility and suggest that, in the context of Nef action, the role of this motif is changed from contacting the endocytic machinery, as it is the case under physiological conditions, to interacting (directly or indirectly) with Nef (Rossi *et al.*, 1996). HIV-1 Nef, in turn, uses its own dileucine-based sorting signal to bring CD4 in contact with the endocytosis machinery (Craig *et al.*, 1998). Colocalization experiments in intact cells have identified clathrin-coated pits as the site of Nef-mediated internalization of CD4 (Mangasarian *et al.*, 1997; Greenberg *et al.*, 1997). Nef was found to increase the density of coated pits as well as their rate of internalization by interacting with the medium (μ) chain of the clathrin adaptor complex. The cellular function of adaptor complexes (AP) is to promote the assembly of clathrin-coated pits at the cell surface (AP-2) and the TGN (AP-1 and AP-3) and to recruit membrane protein bearing exposed tyrosine or dileucine-based sorting signals (Kirchhausen *et al.*, 1997). HIV-1, HIV-2, and SIV Nef mediate CD4 internalization and lysosomal routing by interacting with components of the AP complex. Data obtained in the yeast two-hybrid system suggest that HIV-1 Nef preferentially interacts with the μ-chain of the AP-1 (μ1) (Le Gall *et al.*, 1998; Piguet *et al.*, 1998) and AP-3 (μ3) complexes (J. Guatelli, personal communication). Cell-free binding assays using GST fusion proteins as well as colocalization experiments indicate that HIV-1 Nef also interacts weakly with μ2 (Le Gall *et al.*, 1998; Greenberg *et al.*, 1997). The weak affinity of Nef with μ2 might suggest that the interaction is indirect. A possible candidate for bridging Nef to the AP complex is the recently cloned Nef binding protein 1 (NBP1) (Lu *et al.*, 1998). NBP1 shares sequence similarity with the vacuolar ATPase, which is known to interact with the AP-2 complex (Myers and Forgac, 1993). NBP1 could therefore connect Nef to the endocytosis machinery. Mutations that disrupt the putative Nef-NBP1-interacting domain (DDPE peptide at position 174; Fig. 5) prevent CD4 downregulation by Nef (Lu *et al.*, 1998), although this observation is subject to controversy (Mangasarian *et al.*, 1999). Nef proteins from HIV-2 and SIV appear less selective than HIV-1 Nef; they interact with both the μ1 and μ2 chains with similar affinity (Le Gall *et al.*, 1998; Piguet *et al.*, 1998).

Other mechanistic differences between HIV-1 and SIV Nef appear to lie in the endocytic signal used to contact the AP complex. SIV and HIV-2 rely on a tyrosine-based endocytic signal present in the N-terminal region of Nef (Piguet *et al.*, 1998), while HIV-1 Nef, which contains no such signal uses a dileucine motif at position 164 to contact AP-1, AP-2 (Craig *et al.*, 1998; Bresnahan *et al.*, 1998; Greenberg *et al.*, 1998a), and AP-3 (J. Guatelli, personal communication). This ENNSLL motif fits the dileucine-based sorting signal consensus and is located within the exposed loop of HIV-1 Nef (Craig *et al.*, 1998; Fig. 5). While Nef interaction with the AP complex provides an attractive mechanism to account for the accelerated endocytosis of CD4 or the rerouting of TGN-associated CD4, it may not account for another activity of Nef that inhibits CD4 recycling. Indeed, it was recently demonstrated that both the HIV-1 and SIV Nef proteins can prevent recycling of CD4 molecules from early endosomes to the plasma membrane in an AP-2-independent fashion (Piguet *et al.*, 1998). Nef proteins bearing a mutation at two conserved acidic residues (154-EE) were shown to operate similarly to wild-type Nef on CD4 endocytosis but have only 25% activity in overall CD4 downregulation (Piguet *et al.*, 1999). Mutation of the diacidic motif was further shown to prevent the recycling of CD4–Nef chimeric proteins and lead to localization of the chimera in lysosomes. Interestingly, lysosomal localization correlated with the ability of the CD4–Nef chimera to interact with ß-COP, a component of the COP-1 coatomer complex that assembles with early endosomes (Whitney *et al.*, 1995) previously found to interact with Nef (Benichou *et al.*, 1994). Although attractive, the hypothesis that Nef–ß-COP complexes can redirect CD4-containing early endosomes to the lysosome compartment remains to be verified (Fig. 6).

XIV. Downregulation of MHC Class I

Nef has also been reported to downregulate the surface expression of the major histocompatibility complex class I (MHC I) (Schwartz *et al.*, 1996). Similarly to CD4, sequences in the HLA cytoplasmic domain confer susceptibility to Nef. In particular, a tyrosine residue at position 320 present in all HLA-A and -B alleles is required for Nef-mediated downregulation (Le Gall *et al.*, 1998; Greenberg *et al.*, 1998b). Interestingly, HLA-C molecules, which contain a cysteine in place of the tyrosine residue at position 320, are not susceptible to Nef-mediated downregulation (Le Gall *et al.*, 1998). In fibroblastic and epithelial cell lines transfected with HIV-1 Nef, the major effect on MHC I was an accumulation in the TGN, where it colocalized with the AP-1 complex (Le Gall *et al.*, 1998; Greenberg *et al.*, 1998b). In addition, Nef molecules unable to colocalize with the AP-2 complex were functional in MHC I downregulation (Greenberg *et al.*, 1998b). Taken together, these data suggest that while Nef might downregulate CD4 by promoting its internalization from the cell surface in an AP-2-dependent manner, a larger part of the effect on MHC I might be due to retention in the TGN followed by routing to the lysosome machinery.

Evidence is accumulating that Nef prevents the stable cell-surface expression of CD4 and MHC I through two different mechanisms that require distinct domains of Nef. Most importantly, the dileucine motif at position 154 essential for CD4 downregulation is dispensable for MHC internalization (Mangasarian *et al.*, 1999), indicating that Nef interaction with APs is not necessary for MHC endocytosis. Instead, Nef binding may expose endocytic signals in MHC I that are normally masked (Le Gall *et al.*, 1998). Nef regions important for MHC I downregulation lie in the N-terminal part of the molecule and include a 4-residue acidic stretch (62-EEEE) and a proline-rich motif in the core region encompassing residues 69 to 78 (Greenberg *et al.*, 1998b; Mangasarian *et al.*, 1999) (Fig. 6). Interestingly, this proline-rich region has been implicated in interactions with the SH3 domain of cellular Src family kinases (Lee *et al.*, 1995; Baur *et al.*, 1997; Saksela *et al.*, 1995; Sawai *et al.*, 1994). Although this may suggest that downregulation of MHC 1 involves the interaction between Nef and an SH3-containing protein, the presence of an associated tyrosine kinase activity is not required (Mangasarian *et al.*, 1999). The effect of Nef on MHC I surface expression is thought to protect the infected cell from CTL-mediated lysis (Le Gall *et al.*, 1997; Collins *et al.*, 1998).

XV. Enhancement of Viral Infectivity

One of the early functions attributed to Nef was that of a transcriptional repressor of HIV-1 LTR (Ahmad and Venkatesan, 1988; Niederman *et al.*, 1989, 1991). This activity was tentatively attributed to the negative effect of Nef on the binding of cellular factors implicated in HIV-1 regulation (Niederman *et al.*, 1993; Guy *et al.*, 1990). However, this activity, which gives Nef its name (*ne*gative *f*actor), was rapidly challenged (Hammes *et al.*, 1989; Kim *et al.*, 1989). Nef has instead emerged as an important positive factor for viral infectivity. The initial confusion with regard to the Nef effect on viral replication might have been partially due to the model systems used. Indeed, the use of HIV-1-permissive cell lines with high basal levels of activation and high multiplicity of infection may have masked the effect of Nef (Spina *et al.*, 1994). The positive effect of Nef on viral replication is rather manifested during low-multiplicity infection in quiescent primary peripheral blood mononuclear cells and monocyte/macrophages (Miller *et al.*, 1994; Terwilliger *et al.*, 1991). This positive effect of Nef on viral replication appears to be a natural activity of the protein since nef genes isolated from infected individuals were also shown to accelerate virus production (de Ronde *et al.*, 1992). End-point titration and single-cycle infection experiments demonstrated that Nef accelerates viral spread by enhancing the infectivity of the viral particles and thus promoting the early stages of viral establishment in the target cell (Chowers *et al.*, 1994). The mechanism by which Nef enhances viral infectivity is not fully understood. However, trans-complementation of Nef-defective viruses has shown that Nef is required in the producer cell and cannot restore a wild-type phenotype to Nef-defective viruses when present in the target cell (Pandori *et al.*, 1996). Nef does not modify the density or fusion activity of the env proteins on virions (Miller *et al.*, 1995), neither does it alter the ability of the virus to bind or enter target cells (Chowers *et al.*, 1995). It is thus likely that Nef operates at a postentry step in the viral life cycle. Indeed, virions produced in Nef-positive cells were shown to more efficiently initiate reverse transcription (Chowers *et al.*, 1995; Aiken and Trono, 1995; Schwartz *et al.*, 1995a,b). Interestingly, this was not due to enhanced processivity of the reverse transcriptase (Chowers *et al.*, 1995). How Nef influences virion formation or maturation to enhance infectivity is unclear. Although provocative, the finding that Nef is incorporated in virions (Pandori *et al.*, 1996; Fackler *et al.*, 1996; Bukovsky *et al.*, 1997) where it is cleaved by the viral protease (Pandori *et al.*, 1996) has not been correlated with enhanced infectivity (Miller *et al.*, 1997; Pandori *et al.*, 1998; Chen *et al.*, 1998; Welker *et al.*, 1996). An intriguing possiblity is that Nef-mediated enhancement of viral infectivity is mechanistically related to its ability to interact with the cellular endocytosis machinery. Although the positive effect of Nef on viral infectivity is not the direct consequence of the removal of cell-surface CD4 (Hua *et al.,* 1997; Pandori *et al.,* 1998), the 164LL motif essential for interactions between HIV-1 Nef and the AP complex is also necessary for optimal virus infectivity (Craig *et al.,* 1998). Whether Nef modifies the cell surface representation of cellular proteins important for viral assembly and whether this contributes to enhanced viral infectivity still needs to be addressed.

XVI. Acceleration of Disease Progression

In the absence of Nef, progression to disease following HIV-1 or SIV infection is slower in both human and animal models (Collette, 1997). This property of Nef cannot be attributed to a single biological activity. It rather results from a combination of effects the expression of Nef has on HIV- and SIV-susceptible cells. Particular attention has been placed on the ability of Nef to interfere with T-cell activation. Both the association of Nef with cellular protein kinases and its ability to downregulate key immunological surface receptors such as CD4, MHC I, and IL-2 R is now thought to bring about a general inability of the immune system to control the spread of infection. Early evidence that the presence of Nef influences the course of disease progression came from the finding that patients infected with HIV-1 that remained free of symptoms and maintained normal CD4 lymphocyte counts for more than 10 years mostly harbored virus sequences with deletions in the *nef* gene (Kirchhoff *et al.,* 1995; Deacon *et al.,* 1995; Salvi *et al.,* 1998). Although the nef deletions also removed potentially important sequences in the 3′ LTR, examples of long-term survivors bearing virus with open but mutated nef open reading frames pointed to Nef rather than the LTR as the disease determinant (Premkumar *et al.,* 1996). These studies correlated well with previous findings in rhesus monkeys infected with the SIVmac239 molecular clone bearing a premature stop codon in *nef.* The study showed that the mutant virus rapidly reverted to restore a functional *nef* open reading frame paralleled by the appearance of higher virus loads and accelerated disease progression (Kestler *et al.,* 1991). Similar results were obtained in the SCID-Hu mouse model as well as in transgenic mice where Nef was shown to cause immunodeficiency independent of other viral factors (Jamieson *et al.,* 1994; Lindemann *et al.,* 1994). One possible explanation for the immunodeficiency observed in Nef transgenic mice is that Nef-mediated downregulation of CD4, MHC I, or IL-2 receptor molecules (Greenway *et al.,* 1994) negatively affects the immune response. For instance, Nef-transgenic mouse lines exhibit a decrease in the number of CD4+ cells and impaired response to T-cell receptor (TcR)-mediated activation signals, suggesting that Nef perturbs the development and possibly the selection of CD4+ T cells in the thymus (Brady *et al.,* 1993; Skowronski *et al.,* 1993; Lindemann *et al.,* 1994). Independent of its action of surface receptors, Nef has been shown to impair the immune system in even more profound ways. This area of research has produced the apparently conflicting findings that Nef can both block T-cell activation by mitogenic signals and lead to constitutive T-cell activation. Early evidence indicated that T cells expressing Nef fail to properly respond to mitogenic signals delivered with phorbol 12-myristate 13-acetate (PMA) or by cross-linking of the CD3 chains of the TcR (Collette *et al.,* 1996; Bandres and Ratner, 1994). This led to the failure of the stimulated T-cell to activate a number of proximal as well as distal mitogenic responses such as induction of NF-κB or interleukin 2 (IL-2) mRNA or cell-surface expression of the early activation antigen CD69 (Iafrate *et al.,* 1997; Luria *et al.,* 1991; Niederman *et al.,* 1992). Nef also interferes with the signaling of other stimuli through cell-surface receptors such as a lack of proliferative response to platelet-derived growth factor (PDGF) (De and Marsh, 1994; Graziani *et al.,* 1996) and reduced proliferate response to IL-2 stimulation in peripheral blood mononuclear cells (PBMCs) treated with recombinant Nef (Greenway *et al.,* 1995). In apparent contrast, other studies in transgenic mice have shown that Nef-expressing cells are hypersensitive to stimuli through the TcR as measured by intracellular calcium flux or phosphorylation of key T-cell activation kinases such as MAPK (Hanna *et al.,* 1998; Skowronski *et al.,* 1993). The ability of Nef to induce T-cell activation was linked to SIV pathogenesis in the rhesus and pigtailed monkey model. The SIVsmmPBj14 isolate induces an acute and lethal disease in monkeys. This correlates with the unusual property to replicate in resting PBMCs and to induce lymphocyte proliferation (Fultz, 1991). When the *nef* gene of the SIVmac239 isolate was made to resemble that of PBj14 by substituting a YERL for a RQRL motif, starting at position 17, reminiscent of a consensus sequences for SH2 binding domains ("YE" Nef), T lymphocyte activation as well as development of an acute disease symptomatic of PBj infection were observed (Du *et al.,* 1995). Similarly, SIVmac239 variants selected *in vivo* for high virulence showed a characteristic R to Y substitution at position 17 in Nef (Kirchhoff *et al.,* 1999). The YE Nef was shown to readily insert itself into the T-cell activation pathway. Indeed, using the YERL sequence as an immunoreceptor tyrosine-based activation motif (ITAM), YE Nef was shown to become a substrate for tyrosine phosphorylation by Lck, thus allowing it to interact with the T-cell-specific ZAP-70 tyrosine kinase and activate the downstream activation factor NFAT (Luo and Peterlin, 1997). While it is tempting to speculate that the ability of Nef to induce T-cell activation is alone responsible for acute disease progression, a number of questions remain open. For instance, the ability of SIVmac239 YE Nef to activate T cells was shown to be dependent on the close proximity of macrophages (Du *et al.,* 1995). Since SIVmac239 is capable of infecting macrophages, it is conceivable that the major effect of YE Nef is to induce the expression of adhesion molecules and/or cytokines in monocyte/macrophages which in turn activate the surrounding T cells. This model suggests a scenario where a slow initial infection of macrophages would be amplified by priming T cells for efficient entry and replication of SIV. Interestingly, PBj isolates with Nef proteins bearing an arginine instead of the tyrosine at position 17 lost the ability to activate T cells concomitant with a loss of macrophage tropism (Saucier *et al.*, 1998). Additional evidence that YE Nef alone might not account for the full array of acute disease in monkeys comes from two studies that employed SIV–HIV chimeric viruses (SHIVs). Indeed, SHIV viruses containing either the PBj or mac239 YE Nef in addition to SIV Gag, Pol, and LTR sequences did not cause the acute disease characteristic of SIVsmmPBj14 (Stephens *et al.*, 1998; Shibata *et al.*, 1997).

Nevertheless, it is now clear that Nef has the ability to interfere with intracellular signals that regulate T-cell activation. An attractive explanation for this property is that Nef physically interacts with and modulates the activity of cellular kinases involved in T-cell signaling pathways. Indeed, interactions of HIV-1 Nef with Src-family kinases involved in early transduction of activation signals such as Lyn, Hck (Saksela *et al.*, 1995), and Lck (Collette *et al.*, 1996; Greenway *et al.*, 1996; Dutartre *et al.*, 1998) have been demonstrated. Nef also interacts with downstream factors such as the Nef-associated kinase (NAK), a putative member of the p21-activated protein kinases (PAK) (Sawai *et al.*, 1994). The exact physiological relevance of these various interactions is not clear. However, recent data may suggest a model by which interaction of Nef with two sets of protein kinases would both inhibit stimulation through cell-surface receptors and induce the artificial activation of T-cells or monocyte/macrophages. The association of Nef with proximal mediators of activation such as Lck and Hck would lead to dissociation of the kinases from their surface receptors, thus disconnecting a key step in transduction of extracellular activation signals (Greenway *et al.*, 1995; Iafrate *et al.*, 1997). A similar effect might be achieved by SIV-mac239 Nef, which was shown to interact with the TcR ζ-chain (Howe *et al.*, 1998). The fact that inhibition of CD3 signaling is dependent on the presence of the PXXP motif at position 72 in HIV-1 Nef, which binds the SH3 domain of src family kinases, is further evidence in favor of that model (Iafrate *et al.*, 1997). Concomitantly, Nef could mediate T-cell activation on its own terms by selectively targeting downstream members of the signaling pathway. This is evidenced by the ability of Nef to activate MAP kinase (Hanna *et al.*, 1998) and to stimulate the production of cytokines from macrophages (Du *et al.*, 1995). Association of Nef with NAK could also contribute in activation. This hypothesis is supported by the fact that, at least for SIVmac239 Nef, association with NAK is not correlated with inhibition of CD3 signaling (Iafrate *et al.*, 1997) and that binding of HIV-1 Nef to NAK activates the kinase (Lu *et al.*, 1996). However, there is evidence that Nef does not bind to NAK or other serine kinases directly but through Nef–Lck or Nef–Hck complexes (Manninen *et al.*, 1998; Baur *et al.*, 1997). Taken together, these data suggest that Nef modulates its effect on T-cell activation through differential interactions with cellular kinases. How Nef regulates these interactions is not clear at present although it was suggested that this might depend on the subcellular localization of Nef (Baur *et al.*, 1994).

XVII. HIV-1-Specific Vpu Protein

Vpu is an 81-amino-acid type-1 integral membrane protein composed of three discrete α-helices. (Strebel *et al.*, 1988; Cohen *et al.*, 1988). The N-terminal helix constitutes the transmembrane anchor and is followed by a cytoplasmic tail containing two amphipathic α-helices (Willbold *et al.*, 1997; Federau *et al.*, 1996). The Rev-dependent bicistronic mRNA that encodes Vpu also contains the downstream Env ORF, which is translated by leaky scanning of the Vpu initiation codon (Schwartz *et al.*, 1990). The *vpu* gene is not always functional due to the presence of mutated initiation codons or internal deletions (Korber *et al.*, 1997), suggesting a mechanism by which Vpu expression is regulated by the virus (Schubert *et al.*, 1999). Although the *vpu* gene is only found in HIV-1 strains, the envelope protein of certain isolates of HIV-2 have been shown to assume some of the functionality of the HIV-1 Vpu protein (Ritter *et al.*, 1996; Bour *et al.*, 1996). The Vpu protein has two main roles in the viral life cycle: it promotes the efficient release of viral particles from the cell surface and it induces the degradation of CD4, and possibly other transmembrane proteins, in the endoplasmic reticulum (ER).

XVIII. Vpu-Mediated CD4 Degradation

As critical levels of Rev are achieved in the HIV-infected cell, the viral genome transcribes a number of genes in addition to the early *tat, rev,* and *nef* (Klotman *et al.*, 1991). Among these late genes, the *env* and *vpu* ORFs are of particular significance to the viral effort to downregulate cell-surface CD4. The gp160 envelope glycoprotein precursor is a major player in CD4 downmodulation that can, in most instances, quantitatively block the bulk of newly synthesized CD4 in the endoplasmic reticulum (ER) (Crise *et al.*, 1990; Jabbar and Nayak, 1990; Bour *et al.*, 1995). This strategy has, however, too principal shortcomings. First, in contrast to Nef, Env is unable to remove preexisting CD4 molecules that have reached the cell surface. Second, the formation of CD4–gp160 complexes in the ER blocks the transport and maturation of not only CD4 but of the envelope protein itself (Bour *et al.*,

1991). In cases where equimolar amounts of CD4 and Env are synthesized, this could lead to the production of noninfectious virions devoid of envelope protein (Buonocore and Rose, 1990, 1993). A first role of the HIV-1 Vpu protein is to induce the degradation of CD4 molecules trapped in intracellular complexes with Env, thus allowing gp160 to resume transport toward the cell surface (Willey, 1992a). In Vpu-expressing cells CD4 is rapidly degraded and its half-life drops from 12 h to approximately 15 min (Willey, 1992b). Interestingly, the rate of Vpu-mediated degradation slowed when CD4 was expressed in the absence of Env or in the presence of a CD4 binding-deficient envelope variant (Willey, 1992a,b). However, the contribution of Env in the mechanism of CD4 degradation was limited to its ability to trap CD4 in the ER and optimal rates of degradation could be achieved by replacing Env with the ER-retention drug brefeldin A (Willey, 1992b). The importance of ER localization for CD4 susceptibility to Vpu-mediated degradation suggests that cellular factors essential for CD4 catalysis are located in the ER and/or that the rate-limiting step in the mechanism of degradation is for Vpu to find and target CD4 (Chen *et al.,* 1993). In agreement with the latter, coimmunoprecipitation experiments showed that CD4 and Vpu physically interact in the ER of cells and that this interaction is essential for targeting CD4 to the degradation pathway (Bour *et al.,* 1995). Using chimeric constructs, several groups have shown a requirement for the CD4 cytoplasmic doman (Willey *et al.,* 1994) and a possible contribution of the transmembrane domain of CD4 for susceptibility to Vpu-mediated degradation (Raja *et al.,* 1994; Buonocore *et al.,* 1994). Mutagenesis studies further confirmed the important role of the CD4 cytoplasmic tail and delineated a Vpu susceptibility domain extending from residues 416 to 418 (EKKT) (Yao *et al.,* 1995; Vincent *et al.,* 1993; Lenburg and Landau, 1993). This Vpu susceptibility domain is also required for CD4–Vpu interactions and is therefore likely to represent the Vpu binding site on CD4 (Bour *et al.,* 1995). Interestingly, the Vpu binding site on CD4 does not include the dileucine motif required for Nef action nor the cysteine residues involved in CD4–Lck interactions. The domains of Vpu required for CD4 binding are less discrete, suggesting that conformational rather than linear structures are involved. Although mutations that disturb the structure of either cytoplasmic α-helix disrupt the ability of Vpu to induce CD4 degradation, the membrane-proximal helix is thought to mediate the direct interactions with CD4 (Margottin *et al.,* 1996; Tiganos *et al.,* 1997). The Vpu cytoplasmic α-helices are separated by a short flexible region that contains two highly conserved serine residues phosphorylated *in vivo* by casein kinase II (Schubert and Strebel, 1994). These serine residues at positions 52 and 56 are critically important for the activity of Vpu on CD4 degradation (Paul and Jabbar, 1997; Schubert and Strebel, 1994). However, phosphorylation-defective mutants of Vpu, while inactive with regard to CD4 degradation, retained the capacity to interact with the CD4 cytoplasmic tail (Bour *et al.,* 1995). This finding led to the hypothesis that Vpu binding to CD4 was necessary but not sufficient to induce degradation (Bour *et al.,* 1995). The role of the Vpu phosphoserine residues in the induction of CD4 degradation was recently elucidated. Indeed, we have shown by yeast two-hybrid assay and direct coimmunoprecipitation that Vpu interacts with the human β-transducin repeat-containing protein or ßTrCP (Margottin *et al.,* 1998). Interestingly, Vpu variants mutated at serines 52 and 56 were unable to interact with ßTrCP, providing a mechanistic explanation for the requirement for Vpu phosphorylation, and strongly suggested that ßTrCP was directly involved in the catalysis of Vpu-targeted CD4 molecules (Margottin *et al.,* 1998). Structurally, ßTrCP shows a modular organization. Similarly to its *Xenopus laevis* homolog (Spevak *et al.,* 1993), human ßTrCP contains seven C-terminal WD repeats, a structure known to mediate protein–protein interactions (Neer *et al.,* 1994). Accordingly, the WD repeats of h-ßTrCP were shown to mediate interactions with Vpu in a phosphoserine-dependent fashion (Margottin *et al.,* 1998). ßTrCP also contains an F-box domain recently identified as a connector between target proteins and the ubiquitin-dependent proteolytic machinery (Bai *et al.,* 1996). This domain was of particular significance since there is evidence for the involvement of the ubiquitin-proteasome machinery Vpu-mediated CD4 degradation (Schubert *et al.,* 1998; Fujita *et al.,* 1997). Although the molecular mechanisms by which Vpu targets CD4 for degradation are now reasonably well defined (Fig. 7), how CD4 is ultimately brought into contact with the proteasome is still unclear. A number of catalytic pathways involving WD–F-box proteins and proteasome degradation have recently been deciphered that may at least partially resemble that of Vpu-mediated CD4 degradation. In the yeast model, for instance, progression of the cell cycle from G1 to S is regulated by activation of the cycline-dependent kinase Cdc28 (Patton *et al.,* 1998). This activation requires the WD–F-box-containing protein Cdc4 to degrade the Cdc28-bound Sic1 inhibitor (Fig. 7). Similar to ßTrCP, Cdc4 interacts with Sic1 through its WD domain while the F-box domain recruits an E3 ubiquitin–ligase complex containing Cdc53 through interactions with the SkpI protein. Similarly, ßTrCP itself was shown to interact with SkpI both in the yeast two-hybrid assay (Margottin *et al.,* 1998) and by direct coimmunoprecipitation (Bour, unpublished data). Recent studies have confirmed the central role of TrCP in the regulated degradation of cellular proteins. Indeed, the TrCP-containing *S*kpI, *C*dc53, *F*-box protein (SCF) E3 complex was shown to mediate the ubiquitination and proteasome targeting of both ß-cetenin and the NFκB inhibitor IκBα (Yaron *et al.,* 1998; Winston *et al.,* 1999; Spencer *et al.,* 1999; Hatakeyama *et al.,* 1999). Interestingly, the signal for recognition of both cellular substrates by ßTrCP consists of a pair of conserved phosphorylated serine residues similar to those required in Vpu (Margottin *et al.,* 1998). These serine residues are arranged in a consensus motif present in all three proteins (D*S*GØX*S*). Serine phosphorylation plays the major

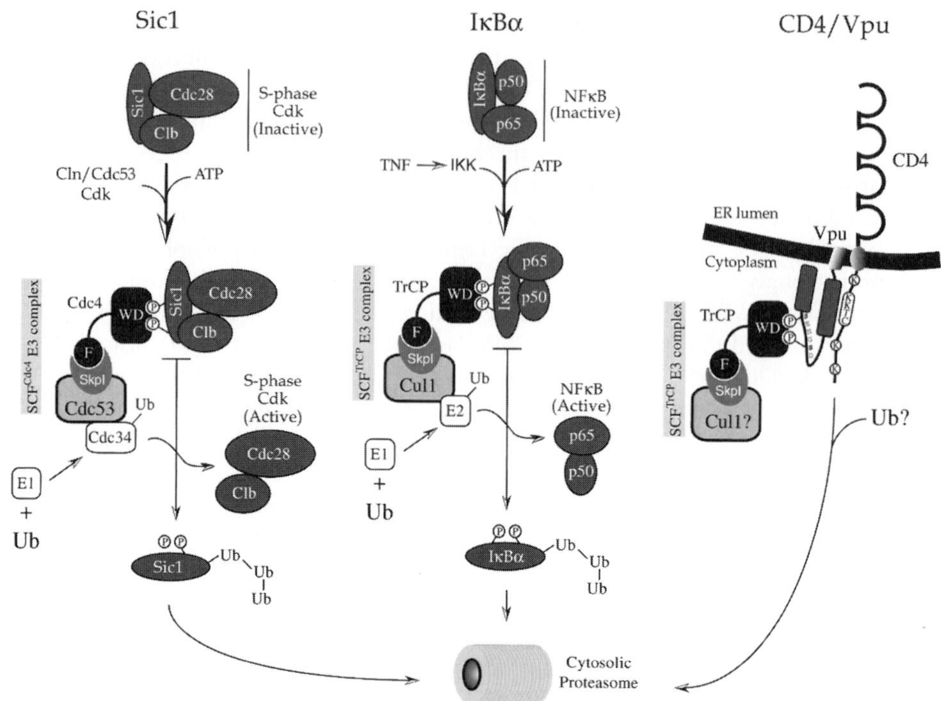

FIGURE 7 Mechanism of protein degradation mediated by the TrCP–SCF complex.

regulatory role in the stability of SCF target proteins. ß-Catenin, a member of a transcriptional activation complex involved in development and tumor progression, is constitutively phosphorylated by the glycogen synthetase kinase 3b and therefore constitutively degraded by the ßTrCP–SCF–E3 complex (Peifer, 1997). Signaling through the Wnt pathway inactivates GSK3ß, leading to stabilization of ß-catenin (Cadigan and Nusse, 1997). Conversely, IκBα becomes phosphorylated and unstable only following activation of the IκB–kinase complex (IKK) by stimuli such as TNFα treatment (Hochstrasser, 1996). In the case of Vpu, however, phosphorylation is constitutively mediated by the ubiquitous casein kinase II (Schubert and Strebel, 1994). Since there is no known phosphatase that mediates dephosphorylation of Vpu, it is reasonable to assume that CD4 degradation is operative throughout the viral life cycle. A number of questions remain to be answered before the mechanism of Vpu-mediated degradation of CD4 is fully elucidated. While in the case of Sic1, ß-catenin and IκBα recruitment of the E3-ubiquitin-ligase activity leads to polyubiquitination and degradation by the proteasome there is no firm evidence that CD4 is ubiquitinated in the course of its degradation by Vpu (Fujita *et al.,* 1997; Schubert *et al.,* 1998). It is also not clear at present whether, similar to other models (Wiertz *et al.,*

1996), extraction of CD4 from the ER membrane is required for complete degradation.

XIX. Enhancement of Viral Particle Release

In addition to its destabilizing effect on CD4, the Vpu protein mediates the efficient release of viral particles from HIV-1-infected cells (Klimkait *et al.,* 1990; Terwilliger *et al.,* 1989; Strebel *et al.,* 1989). The two biological activities of Vpu appear to be mechanistically and structurally distinct. Indeed, the particle release activity of Vpu is independent of CD4, the envelope glycoprotein, and, to a large extent, of the presence of serine residues 52 and 56 (Friborg *et al.,* 1995; Yao *et al.,* 1992; Schubert and Strebel, 1994; Geraghty and Panganiban, 1993). In addition, while the determinants for CD4 degradation have all localized to the cytoplasmic tail of Vpu, the transmembrane domain has been shown to play an essential role for the particle release activity (Paul *et al.,* 1998; Schubert *et al.,* 1996). The ability of Vpu to enhance viral production relies on the integrity of the transmembrane domain and its ability to mediate homooligomerization of Vpu (Paul *et al.,* 1998; Schubert *et al.,* 1996; Maldarelli *et al.,* 1993; Moore *et al.,* 1998). It is still debated whether Vpu enhances virus production through a global modification of the cellular environment or through discreet interactions with cellular or viral factors. On the one hand, the finding that Vpu form ion-conductive channels at the cell surface argues in favor of the former possibility (Ewart *et al.,* 1996; Gonzalez and Carrasco, 1998; Grice *et al.,* 1997; Schubert *et al.,* 1996a,b). On the other hand, the exact nature of the membrane pore formed by Vpu and its effect on viral assembly are still unclear (Coady *et al.,* 1998; Lamb and Pinto, 1997). In addition, interactions between Vpu and a novel cellular protein (Vpu-binding protein or UBP) have now been shown to be involved in the mechanism of viral particle production (Callahan *et al.,* 1998). Interestingly, UBP also interacts with the 55-kDa Gag precursor of HIV-1. However, the binding of UBP to Vpu or Gag appears to be competitive rather that cooperative. Moreover, overexpression of UBP abrogates the ability of Vpu to promote viral particle release, suggesting that UBP is a negative factor for virus assembly which needs to be displaced from Gag by Vpu (Callahan *et al.,* 1998). Alternatively, the association followed by Vpu-mediated dissociation of UBP from Gag might generate conformational changes in Gag important for virus formation and release (Callahan *et al.,* 1998).

Although the *vpu* gene is unique to HIV-1, the activity Vpu provides for enhanced viral particle release is not. Indeed, the envelope proteins of both HIV-2 ROD10 and ST2 were shown to promote viral particle release in a manner similar to that of Vpu (Bour *et al.,* 1996; Ritter *et al.,* 1996). The finding that the Env protein of HIV-2 ROD could also enhance the particle release of HIV-1 strongly suggests that the two activities are mechanistically and perhaps evolutionally related (Bour and Strebel, 1996). Accordingly, the transmembrane subunit of HIV-2, which resembles Vpu structurally, was shown to be important for efficient particle release (Bour and Strebel, 1996). Further similarities between Vpu and HIV-2 ROD Env lie in the fact that, in both cases, the cytoplasmic tail is largely dispensable for enhancement of particle release (Bour *et al.,* 1999a). The absence of a CD4 degradation activity in the ROD10 Env further suggests that this additional function may have evolved in Vpu from the ancestral particle release activity in response to increased affinity between HIV-1 Env and CD4 (Willey *et al.,* 1992a,b; Bour and Strebel, 1996).

As details of Vpu's action on CD4 degradation and particle release emerge, the question arises as to why such apparently unrelated activities have evolved within a single protein. One possible explanation relates to the ability of Vpu to liberate envelope protein precursors trapped in CD4 intracellular complexes. As the rate of viral particle production augments, the action of Vpu would guarantee that enough mature envelope proteins are available for incorporation into virions. Another possibility is that the presence of CD4 at the cell surface actively interferes with the ability of Vpu to promote viral particle release. Our recent findings that CD4 and Vpu can physically interact at the plasma membrane and that the presence of cell-surface CD4 leads to decreased viral particle release lends credence to this hypothesis (Bour *et al.,* 1999b).

Acknowledgments

We thank John Guatelli, Sundararajan Venkatesan, Malcolm Martin, and Eric Freed for discussions.

References

Agostini, I., Navarro, J. M., Rey, F., Bouhamdan, M., Spire, B., Vigne, R., and Sire, J. (1996). The human immunodeficiency virus type 1 Vpr transactivator: cooperation with promoter-bound activator domains and binding to TFIIB. *J Mol Biol.* **261,** 599–606.

Ahmad, N., and Venkatesan, S. (1988). Nef protein of HIV-1 is a transcriptional repressor of HIV-1 LTR (published erratum in Science 1988 242:242). *Science* **241,** 1481–1485.

Aiken, C., and Trono, D. (1995). Nef stimulates human immunodeficiency virus type 1 proviral DNA synthesis. *J. Virol.* **69,** 5048–5056.

Aiken, C., Konner, J., Landau, N. R., Lenburg, M. E., and Trono, D. (1994). Nef induces CD4 endocytosis: Requirement for a critical dileucine motif in the membrane-proximal CD4 cytoplasmic domain. *Cell* **76,** 853–864.

Aldrovandi, G. M., Gao, L., Bristol, G., and Zack, J. A. (1998). Regions of human immunodeficiency virus type 1 nef required for function *in vivo. J. Virol.* **72,** 7032–7039.

Anderson, S., Shugars, D. C., Swanstrom, R., and Garcia, J. V. (1993). Nef from primary isolates of human immunodeficiency virus type 1 suppresses surface CD4 expression in human and mouse T cells. *J. Virol.* **67,** 4923–4931.

Anderson, S. J., Lenburg, M., Landau, N. R., and Garcia, J. V. (1994). The cytoplasmic domain of CD4 is sufficient for its down-regulation from the cell surface by human immunodeficiency virus type 1 Nef. *J. Virol.* **68,** 3092–3101. [published erratum appears in *J. Virol* (1994) **68,** 4705].

Bai, C., Sen, P., Hofmann, K., Ma, L., Goebl, M., Harper, J. W., and Elledge, S. J. (1996). SKP1 connects cell cycle regulators to the ubiquitin proteolysis machinery through a novel motif, the F-box. *Cell* **86,** 263–274.

Balotta, C., Lusso, P., Crowley, R., Gallo, R. C., and Franchini, G. (1993). Antisense phosphothioate oligodeoxynucleotides targeted to the vpr gene inhibit human immunodeficiency virus type 1 replication in primary human macrophages. *J. Virol.* **67,** 4409–4414.

Bandres, J. C., and Ratner, L. (1994). Human immunodeficiency virus type 1 Nef protein down-regulates transcription factors NF-kappa B and AP-1 in human T cells *in vitro* after T-cell receptor stimulation. *J. Virol.* **68,** 3243–3249.

Baraz, L., Friedler, A., Blumenzweig, I., Nussinuv, O., Chen, N., Steinitz, M., Gilon, C., and Kotler, M. (1998). Human immunodeficiency virus type 1 Vif-derived peptides inhibit the viral protease and arrest virus production. *FEBS Lett.* **441,** 419–426.

Bartz, S. R., Rogel, M. E., and Emerman, M. (1996). Human immunodeficiency virus type 1 cell cycle control: Vpr is cytostatic and mediates G2 accumulation by a mechanism which differs from DNA damage checkpoint control. *J. Virol.* **70,** 2324–2331.

Baur, A. S., Sass, G., Laffert, B., Willbold, D., Cheng-Mayer, C., and Peterlin, B. M. (1997). The N-terminus of Nef from HIV-1/SIV associates with a protein complex containing Lck and a serine kinase. *Immunity* **6,** 283–291.

Baur, A. S., Sawai, E. T., Dazin, P., Fantl, W. J., Cheng-Mayer, C., and Peterlin, B. M. (1994). HIV-1 Nef leads to inhibition or activation of T cells depending on its intracellular localization. *Immunity* **1,** 373–384.

Benichou, S., Bomsel, M., Bodeus, M., Durand, H., Doute, M., Letourneur, F., Camonis, J., and Benarous, R. (1994). Physical interaction of the HIV-1 Nef protein with beta-COP, a component of non-clathrin-coated vesicles essential for membrane traffic. *J. Biol. Chem.* **269,** 30073–30076.

Benson, R. E., Sanfridson, A., Ottinger, J. S., Doyle, C., and Cullen, B. R. (1993). Downregulation of cell-surface CD4 expression by simian immunodeficiency virus Nef prevents viral super infection. *J. Exp. Med.* **177,** 1561–1566.

Blanc, D., Patience, C., Schulz, T. F., Weiss, R., and Spire, B. (1993). Transcomplementation of VIF- HIV-1 mutants in CEM cells suggests that VIF affects late steps of the viral life cycle. *Virology* **193,** 186–192.

Borman, A. M., Quillent, C., Charneau, P., Dauguet, C., and Clavel, F. (1995). Human immunodeficiency virus type 1 Vif- mutant particles from restrictive cells: Role of Vif in correct particle assembly and infectivity. *J. Virol.* **69,** 2058–2067.

BouHamdan, M., Xue, Y., Baudat, Y., Hu, B., Sire, J., Pomerantz, R. J., and Duan, L. X. (1998). Diversity of HIV-1 Vpr interactions involves usage of the WXXF motif of host cell proteins. *J. Biol. Chem.* **273,** 8009–8016.

Bour, S., and Strebel, K. (1996). The human immunodeficiency virus (HIV) type 2 envelope protein is a functional complement to HIV type 1 Vpu that enhances particle release of heterologous retroviruses. *J. Virol.* **70,** 8285–8300.

Bour, S., Boulerice, F., and Wainberg, M. A. (1991). Inhibition of gp160 and CD4 maturation in U937 cells after both defective and productive infections by human immunodeficiency virus type 1. *J. Virol.* **65,** 6387–6396.

Bour, S., Geleziunas, R., and Wainberg, M. A. (1995). The human immunodeficiency virus type 1 (HIV-1) CD4 receptor and its central role in the promotion of HIV-1 infection. *Microbiol. Rev.* **59,** 63–93.

Bour, S., Schubert, U., Peden, K., and Strebel, K. (1996). The envelope glycoprotein of human immunodeficiency virus type 2 enhances viral particle release: A Vpu-like factor? *J. Virol.* **70**, 820–829.

Bour, S., Schubert, U., and Strebel, K. (1995). The human immunodeficiency virus type 1 Vpu protein specifically binds to the cytoplasmic domain of CD4: Implications for the mechanism of degradation. *J. Virol.* **69**, 1510–1520.

Bour, S. P., Aberham, C., Perrin, C., and Strebel, K. (1999a). Lack of effect of cytoplasmic tail truncations of human immunodeficiency virus type 2 ROD env particle release activity. *J. Virol.* **73**, 778–782.

Bour, S., Perrin, C., and Strebel, K. (1999b). Cell surface CD4 inhibits HIV-1 particle release by interfering with Vpu activity. *J. Biol. Chem.* **274**, 33800–33806.

Bouyac, M., Courcoul, M., Bertoia, G., Baudat, Y., Gabuzda, D., Blanc, D., Chazal, N., Boulanger, P., Sire, J., Vigne, R., and Spire, B. (1997). Human immunodeficiency virus type 1 Vif protein binds to the Pr55Gag precursor. *J. Virol.* **71**, 9358–9365.

Bouyac, M., Rey, F., Nascimbeni, M., Courcoul, M., Sire, J., Blanc, D., Clavel, F., Vigne, R., and Spire, B. (1997). Phenotypically Vif- human immunodeficiency virus type 1 is produced by chronically infected restrictive cells. *J. Virol.* **71**, 2473–2477.

Brady, H. J., Pennington, D. J., Miles, C. G., and Dzierzak, E. A. (1993). CD4 cell surface downregulation in HIV-1 Nef transgenic mice is a consequence of intracellular sequestration. *EMBO J.* **12**, 4923–4932.

Bresnahan, P. A., Yonemoto, W., Ferrell, S., Williams-Herman, D., Geleziunas, R., and Greene, W. C. (1998). A dileucine motif in HIV-1 Nef acts as an internalization signal for CD4 downregulation and binds the AP-1 clathrin adaptor. *Curr. Biol.* **8**, 1235–1238.

Bukovsky, A. A., Dorfman, T., Weimann, A., and Gottlinger, H. G. (1997). Nef association with human immunodeficiency virus type 1 virions and cleavage by the viral protease. *J. Virol.* **71**, 1013–1018.

Bukrinsky, M. I., and Haffar, O. K. (1997). HIV-1 nuclear import: In search of a leader. *Front Biosci.* **2**, 578–587.

Bukrinsky, M. I., Haggerty, S., Dempsey, M. P., Sharova, N., Adzhubel, A., Spitz, L., Lewis, P., Goldfarb, D., Emerman, M., and Stephenson, M. (1993). A nuclear localization signal within HIV-1 matrix protein that governs infection of non-dividing cells. *Nature* **365**, 666–669.

Buonocore, L., and Rose, J. K. (1990). Prevention of HIV-1 glycoprotein transport by soluble CD4 retained in the endoplasmic reticulum. *Nature* **345**, 625–628.

Buonocore, L., and Rose, J. K. (1993). Blockade of human immunodeficiency virus type 1 production in CD4+ T cells by an intracellular CD4 expressed under control of the viral long terminal repeat. *Proc. Natl. Acad. Sci. USA* **90**, 2695–2699.

Buonocore, L., Turi, T. G., Crise, B., and Rose, J. K. (1994). Stimulation of heterologous protein degradation by the Vpu protein of HIV-1 requires the transmembrane and cytoplasmic domains of CD4. *Virology* **204**, 482–486.

Cadigan, K. M., and Nusse, R. (1997). Wnt signaling: a common theme in animal development. *Genes Dev.* **11**, 3286–3305.

Callahan, M. A., Handley, M. A., Lee, Y. H., Talbot, K. J., Harper, J. W., and Panganiban, A. T. (1998). Functional interaction of human immunodeficiency virus type 1 Vpu and Gag with a novel member of the tetratricopeptide repeat protein family. *J. Virol.* **72**, 5189–5197. [Erratum appears in *J. Virol.* **72**, 8461]

Camaur, D., and Trono, D. (1996). Characterization of human immunodeficiency virus type 1 Vif particle incorporation. *J. Virol.* **70**, 6106–6111.

Chen, M. Y., Maldarelli, F., Karczewski, M. K., Willey, R. L., and Strebel, K. (1993). Human immunodeficiency virus type 1 Vpu protein induces degradation of CD4 *in vitro:* The cytoplasmic domain of CD4 contributes to Vpu sensitivity. *J. Virol.* **67**, 3877–3884.

Chen, Y. L., Trono, D., and Camaur, D. (1998). The proteolytic cleavage of human immunodeficiency virus type 1 Nef does not correlate with its ability to stimulate virion infectivity. *J. Virol.* **72**, 3178–3184.

Chowers, M. Y., Pandori, M. W., Spina, C. A., Richman, D. D., and Guatelli, J. C. (1995). The growth advantage conferred by HIV-1 nef is determined at the level of viral DNA formation and is independent of CD4 downregulation. *Virology* **212,** 451–457.

Chowers, M. Y., Spina, C. A., Kwoh, T. J., Fitch, N. J., Richman, D. D., and Guatelli, J. C. (1994). Optimal infectivity *in vitro* of human immunodeficiency virus type 1 requires an intact nef gene. *J. Virol.* **68,** 2906–2914.

Coady, M. J., Daniel, N. G., Tiganos, E., Allain, B., Friborg, J., Lapointe, J. Y., and Cohen, E. A. (1998). Effects of Vpu expression on *Xenopus* oocyte membrane conductance. *Virology* **244,** 39–49.

Cohen, E. A., Dehni, G., Sodroski, J. G., and Haseltine, W. A. (1990). Human Immunodeficiency Virus vpr product is a virion-associated regulatory protein. *J. Virol.* **64,** 3097–3099.

Cohen, E. A., Terwilliger, E. F., Sodroski, J. G., and Haseltine, W. A. (1988). Identification of a protein encoded by the vpu gene of HIV-1. *Nature* **334,** 532–534.

Coleman, T. R., and Dunphy, W. G. (1994). Cdc2 regulatory factors. *Curr. Opin. Cell. Biol.* **6,** 877–882.

Collette, Y. (1997). *Res. Virol.* **148,** 23–30.

Collette, Y., Chang, H. L., Cerdan, C., Chambost, H., Algarte, M., Mawas, C., Imbert, J., Burny, A., and Olive, D. (1996). Specific Th1 cytokine down-regulation associated with primary clinically derived human immunodeficiency virus type 1 Nef gene-induced expression. *J. Immunol.* **156,** 360–370.

Collette, Y., Dutartre, H., Benziane, A., Ramos, M., Benarous, R., Harris, M., and Olive, D. (1996). Physical and functional interaction of Nef with Lck: HIV-1 Nef-induced T-cell signaling defects. *J. Biol. Chem.* **271,** 6333–6341.

Collette, Y., Mawas, C., and Olive, D. (1996). Evidence for intact CD28 signaling in T cell hyporesponsiveness induced by the HIV-1 nef gene. *Eur. J. Immunol.* **26,** 1788–1793.

Collins, K. L., Chen, B. K., Kalams, S. A., Walker, B. D., and Baltimore, D. (1998). HIV-1 Nef protein protects infected primary cells against killing by cytotoxic T lymphocytes. *Nature* **391,** 397–401.

Connor, R. I., Chen, B. K., Choe, S., and Landau, N. R. (1995). Vpr is required for efficient replication of human immunodeficiency virus type-1 in mononuclear phagocytes. *Virology* **206,** 935–944.

Conti, L., Rainaldi, G., Matarrese, P., Varano, B., Rivabene, R., Columba, S., Sato, A., Belardelli, F., Malorni, W., and Gessani, S. (1998). The HIV-1 vpr protein acts as a negative regulator of apoptosis in a human lymphoblastoid T cell line: Possible implications for the pathogenesis of AIDS. *J. Exp. Med.* **187,** 403–413.

Courcoul, M., Patience, C., Rey, F., Blanc, D., Harmache, A., Sire, J., Vigne, R., and Spire, B. (1995). Peripheral blood mononuclear cells produce normal amounts of defective Vif-human immunodeficiency virus type 1 particles which are restricted for the preretrotranscription steps. *J. Virol.* **69,** 2068–2074.

Craig, H. M., Pandori, M. W., and Guatelli, J. C. (1998). Interaction of HIV-1 nef with the cellular dileucine-based sorting pathway is required for CD4 down-regulation and optimal viral infectivity. *Proc. Natl. Acad. Sci. USA* **95,** 11229–11234.

Crise, B., Buonocore, L., and Rose, J. K. (1990). CD4 is retained in the endoplasmic reticulum by the human immunodeficiency virus type 1 glycoprotein precursor. *J. Virol.* **64,** 5585–5593.

de Rocquigny, H., Petitjean, P., Tanchou, V., Decimo, D., Drouot, L., Delaunay, T., Darlix, J. L., and Roques, B. P. (1997). The zinc fingers of HIV nucleocapsid protein NCp7 direct interactions with the viral regulatory protein Vpr. *J. Biol. Chem.* **272,** 30753–30759.

de Ronde, A., Klaver, B., Keulen, W., Smit, L., and Goudsmit, J. (1992). Natural HIV-1 NEF accelerates virus replication in primary human lymphocytes. *Virology* **188,** 391–395.

De, S. K., and Marsh, J. W. (1994). HIV-1 Nef inhibits a common activation pathway in NIH-3T3 cells. *J. Biol. Chem.* **269,** 6656–6660.

Deacon, N. J., Tsykin, A., Solomon, A., Smith, K., Ludford-Menting, M., Hooker, D. J., McPhee, D. A., Greenway, A. L., Ellett, A., Chatfield, C., Lawson, V. A., Crowe, S.,

Maerz, A., Sonza, S., Learmont, J., Sullivan, J. S., Cunningham, A., Dwyer, D., Dowton, D., and Mills, J. (1995). Genomic structure of an attenuated quasi species of HIV-1 from a blood transfusion donor and recipients. *Science* **270,** 988–991.

Dettenhofer, M., and Yu, X. F. (1999). Highly purified human immunodeficiency virus type 1 reveals a virtual absence of Vif in virions. *J. Virol.* **73,** 1460–1467.

Di Marzio, P., Choe, S., Ebright, M., Knoblauch, R., and Landau, N. R. (1995). Mutational analysis of cell cycle arrest, nuclear localization and virion packaging of human immunodeficiency virus type 1 Vpr. *J. Virol.* **69,** 7909–7916.

Du, Z., Lang, S. M., Sasseville, V. G., Lackner, A. A., Ilyinskii, P. O., Daniel, M. D., Jung, J. U., and Desrosiers, R. C. (1995). Identification of a nef allele that causes lymphocyte activation and acute disease in macaque monkeys. *Cell* **82,** 665–675.

Dutartre, H., Harris, M., Olive, D., and Collette, Y. (1998). The human immunodeficiency virus type 1 Nef protein binds the Src- related tyrosine kinase Lck SH2 domain through a novel phosphotyrosine independent mechanism. *Virology* **247,** 200–211.

Ewart, G. D., Sutherland, T., Gage, P. W., and Cox, G. B. (1996). The Vpu protein of human immunodeficiency virus type 1 forms cation-selective ion channels. *J. Virol.* **70,** 7108–7115.

Fackler, O. T., Kremmer, E., and Mueller-Lantzsch, N. (1996). Evidence for the association of Nef protein with HIV-2 virions. *Virus Res.* **46,** 105–110.

Fan, L., and Peden, K. (1992). Cell-free transmission of Vif mutants of HIV-1. *Virology* **190,** 19–29.

Federau, T., Schubert, U., Flossdorf, J., Henklein, P., Schomburg, D., and Wray, V. (1996). Solution structure of the cytoplasmic domain of the human immunodeficiency virus type 1 encoded virus protein U (Vpu). *Int. J. Pept. Protein Res.* **47,** 297–310.

Felzien, L. K., Woffendin, C., Hottiger, M. O., Subbramanian, R. A., Cohen, E. A., and Nabel, G. J. (1998). HIV transcriptional activation by the accessory protein, VPR, is mediated by the p300 co-activator. *Proc. Natl. Acad. Sci. USA* **95,** 5281–5286.

Fisher, A. G., Ensoli, B., Ivanoff, L., Chamberlain, M., Petteway, S., Ratner, L., Gallo, R. C., and Wong-Staal, F. (1987). The sor gene of HIV-1 is required for efficient virus transmission *in vitro*. *Science* **237,** 888–893.

Fletcher, T. M., III, Brichacek, B., Sharova, N., Newman, M. A., Stivahtis, G., Sharp, P. M., Emerman, M., Hahn, B. H., and Stevenson, M. (1996). Nuclear import and cell cycle arrest functions of the HIV-1 Vpr protein are encoded by two separate genes in HIV-2/SIV(SM). *EMBO J.* **15,** 6155–6165.

Forget, J., Yao, X. J., Mercier, J., and Cohen, E. A. (1998). Human immunodeficiency virus type 1 vpr protein transactivation function: Mechanism and identification of domains involved. *J. Mol. Biol.* **284,** 915–923.

Fouchier, R. A., Meyer, B. E., Simon, J. H., Fischer, U., Albright, A. V., Gonzalez-Scarano, F., and Malim, M. H. (1998). Interaction of the human immunodeficiency virus type 1 Vpr protein with the nuclear pore complex. *J. Virol.* **72,** 6004–6013.

Fouchier, R. A., Simon, J. H., Jaffe, A. B., and Malim, M. H. (1996). Human immunodeficiency virus type 1 Vif does not influence expression or virion incorporation of gag-, pol-, and env-encoded proteins. *J. Virol.* **70,** 8263–8269.

Freed, E. O., Englund, G., Maldarelli, F., and Martin, M. A. (1997). Phosphorylation of residue 131 of HIV-1 matrix is not required for macrophage infection. *Cell* **88,** 171–173.

Friborg, J., Ladha, A., Gottlinger, H., Haseltine, W. A., and Cohen, E. A. (1995). Functional analysis of the phosphorylation sites on the human immunodeficiency virus type 1 Vpu protein. *J. Acquir. Immune Defic. Syndr. Hum. Retrovirol.* **8,** 10–22.

Friedler, A., Blumenzweig, I., Baraz, L., Steinitz, M., Kotler, M., and Gilon, C. (1999). Peptides derived from HIV-1 Vif: A non-substrate based novel type of HIV-1 protease inhibitors. *J. Mol. Biol.* **287,** 93–101.

Fujita, K., Omura, S., and Silver, J. (1997). Rapid degradation of CD4 in cells expressing human immunodeficiency virus type 1 Env and Vpu is blocked by proteasome inhibitors. *J. Gen. Virol.* **78,** 619–625. [published erratum appears in *J. Gen. Virol.* **78,** 2129–2130]

Fukumori, T., Akari, H., Iida, S., Hata, S., Kagawa, S., Aida, Y., Koyama, A. H., and Adachi, A. (1998). The HIV-1 Vpr displays strong anti-apoptotic activity. *FEBS Lett.* **432,** 17–20.

Fultz, P. N. (1991). Replication of an acutely lethal simian immunodeficiency virus activates and induces proliferation of lymphocytes. *J. Virol.* **65,** 4902–4909.

Gabuzda, D. H., Lawrence, K., Langhoff, E., Terwilliger, E., Dorfman, T., Haseltine, W. A., and Sodroski, J. (1992). Role of vif in replication of human immunodeficiency virus type 1 in CD4+ T lymphocytes. *J. Virol.* **66,** 6489–6495.

Gabuzda, D. H., Li, H., Lawrence, K., Vasir, B. S., Crawford, K., and Langhoff, E. (1994). Essential role of vif in establishing productive HIV-1 infection in peripheral blood T lymphocytes and monocyte/macrophages. *J. Acquir. Immune Defic. Syndr.* **7,** 908–915.

Gallay, P., Hope, T., Chin, D., and Trono, D. (1997). HIV-1 infection of nondividing cells through the recognition of integrase by the importin/karyopherin pathway. *Proc. Natl. Acad. Sci. USA* **94,** 9825–9830.

Gallay, P., Stitt, V., Mundy, C., Oettinger, M., and Trono, D. (1996). Role of the karyopherin pathway in human immunodeficiency virus type 1 nuclear import. *J Virol.* **70,** 1027–1032.

Garcia, J. V., Alfano, J., and Miller, A. D. (1993). The negative effect of human immunodeficiency virus type 1 Nef on cell surface CD4 expression is not species specific and requires the cytoplasmic domain of CD4. *J. Virol.* **67,** 1511–5116.

Garcia, J. V., and Miller, A. D. (1991). Serine phosphorylation-independent downregulation of cell-surface CD4 by nef. *Nature* **350,** 508–511.

Garrett, E. D., Tiley, L. S., and Cullen, B. R. (1991). Rev activates expression of the human immunodeficiency virus type 1 vif and vpr gene products. *J. Virol.* **65,** 1653–1657.

Geleziunas, R., Bour, S., and Wainberg, M. A. (1994). Human immunodeficiency virus type 1-associated CD4 downmodulation. *Adv. Virus Res.* **44,** 203–266.

Geraghty, R. J., and Panganiban, A. T. (1993). Human immunodeficiency virus type 1 Vpu has a CD4- and an envelope glycoprotein-independent function. *J. Virol.* **67,** 4190–4194.

Gerritsen, M. E., Williams, A. J., Neish, A. S., Moore, S., Shi, Y., and Collins, T. (1997). CREB-binding protein/p300 are transcriptional coactivators of p65. *Proc. Natl. Acad. Sci. USA* **94,** 2927–2932.

Goh, W. C., Rogel, M. E., Kinsey, C. M., Michael, S. F., Fultz, P. N., Nowak, M. A., Hahn, B. H., and Emerman, M. (1998). HIV-1 Vpr increases viral expression by manipulation of the cell cycle: A mechanism for selection of Vpr *in vivo*. *Nat. Med.* **4,** 65–71.

Goncalves, J., Jallepalli, P., and Gabuzda, D. H. (1994). Subcellular localization of the Vif protein of human immunodeficiency virus type 1. *J. Virol.* **68,** 704–712.

Goncalves, J., Korin, Y., Zack, J., and Gabuzda, D. (1996). Role of Vif in human immunodeficiency virus type 1 reverse transcription. *J. Virol.* **70,** 8701–8709.

Goncalves, J., Shi, B., Yang, X., and Gabuzda, D. (1995). Biological activity of human immunodeficiency virus type 1 Vif requires membrane targeting by C-terminal basic domains. *J. Virol.* **69,** 7196–7204.

Gonzalez, M. E., and Carrasco, L. (1998). The human immunodeficiency virus type 1 Vpu protein enhances membrane permeability. *Biochemistry* **37,** 13710–13719.

Graziani, A., Galimi, F., Medico, E., Cottone, E., Gramaglia, D., Boccaccio, C., and Comoglio, P. M. (1996). The HIV-1 nef protein interferes with phosphatidylinositol 3-kinase activation 1. *J. Biol. Chem.* **271,** 6590–6593.

Greenberg, M. E., Bronson, S., Lock, M., Neumann, M., Pavlakis, G. N., and Skowronski, J. (1997). Co-localization of HIV-1 Nef with the AP-2 adaptor protein complex correlates with Nef-induced CD4 down-regulation. *EMBO J.* **16,** 6964–6964.

Greenberg, M., DeTulleo, L., Rapoport, I., Skowronski, J., and Kirchhausen, T. (1998a). A dileucine motif in HIV-1 Nef is essential for sorting into clathrin-coated pits and for downregulation of CD4. *Curr. Biol.* **8,** 1239–1242.

Greenberg, M. E., Iafrate, A. J., and Skowronski, J. (1998b). The SH3 domain-binding surface and an acidic motif in HIV-1 Nef regulate trafficking of class I MHC complexes. *EMBO J.* **17,** 2777–2789.

Greenway, A., Azad, A., and McPhee, D. (1995). Human immunodeficiency virus type 1 Nef protein inhibits activation pathways in peripheral blood mononuclear cells and T-cell lines. *J. Virol.* **69,** 1842–1850.

Greenway, A., Azad, A., Mills, J., and McPhee, D. (1996). Human immunodeficiency virus type 1 Nef binds directly to Lck and mitogen-activated protein kinase, inhibiting kinase activity. *J. Virol.* **70,** 6701–6708.

Greenway, A. L., McPhee, D. A., Grgacic, E., Hewish, D., Lucantoni, A., Macreadie, I., and Azad, A. (1994). Nef 27, but not the Nef 25 isoform of human immunodeficiency virus-type 1 pNL4.3 down-regulates surface CD4 and IL-2R expression in peripheral blood mononuclear cells and transformed T cells. *Virology* **198,** 245–256.

Grice, A. L., Kerr, I. D., and Sansom, M. S. (1997). Ion channels formed by HIV-1 Vpu: A modelling and simulation study. *FEBS Lett.* **405,** 299–304.

Grzesiek, S., Bax, A., Clore, G. M., Gronenborn, A. M., Hu, J. S., Kaufman, J., Palmer, I., Stahl, S. J., and Wingfield, P. T. (1996). The solution structure of HIV-1 Nef reveals an unexpected fold and permits delineation of the binding surface for the SH3 domain of Hck tyrosine protein kinase. *Nat. Struct. Biol.* **3,** 340–345.

Grzesiek, S., Stahl, S. J., Wingfield, P. T., and Bax, A. (1996). The CD4 determinant for downregulation by HIV-1 Nef directly binds to Nef: Mapping of the Nef binding surface by NMR. *Biochemistry* **35,** 10256–10261.

Guatelli, J. C., Gingeras, T. R., and Richman, D. D. (1990). Alternative splice acceptor utilization during human immunodeficiency virus type 1 infection of cultured cells. *J. Virol.* **64,** 4093–4098.

Gurland, G., and Gundersen, G. G. (1995). Stable, detyrosinated microtubules function to localize vimentin intermediate filaments in fibroblasts. *J. Cell Biol.* **131,** 1275–1290.

Guy, B., Acres, B., Kieny, M. P., and Lecocq, J. P. (1990). DNA binding factors that bind to the negative regulatory element of the human immunodeficiency virus-1: Regulation by nef. *J. Acquir. Immune Defic. Syndr.* **3,** 797–809.

Guy, B., Geist, M., Dott, K., Spehner, D., Kieny, M. P., and Lecocq, J. P. (1991). A specific inhibitor of cysteine proteases impairs a Vif-dependent modification of human immunodeficiency virus type 1 Env protein. *J. Virol.* **65,** 1325–1331.

Hammes, S. R., Dixon, E. P., Malim, M. H., Cullen, B. R., and Greene, W. C. (1989). Nef protein of human immunodeficiency virus type 1: Evidence against its role as a transcriptional inhibitor. *Proc. Natl. Acad. Sci. USA* **86,** 9549–9553.

Hanna, Z., Kay, D. G., Rebai, N., Guimond, A., Jothy, S., and Jolicoeur, P. (1998). Nef harbors a major determinant of pathogenicity for an AIDS-like disease induced by HIV-1 in transgenic mice. *Cell* **95,** 163–175.

Harris, M. P., and Neil, J. C. (1994). Myristoylation-dependent binding of HIV-1 Nef to CD4. *J. Mol. Biol.* **241,** 136–142.

Hatakeyama, S., Kitagawa, M., Nakayama, K., Shirane, M., Matsumoto, M., Hattori, K., Higashi, H., Nakano, H., Okumura, K., Onoe, K., Good, R. A., and Nakayama, K. (1999). Ubiquitin-dependent degradation of IkappaBalpha is mediated by a ubiquitin ligase Skp1/Cul 1/F-box protein FWD1. *Proc. Natl. Acad. Sci. USA* **96,** 3859–3863.

He, J., Choe, S., Walker, R., Di Marzio, P., Morgan, D. O., and Landau, N. R. (1995). Human immunodeficiency virus type 1 viral protein R (Vpr) arrests cells in the G2 phase of the cell cycle by inhibiting p34cdc2 activity. *J. Virol.* **69,** 6705–6711.

Heery, D. M., Kalkhoven, E., Hoare, S., and Parker, M. G. (1997). A signature motif in transcriptional co-activators mediates binding tonuclear receptors. *Nature* **387,** 733–736.

Heinzinger, N. K., Bukinsky, M. I., Haggerty, S. A., Ragland, A. M., Kewalramani, V., Lee, M. A., Gendelman, H. E., Ratner, L., Stevenson, M., and Emerman, M. (1994). The Vpr protein of human immunodeficiency virus type 1 influences nuclear localization of viral nucleic acids in nondividing host cells. *Proc. Natl. Acad. Sci. USA* **91,** 7311–7315.

Hochstrasser, M. (1996). Protein degradation or regulation: Ub the judge. *Cell* **84,** 813–815.

Hoglund, S., Ohagen, A., Lawrence, K., and Gabuzda, D. (1994). Role of vif during packing of the core of HIV-1. *Virology* **201,** 349–355.

Hottiger, M. O., Felzien, L. K., and Nabel, G. J. (1998). Modulation of cytokine-induced HIV gene expression by competitivebinding of transcription factors to the coactivator p300. *EMBO J.* **17,** 3124–3134.

Howe, A. Y., Jung, J. U., and Desrosiers, R. C. (1998). Zeta chain of the T-cell receptor interacts with nef of simian immunodeficiency virus and human immunodeficiency virus type 2. *J. Virol.* **72,** 9827–9834.

Hua, J., Blair, W., Truant, R., and Cullen, B. R. (1997). Identification of regions in HIV-1 Nef required for efficient downregulation of cell surface CD4. *Virology* **231,** 231–238.

Huet, T., Cheynier, R., Meyerhans, A., Roelants, G., and Wain-Hobson, S. (1990). Genetic organization of a chimpanzee lentivirus related to HIV-1. *Nature* **345,** 356–359.

Humphries, E. H., and Temin, H. M. (1974). Requirement for cell division for initiation of transcription of Rous sarcoma virus RNA. *J. Virol.* **14,** 531–546.

Huvent, I., Hong, S. S., Fournier, C., Gay, B., Tournier, J., Carriere, C., Courcoul, M., Vigne, R., Spire, B., and Boulanger, P. (1998). Interaction and co-encapsidation of human immunodeficiency virus type 1 Gag and Vif recombinant proteins. *J. Gen. Virol.* **79,** 1069–1081.

Iafrate, A. J., Bronson, S., and Skowronski, J. (1997). Separable functions of Nef disrupt two aspects of T cell receptor machinery: CD4 expression and CD3 signaling. *EMBO J.* **16,** 673–684.

Jabbar, M. A., and Nayak, D. P. (1990). Intracellular interaction of human immunodeficiency virus type 1 (ARV-2) envelope glycoprotein gp160 with CD4 blocks the movement and maturation of CD4 to the plasma membrane. *J. Virol.* **64,** 6297–6304.

Jamieson, B. D., Aldrovandi, G. M., Planelles, V., Jowett, J. B., Gao, L., Bloch, L. M., Chen, I. S., and Zack, J. A. (1994). Requirement of human immunodeficiency virus type 1 nef for *in vivo* replication and pathogenicity. *J. Virol.* **68,** 3478–3485.

Jenkins, Y., McEntee, M., Weis, K., and Greene, W. C. (1998). Characterization of HIV-1 vpr nuclear import: Analysis of signals and pathways. *J. Cell Biol.* **143,** 875–885.

Jowett, J. B., Planelles, V., Poon, B., Shah, N. P., Chen, M. L., and Chen, I. S. (1995). The human immunodeficiency virus type 1 vpr gene arrests infected T cells in the G2+ M phase of the cell cycle. *J. Virol.* **69,** 6304–6313.

Kappes, J. C. (1995). Viral protein X. *Curr. Top. Microbiol. Immunol.* **193,** 121–132.

Karczewski, M. K., and Strebel, K. (1996). Cytoskeleton association and virion incorporation of the human immunodeficiency virus type 1 Vif protein. *J. Virol.* **70,** 494–507.

Kestler, H. W., Ringler, D. J., Mori, K., Panicali, D. L., Sehgal, P. K., Daniel, M. D., and Desrosiers, R. C. (1991). Importance of the nef gene for maintenance of high virus loads and for development of AIDS. *Cell* **65,** 651–662.

Kim, S., Ikeuchi, K., Byrn, R., Groopman, J., and Baltimore, D. (1989). Lack of a negative influence on viral growth by the nef gene of human immunodeficiency virus type 1. *Proc. Natl. Acad. Sci. USA* **86,** 9544–9548.

Kino, T., Gragerov, A., Kopp, J. B., Stauber, R. H., Pavlakis, G. N., and Chrousos, G. P. (1999). The HIV-1 virion-associated protein vpr is a coactivator of the human glucocorticoid receptor. *J. Exp. Med.* **189,** 51–62.

Kirchhausen, T., Bonifacino, J. S., and Riezman, H. (1997). Linking cargo to vesicle formation: Receptor tail interactions with coat proteins. *Curr. Opin. Cell. Biol.* **9,** 488–495.

Kirchhoff, F., Carl, S., Sopper, S., Sauermann, U., Matz-Rensing, K., and Stahl-Hennig, C. (1999). Selection of the R17Y substitution in SIVmac239 nef coincided with a dramatic increase in plasma viremia and rapid progression to death. *Virology* **254,** 61–70.

Kirchhoff, F., Greenough, T. C., Brettler, D. B., Sullivan, J. L., and Desrosiers, R. C. (1995). Brief report: Absence of intact nef sequences in a long-term survivor with nonprogressive HIV-1 infection. *N. Engl. J. Med.* **332,** 228–232.

Kishi, M., Nishino, Y., Sumiya, M., Ohki, K., Kimura, T., Goto, T., Nakai, M., Kakinuma, M., and Ikuta, K. (1992). Cells surviving infection by human immunodeficiency virus type 1: vif or vpu mutants produce non-infectious or markedly less cytopathic viruses. *J. Gen. Virol.* **73,** 77–87.

Klimkait, T., Strebel, K., Hoggan, M. D., Martin, M. A., and Orenstein, J. M. (1990). The human immunodeficiency virus type 1-specific protein vpu is required for efficient virus maturation and release. *J. Virol.* **64,** 621–629.

Klotman, M. E., Kim, S., Buchbinder, A., DeRossi, A., Baltimore, D., and Wong-Staal, F. (1991). Kinetics of expression of multiply spliced RNA in early human immunodeficiency virus type 1 infection of lymphocytes and monocytes. *Proc. Natl. Acad. Sci. USA* **88,** 5011–5015. [published erratum appears in *Proc. Natl. Acad. Sci. USA* **89,** 1148]

Kondo, E., and Gottlinger, H. G. (1996). A conserved LXXLF sequence is the major determinant in p6gag required for the incorporation of human immunodeficiency virus type 1 Vpr. *J. Virol.* **70,** 159–164.

Kondo, E., Mammano, F., Cohen, E. A., and Gottlinger, H. G. (1995). The p6gag domain of human immunodeficiency virus type 1 is sufficient for the incorporation of Vpr into heterologous viral particles. *J. Virol.* **69,** 2759–2764.

Korber, B., Foley, B., Leitner, T., McCutchan, F., Hahn, B., Mellors, J. W., Myers, G., and Kuiken, C. (eds.). 1997. Human retroviruses and AIDS. Los Alamos National Laboratory, N.M.

Kotler, M., Simm, M., Zhao, Y. S., Sova, P., Chao, W., Ohnona, S. F., Roller, R., Krachmarov, C., Potash, M. J., and Volsky, D. J. (1997). Human immunodeficiency virus type 1 (HIV-1) protein Vif inhibits the activity of HIV-1 protease in bacteria and *in vitro*. J. Virol. **71,** 5774–5781.

Lamb, R. A., and Pinto, L. H. (1997). Do Vpu and Vpr of human immunodeficiency virus type 1 and NB of influenza B virus have ion channel activities in the viral life cycles? *Virology* **229,** 1–11.

Lavallee, C., Yao, X. J., Ladha, A., Gottlinger, H., Haseltine, W. A., and Cohen, E. A. (1994). Requirement of the Pr55gag precursor for incorporation of the Vpr product into human immunodeficiency virus type 1 viral particles. *J. Virol.* **68,** 1926–1934.

Le Gall, S., Erdtmann, L., Benichou, S., Berlioz-Torrent, C., Liu, L., Benarous, R., Heard, J. M., and Schwartz, O. (1998). Nef interacts with the mu subunit of clathrin adaptor complexes and reveals a cryptic sorting signal in MHC I molecules. *Immunity* **8,** 483–495.

Le Gall, S., Prevost, M. C., Heard, J. M., and Schwartz, O. (1997). Human immunodeficiency virus type I Nef independently affects virion incorporation of major histocompatibility complex class I molecules and virus infectivity. *Virology* **229,** 295–301.

Lee, C. H., Leung, B., Lemmon, M. A., Zheng, J., Cowburn, D., Kuriyan, J., and Saksela, K. (1995). A single amino acid in the SH3 domain of Hck determines its high affinity and specificity in binding to HIV-1 Nef protein. *EMBO J.* **14,** 5006–5015.

Lee, C. H., Saksela, K., Mirza, U. A., Chait, B. T., and Kuriyan, J. (1996). Crystal structure of the conserved core of HIV-1 Nef complexed with a Src family SH3 domain. *Cell* **85,** 931–942.

Lenburg, M. E., and Landau, N. R. (1993). Vpu-induced degradation of CD4: Requirement for specific amino acid residues in the cytoplasmic domain of CD4. *J. Virol.* **67,** 7238–7245.

Li, M. S., Garcia-Asua, G., Bhattacharyya, U., Mascagni, P., Austen, B. M., and Roberts, M. M. (1996). The Vpr protein of human immunodeficiency virus type 1 binds to nucleocapsid protein p7 *in vitro*. *Biochem. Biophys. Res. Commun.* **218,** 352–355.

Lindemann, D., Wilhelm, R., Renard, P., Althage, A., Zinkernagel, R., and Mous, J. (1994). Severe immunodeficiency associated with a human immunodeficiency virus 1 NEF/3′-long terminal repeat transgene. *J. Exp. Med.* **179,** 797–807.

Liu, H., Wu, X., Newman, M., Shaw, G. M., Hahn, B. H., and Kappes, J. C. (1995). The Vif protein of human and simian immunodeficiency viruses is packaged into virions and associates with viral core structures. *J. Virol.* **69,** 7630–7638.

Lu, X., Wu, X., Plemenitas, A., Yu, H., Sawai, E. T., Abo, A., and Peterlin, B. M. (1996). CDC42 and Rac1 are implicated in the activation of the Nef-associated kinase and replication of HIV-1. *Curr. Biol.* **6,** 1677–1684.

Lu, X., Yu, H., Liu, S. H., Brodsky, F. M., and Peterlin, B. M. (1998). Interactions between HIV1 Nef and vacuolar ATPase facilitate the internalization of CD4. *Immunity* **8,** 647–656.

Lu, Y. L., Bennett, R. P., Wills, J. W., Gorelick, R., and Ratner, L. (1995). A leucine triplet repeat sequence (LXX)4 in p6gag is important for Vpr incorporation into human immunodeficiency virus type 1 particles. *J. Virol.* **69,** 6873–6879.

Lu, Y. L., Spearman, P., and Ratner, L. (1993). Human immunodeficiency virus type 1 viral protein R localization in infected cells and virions. *J. Virol.* **67,** 6542–6550.

Luo, W., and Peterlin, B. M. (1997). Activation of the T-cell receptor signaling pathway by Nef from an aggressive strain of simian immunodeficiency virus. *J. Virol.* **71,** 9531–9537.

Luria, S., Chambers, I., and Berg, P. (1991). Expression of the type 1 human immunodeficiency virus Nef protein in T cells prevents antigen receptor-mediated induction of interleukin 2 mRNA. *Proc. Natl. Acad. Sci. USA* **88,** 5326–5330.

Ma, X. Y., Sova, P., Chao, W., and Volsky, D. J. (1994). Cysteine residues in the Vif protein of human immunodeficiency virus type 1 are essential for viral infectivity. *J. Virol.* **68,** 1714–1720.

Macreadie, I. G., Arunagiri, C. K., Hewish, D. R., White, J. F., and Azad, A. A. (1996). Extracellular addition of a domain of HIV-1 Vpr containing the amino acid sequence motif H(S/F)RIG causes cell membrane permeabilization and death. *Mol. Microbiol.* **19,** 1185–1192.

Madani, N., and Kabat, D. (1998). An endogenous inhibitor of human immunodeficiency virus in human lymphocytes is overcome by the viral Vif protein. *J. Virol.* **72,** 10251–10255.

Mahalingam, S., Ayyavoo, V., Patel, M., Kieber-Emmons, T., and Weiner, D. B. (1997). Nuclear import, virion incorporation, and cell cycle arrest/differentiation are mediated by distinct functional domains of human immunodeficiency virus type 1 Vpr. *J. Virol.* **71,** 6339–6347.

Mahalingam, S., Collman, R. G., Patel, M., Monken, C. E., and Srinivasan, A. (1995a). Functional analysis of HIV-1 Vpr: Identification of determinants essential for subcellular localization. *Virology* **212,** 331–339.

Mahalingam, S., Khan, S. A., Jabbar, M. A., Monken, C. E., Collman, R. G., and Srinivasan, A. (1995b). Identification of residues in the N-terminal acidic domain of HIV-1 Vpr essential for virion incorporation. *Virology* **207,** 297–302.

Mahalingam, S., Khan, S. A., Murali, R., Jabbar, M. A., Monken, C. E., Collman, R. G., and Srinivasan, A. (1995c). Mutagenesis of the putative alpha-helical domain of the Vpr protein of human immunodeficiency virus type 1: Effect on stability and virion incorporation. *Proc. Natl. Acad. Sci. USA* **92,** 3794–3798.

Maldarelli, F., Chen, M. Y., Willey, R. L., and Strebel, K. (1993). Human immunodeficiency virus type 1 Vpu protein is an oligomeric type I integral membrane protein. *J. Virol.* **67,** 5056–5061.

Mangasarian, A., Foti, M., Aiken, C., Chin, D., Carpentier, J. L., and Trono, D. (1997). The HIV-1 Nef protein acts as a connector with sorting pathways in the Golgi and at the plasma membrane. *Immunity* **6,** 67–77.

Mangasarian, A., Piguet, V., Wang, J. K., Chen, Y. L., and Trono, D. (1999). Nef-induced CD4 and major histocompatibility complex class I (MHC-I) down-regulation are governed by distinct determinants: N-terminal alpha helix and proline repeat of Nef selectively regulate MHC-I trafficking. *J. Virol.* **73,** 1964–1973.

Manninen, A., Hiipakka, M., Vihinen, M., Lu, W., Mayer, B. J., and Saksela, K. (1998). SH3-Domain binding function of HIV-1 Nef is required for association with a PAK-related kinase. *Virology* **250,** 273–282.

Margottin, F., Benichou, S., Durand, H., Richard, V., Liu, L. X., Gomas, E., and Benarous, R. (1996). Interaction between the cytoplasmic domains of HIV-1 Vpu and CD4: Role of Vpu residues involved in CD4 interaction and *in vitro* CD4 degradation. *Virology* **223**, 381–386.

Margottin, F., Bour, S. P., Durand, H., Selig, L., Benichou, S., Richard, V., Thomas, D., Strebel, K., and Benarous, R. (1998). A novel human WD protein, h-beta TrCp, that interacts with HIV-1 Vpu connects CD4 to the ER degradation pathway through an F-box motif. *Mol. Cell.* **1**, 565–574.

Mariani, R., and Skowronski, J. (1993). CD4 down-regulation by nef alleles isolated from human immunodeficiency virus type 1-infected individuals. *Proc. Natl. Acad. Sci. USA* **90**, 5549–5553.

Marsh, M., and Pelchen-Matthews, A. (1996). Endocytic and exocytic regulation of CD4 expression and function. *Curr. Top. Microbiol. Immunol.* **205**, 107–135.

Miller, M. D., Warmerdam, M. T., Ferrell, S. S., Benitez, R., and Greene, W. C. (1997). Intravirion generation of the C-terminal core domain of HIV-1 Nef by the HIV-1 protease is insufficient to enhance viral infectivity. *Virology* **234**, 215–225.

Miller, M. D., Warmerdam, M. T., Gaston, I., Greene, W. C., and Feinberg, M. B. (1994). The human immunodeficiency virus-1 nef gene product: A positive factor for viral infection and replication in primary lymphocytes and macrophages. *J. Exp. Med.* **179**, 101–113.

Miller, M. D., Warmerdam, M. T., Page, K. A., Feinberg, M. B., and Greene, W. C. (1995). Expression of the human immunodeficiency virus type 1 (HIV-1) nef gene during HIV-1 production increases progeny particle infectivity independently of gp160 or viral entry. *J. Virol.* **69**, 579–584.

Moore, P. B., Zhong, Q., Husslein, T., and Klein, M. L. (1998). Simulation of the HIV-1 Vpu transmembrane domain as a pentameric bundle. *FEBS Lett.* **431**, 143–148.

Munis, J. R., Kornbluth, R. S., Guatelli, J. C., and Richman, D. D. (1992). Ordered appearance of human immunodeficiency virus type 1 nucleic acids following high multiplicity infection of macrophages. *J. Gen. Virol.* **73**, 1899–1906.

Myers, M., and Forgac, M. (1993). The coated vesicle vacuolar (H+)-ATPase associates with and is phosphorylated by the 50-kDa polypeptide of the clathrin assembly protein AP-2. *J. Biol. Chem.* **268**, 9184–9186.

Nascimbeni, M., Bouyac, M., Rey, F., Spire, B., and Clavel, F. (1998). The replicative impairment of Vif- mutants of human immunodeficiency virus type 1 correlates with an overall defect in viral DNA synthesis. *J. Gen. Virol.* **79**, 1945–1950.

Neer, E. J., Schmidt, C. J., Nambudripad, R., and Smith, T. F. (1994). The ancient regulatory-protein family of WD-repeat proteins. *Nature* **371**, 297–300. [published erratum appears in *Nature* **371**, 812]

Niederman, T. M., Garcia, J. V., Hastings, W. R., Luria, S., and Ratner, L. (1992). Human immunodeficiency virus type 1 Nef protein inhibits NF-kappa B induction in human T cells. *J. Virol.* **66**, 6213–6219.

Niederman, T. M., Hastings, W. R., Luria, S., Bandres, J. C., and Ratner, L. (1993). HIV-1 Nef protein inhibits the recruitment of AP-1 DNA-binding activity in human T-cells. *Virology* **194**, 338–344.

Niederman, T. M., Hu, W., and Ratner, L. (1991). Simian immunodeficiency virus negative factor suppresses the level of viral mRNA in COS cells. *J. Virol.* **65**, 3538–3546.

Niederman, T. M., Thielan, B. J., and Ratner, L. (1989). Human immunodeficiency virus type 1 negative factor is a transcriptional silencer. *Proc. Natl. Acad. Sci. USA* **86**, 1128–1132.

Oberste, M. S., and Gonda, M. A. (1992). Conservation of amino-acid sequence motifs in lentivirus Vif proteins. *Virus Genes* **6**, 95–102.

Ochsenbauer, C., Wilk, T., and Bosch, V. (1997). Analysis of vif-defective human immunodeficiency virus type 1 (HIV-1) virions synthesized in 'non-permissive' T lymphoid cells stably infected with selectable HIV-1. *J. Gen. Virol.* **78**, 627–635.

Pancio, H. A., and Ratner, L. (1998). Human immunodeficiency virus type 2 Vpx-Gag interaction. *J. Virol.* **72**, 5271–5275.

Pandori, M. W., Fitch, N. J., Craig, H. M., Richman, D. D., Spina, C. A., and Guatelli, J. C. (1996). Producer-cell modification of human immunodeficiency virus type 1: Nef is a virion protein. *J. Virol.* **70,** 4283–4290.

Pandori, M., Craig, H., Moutouh, L., Corbeil, J., and Guatelli, J. (1998). Virological importance of the protease-cleavage site in human immunodeficiency virus type 1 Nef is independent of both intravirion processing and CD4 down-regulation. *Virology* **251,** 302–316.

Patton, E. E., Willems, A. R., and Tyers, M. (1998). Combinatorial control in ubiquitin-dependent proteolysis: Don't Skp the F-box hypothesis. *Trends Genet.* **14,** 236–243.

Paul, M., and Jabbar, M. A. (1997). Phosphorylation of both phosphoacceptor sites in the HIV-1 Vpu cytoplasmic domain is essential for Vpu-mediated ER degradation of CD4. *Virology* **232,** 207–216.

Paul, M., Mazumder, S., Raja, N., and Jabbar, M. A. (1998). Mutational analysis of the human immunodeficiency virus type 1 Vpu transmembrane domain that promotes the enhanced release of virus-like particles from the plasma membrane of mammalian cells. *J. Virol.* **72,** 1270–1279.

Paxton, W., Connor, R. I., and Landau, N. R. (1993). Incorporation of Vpr into human immunodeficiency virus type 1 virions: Requirement for the p6 region of gag and mutational analysis. *J. Virol.* **67,** 7229–7237.

Peifer, M. (1997). Beta-catenin as oncogene: The smoking gun. *Science* **275,** 1752–1753.

Pelchen-Matthews, A., Boulet, I., Littman, D. R., Fagard, R., and Marsh, M. (1992). The protein tyrosine kinase p56lck inhibits CD4 endocytosis by preventing entry of CD4 into coated pits. *J. Cell Biol.* **117,** 279–290.

Pelchen-Matthews, A., Parsons, I. J., and Marsh, M. (1993). Phorbol ester-induced downregulation of CD4 is a multistep process involving dissociation from p56lck, increased association with clathrin- coated pits, and altered endosomal sorting. *J. Exp. Med.* **178,** 1209–1222.

Pemberton, L. F., Blobel, G., and Rosenblum, J. S. (1998). Transport routes through the nuclear pore complex. *Curr. Opin. Cell Biol.* **10,** 392–399.

Perkins, N. D., Felzien, L. K., Betts, J. C., Leung, K., Beach, D. H., and Nabel, G. J. (1997). Regulation of NF-kappaB by cyclin-dependent kinases associated with the p300 coactivator. *Science* **275,** 523–527.

Piguet, V., Chen, Y. L., Mangasarian, A., Foti, M., Carpentier, J. L., and Trono, D. (1998). Mechanism of Nef-induced CD4 endocytosis: Nef connects CD4 with the mu chain of adaptor complexes. *EMBO J.* **17,** 2472–2478.

Piguet, V., Gu, F., Foti, M., Demaurex, N., Gruenberg, J., Carpentier, J. L., and Trono, D. (1999). Nef-induced CD4 degradation: A diacidic-based motif in Nef functions as a lysosomal targeting signal through the binding of beta-COP in endosomes. *Cell* **97,** 63–73.

Planelles, V., Jowett, J. B., Li, Q. X., Xie, Y., Hahn, B., and Chen, I. S. (1996). Vpr-induced cell cycle arrest is conserved among primate lentiviruses. *J. Virol.* **70,** 2516–2524.

Poon, B., Grovit-Ferbas, K., Stewart, S. A., and Chen, I. S. Y. (1998). Cell cycle arrest by Vpr in HIV-1 virions and insensitivity to antiretroviral agents. *Science* **281,** 266–269.

Popov, S., Rexach, M., Ratner, L., Blobel, G., and Bukrinsky, M. (1998). Viral protein R regulates docking of the HIV-1 preintegration complex to the nuclear pore complex. *J. Biol. Chem.* **273,** 13347–13352.

Popov, S., Rexach, M., Zybarth, G., Reiling, N., Lee, M. A., Ratner, L., Lane, C. M., Moore, M. S., Blobel, G., and Bukrinsky, M. (1998). Viral protein R regulates nuclear import of the HIV-1 pre-integration complex. *EMBO J.* **17,** 909–917.

Potash, M. J., Bentsman, G., Muir, T., Krachmarov, C., Sova, P., and Volsky, D. J. (1998). Peptide inhibitors of HIV-1 protease and viral infection of peripheral blood lymphocytes based on HIV-1 Vif. *Proc. Natl. Acad. Sci. USA* **95,** 13865–13868.

Premkumar, D. R., Ma, X. Z., Maitra, R. K., Chakrabarti, B. K., Salkowitz, J., Yen-Lieberman, B., Hirsch, M. S., and Kestler, H. W. (1996). The nef gene from a long-term HIV type 1 nonprogressor. *AIDS Res. Hum. Retroviruses* **12,** 337–345.

Raja, N. U., Vincent, M. J., and Abdul Jabbar, M. (1994). Vpu-mediated proteolysis of gp160/CD4 chimeric envelope glycoproteins in the endoplasmic reticulum: Requirement of both the anchor and cytoplasmic domains of CD4. *Virology* **204,** 357–366.

Ravi, R., Mookerjee, B., van Hensbergen, Y., Bedi, G. C., Giordano, A., El-Deiry, W. S., Fuchs, E. J., and Bedi, A. (1998). p53-mediated repression of nuclear factor-kappaB RelA via thetranscriptional integrator p300. *Cancer Res.* **58,** 4531–4536.

Re, F., and Luban, J. (1997). HIV-1 Vpr: G2 cell cycle arrest, macrophages and nuclear transport. *Prog. Cell Cycle Res.* **3,** 21–27.

Re, F., Braaten, D., Franke, E. K., and Luban, J. (1995). Human immunodeficiency virus type 1 Vpr arrests the cell cycle in G2 by inhibiting the activation of p34cdc2-cyclin B. *J. Virol.* **69,** 6859–6864.

Reddy, T. R., Kraus, G., Yamada, O., Looney, D. J., Suhasini, M., and Wong-Staal, F. (1995). Comparative analyses of human immunodeficiency virus type 1 (HIV-1) and HIV-2 Vif mutants. *J. Virol.* **69,** 3549–3553.

Refaeli, Y., Levy, D. N., and Weiner, D. B. (1995). The glucocorticoid receptor type II complex is a target of the HIV-1 vpr gene product. *Proc. Natl. Acad. Sci. USA* **92,** 3621–3625.

Rhee, S. S., and Marsh, J. W. (1994). Human immunodeficiency virus type 1 Nef-induced down-modulation of CD4 is due to rapid internalization and degradation of surface CD4. *J. Virol.* **68,** 5156–5163.

Ritter, G. D., Jr., Yamshchikov, G., Cohen, S. J., and Mulligan, M. J. (1996). Human immunodeficiency virus type 2 glycoprotein enhancement of particle budding: Role of the cytoplasmic domain. *J. Virol.* **70,** 2669–2673.

Rogel, M. E., Wu, L. I., and Emerman, M. (1995). The human immunodeficiency virus type 1 vpr gene prevents cell proliferation during chronic infection. *J. Virol.* **69,** 882–888.

Rossi, F., Gallina, A., and Milanesi, G. (1996). Nef-CD4 physical interaction sensed with the yeast two-hybrid system. *Virology* **217,** 397–403.

Saib, A., Puvion-Dutilleul, F., Schmid, M., Peries, J., and De The, H. (1997). Nuclear tareting of incoming human foamy virus gag proteins involves a centriolar step. *J. Virol.* **71,** 1155–1161.

Sakai, H., Shibata, R., Sakuragi, J., Sakuragi, S., Kawamura, M., and Adachi, A. (1993). Cell-dependent requirement of human immunodeficiency virus type 1 Vif protein for maturation of virus particles. *J. Virol.* **67,** 1663–1666.

Saksela, K., Cheng, G., and Baltimore, D. (1995). Proline-rich (PxxP) motifs in HIV-1 Nef bind to SH3 domains of a subset of Src kinases and are required for the enhanced growth of Nef+ viruses but not for down-regulation of CD4. *EMBO J.* **14,** 484–491.

Salvi, R., Garbuglia, A. R., Di Caro, A., Pulciani, S., Montella, F., and Benedetto, A. (1998). Grossly defective nef gene sequences in a human immunodeficiency virus type 1-seropositive long-term nonprogressor. *J. Virol.* **72,** 3646–3657.

Sanfridson, A., Cullen, B. R., and Doyle, C. (1994). The simian immunodeficiency virus Nef protein promotes degradation of CD4 in human T cells. *J. Biol. Chem.* **269,** 3917–3920.

Saucier, M., Hodge, S., Dewhurst, S., Gibson, T., Gibson, J. P., McClure, H. M., and Novembre, F. J. (1998). The tyrosine-17 residue of Nef in SIVsmmPBj14 is required for acute pathogenesis and contributes to replication in macrophages. *Virology* **244,** 261–272.

Sawai, E. T., Baur, A., Struble, H., Peterlin, B. M., Levy, J. A., and Cheng-Mayer, C. (1994). Human immunodeficiency virus type 1 Nef associates with a cellular serine kinase in T lymphocytes. *Proc. Natl. Acad. Sci. USA* **91,** 1539–1543.

Schubert, U., and Strebel, K. (1994). Differential activities of the human immunodeficiency virus type 1-encoded Vpu protein are regulated by phosphorylation and occur in different cellular compartments. *J. Virol.* **68,** 2260–2271.

Schubert, U., Anton, L. C., Bacik, I., Cox, J. H., Bour, S., Bennink, J. R., Orlowski, M., Strebel, K., and Yewdell, J. W. (1998). CD4 glycoprotein degradation induced by human immunodeficiency virus type 1 Vpu protein requires the function of proteasomes and the ubiquitin-conjugating pathway. *J. Virol.* **72,** 2280–2288.

Schubert, U., Bour, S., Ferrer-Montiel, A. V., Montal, M., Maldarell, F., and Strebel, K. (1996a). The two biological activities of human immunodeficiency virus type 1 Vpu protein involve two separable structural domains. *J. Virol.* **70**, 809–819.

Schubert, U., Bour, S., Willey, R. L., and Strebel, K. (1999). Regulation of virus release by the macrophage-tropic human immunodeficiency virus type 1 AD8 isolate is redundant and can be controlled by either Vpu or Env. *J. Virol.* **73**, 887–896.

Schubert, U., Ferrer-Montiel, A. V., Oblatt-Montal, M., Henklein, P., Strebel, K., and Montal, M. (1996b). Identification of an ion channel activity of the Vpu transmembrane domain and its involvement in the regulation of virus release from HIV− 1-infected cells. *FEBS Lett.* **398**, 12–18.

Schubert, U., Henklein, P., Boldyreff, B., Wingender, E., Strebel, K., and Porstmann, T. (1994). The human immunodeficiency virus type 1 encoded Vpu protein is phosphorylated by casein kinase-2 (CK-2) at positions Ser52 and Ser56 within a predicted alpha-helix-turn-alpha-helix-motif. *J. Mol. Biol.* **236**, 16–25.

Schuler, W., Wecker, K., de Rocquigny, H., Baudat, Y., Sire, J., and Roques, B. P. (1999). NMR structure of the (52-96) C-terminal domain of the HIV-1 regulatory protein Vpr: Molecular insights into its biological functions. *J. Mol. Biol.* **285**, 2105–2117.

Schwartz, O., Dautry-Varsat, A., Goud, B., Marechal, V., Subtil, A., Heard, J. M., and Danos, O. (1995a). Human immunodeficiency virus type 1 Nef induces accumulation of CD4 in early endosomes. *J. Virol.* **69**, 528–533.

Schwartz, O., Marechal, V., Danos, O., and Heard, J. M. (1995b). Human immunodeficiency virus type 1 Nef increases the efficiency of reverse transcription in the infected cell. *J. Virol.* **69**, 4053–4059.

Schwartz, O., Marechal, V., Le Gall, S., Lemonnier, F., and Heard, J. M. (1996). Endocytosis of major histocompatibility complex class I molecules is induced by the HIV-1 Nef protein. *Nat. Med.* **2**, 338–342.

Schwartz, S., Felber, B. K., Fenyo, E. M., and Pavlakis, G. N. (1990). Env and Vpu proteins of human immunodeficiency virus type 1 are produced from multiple bicistronic mRNAs. *J. Virol.* **64**, 5448–5456.

Schwartz, S., Felber, B. K., and Pavlakis, G. N. (1991). Expression of human immunodeficiency virus type 1 vif and vpr mRNAs is Rev-dependent and regulated by splicing. *Virology* **183**, 677–686.

Selig, L., Pages, J. C., Tanchou, V., Preveral, S., Berlioz-Torrent, C., Liu, L. X., Erdtmann, L., Darlix, J., Benarous, R., and Benichou, S. (1999). Interaction with the p6 domain of the gag precursor mediates incorporation into virions of Vpr and Vpx proteins from primate lentiviruses. *J. Virol.* **73**, 592–600.

Sharp, P. M., Bailes, E., Stevenson, M., Emerman, M., and Hahn, B. H. (1996). Gene acquisition in HIV and SIV. *Nature* **383**, 586–587.

Shibata, R., Maldarelli, F., Siemon, C., Matano, T., Parta, M., Miller, G., Fredrickson, T., and Martin, M. A. (1997). Infection and pathogenicity of chimeric simian-human immunodeficiency viruses in macaques: Determinants of high virus loads and CD4 cell killing. *J. Infect. Dis.* **176**, 362–376.

Simm, M., Shahabuddin, M., Chao, W., Allan, J. S., and Volsky, D. J. (1995). Aberrant Gag protein composition of a human immunodeficiency virus type 1 vif mutant produced in primary lymphocytes. *J. Virol.* **69**, 4582–4586.

Simon, J. H., Fouchier, R. A., Southerling, T. E., Guerra, C. B., Grant, C. K., and Malim, M. H. (1997). The Vif and Gag proteins of human immunodeficiency virus type 1 colocalize in infected human T cells. *J. Virol.* **71**, 5259–5267.

Simon, J. H., Gaddis, N. C., Fouchier, R. A., and Malim, M. H. (1998a). Evidence for a newly discovered cellular anti-HIV-1 phenotype. *Nat. Med.* **4**, 1397–1400.

Simon, J. H., Miller, D. L., Fouchier, R. A., and Malim, M. H. (1998b). Virion incorporation of human immunodeficiency virus type-1 Vif is determined by intracellular expression level and may not be necessary for function. *Virology* **248**, 182–187.

Simon, J. H., Miller, D. L., Fouchier, R. A., Soares, M. A., Peden, K. W., and Malim, M. H. (1998c). The regulation of primate immunodeficiency virus infectivity by Vif is cell species restricted: A role for Vif in determining virus host range and cross-species transmission. *EMBO J.* **17**, 1259–1267.

Simon, J. H., Southerling, T. E., Peterson, J. C., Meyer, B. E., and Malim, M. H. (1995). Complementation of vif-defective human immunodeficiency virus type 1 by primate, but not nonprimate, lentivirus vif genes. *J. Virol.* **69**, 4166–4172.

Simon, J. H. M., Sheehy, A. M., Carpenter, E. A., Fouchier, R. A. M., and Malim, M. H. (1999). Mutational analysis of the human immunodeficiency virus type 1 vif protein. *J. Virol.* **73**, 2675–2681.

Skowronski, J., Parks, D., and Mariani, R. (1993). Altered T cell activation and development in transgenic mice expressing the HIV-1 nef gene. *EMBO J.* **12**, 703–713.

Sodeik, B., Ebersold, M. W., and Helenius, A. (1997). Microtubule-mediated transport of incoming herpes simplex virus 1 capsides to the nucleus. J. Cell Biol. **136**, 1007–1021.

Sova, P., and Volsky, D. J. (1993). Efficiency of viral DNA synthesis during infection of permissive and nonpermissive cells with vif-negative human immunodeficiency virus type 1. *J. Virol.* **67**, 6322–6326.

Sova, P., Chao, W., and Volsky, D. J. (1997). The redox state of cysteines in human immunodeficiency virus type 1 Vif in infected cells and in virions. *Biochem. Biophys. Res. Commun.* **240**, 257–260.

Spencer, E., Jiang, J., and Chen, Z. J. (1999). Signal-induced ubiquitination of IkappaBalpha by the F-box protein Slimb/beta-TrCP. *Genes Dev.* **13**, 284–294.

Spevak, W., Keiper, B. D., Stratowa, C., and Castanon, M. J. (1993). Saccharomyces cerevisiae cdc15 mutants arrested at a late stage in anaphase are rescued by *Xenopus* cDNAs encoding N-ras or a protein with beta-transducin repeats. *Mol. Cell. Biol.* **13**, 4953–4966. [published erratum appears in *Mol. Cell. Biol.* **13**, 7199]

Spina, C. A., Kwoh, T. J., Chowers, M. Y., Guatelli, J. C., and Richman, D. D. (1994). The importance of nef in the induction of human immunodeficiency virus type 1 replication from primary quiescent CD4 lymphocytes. *J. Exp. Med.* **179**, 115–123.

Stephens, E. B., Mukherjee, S., Liu, Z. Q., Sheffer, D., Lamb-Wharton, R., Leung, K., Zhuge, W., Joag, S. V., Li, Z., Foresman, L., Adany, I., and Narayan, O. (1998). Simian-human immunodeficiency virus (SHIV) containing the nef/long terminal repeat region of the highly virulent SIVsmmPBj14 causes PBj-like activation of cultured resting peripheral blood mononuclear cells, but the chimera showed No increase in virulence. *J. Virol.* **72**, 5207–5214.

Strebel, K., Klimkait, T., Maldarelli, F., and Martin, M. A. (1989). Molecular and biochemical analyses of human immunodeficiency virus type 1 vpu protein. *J. Virol.* **63**, 3784–3791.

Strebel, K., Klimkait, T., and Martin, M. A. (1988). A novel gene of HIV-1, vpu, and its 16-kilodalton product. *Science* **241**, 1221–1223.

Strebel, K., Daugherty, D., Clouse, K., Cohen, D., Folks, T, and Martin, M. A. (1987). The HIV 'A' (sor) gene product is essential for virus infectivity. *Nature* **328**, 728–730.

Subbramanian, R. A., Kessous-Elbaz, A., Lodge, R., Forget, J., Yao, X. J., Bergeron, D., and Cohen, E. A. (1998a). Human immunodeficiency virus type 1 Vpr is a positive regulator of viral transcription and infectivity in primary human macrophages. *J. Exp. Med.* **187**, 1103–1111.

Subbramanian, R. A., Yao, X. J., Dilhuydy, H., Rougeau, N., Bergeron, D., Robitaille, Y., and Cohen, E. A. (1998b). Human immunodeficiency virus type 1 Vpr localization: Nuclear transport of a viral protein modulated by a putative amphipathic helical structure and its relevance to biological activity. *J. Mol. Biol.* **278**, 13–30.

Terwilliger, E. F., Cohen, E. A., Lu, Y. C., Sodroski, J. G., and Haseltine, W. A. (1989). Functional role of human immunodeficiency virus type 1 vpu. *Proc. Natl. Acad. Sci. USA* **86**, 5163–5167.

Terwilliger, E. F., Langhoff, E., Gabuzda, D., Zazopoulos, E., and Haseltine, W. A. (1991). Allelic variation in the effects of the nef gene on replication of human immunodeficiency virus type 1. *Proc. Natl. Acad. Sci. USA* **88,** 10971–10975.

Tiganos, E., Yao, X. J., Friborg, J., Daniel, N., and Cohen, E. A. (1997). Putative alpha-helical structures in the human immunodeficiency virus type 1 Vpu protein and CD4 are involved in binding and degradation of the CD4 molecule. *J. Virol.* **71,** 4452–4460.

Tristem, M., Marshall, C., Karpas, A., Petrik, J., and Hill, F. (1990). Origin of vpx in lentiviruses. *Nature* **347,** 341–342.

Tristem, M., Purvis, A., and Quicke, D. L. (1998). Complex evolutionary history of primate lentiviral vpr genes. *Virology* **240,** 232–237.

Tung, H. Y., De Rocquigny, H., Zhao, L. J., Cayla, X., Roques, B. P., and Ozon, R. (1997). Direct activation of protein phosphatase-2A0 by HIV-1 encoded protein complex NCp7:vpr. *FEBS Lett.* **401,** 197–201.

Vincent, M. J., Raja, N. U., and Jabbar, M. A. (1993). Human immunodeficiency virus type 1 Vpu protein induces degradation of chimeric envelope glycoproteins bearing the cytoplasmic and anchor domains of CD4: Role of the cytoplasmic domain in Vpu-induced degradation in the endoplasmic reticulum. *J. Virol.* **67,** 5538–5549.

Vodicka, M. A., Koepp, D. M., Silver, P. A., and Emerman, M. (1998). HIV-1 Vpr interacts with the nuclear transport pathway to promote macrophage infection. *Genes Dev.* **12,** 175–185.

von Schwedler, U., Song, J., Aiken, C., and Trono, D. (1993). Vif is crucial for human immunodeficiency virus type 1 proviral DNA synthesis in infected cells. *J. Virol.* **67,** 4945–4955.

Wang, L., Mukherjee, S., Jia, F., Narayan, O., and Zhao, L. J. (1995). Interaction of virion protein Vpr of human immunodeficiency virus type 1 with cellular transcription factor Sp1 and trans-activation of viral long terminal repeat. *J. Biol. Chem.* **270,** 25564–25569.

Wang, L., Mukherjee, S., Narayan, O., and Zhao, L. J. (1996). Characterization of a leucine-zipper-like domain in Vpr protein of human immunodeficiency virus type 1. *Gene* **178,** 7–13.

Welker, R., Kottler, H., Kalbitzer, H. R., and Krausslich, H. G. (1996). Human immunodeficiency virus type 1 Nef protein is incorporated into virus particles and specifically cleaved by the viral proteinase. *Virology* **219,** 228–236.

Whitney, J. A., Gomez, M., Sheff, D., Kreis, T. E., and Mellman, I. (1995). Cytoplasmic coat proteins involved in endosome function. *Cell* **83,** 703–713.

Wiertz, E. J., Jones, T. R., Sun, L., Bogyo, M., Geuze, H. J., and Ploegh, H. L. (1996). The human cytomegalovirus US11 gene product dislocates MHC class I heavy chains from the endoplasmic reticulum to the cytosol. *Cell* **84,** 769–779.

Willbold, D., Hoffmann, S., and Rosch, P. (1997). Secondary structure and tertiary fold of the human immunodeficiency virus protein U (Vpu) cytoplasmic domain in solution. *Eur. J. Biochem.* **245,** 581–588.

Willey, R. L., Buckler-White, A., and Strebel, K. (1994). Sequences present in the cytoplasmic domain of CD4 are necessary and sufficient to confer sensitivity to the human immunodeficiency virus type 1 Vpu protein. *J. Virol.* **68,** 1207–1212.

Willey, R. L., Maldarelli, F., Martin, M. A., and Strebel, K. (1992a). Human immunodeficiency virus type 1 Vpu protein regulates the formation of intracellular gp160-CD4 complexes. *J. Virol.* **66,** 226–234.

Willey, R. L., Maldarelli, F., Martin, M. A., and Strebel, K. (1992b). Human immunodeficiency virus type 1 Vpu protein induces rapid degradation of CD4. *J. Virol.* **66,** 7193–7200.

Winston, J. T., Strack, P., Beer-Romero, P., Chu, C. Y., Elledge, S. J., and Harper, J. W. (1999). The SCFbeta-TRCP-ubiquitin ligase complex associates specifically with phosphorylated destruction motifs in IkappaBalpha and beta-catenin and stimulates IkappaBalpha ubiquitination *in vitro*. *Genes Dev.* **13,** 270–283.

Wu, X., Conway, J. A., Kim, J., and Kappes, J. C. (1994). Localization of the Vpx packaging signal within the C terminus of the human immunodeficiency virus type 2 Gag precursor protein. *J. Virol.* **68,** 6161–6169.

Yang, X., and Gabuzda, D. (1998). Mitogen-activated protein kinase phosphorylates and regulates the HIV-1 Vif protein. *J. Biol. Chem.* **273,** 29879–29887.

Yang, X., Goncalves, J., and Gabuzda, D. (1996). Phosphorylation of Vif and its role in HIV-1 replication. *J. Biol. Chem.* **271,** 10121–10129.

Yao, X. J., Friborg, J., Checroune, F., Gratton, S., Boisvert, F., Sekaly, R. P., and Cohen, E. A. (1995). Degradation of CD4 induced by human immunodeficiency virus type 1 Vpu protein: A predicted alpha-helix structure in the proximal cytoplasmic region of CD4 contributes to Vpu sensitivity. *Virology* **209,** 615–623.

Yao, X. J., Gottlinger, H., Haseltine, W. A., and Cohen, E. A. (1992). Envelope glycoprotein and CD4 indepedence of vpu-facilitated human immunodeficiency virus type 1 capsid export. *J. Virol.* **66,** 5119–5126.

Yao, X. J., Subbramanian, R. A., Rougeau, N., Boisvert, F., Bergeron, D., and Cohen, E. A. (1995). Mutagenic analysis of human immunodeficiency virus type 1 Vpr: Role of a predicted N-terminal alpha-helical structure in Vpr nuclear localization and virion incorporation. *J. Virol.* **69,** 7032–7044.

Yaron, A., Hatzubai, A., Davis, M., Lavon, I., Amit, S., Manning, A., M., Andersen, J. S., Mann, M., Mercurio, F., and Ben-Neriah, Y. (1998). *Nature* **396,** 590–594.

Zhang, S., Pointer, D., Singer, G., Feng, Y., Park, K., and Zhao, L. J. (1998). Direct binding to nucleic acids by Vpr of human immunodeficiency virus type 1. *Gene* **212,** 157–166.

Zhao, L. J., Mukherjee, S., and Narayan, O. (1994a). Biochemical mechanism of HIV-I Vpr function: Specific interaction with a cellular protein. *J. Biol. Chem.* **269,** 15577–15582.

Zhao, L. J., Wang, L., Mukherjee, S., and Narayan, O. (1994b). Biochemical mechanism of HIV-1 Vpr function: Oligomerization mediated by the N-terminal domain. *J. Biol. Chem.* **269,** 32131–32137.

Carine Van Lint

Département de Biologie Moléculaire
Laboratoire de Chimie Biologique
Institut de Biologie et de Médecine Moléculaires
Université Libre de Bruxelles
6041 Gosselies, Belgium

Role of Chromatin in HIV-1 Transcriptional Regulation

I. Introduction

In eukaryotes, the chromatin environment of a given gene is critical for its transcriptional regulation. Chromatin structures can inhibit the binding and function of the numerous proteins that collaborate to produce appropriate levels of transcription. In addition, recent studies have identified a large group of proteins whose primary function is to help activate or repress transcription by altering chromatin (Workman and Kingston, 1998). The nucleosome core is the target of many of these activities. Thus, the nucleosomal state of any given region of chromatin plays a central role in determining its transcriptional competence.

Retroviruses in general, and HIV (human immunodeficiency virus) in particular, are confronted with a unique problem in terms of transcriptional regulation and packaging into chromatin. They can integrate into many different sites within the host-cell genome, each site with its own properties

Advances in Pharmacology, Volume 48

1054-3589/00 $35.00

that can influence the degree of viral expression. It is likely that the virus has evolved mechanisms to circumvent this potential problem. Two distinct mechanisms have been proposed:

1. The virus might be able to select a restricted set of integration sites that are conducive to a productive infection. It is thought that the preintegration complex is targeted to specific regions of chromatin by a protein–protein interaction between one of its components and a component of chromatin or of the transcriptional apparatus (see Section IV).
2. Alternatively, the virus might integrate randomly into the cell genome and then organize the local chromatin environment at the site of integration.

Evidence has accumulated that the HIV-1 retrovirus is in fact able to determine its local chromatin organization, independent of the site of integration. However, the existence of such a mechanism does not exclude target site selection and both mechanisms likely cooperate to favor optimal viral expression.

II. HIV-1 Transcriptional Regulatory Elements

The rate of replication of HIV-1 in infected individuals directly determines the rate of the decrease of $CD4^+$ T lymphocytes and therefore the rate of progression to immunodeficiency. This replication rate is controlled primarily at the level of transcription initiation and elongation. The main HIV-1 transcriptional *cis*-regulatory elements are located in the viral long terminal repeats (LTR), present at both extremities of the integrated DNA, which have been divided into three regions, named U3, R, and U5 in reference to their respective origin in the viral RNA genome. Transcription initiates in the 5′ LTR at the U3/R junction and transcripts are polyadenylated at the R/U5 junction in the 3′ LTR.

The 5′ LTR has been extensively characterized *in vitro* and is divided into four functional domains that control basal and activated transcription (Fig. 1). Binding sites for several transcription factors have been identified in each of these domains using *in vitro* footprinting and gel-retardation assays (reviewed in Cullen, 1991; Vaishnav and Wong-Staal, 1991; Gaynor, 1992; Jones and Peterlin, 1994). From the 5′ end to the 3′ end, the four domains are as follows (Fig. 1). First, the negative regulatory element (NRE), a silencer, contains two binding sites for the AP-1 and/or the COUP cellular

FIGURE 1 Transcription elements in the 5′ LTR of HIV-1. The U3, R, and U5 regions of the LTR and binding sites for several transcription factors as well as other sequence elements are indicated (see text for further details). Nucleotide (nt) +1 is the start of U3 in the 5′ LTR. The arrow at the U3/R junction denotes the start site of transcription.

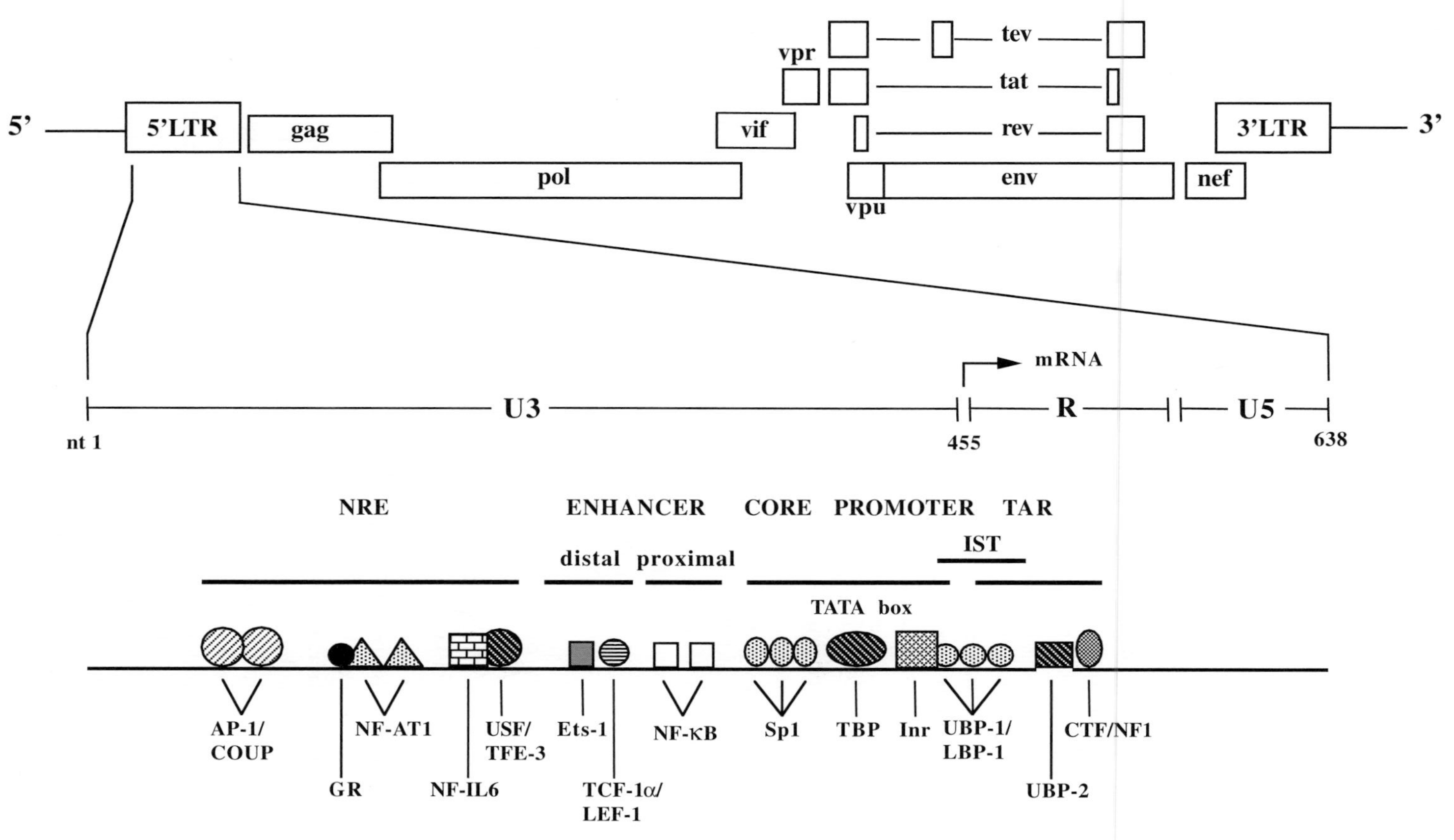
5'
5'LTR
gag
pol
vif
vpr
tev
tat
rev
env
vpu
nef
3'LTR
3'
mRNA
U3
R
U5
nt 1
455
638
NRE
ENHANCER
CORE PROMOTER
TAR
distal
proximal
IST
TATA box
AP-1/
COUP
GR
NF-AT1
NF-IL6
USF/
TFE-3
Ets-1
TCF-1α/
LEF-1
NF-κB
Sp1
TBP
Inr
UBP-1/
LBP-1
UBP-2
CTF/NF1

transcription factors as well as binding sites for the glucocorticoid receptor (GR), NF-AT1, NF-IL6 (Tesmer *et al.,* 1993), and USF/TFE-3. Second, the enhancer is composed of a distal region and a proximal region. The distal enhancer region contains DNA-binding sites for the cellular transcription factors TCF-1α/LEF-1 and Ets-1 (Holzmeister *et al.,* 1993; Sieweke *et al.,* 1998), and the proximal enhancer region contains two binding sites for NF-κB. Third, the core promoter contains three Sp1 binding sites, a TATA box, and an initiator (Inr) element close to the transcription initiation site. Fourth, the 5′ untranslated leader region contains the inducer of short transcripts (IST) element and binding sites for UBP-1/LBP-1, UBP-2, and CTF/NF1 (Jones, 1989; Greene, 1990; Pavlakis and Felber, 1990; Cullen, 1991; Vaishnav and Wong-Staal, 1991; Gaynor, 1992; Jones and Peterlin, 1994). Furthermore, the latter region corresponds to the trans-activating response (TAR) element whose RNA forms a stable stem–loop structure interacting with the viral trans-activator Tat in a manner critical for high activity of the HIV-1 promoter (Rosen, 1991; Frankel, 1992; Karn and Graeble, 1992; Cullen, 1993; Jones and Peterlin, 1994; Cullen, 1998).

The NF-κB binding sites of the proximal enhancer region confer a high rate of transcription to the promoter under conditions of cellular activation. These sites bind inducible factors whose activity in T cells and monocytes/macrophages is induced by factors such as phorbol ester TPA or cytokine TNF (Nabel and Baltimore, 1987; Osborn *et al.,* 1989; Siebenlist *et al.,* 1994). Since HIV expression in latently infected cells can be induced by the same agents, the NF-κB factors have been proposed as the mediators of HIV-1 reactivation from latency (Griffin *et al.,* 1989; Poli and Fauci, 1992). However, the absence of NF-κB in basal conditions is not sufficient to fully explain latency. Indeed, if this was true, HIV-1 mutants with no NF-κB binding sites in their LTR would be expected to always be latent. The fact that this has not been observed (Leonard *et al.,* 1989; Ross *et al.,* 1991) suggests that other factor(s) are responsible for suppressing basal HIV-1 transcription in latently infected cells. The results presented in this chapter are consistent with a model in which chromatin is the primary cause for the suppression of HIV-1 promoter transcription under basal conditions.

Until the early 1990s, the molecular mechanisms of HIV-1 transcriptional regulation were studied essentially using conventional *in vitro* gene expression methodologies, including transient transfections in cell lines, *in vitro* transcription assays, and *in vitro* binding studies. Thus, little was known about the relevance *in vivo* of the interactions between transcription factors and viral elements observed *in vitro*. However, when integrated into cellular genomic DNA, the HIV-1 provirus is organized into nucleosomes, and it is probable that the interactions between cis-acting DNA elements and trans-regulatory proteins are modulated by this nucleosomal organization. Recent investigations on the integrated provirus and on *in vitro* reconstituted chromatin templates have indeed revealed that the mechanisms of HIV-1 transcription are considerably more complex than has been apparent from *in vitro* studies of promoter activity. This chapter highlights recent developments in understanding HIV-1 transcription within its natural context, represented by the provirus integrated in the cellular genome and packaged into chromatin.

III. Chromatin Is an Integral Component of the Transcriptional Regulatory Apparatus

The eukaryotic genome is compacted with histones and other proteins to form chromatin (van Holde, 1989), which allows for efficient storage of genetic information. However, this packaging also prevents the transcription machinery from gaining access to the DNA template (Paranjape *et al.*, 1994).

The repeating subunit of chromatin is the nucleosome core, composed of about 146 bp of DNA tightly wrapped, in 1.65 turns, in a left-handed superhelix around a central histone octamer, which contains two molecules of each of the four core histones: H2A, H2B, H3, and H4. Two adjacent nucleosome cores are separated by a region of linker DNA (14–44 bp) that is associated with a single molecule of histone H1. Histone H1 binds primarily to DNA in the nucleosome at the pseudodyad and at the linker DNA as it enters and leaves the nucleosome core particle (Zhou *et al.*, 1998). A subnucleosome particle containing the nucleosome core, histone H1, and 168 bp of DNA (i.e., the nucleosome core and ~20 bp interacting with histone H1) has been termed a chromatosome. Numerous details of nucleosome structure have been revealed over the past $2\frac{1}{2}$ decades (for review see Pruss *et al.*, 1995). These discoveries culminated with the recent determination of a high-resolution crystal structure of the nucleosome core (Luger *et al.*, 1997). The core octamer has a tripartite structure with a central $(H3–H4)_2$ tetramer that interacts with two H2A–H2B dimers. The core histones contain two domains: a globular, highly α-helical carboxyl (C)-terminal domain, which forms the core of the nucleosome and is involved in histone–histone interactions and DNA binding, and a less structured, highly basic amino (N)-terminal tail domain, located outside of the core particle and interacting with DNA at the outside of the superhelical turns, with other DNA sequences, or with other proteins. The histone N-terminal tails are accessible to various nonhistone proteins, such as proteases, and they are subject to multiple posttranslational modifications, including phosphorylation, methylation, acetylation, poly-ADP-ribosylation, and ubiquitination. All four histones are subject to reversible acetylation and this modification has been studied extensively (see below).

The packaging of genes into chromatin is increasingly recognized as an important component in the regulation of transcription initiation and elongation (Workman and Kingston, 1998). Packaging of DNA modifies its accessibility and geometry in a manner that interferes with these processes (Felsenfeld, 1992; Wolffe, 1995; Workman and Kingston, 1998). First, the close contact between the histone octamer and DNA can inhibit the binding of transcription factors by steric hindrance. Second, nucleosomal DNA is conformationally constrained, both at the level of DNA bending around the histone octamer (80 bp/turn) and at the level of the helical pitch of DNA. Third, the posttranslational modifications of the histone tails can potentially alter the interaction between DNA and the histone octamer and thus modulate transcription factor access to DNA, as has been demonstrated for TFIIIA (Lee *et al.,* 1993).

Several functional studies *in vivo* and *in vitro* have illustrated the regulatory roles of chromatin in transcription. Mutant strains of yeast, in which the synthesis of a specific histone (H4 or H3) can be suppressed, have shown that several genes are induced when histone synthesis is blocked (Han and Grunstein, 1988a; Han *et al.,* 1988b). In addition, chromatin changes have been observed in the promoter of several highly inducible genes during the process of transcriptional activation (Cartwright and Elgin, 1986; Elgin, 1988; Gross and Garrard, 1988). These observations paint a new picture of transcriptional regulation in which chromatin plays an active role by modulating the binding and activity of transcription factors. Chromatin is heterogeneous in the nucleus: transcriptionally active genes are characterized by a more diffuse chromatin structure (active chromatin or euchromatin), whereas inactive genes are packaged in a highly condensed chromatin configuration (inactive chromatin or heterochromatin) (Weintraub and Groudine, 1976). Active chromatin is generally characterized by (1) a general sensitivity (several kb) of its DNA to several nucleases; (2) a more relaxed structure when compared to the condensed nature of inactive chromatin; (3) the presence of nuclease-hypersensitive sites (100–200 bp), which are superimposed on its general nuclease sensitivity; (4) an enrichment in chromosomal nonhistone proteins; and (5) the presence of posttranslational modifications on its histones (acetylation, phosphorylation, methylation) and a relative hypomethylation of its DNA.

In order for the transcription machinery to gain access to DNA, the compacted chromatin structure needs to be altered. At least two different, yet highly conserved, mechanisms are used by eukaryotic cells to alter chromatin structure: (1) chromatin remodeling activities/factors and (2) posttranslational modifications of chromatin components, in particular, histone acetylation.

Several ATP-dependent nucleosome remodeling complexes have been purified from different organisms. These include the SWI/SNF and RSC complexes from yeast; the NURF, CHRAC, ACF, and Brahma complexes from *Drosophila;* and the mammalian SWI/SNF complex from human (reviewed in Tsukiyama and Wu, 1997; Kadonaga, 1998; Varga-Weisz and Becker, 1998). Although these protein complexes have different components and different properties, all contain a related subunit that possesses a conserved SWI2/SNF2-type helicase/ATPase domain (reviewed in Eisen *et al.*, 1995). It has been postulated that this subunit might function as a processive, ATP-driven motor to disrupt DNA–histone interactions (Pazin and Kadonaga, 1997a).

Acetylation of internal lysine residues of core histone N-terminal domains has been found correlatively associated with gene transcription in eukaryotes for more than 3 decades. Histone acetylation levels are the result of a dynamic equilibrium between competing nuclear enzymes: histone acetyltransferases (HATs) and histone deacetylases (HDACs). The acetylation reaction consists in the transfer of an acetyl group from acetylcoenzyme A onto the ε-amino group of the lysine residue, resulting in the neutralization of one positive charge on the histone protein. The mechanism by which acetylation of core histones influences transcriptional activation remains uncertain. It has, nonetheless, been proposed that acetylation of the amino-terminal lysine-rich histone tails promotes destabilization of histone–DNA interactions in the nucleosome, resulting in increased accessibility of the chromatin to the transcription machinery. In the past 2 years, a virtual explosion has occurred in the discovery of enzymes involved in the reversible acetylation/deacetylation of histones (reviewed in Davie, 1998; Kuo and Allis, 1998; Mizzen and Allis, 1998; Workman and Kingston, 1998). The first HAT has been identified as a homolog of the yeast transcriptional coactivator Gcn5 (Brownell *et al.,* 1996). More recently, a number of transcriptional coactivators [TAFII250, TIP60, p300/CBP, and p300/CBP-associated co-factors (P/CAF, P/CIP-ACTR, and SRC-1)] were shown to possess intrinsic HAT activity that is critical for their function *in vivo* (reviewed in Mizzen and Allis, 1998; Struhl, 1998). Targeting of yeast HAT complexes to nucleosomes within the vicinity of the E4 promoter results in transcriptional activation *in vitro* (Utley *et al.,* 1998). Similarly, three human histone deacetylases (HDAC1, HDAC2, and HDAC3) have been cloned and found to be homologous to the yeast transcriptional regulator Rpd3 (Taunton *et al.,* 1996; Yang *et al.,* 1996, 1997; Emiliani *et al.,* 1998b). Rpd3 interacts with transcription factors and modulates transcription of the target genes either positively or negatively, depending on the gene examined (Vidal and Gaber, 1991). Transcriptional repressors (such as Mad/Max; unliganted retinoid and thyroid hormone receptors, YY1) have been shown to repress transcription by recruiting HDACs to specific regulatory regions (reviewed in Grunstein, 1997; Wade *et al.,* 1997; Wolffe, 1997; Struhl, 1998). Collectively, the identification of HATs and HDACs as previously known transcriptional regulators provides strong evidence that histone acetylation/deacetylation plays a causative role in transcriptional regulation. However, the mechanism by which histone acetylation regulates transcription is still not clear. Importantly, a global increase in core histone acetylation does not necessarily induce widespread transcription. Indeed, using RNA differential display, it has been demonstrated that the expression of only ~2% of cellular genes is modulated positively or negatively in response to histone hyperacetylation (Van Lint *et al.,* 1996b). Furthermore, loss of Rpd3 function in yeast and *Drosophila* results in enhanced heterochromatin silencing (De Rubertis *et al.,* 1996; Rundlett *et al.,* 1996; Vannier *et al.,* 1996). Thus, in some contexts, increased histone acetylation actually correlates with decreased transcription (reviewed in Pazin and Kadonaga, 1997b), suggesting a complex relationship between acetyltransferase function and transcriptional activity. This complexity is further revealed in several recent studies showing that histones are not the sole functional substrates for HATs (Gu and Roeder, 1997; Imhof *et al.,* 1997; Boyes *et al.,* 1998; Li *et al.,* 1998; Munshi *et al.,* 1998; Waltzer and Bienz, 1998; Zhang and Bieker, 1998). For example, in addition to histones, p300 acetylates the tumor suppressor protein p53, enhancing its *in vitro* DNA-binding activity (Gu and Roeder, 1997). These latter studies suggest that internal lysine acetylation of multiple proteins exists as a rapid and reversible regulatory mechanism much like protein phosphorylation.

IV. HIV-1 Retroviral Integration and Chromatin

The molecular mechanism underlying the specific integration of a provirus into the host cell chromatin is still poorly understood. Although retroviral integration is clearly not sequence specific, it is not a random phenomenon either (reviewed in Sandmeyer *et al.,* 1990). The following facts are generally accepted: (1) The cellular genome contains highly preferred sites for retroviral integration, which in the case of Rous sarcoma virus were estimated at 500–1000 sites per cell (Shih *et al.,* 1988). (2) Retroviruses integrate preferentially in regions of chromosomal DNA that are being actively transcribed or replicated (near open chromatin structures, such as DNase I-hypersensitive sites) (Robinson and Gagnon, 1986; Vijaya *et al.,* 1986; Rasmussen and Gilboa, 1987; Rohdewohld *et al.,* 1987; Scherdin *et al.,* 1990). Correspondingly, it has been shown that the transcriptional activity (both basal and Tat-transactivated) of the HIV-1 promoter is stimulated by an origin of replication (Kessler and Mathews, 1991; Jeang *et al.,* 1993; Nahreini and Mathews, 1995). (3) Intracellularly, retrovirus integration studies have shown a strong preference by their integrase for nucleosomal rather than naked DNA (Pryciak and Varmus, 1992; Pruss *et al.,* 1994). (4) The site of integration of a given provirus into the cell genome can influence its transcriptional activity (Conklin and Groudine, 1986; Akroyd *et al.,* 1987; Wyke *et al.,* 1989). (5) Expression of genes introduced into cells by retroviral infection is more efficient than that of genes introduced into cells by DNA transfection by a factor of 10- to 50-fold (Hwang and Gilboa, 1984; Rasmussen and Gilboa, 1987).

These observations suggest that retroviruses have evolved mechanisms to specifically direct their genome to sites in cellular DNA which possess transcriptional "potential" (transcribed or inducible). Host transcription factors that bind DNA in a site-specific manner are thought to mediate target site selection by related retrotransposable elements (Sandmeyer *et al.,* 1990). This model is supported by studies using the Ty3 system, a retrotransposon of *Saccharomyces cerevisiae,* which have shown that this retroelement always integrates its genome within the first few base pairs of the transcription initiation site for genes transcribed by RNA polymerase III (Chalker and Sandmeyer, 1990, 1992). Ty3 is targeted to these genes by a direct interaction between the preintegration complex and factors bound at the TATA box of pol III promoters (Chalker and Sandmeyer, 1993; Kirchner *et al.,* 1995). This links, for the first time, a discrete genomic function with the preferential insertion of a retrotransposon.

In the case of HIV-1, Goff and collaborators, using the two-hybrid system, have demonstrated a direct interaction between the HIV-1 integrase protein (IN), a crucial component of the HIV preintegration complex, and the human homolog of yeast chromatin-associated protein SNF5 (Kalpana *et al.,* 1994). The human gene, termed Ini1 (for integrase interactor 1), may encode a nuclear factor that promotes integration and targets incoming viral DNA to active genes. Indeed, the yeast transcriptional activator SNF5 is a component of the multiprotein SWI/SNF complex, which activates transcription by remodeling the chromatin (Wallrath *et al.,* 1994) and Ini1 is part of the analogous mammalian SWI/SNF complex (Wang *et al.,* 1996). Addition of complexes containing Ini1 to nucleosomal DNA results in remodeling of the chromatin *in vitro* (Wang *et al.,* 1996). Because Ini1 directly interacts with HIV IN, is capable of binding to DNA (Morozov *et al.,* 1998), and is involved in chromatin remodeling, Ini1 may target the retroviral integration machinery to open chromatin regions.

Another study examining the flanking sites of eight HIV-1 proviral integrations revealed that seven of the eight proviruses had integrated either directly in or close to (less than 200 bp) an L1H or Alu repetitive element (Stevens and Griffith, 1994), further suggesting that preferential integration sites exist for HIV-1. Recently, Katz and co-workers (Katz *et al.,* 1998) described preferred *in vitro* integration sites for avian sarcoma virus and HIV-1 integrases within the stems of plasmid DNA cruciform structures. The preferred sites are adjacent to the loops in the cruciform and are strand specific (Katz *et al.,* 1998).

A recent study provides evidence that the transcriptional activity of the HIV-1 promoter can be greatly influenced by the site of proviral insertion (Nahreini and Mathews, 1997). To investigate the viral promoter activity in the context of human chromosomes, these authors have used recombinant adeno-associated virus-2 (AAV-2) vectors to transduce a reporter gene driven by the HIV-1 LTR into chromosomal DNA (at a random or specific site).

Basal promoter activity varied substantially among the isolated cell clones, and its responsiveness to Tat was also variable (Nahreini and Mathews, 1997).

V. DNase I-Hypersensitive Sites in the HIV-1 Genome

DNase I digestion of chromatin in purified nuclei shows that different regions of chromatin are differently susceptible to endonuclease cleavage. Small regions of the genome are exquisitely sensitive to digestion by nucleases and are called nuclease-hypersensitive sites. Such sites are thought to represent nucleosome-free or -disrupted regions of chromatin which are bound by *trans*-acting factors, and they are generally found associated with regions of the genome that are important for the regulation of gene expression (such as enhancers, promoters, silencers, and origins of replication) (Gross and Garrard, 1988). The identification of a DNase I-hypersensitive site in a gene usually points to the presence of an underlying regulatory element. This property has been used in several systems, including the discovery of the locus control (LCR) region in the β-globin gene, originally identified as four nuclease-sensitive sites located 10–20 kb upstream of the gene cluster (Wolffe, 1995). However, this situation is not static and, in a growing number of cases, changes in nuclease sensitivity secondary to the disruption of specific nucleosomes have been observed during transcriptional activation (Wallrath *et al.,* 1994). Consequently, understanding the mechanisms responsible for both the positioning and the disruption of specific nucleosomes is a crucial part of understanding the mechanisms of transcriptional regulation of the corresponding gene.

The chromatin structure of the complete HIV-1 genome, integrated in chronically infected cell lines, has been analyzed for the presence of DNase I-hypersensitive sites using the indirect end-labeling technique (Verdin, 1991). Three well-characterized chronically infected cell lines were used: ACH2 (Clouse *et al.,* 1989) and 8E5 (Folks *et al.,* 1986) (both derived from the CEM cell line, a CD4+ T-lymphoid cell line) and U1 (Folks *et al.,* 1987, 1988) (derived from the U937 cell line, a monocyte/macrophage cell line). Two of these cell lines (ACH2 and U1) express little viral RNA in basal conditions, presumably as a consequence of a block at the level of transcription (Pomerantz *et al.,* 1990). HIV-1 transcription in these cells can be induced up to 100-fold by treatment with cytokines or phorbol esters (reviewed in Rosenberg and Fauci, 1990). Therefore, these cell lines have been used as model systems of postintegration latency and allow analysis of the virus chromatin under two distinct functional states: low and high rates of transcription. The 8E5 cell line constitutively produces virus (Folks *et al.,* 1986).

Verdin has demonstrated the presence of five major hypersensitive sites in the U1 cell line and four major sites in the ACH2 and 8E5 cell lines (Verdin, 1991) (Fig. 2). Some of these sites are associated with previously identified cis-acting regulatory elements located in the 5′ and 3′ LTRs. Other additional sites have been identified in the HIV-1 provirus downstream of the 5′ LTR and in the *pol* gene, pointing to the presence of potentially new regulatory elements in these regions (Verdin, 1991).

Under basal conditions, two major DNase I-hypersensitive sites are present in the 5′ LTR: HS2 (nt 223–325) and HS3 (nt 390–449), which map to the HIV-1 promoter in the U3 region. Site HS4 (nt 656–720) is located immediately downstream of the 5′ LTR in a region overlapping the primer binding site (Verdin, 1991). It has recently been demonstrated that a new positive transcriptional regulatory element is associated with DNase I-hypersensitive site HS4 (see Section VIII,A). A single major DNase I-hypersensitive site, HS7 (nt 4534–4733), is present in the 8-kb region located between the two LTRs and maps to a part of the pol gene coding for the integrase (Verdin, 1991) (Fig. 2). A new transcriptional regulatory element is associated with HS7 (see Section VIII,B). This site is present only in the chronically infected U1 cell line of monocytic origin and not in two cell lines of lymphoid origin (ACH2 and 8E5), suggesting a cellular specificity associated with this intragenic element. The 3′ LTR, where viral transcripts are polyadenylated, exhibits a pattern of DNase I-hypersensitivity that is different from the 5′ LTR, with only a single major hypersensitive site (HS8) mapping to nt 9322–9489 (Fig. 2). The 5′ and 3′ LTRs in retroviruses exert different functions despite their identical nucleotide sequence: the 5′ LTR acts as a promoter, whereas the 3′ LTR functions as a polyadenylation site for viral transcripts. The significant structural difference between the two LTRs in terms of DNase I sensitivity could represent the molecular basis of promoter occlusion suggested by Cullen *et al.* (1984). During HIV-1 transcriptional activation by TPA or TNF, an increase in sensitivity appears in the 5′ LTR between HS3 and HS4 (Verdin, 1991). The 8E5 cell line, which constitutively expresses HIV-1, presents the same pattern of hypersensitivity as the ACH2 cells except that the region between HS3 and HS4 is hypersensitive to DNase I digestion in basal conditions (Verdin, 1991).

VI. Nucleosomes Are Precisely Positioned in the 5′ LTR

Chromatin analysis of five different chronically infected cell lines has been performed with micrococcal nuclease (MCnuc), which cuts DNA in nucleosome-free regions and in linker regions separating nucleosomes (Verdin *et al.*, 1993). This analysis found a pattern of discrete cutting for MCnuc in the 5′ LTR region. This pattern is consistent with the presence of precisely

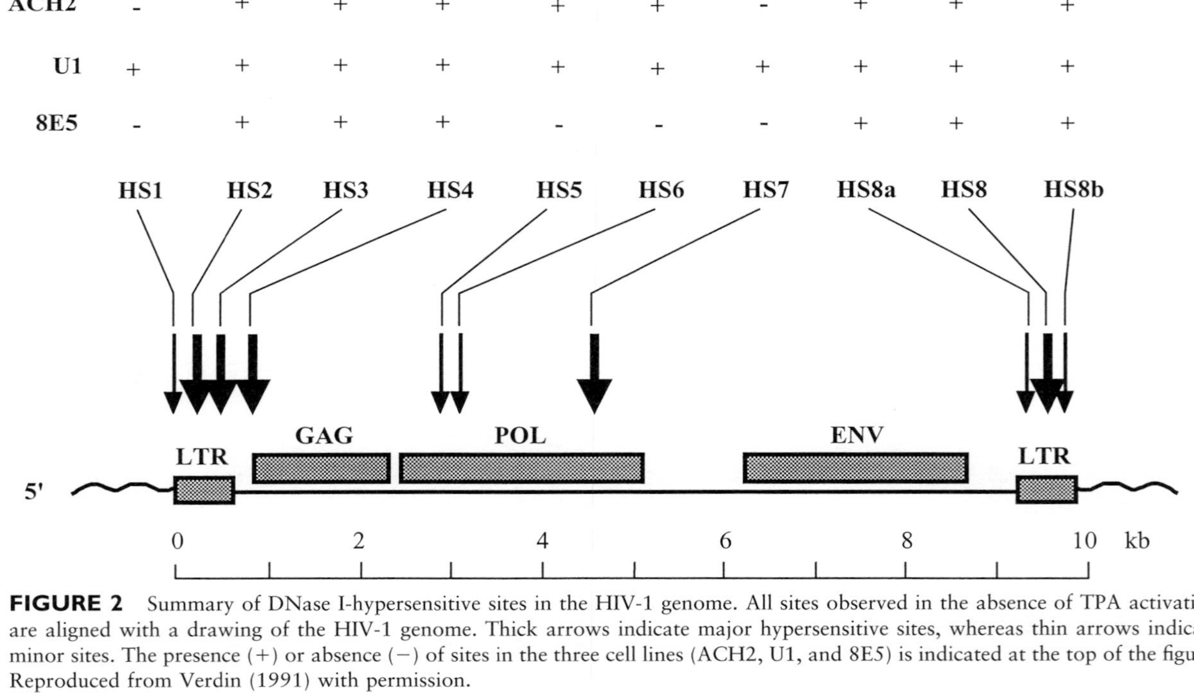

FIGURE 2 Summary of DNase I-hypersensitive sites in the HIV-1 genome. All sites observed in the absence of TPA activation are aligned with a drawing of the HIV-1 genome. Thick arrows indicate major hypersensitive sites, whereas thin arrows indicate minor sites. The presence (+) or absence (−) of sites in the three cell lines (ACH2, U1, and 8E5) is indicated at the top of the figure. Reproduced from Verdin (1991) with permission.

positioned nucleosomes in all of the integrated copies of HIV-1 examined (Verdin *et al.,* 1993). Remarkably, the region separating HS3 and HS4 in R/U5 also becomes hypersensitive to MCnuc digestion after TPA or TNF treatment (Verdin *et al.,* 1993), as previously found for DNase I (Verdin, 1991). These observations support the hypothesis that a chromatin disruption takes place in the R/U5 region during transcriptional activation. Restriction enzymes have been used to confirm the results obtained with the other nucleases. Most restriction enzymes cut in nucleosome-free and in linker regions (Archer *et al.,* 1991), except when their recognition sequence is occupied by a DNA-binding factor (Archer *et al.,* 1991). Multiple enzymes, whose target sequences span the 5′ LTR region, have been used to digest purified nuclei and the digestion products were analyzed by Southern blotting, using the indirect end-labeling technique. These experiments demonstrated the presence of two broad regions of increased accessibility (Verdin *et al,*. 1993), confirming the MCnuc- and DNase I-hypersensitive sites.

Taken together, these experimental results have been used to build a model for the nucleosomal organization of the 5′ LTR (Fig. 3) (Verdin *et al.,* 1993). Under basal conditions, when the provirus is transcriptionally silent, two nucleosomes, nuc-0 (nt 40–200) and nuc-1 (nt 465–610), are precisely positioned in the 5′ LTR. These two nucleosomes define two large open chromatin regions. The first region is associated with the promoter/enhancer elements in the U3 region and spans two distinct DNase I-hypersensitive sites (HS2 and HS3). Genomic footprinting studies have shown that in the silent LTR, several critical protein–DNA interactions are still preserved in this latter open chromatin region (Demarchi *et al.,* 1992, 1993). The second open region is associated with an area that overlaps the primer binding site immediately downstream of the 5′ LTR and spans a DNase I-hypersensitive site called HS4 (nt 610–720). HS4 corresponds to a new regulatory domain (see Section VIII,A). Three precisely positioned nucleosomes (nuc-2, nuc-3, and nuc-4) organize the rest of the 5′ region into chromatin. During transcriptional activation of the HIV-1 promoter, only one change occurs in this organization: a single nucleosome, nuc-1, located immediately after the transcription start site, is specifically disrupted (Fig. 3) (see Section VII). This is the only alteration observed over the complete HIV-1 genome upon transcriptional activation. These results suggest that the translational positioning of nucleosomes in the 5′ LTR is an intrinsic property of the LTR, as the same positions were observed for five distinct chromosomal sites of HIV-1 integration (Verdin *et al.,* 1993).

VII. Disruption of a Single Nucleosome at the Transcription Start Site during HIV-1 Transcriptional Activation

The R/U5 region of HIV-1 separating HS3 and HS4 spans ~200 nt and is the site where a change in nuclease accessibility is observed during

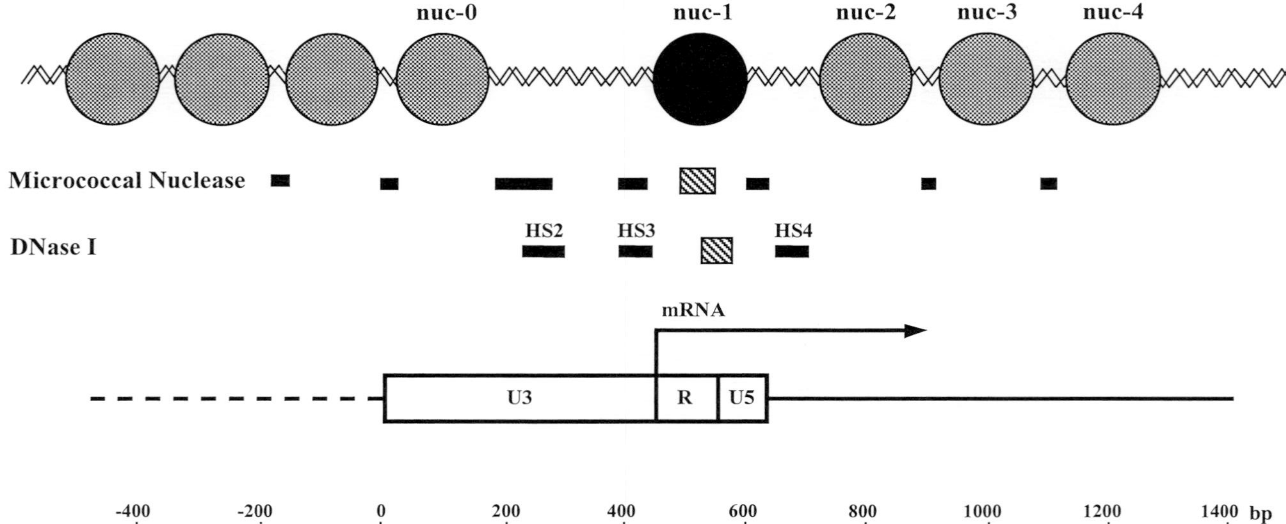

FIGURE 3 Model for the chromatin organization of the 5′ region of the HIV-1 genome. Sites of cutting by DNase I and micrococcal nuclease in purified nuclei are indicated by solid bars and are aligned with the U3, R, and U5 regions of the 5′ LTR. Hypersensitive sites HS2, HS3, and HS4 are shown. The hatched bar indicates the region which becomes hypersensitive to nucleases during transcription activation. The assignment of nucleosome positions in this region based on nuclease digestion is shown above. Taken with modification from Verdin *et al.* (1993) with permission.

transcription activation. To further study this region, DNA samples from nuclease treatment of purified nuclei or following dimethylsulfate (DMS) treatment of intact cells have been analyzed by ligation-mediated PCR (LM-PCR) *in vivo* footprinting (Verdin *et al.,* 1993). This method, based on PCR amplification of nuclease digestion products, permits the examination *in vivo* of a DNA region of known sequence with a resolution of a single nucleotide (Mueller and Wold, 1989; Pfeifer *et al.,* 1990; Pfeifer and Riggs, 1991). In basal conditions, MCnuc digestion demonstrates the presence of a large footprint extending from nt 465 to nt 610 when compared with naked DNA that has been digested *in vitro.* After digestion with DNase I in basal conditions, a periodic pattern of digestion was observed (period ~10 bp) when compared with naked DNA digested *in vitro.* Such a periodic decrease in digestion has been previously reported for nucleosomal DNA and is thought to reflect the deposition of DNA on the surface of the histone octamer (Wolffe, 1995), hindering access to the nuclease with a periodicity equal to the helical repeat of DNA. Modification with DMS showed no significant difference between *in vivo* and *in vitro,* as expected for nucleosomal DNA (Verdin *et al.,* 1993). Following TPA treatment, digestion with MCnuc revealed the disappearance of the nucleosomal footprint and a pattern of digestion indistinguishable from naked DNA. Digestion with DNase I showed the disappearance of the periodic DNA protection pattern and also revealed a profile of digested template essentially identical to naked DNA (Verdin *et al.,* 1993). These results confirm the model regarding the presence of nuc-1 in the R/U5 region under basal conditions in the R/U5 region and the disruption of this nucleosome during transcriptional activation (Fig. 3). The term "disruption" refers to changes in the contacts between DNA and the histone octamer. Indeed, further studies are necessary in order to distinguish between a partial or total displacement of the nucleosome from DNA and a conformational change. The disruption of nuc-1 is independent of DNA replication since it is completed in less than 20 min (Verdin *et al.,* 1993). The disruption of nuc-1 is also independent of RNA polymerase II (RNAPII) transcription since it is insensitive to α-amanitin, an inhibitor of RNAPII elongation (Verdin *et al.,* 1993). The latter result suggests that the disruption is not a passive phenomenon secondary to increased polymerase trafficking following transcriptional activation.

These observations are consistent with a model in which the repression of HIV-1 transcription in latently infected cell lines is mediated by nuc-1. The position of nuc-1 immediately after the transcription start site and the fact that a block to transcriptional elongation colocalizes to this region (Kao *et al.,* 1987) suggest that nuc-1 might be playing a direct role in the inhibition of HIV-1 transcription during latency. Nuc-1 could mediate such an inhibition either by enhancing sequence-specific pausing (Izban and Luse, 1991) or by blocking the binding of a transcription factor (Garcia *et al.,* 1987; Jones *et al.,* 1988; Wu *et al.,* 1988; Lee *et al.,* 1993).

Additional evidence for a role of chromatin in HIV-1 transcription comes from studies on the activation of the promoter by ultraviolet light (UV) and other agents that produce bulky DNA lesions, such as mitomycin C and 4-nitro-quinoline-*N*-oxide (Valerie *et al.*, 1988; Stanley *et al.*, 1989; Valerie and Rosenberg, 1990). However, in contrast to the local chromatin decondensation event limited to nuc-1, these studies have proposed that HIV-1 activation by UV light results from a global decondensation of a chromatin domain encompassing the entire viral genome (Valerie and Rosenberg, 1990). Interestingly, this activation, in stably transfected human cells, does not require the NF-κB binding elements (Valerie *et al.*, 1995).

VIII. Two Novel Regulatory Regions Are Associated with DNase I-Hypersensitive Sites in the HIV-1 Provirus

Structural studies of the chromatin organization of integrated HIV-1 have identified two regions (HS4 and HS7) that are likely to play a role in controlling viral transcription by virtue of their accessibility in chromatin (Verdin, 1991).

A. Characterization of the DNase I-Hypersensitive Site HS4 Associated with the 5′ Untranslated Leader Region of HIV-1

The transcriptionally active HIV-1 promoter is characterized by a large open chromatin region encompassing both the promoter/enhancer region (HS2 + HS3) and a 255-nt region downstream of the transcription start site (nt 465–720) (nuc-1 region and HS4). *In vivo* and *in vitro* footprint analysis of the latter region has identified new recognition sites for several constitutive and inducible transcription factors (Roebuck *et al.*, 1993, 1996; El Kharroubi and Verdin, 1994; Mallardo *et al.*, 1996; Rabbi *et al.*, 1997a,b, 1998): an NF-κB binding site and three AP-1 binding sites (I, II, and III) which lie within the region protected by nuc-1 in basal conditions; an AP3-like motif (AP3-L); a motif interacting with a nuclear factor called downstream binding factor (DBF) site; and juxtaposed Sp1 binding sites (Fig. 4). It has recently been shown that the DBF site is an interferon-regulatory factor (IRF) binding site and that the AP3-L motif binds the T-cell-specific factor NF-AT (Van Lint *et al.*, 1997). Mutations that abolish the binding of each factor to its cognate site have been introduced in an infectious HIV-1 molecular clone to study their effect on HIV-1 transcription and replication. Individual mutation of the DBF or AP3-L site as well as the double mutation AP-1(III)/AP3-L do not affect HIV-1 replication as compared to that of the wild-type virus. In contrast, proviruses carrying

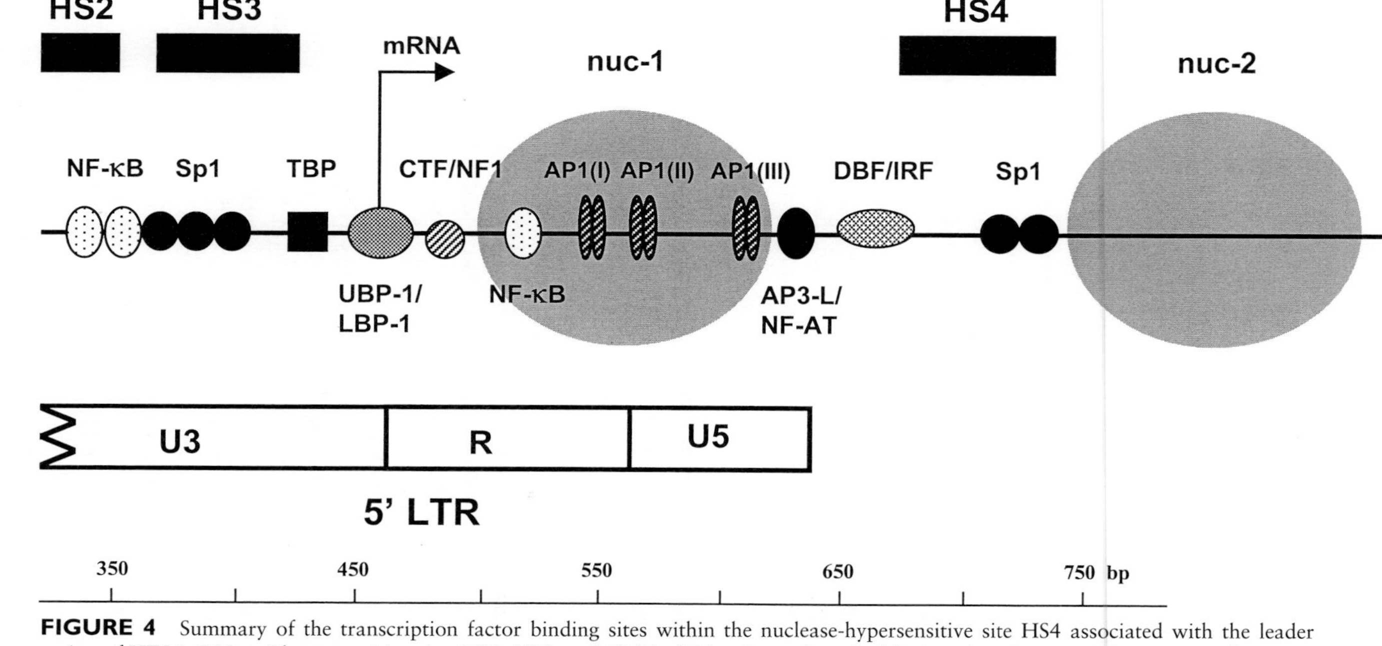

FIGURE 4 Summary of the transcription factor binding sites within the nuclease-hypersensitive site HS4 associated with the leader region of HIV-1. DNase I-hypersensitive sites HS2, HS3, and HS4 (solid bars) are aligned with *cis*-acting elements. The position of putative nucleosomes are indicated by ovals encompassing 146 bp. The location of the transcription initiation site at the U3/R junction is indicated by an arrow. Taken with modification from Van Lint *et al.* (1997) with permission.

mutations in the Sp1 sites are totally defective in terms of replication. Virus production occurs with slightly delayed kinetics for viruses containing combined mutations in the AP-1(III), AP3-L, and DBF sites and in the AP3-L and DBF-sites, whereas viruses mutated in the AP-1(I,II,III) and AP3-L sites and in the AP-1(I,II,III), AP3-L, and DBF sites exhibit a severely defective replicative phenotype (Van Lint *et al.,* 1997). No RNA-packaging defect could be measured for any of the mutant viruses as determined by quantification of their genomic RNA. Measurement of the transcriptional activity of the HIV-1 promoter after transient transfection of the HIV-1 provirus DNA or of long terminal repeat-luciferase constructs shows a positive correlation between the transcriptional and the replication defects for most mutants (Van Lint *et al.,* 1997). These results demonstrate an important role in HIV-1 infectivity of the nucleosome-free region located downstream of the 5′LTR.

In good agreement with these results, an LTR containing mutations in most of the HS4 binding sites [AP-1(III), AP3-L, DBF, and Sp1] and stably integrated into HeLa cells shows reduced promoter activity accompanied by the disappearance of nuclease-hypersensitive site HS4 (El Kharroubi and Martin, 1996). These experiments demonstrate that these sites collectively contribute to the establishment of a nucleosome-free region corresponding to HS4.

B. Characterization of the DNase I-Hypersensitive Site HS7 Associated with the HIV-1 *pol* Gene

High-resolution mapping of the major constitutive DNase I-hypersensitive site associated with the *pol* gene (HS7) defines a nucleosome-free region centered around nucleotides (nt) 4490–4766 (Van Lint *et al.,* 1994). A 500-bp fragment encompassing this hypersensitive site (nt 4481–4982) exhibits transcription enhancing activity after transient transfection in U937 and CEM cells, when it is cloned in its natural position with respect to the HIV-1 promoter after transient transfection in U937 and CEM cells. Five distinct binding sites for nuclear proteins were identified within this positive regulatory element by *in vitro* binding studies (Fig. 5) (Van Lint *et al.,* 1994). Site B (nt 4519–4545) specifically binds four distinct nuclear protein complexes: a ubiquitous factor, a T-cell-specific factor, a B-cell-specific factor, and the monocyte/macrophage and B-cell-specific transcription factor PU.1/Spi-1. In most HIV-1 isolates where this PU box is not conserved, it is replaced by a binding site for the related factor Ets-1. Factors binding to site C (nt 4681–4701) have a DNA-binding specificity similar to that of factors binding to site B, except for PU.1/Spi-1 (Fig. 5). A binding site for Sp1 is located at nt 4623–4631. Site D (nt 4816–4851) specifically binds a ubiquitously expressed factor (Van Lint *et al.,* 1994). These data identify a new transcriptional regulatory element associated with a nuclease hypersensitive site in the *pol* gene of HIV-1 and suggest that its activity may be

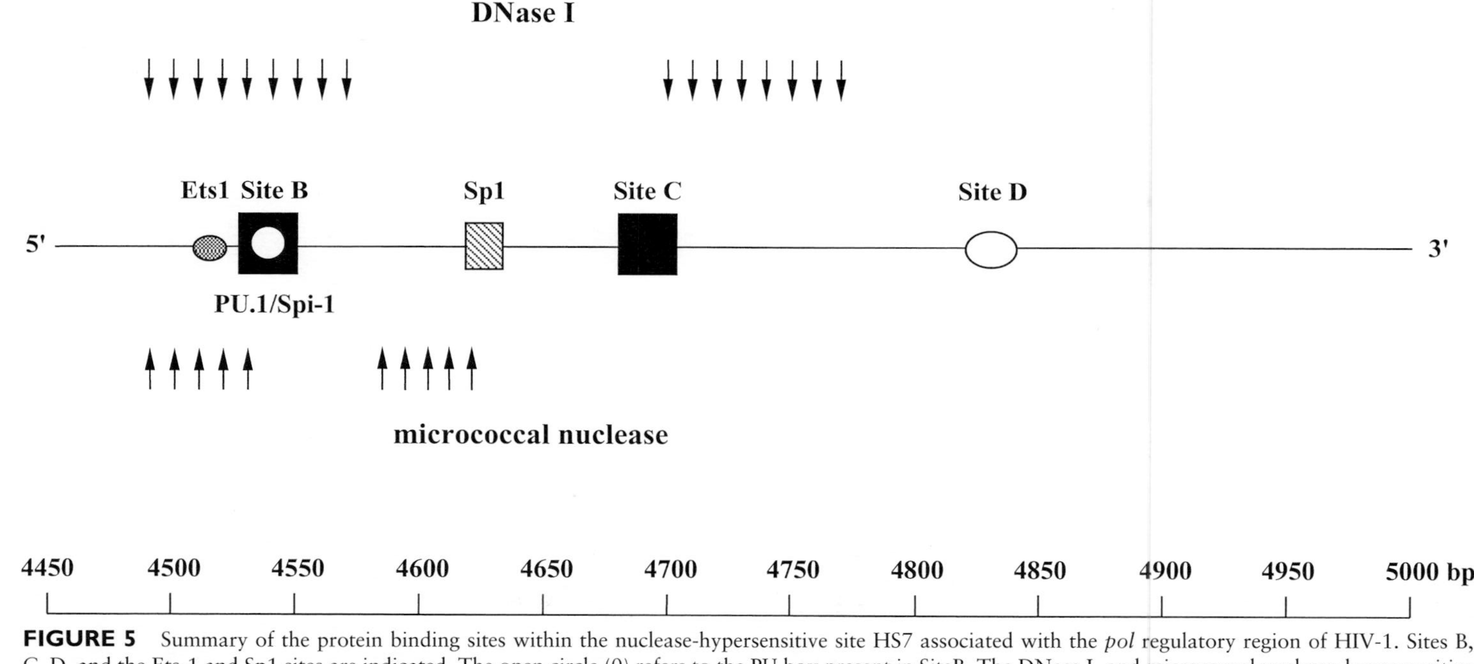

FIGURE 5 Summary of the protein binding sites within the nuclease-hypersensitive site HS7 associated with the *pol* regulatory region of HIV-1. Sites B, C, D, and the Ets-1 and Sp1 sites are indicated. The open circle (0) refers to the PU box present in SiteB. The DNase I- and micrococcal nuclease-hypersensitive sites identified in the *pol* gene are indicated by arrows. Taken with modification from Van Lint *et al.* (1994) with permission.

controlled by a complex interplay of cis-regulatory elements (Van Lint *et al.*, 1994).

The positive regulatory element associated with HS7 could play an important role in the macrophagetropism of HIV-1. Previous studies have demonstrated that HIV-1 macrophagetropism is primarily determined at the level of the entry, by regions of the surface envelope glycoprotein gp120, including the V3 loop region (Cheng-Mayer *et al.*, 1990; O'Brien *et al.*, 1990; York-Higgins *et al.*, 1990; Hwang *et al.*, 1991; Shioda *et al.*, 1991; Westervelt *et al.*, 1991; Westervelt *et al.*, 1992b). However, it is also clear that other determinants of tropism may manifest themselves after viral entry, as illustrated by the *vpr* and *vpu* genes, which are required for efficient viral replication in primary monocytes/macrophages (Westervelt *et al.*, 1992a; Balotta *et al.*, 1993). The presence in the HIV-1 *pol* gene of a regulatory region containing a PU.1 site and associated with a monocyte-specific DNase I-hypersensitive site suggests that an additional determinant of macrophagetropism might be present at the level of transcription. Whereas a large number of studies have examined the lymphoid specificity of the HIV-1 promoter (reviewed in Jones, 1989; Greene, 1990; Pavlakis and Felber, 1990; Cullen, 1991; Vaishnav and Wong-Staal, 1991; Gaynor, 1992; Jones and Peterlin, 1994), little is known on its monocyte/macrophage specificity or other cis-acting elements responsible for this specificity. Tropism of other viruses for their target tissues is determined, in part, at the transcriptional level. In murine leukemia viruses, a switch in tropism to T cells was found to be caused by alterations in the U3 region of the LTR (Celander and Haseltine, 1984). The enhancer element of lymphotropic papovavirus, which contains a PU box required for its activity, contributes to the restricted tropism of the virus for primate B-lymphocytes (Erselius *et al.*, 1990). Importantly, the PU.1/Spi-1 binding site situated in the equine infectious anemia virus (EIAV) LTR is a primary determinant of promoter activity in transfected monocyte cell lines, suggesting that it may be relevant to the macrophagetropism of the EIAV lentivirus (Carvalho and Derse, 1993). Most studies on the transcriptional regulatory mechanisms of retroviruses have focused on the 5′ LTR. However, studies by several groups have identified elements located outside of the LTR that play a significant role in determining the transcriptional rate and tissue specificity of retroviral promoters (Payvar *et al.*, 1983; Arrigo *et al.*, 1987; Miller *et al.*, 1992; Ryden *et al.*, 1993; Verdin and Van Lint, 1995). The element described here in the *pol* gene of HIV-1 offers another example of this concept (Verdin and Van Lint, 1995).

The biological role of the *pol* hypersensitive region in the viral life cycle and macrophagetropism remains to be assessed. It is interesting to note that, in addition to the binding sites identified in HS7, three AP-1 sites (Van Lint *et al.*, 1991) and a glucocorticoid response element (Soudeyns *et al.*, 1993) are present upstream and downstream of HS7, respectively.

Thus, analysis of the chromatin accessibility of integrated HIV-1 genome has identified two new regions that may participate in the proper control of HIV-1 transcription, suggesting that this control may be more complex than originally thought.

IX. Role of Histone Acetylation/Deacetylation in HIV-1 Transcriptional Regulation

A single nucleosome, called nuc-1, is precisely positioned immediately after the transcription start site in cell lines in which the HIV-1 promoter is silent, and nuc-1 is disrupted during transcriptional activation (Verdin *et al.,* 1993). Therefore, chromatin modification(s) might result in HIV-1 promoter activation. Consistently, the silent, integrated LTR can be strongly activated by drugs inducing a chromatin modification (histone acetylation) (Bohan *et al.,* 1987, 1989; Golub *et al.,* 1991; Laughlin *et al.,* 1993, 1995; Van Lint *et al.,* 1996a,b).

Sodium butyrate, a noncompetitive inhibitor of histone deacetylase(s), induces HIV-1 expression in chronically infected cell lines (Bohan *et al.,* 1987, 1989; Golub *et al.,* 1991; Laughlin *et al.,* 1993, 1995) and some controversy has existed regarding the mechanism of this induction and the mechanism of action of *n*-butyrate in general (Kruh, 1982; Kruh *et al.,* 1992; Bohan *et al.,* 1989; Golub *et al.,* 1991). A variety of biological phenomena induced by *n*-butyrate, such as differentiation and cell cycle arrest, have been ascribed to a global increase in histone acetylation caused by the inhibition of histone deacetylase(s) (Kruh, 1982; Kruh *et al.,* 1992). However, at high concentrations, *n*-butyrate also causes nonspecific effects on other enzymes, cell membranes, and cytoskeleton (Kruh, 1982; Kruh *et al.,* 1992). In the case of the HIV-1 promoter, the *n*-butyrate-mediated induction has been mapped to cis-acting elements in the LTR (Bohan *et al.,* 1989; Golub *et al.,* 1991). In contrast, a study using linker-scanning mutants spanning the complete LTR has shown that the induction of LTR transcription could not be mapped to any specific DNA sequence (Laughlin *et al.,* 1993).

Van Lint and co-workers (Van Lint *et al.,* 1996a,b) have used two new specific inhibitors of histone decetylase(s), trapoxin (TPX) and trichostatin A (TSA), to examine the effect of histone hyperacetylation on HIV-1 gene expression. Both drugs inhibit histone deacetylase at nanomolar concentrations *in vitro* and *in vivo* reversibly (TSA) (Yoshida *et al.,* 1990) and irreversibly (TPX) (Kijima *et al.,* 1993). TPX and TSA block cell-cycle progression in normal fibroblasts, induce phenotypic reversion in sis-transformed fibroblasts cells, and lack the pleiotropic effects observed with *n*-butyrate (Yoshida *et al.,* 1990; Kijima *et al.,* 1993). Purification of histone deacetylase from a mutant cell line selected for its resistance to TSA has shown that the purified enzyme is also resistant to the drug *in vitro* (Kijima *et al.,* 1993).

This cell line is also resistant to the biological effects of TPX, suggesting that the two drugs inhibit histone deacetylase activity by binding to a common domain of the enzyme (Kijima *et al.*, 1993).

Several latently HIV-1-infected cell lines have been treated with either TSA or TPX and a marked increase in virus production was detected in cell supernatants (Van Lint *et al.*, 1996a). This activation of virus production occurs at the transcriptional level. A dose–response study shows that maximal induction occurs at 10 n*M* TPX and 1 m*M* TSA. Using Triton-acid-urea gels, Van Lint and collaborators (Van Lint *et al.*, 1996a) have demonstrated a strong positive correlation between the degree of histone hyperacetylation and the degree of transcriptional activation. It has been proposed that activation of the transcription factor NF-κB is a crucial event mediating the transcriptional activation of the HIV-1 promoter from latency (Nabel and Baltimore, 1987; Griffin *et al.*, 1989; Osborn *et al.*, 1989; Poli and Fauci, 1992). However, no induction of NF-κB activity is found in response to either drug (Van Lint *et al.*, 1996a). Using RNA differential display to screen a large number of genes, Van Lint and colleagues have shown that the expression of a small fraction of cellular genes (~2% of all genes) is changed in response to histone hyperacetylation (Van Lint *et al.*, 1996b). This suggests that activation of HIV-1 transcription in response to TSA or TPX demonstrates a specificity and does not result from a global derepression of many cellular genes. Remarkably, despite the global histone hyperacetylation observed following drug treatment, the only detectable modification at the level of HIV-1 chromatin is the disruption of nuc-1 in the 5′ LTR (Van Lint *et al.*, 1996a), whereas other nucleosomes remain unaffected. Therefore, the integrated HIV-1 promoter belongs to a family of genes whose transcriptional activation is modulated by histone acetylation/deacetylation. These experiments demonstrate that independent activation signals for HIV-1 transcription result in the same chromatin modification (nuc-1 disruption) and suggest that this modification is a necessary step for viral transcriptional activation.

As mentioned in Section III, the current view postulates that recruitment of coactivators bearing HAT activity by promoter-bound transcription factors results in histone acetylation of nearby nucleosomes, thus enhancing access of the transcriptional machinery to DNA (Grunstein, 1997; Mizzen and Allis, 1998; Struhl, 1998). Conversely, some transcriptional repressors can recruit histone deacetylases that inhibit transcription by deacetylating chromatin (Pazin and Kadonaga, 1997b; Struhl, 1998). Therefore, the observations described above are consistent with the following model for the suppression of HIV-1 expression under basal conditions: the viral promoter is poised for transcription but repressed by the presence of a single nucleosome, nuc-1, positioned immediately after the transcription start site. The fact that inhibition of histone deacetylase activity is sufficient for transcriptional activation implies that nuc-1 is constitutively deacetylated by targeted histone deacetylase(s). This model is supported by recent data showing that histone deacetylases are targeted to specific promoter regions by virtue of their interactions with specific transcription factors (Ayer *et al.,* 1995; Schreiber-Agus *et al.,* 1995; Hassig *et al.,* 1997; Laherty *et al.,* 1997). In the HIV-1 promoter, several E-box binding sites have been identified (Giacca *et al.,* 1992; Zhang *et al.,* 1992; Ou *et al.,* 1994; di Fagagna *et al.,* 1995). Furthermore, a YY-1 binding site around the transcription start site has been shown to be involved in the suppression of viral expression (Margolis *et al.,* 1994). Since E-box binding proteins (Hassig *et al.,* 1997; Laherty *et al.,* 1997) and YY-1 (Yang *et al.,* 1996) have been shown to interact with histone deacetylases, these factors and possibly other transcription factors represent good candidates for specifically targeting histone deacetylases to nuc-1, causing transcription suppression.

The functional role of nuc-1 in latency remains to be determined. Nuc-1 could block the binding of a transcription factor necessary for the assembly of the initiation complex. Alternatively, nuc-1, by its position immediately after the transcription start site, could impede the progression of RNAPII (by accentuating a natural pausing site), resulting in inefficient elongation (Laspia *et al.,* 1989, 1990; Feinberg *et al.,* 1991) and in the accumulation of short attenuated transcripts detected *in vivo* (Kao *et al.,* 1987).

The mechanism of nuc-1 disruption in response to TNF or to histone hyperacetylation is unclear at the present time. A potential mechanism is that TPA and TNF could induce a transcription factor that would enter in competition with nuc-1 histones for binding to DNA. Such a role could be played by NF-κB, AP-1, and/or NF-AT since binding sites for these transcription factors lie within the region covered by nuc-1 (Fig. 4) and since these factors are induced in response to T-cell activation signals, as is the disruption of nuc-1. The possible role of the viral trans-activator Tat in nuc-1 disruption is discussed in the following section.

X. Role of Tat in HIV-1 Promoter Chromatin Remodeling and Transcriptional Activation

The HIV-1 5′ LTR region contains a number of sites for cellular transcription factors. These factors help control the rate of RNAPII transcription initiation from the integrated provirus, and their abundance in different cell types or at different times likely determines whether a provirus is quiescent or actively replicating. Despite the importance of these factors, HIV-1 transcription is strongly dependent on the viral trans-activator Tat. In the absence of Tat, most viral transcripts terminate prematurely at random locations within ~200 nt of the transcription start site. Tat causes transcribing polymerases to become sufficiently processive to completely transcribe the ~9 kb viral genome. Under some conditions, Tat may also enhance the rate of transcription initiation (reviewed in Jones and Peterlin, 1994; Jeang, 1996). *In vivo* studies with integrated Tat-defective provirus have demonstrated a predominant effect at the level of transcriptional elongation (Feinberg *et al.,* 1991).

Tat increases production of viral mRNAs ~100-fold and is essential for viral replication. Full-length Tat is encoded by two exons on a spliced transcript and is 86–101 aa in length, depending on the viral isolate. Two distinct forms of Tat are detected during HIV-1 infection. A 101–aa protein (Tat101), encoded by both exons (multiply spliced mRNA of ~2000 nt), is expressed in a Rev-independent fashion early and throughout the infectious cycle. A second form of Tat, comprised of 72 aa (Tat72), is translated from an intermediate-size mRNA (~4000 nt). The transport of this mRNA is Rev-dependent and consequently, expression of this form of Tat only occurs later in the virus life cycle, when Rev has accumulated to significant levels (Kim *et al.,* 1989). A stop codon, located in the intron adjacent to the first exon terminates translation after the first 72 aa. The biological reason for the existence of these two forms of Tat is still unclear.

Unlike typical transcriptional activators, Tat does not bind to a DNA site but to an RNA hairpin structure known as TAR (trans-activating response element), present at the 5′ end of all nascent viral transcripts (Berkhout *et al.,* 1989; Dingwall *et al.,* 1990). The 5′ TAR RNA hairpin is formed by base pairing of sequences between positions +1 and +57 relative to the transcription start site. TAR presents two critical sequence elements: the apical 6-nt loop and the 3-nt U-rich bulge. Whereas Tat binds to the bulge, the essential TAR loop has been proposed to contribute to the trans-activation mechanism by binding a cellular cofactor. Functional data suggest that the loop-binding factor is encoded by human chromosome 12. In addition, cofactors may interact with the Tat protein itself.

It is not clear how Tat acts to enhance the processivity of transcribing RNAPII complexes, but recent experiments suggest that Tat may assemble into transcription complexes and recruit or activate factors that phosphorylate the RNAPII C-terminal domain (CTD). Two kinases have been implicated in transcriptional activation of the HIV-1 LTR promoter by Tat/TAR; the cyclin-dependent kinases CDK7 and CDK9 (reviewed in Jones, 1997; Cullen, 1998; Jeang, 1998; Yankulov and Bentley, 1998). CDK7 is associated with cyclin H in the CAK and TFIIH complexes (Jones, 1997). CDK9 (TAK or PITARLE) has recently been shown to interact with cyclin T in the human P-TEFb complex (Wei *et al.,* 1998). These findings support a model in which Tat enhances phosphorylation of the CTD, a process known to occur as RNAPII converts from an initiating to an elongating enzyme. Recent data from Wei *et al.* (1998) demonstrate that the Tat activation domain directly interacts with a novel cyclin-related protein, termed cyclin T, that, in turn, binds specifically to CDK9. Importantly, the Tat-cyclin T complex binds to TAR with higher affinity and specifity than Tat alone, and cyclin T confers recognition of the loop as well as the bulge (Wei *et al.*, 1998). It has been proposed that Tat recruits cyclin T-CDK9 complexes to RNAPII through cooperative binding to nascent TAR RNA, thereby relieving the block to elongation in a Tat-dependent fashion (reviewed in Cullen, 1998). Cyclin T is a probable candidate for the loop-binding factor because it is encoded on human chromosome 12 and leads to protection of the TAR loop in RNA probing assays in the presence of Tat (Wei *et al.*, 1998).

A potentially repressive nucleosome (nuc-1) positioned immediately after the transcription start site in the HIV-1 promoter is disrupted during transcriptional activation (Verdin *et al.*, 1993) and histone hyperacetylation causes HIV-1 transcriptional activation and nuc-1 disruption (Van Lint *et al.*, 1996a,b). These results suggest that nuc-1 might play a direct role in transcription inhibition. Since Tat binds to TAR in a region close to nuc-1 and since Tat activity is critical for transcriptional elongation through a region corresponding to nuc-1 (Kao *et al.*, 1987; Laspia *et al.*, 1989, 1990; Feinberg *et al.*, 1991; Marciniak and Sharp, 1991; Jones and Peterlin, 1994), these results point to a possible role for Tat or a Tat cofactor in the disruption of nuc-1 through a posttranslational modification such as histone acetylation.

Several recent findings are consistent with this model. First, in stably transfected cell lines containing the integrated HIV-1 LTR, chromatin remodeling occurs in the region of nuc-1 upon activation of the promoter by Tat, but not by other stimuli acting through the upstream enhancer sequence (El Kharroubi *et al.*, 1998). This observation suggests that Tat could mediate the transition from repressive to active chromatin structure. Second, disruption of the Tat–TAR axis is responsible for postintegration latency in the chronically infected cell lines U1 and ACH2 (Emiliani *et al.*, 1996, 1998a). In ACH2 cells, the proviral TAR sequence is mutated so that its responsiveness to Tat is diminished, resulting in a latent phenotype when reintroduced into cells as a provirus (Emiliani *et al.*, 1996). Sequence analysis of Tat cDNAs from the U1 cell line have identified two distinct forms of Tat, in agreement with the fact that this cell line contains two integrated HIV-1 proviruses. Both *tat* cDNAs contain mutations and are defective in terms of transcriptional activation in transient-transfection experiments (Emiliani *et al.*, 1998a). These observations signify that the presence of nuc-1 on the HIV-1 promoter is associated with the absence of functional Tat activity and that Tat may play a significant role in chromatin remodeling. Third, Tat has been recently shown to interact with the histone acetyltransferases p300/CBP (Benkirane *et al.*, 1998; Hottiger and Nabel, 1998; Marzio *et al.*, 1998). Tat forms intracellularly a ternary complex with p300/CBP and P/CAF and these HATs function as coactivators for Tat-mediated transcription of chromatinized LTRs (Benkirane *et al.*, 1998; Hottiger and Nabel, 1998; Marzio *et al.*, 1998). *In vivo* quantitative chromatin cross-linking experiments have demonstrated that Tat-mediated activation of the integrated LTR is concomitant with the recruitment of p300 and CBP specifically to the promoter region (Marzio *et al.*, 1998). Fourth, a protein called TIP60 (Tat interactive protein 60), which binds Tat and is critical for trans-activation by Tat (Kamine *et al.*, 1996), was demonstrated to possess HAT activity (Yamamoto and Horikoshi, 1997).

Further work is necessary to test the hypothesis that Tat activity on the HIV-1 promoter is mediated by its interaction with a histone-modifying enzyme. Such interaction might be critical early in the virus life cycle when the provirus has become packaged into chromatin after integration. Indeed, a chromatin remodeling activity of Tat might be critical at this time to modify the chromatin environment in the HIV-1 promoter, in particular at the level of nuc-1. Once the promoter has been remodeled through acetylation and nuc-1 is modified, Tat action might be mostly mediated via its interaction with CDK9. This model is in agreement with the observation that Tat101, which is expressed early in the virus life cycle, is a more efficient transcriptional activator than Tat72 (Tong-Starksen *et al.*, 1993) and with the observation that the second coding exon of Tat plays a role in the optimal trans-activation of integrated (but not unintegrated) LTRs (Jeang *et al.*, 1993). These results further suggest a possible role for Tat in chromatin remodeling.

XI. Study of HIV-1 Transcription using *in vitro* Chromatin-Reconstituted Templates

Nucleosomes assembled *in vitro* on DNA templates inhibit the formation of RNAPII preinitiation complexes in cell-free extracts. In many instances, the presence of transcription factors during nucleosome assembly has been shown to prevent the repression of transcription (reviewed in Owen-Hughes and Workman, 1994). The critical role played by chromatin in HIV-1 transcriptional regulation is supported by studies using the LTR reconstituted *in vitro* with chromatin. *In vitro* packaging of the LTR with *Drosophila* nucleosome assembly extracts strongly represses transcription (Sheridan *et al.*, 1995). This chromatin-mediated repression can be counteracted by preincubation of the DNA prior to incorporation into chromatin with LEF-1, Ets-1, TFE-3 (these factors bind in close proximity near nuc-0), and purified Sp1 (Sheridan *et al.*, 1995). While each of the three distal enhancer factors (LEF-1, Ets-1, TFE-3) can be used interchangeably, presence of Sp1 during nucleosome assembly is essential to obtain a chromatin template that can be transcribed. This suggests that these factors are important for the establishment of the nucleosome-free region corresponding to DNase I-hypersensitive sites HS2 and HS3 observed *in vivo* (Verdin *et al.*, 1991, 1993). Nucleosome assembly in the presence of Sp1 only does not result in high transcription levels. However, addition of the upstream binding transcription factor LEF-1 (a T cell-enriched high-mobility-group transcription factor) during assembly results in transcription levels a few hundredfold above the chromatin repressed template levels (Sheridan *et al.,* 1995). This effect is enhanced in the presence of Ets-1, which binds close to LEF-1 (Sheridan *et al.,* 1995). Interestingly, in the absence of nucleosomal reconstitution, the presence of LEF-1 has no effect on transcription on naked DNA templates (Sheridan *et al.,* 1995), indicating that HIV-1 transcription activation by LEF-1 *in vitro* is a chromatin-dependent process. Since insertion of a linker in that region greatly inhibits virus replication (Kim *et al.,* 1993), it is probable that LEF-1 is essential for virus propagation by preventing nucleosomal repression. In addition, nucleosomal derepression by LEF-1 is not the simple consequence of the bending of DNA by the LEF-1 HMG domain, since it requires a functional trans-activation domain in addition to the HMG domain (Sheridan *et al.,* 1995).

While the factors Sp1, LEF-1, Ets-1, and TFE-3 can prevent the assembly of a repressive chromatin structure, they cannot activate transcription of previously packaged templates. Remarkably, the p65 subunit of the NF-κB transcription factor, and not p50, can activate transcription of nucleosome-assembled LTR synergistically with Sp1 and distal enhancer-binding factors (LEF-1, Ets-1, TFE-3) (Pazin *et al.,* 1996). This trans-activation is observed with chromatin, but not with nonchromatin templates. Furthermore, binding of NF-κB (either p50 or p65) with Sp1 to the HIV-1 promoter causes rearrangement of the chromatin to a structure similar to that of the uninduced integrated provirus *in vivo* (including the precise positioning of nuc-1) (Pazin *et al.,* 1996). These observations therefore suggest that p50 and Sp1 contribute to the establishment of the large open chromatin region corresponding to HS2 + HS3 in the uninduced provirus in resting T cells and that p65 activates transcription by recruitment of the RNAPII transcriptional machinery to the chromatin-repressed basal promoter. However, these authors did not detect a significant difference between the chromatin structure of transcriptionally active templates that were bound by the p65 subunit of NF-κB and Sp1 and those, transcriptionally inactive, bound by the p50 subunit and Sp1.

A recent study has confirmed the observations from the Kadonaga and Jones laboratories (Sheridan *et al.,* 1995; Pazin *et al.,* 1996) and demonstrated that the Sp1 + NF-κB-mediated reconfiguration of the LTR chromatin organization is ATP-dependent (Widlak *et al.,* 1997). While this *in vitro* reconstituted system reproduces several of the characteristics of the *in vivo* situation, it is unresponsive to Tat stimulation (Widlak *et al.,* 1997).

Another recent study suggests that the hypersensitive areas detected *in vivo* do not reflect the absence of nucleosomes but rather occur as a consequence of factor binding (Sp1, NF-κB, LEF-1, Ets-1, and TFE-3) to nucleosomal DNA and the formation of a ternary complex between DNA, histones, and transcription factors (Steger and Workman, 1997). It will be informative to determine *in vivo* whether nucleosomes stay associated with DNA in a reconfigured conformation or whether they are truly displaced from the HIV-1 promoter.

Specific inhibitors of histone deacetylase(s), such as TSA and TPX, are potent inducers of HIV-1 transcription in latently infected T-cell lines, and the activation is accompanied by the loss or rearrangement of nuc-1 (Van Lint *et al.,* 1996a,b). Recent *in vitro* studies examining a chromatin-reconstituted HIV-1 template also support a role for acetylation in LTR-directed transcription (Sheridan *et al.,* 1997; Steger *et al.,* 1998). Experiments from the Jones laboratory have shown that TSA strongly induces HIV-1 transcription on nucleosomal DNA *in vitro,* concomitant with an enhancer-dependent increase in the level of acetylated histones (Sheridan *et al.,* 1997). They conclude that HIV-1 enhancer complexes greatly facilitate transcription reinitiation on chromatin *in vitro* and act at a limiting step to promote the acetylation of histones and/or other, as yet undefined, regulatory transcription factors required for HIV-1 enhancer activity (Sheridan *et al.,* 1997). TSA treatment, however, did not detectably alter enhancer factor binding or the positioning of nuc-1 on the majority of the chromatin templates. These results provide general support for the proposal that recruitment of coactivator complexes containing associated HAT activities is a critical step in enhancer function.

Another study by Steger *et al.* (1998) has demonstrated that, *in vitro,* HAT activities acetylating either histone H3 or H4 stimulate HIV-1 transcription in a chromatin-specific fashion. Acetylation of only histone proteins mediates enhanced transcription, suggesting that these HAT activities facilitate transcription at least in part by modifying histones. In addition, HATs increase accessibility of HIV-1 chromatin in the absence of transcription, suggesting that histone acetylation leads to nucleosome remodeling (Steger *et al.,* 1998).

These experimental systems should become powerful tools to further dissect biochemically the role of chromatin in HIV-1 transcription.

XII. Conclusions

The studies reviewed here of HIV-1 chromatin structure *in vitro* and *in vivo* have led to the unequivocal conclusion that nucleosomal organization of the provirus plays an active role in its transcription regulation. These studies provide fundamental new insights into the process of HIV-1 transcriptional latency and reactivation and ultimately should contribute to an increased understanding of AIDS pathogenesis.

Two new regulatory regions, associated with nuclease-hypersensitive sites in the integrated proviral DNA, are located downstream of the 5′ LTR and in the *pol* gene, respectively. They contain DNA-binding sites for several transcription factors. Further experiments will be necessary to investigate the physiological role of these sites in HIV-1 transcription and replication and their role in the local establishment of an open chromatin configuration in these two regions. Demonstration of positive regulatory elements in the transcribed region of the HIV-1 genome introduces an additional factor into an already-complex network of regulators affecting the degree of viral gene expression.

A potentially repressive nucleosome (nuc-1), located near the transcription start site under basal conditions, is specifically disrupted during transcriptional activation of the HIV-1 promoter by TNF or inhibitors of histone deacetylase(s). To fully understand the mechanism underlying the induction of HIV-1 transcription in stimulated T cells, it will be important to determine how this nucleosome is positioned and disrupted, as well as the exact nature of its disruption. Moreover, it will be of great interest to further explore the mechanism of Tat trans-activation in relation to chromatin modification. Another major task in chromatin studies is to faithfully reproduce the complex regulation found in the provirus on templates constructed *in vitro*. Most challenging is to determine the significance of the data, obtained in model cell cultures and in cell-free chromatin assembly reactions, to the HIV-1 life cycle in human patients.

The HIV-1 provirus belongs to a family of genes whose transcriptional activation is modulated by histone acetylation/deacetylation. Therefore, it will be important to identify the histone deacetylase and acetyltransferase complexes involved in positioning and selective reconfiguration of nuc-1. Beyond the role of these enzymes in chromatin remodeling, it will be important to determine whether regulatory proteins are also relevant targets for acetylation, in addition to the histones.

It is likely that specific promoters will prove to be uniquely dependent on the activity of specific histone deacetylases. Such specificity could be conferred by specific interactions of a given histone deacetylase with a distinct set of DNA-binding factors. In addition to the specific targeting of their enzymatic activities, the histone deacetylases have specificity in their choice of substrates: both in terms of which histone protein they target (H4 vs H3 vs H2A vs H2B) and in terms of which specific lysine residues are affected on each histone N-terminal tail. For the HIV-1 promoter, the next challenge is to identify how and which deacetylase(s) is targeted to specifically modify nuc-1 and therefore inhibit transcription. In order to examine the nucleosomal and acetylation state of nuc-1 under basal and stimulated conditions (− and + Tat or other inducers), it is necessary to perform immunoprecipitation experiments using antisera specific for each of the histones and for acetylated vs unacetylated histones.

These aspects of HIV-1 biology should define new targets for drug design and therapeutic intervention aimed at interfering with HIV-1 replication by maintaining cells in the latent state or by preventing the provirus to remain latent and thus forcing virus expression and sensitivity to viral inhibitors (nucleoside analogs and protease inhibitors). In particular, the enzymatic pathways that control reversible histone acetylation could be modulated through pharmacological intervention. They constitute targets totally inexplored to date and very specific considering the differential role of chromatin modifications in the expression of eukaryotic genes. Additionally, since chromatin influences the transcriptional regulation of many genes besides HIV-1, the observations made in this system will bear relevance to other human diseases in which the expression of genes is similarly altered.

It is hoped that, in the coming years, answers to these and other questions will provide many more exciting insights into the mechanisms that control HIV-1 transcription.

Acknowledgments

The author thanks Arsène Burny, Eric Verdin, and Karen Willard-Gallo for critical reading of the manuscript and apologizes to those whose work could not be cited directly because of space limitations. Carine Van Lint is a "Chercheur Qualifié" of the "Fonds National de la Recherche Scientifique" (F.N.R.S., Belgium). The author acknowledges grant support from the F.N.R.S., Televie, Free University of Brussels (ARC), Internationale Brachet Stiftung (I.B.S.), CGRI-INSERM cooperation, and Theyskens-Mineur Foundation.

References

Akroyd, J., Fincham, V. J., Green, A. R., Levantis, P., Searle, S., and Wyke, J. A. (1987). Transcription of Rous sarcoma proviruses in rat cells is determined by chromosomal position effects that fluctuate and can operate over long distances. *Oncogene* **1**, 347–354.

Archer, T. K., Cordingley, M. G., Wolford, R. G., and Hager, G. L. (1991). Transcription factor access is mediated by accurately positioned nucleosomes on the mouse mammary tumor virus promoter. *Mol. Cell. Biol.* **11**, 688–698.

Arrigo, S., Yun, M., and Beemon, K. (1987). cis-acting regulatory elements within gag genes of avian retroviruses. *Mol. Cell. Biol.* **7**, 388–397.

Ayer, D. E., Lawrence, Q. A., and Eisenman, R. N. (1995). Mad-Max transcriptional repression is mediated by ternary complex formation with mammalian homologs of yeast repressor Sin3. *Cell* **80**, 767–776.

Balotta, C., Lusso, P., Crowley, R., Gallo, R. C., and Franchini, G. (1993). Antisense phosphorothioate oligodeoxynucleotides targeted to the vpr gene inhibit human immunodeficiency virus type 1 replication in primary human macrophages. *J. Virol.* **67**, 4409–4414.

Benkirane, M., Chun, R. F., Xiao, H., Ogryzko, V. V., Howard, B. H., Nakatani, Y., and Jeang, K. T. (1998). Activation of integrated provirus requires histone acetyltransferase: p300 and P/CAF are coactivators for HIV-1 Tat. *J. Biol. Chem.* **273**, 24898–24905.

Berkhout, B., Silverman, R. H., and Jeang, K. T. (1989). Tat trans-activates the human immunodeficiency virus through a nascent RNA target. *Cell* **59**, 273–282.

Bohan, C., York, D., and Srinivasan, A. (1987). Sodium butyrate activates human immunodeficiency virus long terminal repeat-directed expression. *Biochem. Biophys. Res. Commun.* **148**, 899–905.

Bohan, C. A., Robinson, R. A., Luciw, P. A., and Srinivasan, A. (1989). Mutational analysis of sodium butyrate inducible elements in the human immunodeficiency virus type I long terminal repeat. *Virology* **172,** 573–583.

Boyes, J., Byfield, P., Nakatani, Y., and Ogryzko, V. (1998). Regulation of activity of the transcription factor GATA-1 by acetylation. *Nature* **396,** 594–598.

Brownell, J. E., Zhou, J., Ranalli, T., Kobayashi, R., Edmondson, D. G., Roth, S. Y., and Allis, C. D. (1996). Tetrahymena histone acetyltransferase A: A homolog to yeast Gcn5p linking histone acetylation to gene activation. *Cell* **84,** 843–851.

Cartwright, I. L., and Elgin, S. C. (1986). Nucleosomal instability and induction of new upstream protein-DNA associations accompany activation of four small heat shock protein genes in Drosophila melanogaster. *Mol. Cell. Biol.* **6,** 779–791.

Carvalho, M., and Derse, D. (1993). The PU.1/Spi-1 proto-oncogene is a transcriptional regulator of a lentivirus promoter. *J. Virol.* **67,** 3885–3890.

Celander, D., and Haseltine, W. A. (1984). Tissue-specific transcription preference as a determinant of cell tropism and leukaemogenic potential of murine retroviruses. *Nature* **312,** 159–162.

Chalker, D. L., and Sandmeyer, S. B. (1990). Transfer RNA genes are genomic targets for de novo transposition of the yeast retrotransposon Ty3. *Genetics* **126,** 837–850.

Chalker, D. L., and Sandmeyer, S. B. (1992). Ty3 integrates within the region of RNA polymerase III transcription initiation. *Genes Dev.* **6,** 117–128.

Chalker, D. L., and Sandmeyer, S. B. (1993). Sites of RNA polymerase III transcription initiation and Ty3 integration at the U6 gene are positioned by the TATA box. *Proc. Natl. Acad. Sci. USA* **90,** 4927–4931.

Cheng-Mayer, C., Quiroga, M., Tung, J. W., Dina, D., and Levy, J. A. (1990). Viral determinants of human immunodeficiency virus type 1 T-cell or macrophage tropism, cytopathogenicity, and CD4 antigen modulation. *J. Virol.* **64,** 4390–4398.

Clouse, K. A., Powell, D., Washington, I., Poli, G., Strebel, K., Farrar, W., Barstad, P., Kovacs, J., Fauci, A. S., and Folks, T. M. (1989). Monokine regulation of human immunodeficiency virus-1 expression in a chronically infected human T cell clone. *J. Immunol.* **142,** 431–438.

Conklin, K. F., and Groudine, M. (1986). Varied interactions between proviruses and adjacent host chromatin. *Mol. Cell. Biol.* **6,** 3999–3407.

Cullen, B. R. (1991). Regulation of human immunodeficiency virus replication. *Annu. Rev. Microbiol.* **45,** 219–250, 219–250.

Cullen, B. R. (1993). Does HIV-1 Tat induce a change in viral initiation rights? *Cell* **73,** 417–420.

Cullen, B. R. (1998). HIV-1 auxiliary proteins: Making connections in a dying cell. *Cell* **93,** 685–692.

Cullen, B. R., Lomedico, P. T., and Ju, G. (1984). Transcriptional interference in avian retroviruses—Implications for the promoter insertion model of leukaemogenesis. *Nature* **307,** 241–245.

Davie, J. R. (1998). Covalent modifications of histones: Expression from chromatin templates. *Curr. Opin. Genet. Dev.* **8,** 173–178.

De Rubertis, F., Kadosh, D., Henchoz, S., Pauli, D., Reuter, G., Struhl, K., and Spierer, P. (1996). The histone deacetylase RPD3 counteracts genomic silencing in *Drosophila* and yeast. *Nature* **384,** 589–591.

Demarchi, F., D'Agaro, P., Falaschi, A., and Giacca, M. (1992). Probing protein-DNA interactions at the long terminal repeat of human immunodeficiency virus type 1 by *in vivo* footprinting. *J. Virol.* **66,** 2514–2518.

Demarchi, F., D'Agaro, P., Falaschi, A., and Giacca, M. (1993). *In vivo* footprinting analysis of constitutive and inducible protein–DNA interactions at the long terminal repeat of human immunodeficiency virus type 1. *J. Virol.* **67,** 7450–7460.

di Fagagna, F., Marzio, G., Gutierrez, M. I., Kang, L. Y., Falaschi, A., and Giacca, M. (1995). Molecular and functional interactions of transcription factor USF with the long terminal repeat of human immunodeficiency virus type 1. *J. Virol.* **69,** 2765–2775.

Dingwall, C., Ernberg, I., Gait, M. J., Green, S. M., Heaphy, S., Karn, J., Lowe, A. D., Singh, M., and Skinner, M. A. (1990). HIV-1 tat protein stimulates transcription by binding to a U-rich bulge in the stem of the TAR RNA structure. *EMBO J.* **9,** 4145–4153.

Eisen, J. A., Sweder, K. S., and Hanawalt, P. C. (1995). Evolution of the SNF2 family of proteins: Subfamilies with distinct sequences and functions. *Nucleic Acids. Res.* **23,** 2715–2723.

El Kharroubi, A., and Martin, M. A. (1996). cis-acting sequences located downstream of the human immunodeficiency virus type 1 promoter affect its chromatin structure and transcriptional activity. *Mol. Cell. Biol.* **16,** 2958–2966.

El Kharroubi, A., and Verdin, E. (1994). Protein–DNA interactions within DNase I-hypersensitive sites located downstream of the HIV-1 promoter. *J. Biol. Chem.* **269,** 19916–19924.

El Kharroubi, A., Piras, G., Zensen, R., and Martin, M. A. (1998). Transcriptional activation of the integrated chromatin-associated human immunodeficiency virus type 1 promoter. *Mol. Cell. Biol.* **18,** 2535–2544.

Elgin, S. C. (1988). The formation and function of DNase I hypersensitive sites in the process of gene activation. *J. Biol. Chem.* **263,** 19259–19262.

Emiliani, S., Fischle, W., Ott, M., Van Lint, C., Amella, C. A., and Verdin, E. (1998a). Mutations in the tat gene are responsible for human immunodeficiency virus type 1 postintegration latency in the U1 cell line. *J. Virol.* **72,** 1666–1670.

Emiliani, S., Fischle, W., Van Lint, C., Al-Abed, Y., and Verdin, E. (1998b). Characterization of a human RPD3 ortholog, HDAC3. *Proc. Natl. Acad. Sci. USA* **95,** 2795–2800.

Emiliani, S., Van Lint, C., Fischle, W., Paras, P. J., Ott, M., Brady, J., and Verdin, E. (1996). A point mutation in the HIV-1 Tat responsive element is associated with postintegration latency. *Proc. Natl. Acad. Sci. USA* **93,** 6377–6381.

Erselius, J. R., Jostes, B., Hatzopoulos, A. K., Mosthaf, L., and Gruss, P. (1990). Cell-type-specific control elements of the lymphotropic papovavirus enhancer. *J. Virol.* **64,** 1657–1666.

Feinberg, M. B., Baltimore, D., and Frankel, A. D. (1991). The role of Tat in the human immunodeficiency virus life cycle indicates a primary effect on transcriptional elongation. *Proc. Natl. Acad. Sci. USA* **88,** 4045–4049.

Felsenfeld, G. (1992). Chromatin as an essential part of the transcriptional mechanism. *Nature* **355,** 219–224.

Folks, T. M., Justement, J., Kinter, A., Dinarello, C. A., and Fauci, A. S. (1987). Cytokine-induced expression of HIV-1 in a chronically infected promonocyte cell line. *Science* **238,** 800–802.

Folks, T. M., Justement, J., Kinter, A., Schnittman, S., Orenstein, J., Poli, G., and Fauci, A.S. (1988). Characterization of a promonocyte clone chronically infected with HIV and inducible by 13-phorbol-12-myristate acetate. *J. Immunol.* **140,** 1117–1122.

Folks, T. M., Powell, D., Lightfoote, M., Koenig, S., Fauci, A. S., Benn, S., Rabson, A., Daugherty, D., Gendelman, H. E., and Hoggan, M. D. (1986). Biological and biochemical characterization of a cloned Leu-3-cell surviving infection with the acquired immune deficiency syndrome retrovirus. *J. Exp. Med.* **164,** 280–290.

Frankel, A. D. (1992). Activation of HIV transcription by Tat. *Curr. Opin. Genet. Dev.* **2,** 293–298.

Garcia, J. A., Wu, F. K., Mitsuyasu, R., and Gaynor, R. B. (1987). Interactions of cellular proteins involved in the transcriptional regulation of the human immunodeficiency virus. *EMBO J.* **6,** 3761–3770.

Gaynor, R. (1992). Cellular transcription factors involved in the regulation of HIV-1 gene expression [published erratum appears in AIDS 1992 Jun;6(6):following 606]. *AIDS* **6,** 347–363.

Giacca, M., Gutierrez, M. I., Menzo, S., Di Fagagna, F. D., and Falaschi, A. (1992). A human binding site for transcription factor USF/MLTF mimics the negative regulatory element of human immunodeficiency virus type 1. *Virology* **186,** 133–147.

Golub, E. I., Li, G. R., and Volsky, D. J. (1991). Induction of dormant HIV-1 by sodium butyrate: Involvement of the TATA box in the activation of the HIV-1 promoter. *AIDS* **5**, 663–668.

Greene, W. C. (1990). Regulation of HIV-1 gene expression. *Annu. Rev. Immunol.* **8**, 453–475.

Griffin, G. E., Leung, K., Folks, T. M., Kunkel, S., and Nabel, G. J. (1989). Activation of HIV gene expression during monocyte differentiation by induction of NF-kappa B. *Nature* **339**, 70–73.

Gross, D. S., and Garrard, W. T. (1988). Nuclease hypersensitive sites in chromatin. *Annu. Rev. Biochem.* **57**, 159–197.

Grunstein, M. (1997). Histone acetylation in chromatin structure and transcription. *Nature* **389**, 349–352.

Gu, W., and Roeder, R. G. (1997). Activation of p53 sequence-specific DNA binding by acetylation of the p53 C-terminal domain. *Cell* **90**, 595–606.

Han, M., and Grunstein, M. (1988a). Nucleosome loss activates yeast downstream promoters *in vivo*. *Cell* **55**, 1137–1145.

Han, M., Kim, U. J., Kayne, P., and Grunstein, M. (1988b). Depletion of histone H4 and nucleosomes activates the PHO5 gene in *Saccharomyces cerevisiae*. *EMBO J.* **7**, 2221–2228.

Hassig, C. A., Fleischer, T. C., Billin, A. N., Schreiber, S. L., and Ayer, D. E. (1997). Histone deacetylase activity is required for full transcriptional repression by mSin3A. *Cell* **89**, 341–347.

Holzmeister, J., Ludewig, B., Pauli, G., and Simon, D. (1993). Sequence specific binding of the transcription factor c-Ets1 to the human immunodeficiency virus type I long terminal repeat. *Biochem. Biophys. Res. Commun.* **197**, 1229–1233.

Hottiger, M. O., and Nabel, G. J. (1998). Interaction of human immunodeficiency virus type 1 Tat with the transcriptional coactivators p300 and CREB binding protein. *J. Virol.* **72**, 8252–8256.

Hwang, L. H., and Gilboa, E. (1984). Expression of genes introduced into cells by retroviral infection is more efficient than that of genes introduced into cells by DNA transfection. *J. Virol.* **50**, 417–424.

Hwang, S. S., Boyle, T. J., Lyerly, H. K., and Cullen, B. R. (1991). Identification of the envelope V3 loop as the primary determinant of cell tropism in HIV-1. *Science* **253**, 71–74.

Imhof, A., Yang, X. J., Ogryzko, V. V., Nakatani, Y., Wolffe, A. P., and Ge, H. (1997). Acetylation of general transcription factors by histone acetyltransferases. *Curr. Biol.* **7**, 689–692.

Izban, M. G., and Luse, D. S. (1991). Transcription on nucleosomal templates by RNA polymerase II *in vitro:* Inhibition of elongation with enhancement of sequence-specific pausing. *Genes Dev.* **5**, 683–696.

Jeang K. T. (1996). HIV-1 Tat structure and function. *In* "Human Retroviruses and AIDS 1996" (G. Meyers, L. E. Henderson, B. Korber, K. T. Jeang, and S. Wain-Obson, Eds.) pp. 3–18. Los Alamos National Laboratory, Los Alamos, NM.

Jeang, K. T. (1998). Tat, Tat-associated kinase, and transcription. *J. Biomed. Sci.* **5**, 24–27.

Jeang, K. T., Berkhout, B., and Dropulic, B. (1993). Effects of integration and replication on transcription of the HIV-1 long terminal repeat. *J. Biol. Chem.* **268**, 24940–24949.

Jones, K. A. (1989). HIV trans-activation and transcription control mechanisms. *New Biol.* **1**, 127–135.

Jones, K. A. (1997). Taking a new TAK on tat transactivation. *Genes Dev.* **11**, 2593–2599.

Jones, K. A., and Peterlin, B. M. (1994). Control of RNA initiation and elongation at the HIV-1 promoter. *Annu. Rev. Biochem.* **63**, 717–743.

Jones, K. A., Luciw, P. A., and Duchange, N. (1988). Structural arrangements of transcription control domains within the 5′- untranslated leader regions of the HIV-1 and HIV-2 promoters. *Genes Dev.* **2**, 1101–1114.

Kadonaga, J. T. (1998). Eukaryotic transcription: An interlaced network of transcription factors and chromatin-modifying machines. *Cell* **92,** 307–313.

Kalpana, G. V., Marmon, S., Wang, W., Crabtree, G. R., and Goff, S. P. (1994). Binding and stimulation of HIV-1 integrase by a human homolog of yeast transcription factor SNF5. *Science* **266,** 2002–2006.

Kamine, J., Elangovan, B., Subramanian, T., Coleman, D., and Chinnadurai, G. (1996). Identification of a cellular protein that specifically interacts with the essential cysteine region of the HIV-1 Tat transactivator. *Virology* **216,** 357–366.

Kao, S. Y., Calman, A. F., Luciw, P. A., and Peterlin, B. M. (1987). Anti-termination of transcription within the long terminal repeat of HIV-1 by tat gene product. *Nature* **330,** 489–493.

Karn, J., and Graeble, M. A. (1992). New insights into the mechanism of HIV-1 trans-activation. *Trends Genet.* **8,** 365–368.

Katz, R. A., Gravuer, K., and Skalka, A. M. (1998). A preferred target DNA structure for retroviral integrase *in vitro*. *J. Biol. Chem.* **273,** 24190–24195.

Kessler, M., and Mathews, M. B. (1991). Tat transactivation of the human immunodeficiency virus type 1 promoter is influenced by basal promoter activity and the simian virus 40 origin of DNA replication [published erratum appears in *Proc. Natl. Acad. Sci. USA* (1993) Oct **90**(19):9233]. *Proc. Natl. Acad. Sci. USA* **88,** 10018–10022.

Kijima, M., Yoshida, M., Sugita, K., Horinouchi, S., and Beppu, T. (1993). Trapoxin, an antitumor cyclic tetrapeptide, is an irreversible inhibitor of mammalian histone deacetylase. *J. Biol. Chem.* **268,** 22429–22435.

Kim, J. Y., Gonzalez-Scarano, F., Zeichner, S. L., and Alwine, J. C. (1993). Replication of type 1 human immunodeficiency viruses containing linker substitution mutations in the -201 to -130 region of the long terminal repeat. *J. Virol.* **67,** 1658–1662.

Kim, S. Y., Byrn, R., Groopman, J., and Baltimore, D. (1989). Temporal aspects of DNA and RNA synthesis during human immunodeficiency virus infection: Evidence for differential gene expression. *J. Virol.* **63,** 3708–3713.

Kirchner, J., Connolly, C. M., and Sandmeyer, S. B. (1995). Requirement of RNA polymerase III transcription factors for *in vitro* position-specific integration of a retroviruslike element. Science **267,** 1488–1491.

Kruh, J. (1982). Effects of sodium butyrate, a new pharmacological agent, on cells in culture. *Mol. Cell. Biochem.* **42,** 65–82.

Kruh, J., Defer, N., and Tichonicky, L. (1992). Molecular and cellular action of butyrate. *C. R. Seances. Soc. Biol. Fil.* **186,** 12–25.

Kuo, M. H., and Allis, C. D. (1998). Roles of histone acetyltransferases and deacetylases in gene regulation. *Bioessays* **20,** 615–626.

Laherty, C. D., Yang, W. M., Sun, J. M., Davie, J. R., Seto, E., and Eisenman, R. N. (1997). Histone deacetylases associated with the mSin3 corepressor mediate mad transcriptional repression. *Cell* **89,** 349–356.

Laspia, M. F., Rice, A. P., and Mathews, M. B. (1989). HIV-1 Tat protein increases transcriptional initiation and stabilizes elongation. *Cell* **59,** 283–292.

Laspia, M. F., Rice, A. P., and Mathews, M. B. (1990). Synergy between HIV-1 Tat and adenovirus E1A is principally due to stabilization of transcriptional elongation. *Genes Dev.* **4,** 2397–2408.

Laughlin, M. A., Chang, G. Y., Oakes, J. W., Gonzalez-Scarano, F., and Pomerantz, R. J. (1995). Sodium butyrate stimulation of HIV-1 gene expression: a novel mechanism of induction independent of NF-kappa B. *J. Acquir. Immune Defic. Syndr. Hum. Retrovirol.* **9,** 332–339.

Laughlin, M. A., Zeichner, S., Kolson, D., Alwine, J. C., Seshamma, T., Pomerantz, R. J., and Gonzalez-Scarano, F. (1993). Sodium butyrate treatment of cells latently infected with HIV-1 results in the expression of unspliced viral RNA. *Virology* **196,** 496–505.

Lee, D. Y., Hayes, J. J., Pruss, D., and Wolffe, A. P. (1993). A positive role for histone acetylation in transcription factor access to nucleosomal DNA. *Cell* **72,** 73–84.

Leonard, J., Parrott, C., Buckler-White, A. J., Turner, W., Ross, E. K., Martin, M. A., and Rabson, A. B. (1989). The NF-kappa B binding sites in the human immunodeficiency virus type 1 long terminal repeat are not required for virus infectivity. *J. Virol.* **63,** 4919–4924.

Li, Q., Herrler, M., Landsberger, N., Kaludov, N., Ogryzko, V. V., Nakatani, Y., and Wolffe, A. P. (1998). *Xenopus* NF-Y pre-sets chromatin to potentiate p300 and acetylation-responsive transcription from the Xenopus hsp70 promoter *in vivo*. *EMBO J.* **17,** 6300–6315.

Luger, K., Mader, A. W., Richmond, R. K., Sargent, D. F., and Richmond, T. J. (1997). Crystal structure of the nucleosome core particle at 2.8 A resolution. *Nature* **389,** 251–260.

Mallardo, M., Dragonetti, E., Baldassarre, F., Ambrosino, C., Scala, G., and Quinto, I. (1996). An NF-kappaB site in the 5′-untranslated leader region of the human immunodeficiency virus type 1 enhances the viral expression in response to NF-kappaB-activating stimuli. *J. Biol. Chem.* **271,** 20820–20827.

Marciniak, R. A., and Sharp, P. A. (1991). HIV-1 Tat protein promotes formation of more-processive elongation complexes. *EMBO J.* **10,** 4189–4196.

Margolis, D. M., Somasundaran, M., and Green, M. R. (1994). Human transcription factor YY1 represses human immunodeficiency virus type 1 transcription and virion production. *J. Virol.* **68,** 905–910.

Marzio, G., Tyagi, M., Gutierrez, M. I., and Giacca, M. (1998). HIV-1 tat transactivator recruits p300 and CREB-binding protein histone acetyltransferases to the viral promoter. *Proc. Natl. Acad. Sci. USA* **95,** 13519–13524.

Miller, C. L., Garner, R., and Paetkau, V. (1992). An activation-dependent, T-lymphocyte-specific transcriptional activator in the mouse mammary tumor virus env gene. *Mol. Cell. Biol.* **12,** 3262–3272.

Mizzen, C. A., and Allis, C. D. (1998). Linking histone acetylation to transcriptional regulation. *Cell Mol. Life Sci.* **54,** 6–20.

Morozov, A., Yung, E., and Kalpana, G. V. (1998). Structure–function analysis of integrase interactor 1/hSNF5L1 reveals differential properties of two repeat motifs present in the highly conserved region. *Proc. Natl. Acad. Sci. USA* **95,** 1120–1125.

Mueller, P. R., and Wold, B. (1989). In vivo footprinting of a muscle specific enhancer by ligation mediated PCR [published erratum appears in *Science* **248**(4957), 802]. Science **246,** 780–786.

Munshi, N., Merika, M., Yie, J., Senger, K., Chen, G., and Thanos, D. (1998). Acetylation of HMG I(Y) by CBP turns off IFN beta expression by disrupting the enhanceosome. *Mol. Cell* **2,** 457–467.

Nabel, G., and Baltimore, D. (1987). An inducible transcription factor activates expression of human immunodeficiency virus in T cells [published erratum appears in *Nature* **344**(6262), 178]. *Nature* **326,** 711–713.

Nahreini, P., and Mathews, M. B. (1995). Effects of the simian virus 40 origin of replication on transcription from the human immunodeficiency virus type 1 promoter. *J. Virol.* **69,** 1296–1301.

Nahreini, P., and Mathews, M. B. (1997). Transduction of the human immunodeficiency virus type 1 promoter into human chromosomal DNA by adeno-associated virus: Effects on promoter activity. *Virology* **234,** 42–50.

O'Brien, W. A., Koyanagi, Y., Namazie, A., Zhao, J. Q., Diagne, A., Idler, K., Zack, J. A., and Chen, I. S. (1990). HIV-1 tropism for mononuclear phagocytes can be determined by regions of gp120 outside the CD4-binding domain. *Nature* **348,** 69–73.

Osborn, L., Kunkel, S., and Nabel, G. J. (1989). Tumor necrosis factor alpha and interleukin 1 stimulate the human immunodeficiency virus enhancer by activation of the nuclear factor kappa B. *Proc. Natl. Acad. Sci. USA* **86,** 2336–2340.

Ou, S. H., Garcia-Martinez, L. F., Paulssen, E. J., and Gaynor, R. B. (1994). Role of flanking E box motifs in human immunodeficiency virus type 1 TATA element function. *J. Virol.* **68**, 7188–7199.

Owen-Hughes, T., and Workman, J. L. (1994). Experimental analysis of chromatin function in transcription control. *Crit. Rev. Eukaryot. Gene Expr.* **4**, 403–441.

Paranjape, S. M., Kamakaka, R. T., and Kadonaga, J. T. (1994). Role of chromatin structure in the regulation of transcription by RNA polymerase II. *Annu. Rev. Biochem.* **63**, 265–297.

Pavlakis, G. N., and Felber, B. K. (1990). Regulation of expression of human immunodeficiency virus. *New Biol.* **2**, 20–31.

Payvar, F., DeFranco, D., Firestone, G. L., Edgar, B., Wrange, O., Okret, S., Gustafsson, J. A., and Yamamoto, K. R. (1983). Sequence-specific binding of glucocorticoid receptor of MTV DNA at sites within and upstream of the transcribed region. *Cell* **35**, 381–392.

Pazin, M. J., and Kadonaga, J. T. (1997a). SWI2/SNF2 and related proteins: ATP-driven motors that disrupt protein–DNA interactions? *Cell* **88**, 737–740.

Pazin, M. J., and Kadonaga, J. T. (1997b). What's up and down with histone deacetylation and transcription? *Cell* **89**, 325–328.

Pazin, M. J., Sheridan, P. L., Cannon, K., Cao, Z., Keck, J. G., Kadonaga, J. T., and Jones, K. A. (1996). NF-kappa B-mediated chromatin reconfiguration and transcriptional activation of the HIV-1 enhancer *in vitro*. *Genes Dev.* **10**, 37–49.

Pfeifer, G. P., and Riggs, A. D. (1991). Chromatin differences between active and inactive X chromosomes revealed by genomic footprinting of permeabilized cells using DNase I and ligation-mediated PCR. *Genes Dev.* **5**, 1102–1113.

Pfeifer, G. P., Tanguay, R. L., Steigerwald, S. D., and Riggs, A. D. (1990). *In vivo* footprint and methylation analysis by PCR-aided genomic sequencing: Comparison of active and inactive X chromosomal DNA at the CpG island and promoter of human PGK-1. *Genes Dev.* **4**, 1277–1287.

Poli, G., and Fauci, A. S. (1992). The effect of cytokines and pharmacologic agents on chronic HIV infection. *AIDS Res. Hum. Retroviruses* **8**, 191–197.

Pomerantz, R. J., Trono, D., Feinberg, M. B., and Baltimore, D. (1990). Cells nonproductively infected with HIV-1 exhibit an aberrant pattern of viral RNA expression: A molecular model for latency. *Cell* **61**, 1271–1276.

Pruss, D., Bushman, F. D., and Wolffe, A. P. (1994). Human immunodeficiency virus integrase directs integration to sites of severe DNA distortion within the nucleosome core. *Proc. Natl. Acad. Sci. USA* **91**, 5913–5917.

Pruss, D., Hayes, J. J., and Wolffe, A. P. (1995). Nucleosomal anatomy—Where are the histones? *Bioessays* **17**, 161–170.

Pryciak, P. M., and Varmus, H. E. (1992). Nucleosomes, DNA-binding proteins, and DNA sequence modulate retroviral integration target site selection. *Cell* **69**, 769–780.

Rabbi, M. F., Al-Harthi, L., and Roebuck, K. A. (1997a). TNFalpha cooperates with the protein kinase A pathway to synergistically increase HIV-1 LTR transcription via downstream TRE-like cAMP response elements. *Virology* **237**, 422–429.

Rabbi, M. F., Saifuddin, M., Gu, D. S., Kagnoff, M. F., and Roebuck, K. A. (1997b). U5 region of the human immunodeficiency virus type 1 long terminal repeat contains TRE-like cAMP-responsive elements that bind both AP-1 and CREB/ATF proteins. *Virology* **233**, 235–245.

Rabbi, M. F., Al-Harthi, L., Saifuddin, M., and Roebuck, K. A. (1998). The cAMP-dependent protein kinase A and protein kinase C-beta pathways synergistically interact to activate HIV-1 transcription in latently infected cells of monocyte/macrophage lineage. *Virology* **245**, 257–269.

Rasmussen, J. A., and Gilboa, E. (1987). Significance of DNase I-hypersensitive sites in the long terminal repeats of a Moloney murine leukemia virus vector. *J. Virol.* **61**, 1368–1374.

Robinson, H. L., and Gagnon, G. C. (1986). Patterns of proviral insertion and deletion in avian leukosis virus- induced lymphomas. *J. Virol.* **57**, 28–36.

Roebuck, K. A., Brenner, D. A., and Kagnoff, M. F. (1993). Identification of c-fos-responsive elements downstream of TAR in the long terminal repeat of human immunodeficiency virus type-1. *J. Clin. Invest.* **92,** 1336–1348.

Roebuck, K. A., Gu, D. S., and Kagnoff, M. F. (1996). Activating protein-1 cooperates with phorbol ester activation signals to increase HIV-1 expression. *AIDS* **10,** 819–826.

Rohdewohld, H., Weiher, H., Reik, W., Jaenisch, R., and Breindl, M. (1987). Retrovirus integration and chromatin structure: Moloney murine leukemia proviral integration sites map near DNase I-hypersensitive sites. *J. Virol.* **61,** 336–343.

Rosen, C. A. (1991). Regulation of HIV gene expression by RNA–protein interactions. *Trends Genet.* **7,** 9–14.

Rosenberg, Z. F., and Fauci, A. S. (1990). Immunopathogenic mechanisms of HIV infection: Cytokine induction of HIV expression. *Immunol. Today* **11,** 176–180.

Ross, E. K., Buckler-White, A. J., Rabson, A. B., Englund, G., and Martin, M. A. (1991). Contribution of NF-kappa B and Sp1 binding motifs to the replicative capacity of human immunodeficiency virus type 1: Distinct patterns of viral growth are determined by T-cell types. *J. Virol.* **65,** 4350–4358.

Rundlett, S. E., Carmen, A. A., Kobayashi, R., Bavykin, S., Turner, B. M., and Grunstein, M. (1996). HDA1 and RPD3 are members of distinct yeast histone deacetylase complexes that regulate silencing and transcription. *Proc. Natl. Acad. Sci. USA* **93,** 14503–14508.

Ryden, T. A., de Mars, M., and Beemon, K. (1993). Mutation of the C/EBP binding sites in the Rous sarcoma virus long terminal repeat and gag enhancers. *J. Virol.* **67,** 2862–2870.

Sandmeyer, S. B., Hansen, L. J., and Chalker, D. L. (1990). Integration specificity of retrotransposons and retroviruses. *Annu. Rev. Genet.* **24,** 491–518.

Scherdin, U., Rhodes, K., and Breindl, M. (1990). Transcriptionally active genome regions are preferred targets for retrovirus integration. *J. Virol.* **64,** 907–912.

Schreiber-Agus, N., Chin, L., Chen, K., Torres, R., Rao, G., Guida, P., Skoultchi, A. I., and DePinho, R. A. (1995). An amino-terminal domain of Mxil mediates anti-Myc oncogenic activity and interacts with a homolog of the yeast transcriptional repressor SIN3. *Cell* **80,** 777–786.

Sheridan, P. L., Mayall, T. P., Verdin, E., and Jones, K. A. (1997). Histone acetyltransferases regulate HIV-1 enhancer activity *in vitro*. *Genes Dev.* **11,** 3327–3340.

Sheridan, P. L., Sheline, C. T., Cannon, K., Voz, M. L., Pazin, M. J., Kadonaga, J. T., and Jones, K. A. (1995). Activation of the HIV-1 enhancer by the LEF-1 HMG protein on nucleosome-assembled DNA *in vitro*. *Genes Dev.* **9,** 2090–2104.

Shih, C. C., Stoye, J. P., and Coffin, J. M. (1988). Highly preferred targets for retrovirus integration. *Cell* **53,** 531–537.

Shioda, T., Levy, J. A., and Cheng-Mayer, C. (1991). Macrophage and T cell-line tropisms of HIV-1 are determined by specific regions of the envelope gp120 gene. *Nature* **349,** 167–169.

Siebenlist, U., Franzoso, G., and Brown, K. (1994). Structure, regulation and function of NF-kappa B. *Annu. Rev. Cell Biol.* **10,** 405–455.

Sieweke, M. H., Tekotte, H., Jarosch, U., and Graf, T. (1998). Cooperative interaction of ets-1 with USF-1 required for HIV-1 enhancer activity in T cells. *EMBO J.* **17,** 1728–1739.

Soudeyns, H., Geleziunas, R., Shyamala, G., Hiscott, J., and Wainberg, M. A. (1993). Identification of a novel glucocorticoid response element within the genome of the human immunodeficiency virus type 1. *Virology* **194,** 758–768.

Stanley, S. K., Folks, T. M., and Fauci, A. S. (1989). Induction of expression of human immunodeficiency virus in a chronically infected promonocytic cell line by ultraviolet irradiation. *AIDS Res. Hum. Retroviruses* **5,** 375–384.

Steger, D. J., and Workman, J. L. (1997). Stable co-occupancy of transcription factors and histones at the HIV-1 enhancer. *EMBO J.* **16,** 2463–2472.

Steger, D. J., Eberharter, A., John, S., Grant, P. A., and Workman, J. L. (1998). Purified histone acetyltransferase complexes stimulate HIV-1 transcription from preassembled nucleosomal arrays. *Proc. Natl. Acad. Sci. USA* **95,** 12924–12929.

Stevens, S. W., and Griffith, J. D. (1994). Human immunodeficiency virus type 1 may preferentially integrate into chromatin occupied by L1Hs repetitive elements. *Proc. Natl. Acad. Sci. USA* **91**, 5557–5561.

Struhl, K. (1998). Histone acetylation and transcriptional regulatory mechanisms. *Genes Dev.* **12**, 599–606.

Taunton, J., Hassig, C. A., and Schreiber, S. L. (1996). A mammalian histone deacetylase related to the yeast transcriptional regulator Rpd3p. *Science* **272**, 408–411.

Tesmer, V. M., Rajadhyaksha, A., Babin, J., and Bina, M. (1993). NF-IL6-mediated transcriptional activation of the long terminal repeat of the human immunodeficiency virus type 1. *Proc. Natl. Acad. Sci. USA* **90**, 7298–7302.

Tong-Starksen, S. E., Baur, A., Lu, X. B., Peck, E., and Peterlin, B. M. (1993). Second exon of Tat of HIV-2 is required for optimal trans-activation of HIV-1 and HIV-2 LTRs. *Virology* **195**, 826–830.

Tsukiyama, T., and Wu, C. (1997). Chromatin remodeling and transcription. *Curr. Opin. Genet. Dev.* **7**, 182–191.

Utley, R. T., Ikeda, K., Grant, P. A., Cote, J., Steger, D. J., Eberharter, A., John, S., and Workman, J. L. (1998). Transcriptional activators direct histone acetyltransferase complexes to nucleosomes. *Nature* **394**, 498–502.

Vaishnav, Y. N., and Wong-Staal, F. (1991). The biochemistry of AIDS. *Annu. Rev. Biochem.* **60**, 577–630.

Valerie, K., and Rosenberg, M. (1990). Chromatin structure implicated in activation of HIV-1 gene expression by ultraviolet light. *New Biol.* **2**, 712–718.

Valerie, K., Delers, A., Bruck, C., Thiriart, C., Rosenberg, H., Debouck, C., and Rosenberg, M. (1988). Activation of human immunodeficiency virus type 1 by DNA damage in human cells. *Nature* **333**, 78–81.

Valerie, K., Singhal, A., Kirkham, J. C., Laster, W. S., and Rosenberg, M. (1995). Activation of human immunodeficiency virus gene expression by ultraviolet light in stably transfected human cells does not require the enhancer element. *Biochemistry* **34**, 15760–15767.

van Holde, K. E. (1989). "Chromatin" (A. Rich, Ed.) Springer-Verlag, New York.

Van Lint, C., Amella, C. A., Emiliani, S., John, M., Jie, T., and Verdin, E. (1997). Transcription factor binding sites downstream of the human immunodeficiency virus type 1 transcription start site are important for virus infectivity. *J. Virol.* **71**, 6113–6127.

Van Lint, C., Burny, A., and Verdin, E. (1991). The intragenic enhancer of human immunodeficiency virus type 1 contains functional AP-1 binding sites. *J. Virol.* **65**, 7066–7072.

Van Lint, C., Emiliani, S., Ott, M., and Verdin, E. (1996a). Transcriptional activation and chromatin remodeling of the HIV-1 promoter in response to histone acetylation. *EMBO J.* **15**, 1112–1120.

Van Lint, C., Emiliani, S., and Verdin, E. (1996b). The expression of a small fraction of cellular genes is changed in response to histone hyperacetylation. *Gene Expr.* **5**, 245–253.

Van Lint, C., Ghysdael, J., Paras, P. J., Burny, A., and Verdin, E. (1994). A transcriptional regulatory element is associated with a nuclease-hypersensitive site in the pol gene of human immunodeficiency virus type 1. *J. Virol.* **68**, 2632–2648.

Vannier, D., Balderes, D., and Shore, D. (1996). Evidence that the transcriptional regulators SIN3 and RPD3, and a novel gene (SDS3) with similar functions, are involved in transcriptional silencing in *S. cerevisiae*. *Genetics* **144**, 1343–1353.

Varga-Weisz, P. D., and Becker, P. B. (1998). Chromatin-remodeling factors: Machines that regulate? *Curr. Opin. Cell Biol.* **10**, 346–353.

Verdin, E. (1991). DNase I-hypersensitive sites are associated with both long terminal repeats and with the intragenic enhancer of integrated human immunodeficiency virus type 1. *J. Virol.* **65**, 6790–6799.

Verdin, E., and Van Lint, C. (1995). Internal transcriptional regulatory elements in HIV-1 and other retroviruses. *Cell Mol. Biol.* **41**, 365–369.

Verdin, E., Paras, P. J., and Van Lint, C. (1993). Chromatin disruption in the promoter of human immunodeficiency virus type 1 during transcriptional activation [published erratum appears in *EMBO J.* **12**(12), 4900]. *EMBO J.* **12**, 3249–3259.

Vidal, M., and Gaber, R. F. (1991). RPD3 encodes a second factor required to achieve maximum positive and negative transcriptional states in *Saccharomyces cerevisiae. Mol. Cell Biol.* **11**, 6317–6327.

Vijaya, S., Steffen, D. L., and Robinson, H. L. (1986). Acceptor sites for retroviral integrations map near DNase I-hypersensitive sites in chromatin. *J. Virol.* **60**, 683–692.

Wade, P. A., Pruss, D., and Wolffe, A. P. (1997). Histone acetylation: Chromatin in action. *Trends Biochem. Sci.* **22**, 128–132.

Wallrath, L. L., Lu, Q., Granok, H., and Elgin, S. C. (1994). Architectural variations of inducible eukaryotic promoters: Preset and remodeling chromatin structures. *Bioessays* **16**, 165–170.

Waltzer, L., and Bienz, M. (1998). Drosophila CBP represses the transcription factor TCF to antagonize Wingless signalling. *Nature* **395**, 521–525.

Wang, W., Cote, J., Xue, Y., Zhou, S., Khavari, P. A., Biggar, S. R., Muchardt, C., Kalpana, G. V., Goff, S. P., Yaniv, M., Workman, J. L., and Crabtree, G. R. (1996). Purification and biochemical heterogeneity of the mammalian SWI-SNF complex. *EMBO J.* **15**, 5370–5382.

Wei, P., Garber, M. E., Fang, S. M., Fischer, W. H., and Jones, K. A. (1998). A novel CDK9-associated C-type cyclin interacts directly with HIV-1 Tat and mediates its high-affinity, loop-specific binding to TAR RNA. *Cell* **92**, 451–462.

Weintraub, H., and Groudine, M. (1976). Chromosomal subunits in active genes have an altered conformation. *Science* **193**, 848–856.

Westervelt, P., Gendelman, H. E., and Ratner, L. (1991). Identification of a determinant within the human immunodeficiency virus 1 surface envelope glycoprotein critical for productive infection of primary monocytes. *Proc. Natl. Acad. Sci. USA* **88**, 3097–3101.

Westervelt, P., Henkel, T., Trowbridge, D. B., Orenstein, J., Heuser, J., Gendelman, H. E., and Ratner, L. (1992a). Dual regulation of silent and productive infection in monocytes by distinct human immunodeficiency virus type 1 determinants. *J. Virol.* **66**, 3925–3931.

Westervelt, P., Trowbridge, D. B., Epstein, L. G., Blumberg, B. M., Li, Y., Hahn, B. H., Shaw, G. M., Price, R. W., and Ratner, L. (1992b). Macrophage tropism determinants of human immunodeficiency virus type 1 *in vivo*. *J. Virol.* **66**, 2577–2582.

Widlak, P., Gaynor, R. B., and Garrard, W. T. (1997). In vitro chromatin assembly of the HIV-1 promoter. ATP-dependent polar repositioning of nucleosomes by Sp1 and NFkappaB. *J. Biol. Chem.* **272**, 17654–17661.

Wolffe, A. P. (1995). "Chromatin—Structure and Function." Academic Press, San Diego.

Wolffe, A. P. (1997). Transcriptional control: Sinful repression. *Nature* **387**, 16–17.

Workman, J. L., and Kingston, R. E. (1998). Alteration of nucleosome structure as a mechanism of transcriptional regulation. *Annu. Rev. Biochem.* **67**, 545–579.

Wu, F. K., Garcia, J. A., Harrich, D., and Gaynor, R. B. (1988). Purification of the human immunodeficiency virus type 1 enhancer and TAR binding proteins EBP-1 and UBP-1. *EMBO J.* **7**, 2117–2130.

Wyke, J. A., Akroyd, J., Gillespie, D. A., Green, A. R., and Poole, C. (1989). Proviral position effects: possible probes for genes that suppress transcription. *Ciba. Found. Symp.* **142**, 117–127.

Yamamoto, T., and Horikoshi, M. (1997). Novel substrate specificity of the histone acetyltransferase activity of HIV-1-Tat interactive protein Tip60. *J. Biol. Chem.* **272**, 30595–30598.

Yang, W. M., Inouye, C., Zeng, Y., Bearss, D., and Seto, E. (1996). Transcriptional repression by YY1 is mediated by interaction with a mammalian homolog of the yeast global regulator RPD3. *Proc. Natl. Acad. Sci. USA* **93**, 12845–12850.

Yang, W. M., Yao, Y. L., Sun, J. M., Davie, J. R., and Seto, E. (1997). Isolation and characterization of cDNAs corresponding to an additional member of the human histone deacetylase gene family. *J. Biol. Chem.* **272**, 28001–28007.

Yankulov, K., and Bentley, D. (1998). Transcriptional control: Tat cofactors and transcriptional elongation. *Curr. Biol.* **8**, R447–R449.

York-Higgins, D., Cheng-Mayer, C., Bauer, D., Levy, J. A., and Dina, D. (1990). Human immunodeficiency virus type 1 cellular host range, replication, and cytopathicity are linked to the envelope region of the viral genome. *J. Virol.* **64**, 4016–4020.

Yoshida, M., Kijima, M., Akita, M., and Beppu, T. (1990). Potent and specific inhibition of mammalian histone deacetylase both *in vivo* and *in vitro* by trichostatin A. *J. Biol. Chem.* **265**, 17174–17179.

Zhang, W., and Bieker, J. J. (1998). Acetylation and modulation of erythroid Kruppel-like factor (EKLF) activity by interaction with histone acetyltransferases. *Proc. Natl. Acad. Sci. USA* **95**, 9855–9860.

Zhang, Y., Doyle, K., and Bina, M. (1992). Interactions of HTF4 with E-box motifs in the long terminal repeat of human immunodeficiency virus type 1. *J. Virol.* **66**, 5631–5634.

Zhou, Y. B., Gerchman, S. E., Ramakrishnan, V., Travers, A., and Muyldermans, S. (1998). Position and orientation of the globular domain of linker histone H5 on the nucleosome. *Nature* **395**, 402–405.

Arnold B. Rabson and Hsin-Ching Lin

Department of Molecular Genetics and Microbiology
Center for Advanced Biotechnology and Medicine and
Cancer Institute of New Jersey
University of Medicine and Dentistry of New Jersey
Robert Wood Johnson Medical School
Piscataway, New Jersey 08854

NF-κB and HIV: Linking Viral and Immune Activation

I. Introduction

Human immunodeficiency virus (HIV) replication is critically dependent on the regulated transcription of HIV RNAs, including both the full-length viral RNA that is packaged as the viral genome in progeny virions and the array of unspliced and spliced mRNAs that encode the different viral proteins (reviewed in Antoni, Stein, and Rabson, 1994). The regulation of HIV RNA transcription is dependent on the functions of both the viral Tat transactivator protein (see Chapter 10) as well as the activities of a series of cellular transcriptional factors which interact with viral promoter and enhancer elements, located predominantly within the viral long terminal repeat (LTR). While a large number of different cellular factors have been suggested to participate in the regulation of HIV transcription, either through direct interactions with HIV DNA or RNA, or through modulating the effects of Tat, only a handful of cellular regulatory factors have been clearly

Advances in Pharmacology, Volume 48

1054-3589/00 $35.00

shown to have important effects on viral replication and therefore could serve as possible targets for inhibition of HIV infection. This chapter focuses on one family of cellular transcription factors known to have important roles in regulating HIV gene expression, the NF-κB/Rel proteins. Studies of the interactions of NF-κB and HIV have provided important understandings of both HIV biology and gene regulation as well as provided the impetus for intensive study of NF-κB itself. These latter studies of the molecular regulation of NF-κB have had profound impacts on our understanding of such fundamental pathological processes as inflammation and oncogenesis. Thus, the study of NF-κB in HIV biology has had tremendous implications for therapeutic approaches to other diseases and provides yet another example of how studies in one area of biology can have unanticipated implications for other aspects of human health and disease.

II. Historical Perspectives

The origins of our current understandings of the NF-κB/Rel proteins and their role in the regulation of HIV gene expression developed during the 1980s out of studies on newly obtained molecular clones of HIV and also out of apparently unrelated studies on lymphocyte activation and retroviral oncogenesis. The identification in retroviral LTRs of "enhancers," cis-acting DNA elements that induce transcription from promoters in a distance- and orientation-independent manner (Levinson *et al.,* 1982), led to a search for such elements in the HIV LTR. A large enhancer region (Rosen, Sodroski, and Haseltine, 1985) was further refined and identified as a 10- to 12-base pair, tandemly repeated, conditional enhancer, responsive to activation by mimetics of T-cell activation such as phorbol esters (Kaufman *et al.,* 1987; Nabel and Baltimore, 1987; Siekevitz *et al.,* 1987; Tong-Starksen, Luciw, and Peterlin, 1987). Nabel and Baltimore noted that the HIV-enhancer element was essentially identical to a transcription factor binding site present in the enhancer of the immunoglobulin kappa light-chain gene (Nabel and Baltimore, 1987), which bound to a transcription factor known as NF-κB (*n*uclear *f*actor binding to the *B* site in the κ immunoglobulin gene enhancer). Taking advantage of related work in the Baltimore laboratory (Sen and Baltimore, 1986), these investigators were able to show that NF-κB bound to the HIV LTR-enhancer element and furthermore were able to demonstrate that NF-κB binding was induced by T-cell activation stimuli. These critical experiments linked HIV gene regulation with fundamental events in the process of T-cell activation and immediately suggested important hypotheses about the role of immune activation and the HIV-enhancer sequences in the pathogenesis of AIDS.

Following the discovery that the HIV-enhancer sequences were NF-κB binding sites, numerous studies (reviewed below) have examined the functions of the HIV LTR NF-κB binding sites in the activation of viral gene expression by many diverse stimuli. These studies suggested possible roles for NF-κB in HIV replication and AIDS progression *in vivo,* as a wide variety of interesting stimuli, and potential cofactors for AIDS progression, were shown to activate HIV expression through NF-κB. Progress at the end of the 1980s in understanding NF-κB and HIV interactions was limited by the lack of biochemical reagents to further these studies. While the Baltimore lab presented a series of elegant studies demonstrating mechanisms of NF-κB activation (Baeuerle and Baltimore, 1988a; Baeuerle and Baltimore, 1988b), it was not until the molecular cloning of the first NF-κB subunit (Ghosh *et al.,* 1990; Kieran *et al.,* 1990) that detailed molecular analyses could be performed. Analysis of the sequence of the NF-κB-1 gene generated the striking observation that the gene was closely related to a previously identified retroviral oncogene, *v-rel,* and its cellular homolog, *c-rel.* Further studies elicited the molecular cloning of the family of NF-κB-Rel transcription factors and have elucidated the molecular mechanisms of their control of transcription. Nonetheless, despite the elegant molecular biology and profound understandings of this fascinating cellular regulatory pathway, the actual role of NF-κB in the progression of HIV infection and AIDS has continued to be controversial, even up to the present time.

III. The NF-κB/Rel Transcription Factors

The nature, identity, and regulation of the NF-κB/Rel family of transcription factors have been unfolded over the past several years. It is clear that the NF-κB/Rel gene regulation pathway represents one of the most important inducible gene expression systems in the cell, regulating literally hundreds of genes involved in immune activation and more general cellular and organismal responses to stress. NF-κB did not evolve to support HIV replication; instead, HIV has cleverly coopted this fundamental cellular regulatory pathway to support its own intimate interactions with the resting and activated host immune system. There are numerous comprehensive reviews of NF-κB effects on immune activation (Baldwin, 1996; Gerondakis *et al.,* 1998), the structure and function of NF-κB proteins, the IκB inhibitors of NF-κB (Ghosh, May, and Kopp, 1998; Siebenlist, Franzoso, and Brown, 1994), and, more recently, the enzymatic activities involved in the regulation of NF-κB and IκB function (May and Ghosh, 1998; Zandi and Karin, 1999). In light of these many elegant reviews, only the most salient points for understanding the effects of NF-κB on HIV infection are discussed in this chapter.

A. Structure of the NF-κB/Rel Proteins

The mammalian NF-κB/Rel family of transcription factors are encoded by five genes whose gene products are illustrated in Fig. 1A. The products

A

c-Rel — DNA BD, NLS, RHD, TAD, TAD

v-Rel — RevT, DNA BD, NLS, RHD, TAD, RevT

RelA — DNA BD, NLS, RHD, TAD

RelB — LeuB, DNA BD, NLS, RHD, TAD

NF-κB-1/p105 — DNA BD, NLS, RHD, Gly, ankyrins

NF-κB-1/p50 — DNA BD, NLS, RHD, Gly

NF-κB-2/p100 — DNA BD, NLS, RHD, Gly, ankyrins

NF-κB-2/p52 — DNA BD, NLS, RHD, Gly

B

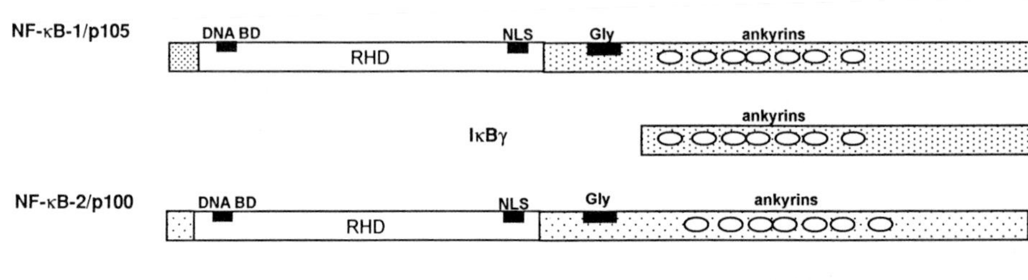

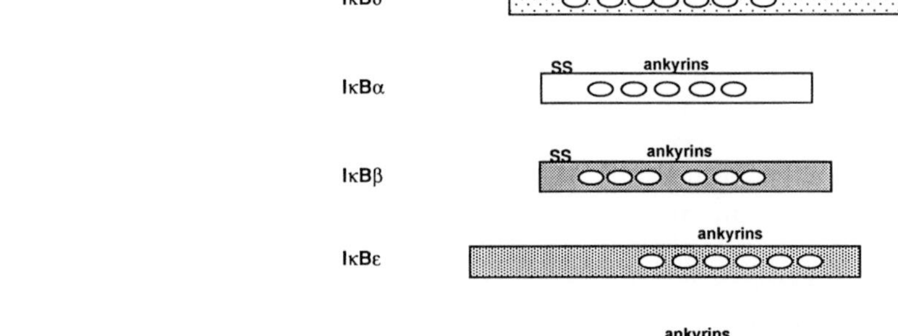

Bcl-3 — ankyrins

of these genes form an array of homo- and heterodimeric forms that account for the detailed contributions of each of these proteins to transcriptional regulation. The amino-terminal portion of the different NF-κB/Rel protein products consists of an approximately 300-amino-acid region referred to as the Rel homology domain (RHD). The RHD contains conserved amino acid sequences involved in important functions common to all NF-κB/Rel proteins, including DNA binding, dimerization, and nuclear localization. The carboxy-terminal sequences of each NF-κB/Rel protein differ, reflecting the unique biological activities of different family members.

The *c-rel, relA,* and *relB* gene products have features of "classic" transcription factors, with DNA binding domains in their RHDs and with identifiable transcriptional activation domains within their differing C-terminal sequences. c-Rel is the cellular homolog of the first NF-κB protein to be identified, the *v-rel* oncogene from the reticulendotheliosis virus of turkeys. c-Rel has two transactivation domains in its C-terminus and has been reported to effectively activate expression of the HIV LTR in one study (McDonnell *et al.,* 1992). V-Rel is deleted in the most C-terminal activation domain, a feature which somewhat decreases its transcriptional activation properties for some promoters (McDonnell *et al.,* 1992). In fact, v-Rel will actually compete with the activating c-Rel protein, or with classic inducible NF-κB dimers, and will thereby reduce activation of the HIV LTR (Ballard *et al.,* 1990; McDonnell *et al.,* 1992). c-Rel itself is less activating than inducible NF-κB and thus may compete with NF-κB dimers and also limit HIV LTR activation (Doerre *et al.,* 1993). Thus c-Rel may either activate LTR expression or limit activation induced by more potent activating dimers. The *relA* gene encodes a 65-kDa protein, variously referred to as RelA or p65 NF-κB. RelA was identified during the initial characterization of NF-κB as participating in the inducible activation of the HIV LTR through its enhancer sequences. RelA contains a potent transcriptional activation domain in its C-terminus containing a series of modular, acidic sequences

FIGURE 1 Structures of the mammalian NF-κB and IκB proteins. (A) The NF-κB proteins. All family members contain the Rel homology domain (RHD) with DNA binding domains (DNA BD) and nuclear localization sequences (NLS). Members of the Rel subfamily contain C-terminal transactivation domains (TAD). v-Rel is derived from the reticuloendotheliosis virus (Rev-T) and has Rev-T fusion sequences at its amino- and carboxy-termini. RelB contains an amino terminal leucine zipper motif (Leu). The full-length members of the NF-κB subfamily contain carboxy-terminal ankyrin motifs (ankyrins) as well as a glycine rich region (Gly) involved in proteolytic processing. (B). The IκB proteins. The IκB proteins contain ankyrin motifs. The full-length NF-κB-1 and NF-κB-2 proteins also function as IκBs, and the C-terminal regions of each of these proteins may also function as an independent IκB (IκBγ derived from NF-κB-1/p105, and IκBδ, derived from NF-κB-2/p100). The locations of critical regulatory serine residues (S) in IκBα and IκBβ are shown. Bcl-3 is similar in structure to the IκBs; however, it may actually potentiate NF-κB function.

contributing the activation properties (Blair *et al.*, 1994; Schmitz *et al.*, 1994). As is described in more detail below (Section III), NF-κB dimers containing RelA are potent activators of HIV gene expression (Kretzschmar *et al.*, 1992; Liu *et al.*, 1992; Nolan *et al.*, 1991; Ruben *et al.*, 1992; Schmid *et al.*, 1991). Like RelA and c-Rel, RelB also contains potent transcriptional activation domains in its C-terminal region. RelB is somewhat unique, as it is unable to homodimerize; however, RelB-containing heterodimers function as activators of NF-κB target gene expression and can activate promoters containing the HIV NF-κB motifs (Bours *et al.*, 1994). Interestingly, relatively little is known about the specific effects of RelB on HIV transcription, although p50/RelB homodimers have been suggested to play a role in HIV transcription in monocytic cells (Lewin *et al.*, 1997).

The second family of NF-κB genes includes *NFKB1* and *NFKB2*. Both of these genes encode two different gene products. Full-length products of these genes (p105 NF-κB-1 and p100 NF-κB-2) contain the RHD, a glycine-rich "hinge" region, and a C-terminal region containing a series of repeated ankyrin motifs. Shorter proteins (p50 NF-κB-1 and p52 NF-κB-2) are also synthesized by the *NFkB1* and *NFKB2* genes. While the mechanisms regulating the synthesis of these processed forms is still controversial, there are data to suggest that at least some p50 is synthesized directly from the *NFKB1* transcript through cotranslational processing by the 26S proteasome, without prior synthesis of p105 NF-κB-2 (Lin, DiMartino, and Greene, 1998). On the other hand, at least in some circumstances (Belich *et al.*, 1999), it is likely that p105 and p100 do serve as true precursor forms from which the p50 and p52 proteins are derived by proteolytic processing, again likely involving the 26S proteasome. p105 NF-κB-1 and p100 NF-κB-2 both contain the ankyrin motifs that are characteristic of the IκB inhibitors of NF-κB (see below), and therefore, not surprisingly, each of these full-length proteins has been reported to function as an inhibitor of NF-κB transcriptional activation. p100 NF-κB-2, in particular, has been reported to inhibit HIV LTR activation (Harhaj *et al.*, 1996). Regulated processing of these proteins into the p50 NF-κB-1 and p52 NF-κB-2 proteins could therefore play a role in the regulation of HIV in infected cells.

p50 NF-κB-1 and p52NF-κB-2 contain only the N-terminal RHD sequences with adjacent carboxy-terminal glycine-rich sequences and do not have a clearly defined transcriptional activation domain. Instead, they participate in transcriptional activation through heterodimerization with members of the Rel subfamily of proteins. p50 NF-κB-1 is a component of the p50/p65 (RelA) heterodimer that is the classic form of NF-κB, originally identified as a cytokine-responsive transcriptional activator in T cells. Thus, p50/p65 heterodimers serve as important activators of HIV LTR transcription. p52 NF-κB-2 also heterodimerizes with RelA, and this heterodimer also will function as a strong activator of HIV LTR transcription (Schmid *et al.*, 1991). The relative contributions of p50/p65 and p52/p65 heterodimers to HIV transcriptional activation probably vary in different cell types and in response to different cellular activation stimuli (Fujita *et al.*, 1992; Kretzschmar *et al.*, 1992; Liu *et al.*, 1992; Schmid *et al.*, 1991); however, at least in some settings, p52/p65 heterodimers appear to be better activators of HIV than the classic p50/p65 heterodimers (Liu *et al.*, 1992; Schmid *et al.*, 1991). In contrast, both p50 and p52 homodimers, lacking transcriptional activation domains, have no independent transcriptional activation properties on the HIV LTR promoter and, in fact, will compete with the strongly activating NF-κB dimers for binding to the LTR κB motifs, thus functionally serving as repressors of LTR activation (Fujita *et al.*, 1992; Schmid *et al.*, 1991).

B. IκB Proteins

Transcriptional activation of HIV and other target genes by NF-κB/Rel proteins is carefully regulated by the cell through an elegant series of mechanisms involving both inhibition and activation of NF-κB activity. The primary focus of this regulatory network are the IκB proteins (Fig. 1B). The IκB proteins bind to NF-κB/Rel dimers and block the nuclear localization and DNA binding activities of NF-κB, thus inhibiting NF-κB-mediated transcriptional activation at two distinct levels, nuclear translocation and direct interaction with NF-κB target gene DNA sequences in the nucleus. The IκBs, like p105 NF-κB-1 and p100 NF-κB-2, contain a series of repeated ankyrin motifs. IκBα, IκBβ, and IκBε all contain a central core of ankyrin motifs flanked by amino- and carboxy-terminal regulatory sequences. IκBγ and the putative IκBδ are composed of the ankyrin motif regions of p105 NF-κB-1 and p100 NF-κB-2, respectively. In fact, as noted above, the full-length p105 and p100 molecules also serve as potent inhibitors of NF-κB activity and thus can rightfully be described as having IκB-like activities.

The mechanisms by which IκBα and IκBβ inhibit NF-κB function have been studied most intensively. Two serine residues in the amino terminus of IκBα and IκBβ are critically important for regulating the inhibitory function of IκB. Phosphorylation of these residues is the first step in the loss of IκB inhibition of NF-κB, which leads to NF-κB activation (Brockman *et al.*, 1995; Brown *et al.*, 1995; DiDonato *et al.*, 1996; Traenckner *et al.*, 1995; Whiteside *et al.*, 1995). The recently determined X-ray crystallographic structures of IκBα in complex with truncated p50/p65 dimers has brought new light to understanding the functions of the IκBs (Huxford *et al.*, 1998; Jacobs and Harrison, 1998). Ankyrin motifs from the IκBα central region contact NF-κB residues involved in both nuclear localization and DNA binding, accounting for the dual activities of IκB on these two processes.

C. Regulation of NF-κB Activity

NF-κB activation is the result of the loss of IκB inhibition of NF-κB. A large array of signals, including a number of immune activation signals, cytokines, cellular stresses, and bacterial and viral infections, have been shown to potently activate NF-κB in human monocytes and T cells (reviewed in Antoni, Stein, and Rabson, 1994). In its simplest form, NF-κB activation results from specific phosphorylation of IκBα or IκBβ on two regulatory serines in the amino terminal portion of the molecule (serine 32 and 36 in IκBα, see Fig. 1B and Fig. 2A). This phosphorylation event serves as a signal to induce the ubiquitination of IκB. Ubiquitinated IκB is recognized by the 26S proteasome and undergoes proteolytic degradation, freeing NF-κB dimers to translocate to the nucleus, where they can activate transcription. This provides a fundamental paradigm for HIV activation by NF-κB, which has been invoked to explain multiple aspects of HIV pathogenesis. In this paradigm, a particular extracellular signal induces IκB phosphorylation and degradation, leading to NF-κB activation, as manifested by NF-κB nuclear localization, binding of activating NF-κB dimers to the HIV LTR DNA, and activation of HIV transcription (Antoni, Stein, and Rabson, 1994; Bednarik and Folks, 1992; Fauci, 1988). Site-directed mutations of the critical regulatory serines in IκB to alanine residues results in an IκB that cannot be phosphorylated or degraded (Brockman *et al.*, 1995; Brown *et al.*, 1995; DiDonato *et al.*, 1996; Traenckner *et al.*, 1995), thus providing a continuous and potent inhibitor of NF-κB. This molecule provides a useful tool for probing the role of NF-κB activation in biological processes such as HIV infection (see Section VI). It should be noted that IκB molecules can also be found in the nucleus, where they can interact with NF-κB dimers to inhibit DNA binding (Arenzana-Seisdedos *et al.*, 1997; Luque and Gélinas, 1998). This may represent an important mechanism for turning off transcription of an NF-κB target gene, following its activation. Thus, the mechanisms responsible for IκB degradation and NF-κB activation are directly relevant to understanding the biology of HIV infection.

The details of NF-κB activation are being elucidated at a rapid rate. While subtle differences in the activation pathways undoubtedly occur for

---

FIGURE 2 Activation of NF-κB transcription factors. (A) Regulation of NF-κB by IκB processing. Cytokine signals through the TNF or IL-1 receptors activate a kinase cascade, including the IκB kinase complex (IKK), leading to the phosphorylation, ubiquitination, and proteasome-mediated degradation of IκB molecules and the nuclear translocation of activating NF-κB dimers. Nuclear NF-κB can then activate transcription of integrated HIV proviruses. Nuclear IκB may regulate nuclear NF-κB function. (B) Regulation of NF-κB by full-length NF-κB-1 and NF-κB-2 proteins. NF-κB-1/p105 (shown) and NF-κB-2/p100 can both serve as IκB-like molecules, retaining activating NF-κB subunits in the cytoplasm. Cytokine and T-cell activation stimuli can activate the Cot/Tpl-2 kinase, which can phosphorylate p105, leading to its ubiquitination and proteasome-mediated processing to p50. Alternatively, p50 can be formed through direct, proteasome-mediated, cotranslational processing of NF-κB-1 translation products. p50 can then participate in activating NF-κB heterodimers and translocate to the nucleus to activate HIV gene expression.

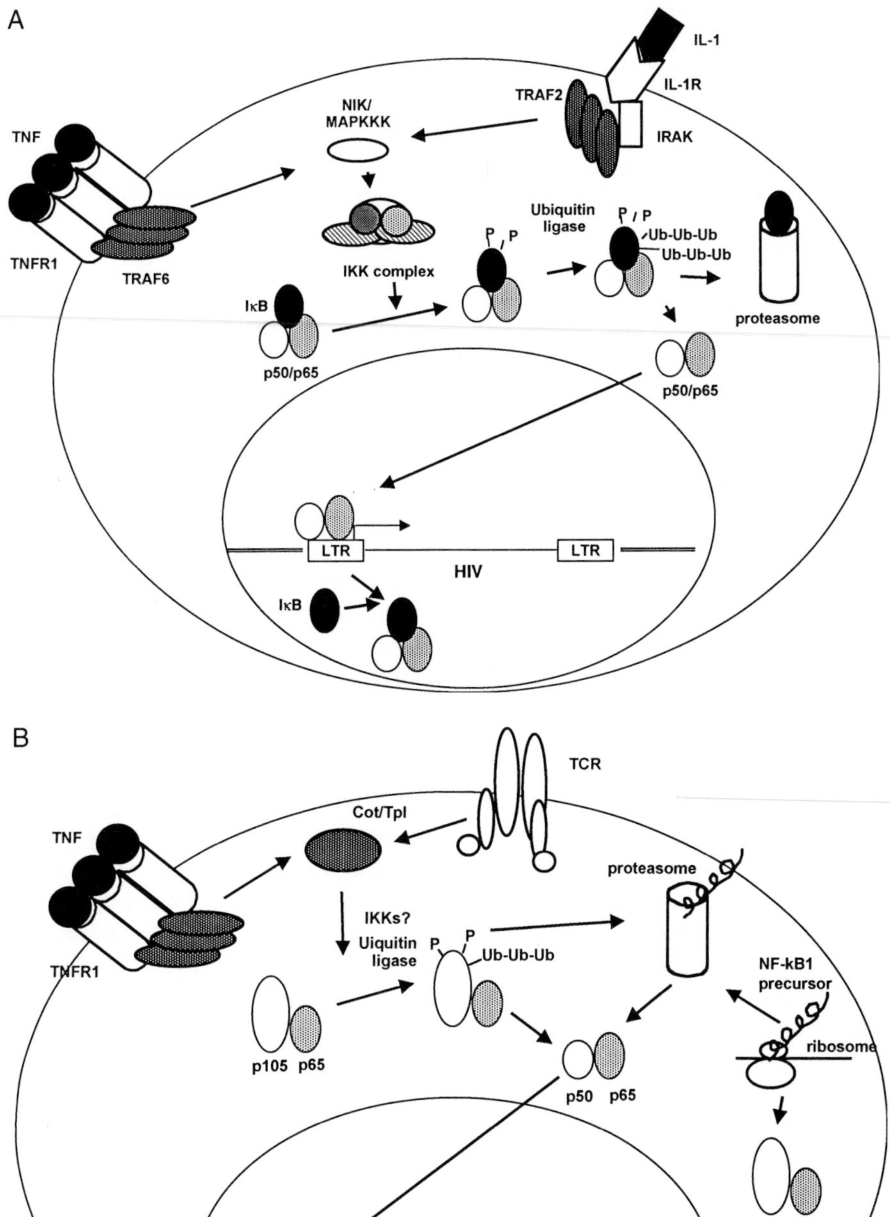

A

IL-1

IL-1R

TRAF2

NIK/
MAPKKK

IRAK

TNF

TNFR1

TRAF6

Ubiquitin
ligase

P P

P / P

Ub-Ub-Ub
Ub-Ub-Ub

IKK complex

IκB

proteasome

p50/p65

p50/p65

LTR

LTR

HIV

IκB

B

TCR

Cot/Tpl

TNF

proteasome

IKKs?

Uiquitin
ligase

P P

Ub-Ub-Ub

TNFR1

NF-kB1
precursor

ribosome

p105 p65

p50 p65

p105 p65

LTR

LTR

HIV

the multiple different activating stimuli, the activation pathways seen following treatment with two cytokines, IL-1 and TNF-α, provide relevant and illustrative models (Fig. 2), as both of these cytokines have been implicated as activators of HIV expression (Duh *et al.*, 1989; Folks *et al.*, 1986; Folks *et al.*, 1987; Griffin *et al.*, 1989; Osborn, Kunkel, and Nabel, 1989). Both IL-1 and TNF-α interact with their receptors on the surface of the target cell. This triggers a cascade of protein phosphorylation events, leading to activation of an IκB kinase complex (DiDonato *et al.*, 1997; Mercurio *et al.*, 1997; Regnier *et al.*, 1997; Zandi *et al.*, 1997) and ultimately to phosphorylation, ubiquitination, and degradation of IκBα (reviewed in May and Ghosh, 1998; Zandi and Karin, 1999). Binding of TNF to TNF receptor components at the plasma membrane leads to recruitment and oligomerization of TNF receptor associated factor 6 (TRAF6), which transduces activation signals into the cell, ultimately leading to phosphorylation and activation of the IκB kinases (IKKs) (Baud *et al.*, 1999). Similarly, IL-1 binding to its receptor leads to activation of an IL-1 receptor associated kinase, IRAK. IRAK recruits TRAF2 to the membrane, and oligomerized TRAF2, similar to TRAF 6, serves as a signal for activation of a further kinase cascade leading to IKK phosphorylation and activation (Baud *et al.*, 1999). The nature of the kinases that transduce signals between the TRAFs and the IKKs remains controversial. The IKKs are phosphorylated on a pair of serine residues in a sequence that has homology to the consensus MEKK phosphorylation site, suggesting that the IKKs are targets of MEKK-related kinases (Nemoto, DiDonato, and Lin, 1998). There is evidence that the NF-κB-inducing kinase (NIK), a MAP kinase kinase kinase, lies proximal to IKK activation by both the IL-1 and TNFα pathways (DiDonato *et al.*, 1997; Mercurio *et al.*, 1997; Regnier *et al.*, 1997; Zandi *et al.*, 1997). NIK has been shown to interact with the receptor-associated TRAF molecules as well as with purified IKKs, presumably transducing cytokine-induced activating signals. NIK itself is likely to be a target of more proximal kinases; for example, TAK1, another MAPKKK, appears to link TRAF6 to the phosphorylation of NIK during IL-1 signal transduction (Ninomiya-Tsuji *et al.*, 1999).

In addition to the cytokine-induced, NIK-IKK phosphorylation cascade, other kinases are likely to transduce cellular stress-induced signals to the IKKs or directly to IκB itself (Zandi and Karin, 1999). MEKK1 is a component of the IKK complexes and will phosphorylate IKKs in response to cytokines and possibly cellular stress (Nakano *et al.*, 1998). Yet another cellular kinase, PKR, may also induce NF-κB activity, through an incompletely defined pathway (Kumar *et al.*, 1995). PKR is activated by double-stranded RNA during viral infections, as part of the interferon response, and has been suggested to play a role in HIV infection (DeLuca *et al.*, 1996). This could represent an additional pathway for NF-κB induction that could play a role in HIV infection.

The IκB-like activities of p105 NF-κB-1 may also be regulated (Fig. 2B). Recent studies have shown that the Cot/Tpl2 kinase protooncogene can bind to, and phosphorylate, p105, targeting it for ubiquitination and processing to p50 (Belich *et al.*, 1999) through proteolysis of the C-terminal ankyrin domain by the 26S proteasome as previously described (Palombella *et al.*, 1994). TNF-α treatment, as well as signaling through the CD3/CD28 T-cell activation costimulatory pathway (Lin *et al.*, 1999), appears to activate Cot/Tpl2 kinase activity, which likely can lead to proteolysis of p105 as well as activation of NIK to induce IκBα degradation. Whether a similar pathway plays a role in the processing of p100 NF-κB-2 to p52 is not yet described.

Thus, the activation of NF-κB is tightly regulated by immune-, cytokine-, and stress-activation pathways in CD4+ T cells and monocytes/macrophages, the cellular targets of HIV infection. As is discussed in the following sections, activation of NF-κB can activate HIV transcription and replication and therefore can be reasonably hypothesized to play a role in HIV pathogenesis. Furthermore, the plethora of regulatory cascades suggest that pharmacologic approaches may be developed to modulate NF-κB activity, although the role that this modulation could play in the treatment of HIV infection is unclear.

IV. NF-κB and the Regulation of HIV Transcription

Following the identification of NF-κB as the transcription factor mediating inducible activation of HIV LTR transcription, a large number of studies have been devoted to exploring the roles of NF-κB in HIV gene regulation. This section focuses on the locations of the cis-acting HIV sequences mediating NF-κB activation and the mechanisms by which NF-κB may activate HIV transcription. Further sections address the roles of NF-κB in HIV replication and pathogenesis.

A. Conservation of NF-κB Binding Sites in Primate Lentiviral LTR-Enhancers

The presence of NF-κB binding sites in the enhancer sequences in the U3 region of the viral LTR is a fundamental characteristic of all the primate immunodeficiency lentiviruses. The locations of the U3 NF-κB and Sp1 binding sites of several HIV and SIV LTRs are shown in Fig. 3. These viruses all contain at least one functional NF-κB binding site directly 5′ to a series (two to four) of Sp1 binding sites. The classic HIV-1 subtype B viruses, present in Europe and North America, and studied most extensively in laboratories, contain two functional NF-κB sites located from −80 bp to −105 bp 5′ to the initiation site of RNA synthesis at the border of the U3

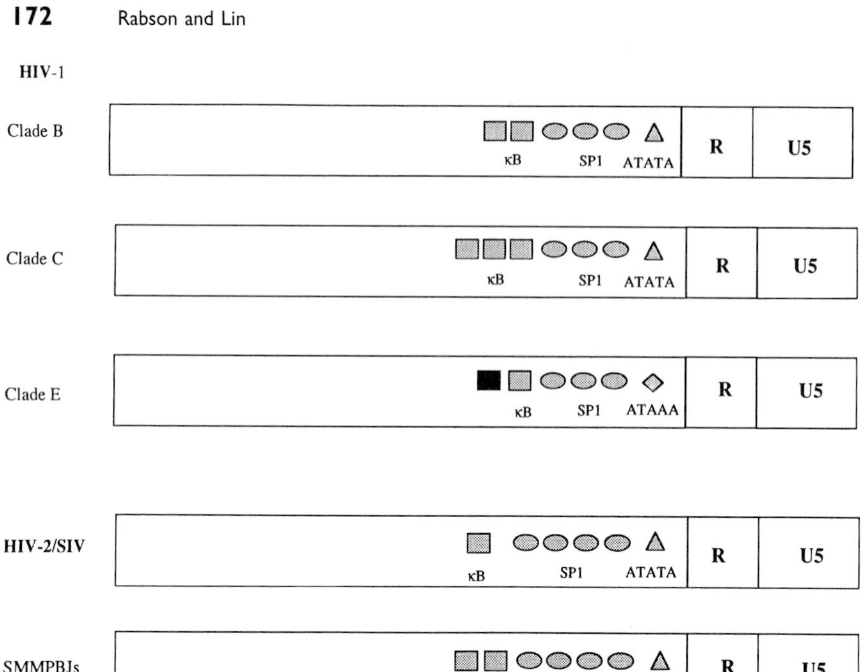

FIGURE 3 NF-κB and Sp1 binding sites in HIV and SIV LTRs. The locations of the U3 region NF-κB binding sites (κB), Sp1 binding sites (SP1), and TATA boxes are shown schematically. The position of a mutated NF-κB site in the Clade E LTR is denoted by the black box.

and R regions of the LTR. These two NF-κB sites lie directly 5′ to three tandem Sp1 binding sites in the LTR. Although there is some variation in the exact position of the sites, two similar NF-κB sites are present in the LTRs of most described HIV-1 isolates, including isolates from subtypes D, F, G, H, and J, as well as SIVcpz, from the chimpanzee (Korber *et al.,* 1998). Other HIV-1 subtypes, encountered in Africa and Asia, contain variant numbers of NF-κB and Sp1 binding sites. For example, subtype C viruses generally contain three NF-κB sites, whereas subtype E viruses contain one functional 3′ NF-κB site (Montano *et al.,* 1997), having instead a point mutation in the 5′ upstream site that reduces its activity (Montano *et al.,* 1997). Subtype O viruses appear to contain a functional 3′ NF-κB site and either intact or variant 5′ sites displaced upstream by a spacer element (Korber *et al.,* 1998). The LTRs of HIV-2 and most SIV isolates have a similar organization to each other, containing a single NF-κB site upstream of four Sp1 binding sites (Korber *et al.,* 1998). Despite these variations, it is striking that the NF-κB and Sp1 sites reside in an overall conserved and highly characteristic position 5′ to the TATA box, which specifies the start site for transcription of viral RNA by the host cell RNA polymerase II. As described elsewhere (see Chapter 2, this volume), the TAR region, which is the target for the HIV and SIV Tat transactivator proteins, encompasses the 5′ region of the nascent RNA transcript. Thus, there is a highly conserved spatial orientation with respect to the NF-κB and Sp1 sites, the TATA box, and the TAR region in all of the primate lentiviruses. This overall structural conservation strongly suggests important functional roles and interactions between these various sites that may contribute to the unique biology of these viruses.

B. Additional NF-κB Binding Sites in HIV Proviruses

The NF-κB sites in the U3 region of the HIV LTRs were originally identified as enhancer elements for HIV transcription. By definition, enhancer elements can work at a distance, in a position- and orientation-independent manner. Thus, it is formally possible for DNA sequences located 3′ to the site of transcription initiation to serve as transcriptional enhancers, and such internal enhancer elements have been defined for a number of retroviruses, including HIV (Verdin *et al.,* 1990). In this regard, both functional assays and DNA sequence analyses have identified a number of additional putative or actual NF-κB binding sites within the proviral DNA, downstream of the transcriptional start site in the 5′ LTR (Fig. 4). One such site actually overlaps the initiator element in the R region of the LTR at positions +1 to +10 with respect to the RNA transcription start site (Montano *et al.,* 1996). This site binds to p50 and p52 homodimers. A second NF-κB site has been identified within the TAR region of the LTR R region mapping to positions +31 to +40 with respect to the transcription initiation site (Mallardo *et al.,* 1996; Montano *et al.,* 1996). Although its nucleotide sequence varies at its 3′ end from the classic consensus κB motifs, this site has all the attributes of a functional NF-κB site, including responsiveness to NF-κB inducers and activation by p50/p65 heterodimers. This site seems to cooperate with the U3 sites in inducing full LTR activation by NF-κB-inducing stimuli. Additional putative NF-κB binding sites may be identified within the body of the HIV genome. For example, in the original identification of an intragenic enhancer in HIV, it was suggested that an NF-κB site might be present within the *vif* gene at position 5288 on the proviral map (Verdin *et al.,* 1990). As part of the preparation of this chapter, we have also performed a computer search for internal transcription factor binding sites, which identified three additional putative NF-κB sites as shown in Fig. 4. These putative sites are located within the *pol* and *nef* genes. Whether any of these non-LTR, putative κB motifs actually contribute to HIV gene expression in the context of the infectious virus will have to be experimentally tested.

C. NF-κB Activation of HIV LTR Transcription

The role that NF-κB plays in the activation of HIV LTR transcription has been studied in many different contexts ranging from *in vitro* transcrip-

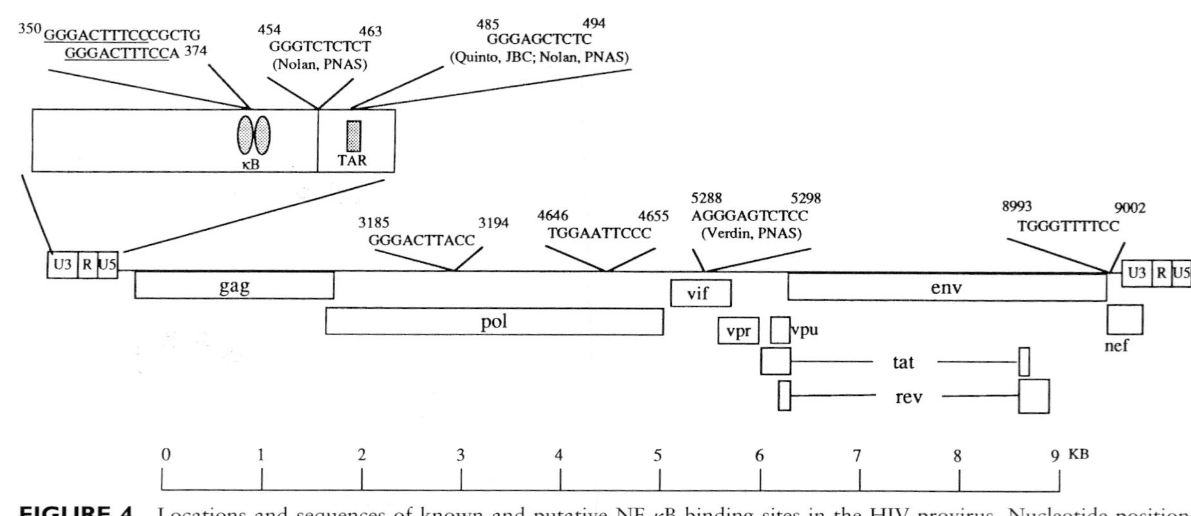

FIGURE 4 Locations and sequences of known and putative NF-κB binding sites in the HIV provirus. Nucleotide positions are shown on the basis of the HIV proviral map with position 1 corresponding to the first nucleotide at the 5′ end of the HXB-2 provirus (Korber *et al.,* 1998). All of the NF-κB sites indicated in the 5′ LTR are also present in the 3′ LTR, but are not shown due to spatial considerations. As described in the text, all four LTR sites have been shown to bind to NF-κB dimers. The function of the internal sites remains unproven.

tion and transient transfections, to analysis of productive and latent HIV infection (see below). Transient transfections of LTR-directed, reporter gene plasmids with mutated NF-κB binding sites have consistently shown a role for the NF-κB binding sites in the enhancement of HIV LTR transcription by NF-κB-inducing stimuli (Duh *et al.*, 1989; Griffin *et al.*, 1989; Kaufman *et al.*, 1987; Nabel and Baltimore, 1987; Osborn, Kunkel, and Nabel, 1989; Siekevitz *et al.*, 1987; Tong-Starksen, Luciw, and Peterlin, 1987; and reviewed in Antoni, Stein, and Rabson, 1994). Other experiments have demonstrated the ability of purified NF-κB proteins to activate the *in vitro* transcription of HIV LTR promoter DNA using reconstituted transcription systems (Fujita *et al.*, 1992; Kretzschmar *et al.*, 1992).

More recent studies have examined the role of NF-κB in activating transcription of the HIV LTR in the context of chromatin. This has been done by examining the LTRs of cell lines containing integrated LTR-reporter gene plasmids or integrated proviruses (see below), as well as in recently developed *in vitro* transcription systems using chromatin-packaged DNA templates. The results of these types of experiments have also suggested important roles for NF-κB in the activation of LTR transcription. NF-κB-inducing stimuli, as well as NF-κB expression plasmids, will activate transcription from integrated HIV LTRs, though NF-κB induction alone, at least in some cells, may not be sufficient to activate transcription of a fully silent integrated promoter (El Kharroubi *et al.*, 1998). RelA, but not p50 NF-κB-1, was shown to strongly activate the *in vitro* transcription of chromatin-assembled HIV LTR DNA templates (Pazin *et al.*, 1996) by counteracting the transcriptional repression induced by nucleosome packaging.

What are the mechanisms by which NF-κB activates HIV transcription? Fundamentally, these are the same mechanisms by which NF-κB activates transcription of the multitude of normal cellular targets for its transcriptional regulation, and the elucidation of the details of these mechanisms is still very much in progress. The first step is binding of activating NF-κB dimers, particularly the p50/p65 or p52/65 heterodimers, to the κB motifs in the LTR, through which the potent transcriptional activation domain of p65 RelA is brought to the LTR promoter. RelA-containing dimers alone can mediate transcriptional activation through their effects both on the basal transcription factors and on the organization of chromatin at the target promoter (Fig. 5). Protein–protein interaction studies have shown that the basal transcription factors, TATA binding protein (TBP), and TFIIB bind to c-Rel and RelA (Kerr *et al.*, 1993; Paul, Baeuerle, and Schmitz, 1997; Xu *et al.*, 1993) and have implicated the Rel transactivation domains in these interactions (Xu *et al.*, 1993; Paul, Baeuerle, and Schmitz, 1997). These interactions appeared to be physiologically significant in that they increased expression of reporter genes under the control of κB binding motifs. In *in vitro* transcription assays, RelA-dependent transcriptional activation of HIV LTR DNA also required the transcriptional coactivator USA

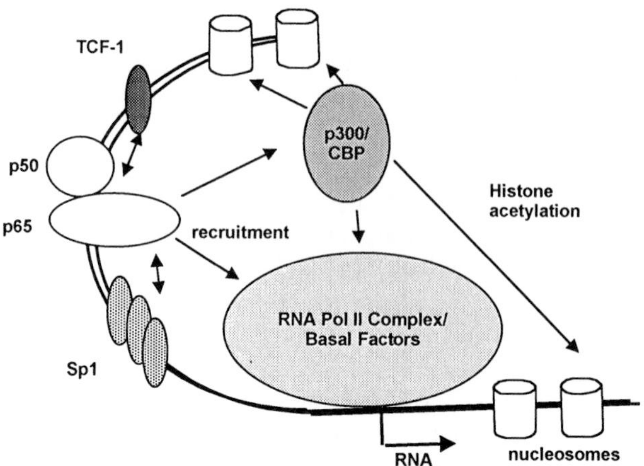

FIGURE 5 Mechanisms of activation of HIV transcription by NF-κB. The interactions of p50/p65 NF-κB dimers with adjacent Sp1 and TCF-1 transcription factors are shown. The ability of p65 (RelA) to recruit constituents of the basal transcription complex is indicated. The interactions of p65 with transcriptional coactivators (p300/CBP) and the resultant acetylation of nucleosome-associated histones are shown, representing another mechanism by which NF-κB likely activates HIV transcription.

(Guermah, Malik, and Roeder, 1998). In an interesting variant of this theme, the binding of p50 or p52 homodimers to the κB site at the initiator region also appears to recruit basal transcription factors to the promoter through enhancing binding of TFII-I (Montano *et al.,* 1996). TFII-I binding was proposed to recruit additional basal factors to the promoter to enhance transcription. Thus, transcriptional activation of HIV by NF-κB occurs, at least in part, by recruitment of proteins involved in the functions of the basal transcriptional apparatus.

It has recently become apparent that an important component of transcriptional activation is the ability to enhance the accessibility of promoter DNA packaged in chromatin to transcription by the holo-RNA polymerase II enzyme through enzymatic modification and remodeling of chromatin (Workman and Kingston, 1998). NF-κB-mediated activation of HIV expression also involves these mechanisms (Fig. 5; also see chapter 4, this volume). This is not surprising, as integrated HIV proviral DNA is packaged into chromatin in the same manner as cellular DNA. Changes in chromatin structure have been identified in conjunction with NF-κB binding both *in vitro* and *in vivo*. NF-κB has been clearly shown to be able to bind to nucleosome-packaged LTR DNA (Pazin *et al.,* 1996; Steger and Workman, 1997); however, the positioning of nucleosomes over HIV LTR DNA was altered during *in vitro* chromatin reconstitution by the addition of Sp1 and the p50 or p65 NF-κB subunits (Pazin *et al.,* 1996; Widlak, Gaynor, and

Garrard, 1997). The presence of these factors leads to a nucleosome-free region over this area of the promoter, similar to that observed *in vivo* (Verdin, Paras, and Van Lint, 1993); thus these factors may help to establish the chromatin organization of the LTR promoter. The substitution of activating p65-containing dimers in place of the nonactivating p50 homodimers counteracts nucleosome-induced transcriptional repression and leads to induction of transcription from the chromatin-assembled template.

The ability of RelA-containing NF-κB dimers to activate transcription of HIV LTRs packaged in chromatin, either *in vitro* or *in vivo,* suggests the possibility that the NF-κB dimers may recruit to the promoter some of the recently identified factors that modify and/or remodel chromatin to allow enhanced transcriptional initiation and elongation (reviewed in Workman and Kingston, 1998). RelA has been shown to be one of many transcriptional activators that recruit the p300 and CBP transcriptional coactivators to promoters (Gerritsen *et al.,* 1997; Neurath *et al.,* 1996; Perkins *et al.,* 1997). p300 and CBP both function as histone acetylases and recruit a histone acetylase, P/CAF, leading to acetylation of the N-terminal tails of histones H3 and H4. Histone acetylation is hypothesized to result in decreased affinity of histones for DNA and increased accessibility of the target gene to other transcriptional regulators, both additional activators and components of the transcriptional machinery. This suggests that altered histone acetylation with increased accessibility of the HIV LTR for RNA polymerase II and associated proteins may be a fundamental mechanism by which NF-κB activates HIV transcription. Consistent with this model is the observation that p300 levels appear to be limiting for the ability of RelA to activate HIV transcription (Hottiger, Felzien, and Nabel, 1998). Activation of other nuclear transcription pathways that also employ p300, such as the STAT transcriptional activation pathway, resulted in a competition for available p300, leading to decreased RelA-mediated LTR activation. Further support for the important role of histone acetylation in controlling the transcription of integrated HIV proviruses comes from the fact that inducers of increased histone acetylation, such as sodium butyrate and trichostatin A, which inhibit histone deacetylases, function as NF-κB-independent activators of HIV transcription (Antoni *et al.,* 1994; El Kharroubi *et al.,* 1998; Laughlin *et al.,* 1995; Sheridan *et al.,* 1997). Most recently, *in vitro* transcription studies with chromatin-assembled HIV LTR DNA have demonstrated a marked effect of histone deacetylase inhibitors, as well as histone acetylases, in inducing transcription from this template, but not from naked HIV LTR DNA (Sheridan *et al.,* 1997; Steger *et al.,* 1998). Furthermore, the effects of RelA in inducing transcription in this same system also correlated with increased histone acetylation, suggesting the possible recruitment of transcriptional coactivators with histone acetylase activity by RelA (Sheridan *et al.,* 1997). These results all suggest that an important consequence of RelA interaction with the HIV LTR enhancer is the recruitment of coactivators to the promoter that would enhance histone acetylation and relieve chromatin-induced repression of transcription.

The important role of the association of NF-κB/Rel proteins with basal transcription factors and transcriptional coactivators in the activation of HIV transcription provides yet another level for cellular regulation of HIV gene expression. Phosphorylation of RelA can affect its ability to transcriptionally activate target genes, such as the HIV LTR, presumably by altering protein–protein interactions required for transcriptional activation. Phosphorylation of the C-terminal transactivation domain of RelA has been proposed to be one important level of control. TNF-induced phosphorylation of serine 529 in RelA leads to enhanced transcriptional activation independent of induction of DNA binding or nuclear translocation (Wang and Baldwin, 1998). Ras-mediated signal transduction pathways may induce a similar activation of RelA. (Finco *et al.,* 1997). Conversely, inhibition of cyclin-dependent kinases found in complex with RelA and p300 resulted in increased transcriptional activation of NF-κB targets (Perkins *et al.,* 1997), suggesting that, in this context, RelA phosphorylation might have inhibitory effects. Whether such phosphorylation events directly affect interactions with transcriptional coactivators is not yet known, but this is one likely explanation for these observations.

As noted above, the HIV LTR NF-κB sites are positioned in a characteristic position and orientation with respect to the adjacent 3′ Sp1 binding sites as well as to the more distal 5′ TCF-1 binding sites. As transcription factors are known to exhibit interactions with each other on promoter DNAs, this raised the question as to the nature of the interactions between NF-κB and these two different factors in the regulation of HIV transcription. The earliest evidence of interactions of NF-κB with the LTR Sp1 binding sites came from exonuclease footprinting studies of protein binding to the NF-κB and Sp1 binding sites (Gimble *et al.,* 1988). Induction of NF-κB by herpes virus infection resulted in expansion of the footprint over both the NF-κB and Sp1 binding sites, suggesting that NF-κB binding also enhanced Sp1 binding to the LTR. Definitive evidence of functionally important interactions between NF-κB and Sp1 was provided by the studies of Perkins *et al.* (1993, 1994). These authors showed cooperative binding of NF-κB and Sp1 to their respective LTR binding sites in response to NF-κB inducers, dependent on the orientation and spacing of the two binding sites on the DNA. They further showed synergistic transcriptional activation of the LTR by NF-κB and Sp1 in transfection assays (Perkins *et al.,* 1993). In subsequent studies, these authors demonstrated direct protein–protein interactions between the DNA binding domains of SP1 and RelA, leading to the cooperative DNA binding to the HIV LTR sites (Perkins *et al.,* 1994). In *in vitro* transcription studies, the functional interactions RelA and Sp1 were particularly apparent on chromatin-packaged template DNA, where these two factors led to marked synergistic activation of HIV LTR transcription (Pazin *et al.*, 1996). These results provide a satisfying functional explanation for the highly conserved juxtaposition of NF-κB and Sp1 binding sites in primate lentiviral LTRs.

NF-κB has also been reported to synergize with several other transcription factors in the activation of HIV transcription. Binding sites for LEF-1/TCF-1 and members of the Ets family of transcription factors are located 5′ to the NF-κB sites in a second enhancer unit mapping from −131 to approximately −150 bp with respect to the RNA start site (Sheridan *et al.*, 1995). Ets family binding sites also play an important role in the regulation of HIV-2 gene expression (Leiden *et al.*, 1992; Markovitz *et al.*, 1992) and synergize with NF-κB in activation of this virus. By analogy with the interactions of Sp1 and NF-κB, this raises the possibility that these upstream transcriptional activators may also physically and functionally interact with NF-κB proteins. In fact, several Ets proteins will bind to the Rel homology domain of p50 NF-κB-1, and this interaction was required for synergistic activation of HIV-1 or HIV-2 LTR transcription (Bassuk, Anandappa, and Leiden, 1997). A potential role for Ets-related transcription factors in synergizing with NF-κB is particularly interesting, as Ets protein-mediated transcriptional activation can also be regulated by T-cell activation stimuli. Thus, Ets proteins and NF-κB transcription factors can participate in coupling HIV transcriptional activation with the processes of T-cell or monocyte activation. Functional cooperation and protein–protein interactions have also been observed between NF-κB proteins and C/EBP transcription factors (Ruocco *et al.*, 1996). C/EBP isoforms, including the C/EBPβ or NF-IL-6 protein (which mediates downstream responses to the cytokine, IL-6) bind to a region of the LTR mapping from −174 to −166 bp with respect to the transcription initiation site. Complexes of p50 NF-κB dimers with C/EBPβ can be formed and these proteins can synergize in the activation of LTR transcription. Thus, the repertoire of NF-κB-regulated effects on HIV transcription can be expanded, and the kinds of control refined, on the basis of additional interactions between NF-κB and other regulatory transcription factors.

It is important to note that NF-κB dimers are apparently not the only DNA binding proteins that can interact with the HIV enhancer κB motifs. An additional Ets family member binding site has been identified within the 5′ NF-κB site (Lodie *et al.*, 1998) and has been reported to play a role with NF-κB in activating HIV LTR transcription in macrophages (Lodie *et al.*, 1998). NFAT-1 is a T-cell transcription factor distantly related to the NF-κB/Rel proteins (Rao, Luo, and Hogan, 1997). NFAT-1 can also apparently bind to the HIV LTR NF-κB sites and block the binding and activation induced by classic NF-κB dimers (Macian and Rao, 1999).

A large zinc finger protein, PRDII-BF1 (also known as EP-1 and MBP-1), has been repeatedly cloned by Southwestern screening of lambda cDNA expression libraries with a radiolabeled HIV κB motif DNA sequence (Baldwin *et al.*, 1990; Fan and Maniatis, 1990; Nakamura *et al.*, 1990; Ron, Brasier, and Haebner, 1991). There has been surprisingly little study of the functions of this protein in regulating HIV expression, possibly because of difficulties in expressing and studying the large protein product. The most informative work has come from transient transfection studies (Seeler *et al.*, 1994), which demonstrated that PRDII-BF1 activated HIV LTR expression between 4- and 10-fold, both in the absence and presence of Tat, thus providing an alternative activation pathway for HIV activation through the LTR NF-κB binding sites. It would seem likely that this protein may have important and as yet uncharacterized roles in HIV infection.

V. Other Roles for NF-κB in HIV Infection

A. Activation of NF-κB by HIV Infection

Growing evidence suggests that alterations in NF-κB activity can play a number of roles in HIV-cell interactions that ultimately could have direct effects on HIV replication. Many viral infections, including infections by a different human retrovirus, the human T-cell leukemia virus, as well as human herpes virus infections, such as cytomegalovirus and herpes simplex virus, induce cellular NF-κB as part of the cellular response to viral infection (reviewed in Mosialos, 1997). These observations lead to the hypothesis that HIV infection may itself activate NF-κB, generating an autocrine stimulatory loop that could further enhance HIV gene expression. Experiments from a number of laboratories have suggested that such an autocrine loop accompanies HIV infection of monocyte/macrophage cell lines. Chronic HIV infection of U937 promonocytic cells (Bachelerie *et al.*, 1991), as well as the PLB-985 monomyeloblastic cell line (Roulston *et al.*, 1992), is associated with NF-κB induction.

A number of possible mechanisms have been proposed to account for HIV induction of NF-κB. HIV infection can lead to IκB kinase activation with phosphorylation and degradation of IκBα (Asin *et al.*, 1999; McElhinny *et al.*, 1995). The binding of HIV or of the HIV gp120 envelope protein to CD4 on the surface of a cell has been shown to result in NF-κB induction (Benkirane, Jeang, and Devaux, 1994). This effect is dependent on the activation of the CD4-associated Lck tyrosine kinase and the subsequent activation of downstream kinase pathways (Briant *et al.*, 1998; Popik, Hesselgesser, and Pitha, 1998). This suggested that HIV may prepare target cells for its transcription through interactions with CD4 during the process of viral binding and entry. In fact, HIV replication was decreased in cells containing a truncated CD4, which did not activate NF-κB upon viral binding (Benkirane, Jeang, and Devaux, 1994). HIV Tat has been reported to induce NF-κB activation through a mechanism involving the activity of PKR, a kinase known to lead to IκB degradation (Demarchi, Guitierrez, and Giacca, 1999). Therefore, Tat could also contribute to the autocrine activation loop. In a third possible mechanism, HIV infection also leads to increased proteolytic processing of the inhibitory p105 NF-κB-1 precursor (McElhinny *et al.*, 1995; Riviere *et al.*, 1991), with the generation of increased levels of p50, which in turn can heterodimerize with p65 to activate HIV transcription. Surprisingly, the HIV protease itself appears to be capable of processing p105, which may further enhance NF-κB binding activity in infected cells (Riviere *et al.*, 1991).

What are the possible consequences of HIV-induced activation of NF-κB for HIV infection of monocyte/macrophages? Infection of these cells is known to be persistent, with relatively limited cytotoxicity. Using *in vivo* footprinting, Jacque *et al.* demonstrated that persistent HIV expression in U937 cells was dependent on occupancy of the LTR NF-κB binding sites by NF-κB p50/p65 heterodimers (Jacque *et al.*, 1996). Thus, the autocrine activation of NF-κB is likely to be critical for persistent viral transcription. Furthermore, as NF-κB induces expression of a number of genes with potent antiapoptotic effects, it is likely that HIV-induced, NF-κB activation in macrophages may contribute to the lack of HIV-induced cytotoxicity that is commonly seen in these cells as compared with infected T cells. This would also contribute to persistent, productive infection, as infected cells would not die, but instead would continue to produce virus. The role of HIV-induced NF-κB in inhibiting apoptosis has been confirmed in monocyte/macrophage-derived cell lines (DeLuca *et al.*, 1998). It will clearly be important to extend these experimental approaches to primary human monocytes and macrophages to further understand the role of HIV-induction of NF-κB in viral infection.

B. NF-κB-Independent Effects of IκB on HIV Infection

IκBα appears to have a distinct effect on HIV gene expression that is independent of NF-κB-mediated transcriptional activation, but instead inhibits HIV RNA transport by interfering with the Rev regulatory pathway (see Chapter 8, this volume). The first hint that IκBα might affect RNA transport came from the observation that IκBα contains a nuclear export signal (NES) in its C-terminus, related to NES present in Rev (Fisher *et al.*, 1995; Wen *et al.*, 1995). This raised the question as to whether these two proteins might use similar nuclear export pathways. Wu *et al.* demonstrated that IκBα potently inhibited HIV production following transient transfection of HIV proviruses and went on to show IκBα specifically inhibited Rev-mediated, RRE-dependent expression of HIV gp160 envelope protein (Wu *et *al.,* 1995). The fact that RelA cotransfection partially relieved the inhibitory effects of IκBα raised the possibility that a cellular NF-κB target gene might play a role in Rev-mediated transport of HIV RNAs. Alternatively, the fact that Rev and IκBα share a NES suggested the possibility that these two proteins could compete with each other for the nuclear export pathways required for Rev-regulated nuclear export of HIV RNAs. Distinguishing these possibilities has been difficult. Further mutagenesis suggested that the N-terminus of IκBα was required to inhibit Rev function and that this effect was independent of both the IκB NES as well as the ability of IκBα to inhibit NF-κB (Wu *et al.,* 1997). These data would suggest that neither of the alternative hypotheses explain IκB's ability to inhibit Rev function. In contrast, others (Bachelerie *et al.,* 1997) have shown that mutations in the IκBα NES do block the ability of IκB to inhibit Rev function. Further studies will be required to elucidate this problem as well as to determine the physiologic role of IκB molecules in the regulation of HIV replication and pathogenesis.

VI. NF-κB and HIV Replication

While NF-κB can clearly modulate HIV transcription in model systems, as described in Section IV, and apparently can even affect other aspects of HIV replication, such as Rev function (Section V), the ultimate issue of the role of NF-κB in the pathogenesis of AIDS remains unclear. Certainly, a number of important hypotheses can be formulated. While the ultimate mediator of high levels of HIV gene expression is the interaction of the HIV Tat protein with TAR RNA, this interaction is dependent on the initial interactions of host cellular transcription factors such as NF-κB and Sp1 with the LTR promoter elements. In an initial infection, these cellular factors, which recruit RNA polymerase II to the integrated HIV LTR promoter, are required to initiate transcription in the absence of both preexisting Tat protein and TAR RNA, as would be the case in a newly infected cell. Thus, NF-κB may play a critical role in the initiation of viral transcription and a productive viral infection. In cells in which integrated HIV proviruses are not being actively transcribed (i.e., latently infected cells), NF-κB interactions with the HIV promoter could provide a mechanism to generate the initial HIV RNA transcripts containing TAR as well as sufficient levels of Tat protein to sustain HIV gene expression. In both cases, NF-κB would provide a link between the state of T-cell or monocyte/macrophage activation and the efficiency of HIV transcription and virus production. Activated cells, containing high levels of nuclear NF-κB would be more likely to express HIV RNA, proteins, and virus particles.

As a first step toward understanding any real functional roles of NF-κB in the virus and the host, it was critical to move from studies of models using *in vitro* transcription and transfection of LTR-reporter genes to studies of viral replication in target cells. In this regard, tissue culture model systems exist for the study of productive and cytopathic HIV infection as well as for the study of chronic, low-level or latent HIV infection. The availability of infectious molecular clones of HIV proviral DNA allowed the construction of HIV proviruses containing mutations in the LTR NF-κB sites, allowing direct tests of the role of these sequences in both productive and latent infection.

Infectious molecular clones of HIV have been generated containing either deletions or point mutations inactivating the function of both HIV LTR NF-κB sites. These viruses were found to replicate efficiently in both CD4+ T-cell lines and PHA-stimulated, human peripheral blood mononuclear cells (in which T cells represent the predominant dividing cell type) (Leonard *et al.,* 1989). Similar replication of viruses missing LTR NF-κB binding sites was seen using reconstructed proviruses from which both NF-κB and Sp1 sites were deleted and replaced with different combinations of functional and/or mutated NF-κB and Sp1 binding sites (Ross *et al.,* 1991). In these studies, some decrease in productive HIV replication was observed with the NF-κB mutated viruses in MT4 cells, an HTLV-1 transformed cell containing very high levels of endogenous NF-κB; however, little effect was seen in other cell lines with lower NF-κB levels (Leonard *et al.,* 1989; Parrott *et al.,* 1991; Ross *et al.,* 1991). These data clearly indicate that in these cell types, the canonical HIV LTR NF-κB sites were not required for viral replication. Using a replication competent HIV vector containing the luciferase reporter gene, Chen *et al.* addressed similar questions and by studying levels of reporter gene expression were able to quantitatively compare the levels of HIV gene expression in a single round of infection with the cellular levels of NF-κB (Chen, Feinberg, and Baltimore, 1997). These investigators also observed no absolute requirement for the LTR NF-κB binding sites. In cell lines containing high levels of NF-κB, they were able to demonstrate a strong enhancement of HIV gene expression by the LTR NF-κB binding sites, suggesting that NF-κB interactions with the U3 LTR binding sites could contribute significantly to the efficiency of virus production. Using markers that could be studied on a single cell basis, these investigators showed that the increased production of the wild-type virus, as compared with the NF-κB mutated virus, resulted in more rapid spread through the infected culture, again suggesting an evolutionary advantage for viruses containing the LTR NF-κB sites and providing a rationale for the extraordinary conservation of these sites in HIV-1. In contrast to these two studies, Alcami *et al.* presented evidence that infection of primary CD4+ T cells, as well as of PHA-stimulated, peripheral blood lymphocytes, was absolutely dependent on the LTR NF-κB sites (Alcami *et al.,* 1995). Whether this fundamentally different observation reflects differences in the properties of the target cells studied or of the sequence of the proviral DNAs used is not clear.

On balance, these studies suggest that the U3 LTR NF-κB binding sites of HIV-1 are not absolutely required for HIV infection of CD4+ T cells, but likely play an important role in enhancing productive infection and, by implication, possibly play a role in disease pathogenesis. This lack of an absolute requirement of the NF-κB sites could reflect several aspects of HIV gene regulation. As noted above, the HIV LTR is a compound promoter element in which different transcription factors can compensate for each other. Numerous studies have suggested that the interactions of Tat and TAR (see Chapter 6, this volume) are the most important for supporting high-level HIV gene expression required for productive infection. As long as the U3 LTR elements direct sufficient transcription for the synthesis of Tat and TAR RNA targets, high levels of viral gene expression can be generated. Furthermore, as also described above, there are several additional putative or demonstrated NF-κB binding sites scattered throughout the HIV provirus that could serve as "intragenic" enhancer elements. The site present in the TAR region (Mallardo *et al.,* 1996), with its proximity to the basal promoter, is particularly likely to provide compensatory effects for any deletion of the U3 NF-κB sites.

Somewhat surprisingly, the effects of the HIV LTR NF-κB sites on infection of cultured monocyte/macrophages have been relatively poorly studied. Mutations of the NF-κB sites appeared to have a profound effect in reducing viral replication in the U937 cell line (Jacque *et al.,* 1996; Lu *et al.,* 1989). These cell-line studies, coupled with the induction of NF-κB during monocyte/macrophage infection, strongly suggest that NF-κB could play a very significant role in HIV infection of monocytes. In this regard, studies to directly address the effects of NF-κB in primary monocytic cells will be critical.

The role of LTR NF-κB binding sites in SIV infections, both in tissue culture and in animal models, has also been evaluated. While the SIV LTR is clearly quite different from the HIV LTR, apparently including additional important enhancer elements (Pohlmann *et al.,* 1998), studies of SIV do allow an assessment of possible effects on the pathogenesis of the SIV-induced AIDS in monkeys (see Section VII). Tissue culture experiments have clearly demonstrated that the single NF-κB site in SIV mac239 (from macaques) is not required for infection of CD4+ T-cell lines or primary macaque PBMCs in culture (Bellas, Hopkins, and Li, 1993; Ilyinskii and Desrosiers, 1996; Zhang, Novembre, and Rabson, 1997), nor does its mutation or deletion appreciably delay or reduce viral replication in these cells. In contrast, the NF-κB site does appear to play a role in SIV replication in monocytes/macrophages (Bellas, Hopkins, and Li, 1993; Ilyinskii and Desrosiers, 1996). Mutation of this site in the context of a provirus with a monocytotropic envelope results in reduced replication in macaque macrophages, suggesting that NF-κB may play a more important role in replication in these cells. This latter observation again raises the important question of the role of the HIV LTR NF-κB sites in infection of primary human monocyte/macrophages by monocytotropic HIVs.

Studies of HIV and SIV clones with mutations in the U3 LTR NF-κB binding sites address only the roles of these LTR sites on viral replication and do not address more global effects of cellular NF-κB and IκB proteins on HIV infection. Other effects of NF-κB on HIV replication would include those mediated by NF-κB interactions with intragenic enhancer elements, the effects of IκB on Rev function in viral RNA processing, and the role of NF-κB in regulating the expression of cellular genes that may, in turn, play a role in modulating HIV infection (such as chemokines and their receptors, other cytokines that could modulate HIV infection, and antiapoptotic genes). One approach to studying the more global roles of NF-κB in HIV infection involves the use of NF-κB inhibitors. While a number of compounds have anti-NF-κB effects (see Section VIII), these agents generally have many other effects on treated cells, such as broad-based antioxidant activity; thus their effects on HIV replication are not necessarily mediated through NF-κB. In contrast, mutated forms of IκBα (deletions or point mutations of the critical regulatory serine residues, see Section III,B) have been developed that block IκB phosphorylation and degradation, therefore leading to sequestration of NF-κB in the cytoplasm of cells and blocking NF-κB function in transcriptional activation. Based on mutagenesis experiments (Wu *et al.*, 1997), these superrepressor IκBs (SRIκBs) would also be expected to inhibit HIV Rev function; thus this experimental approach would not yield a simple interpretation of the effects of NF-κB transcriptional regulation on HIV infection. Nonetheless, CD4+ T-cell lines have been developed that inducibly express SRIκBα molecules. Expression of SRIκBα resulted in a dramatic inhibition of HIV infection in these cells, consistent with inhibitory effects on both HIV transcription and RNA processing as well as with possible effects on the synthesis of cytokines that could stimulate HIV replication and NF-κB-inhibitory effects that could enhance HIV induced apoptosis prior to synthesis of progeny virions (Kwon *et al.*, 1998). Quinto *et al.* have built upon this approach to design an HIV vector carrying the SRIκBα gene in place of the HIV *nef* gene (Quinto *et al.*, 1999). This construct resulted in the production of a highly attenuated HIV that both loses the function of Nef and carries a stable HIV inhibitory activity. Such an attenuated virus has been proposed as a prototype for the development of live, attenuated HIV vaccines.

Another interesting line of investigation has also pointed to an effect of NF-κB on HIV infection, again without specifically defining a direct effect on HIV transcription as compared to effects on cellular targets of NF-κB. Qian *et al.* used mutagenesis of HIV target cells to identify CD4+ T-cell clones of the CEM cell line that were resistant to HIV infection (Qian *et al.*, 1994). Two clones demonstrated reduced activation of the HIV LTR, and by electrophoretic mobility shift assays, demonstrated specific reduction in inducible NF-κB activity. These investigators suggested that an overall reduction in cellular NF-κB activity could lead to inhibition of HIV infection. As CEM cell subclones were among the cell lines in which HIVs containing mutations in the U3 LTR NF-κB sites replicated efficiently (Leonard *et al.*, 1989), these results suggest that NF-κB-induced cellular target genes could be important for enhancing HIV replication in addition to direct effects of NF-κB on HIV transcription.

The studies described above have focused primarily on the effects of NF-κB transcriptional regulation of HIV in models of productive and cytopathic HIV infection and have suggested that NF-κB plays a role in enhancing replication, but is not absolutely required for viral growth. In contrast, there is abundant evidence that NF-κB plays a central and critical role in another aspect of the viral life cycle, the activation of transcription of a latently integrated HIV provirus. Studies in support of this hypothesis had their origin in the isolation of a series of T-cell and monocytic cell lines (ACH-2, U1, J1, and OM10.1 cells) that had survived cytopathic HIV infection. These cells were themselves chronically infected with HIV and contained integrated HIV proviruses and yet were producing either only very low levels of HIV or were silently (or latently) infected (reviewed in Butera and Folks, 1992). The seminal observations were that a series of immune activation stimuli, including TNF-α and IL-1, would potently activate HIV transcription and virus production from these lines (Folks *et al.*, 1986, 1987, 1989), immediately suggesting the model that immune activation could also enhance HIV production and replication *in vivo*. With the observation that PMA (also an inducer of HIV in these chronically infected cell lines) activated HIV LTR expression in transfection experiments through NF-κB induction (Nabel and Baltimore, 1987), it was quickly shown that cytokine induction of latent HIV proviruses also was associated with NF-κB induction (Duh *et al.*, 1989; Griffin *et al.*, 1989). These experiments were inherently correlative. TNF and IL-1 induced HIV LTR activity in transfections through NF-κB, and treatment of the latently infected cells with cytokines also induced NF-κB nuclear activity and transcription of the latent HIV provirus. Confirmation of the mechanistic role of the HIV LTR binding sites in the cytokine-induced activation of HIV transcription from latently infected cells came from experiments utilizing the mutated HIV infectious molecular clones described above. An HIV clone containing mutations in the LTR U3 NF-κB binding sites was used to generate latently infected, clonal, CD4+ T cells (Antoni *et al.*, 1994). Transcription of the silent HIV provirus in these cells was no longer inducible by TNF-α, proving the role of NF-κB interaction with the LTR enhancer binding sites was responsible for cytokine activation of latent HIV. It is important to note that the HIV transcription in NF-κB deleted, latently infected cells was still activated by nonspecific transcriptional activators such as sodium butyrate and hexamethylene bis-acetamide, also shown by others to induce HIV transcription in an NF-κB-independent manner (Laughlin *et al.*, 1995; Vlach and Pitha, 1993).

Studies of integrated proviruses have also provided clues as to the mechanisms of NF-κB activation. Even in the untranscribed state, there is evidence of establishment of a nucleosome-free region (Verdin, Paras, and Van Lint, 1993) and transcription factor binding over the LTR NF-κB and Sp1 binding sites (Demarchi *et al.*, 1993; Verdin, Paras, and Van Lint, 1993). Activation of HIV transcription through NF-κB induction is associated with rapid disruption of a nucleosome positioned directly 3′ to the transcription initiation site (Verdin, Paras, and Van Lint, 1993), similar in kinetics to the activation of NF-κB. This disruption was not apparently dependent on the actual process of RNA transcription, as it was not inhibited by α-amanitin. Instead, it likely results from effects of the binding of activating NF-κB dimers to the promoter, with the recruitment of transcriptional coactivators as well as components of the transcriptional machinery [possibly including recently described chromatin remodeling activities (LeRoy *et al.*, 1998)] to the promoter. Interestingly, treatment of cells with histone deacetylase inhibitors leads to the same pattern of nucleosome disruption (Van Lint *et al.*, 1996), providing further indirect evidence that the effects of NF-κB induction may be mediated through effects on nucleosome structure, possibly mediated by the recruitment of transcriptional coactivators and alterations in histone modification and chromatin structure.

VII. NF-κB and HIV Pathogenesis

The most important, and ultimately most difficult, question to answer is the role of NF-κB in the pathogenesis of AIDS. It is easy to formulate hypotheses in this regard. In particular, as NF-κB can enhance HIV replication and can activate expression of HIV from latently infected cells, it is attractive to postulate that stimuli that induce NF-κB would lead to increased HIV replication and ultimately to more rapid progression from asymptomatic HIV infection to AIDS. A large array of NF-κB-inducing stimuli that may occur *in vivo* in AIDS patients, such as inflammatory cytokines and intercurrent infections, could contribute to the progression of HIV infection in this way (reviewed in Antoni, Stein, and Rabson, 1994; Fauci, 1993; Vicenzi and Poli, 1994). Indirect data in support of this hypothesis comes from several sources. Levels of *in vivo* HIV production are strong predictors of progression to AIDS (Mellors *et al.*, 1996). Immune activation stimuli such as vaccinations have also been shown to increase HIV viral load (Mellors *et al.*, 1996; O'Brien *et al.*, 1995; Staprans *et al.*, 1995), although this observation is controversial (Fowke *et al.*, 1997). Similarly, it could be hypothesized that intercurrent infections, for example, active infection with cytomegalovirus, which also induce cytokines and likely NF-κB activation, could also predispose to HIV disease progression (reviewed in Webster, 1991). Thus, it is appealing to attribute this enhanced HIV replication to NF-κB activation in infected target cells. It should be noted, however, that several different facets of T-cell activation can also contribute to enhanced HIV replication. While HIV replication is not completely dependent on cell division, it is clear that productive HIV replication in primary T cells in culture is markedly enhanced in dividing cells (Stevenson *et al.*, 1990). Furthermore, T-cell activation is also associated with induction of cyclin T (Garriga *et al.*, 1998), a necessary cofactor for Tat activation of transcription (see Chapter 6, this volume). Thus, NF-κB may only be one of several mechanisms by which HIV replicates more efficiently in activated T cells.

It is of course not possible to directly experimentally test the role of NF-κB in the pathogenesis of HIV in humans. There are several indirect methods for assessing the possible contributions of NF-κB to the development of human AIDS. These include the characterization of variants of HIV that have altered NF-κB binding sites, the study of the SIV model system, and the possible therapeutic use of NF-κB inhibitors in HIV-infected patients.

Nucleotide sequencing studies have addressed HIV LTR sequence variability over time and within "quasi-species" sequences in HIV-infected individuals. These studies have identified a relatively common change in LTR DNA sequence, predominantly involving duplications of a DNA sequence (the TCF-1/Ets family binding region) 5′ to the NF-κB binding sites (Estable *et al.*, 1996; Koken *et al.*, 1992; Michael *et al.*, 1994). Alterations in the NF-κB sequences have been relatively uncommon in the subtype B viruses commonly found in North America and Europe. Individual mutations have been detected in patients over time (McNearney *et al.*, 1995; Michael *et al.*, 1994); however, it was not clear that these mutations had effects on pathogenicity. In a study of NF-κB sequences from 478 LTRs derived from 42 patients, only 6% of LTRs had sequence changes predicted to abrogate binding at both NF-κB sites and 95 had alterations which would disrupt binding at one site (Estable *et al.*, 1996). The presence of two NF-κB sites is highly conserved *in vivo* and therefore would be predicted to have some effect on viral replication and therefore possibly pathogenesis. On the other hand, Zhang *et al.* carefully characterized an unusual HIV species from a patient who progressed to AIDS over approximately 11 years postinfection (Zhang *et al.*, 1997b). HIV isolates from this patient were deleted in both U3 LTR NF-κB binding sites. This virus was able to replicate in primary PBMCs and macrophages, but was, however, defective and/or delayed in replication in tissue culture cell lines. The virus did contain a duplication of the upstream LTR TCF-1α binding sites, commonly seen in primary HIV-1 isolates from patients; nonetheless, the LTR exhibited somewhat reduced activity in promoter assays. The reduced promoter activity was still sufficient to support replication of HIV when the NF-κB deleted LTR was transferred into a heterologous infectious clone of HIV, a result quite consistent with the studies of *in vitro* mutated HIVs. The NF-κB binding site deletion was present as early as 7 years prior to the development of AIDS in this patient; thus, the LTR NF-κB sites were clearly not required for the AIDS pathogenesis in this patient.

In an alternative approach, several groups have examined the LTR sequences of long-term survivors of HIV infection (Quinones-Mateu *et al.*, 1998; Rousseau *et al.*, 1997; Visco-Comandi *et al.*, 1999; Zhang *et al.*, 1997a). The hypothesis was that if NF-κB was associated with pathogenesis of AIDS, mutation of the NF-κB sites in the LTR might be more frequent in long-term nonprogressors. These studies confirmed the high degree of conservation of the LTR NF-κB sites and failed to identify a consistent association between NF-κB site alterations and progression rate to AIDS, although a single long-term nonprogressor did exhibit functionally significant alterations in both NF-κB and SP1 binding sites (Zhang *et al.*, 1997a).

As described above (section IV,A), NF-κB site variations are more commonly detected in HIV-1 subtype C and subtype E. Subtype C viruses contain three (or even four) U3 LTR NF-κB binding sites (Montano *et al.*, 1997). Subtype E viruses contain one classic and one variant NF-κB site with markedly reduced binding to NF-κB, in addition to other LTR changes in the TATA box and TAR region (Montano, Nixon, and Essex, 1998). Transfections demonstrated that transcriptional activation by NF-κB proteins and by TNF-α was correlated with the number of functional NF-κB sites (Montano, Nixon, and Essex, 1998). Whether these variant LTR structures are in any way correlated with altered pathogenicity or with the rapid spread of these viruses in certain geographical locations is unknown. A provocative observation in this regard has come from the studies of Verhoef *et al.* (1999). These investigators observed in *in vitro* viral replication studies the conversion of the 5′ NF-κB site into a binding site for a distinct transcriptional activator, the GA binding protein (GABP). GABP strongly synergized with Tat to activate transcription from this promoter and the incorporation of the GABP binding site enhanced viral replication in some cell lines, as compared to wild-type HIV with two NF-κB sites. The same nucleotide alteration is seen in subtype E viruses from Thailand, raising the interesting, though strongly speculative hypothesis that this sequence alteration could contribute to highly pathogenic nature of this virus *in vivo* (Verhoef *et al.*, 1999).

None of these HIV LTR sequencing studies provide any definitive insights into the role of NF-κB in the pathogenesis of human AIDS. At best, these studies suggest that the LTR NF-κB sites are clearly not required for progression to AIDS nor are mutations in these sites associated with delayed progression to AIDS.

SIV-induced, simian AIDS shares many pathogenic features with HIV-induced human AIDS. The roles of NF-κB in the development and progression of SIV-induced AIDS have been studied in two different contexts. A naturally occurring, highly pathogenic strain of SIV, SIVsmmPBj14, induces a severe enteropathy that is fatal within days of inoculation into susceptible animals. These animals die of a disease characterized by high levels of SIVsmmPBj14 replication in the gut-associated lymphoid tissue, associated with lymphocyte activation and proliferation, and high levels of secretion of inflammatory cytokines (reviewed in Fultz and Zack, 1994). SIVsmmPBj14 exhibits a duplication of the SIV LTR NF-κB binding site as compared to the parental SIV LTR. While this NF-κB site contributes to enhanced viral replication in cultured cells (Dollard *et al.,* 1994), studies of the *in vivo* pathogenicity of recombinant viruses suggested that the LTR NF-κB duplication is not a determinant of the unique biological behavior and pathogenicity of SIVsmmPBj14 (Du *et al.,* 1996; Novembre *et al.,* 1993). In fact, NF-κB binding sites were not essential for SIVsmmPBj pathogenicity.

Ilyinskii *et al.* have employed SIV molecular clones containing LTR mutations (see Section IV) to study the role of the SIV LTR NF-κB site in the pathogenesis of SIV-induced AIDS (Ilyinskii *et al.,* 1997). Two monkeys injected intravenously with a virus deleted in the LTR NF-κB site exhibited high levels of SIV viremia, and evidence of progression to AIDS was reported in one animal (Ilyinskii *et al.,* 1997). Additional animals inoculated with viruses deleted or mutated in both the NF-κB site and the LTR Sp1 binding sites also exhibited progression to AIDS. These results suggest that the LTR NF-κB sites certainly are not required for simian AIDS, but, as noted by these authors, it is clear that the SIV LTR contains additional enhancer elements not present in the HIV-1 LTR, which apparently strongly contribute to SIV gene expression (Pohlmann *et al.,* 1998). One provocative implication of the SIV data is that in this system, macrophage infection may not be important for disease progression. The disease-inducing, NF-κB mutated viruses had shown reduced replication in simian macrophages in tissue culture (Bellas, Hopkins, and Li, 1993; Ilyinskii and Desrosiers, 1996). Interestingly, animals inoculated with the NF-κB and/or Sp1 deleted viruses exhibited decreased evidence of two pathologic features of SIV AIDS, encephalitis and pneumonitis. Infection of macrophagelike cells has been long thought to be part of the pathogenesis of CNS AIDS, thus the inability of the mutated viruses to replicate efficiently in monocyte/macrophages could be responsible for these subtle alterations in disease manifestations (Ilyinskii *et al.,* 1997). Thus, while the LTR NF-κB sites themselves may not be required for AIDS, they could play a role in certain pathologic changes that are part of the disease process.

The published results of SIV infections using limited sets of monkeys do not necessarily rule out any contribution of NF-κB to the progression of SIV-induced disease, however. Recent studies have suggested that, similar to the models of progression proposed for HIV infection, chronic immune stimulation of SIV-infected monkeys may enhance viral replication and accelerate the development of simian AIDS (Folks *et al.,* 1997). It would be most interesting to directly test the effects of immune activation stimuli on the rate of progression of simian AIDS induced by SIVs containing NF-κB site mutations. Such an analysis might provide evidence bearing on the hypothesis that NF-κB could play a role in the progression of human AIDS. Nonetheless, it is important to note that there are multiple, important differences between the functional organization of the HIV-1 and SIV LTRs, including the presence of additional, potent enhancer sequences in the SIV LTR which allow SIV replication even without NF-κB and Sp1 binding sites (Pohlmann *et al.,* 1998). Thus, conclusions drawn from *in vivo* SIV studies might not be directly relevant to the role of NF-κB in HIV-1-induced disease pathogenesis. On balance, however, the SIV studies also fail to support a role for the LTR NF-κB sites in the development of AIDS, but do suggest that differences in the disease manifestations could be observed in the absence of the LTR NF-κB sites. It would be interesting to know if the reported case of human AIDS associated with HIV LTR NF-κB site deletions (Zhang *et al.,* 1997b) was associated with any differences in disease manifestations, such as a lack of neurological disease, which is associated with infection of macrophage-related cells in the central nervous system.

VIII. Therapeutic Modulation of NF-κB: A Role in HIV Therapy?

The possible roles of NF-κB in the pathogenesis of AIDS, as well as its central role in the immune and inflammatory responses, have prompted significant efforts to identify and develop clinically useful inhibitors of NF-κB. A safe, effective inhibitor of NF-κB could be a potent anti-inflammatory agent, as many of the initial steps in inflammation are critically dependent on NF-κB induction in leukocytes and endothelial cells. A number of different classes of compounds with NF-κB inhibitory activity have been described. These have been recently reviewed in several contexts (Baeuerle and Baichwal, 1997; Lee and Burckart, 1998); therefore, this chapter focuses on only a few possible inhibitors and, more importantly, on more general issues related to the possible therapeutic utility of NF-κB inhibition for the treatment of HIV infection.

At this time, there are no commercially available, specific NF-κB inhibitors. Several different classes of compounds have demonstrated inhibitory effects on NF-κB activation. Among the first identified were antioxidants, which were shown to inhibit cytokine-induced NF-κB activation (Kalebic *et al.,* 1991; Roederer *et al.,* 1990; Schreck, Rieber, and Baeuerle, 1991), as well as basal HIV LTR transcription in some lymphoid lines (Israel *et al.,* 1992). One antioxidant, PDTC, has been a particularly useful experimental tool to examine the effects of NF-κB inhibition on numerous cellular processes, as it is effective at micromolar amounts and blocks NF-κB induction by multiple stimuli (Schreck *et al.*, 1992). A second antioxidant, N-acetylcysteine (NAC), which serves as a cysteine prodrug to replenish intracellular thiol levels, will inhibit NF-κB activation by cytokines when used at relatively high (millimolar) concentrations (Kalebic *et al.*, 1991; Roederer *et al.*, 1990; Staal *et al.*, 1993). Such high levels are clinically achievable, which prompted clinical trials to assess whether NAC would actually demonstrate any anti-HIV activity in AIDS patients. To date, the results of these trials have been inconclusive (Akerlund *et al.*, 1996; Look *et al.*, 1998). NAC treatment has not resulted in increased intracellular levels of glutathione in AIDS patients (Witschi *et al.*, 1995), nor have significant antiviral effects been reported. Thus, the utility of antioxidant therapy as NF-κB inhibitors for HIV infection remains questionable, particularly in view of the fact that potential immunosuppressive effects are seen *in vitro* at the concentrations required to affect HIV production (Aillet *et al.*, 1994). Certain antioxidant vitamins may serve as possible therapeutic approaches in the developing world. These would have the advantages of being relatively inexpensive and of contributing to overall enhanced nutrition in the patients, although the therapeutic effect on HIV infection per se is still questionable (reviewed in Kotler, 1998). From a theoretical standpoint, one disadvantage to antioxidant therapy for HIV infection is that most antioxidants can also exhibit prooxidant effects at certain concentrations and under certain cellular conditions. For example, in macrophages, NAC has been found to induce NF-κB and HIV gene expression (Nottet *et al.*, 1997). Thus, the *in vivo* effects of these agents may not always be predictable.

Several anti-inflammatory agents have been shown to have NF-κB inhibitory activity, which may in fact be an important component of their anti-inflammatory effects. Glucocorticoids have been shown to inhibit NF-κB activity. The mechanisms responsible for this inhibition are unclear. Dexamethasone has been reported to induce increased transcription of IκBα (Auphan *et al.*, 1995; Scheinman *et al.*, 1995a), block TNF-induced IκBα degradation (Hofmann *et al.*, 1998), and to inhibit nuclear transcriptional activation induced by NF-κB (De Bosscher *et al.*, 1997). The latter effect is likely mediated by both direct physical interactions of the glucocorticoid receptor with RelA (Caldenhoven *et al.*, 1995; Ray and Prefontaine, 1994; Scheinman *et al.*, 1995b) as well as through competition for transcriptional coactivator utilization (Sheppard *et al.*, 1998). The effect of glucocorticoids on NF-κB likely partially explains their potent anti-inflammatory and immunosuppressive effects, although additional direct effects on leukocyte viability and function probably play important roles as well. Glucocorticoids have been used clinically to treat a wide variety of complications of HIV infection; however, there is no direct *in vivo* evidence of a significant antiviral effect. In fact, endogenous cortisol levels may be elevated (Lortholary *et al.*, 1996) in AIDS patients and may play an immunosuppressive role in AIDS patients, reducing cytotoxic T-cell response (Norbiato *et al.*, 1997).

Other anti-inflammatory compounds such as aspirin and sodium salicylate have also been reported to have NF-κB-inhibitory effects. Millimolar concentrations were reported to block phorbol ester plus phytohemagglutinin induction of NF-κB and HIV LTR transcriptional activity and TNF-α activation of the HIV LTR (Kopp and Ghosh, 1994). Recently, this effect was shown to be mediated through a direct inhibition of the activity of the IκB kinases (Yin, Yamato, and Gaynor, 1998). This effect was specific for IKK-β and was observed with micromolar concentrations of the inhibitors. Aspirin was observed to compete with ATP for binding to IKK-β. Interestingly, not all nonsteroidal anti-inflammatory agents exhibited inhibition of IKK activity (Yin, Yamato, and Gaynor, 1998). Although it has not been demonstrated that aspirin itself is a sufficiently potent inhibitor to affect HIV replication at clinically achievable concentrations; nonetheless, this observation provides a proof of principle that it should be possible to identify specific and potent inhibitors of the IKKs.

In addition to these known anti-inflammatory compounds, a number of other classes of molecules have been reported to inhibit NF-κB. A large number of natural products, including curcumin from tumeric (Chan, 1995; Singh and Aggarwal, 1995), emodin from the roots of *Polygonum cuspidatum* (Kumar, Dhawan, and Aggarwal, 1998), caffeic acid phenethyl ester from honey bee hives (Natarajan *et al.,* 1996), and hymenialdisine from marine sponges (Breton and Chabot-Fletcher, 1997), have been shown to inhibit NF-κB transcriptional activation. The precise mechanisms of action and the potential clinical utility of these agents and many other natural products, either as anti-inflammatory agents or as anti-HIV agents, are by and large unknown.

Inhibitors of proteasome function also inhibit NF-κB transcriptional activity by blocking degradation of IκB in response to inductive stimuli. Proteasome inhibitors such as peptide aldehyde inhibitors of the catalytic subunit of the proteasome, are routinely used as experimental inhibitors of NF-κB induction in tissue culture (Palombella *et al.,* 1994; Traenckner, Wilk, and Baeuerle, 1994). Again, the potential clinical utility of proteasome inhibitors is currently under investigation, but the central role of the ubiquitination/proteasome degradation system in regulating the turnover of the majority of cellular proteins suggests that these agents may have significant *in vivo* toxicities.

Several cytokines may have NF-κB inhibitory activity, presumably as a part of their normal functions in regulating immune system function. IL-16 has been shown to repress HIV gene expression (Maciaszek *et al.,* 1997) and may be a component of the CD8+ cell-produced, HIV suppressive factor(s). The effects of cytokines are likely to be quite complicated, however. For example, while IL-10 is clearly able to repress NF-κB induction by several stimuli (Romano *et al.,* 1996; Wang *et al.,* 1995), somewhat surprisingly, IL-10 has been shown to cooperate with TNF-α to enhance activation of HIV from latently infected cells (Finnegan *et al.*, 1996). Thus, the complexity of cytokine-induced regulation of NF-κB and other lymphoid transcriptional pathways is likely to limit the clinical utility of these agents as specific NF-κB inhibitors.

Nucleic acid-based therapeutics targeted against NF-κB have also been shown to have efficacy in animal models of inflammation and cell damage. Antisense phosphorothioate oligonucleotides directed against Rel A demonstrated a striking effect on murine experimental colitis models when administered locally (Neurath *et al.*, 1996). An alternative approach is the use of double-stranded NF-κB binding site oligonucleotides that act as "decoys" that bind to activated nuclear NF-κB and block binding to and activation of endogenous NF-κB target genes. The transcription factor decoy approach has also been used *in vivo* in rats to reduce myocardial damage following cardiac ischemia and reperfusion (Morishita *et al.*, 1997). In an interesting study of effects of synthetic oligonucleotides directed against NF-κB in autoimmune splenocytes, the transcription factor decoy appeared to provide longer lasting inhibition than did antisense oligonucleotides (Khaled *et al.*, 1998).

Although it is likely that effective NF-κB inhibitors will be developed for use in inflammation and other disorders, it remains questionable as to whether such agents would ever have a role in the therapy of HIV infections. First of all, it is still not clear that NF-κB plays a critical role in *in vivo* viral replication and disease pathogenesis. Only a minority of studies suggest that NF-κB is critical for viral replication in T cells, the principal cause of the dramatic immunosuppression associated with HIV infection. NF-κB may play a more important role in monocyte/macrophage replication, although this remains unproven. If this is true, inhibition of NF-κB might have a more important therapeutic effect for HIV encephalopathy, whose pathogenesis appears to involve infection of microglial cells, CNS macrophages. Perhaps more importantly, it is clear that HIV-infected individuals do mount an effective, although ultimately unsuccessful, immune response to HIV, characterized by active antibody and cytotoxic T-cell responses to the virus. Any potent NF-κB inhibitor will inevitably block NF-κB function in B and T lymphocytes and macrophages. This will induce immunosuppressive effects, limiting any effective host response to HIV. This action could have the potentially very unfavorable effect of limiting host immune response to HIV, further accelerating the progression of AIDS. Thus, at this time, it is not clear that inhibition of NF-κB would have any beneficial effects in the therapy of HIV infections and AIDS.

IX. Conclusions

The evolution of primate lentiviruses has consistently selected for the maintenance of at least one NF-κB binding site in the viral LTR enhancer regions. Given the critical role of the NF-κB transcription factors in regulating lymphoid and monocyte function during the immune response, the presence of the viral NF-κB sites intimately links host immune activation with viral gene expression. While the precise roles of NF-κB function in HIV replication and pathogenesis remain to be further elucidated, it is likely that induction of NF-κB contributes to high levels of virus production in acutely and cytopathically infected cells. NF-κB certainly contributes to the induction of HIV production from latently infected cell lines, and with the growing realization of the persistence of latently infected cells *in vivo,* even in the face of aggressive antiviral therapies (Chun *et al.*, 1999), it is highly likely that NF-κB also plays a role in activation of virus from latent reservoirs in HIV-infected individuals. Unfortunately, the intimate link between NF-κB, viral gene expression, and the host response raises serious doubts about the ability to successfully exploit NF-κB inhibition as part of an antiviral therapeutic strategy. Inhibition of NF-κB would be likely to inhibit normal immune responses that may play a critical role in containment of viral infection or in the reconstitution of immune function as a part of long-term, highly active, antiretroviral therapies. The development of novel, highly specific NF-κB inhibitors will allow more direct study of the possible efficacy of NF-κB inhibition in the treatment of HIV infection.

Acknowledgments

We thank Loretta Miller for assistance with the manuscript and references and Drs. Céline Gélinas and Aaron Shatkin for suggestions and helpful discussions. We acknowledge the support of our related research programs by the National Institutes of Health (Research Grants CA68333 and AI30901) and by the New Jersey Commission on Science and Technology.

References

Aillet, F., Gougerot-Pocidalo, M. A., Virelizier, J. L., and Israel, N. (1994). Appraisal of potential therapeutic index of antioxidants on the basis of their *in vitro* effects on HIV replication in monocytes and interleukin 2-induced lymphocyte proliferation. *AIDS Res. Hum. Retroviruses* 10, 405–411.

Akerlund, B., Jarstrand, C., Lindeke, B., Sonnerborg, A., Akerblad, A. C., and Rasool, O. (1996). Effect of *N*-acetylcysteine(NAC) treatment on HIV-1 infection: A double-blind placebo-controlled trial. *Eur. J. Clin. Pharmacol.* 50, 457–461.

Alcami, J., Lain de Lera, T., Folgueira, L., Pefraza, M.-A., Jacqué, J.-M., Bachelerie, F., Noriega, A. R., Hay, R. T., Harrich, D., Gaynor, R. B., Virelizier, J.-L., and Arenzana-Seisdedos, F. (1995). Absolute dependence on κB responsive elements for initiation and Tat-mediated amplification of HIV transcription in blood CD4 T lymphocytes. *EMBO J.* **14**, 1552–1560.

Antoni, B. A., Rabson, A. B., Kinter, A., Bodkin, M., and Poli, G. (1994). NF-kappa B-dependent and -independent pathways of HIV activation in a chronically infected T cell line. *Virology* 202, 684–694.

Antoni, B. A., Stein, S. B., and Rabson, A. B. (1994). Regulation of human immunodeficiency virus infection: implications for pathogenesis. *In* "Advances Virus Research" (K. Mamarosch, F. A. Murphy, and A. J. Shatkin, Eds.), Vol. 43. pp. 53–145. Academic Press, San Diego.

Arenzana-Seisdedos, F., Turpin, P., Rodriguez, M., Thomas, D., Hay, R. T., and Virelizier, J. L. (1997). Nuclear localization of I Kappa B alpha promotes active transport of NF-kappa B from the nucleus to the cytoplasm. *J. Cell Sci.* **110**, 369–378.

Asin, S., Taylor, J. A., Trushin, S., Bren, G., and Paya, C. V. (1999). Ikappakappa mediates NF-kappaB activation in human immunodeficiency virus-infected cells. *J. Virol.* **73**, 3898–3903.

Auphan, N., DiDonato, J. A., Rosette, C., Helmberg, A., and Karin, M. (1995). Immunosuppression by glucocorticoids: Inhibition of NF-κB activity through induction of IκB synthesis. *Science* **270**, 286–290.

Bachelerie, F., Alcami, J., Arenzana-Seisdedos, F., and Virelizier, J.-L. (1991). HIV enhancer activity perpetuated by NF-κB induction on infection of monocytes. *Nature* **350**, 709–712.

Bachelerie, F., Rodriguez, M. S., Dargemont, C., Rousset, D., Thomas, D., Virelizier, J.-L., and Arenzana-Seisdedos, F. (1997). Nuclear export signal of IκBα interferes with the Rev-dependent posttranscriptional regulation of human immunodeficiency virus type 1. *J. Cell Sci.* **110**, 2883–2893.

Baeuerle, P. A., and Baichwal, V. R. (1997). NF-kappa B as a frequent target for immunosuppressive and anti-inflammatory molecules. *Adv. Immunol.* **65**, 111–137.

Baeuerle, P. A., and Baltimore, D. (1988a). Activation of DNA-binding activity in an apparently cytoplasmic precursor of the NF-κB transcription factor. *Cell* **53**, 211–217.

Baeuerle, P. A., and Baltimore, D. (1988b). IκB: a specific inhibitor of the NF-κB transcription factor. *Science* **242**, 540–546.

Baldwin, A. S., LeClair, K. P., Harinder, S., and Sharp, P. A. (1990). A large protein containing zinc finger domains binds to related sequence elements in the enhancers of the class I major histocompatability complex and kappa immunoglobulin genes. *Mol. Cell. Biol.* **10**, 1406–1414.

Baldwin, A. S. J. (1996). The NF-kappa B and I kappa B proteins: new discoveries and insights. *Annu. Rev. Immunol.* **14**, 649–683.

Ballard, D. W., Walker, W. H., Doerre, S., Sista, P., Molitor, J. A., Dixon, E. P., Peffer, N. J., Hannink, M., and Greene, W. C. (1990). The v-rel oncogene encodes a kappa B enhancer binding protein that inhibits NF-kappa B function. *Cell* **63**, 803–814.

Bassuk, A., Anandappa, R., and Leiden, J. M. (1997). Physical interactions between Ets and NF-κB/NFAT proteins play an important role in their cooperative activation of the human immunodeficiency virus enhancer in T cells. *J. Virol.* **71**, 3563–3573.

Baud, V., Liu, Z. G., Bennett, B., Suzuki, N., Xia, Y., and Karin, M. (1999). Signaling by proinflammatory cytokines: oligomerization of TRAF2 and TRAF6 is sufficient for JNK and IKK activation and target gene induction via an amino-terminal effector domain. *Genes Dev.* **15**, 1297–1308.

Bednarik, D. P., and Folks, T. M. (1992). Mechanisms of HIV-1 latency. *AIDS* **6**, 3–16.

Belich, M. P., Salmeron, A., Johnston, L. H., and Ley, S. C. (1999). TPL-2 kinase regulates the proteolysis of the NF-kappaB-inhibitory protein NF-kappaB1 p105. *Nature* **397**, 363–368.

Bellas, R. E., Hopkins, N., and Li, Y. (1993). The NF-kappa B binding site is necessary for efficient replication of simian immunodeficiency virus of macaques in primary macrophages but not in T cells in vitro. *J. Virol.* **67**, 2908–2913.

Benkirane, M., Jeang, K. T., and Devaux, C. (1994). The cytoplasmic domain of CD4 plays a critical role during the early stages of HIV infection in T cells. *EMBO J.* **13**, 5559–5569.

Blair, W. S., Bogard, H. P., Madore, S. J., and Cullen, B. R. (1994). Mutational analysis of the transcription activation domain of RelA: Identification of a highly synergistic minimal acidic activation module. *Mol. Cell. Biol.* **14**, 7226–7234.

Bours, V., Azarenko, V., Dejaradin, E., and Siebenlist, U. (1994). Human RelB (I-Rel) functions as a kappa B site-dependent transactivating member of the family of Rel-related proteins. *Oncogene* **9**, 1699–1702.

Breton, J. J., and Chabot-Fletcher, M. C. (1997). The natural product hymenialdisine inhibits interleukin-8 production in U937 cells by inhibition of nuclear factor-κB. *Pharmacol. Exp. Therap.* **282,** 459–466.

Briant, L., Robert-Hebmann, V., Acquaviva, C., Pelchen-Matthews, A., Marsh, M., and Devaux, C. (1998). The protein tyrosine kinase p56lck is required for triggering NF-kappaB activation upon interaction of human immunodeficiency virus type 1 envelope glycoprotein gp120 with cell surface CD4, *J. Virol.* **72,** 6207–6214.

Brockman, J. A., Scherer, D. C., McKinsey, T. A., Hall, S. M., Qi, X., Lee, W. Y., and Ballard, D. W. (1995). Coupling of a signal response domain in IκBα to multiple pathways for NF-κB activation. *Mol. Cell. Biol.* **15,** 2809–2818.

Brown, K., Gerstberger, S., Carlson, L., Franzoso, G., and Siebenlist, U. (1995). Control of IκB-α proteolysis by site-specific, signal-induced phosphorylation. *Science* **267,** 1485–1488.

Butera, S. T., and Folks, T. M. (1992). Application of latent HIV-1 infected cellular models to therapeutic intervention. *AIDS Res. Hum. Retroviruses* **8,** 991–995.

Caldenhoven, E., Liden, J., Wissink, S., Van de Stolpe, A., Raaijmakers, J., Koenderman, L., Okret, S., Gustafsson, J. A., and Van der Saag, P. T. (1995). Negative cross-talk between RelA and the glucocorticoid receptor: a possible mechanism for the antiinflammatory action of glucocorticoids. *Mol. Endocrinol.* **9,** 401–412.

Chan, M. M. (1995). Inhibition of tumor necrosis factor by curcuim, a phytochemical. *Biochem. Pharmacol.* **49,** 1551–1556.

Chen, B. K., Feinberg, M. B., and Baltimore, D. (1997). The κB sites in the human immunodeficiency virus type 1 long terminal repeat enhance virus replication yet are not absolutely required for viral growth. *J. Virol.* **71,** 5495–5504.

Chun, T. W., Engel, D., Mizell, S. B., Hallahan, C. W., Fischette, M., Park, S., Davey, R. T., Jr., Dybul, M., Kovacs, J. A., Metcalf, J. A., Mican, J. M., Berrey, M. M., Corey, L., Lane, H. C., and Fauci, A. S. (1999). Effect of interleukin-2 on the pool of latently infected, resting CD4+ T cells in HIV-1-infected patients receiving highly active antiretroviral therapy. *Nat. Med.* **5,** 651–655.

De Bosscher, K., Schmitz, M. L., Vanden Berghe, W., Plaisance, S., Fiers, W., and Haegeman, G. (1997). Glucocorticoid-mediated repression of nuclear factor-kappaB-dependent transcription involves direct interference with transactivation. *Proc. Natl. Acad. Sci. USA* **94,** 13504–14509.

DeLuca, C., Kwon, H., Pelletier, N., Wainberg, M. A., and Hiscott, J. (1998). NF-κB protects HIV-1-infected myeloid cells from apoptosis. *Virology* **244,** 27–38.

DeLuca, C., Roulston, A., Koromilas, A., Wainberg, M. A., and Hiscott, J. (1996). Chronic human immunodeficiency virus type 1 infection of myeloid cells disrupts the autoregulatory control of the NF-kappaB/Rel pathway via enhanced IkappaBalpha degradation. *J. Virol.* **70,** 5183–5193.

Demarchi, F., D'Agaro, P., Falaschi, A., and Giacca, M. (1993). *In vivo* footprinting analysis of constitutive and inducible protein–DNA interactions at the long terminal repeat of human immunodeficiency virus type 1, *J. Virol.* **67,** 7450–7460.

Demarchi, F., Guitierrez, M. I., and Giacca, M. (1999). Human immunodeficiency virus type 1 tat protein activates transcription factor NF-kappaB through the cellular interferon-inducible, double-stranded RNA-dependent protein kinase, PKR. *J. Virol.* **73,** 7080–7086.

DiDonato, J., Mercurio, F., Rosette, C., Wu-Li, J., Suyang, H., Ghosh, S., and Karin, M. (1996). Mapping of the inducible IκB phosphorylation sites that signal its ubiquitination and degradation. *Mol. Cell. Biol.* **16,** 1295–1304.

DiDonato, J. A., Hayakawa, M., Rothwarf, M., Zandi, E., and Karin, M. (1997). A cytokine-responsive IκB kinase that activates the transcription factor, NF-κB. *Nature* **388,** 548–554.

Doerre, S., Sista, P., Sun, S. C., Ballard, D. W., and Greene, W. C. (1993). The c-rel protooncogene product represses NF-κB p65-mediated transcriptional activation of the long terminal repeat of type 1 human immunodeficiency virus. *Proc. Natl. Acad. Sci. USA* **90,** 1023–1027.

Dollard, S. C., Gummuluru, S., Tsang, S., Fultz, P. N., and Dewhurst, S. (1994). Enhanced responsiveness to nuclear factor kappa B contributes to the unique phenotype of simian immunodeficiency virus variant SIVsmmPBj14. *J. Virol.* **68,** 7800–7809.

Du, Z., Ilyinskii, P. O., Sasseville, V. G., Newstein, M., Lackner, A. A., and Desrosiers, R. C. (1996). Requirements for lymphocyte activation by unusual strains of simian immunodeficiency virus. *J. Virol.* **70,** 4157–4161.

Duh, E. J., Maury, W. J., Folks, T. M., Fauci, A. S., and Rabson, A. B. (1989). Tumor necrosis factor alpha activates human immunodeficiency virus type 1 through induction of nuclear factor binding to the NF-kappa B sites in the long terminal repeat. *Proc. Natl. Acad. Sci. USA* **86,** 5974–5978.

El Kharroubi, A., Piras, G., Zensen, R., and Martin, M. A. (1998). Transcriptional activation of the integrated chromatin-associated human immunodeficiency virus type 1 promoter. *Mol. Cell. Biol.* **18,** 2535–2544.

Estable, M. C., Bell, B., Merzouki, A., Montaner, J. S. G., O'Shaughnessy, M. V., and Sadowski, I. J. (1996). Human immunodeficiency virus type 1 long terminal repeat variants from 42 patients representing all stages of infection display a wide range of sequence polymorphism and transcription activity. *J. Virol.* **70,** 4053–4062.

Fan, C. M., and Maniatis, T. (1990). A DNA-binding protein containing two widely separated zinc finger motifs that recognize the same DNA sequence. *Genes Dev.* **4,** 29–42.

Fauci, A. S. (1988). The human immunodeficiency virus: Infectivity and mechanisms of pathogenesis. *Science* **239,** 617–622.

Fauci, A. S. (1993). Multifactorial nature of human immunodeficiency virus disease: Implications for therapy. *Science* **262,** 1011–1018.

Finco, T. S., Westwick, J. K., Norris, J. L., Beg, A. A., Der, C. J., and Baldwin, Jr., A. S. (1997). Oncogenic Ha-Ras-induced signaling activates NF-kappaB transcriptional activity, which is required for cellular transformation. *J. Biol. Chem.* **272,** 24113–24116.

Finnegan, A., Roebuck, K. A., Nakai, B. E., Gu, D. S., Rabbi, M. F., Song, S., and Landay, A. L. (1996). IL-10 cooperates with TNF-alpha to activate HIV-1 from latently and acutely infected cells of monocyte/macrophage lineage. *J. Immunol.* **156,** 841–851.

Fisher, U., Huber, J., Boelens, W. C., Mattaj, I. W., and Luhrmann, R. (1995). The HIV-1 rev activation domain is a nuclear export signal that accesses an export pathway used by specific cellular RNAs. *Cell* **82,** 475–483.

Folks, T., Powell, D. M., Lightfoote, M. M., Benn, S., Martin, M. A., and Fauci, A. S. (1986). Induction of HTL VIII/LAV from a nonvirus-producing T-cell line: Implications for latency. *Science* **231,** 600–602.

Folks, T., Rowe, T., Villinger, F., Parekh, B., Mayne, A., Anderson, D., McClure, H., and Ansari, A. A. (1997). Immune stimulation may contribute to enhanced progression of SIV induced disease in rhesus macaques. *J. Med. Primatol.* **26,** 181–189.

Folks, T. M., Clouse, K. A., Justement, J., Rabson, A., Duh, E., Kehrl, J., and Fauci, A. S. (1989). Tumor necrosis factor alpha induces expression of human immunodeficiency virus in a chronically infected T-cell line. *Proc. Natl. Acad. Sci. USA* **86,** 2365–2368.

Folks, T. M., Justement, J., Kinter, A., Dinarello, C. A., and Fauci, A. S. (1987). Cytokine-induced expression of HIV-1 in a chronically infected promonocyte cell line. *Science* **238,** 800–802.

Fowke, K. R., D'Amico, R., Chernoff, D. N., Pottage, J. C. J., Benson, C. A., Sha, B. E., Kessler, H. A., Landay, A. L., and Shearer, G. M. (1997). Immunologic and virologic evaluation after influenza vaccination of HIV-1 infected patients. *AIDS* **11,** 1013–1021.

Fujita, T., Nolan, G. P., Ghosh, S., and Baltimore., D. (1992). Independent modes of transcriptional activation by the p50 and p65 subunits of NF-κB. *Genes Dev.* **6,** 775–787.

Fultz, P. N., and Zack, P. M. (1994). Unique lentivirus-host interactions: SIVsmmPBj14 infection of macaques. *Virus Res.* **32,** 205–225.

Garriga, J., Peng, J., Parreno, M., Price, D. H., Henderson, E. E., and Grana, X. (1998). Upregulation of cycin T1/CDK9 complexes during T cell activation. *Oncogene* **17,** 3093–3102.

Gerondakis, S., Grumont, R., Rourke, I., and Grossmann, M. (1998). The regulation and roles of Rel/NF-kappaB transcription factors during lymphocyte activation. *Curr. Opin. Immunol.* **10,** 353–359.

Gerritsen, M. E., Williams, A. J., Neish, A. S., Moore, S., Shi, Y., and Collins, T. (1997). CREB-binding protein/p300 are transcriptional coactivators of p65. *Proc. Natl. Acad. Sci. USA* **94,** 2927–2932.

Ghosh, S., Gifford, A. M., Riviere, L. R., Tempst, P., Nolan, G. P., and Baltimore, D. (1990). Cloning of the p50 DNA binding subunit of NF-kappa B: Homology to rel and dorsal. *Cell* **62,** 1019–1029.

Ghosh, S., May, M. J., and Kopp, E. B. (1998). NF-kappa B and Rel proteins: evolutionarily conserved mediators of immune responses. *Annu. Rev. Immunol.* **16,** 225–260.

Gimble, J. M., Duh, E., Ostrove, J. M., Gendelman, H. E., Max, E. E., and Rabson, A. B. (1988). Activation of the human immunodeficiency virus long terminal repeat by herpes simplex virus type 1 is associated with induction of a nuclear factor that binds to the NF-kappa B/core enhancer sequence. *J. Virol.* **62,** 4104–4112.

Griffin, G. E., Leung, K., Folks, T. M., Kunkel, S., and Nabel, G. J. (1989). Activation of HIV gene expression during monocyte differentiation by induction of NF-kappa B. *Nature* **339,** 70–73.

Guermah, M., Malik, S., and Roeder, R. G. (1998). Involvement of TFIID and USA components in transcriptional activation of the human immunodeficiency virus promoter by NF-kappaB and Sp1. *Mol. Cell. Biol.* **18,** 3234–3244.

Harhaj, E., Blaney, J., Millhouse, S., and Sun, S. C. (1996). Differential effects of I Kappa B molecules on Tat-mediated transactivation of HIV-1 LTR. *Virology* **216,** 284–287.

Hofmann, T. G., Hehner, S. P., Bacher, S., Droge, W., and Schmitz, M. L. (1998). Various glucocorticoids differ in their ability to induce gene expression, apoptosis, and to repress NF-kappaB-dependent transcription. *FEBS Lett.* **441,** 441–446.

Hottiger, M. O., Felzien, L. K., and Nabel, G. J. (1998). Modulation of cytokine-induced HIV gene expression by competitive binding of transcription factors to the coactivator p300. *EMBO J.* **17,** 3124–3134.

Huxford, T., Huang, D. B., Malek, S., and Ghosh, G. (1998). The crystal structure of the IkappaBalpha/NF-kappaB complex reveals mechanisms of NF-kappaB inactivation. *Cell* **95,** 759–770.

Ilyinskii, P. O., and Desrosiers, R. C. (1996). Efficient transcription and replication of simian immunodeficiency virus in the absence of NF-kappaB and Sp1 binding elements. *J. Virol.* **70,** 3118–3126.

Ilyinskii, P. O., Simon, M. A., Czajak, S. C., Lackner, A. A., and Desrosiers, R. C. (1997). Induction of AIDS by simian immunodeficiency virus lacking NF-κB and SP1 binding elements. *J. Virol.* **71,** 1880–1887.

Israel, N., Gougerot-Pocidalo, M. A., Aillet, F., and Virelizier, J. L. (1992). Redox status of cells influences constitutive or induced NF-kappa B translocation and HIV long terminal repeat activity in human T and monocytic cell lines. *J. Immunol.* **149,** 3386–3393.

Jacobs, M. D., and Harrison, S. C. (1998). Structure of an IkappaBalpha/NF-kappaB complex. *Cell* **95,** 749–758.

Jacque, J. M., Fernandez, B., Arenzana-Seisdedos, F., Thomas, D., Baleux, F., Virelizier, J. L., and Bachelerie, F. (1996). Permanent occupancy of the human immunodeficiency virus type 1 enhancer by NF-kappa B is needed for persistent viral replication in monocytes. *J. Virol.* **70,** 2930–2938.

Kalebic, T., Kinter, A., Poli, G., Anderson, M. E., Meister, A., and Fauci, A. S. (1991). Suppression of human immunodeficiency virus expression in chronically infected monocytic cells by glutathione, glutathione ester, and *N*-acetylcysteine. *Proc. Natl. Acad. Sci. USA* **88**, 986–990.

Kaufman, J. D., Valandra, G., Roderiquez, G., Bushar, G., Giri, C., and Norcross, M. A. (1987). Phorbol ester enhances human immunodeficiency virus-promoted gene expression and acts on a repeated 10-base-pair functional enhancer element. *Mol. Cell. Biol.* **7**, 3759–3766.

Kerr, L. D., Ransone, L. J., Wamsley, P., Schmitt, M. J., Boyer, T. G., Zhou, Q., Berk, A. J., and Verma, I. M. (1993). Association between proto-oncogene Rel and TATA-binding protein mediates transcriptional activation by NF-kappa B. *Nature* **365**, 4120419.

Khaled, A. R., Butfiloski, E. J., Sobel, E. S., and Schiffenbauer, J. (1998). Use of phosphorothioate-modified oligodeoxynucleotides to inhibit NF-kappaB expression and lymphocyte function. *Clin. Immunol. Immunopathol.* **86**, 170–179.

Kieran, M., Blank, V., Logeat, F., Vandekerckhove, J., Lottspeich, F., Le-Bail, O., Urban, M. B., Kourilsky, P., Baeuerle, P. A., and Israel, A. (1990). The DNA binding subunit of NF-kappa B is identical to factor KBF1 and homologous to the rel oncogene product. *Cell* **62**, 1007–1018.

Koken, S. E., van Wamel, J. L., Goudsmit, J., Berkhout, B., and Geelen, J. L. (1992). Natural variants of the HIV-1 long terminal repeat: Analysis of promoters with duplicated DNA regulatory motifs. *Virology* **191**, 968–972.

Kopp, E., and Ghosh, S. (1994). Inhibition of NF-κB by sodium salicylate and asprin. *Science* **265**, 956–958.

Korber, B., Kuiken, C., Foley, B., Hahn, B., McCutchan, F., Mellors, J., and Sodroski, J. (Eds.) (1998). "Human Retroviruses and AIDS 1998". Los Alamos, NM: Los Alamos National Laboratory.

Kotler, D. P. (1998). Antioxidant therapy and HIV infection: 1998. *Am. J. Clin. Nutr.* **67**, 7–9.

Kretzschmar, M., Meisterernst, M., Scheidereit, C., Li, G., and Roeder, R. G. (1992). Transcriptional regulation of the HIV-1 promoter by NF-kappa B *in vitro*. *Genes Dev.* **6**, 761–774.

Kumar, A., Dhawan, S., and Aggarwal, B. B. (1998). Emodin (3-methyl-1,6,8-trihydroxyanthraquinone) inhibits TNF-induced NF-kappaB activation, IkappaB degradation, and expression of cell surface adhesion proteins in human vascular endothelial cells. *Oncogene* **17**, 913–918.

Kumar, A., Haque, J., Lacoste, J., Hiscott, J., and Williams, B. R. (1995). Double-stranded RNS-dependent protein kinase activates transcription factor NF-kappa B by phosphorylating I kappa B. *Proc. Natl. Acad. Sci. USA* **91**, 6288–6292.

Kwon, H., Pelletier, N., DeLuca, C., Genin, P., Cisternas, S., Lin, R., Wainberg, M. A., and Hiscott, J. (1998). Inducible expression of IκBα repressor mutants interferes with NF-κB activity and HIV-1 replication in Jurkat T cells. *J. Biol. Chem.* **273**, 7431–7440.

Laughlin, M. A., Chang, G. Y., Oakes, J. W., González-Scarano, F., and Pomerantz, R. J. (1995). Sodium butyrate stimulation of HIV-1 gene expression: A novel mechanism of induction independent of NF-κB. *J. Acq. Immune Def. Syndr.* **9**, 332–339.

Lee, J. L., and Burckart, G. J. (1998). Nuclear factor kappa B: Important transcription factor and therapeutic target. *J. Clin. Pharmacol.* **38**, 981–983.

Leiden, J. M., Wang, C. Y., Petryniak, B., Markovitz, D. M., Nabel, G. J., and Thompson, C. B. (1992). A novel Ets-related transcription factor, Elf-1 binds to human immunodeficiency virus type 2 regulatory elements that are required for inducible trans activation in T cells. *J. Virol.* **66**, 5890–5897.

Leonard, J., Parrott, C., Buckler-White, A. J., Turner, W., Ross, E. K., Martin, M. A., and Rabson, A. B. (1989). The NF-kappa B binding sites in the human immunodeficiency virus type 1 long terminal repeat are not required for virus infectivity. *J. Virol.* **63**, 4919–4924.

LeRoy, G., Orphanides, G., Lane, W. S., and Reinberg, D. (1998). Requirement of RSF and FACT for transcription of chromatin templates *in vitro*. *Science* **282**, 1900–1904.

Levinson, B., Khoury, G., Woude, G. V., and Gruss, P. (1982). Activation of SV40 genome by 72-base pair tandem repeats of Moloney sarcoma virus. *Nature* **295**, 568–572.

Lewin, S. R., Lambert, P., Deacon, N. J., Mills, J., and Crowe, S. M. (1997). Constitutive expression of p50 homodimer in freshly isolated human monocytes decreases with *in vitro* and *in vivo* differentiation: A possible mechanism influencing human immunodeficiency virus replication in monocytes and mature macrophages. *J. Virol.* **71**, 2114–2119.

Lin, L., DiMartino, G. N., and Greene, W. C. (1998). Cotranslational biogenesis of NF-kappaB p50 by the 26S proteasome. *Cell* **92**, 819–828.

Lin, X., Cunningham, E. T. J., Mu, Y., Geleziunas, R., and Greene, W. C. (1999). The proto-oncogene Cot kinase participates in CD3/CD28 induction of NF-kappaB acting through the NF-kappaB-inducing kinase and the IkappaB kinases. *Immunity* **10**, 271–280.

Liu, J., Perkins, N. D., Schmid, R. M., and Nabel, G. J. (1992). Specific NF-kappa B subunits act in concert with Tat to stimulate human immunodeficiency virus type 1 transcription. *J. Virol.* **66**, 3883–3887.

Lodie, T. A., Reiner, M., Coniglio, S., Viglianti, G., and Fenton, M. J. (1998). Both Pu.1 and nuclear factor-κB mediate lipopolysaccharide-induced HIV long terminal repeat transcription in macrophages. *J. Immunol.* **161**, 268–276.

Look, M. P., Rochstroh, J. K., Rao, G. S., Barton, S., Lemoch, H., Kaiser, R., Kupfer, B., Sudhop, T., Spengler, U., and Sauerbruch, T. (1998). Soldium selenite and *N*-acetylcysteine in antiretroviral-naive HIV-1-infected patients: A randomized, controlled pilot study. *Eur. J. Clin. Invest.* **28**, 389–397.

Lortholary, O., Christeff, N., Casassus, P., Thobie, N., Veyssier, P., Trogoff, B., Torri, O., Brauner, M., Nunez, E. A., and Guillevin, L. (1996). Hypothalamo-pituitaty-adrenal function in human immunodeficiency virus-infected men. *J. Clin. Endocrinol. Metab.* **81**, 791–796.

Lu, Y., Stenzel, M., Sodroski, J. G., and Haseltine, W. A. (1989). Effects of long terminal repeat mutations on human immunodeficiency virus type 1 replication. *J. Virol.* **63**, 4115–4119.

Luque, I., and Gélinas, C. (1998). Distinct domains of IkappaBalpha regulate c-Rel in the cytoplasm and in the nucleus. *Mol. Cell. Biol.* **18**, 1213–1224.

Macian, F., and Rao, A. (1999). Reciprocal modulatory interaction between human immunodeficiency virus type 1 tat and transcription factor NFAT1. *Mol. Cell. Biol.* **19**, 3645–3653.

Maciaszek, J. W., Parada, N. A., Cruikshank, W. W., Center, D. M., Kornfeld, H., and Viglianti, G. A. (1997). IL-16 represses HIV-1 promoter activity. *J. Immunol.* **158**, 5–8.

Mallardo, M., Dragonetti, E., Baldassarre, F., Ambrosino, C., Scala, G., and Quinto, I. (1996). An NF-kappaB site in the 5′-untranslated leader region of the human immunodeficiency virus type 1 enhances the viral expression in response to NF-kappaB-activating stimuli. *J. Biol. Chem.* **271**, 20820–20827.

Markovitz, D. M., Smith, M. J., Hilfinger, J., Hannibal, M. C., Patryniak, B., and Nabel, G. J. (1992). Activation of the human immunodeficiency virus type 2 enhancer is dependent on purine box and kappa B regulatory elements. *J. Virol.* **66**, 5479–5484.

May, M. J., and Ghosh, S. (1998). Signal transduction through NF-kappaB. *Immunol. Today* **19**, 80–88.

McDonnell, P. C., Kumar, S., Rabson, A. B., and Gélinas, C. (1992). Transcriptional activity of rel family proteins. *Oncogene* **7**, 163–170.

McElhinny, J. A., MacMorran, W. S., Bren, G. D., Ten, R. M., Israel, A., and Paya, C. V. (1995). Regulation of IκBα and p105 in monocytes and macrophages persistently infected with human immunodeficiency virus. *J. Virol.* **69**, 1500–1509.

McNearney, T., Hornikova, Z., Templeton, A., Birdwell, A., Arens, M., Markham, R., Saah, A., and Ratner, L. (1995). Nef and LTR sequence variation from sequentially derived human immunodeficiency virus type 1 isolates. *Virology* **208**, 388–398.

Mellors, J. W., Rinaldo, C. R. J., Gupta, P., White, R. M., Todd, J. A., and Kingsley, L. A. (1996). Prognosis in HIV-1 infection predicted by the quantity of virus in plasma. *Science* **272**, 1167–1170.

Mercurio, F., Zhu, H., Murray, B. W., Shevchenko, A., Bennett, B. L., Li, J., Young, D. B., Barbosa, M., Mann, M., Manning, A., and Rao, A. (1997). IKK-1 and IKK-2: Cytokine activated IκB kinases essential for NF-κB activation. *Science* **278,** 860–866.

Michael, N. L., D'Arcy, L., Ehrenberg, P. K., and Redfield, R. R. (1994). Naturally occurring genotypes of the human immunodeficiency virus type 1 long terminal repeat display a wide range of basal and Tat-induced transcriptional activities. *J. Virol.* **68,** 3163–3174.

Montano, M. A., Kripke, K., Norina, C. D., Achacoso, P., Herzenberg, L. A., Roy, A. L., and Nolan, G. P. (1996). NF-κB homodimer binding within the HIV-1 initiator region and interactions with TFII-I. *Proc. Natl. Acad. Sci. USA* **93,** 12376–12381.

Montano, M. A., Nixon, C. P., and Essex, M. (1998). Dysregulation through the NF-κB enhancer and TATA box of the human immunodeficiency virus type 1 subtype E promoter. *J. Virol.* **72,** 8446–8452.

Montano, M. A., Novitsky, V. A., Blackard, J. T., Cho, N. L., Katzenstein, D. A., and Essex, M. (1997). Divergent transcriptional regulation among expanding human immunodeficiency virus type 1 subtypes. *J. Virol.* **71,** 8657–8665.

Morishita, R., Sugimoto, T., Aoki, M., Kida, I., Tomita, N., Moriguchi, A., Maeda, K., Sawa, Y., Kaneda, Y., Higaki, J., and Ogihara, T. (1997). *In vivo* transfection of *cis* element "decoy" against nuclear factor-κB binding site prevents myocardial infarction. *Nature Med.* **3,** 894–899.

Mosialos, G. (1997). The role of Rel/NF-kappa B proteins in viral oncogenesis and the regulation of viral transcription. *Semin. Cancer Biol.* **8,** 121–129.

Nabel, G., and Baltimore, D. (1987). An inducible transcription factor activates expression of human immunodeficiency virus in T cells. *Nature* **326,** 711–713.

Nakamura, T., Donovan, D. M., Hamada, K., Sax, C. M., Norman, B., Flanagan, J. R., Ozato, K., Westphal, H., and Piatigorsky, J. (1990). Regulation of the mouse alpha A-crystallin gene: Isolation of a cDNA encoding a protein that binds to a *cis* sequence motif shared with the major histocompatability complex class I gene and other genes. *Mol. Cell. Biol.* **10,** 3700-3708.

Nakano, H., Shindo, M., Sakon, S., Nishinaka, S., Mihara, M., Yagita, H., and Okumura, K. (1998). Differential regulation of IkappaB kinase alpha and beta by two upstream kinases, NF-kappaB-inducing kinase and mitogen-activated protein kinase/ERK kinase kinase-1. *Proc. Natl. Acad. Sci. USA* **95,** 3537–3542.

Natarajan, K., Singh, S., Burke, T. R., Jr., Grunberger, D., and Aggarwal, B. B. (1996). Caffeic acid phenethyl ester is a potent and specific inhibitor of activation of nuclear transcription factor NF-kappa B. *Proc. Natl. Acad. Sci. USA* **93,** 9090–9095.

Nemoto, S., DiDonato, J. A., and Lin, A. (1998). Coordinate regulation of IkappaB kinases by mitogen-activated protein kinase kinase kinase 1 and NF-kappaB-inducing kinase. *Mol. Cell. Biol.* **18,** 7336–7343.

Neurath, M. F., Petterson, S., Meyer zum Büschenfelde, K.-H., and Strober, W. (1996). Local administration of antisense phosphorothioate oligonucleotides to the p65 subunit of NF-κB abrogates established colitis in mice. *Nature Med.* **2,** 998–1004.

Ninomiya-Tsuji, J., Kishimoto, K., Hiyama, A., Inoue, J., Cao, Z., and Matsumoto, K. (1999). The kinase TAK1 can activated the NIK-IkappaB as well as the MAP kinase cascade in the IL-1 signaliing pathway. *Nature* **398,** 252–256.

Nolan, G. P., Ghosh, S., Liou, H. C., Tempst, P., and Baltimore, D. (1991). DNA binding and I kappa B inhibition of the cloned p65 subunit of NF-kappa B, a rel-related polypeptide. *Cell* **64,** 961–969.

Norbiato, G., Bevilacqua, M., Vago, T., Taddei, A., and Clerici, M. (1997). Glucocorticoids and the immune function in the human immunodeficiency virus infection: A study in hypercortisolemic and cortisol-resistant patients. *J. Clin. Endocrinol. Metab.* **82,** 3260–3263.

Nottet, H. S., Moeleans, I. I., de Vos, N. M., de Graaf, L., Visser, M. R., and Verhoef, J. (1997). *N*-acetyl-L-cysteine-induced up-regulation of HIV-1 gene expression in monocyte-derived macrophages correlates with increased NF-kappaB DNA binding activity. *J. Leukoc. Biol.* **61,** 33–39.

Novembre, F. J., Johnson, P. R., Lewis, M. G., Anderson, D. C., Klumpp, S., McClure, H. M., and Hirsch, V. M. (1993). Multiple determinants contribute to pathogenicity of the acutely lethal simian immunodeficiency virus SIV smmPBj14 variant. *J. Virol.* **67,** 2466–2474.

O'Brien, W. A., Grovit-Ferbas, K., Namzi, A., Ovcak-Derzic, S., Wang, H.-J., Park, J., Yeramian, C., Mao, S.-H., and Zack, J. A. (1995). Human immunodeficiency virus-type 1 replication can be increased in peripheral blood of seropositive patients after influenza vaccination. *Blood* **86,** 1082–1089.

Osborn, L., Kunkel, S., and Nabel, G. J. (1989). Tumor necrosis factor alpha and interleukin 1 stimulate the human immunodeficiency virus enhancer by activation of the nuclear factor kappa B. *Proc. Natl. Acad. Sci. USA* **86,** 2336–2340.

Palombella, V. J., Rando, O. J., Goldberg, A. L., and Maniatis, T. (1994). The ubiquitin-proteasome pathway is required for processing the NF-kappa B 1 precursor proteiin and the activation of NF-kappa B. *Cell* **78,** 773–785.

Parrott, C., Seidner, T., Duh, E., Leonard, J., Theodore, T. S., Buckler-White, A., Martin, M. A., and Rabson, A. B. (1991). Variable role of the long terminal repeat Sp1-binding sites in human immunodeficiency virus replication in T lymphocytes. *J. Virol.* **65,** 1414–1419.

Paul, I., Baeuerle, P. A., and Schmitz, M. L. (1997). Basal transcription factors TBP and TFIIB and the viral coactivator E1A 13S bind with distinct affinities and kinetics to the transactivation domain of NF-kappaB p65. *Nucleic Acids Res.* **25,** 1050–1055.

Pazin, M. J., Sheridan, P. L., Cannon, K., Cao, Z., Keck, J. G., Kadonaga, J. T., and Jones, K. A. (1996). NF-κB-mediated chromatin reconfiguration and transcriptional activation of the HIV-1 enhancer in vitro. *Genes Dev.* **10,** 37–49.

Perkins, N. D., Agranoff, A. B., Pascal, E., and Nabel, G. J. (1994). An interaction between the DNA-binding domains of RelA(p65) and Sp1 mediates human immunodeficiency virus gene activation. *Mol. Cell. Biol.* **14,** 6570–6583.

Perkins, N. D., Edwards, N. L., Duckett, C. S., Agranoff, A. B., Schmid, R. M., and Nabel, G. J. (1993). A cooperative interaction between NF-κB and Sp1 is required for HIV-1 enhancer activation. *EMBO J.* **12,** 3551–3558.

Perkins, N. D., Felzien, L. K., Betts, J. C., Leung, K., Beach, D. H., and Nabel, G. J. (1997). Regulation of NF-kappaB by cyclin-dependent kinases associated with the p300 coactivator. *Science* **275,** 523–527.

Pohlmann, S., Floss, S., Ilyinskii, P. O., Stamminger, T., and Kirchhoff, F. (1998). Sequences just upstream of the simian immunodeficiency virus core enhancer allow efficient replication in the absence of NF-kappaB and Sp1 binding elements *J. Virol.* **72,** 5589–5598.

Popik, W., Hesselgesser, J. E., and Pitha, P. M. (1998). Binding of human immunodeficiency virus type 1 to CD4 and CXCR4 receptors differentially regulates expression of inflammatory genes and activates the MEK/ERK signaling pathway. *J. Virol.* **72,** 6406–6413.

Qian, J., Bours, V., Manischewitz, J., Blackburn, R., Siebenlist, U., and Golding, H. (1994). Chemically selected subclones of the CEM cell line demonstrate resistance to HIV-1 infection resulting from a selective loss of NF-κB DNA binding proteins. *J. Immunol.* **152,** 4183–4191.

Quinones-Mateu, M. E., Mas, A., Lain de Lara, T., Soriano, V., Alcami, J., Lederman, M. M., and Domingo, E. (1998). LTR and tat variability of HIV-1 isolates from patients with divergent rates of disease progression. *Virus Res.* **57,** 11–20.

Quinto, I., Mallardo, M., Baldassarre, F., Scala, G., Englund, G., and Jeang, K.-T. (1999). Potent and stable attenuation of live-HIV-1 by gain of a proteolysis-resistant inhibitor of NF-κB (IκB-αS32/36A) and the implications for vaccine development. *J. Biol. Chem.* **274,** 17567–17572.

Rao, A., Luo, C., and Hogan, P. G. (1997). Transcription factors of the NFAT family: regulation and function. *Annu. Rev. Immunol.* **15,** 707–747.

Ray, A., and Prefontaine, K. E. (1994). Physical association and functional antagonism between the p65 subunit of transcription factor NF-kappa B and the glucocorticoid receptor. *Proc. Natl. Acad. Sci. USA* **18**, 752–756.

Regnier, C. H., Song, H. Y., Gao, X., Goeddel, D. V., Cao, Z., and Rothe, M. (1997). Identification and characterization of an IkappaB kinase. *Cell* **90**, 373–383.

Riviere, Y., Blank, V., Kourilsky, P., and Israel, A. (1991). Processing of the precursor of NF-κB by the HIV-1 protease during acute infection. *Nature* **350**, 625–626.

Roederer, M., Staal, F. J., Raju, P. A., Ela, S. W., Herzenberg, L. A., and Herzenberg, L. A. (1990). Cytokine-stimulated human immunodeficiency virus replication is inhibited by *N*-acetyl-L-cysteine. *Proc. Natl. Acad. Sci. USA* **87**, 4884–4888.

Romano, M. F., Lamberti, A., Petrella, A., Bisogni, R., Tassone, P. F., Formisano, S., Venuta, S., and Turco, M. C. (1996). IL-10 inhibits nuclear factor-kappa B/Rel nuclear activity in CD3-stimulated human peripheral T lymphocytes. *J. Immunol.* **156**, 2119–2123.

Ron, D., Brasier, A. R., and Haebner, J. F. (1991). Angiotensin gene-inducible enhancer binding protein 1: a member of a new family of large nuclear proteins that recognize NF-κB sites through a zinc finger motif. *Mol. Cell. Biol.* **11**, 2887–2895.

Rosen, C. A., Sodroski, J. G., and Haseltine, W. A. (1985). The location of *cis*-acting regulatory sequences in the HTL V-III/LAV LTR. *Cell* **41**, 813–823.

Ross, E. K., Buckler-White, A. J., Rabson, A. B., Englund, G., and Martin, M. A. (1991). Contribution of NF-kappa B and Sp1 binding motifs to the replicative capacity of human immunodeficiency virus type 1: Distinct patterns of viral growth are determined by T-cell types. *J. Virol.* **65**, 4350–4358.

Roulston, A., D'Addario, M., Boulerice, F., Caplan, S., Wainberg, M. A., and Hiscott, J. (1992). Induction of monocytic differentiation and NF-kappaB-like activities by human immunodeficiency virus 1 infection of myelomonoblastic cells. *J. Exp. Med.* **175**, 751–763.

Rousseau, C., Abrams, E., Lee, M., Urbano, R., and King, M.-C. (1997). Long terminal repeat and *nef* gene variants of human immunodeficiency virus type 1 in perinatally infected long-term survivors and rapid progressors. *AIDS Res. Hum. Retroviruses* **13**, 1611–1623.

Ruben, S. M., Narauanan, R., Klement, J. F., Chen, C. H., and Rosen, C. A. (1992). Functional characterization of the NF-κB p65 transcriptional activator and an alternatively spliced derivative. *Mol. Cell. Biol.* **12**, 444–454.

Ruocco, M. R., Chen, X., Ambrosino, C., Dragonetti, E., Liu, W., Mallardo, M., De Falco, G., Palmieri, C., Franzoso, G., Quinto, I., Venuta, S., and Scala, G. (1996). Regulation of HIV-1 long terminal repeats by interaction of C/EBP(NF-IL6) and NF-kappaB/Rel transcription factors. *J. Biol. Chem.* **271**, 22479–22486.

Scheinman, R. I., Cogswell, P. C., Lofquist, A. K., and Baldwin, A. S. Jr. (1995a). Role of transcriptional activation of IκBα in mediation of immunosuppression by glucocorticoids. *Science* **270**, 283–286.

Scheinman, R. I., Gualberto, A., Jewell, C. M., Cidlowski, J. A., and Baldwin, A. S., Jr. (1995b). Characterization of mechanisms involved in transrepression of NF-kappa B by activated glucocorticoid receptors. *Mol. Cell. Biol.* **15**, 943–953.

Schmid, R. M., Perkins, N. D., Duckett, C. S., Andrews, P. C., and Nabel, G. J. (1991). Cloning of an NF-kappa B subunit which stimulates HIV transcription in synergy with p65. *Nature* **352**, 733–736.

Schmitz, M. L., dos Santos Siva, M. A., Altmann, H., Czisch, M., Holak, T. A., and Baeuerle, P. A. (1994). Structural and functional analysis of the NF-kappa B p65 C terminus: An acidic and modular transactivation domain with the potential to adopt an alpha-helical conformation. *J. Biol. Chem.* **269**, 25613–25620.

Schreck, R., Meier, B., Mannel, D. N., Droge, W., and Baeuerle, P. A. (1992). Dithiocarbamates as potent inhibitors of nuclear factor kappa B activation in intact cells. *J. Exp. Med.* **175**, 1181–1194.

Schreck, R., Rieber, P., and Baeuerle, P. A. (1991). Reactive oxygen intermediates as apparently widely used messengers in the activation of the NF-kappa B transcription factor and HIV-1. *EMBO J.* **10**, 2247–2258.

Seeler, J. S., Murchadt, C., Suessle, A., and Gaynor, R. B. (1994). Transcription factor PRDII-BF1 activates human immunodeficiency virus type 1 gene expression. *J. Virol.* **68**, 1002–1009.

Sen, R., and Baltimore, D. (1986). Inducibility of kappa immunoglobulin enhancer-binding protein Nf-kappa B by a posttranslational mechanism. *Cell* **47**, 921–928.

Sheppard, K. A., Phelps, K. M., Williams, A. J., Thanos, D., Glass, C. K., Rosenfeld, M. G., Gerritsen, M. E., and Collins, T. (1998). Nuclear integration of glucocorticoid receptor and nuclear factor-kappaB signaling by the CREB-binding protein and steroid receptor coactivator-1. *J. Biol. Chem.* **273**, 29291–29294.

Sheridan, P. L., Mayall, T. P., Verdin, E., and Jones, K. A. (1997). Histone acetyltranferases regulate HIV-1 enhancer activity *in vitro*. *Genes Dev.* **11**, 3327–3340.

Sheridan, P. L., Sheline, C. T., Cannon, K., Voz, M. L., Pazin, M. J., Kadonaga, J. T., and Jones, K. A. (1995). Activation of the HIV-1 enhancer by the LEF-1 HMG protein on nucleosome-assembled DNA in vitro. *Genes Dev.* **9**, 2090–2104.

Siebenlist, U., Franzoso. G., and Brown, K. (1994). Structure, regulation, and function of NF-kappa B. *Annu. Rev. Cell Biol.* **10**, 405–455.

Siekevitz, M., Josephs, S. F., Dukovich, M., Peffer, N., Wong-Staal, F., and Greene, W. C. (1987). Activation of the HIV-1 LTR by T cell mitogens and the trans-activator protein of HTL V-1. *Science* **238**, 1575–1578.

Singh, S., and Aggarwal, B. B. (1995). Activation of transcription factor NF-kappa B is suppressed by curcumin. *J. Biol. Chem.* **270**, 24995–25000.

Staal, F. J. T., Roederer, M., Raju, P. A., Anderson, M. T., Ela, S. W., Herzenberg, L. A., and Herzenberg, L. A. (1993). Antioxidants inhibit stimulation of HIV transcription. *AIDS Res. Hum. Retroviruses* **9**, 299–306.

Staprans, S. I., Hamilton, B. L., Follansgee, S. E., Elbeik, T., Barbosa, P., Grant, R. M., and Feinberg, M. B. (1995). Activation of virus replication after vaccination of HIV-1 infected individuals. *J. Exp. Med.* **182**, 1727–1737.

Steger, D. J., Eberharter, A., John, S., Grant, P. A., and Workman, J. L. (1998). Purified histone acetyltransferase complexes stimulate HIV-1 transcription from preassembled nucleosomal arrays. *Proc. Natl. Acad. Sci. USA* **95**, 12924–12929.

Steger, D. J., and Workman, J. L. (1997). Stable co-occupancy of transcription factors and histones at the HIV-1 enhancer. *EMBO J.* **16**, 2463–2472.

Stevenson, M., Stanwick, T. L., Dempsey, M. P., and Lamonica, C. A. (1990). HIV-1 replication is controlled at the level of T cell activation and proviral integration. *EMBO J.* **9**, 1551–1560.

Tong-Starksen, S. E., Luciw, P. A., and Peterlin, B. M. (1987), Human immunodeficiency virus long terminal repeat responds to T-cell activation signals. *Proc. Natl. Acad. Sci. USA* **84**, 6845–6849.

Traenckner, B.-M., Pahl, H., Henkel, T., Schmidt, K., Wilk, S., and Baeuerle, P. (1995). Phosphorylation of human IκB-α on serines 32 and 36 controls IkappaB-alpha proteolysis and NF-kappaB activation in reponse to diverse stimuli. *EMBO J.* **14**, 2876–2883.

Traenckner, E. B., Wilk, S., and Baeuerle, P. A. (1994). A proteasome inhibitor prevents activation of NF-kappa B and stabilizes a newly phosphorylated form of I kappa B-alpha that is still bound to NF-kappa B. *EMBO J.* **13**, 5433–5441.

Van Lint, C., Emiliani, S., Ott, M., and Verdin, E. (1996). Transcriptional activation and chromatin remodeling of the HIV-1 promoter in response to histone acetylation. *EMBO J.* **15**, 1112–1120.

Verdin, E., Becker, N., Bex, F., Droogmans, L., and Burny, A. (1990). Identification and characterization of an enhancer in the coding region of the genome of human immunodeficiency virus type 1. *Proc. Natl. Acad. Sci. USA* **87**, 4874–4878.

Verdin, E., Paras, P., Jr., and Van Lint, C. (1993). Chromatin disruption in the promoter of human immunodeficiency virus type 1 during transcriptional activation. *EMBO J.* **12**, 3249–3259.

Verhoef, K., Sanders, R. W., Fontaine, V., Kitajima, S., and Berkhout, B. (1999). Evolution of the human immunodeficiency virus type 1 long terminal repeat promoter by conversion of an NF-κB enhancer element into a GABP binding site. *J. Virol.* **73,** 13331–1340.

Vicenzi, E., and Poli, G. (1994). Regulation of HIV expression by viral genes and cytokines. *J. Leukocyte Biol.* **56,** 328–334.

Visco-Comandi, U., Yun, Z., Vahlne. A., and Sönnerborg, A. (1999). No association of HIV type 1 long terminal repeat sequence pattern with long-term nonprogression and *in vivo* viral replication levels in European subjects. *AIDS Res. Hum. Retroviruses* **15,** 609–617.

Vlach, J., and Pitha, P. M. (1993). Hexamethylene bisacetamide activates the human immunodeficiency virus type 1 provirus by an NF-κB-independent mechanism. *J. Gen. Virol.* **74,** 2401–2408.

Wang, D., and Baldwin, A. S. Jr. (1998). Activation of nuclear factor-kappaB-dependent transcription by tumor necrosis factor-alpha is mediated through phosphorylation of RelA/p65 on serine 529. *J. Biol. Chem.* **273,** 29411–29416.

Wang, P., Wu, P., Siegel, M. I., Egan, R. W., and Billah, M. M. (1995). Interleukin (IL)-10 inhibits nuclear factor kappa B (NF kappa B) activation in human monocytes. IL-10 and IL-4 suppress cytokine synthesis by different mechanisms. *J. Biol. Chem.* **270,** 9558–9563.

Webster, A. (1991). Cytomegalovirus as a possible cofactor in HIV disease progression. *J. Acquir. Immune Defic. Syndr.* **4** (Suppl.), S47–S52.

Wen, W., Meinkoth, J. I., Tsien, R. Y., and Taylor, S. S. (1995). Identification of a signal for rapid export of proteins from the nucleus. *Cell* **82,** 463–473.

Whiteside, S. L., Ernst, M. K., LeBail, O., Laurent-Winter, C., Rice, N., and Israel, A. (1995). N- and C- terminal sequences control degradation of MAD3/IκBα in response to inducers of NF-κB activity. *Mol. Cell. Biol.* **15,** 5339–5345.

Widlak, P., Gaynor, R. B., and Garrard, W. T. (1997). *In vitro* chromatin assembly of the HIV-1 promoter: ATP-dependent polar repositioning of nucleosomes by Sp1 and NFκB. *J. Biol. Chem.* **272,** 17654–17661.

Witschi, A., Junker, E., Schranz, C., Speck, R. F., and Lauterburg, B. H. (1995). Supplementation of *N*-acetylcysteine fails to increase glutathione in lymphocytes and plasma of patients with AIDS. *AIDS Res. Hum. Retroviruses* **11,** 141–143.

Workman, J. L., and Kingston, R. E. (1998). Alteration of nucleosome structure as a mechanism of transcriptional regulation. *Annu. Rev. Biochem.* **67,** 545–579.

Wu, B.-Y., Woffendin, C., Duckett, C. S., Ohno, T., and Nabel, G. J. (1995). Regulation of human retroviral latency by the NF-κB/IκB family: Inhibition of human immunodeficiency virus replication by IκB through a rev-dependent mechanism. *Proc. Natl. Acad. Sci. USA* **92,** 1480–1484.

Wu, B.-Y., Woffendin, C., MacLachlan, I., and Nabel, G. J. (1997). Distinct domains of IκB-α inhibit human immunodeficiency virus type 1 replication through NF-κB and Rev. *J. Virol.* **71,** 3163–3167.

Xu, X., Prorock, C., Ishikawa, H., Maldonado, E., Ito, Y., and Gélinas, C. (1993). Functional interaction of the v-Rel and c-Rel oncoproteins with the TATA-binding protein and association with transcription factor IIB. *Mol. Cell. Biol.* **13,** 6733–6741.

Yin, M.-J., Yamato, Y., and Gaynor, R. B. (1998). The anti-inflammatory agents aspirin and salicylate inhibit the activity of IκB Kinase-β. *Nature* **396,** 77–80.

Zandi, E., and Karin, M. (1999). Bridging the gap: Composition, regulation, and physiological function of the IkappaB kinase complex. *Mol. Cell. Biol.* **19,** 4547–4551.

Zandi, E., Rothwarf, M., Delhase, M., Hayakawa, M., and Karin, M. (1997). The IκB kinase complex (IKK) contains two kinase subunits, IKKα and IKKβ, necessary for IκB phosphorylation and NF-κB activation. *Cell* **91,** 243–252.

Zhang, J., Novembre, F., and Rabson, A. B. (1997). Simian immunodeficiency viruses containing mutations in the long terminal repeat NF-κB or Sp1 binding sites replicate efficiently in T cells and PHA-stimulated PBMCs. *Virus Res.* **49,** 205–213.

Zhang, L., Huang, Y., Yuan, H., Chen, B. K., Ip, J., and Ho, D. D. (1997a). Genotypic and phenotypic characterization of long terminal repeat sequences from long-term survivors of human immunodeficiency virus type 1 infection. *J. Virol.* **71,** 5608–5613.

Zhang, L., Huang, Y., Yuan, H., Chen, B. K., Ip, J., and Ho, D. D. (1997b). Identification of a replication-competent pathogenic human immunodeficiency virus type 1 with a duplication in the TCF-1α region but lacking NF-κB binding sites. *J. Virol.* **71,** 1651–1656.

Anne Gatignol[*,†] and Kuan-Teh Jeang[‡]

*U529 INSERM
Institut Cochin de Génétique Moléculaire
75014 Paris, France
†Molecular Oncology Group
McGill AIDS Centre
Lady Davis Institute for Medical Research
Montréal, QC, H3T 1E2 Canada

‡Molecular Virology Section
Laboratory of Molecular Microbiology
National Institute for Allergy and Infectious Diseases
National Institutes of Health
Bethesda, Maryland 20892-0460

Tat as a Transcriptional Activator and a Potential Therapeutic Target for HIV-1

Therapeutic approaches against AIDS will likely involve multiple drug combinations in order to avoid the development of mutations leading to resistant virus and to decrease the numerous side effects of the presently available drugs. To be most effective, these drugs should reach different targets. HIV gene expression and its strong activation by the viral Tat protein represents one potential drug target (De Clercq, 1995). Recent insights on the mechanism of Tat trans-activation suggest that Tat acts in concert with cellular cofactors to increase viral transcription. We review here the role of Tat and cellular proteins in mediating transcription from the HIV-1 LTR-promoter and anti-Tat therapeutic strategies.

I. Mechanisms of Tat-Mediated Trans-Activation

A. Tat Protein

The 14-kDa Tat protein (*t*rans-*a*ctivator of *t*ranscription) activates the expression of the long terminal repeat (LTR; Fig. 1) through an RNA-

Advances in Pharmacology, Volume 48

1054-3589/00 $35.00

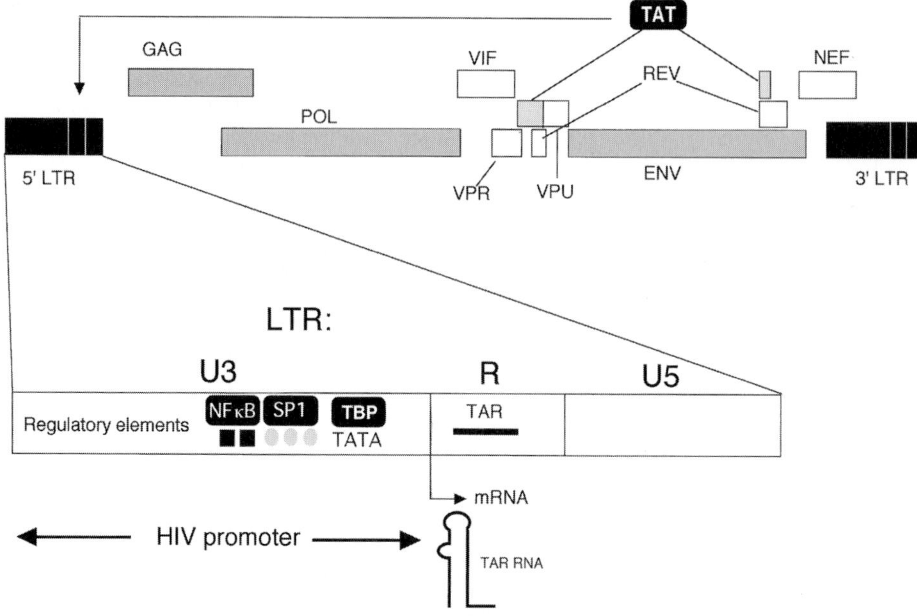

FIGURE 1 HIV-1 schematic and detail's for the long terminal repeat (LTR), the HIV promoter, and TAR RNA.

dependent mechanism. In the presence of Tat transcription from the HIV-1 LTR is more than a 100-fold higher than in its absence. Tat is encoded by two exons (exon 1, aa 1–72, and exon 2, aa 73–101). The protein can be divided into five domains: domain I (aa 1–20) is located in the N-terminus and mutations in this region have little influence on trans-activation. Domain II has seven highly conserved cysteines which are important for Tat function. Domain III (aa 40–48), or the "core", is essential for trans-activation; in this region a single change at lysine 41 has been shown to abolish Tat activity. Domain IV (aa 49–72) contains an arginine-rich stretch which is important in RNA binding and in nuclear localization. Although the first three domains constitute the minimal portion of Tat necessary for transcription (Derse *et al.,* 1991; Southgate and Green, 1995), in the context of a promoter, they function only when linked to region IV or to a heterologous RNA/DNA binding domain (Berkhout *et al.,* 1990; Kamine *et al.,* 1991; Selby and Peterlin, 1990; Southgate *et al.,* 1990; Southgate and Green, 1991). Domain V, including the second exon, is less important for transcription but contributes importantly toward viral infectivity and other Tat functions (reviewed in Rana and Jeang, 1999).

In activating transcription from the HIV-1 LTR, Tat acts through an RNA target called TAR (Trans-Activation Response) located in the R region of the LTR. TAR RNA forms a stable stem-bulge-loop structure. The sequence in the loop, the bulge, as well as the stem structure are necessary for Tat trans-activation (Berkhout and Jeang, 1989, 1991; Berkhout *et al.*, 1989; Delling *et al.*, 1991; Feng and Holland, 1988; Jakobovits *et al.*, 1988; Roy *et al.*, 1990; Selby *et al.*, 1989) and HIV replication (Klaver and Berkhout, 1994; Rounseville *et al.*, 1996; Verhoef *et al.*, 1997). *In vitro,* Tat can directly bind the UCU bulge of TAR (Dingwall *et al.*, 1990); this binding by Tat is enhanced by a recently characterized cellular cofactor, cyclin T (Wei *et al.*, 1998). Functional data on Tat activity indicate that protein–DNA interactions in the LTR (Berkhout *et al.*, 1990; Kamine *et al.*, 1993; Southgate *et al.*, 1990; Southgate and Green, 1995), protein–TAR RNA interactions (Gatignol *et al.*, 1991; Wei *et al.*, 1998; Wu-Baer *et al.*, 1995), and the cellular context (Hart *et al.*, 1989; Newstein *et al.*, 1990) all contribute to optimal trans-activation. In binding to an RNA target, Tat has the potential to influence initiation as well as elongation of transcription (for recent reviews, see Gatignol *et al.*, 1996; Jeang, 1998; Jeang *et al.*, Jones, 1997; Kingsman and Kingsman, 1996; Rana and Jeang, 1999; Yankulov and Bentley, 1998).

B. Tat Mechanism on Transcriptional Initiation and Elongation

Intracellularly, it has been suggested that a part of Tat's function contributes to the formation of transcriptionally competent complexes at the promoter (Jeang *et al.*, 1988; Laspia *et al.*, 1989). Some observations relevant to this aspect of Tat function include (1) the demonstration of a position-dependent effect of TAR on transcription. Thus, when TAR RNA was progressively distanced away from the site of initiation the transcriptional activity of Tat was diminished (Selby *et al.*, 1989). (2) The ability of Tat to function when targeted to the promoter via a DNA binding domain (Berkhout *et al.*, 1990; Daviet *et al.*, 1998; Kamine *et al.*, 1991; Kim and Risser, 1993; Southgate and Green, 1991, 1995). (3) The interaction of Tat with TBP (Kashanchi *et al.*, 1994; Veschambre *et al.*, 1995) and with promoter-upstream factors such as Sp1 (Chun *et al.*, 1998; Jeang *et al.*, 1993; Pagtakhan and Tong-Starksen, 1997). (4) Kinetic evidence that Tat acts early in transcription (Jeang and Berkhout, 1992) through a nascent TAR RNA (Berkhout *et al.*, 1989). (5) The physical presence of Tat protein in preinitiation complexes (Cujec *et al.*, 1997; García-Martinez *et al.*, 1997). A recent study clarified the role of Tat at the promoter as occuring at an event after TBP binding, likely at the step of promoter clearance. Such an activity would facilitate both the processive elongation of RNA polymerase II complexes and subsequent reinitiations at the promoter (Xiao *et al.*, 1997).

Effects of Tat on elongation of transcription have been well-documented (Jones and Peterlin, 1994; Laspia *et al.*, 1989). Several relevant observations include (1) the presence of short LTR-derived transcripts in the absence of

Tat in some cell lines and in cell free transcription assays (Kao *et al.*, 1987; Zhou and Sharp, 1995); (2) nuclear run-on experiments consistent with differential processivity of transcription in the absence of Tat (Blau *et al.*, 1996; Feinberg *et al.*, 1991; Laspia *et al.*, 1989); and (3) the physical association of Tat with the elongation complex (Keen *et al.*, 1996) and elongation factors (Mancebo *et al.*, 1997; Yang *et al.*, 1997; Zhou and Sharp, 1996; Zhu *et al.*, 1997). Cell free transcription assays very efficiently measure the elongation of RNA polymerase II (Keen *et al.*, 1997; Zhou and Sharp, 1995). However, templates in these *in vitro* extracts do not reinitiate and therefore exclude initiation events (Green, 1993; Hahn, 1998; Jeang *et al.*, 1999).

Although promoter-proximal short (~80 nucleotides) transcripts are effectively transcribed from the HIV-1 LTR in several cellular settings, studies show that these prematurely "terminated" RNAs do not appear to be precursors for processively elongated transcripts observed in the presence of Tat. First, Tat-induced RNAs are seen in the absence of short transcripts (Jeang *et al.*, 1993). Second, mutational analyses in the HIV-1 LTR of the inducer of short transcripts (IST) motif, which is distinct from but overlaps TAR, show that IST function and Tat trans-activation are distinct and independent properties (Jeang *et al.*, 1993; Sheldon *et al.*, 1993). One interpretation of these results is that nonprocessive complexes can form in the absence of Tat; however, presence of Tat induces at the promoter a new set of unrelated processive complexes that are competent for synthesis of full-length transcripts (Pendergrast and Hernandez, 1997; Pessler and Hernandez, 1998).

A mechanistic explanation for the action of Tat on processivity of transcription emerged from the recent demonstration of a critical role for the carboxy-terminal domain (CTD) of RNA polymerase II. Tat was found to physically associate with cellular kinases that phosphorylate the RNAP II CTD (Cujec *et al.*, 1997; Herrmann and Rice, 1995; Mancebo *et al.*, 1997; Yang *et al.*, 1996, 1997; Zhu *et al.*, 1997). Phosphorylation of the CTD triggers release from and clearance of the promoter by RNAP II leading to efficient elongation (Dahmus, 1996; Hahn, 1998; McCracken *et al.*, 1997; Reinberg *et al.*, 1998). Promoter clearance also facilitates subsequent reinitiation at the same promoter. Thus an effect of Tat on CTD phosphorylation could account for effects on processivity of elongation as well as transcriptional reinitiations (Fig. 2).

II. Cellular Cofactors for Trans-Activation

The requirement for cellular cofactors in Tat-mediated trans-activation came from two observations: (1) although Tat binds to the bulge of TAR RNA, mutations in the loop and the upper stem of TAR severely impair trans-activation; and (2) trans-activation occurs optimally in primate cells

(Barry *et al.*, 1991; Seigel *et al.*, 1986). It is poorly active in rodents (Hart *et al.*, 1989; Newstein *et al.*, 1990) and does not occur in insect, yeast, or bacterial cells (Daviet *et al.*, 1998; Jeang *et al.*, 1988; Toyama *et al.*, 1992). Likely, then, nonprimate cells lack critical cellular Tat cofactors which are needed for trans-activation. Since trans-activation in mouse and hamster cells can be restored in somatic cells hybrids that have received human chromosome 12, this suggests that the critical cofactor is encoded within this chromosome. When Tat was targeted to the promoter in a manner independent of TAR RNA binding, differences in trans-activation between rodent cells and their counterparts which contained human chromosome 12 were no longer observed. In this setting TAR-loop mutants were transactivated by Tat with the same efficiency in rodent and in primate cells. These data are compatible with human chromosome 12 encoding a loop binding factor that allows efficient intracellular Tat–TAR interaction (Alonso *et al.*, 1992, 1994; Hart *et al.*, 1993; Newstein *et al.*, 1993). Interestingly, genes for three proteins (Sp1, TRBP, and cyclin T) shown to have roles in Tat trans-activation have been mapped to human chromosome 12. Other factors such as histone acetyl transferases have also been suggested to contribute to the activation by Tat of proviral LTRs inside cells.

A. Histone Acetyl Transferases

For optimal activity several critically required Tat-associated cellular cofactors, including RNA polymerase II (RNAPII) carboxyl-terminal domain kinases (TAK) [TFIIH (Cujec *et al.*, 1997; Garcia-Martinez *et al.*, 1997; Parada and Roeder, 1996) and P-TEFb (Garber *et al.*, 1998; Wei *et al.*, 1998; Zhu *et al.*, 1997); reviewed by Jones, 1997] and histone acetylases (HAT) p300/CBP and PCAF (Benkirane *et al.*, 1998; Hottiger and Nabel, 1998; Marzio *et al.*, 1998), have been recently characterized. TAK effects processive transcription of RNAPII from the HIV-1 LTR promoter (Chun and Jeang, 1996; Okamoto *et al.*, 1996; Parada and Roeder, 1996; Yang *et al.*, 1996). By contrast, one mechanistic function of Tat-associated HAT (TAH) is in the activation of chromatinized HIV-1 LTRs, presumably through the acetylation of histones (Benkirane *et al.*, 1998).

Like all retroviruses, an essential step in the life cycle of HIV-1 is the integration of proviral DNA into host-cell chromosomes. HIV-1 proviral DNAs are organized into nucleosomal forms (Sheridan *et al.*, 1997; Van Lint *et al.*, 1996; Verdin *et al.*, 1993). Thus, access to the integrated LTR-promoter is likely to be the first rate-limiting step encountered by Tat in its transcriptional activation (Jeang *et al.*, 1993). The answer as to how Tat gains entry to integrated HIV-1 LTRs has recently been addressed by four separate studies which showed a Tat-associated histone acetyltransferase (TAH) activity (Benkirane *et al.*, 1998; Hottiger and Nabel, 1998; Marzio *et al.*, 1998; Weissman *et al.*, 1998). TAH has been variously characterized

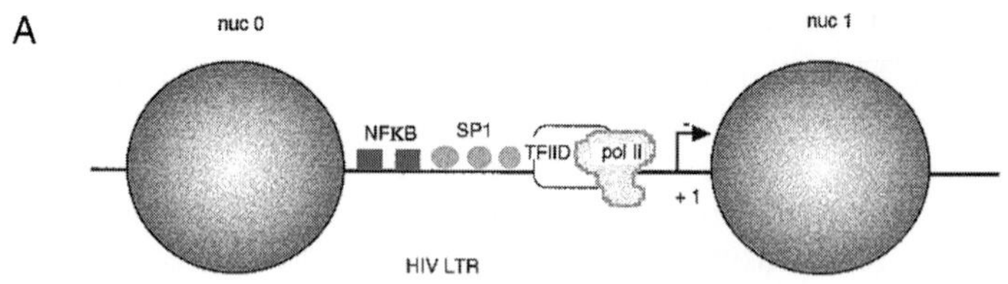
A
nuc 0
nuc 1
NFKB
SP1
TFIID
pol II
+1
HIV LTR

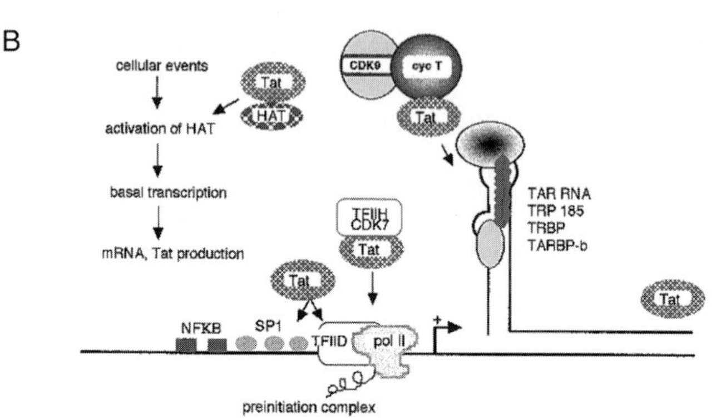
B
cellular events
activation of HAT
basal transcription
mRNA, Tat production
Tat
HAT
CDK9
cyc T
TAR RNA
TRP 185
TRBP
TARBP-b
TFIIH
CDK7
NFKB
SP1
TFIID
pol II
preinitiation complex

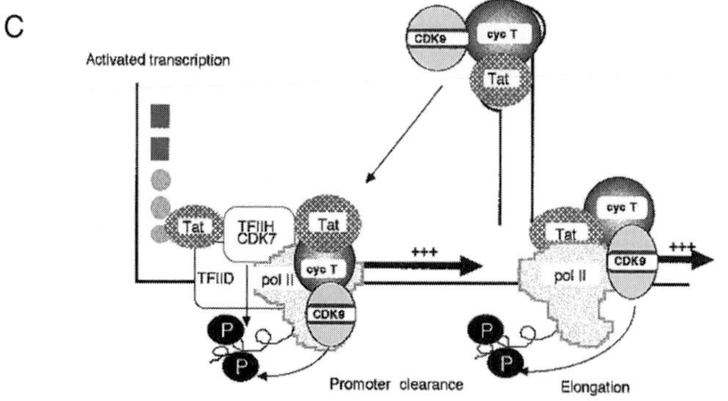
C
Activated transcription
CDK9
cyc T
Tat
TFIIH
CDK7
TFIID
pol II
+++
P
Promoter clearance
Elongation

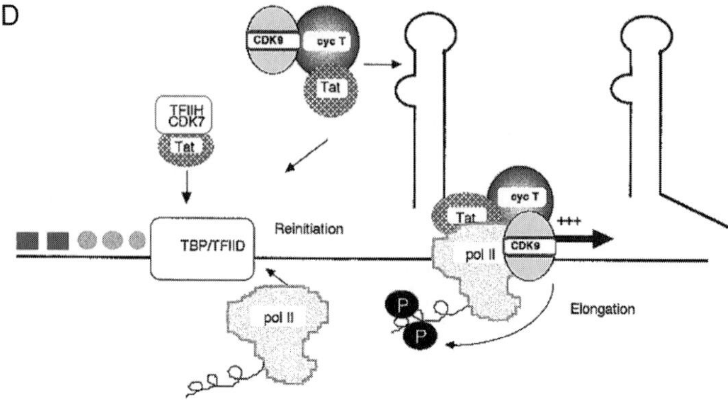
D
CDK9
cyc T
Tat
TFIIH
CDK7
TBP/TFIID
Reinitiation
pol II
+++
P
Elongation

as p300/CBP, PCAF, and/or TAF250. One envisions that only after Tat accesses an integrated LTR can promoter-proximal effects of transcriptional activation then ensue.

B. Kinases and Cyclin T

Recently, two Tat-associated kinases have been characterized. One is CDK7, which belongs to the CAK complex of TFIIH. CDK7/TFIIH has a role in the phosphorylation of the carboxy-terminal domain of the largest subunit of RNA polymerase II during its transition from initiation to promoter clearance (Cujec *et al.,* 1997; Garcia-Martinez *et al.,* 1997; McCracken *et al.,* 1997; Nekhai *et al.,* 1997; Reinberg *et al.,* 1998). Another one is CDK9 (also called TAK), which is a component of the P-TEFb complex. P-TEFb phosphorylates RNAP II CTD during transcriptional elongation (Gold *et al.,* 1998; Mancebo *et al.,* 1997; Yang *et al.,* 1997; Zhu *et al.,* 1997).

Several steps are necessary to progress from the recruitment of a preinitiation complex to transcriptional elongation. These include (1) the formation of an open promoter complex, (2) the initiation of the first phosphodiester bond followed by the incorporation of 7–14 nucleotides, (3) promoter clearance, and (4) promoter escape by a fully competent elongation complex (Holstege *et al.,* 1997; Lis, 1998; Reinberg *et al.,* 1998). Phosphorylation of RNAP II CTD is required for promoter clearance, promoter escape, and elongation. TFIIH has been implicated in triggering polymerase departure, promoter clearance, and rapid reinitiation; its activity in the preinitiation complex prevents the formation of a promoter-proximal paused complex (Kumar *et al.,* 1998).

TFIIH is released from the RNAP II complex after the synthesis of approximately the first 30 nucleotides; thus, it cannot have an effect on later steps of polymerase elongation (Chen and Zhou, 1999; Reinberg *et al.,* 1998). By contrast, CDK9 modulates the processivity of promoter distal RNAP II elongation; however, CDK9 has also been found in the preinitiation complex, which suggests that it might also have a role in promoter clearance

FIGURE 2 *In vivo* model for HIV-1 Tat trans-activation. (A) Integrated HIV-1 LTR into a chromatin structure. Nuc 0 and nuc 1 represent nucleosomal structures (Sheridan *et al.,* 1997; Van Lint *et al.,* 1996). The +1 indicates the transcriptional start site. The (−) indicates a repressed transcription. (B) Induction of basal level of transcription by chromatin disruption due to cellular events and subsequently enhanced by Tat. Basal level of transcription induces a small amount of mRNA translated to produce Tat. The different complexes represented with Tat have been described (see text). (C) Activated transcription in the presence of Tat and formation of a processive transcription complex by CTD phosphorylation of RNA pol II. (D) Increased transcriptional reinitiation rate due to fast promoter clearance and processive elongation.

(Ping and Rana, 1999). The association of CDK9 with Tat is through a ternary complex with cyclin T (Peng *et al.,* 1998; Wei *et al.,* 1998). Tat has been shown to interact with TFIIH holoenzyme, CAK complex, and recombinant CDK7 (Cujec *et al.,* 1997; Garcia-Martinez *et al.,* 1997; Parada and Roeder, 1996). It should be noted that these last protein–protein contacts have been disputed by others and therefore await further confirmation. Nonetheless, these data are compatible with a recruitment of both kinases by Tat with an activity of CDK7 on promoter clearance and CDK9 on both promoter clearance and transcriptional elongation.

Cyclin Ts (T1, T2a, T2b) were isolated based on their affinity for CDK9 (Peng *et al.,* 1998; Wei *et al.,* 1998). Cyclin T1 forms a complex with Tat and facilitates the binding of Tat to TAR RNA. The gene for cyclin T1 has been mapped to human chromosome 12. Functionally, overexpression of human cyclin T1 in rodent cells enhances Tat trans-activation by 7- to 18-fold, suggesting its role as a critical Tat cofactor (Wei *et al.,* 1998). Murine cyclin T binds to Tat more weakly or similarly than its human homolog, but it neither forms a ternary complex with TAR RNA nor restores trans-activation in rodent cells. This difference is explained by a single cysteine (human)-to-tyrosine (mouse) change (Bieniasz *et al.,* 1998; Fujinaga *et al.,* 1999; Garber *et al.,* 1998; Kwak *et al.,* 1999). Murine cells engineered to express human cyclin T1 and receptors remain incapable of supporting HIV-1 replication, suggesting that additional yet-understood virus-relevant human–mouse genetic differences exist (Garber *et al.,* 1998).

C. TRBP (TAR RNA Binding Protein)

Several cellular proteins with contributory roles in Tat-mediated trans-activation have been identified by virtue of their interaction with the LTR, Tat, or TAR RNA. These proteins act either as part of multiprotein complexes or individually with Tat to increase viral expression (Gatignol *et al.,* 1996; Jeang, 1998; Jeang *et al.,* 1999; Jones, 1997; Kingsman and Kingsman, 1996; Rana and Jeang, 1999). TRBP—the first cloned TAR RNA-binding protein (Gatignol *et al.,* 1991; Gatignol and Jeang, 1994)—is a cellular protein that belongs to the family of double-stranded RNA binding proteins with well-defined dsRNA binding domains (dsRBD) (Gatignol *et al.,* 1993; St. Johnston *et al.,* 1992). TRBP has two dsRBDs, but only one is functional for TAR binding (Daviet *et al.,* 2000). A 24-aa peptide defines the TRBP dsRBD and is wholly sufficient for binding TAR RNA. Binding by this peptide occurs with high affinity for the upper stem/loop of TAR and destabilizes the TAR RNA structure (Erard *et al.,* 1998; Gatignol *et al.,* 1993). Available structural data show overlapping binding sites between TRBP and Tat/cyclin T and predict a possible interference between the two complexes for RNA binding (Erard *et al.,* 1998; Wei *et al.,* 1998).

TRBP is encoded by the TARBP2 gene, which also maps to human chromosome 12 (Kozak *et al.,* 1995). From a functional perspective, TRBP acts in synergy with Tat to stimulate expression of the HIV-1 LTR (Gatignol *et al.,* 1991) and is associated with TAR RNA during HIV infection (Gatignol *et al.,* 1993). In murine cells, TRBP increases basal level as well as Tat trans-activation, suggesting a dual mechanism on viral expression (Battisti, Daher, and Gatignol, unpublished). Besides TAR RNA, TRBP also binds to dsRNA-induced, interferon-regulated protein kinase PKR (Blair *et al.,* 1995; Cosentino *et al.,* 1995; Meurs *et al.,* 1990). TRBP prevents the inhibitory effects of PKR on translation and on HIV replication and its expression is inhibited by interferon-α (Benkirane *et al.,* 1997; Park *et al.,* 1994). Furthermore, astrocytes, which express low levels of TRBP, replicate HIV poorly due to inefficient translation of HIV structural proteins (Gorry *et al.,* 1999). This effect is reversed by TRBP overexpression (D. Purcell, personal communication). Current data suggest that TRBP facilitates viral replication by two mechanisms: direct activation of the LTR through TAR binding and inhibition of the host antiviral (e.g., PKR) mechanisms.

D. Other Positive Effectors for Tat Activity

Other proteins have been shown to positively influence HIV LTR expression and trans-activation through Tat or TAR binding. These include components of the transcriptional machinery such as TBP (Kashanchi *et al.,* 1994; Veschambre *et al.,* 1995), Sp1 (Chun *et al.,* 1998; Jeang *et al.,* 1993; Pagtakhan and Tong-Starksen, 1997), and TAF55 (Chiang and Roeder, 1995) that interact directly with Tat. Tat–Sp1 interaction is functionally important for *in vivo* trans-activation and HIV replication through Sp1 phosphorylation (Chun *et al.,* 1998). TAP/p32 is a strong transcriptional activator that binds Tat and TFIIB (Yu *et al.,* 1995). TARBP-b is a TAR-bulge binding protein that increases trans-activation (Reddy *et al.,* 1995). TRP185 binds to the loop of TAR RNA and regulates RNA polymerase II-TAR binding (Wu-Baer *et al.,* 1995). Although these factors have been less studied because they are not deficient in rodent cells, they might contribute significantly to Tat activity *in vivo.* Tat trans-activation and virus expression in the context of the whole virus is likely to be more complex than in *in vitro* assays. Therefore the study of all components of this mechanism will contribute to the discovery of new targets for molecular intervention against the virus.

III. Therapeutic Anti-Tat Strategies

A. Tat Is Essential for HIV Replication

Tat has been long recognized as a potential anti-HIV target since its inactivation results in an abrogration of HIV transcription and replication

(Dayton *et al.*, 1986; De Clercq, 1995). Curiously, several examples of HIV that replicate despite being either Tat defective or TAR defective question the use of Tat as a therapeutic target. However, a close examination of these findings indicate that replication of Tat(-) or TAR(-) viruses occurs only in rather artificial settings (i.e., either with modified viral genomes or in cell types that are not the natural targets for HIV). Even within those parameters, observed virus replication is severely attenuated compared with that expected for wild-type virus. Thus TAR(-) viruses can replicate if (1) Tat is artificially targeted to DNA, (2) in glial cells (Taylor *et al.*, 1992; Thatikunta *et al.*, 1997), (3) in PMA-activated cells (Harrich *et al.*, 1990), or (4) in a setting of coexpression of HCMV activator (Dal Monte *et al.*, 1997). Tat(-) viruses replicate feebly if basal transcription is artificially increased by either use of a CMV enhancer (Chang and Zhang, 1995) or treatment with TNFα (Popik and Pitha, 1993). The artificial nature of these findings is supported by the fact that no naturally occurring or tissue-culture-selected infectious HIV-1 has been isolated that is inactivated for either Tat or TAR (Klaver and Berkhout, 1994; Korber *et al.*, 1997; Verhoef and Berkhout, 1999; Verhoef *et al.*, 1997). Furthermore, *in vivo* analysis of Tat evolution without artifactual constraints demonstrated a positive selection for Tat function in primary cell lines (Neuveut and Jeang, 1996). Therefore, in physiological settings, Tat is absolutely required for viral replication. It stands to reason that anti-Tat would be very useful in combination therapy.

B. Anti-Tat Compounds

Tat has been the target for the development of several active compounds; none has yet reached therapeutic use. Trans-dominant mutants, TAR RNA analogs, antisense, and ribozymes have all been used; efficacy of each has remained weak. A combination of these molecules (Aguilar-Cordova *et al.*, 1995; Chang *et al.*, 1994; Lisziewicz *et al.*, 1995) or fusion with a repressor domain (Fraisier *et al.*, 1998) could provide better results. Early screening of a large number of compounds led to the selection of anti-Tat candidates such as benzodiazepines Ro 5-3335 and Ro 24-7429 (Hsu *et al.*, 1991) and keto/enol epoxy steroids (Michne *et al.*, 1995). Clinical trials for the former have been interrupted because of their toxicity (Cupelli and Hsu, 1995), whereas the latter compounds appear more active on viral replication (Michne *et al.*, 1995). Screening based on Tat–TAR inhibition resulted in the selection of a modified nona-D-arginine compound, ALX-40C, for which clinical trials have started (Anonymous, 1993), and other compounds (Hamy *et al.*, 1997, 1998). Interestingly, however, ALX-40C was recently shown to be an inhibitor of gp120-V3 interactions and an inhibitor of the CXCR4 coreceptor but a poor inhibitor of Tat's transcriptional activity (Doranz *et al.*, 1997; O'Brien *et al.*, 1996). Other Tat inhibitors include carbocyclic adenosine analogs (De Clercq, 1998), nordihydroguaiaretic acids (Hwu *et al.*, 1998), and bisbenzimidazole derivative Hoechst 33258 (Dassonneville *et al.*, 1997). Despite the relative abundance of anti-Tat agents, none can be considered to be useful therapeutically until results are accrued from clinical trials.

C. Future Approaches

Screenings for a large number of anti-Tat compounds have mainly used two types of assays based on a Tat-dependent reporter gene in primate cells (Hsu *et al.*, 1991) and on *in vitro* Tat–TAR interactions (Hamy *et al.*, 1997). It is likely that future approaches will also use different strategies. Molecular modeling has been successively used to design HIV antiprotease drugs and other therapeutic agents. Availability of NMR data on Tat, TAR RNA, and other functional interactions will help to design anti-Tat compounds by modelization (Aboul-ela *et al.*, 1996; Gregoire and Loret, 1996; Long and Crothers, 1999). Another strategy is to use the knowledge of the molecular mechanistic events of trans-activation to identify appropriate targets and then compete with peptide analogs for cellular interactions involved in this mechanism. A very effective technique to find antagonist peptides of known proteins is the screening of phage-displayed libraries of proteins and peptides (Allen *et al.*, 1995; Cortese *et al.*, 1996; Reineke *et al.*, 1999; Wilson and Finlay, 1998). This technology would be useful in the search of inhibitors of protein–protein interactions that occur in trans-activation.

Acknowledgments

Work done in our laboratories was supported by grants from ANRS, FRM, CNRS/ARC, Ensemble contre le SIDA and INSERM/MRC (to A.G.) and from the Office of the Director of NIH (to K.T.J).

References

Aboul-ela, F., Karn, J., and Varani, G. (1996). Structure of HIV-1 TAR RNA in the absence of ligands reveals a novel conformation of the trinucleotide bulge. *Nucleic Acids Res.* **24**, 3974–3981.

Aguilar-Cordova, E., Chinen, J., Donehower, L., Harper, J., Rice, A., Butel, J., and Belmont, J. (1995). Inhibition of HIV-1 by a double transdominant fusion gene. *Gene Ther.* **2**, 181–186.

Allen, J., Walberg, M., Edwards, M., and Elledge, S. (1995). Finding prospective partners in the library: The two-hybrid system and phage display find a match. *Trends Biochem. Sci.* **20**, 511–516.

Alonso, A., Cujec, T. P., and Peterlin, B. M. (1994). Effects of human chromosome 12 on interactions between Tat and TAR of human immunodeficiency virus type 1. *J. Virol.* **68**, 6505–6513.

Alonso, A., Derse, D., and Peterlin, B. (1992). Human chromosome 12 is required for optimal interactions between Tat and TAR of human immunodeficiency virus type 1 in rodent cells. *J. Virol.* **66,** 4617–4621.

Anonymous (1993). HIV Tat inhibitor from Allelix entering clinical Trials. *Antiviral Agents Bull.* **6,** 130–131.

Barry, P. A., Pratt-Lowe, E., Unger, R. E., and Luciw, P. A. (1991). Cellular factors regulate transactivation of human immunodeficiency virus type 1. *J. Virol.* **65,** 1392–1399.

Benkirane, M., Chun, R. F., Xiao, H., Ogryzko, V. V., Howard, B. H., Nakatani, Y., and Jeang, K. T. (1998). Activation of integrated provirus requires histone acetyltransferase: p300 and P/CAF are coactivators for HIV-1 Tat. *J. Biol. Chem.* **273,** 24898–24905.

Benkirane, M., Neuveut, C., Chun, R., Smith, S., Samuel, C., Gatignol, A., and Jeang, K.-T. (1997). Oncogenic potential of TAR RNA-binding protein TRBP and its regulatory interaction with Protein Kinase PKR. *EMBO J.* **16,** 611–624.

Berkhout, B., Gatignol, A., Rabson, A. B., and Jeang, K.-T. (1990). TAR-independent activation of the HIV-1 LTR: Evidence that Tat requires specific regions of the promoter. *Cell* **62,** 757–767.

Berkhout, B., and Jeang, K.-T. (1991). Detailed analysis of TAR RNA: Critical spacing between the bulge and loop recognition domains. *Nucleic Acids Res.* **19,** 6169–6176.

Berkhout, B., and Jeang, K.-T. (1989). Trans-activation of human immunodeficiency virus type 1 is sequence specific for both the single-stranded bulge and loop of the trans-acting-responsive hairpin: A quantitative analysis. *J. Virol.* **63,** 5501–5504.

Berkhout, B., Silverman, R. H., and Jeang, K.-T. (1989). Tat trans-activates the human immunodeficiency virus through a nascent RNA target. *Cell* **59,** 273–282.

Bieniasz, P. D., Grdina, T. A., Bogerd, H. P., and Cullen, B. R. (1998). Recruitment of a protein complex containing Tat and cyclin T1 to TAR governs the species specificity of HIV-1 Tat. *EMBO J.* **17,** 7056–7065.

Blair, E., Roberts, C., Snowden, B., Gatignol, A., Benkirane, M., and Jeang, K.-T. (1995). Expression of TAR RNA-binding protein in baculovirus and co-immunoprecipitation with insect cell protein kinase. *J. Biomed. Sci.* **2,** 322–329.

Blau, J., Xiao, H., McCracken, S., O'Hare, P., Greenblatt, J., and Bentley, D. (1996). Three functional classes of transcriptional activation domain. *Mol. Cell. Biol.* **16,** 2044–2055.

Chang, H., Gendelman, R., Lisziewicz, J., Gallo, R., and Ensoli, B. (1994). Block of HIV-1 infection by a combination of antisensé tat RNA and TAR decoys: A strategy for control of HIV-1. *Gene Ther.* **1,** 208–216.

Chang, L., and Zhang, C. (1995). Infection and replication of Tat- human immunodeficiency viruses: Genetic analyses of LTR and *tat* mutations in primary and long-term human lymphoid cells. *Virology* **211,** 157–169.

Chen, D., and Zhou, Q. (1999). Tat activates human immunodeficiency virus type 1 transcriptional elongation independent of TFIIH kinase. *Mol. Cell. Biol,* **19,** 2863–2871.

Chiang, C. M., and Roeder, R. G. (1995). Cloning of an intrinsic human TFIID subunit that interacts with multiple transcriptional activators. *Science* **267,** 531–536.

Chun, R. F., and Jeang, K. T. (1996). Requirements for RNA polymerase II carboxyl-terminal domain for activated transcription of human retroviruses human T-cell lymphotropic virus I and HIV-1. *J. Biol. Chem.* **271,** 27888–27894.

Chun, R. F., Semmes, O. J., Neuveut, C., and Jeang, K. T. (1998). Modulation of Sp1 phosphorylation by human immunodeficiency virus type 1 Tat. *J. Virol.* **72,** 2615–2629.

Cortese, R., Monaci, P., Luzzago, A., Santini, C., Bartoli, F., Cortese, I., Fortugno, P., Galfre, G., Nicosia, A., and Felici, F. (1996). Selection of biologically active peptides by phage display of random peptide libraries. *Curr. Opin. Biotechnol.* **7,** 616–621.

Cosentino, G. P., Venkatesan, S., Serluca, F. C., Green, S. R., Mathews, M. B., and Sonenberg, N. (1995). Double-stranded-RNA-dependent protein kinase and TAR RNA-binding protein form homo- and heterodimers *in vivo*. *Proc. Natl. Acad. Sci. USA* **92,** 9445–9449.

Cujec, T. P., Cho, H., Maldonado, E., Meyer, J., Reinberg, D., and Peterlin, B. M. (1997). The human immunodeficiency virus transactivator Tat interacts with the RNA polymerase II holoenzyme. *Mol. Cell. Biol.* **17,** 1817–1823.

Cujec, T. P., Okamoto, H., Fujinaga, K., Meyer, J., Chamberlin, H., Morgan, D. O., and Peterlin, B. M. (1997). The HIV transactivator TAT binds to the CDK-activating kinase and activates the phosphorylation of the carboxy-terminal domain of RNA polymerase II. *Genes Dev.* **11,** 2645–2657.

Cupelli, L., and Hsu, M. (1995). The human immunodeficiency virus type 1 Tat antagonist, Ro 5-3335, predominantly inhibits transcription initiation from the viral promoter. *J. Virol.* **69,** 2640–2643.

Dahmus, M. E. (1996). Reversible phosphorylation of the C-terminal domain of RNA polymerase II. *J. Biol. Chem.* **271,** 19009–19012.

Dal Monte, P., Landini, M. P., Sinclair, J., Virelizier, J. L., and Michelson, S. (1997). TAR and Sp1-independent transactivation of HIV long terminal repeat by the Tat protein in the presence of human cytomegalovirus IE1/IE2. *AIDS* **11,** 297–303.

Dassonneville, L., Hamy, F., Colson, P., Houssier, C., and Bailly, C. (1997). Binding of Hoechst 33258 to the TAR RNA of HIV-1: Recognition of a pyrimidine bulge-dependent structure. *Nucleic Acids Res.* **25,** 4487–4492.

Daviet, L., Bois, F., Battisti, P.-L., and Gatignol, A. (1998). Identification of limiting steps for efficient trans-activation of HIV-1 promoter by Tat in *Saccharomyces cerevisiae. J. Biol. Chem.* **273,** 28219.

Daviet, L., Erard, M., Dorin, D., Duarte, M., Vaquero, C., and Gatignol, A. (2000). The analysis of a binding difference between the two dsRNA-binding domains in TRBP reveals the modular function of a KR-helix motif. *Eur. J. Bioch.* **267,** 2419–2431.

Dayton, A. I., Sodroski, J., Rosen, C., Goh, W., and Haseltine, W. (1986). The trans-activator gene of the human T cell lymphotropic virus type III is required for replication. *Cell* **44,** 941–947.

De Clercq, E. (1995). Antiviral therapy for human immunodeficiency virus infections. *Clin. Microbiol. Rev.* **8,** 200–239.

De Clercq, E. (1998). Carbocyclic adenosine analogues as *S*-adenosylhomocysteine hydrolase inhibitors and antiviral agents: Recent advances. *Nucleosides Nucleotides* **17,** 625–634.

Delling, U., Roy, S., Sumner-Smith, M., Barnett, R., Reid, L., Rosen, C. A., and Sonenberg, N. (1991). The number of positively charged amino acids in the basic domain of Tat is critical for trans-activation and complex formation with TAR RNA. *Proc. Natl. Acad. Sci. USA* **88,** 6234–6238.

Derse, D., Carvalho, M., Carroll, R., and Peterlin, B. M. (1991). A minimal lentivirus Tat. *J. Virol.* **65,** 7012–7015.

Dingwall, C., Ernberg, I., Gait, M., Green, S., Heaphy, S., Karn, J., Lowe, A., Singh, M., and Skinner, M. (1990). HIV-1 tat protein stimulates transcription by binding to a U-rich bulge in the stem of the TAR RNA structure. *EMBO J.* **9,** 4145–4153.

Doranz, B. J., Grovit-Ferbas, K., Sharron, M. P., Mao, S. H., Goetz, M. B., Daar, E. S., Doms, R. W., and O'Brien, W. A. (1997). A small-molecule inhibitor directed against the chemokine receptor CXCR4 prevents its use as an HIV-1 coreceptor. *J. Exp. Med.* **186,** 1395–1400.

Erard, M., Barker, D., Amalric, F., Jeang, K.-T., and Gatignol, A. (1998). An Arg/Lys-rich core peptide mimics TRBP binding to the HIV-1 TAR RNA upper-stem/loop. *J. Mol. Biol.* **279,** 1085–1099.

Feinberg, M., Baltimore, D., and Frankel, A. (1991). The role of Tat in the human immunodeficiency virus life cycle indicates a primary effect on transcriptional elongation. *Proc. Natl. Acad. Sci. USA* **88,** 4045–4049.

Feng, S., and Holland, E. C. (1988). HIV-1 tat trans-activation requires the loop sequence within tar. *Nature* **334,** 165–167.

Fraisier, C., Abraham, D., van Oijen, M., Cunliffe, V., Irvine, A., Craig, R., and Dzierzak, E. (1998). Inhibition of Tat-mediated transactivation and HIV replication with Tat mutant and repressor domain fusion proteins. *Gene Ther.* **5**, 946–954.

Fujinaga, K., Taube, R., Wimmer, J., Cujec, T. P., and Peterlin, B. M. (1999). Interactions between human cyclin T, Tat, and the transactivation response element (TAR) are disrupted by a cysteine to tyrosine substitution found in mouse cyclin T. *Proc. Natl. Acad. Sci. USA* **96**, 1285–1290.

Garber, M. E., Wei, P., KewalRamani, V., Mayall, T. P., Herrmann, C., Rice, A. P., Littman, D. R., and Jones, K. (1998). The interaction between HIV-1 Tat and human cyclin T1 requires zinc and a critical cysteine residue that is not conserved in the murine CycT1 protein. *Genes Dev.* **12**, 3512–3527.

García-Martinez, L. F., Ivanov, D., and Gaynor, R. B. (1997). Association of Tat with purified HIV-1 and HIV-2 transcription preinitiation complexes. *J. Biol. Chem.* **272**, 6951–6958.

Garcia-Martinez, L. F., Mavankal, G., Neveu, J. M., Lane, W. S., Ivanov, D., and Gaynor, R. B. (1997). Purification of a Tat-associated kinase reveals a TFIIH complex that modulates HIV-1 transcription. *EMBO J.* **16**, 2836–2850.

Gatignol, A., Buckler, C., and Jeang, K.-T. (1993). Relatedness of an RNA binding motif in HIV-1 TAR RNA binding protein TRBP to human P1/dsI kinase and *Drosophila staufen. Mol. Cell. Biol.* **13**, 2193–2202.

Gatignol, A., Buckler-White, A., Berkhout, B., and Jeang, K.-T. (1991). Characterization of a human TAR RNA-binding protein that activates the HIV-1 LTR. *Science* **251**, 1597–1600.

Gatignol, A., Duarte, M., Daviet, L., Chang, Y.-N., and Jeang, K.-T. (1996). Sequential steps in Tat trans-activation of HIV-1 mediated through cellular DNA, RNA, and protein binding factors. *Gene Expr.* **5**, 217–228.

Gatignol, A., and Jeang, K.-T. (1994). Expression cloning of genes encoding RNA binding proteins. *In* "Methods in Molecular Genetics: Molecular Virology Techniques A" (K. W. Adolph, Ed.), pp. 18–28. Academic Press, San Diego, CA.

Gold, M., Yang, X., Herrmann, C., and Rice, A. (1998). PITALRE, the catalytic subunit of TAK, is required for human immunodeficiency virus Tat transactivation *in vivo. J. Virol.* **72**, 4448–4453.

Gorry, P. R., Howard, J. L., Churchill, M. J., Anderson, J. L., Cunningham, A., Adrian, D., McPhee, D. A., and Purcell, D. F. (1999). Diminished production of human immunodeficiency virus type 1 in astrocytes results from inefficient translation of gag, env, and nef mRNAs despite efficient expression of Tat and Rev. *J. Virol.* **73**, 352–361.

Green, M. R. (1993). Molecular mechanisms of Tat and Rev. *AIDS Res. Rev.* **3**, 41–55.

Gregoire, C., and Loret, E. (1996). Conformational heterogeneity in two regions of TAT results in structural variations of this protein as a function of HIV-1 isolates. *J. Biol. Chem.* **271**, 22641–22646.

Hahn, S. (1998). Activation and the role of reinitiation in the control of transcription by RNA polymerase II. *Cold Spring Harb. Symp. Quant. Biol.* **63**, 181–188.

Hamy, F., Brondani, V., Florsheimer, A., Stark, W., Blommers, M. J., and Kilmkait, T. (1998). A new class of HIV-1 Tat antagonist acting through Tat-TAR inhibition. *Biochemistry* **37**, 5086–5095.

Hamy, F., Felder, E. R., Heizmann, G., Lazdins, J., Aboul-ela, F., Varani, G., Karn, J., and Klimkait, T. (1997). An inhibitor of the Tat/TAR RNA interaction that effectively suppresses HIV-1 replication. *Proc. Natl. Acad. Sci. USA* **94**, 3548–3553.

Harrich, D., Garcia, J., Mitsuyasu, R., and Gaynor, R. (1990). TAR independent activation of the human immunodeficiency virus in phorbol ester stimulated T lymphocytes. *EMBO J.* **9**, 4417–4423.

Hart, C., Galphin, J., Westhafer, M., and Schochetman, G. (1993). TAR loop-dependant human immunodeficiency virus trans activation requires factors encoded on human chromosome 12. *J. Virol.* **67**, 5020–5024.

Hart, C., Ou, C.-Y., Galphin, J., Moore, J., Bacheler, L., Wasmuth, J., Petteway, S., and Schochetman, G. (1989). Human chromosome 12 is required for elevated HIV-1 expression in human–hamster hybrid cells. *Science* **246,** 488–491.

Herrmann, C. H., and Rice, A. P. (1995). Lentivirus Tat proteins specifically associate with a cellular protein kinase, TAK, that hyperphosphorylates the carboxyl-terminal domain of the large subunit of RNA polymerase II: Candidate for a Tat cofactor. *J. Virol.* **69,** 1612–1620.

Holstege, F. C., Fiedler, U., and Timmers, H. T. (1997). Three transitions in the RNA polymerase II transcription complex during initiation. *EMBO J.* **16,** 7468–7480.

Hottiger, M. O., and Nabel, G. J. (1998). Interaction of human immunodeficiency virus type 1 Tat with the transcriptional coactivators p300 and CREB binding protein. *J. Virol.* **72,** 8252–8256.

Hsu, M.-C., Schmutt, A. D., Holly, M., Slice, L. W., Sherman, M. I., Richman, D. D., Potash, M. J., and Volsky, D. J., (1991). Inhibition of HIV replication in acute and chronic infections *in vitro* by a Tat antagonist. *Science* **254,** 1799–1802.

Hwu, J. R., Tseng, W. N., Gnabre, J., Giza, P., and Huang, R. C. (1998). Antiviral activities of methylated nordihydroguaiaretic acids. 1. Synthesis, structure identification, and inhibition of tat-regulated HIV transactivation. *J. Med. Chem.* **41,** 2994–3000.

Jakobovits, A., Smith, D. H., Jakobovits, E. B., and Capon, D. J. (1988). A discrete element 3′ of human immunodeficiency virus 1 (HIV-1) and HIV-2 mRNA initiation sites mediates transcriptional activation by an HIV trans activator. *Mol. Cell. Biol.* **8,** 2555–2561.

Jeang, K., Berkhout, B., and Dropulic, B. (1993). Effects of integration and replication on transcription of the HIV-1 long terminal repeat. *J. Biol. Chem.* **268,** 24940–24949.

Jeang, K.-T. (1998). Tat, Tat-associated kinase, and transcription. *J. Biomed. Sci.* **5,** 24–27.

Jeang, K.-T., and Berkhout, B. (1992). Kinetics of HIV-1 long terminal repeat trans-activation. Use of intragenic ribozyme to assess rate-limiting steps. *J. Biol. Chem.* **267,** 17891–17899.

Jeang, K.-T., Chun, R., Lin, N. H., Gatignol, A., Glabe, C. G., and Fan, H. (1993). *In vitro* and *in vivo* binding of human immunodeficiency virus type 1 tat protein and Sp1 transcription factor. *J. Virol.* **67,** 6224–6233.

Jeang, K.-T., Shank, P. R., and Kumar, A. (1988). Transcriptional activation of homologous viral long terminal repeats by the human immunodeficiency virus type 1 or the human T-cell leukemia virus type I tat proteins occurs in the absence of *de novo* protein synthesis. *Proc. Natl. Acad. Sci. USA* **85,** 8291–8295.

Jeang, K.-T., Xiao, H., and Rich, E. (1999). Multifaceted activities of the human immunodeficiency virus type 1 transactivator of transcription, Tat. *J. Biol. Chem.* **274,** 28837–28840.

Jones, K., and Peterlin, B. (1994). Control of RNA initiation and elongation at the HIV-1 promoter. *Annu. Rev. Biochem.* **63,** 717–743.

Jones, K. A. (1997). Taking a new TAK on tat transactivation. *Genes Dev.* **11,** 2593–2599.

Kamine, J., Subramanian, T., and Chinnadurai, G. (1993). Activation of a heterologous promoter by HIV 1 Tat requires Sp1 and is distinct from the mode of activation by acidic transcriptional activators. *J. Virol.* **67,** 6828–6834.

Kamine, J., Subramanian, T., and Chinnadurai, G. (1991). Sp1-dependent activation of a synthetic promoter by human immunodeficiency virus type 1 Tat protein. *Proc. Natl. Acad. Sci. USA* **88,** 8510–8514.

Kao, S., Calman, A., Luciw, P., and Peterlin, B. (1987). Anti-termination of transcription within the long terminal repeat of HIV-1 by tat gene product. *Nature* **330,** 489–493.

Kashanchi, F., Piras, G., Radonovich, M. F., Duvall, J. F., Fattaey, A., Chiang, C.-M., Roeder, R. G., and Brady, J. N. (1994). Direct interaction of human TFIID with the HIV-1 transactivator Tat. *Nature* **367,** 295–299.

Keen, N. J., Churcher, M. J., and Karn, J. (1997). Transfer of Tat and release of TAR RNA during the activation of the human immunodeficiency virus type-1 transcription elongation complex. *EMBO J.* **16,** 5260–5272.

Keen, N. J., Gait, M. J., and Karn, J. (1996). Human immunodeficiency virus type-1 Tat is an integral component of the activated transcription-elongation complex. *Proc. Natl. Acad. Sci. USA* **93**, 2505–2510.

Kim, Y. S., and Risser, R. (1993). TAR-independent transactivation of the murine cytomegalovirus major immediate-early promoter by the Tat protein. *J. Virol.* **67**, 239–248.

Kingsman, S. M., and Kingsman, A. J. (1996). The regulation of human immunodeficiency virus type-1 gene expression. *Eur. J. Biochem.* **240**, 491–507.

Klaver, B., and Berkhout, B. (1994). Evolution of a disrupted TAR RNA hairpin structure in the HIV-1 virus. *EMBO J.* **13**, 2650–2659.

Korber, B., Hahn, B., Foley, B., Mellors, J., Leitner, T., Myers, G., McCutchan, F., and Kuiken, C. (1997). "A Compilation and Analysis of Nucleic Acid and Amino Acid Sequences" (B. Korber, B. Hahn, B. Foley, J. W. Mellors, T. Leitner, G. Myers, F. McCutchan, and C. Kuiken, Eds.). Los Alamos, NM: Los Alamos National Laboratory.

Kozak, C., Gatignol, A., Graham, K., Jeang, K.-T., and McBride, O. (1995). Genetic mapping in human and mouse of the locus encoding TRBP, a protein that binds the TAR region of the human immunodeficiency virus (HIV-1). *Genomics* **25**, 66–72.

Kumar, K. P., Akoulitchev, S., and Reinberg, D. (1998). Promoter-proximal stalling results from the inability to recruit transcription factor IIH to the transcription complex and is a regulated event. *Proc. Natl. Acad. Sci. USA* **95**, 9767–9772.

Kwak, Y., Ivanov, D., Guo, J., Nee, E., and Gaynor, R. (1999). Role of the human and murine cyclin T proteins in regulating HIV-1 tat-activation. *J. Mol. Biol.* **288**, 57–69.

Laspia, M., Rice, A., and Mathews, M. (1989). HIV-1 tat protein increases transcriptional initiation and stabilizes elongation. *Cell* **59**, 283–292.

Lis, J. (1998). Promoter-associated pausing in promoter architecture and postinitiation transcriptional regulation. *Cold Spring Harb. Symp. Quant. Biol.* **63**, 347–356.

Lisziewicz, J., Sun, D., Trapnell, B., Thomson, M., Chang, H., Ensoli, B., and Peng, B. (1995). An autoregulated dual-function antitat gene for human immunodeficiency virus type 1 gene therapy. *J. Virol.* **69**, 206–212.

Long, K., and Crothers, D. (1999). Characterization of the solution conformations of unbound and tat peptide-bound forms of HIV-1 TAR RNA. *Biochemistry* **38**, 10059–10069.

Mancebo, H. S., Lee, G., Flygare, J., Tomassini, J., Luu, P., Zhu, Y., Peng, J., Blau, C., Hazuda, D., Price, D., and Flores, O. (1997). P-TEFb kinase is required for HIV Tat transcriptional activation *in vivo* and *in vitro*. *Genes Dev.* **11**, 2633–2644.

Marzio, G., Tyagi, M., Gutierrez, M. I., and Giacca, M. (1998). HIV-1 tat transactivator recruits p300 and CREB-binding protein histone acetyltransferases to the viral promoter. *Proc. Natl. Acad. Sci. USA* **95**, 13519–13524.

McCracken, S., Fong, N., Yankulov, K., Ballantyne, S., Pan, G., Greenblatt, J., Patterson, S. D., Wickens, M., and Bentley, D. L. (1997). The C-terminal domain of RNA polymerase II couples mRNA processing to transcription. *Nature* **385**, 357–361.

Meurs, E., Chong, K., Galabru, J., Thomas, N., Kerr, I., Williams, B., and Hovanessian, A. (1990). Molecular cloning and characterization of the human double-stranded RNA-activated protein kinase induced by interferon. *Cell* **62**, 379–390.

Michne, W., Schroeder, J., Bailey, T., Neumann, H., Cooke, D., Young, D., Hughes, J., Kingsley, S., KA, R., Putz, H., Shaw, L., and Dutko, F. (1995). Keto/enol epoxy steroids as HIV-1 Tat inhibitors: Structure–activity relationships and pharmacophore localization. *J. Med. Chem.* **38**, 3197–3206.

Nekhai, S., Shukla, R. R., and Kumar, A. (1997). A human primary T-lymphocyte-derived human immunodeficiency virus type 1 Tat-associated kinase phosphorylates the C-terminal domain of RNA polymerase II and induces CAK activity. *J. Virol.* **71**, 7436–7441.

Neuveut, C., and Jeang, K.-T. (1996). Recombinant human immunodeficiency virus type 1 genomes with *tat* unconstrained by overlapping reading frames reveal residues in Tat important for replication in tissue culture. *J. Virol.* **70**, 5572–5581.

Newstein, M., Lee, I. S., Venturini, D., and Shank, P. (1993). A chimeric human immunodeficiency virus type 1 TAR region which mediates high level trans-activation in both rodent and human cells. *Virology* **197**, 825–828.

Newstein, M., Stanbridge, E., Casey, G., and Shank, P. (1990). Human chromosome 12 encodes a species-specific factor which increases human immunodeficiency virus type 1 tat-mediated trans activation in rodent cells. *J. Virol.* **64**, 4565–4567.

O'Brien, W. A., Sumner-Smith, M., Mao, S. H., Sadeghi, S., Zhao, J. Q., and Chen, I. S. (1996). Anti-human immunodeficiency virus type 1 activity of an oligocationic compound mediated via gp120 V3 interactions. *J. Virol.* **70**, 2825–2831.

Okamoto, H., Sheline, C. T., Corden, J. L., Jones, K. A., and Peterlin, B. M. (1996). Trans-activation by human immunodeficiency virus Tat protein requires the C-terminal domain of RNA polymerase II. *Proc. Natl. Acad. Sci. USA* **93**, 11575–11579.

Pagtakhan, A. S., and Tong-Starksen, S. E. (1997). Interactions between Tat of HIV-2 and transcription factor Sp1. *Virology* **238**, 221–230.

Parada, C. A., and Roeder, R. G. (1996). Enhanced processivity of RNA polymerase II triggered by Tat-induced phosphorylation of its carboxy-terminal domain. *Nature* **384**, 375–378.

Park, H., Davies, M., Langland, J., Chang, H.-W., Nam, Y. S., Tartaglia, J., Paoletti, E., Jacobs, B., Kaufman, R., and Venkatesan, S. (1994). TAR RNA-binding protein is an inhibitor of the interferon-induced protein kinase PKR. *Proc. Natl. Acad. Sci. USA* **91**, 4713–4717.

Pendergrast, P. S., and Hernandez, N. (1997). RNA-targeted activators, but not DNA-targeted activators, repress the synthesis of short transcripts at the human immunodeficiency virus type 1 long terminal repeat. *J. Virol.* **71**, 910–917.

Peng, J., Zhu, Y., Milton, J. T., and Price, D. H. (1998). Identification of multiple cyclin subunits of human P-TEFb. *Genes Dev.* **12**, 755–762.

Pessler, F., and Hernandez, N. (1998). The HIV-1 inducer of short transcripts activates the synthesis of 5,6-dichloro-1-beta-D-benzimidazole-resistant short transcripts *in vitro*. *J. Biol. Chem.* **273**, 5375–5384.

Ping, Y. H., and Rana, T. M. (1999). Tat-associated kinase (P-TEFb): A component of transcription preinitiation and elongation complexes. *J. Biol. Chem.* **274**, 7399–7404.

Popik, W., and Pitha, P. (1993). Role of tumor necrosis factor alpha in activation and replication of the tat-defective human immunodeficiency virus type 1. *J. Virol.* **67**, 1094–1099.

Rana, T., and Jeang, K.-T. (1999). Biochemical and functional interactions between HIV-1 Tat protein and TAR RNA. *Arch. Biochem. Biophys.* **365**, 175–85.

Reddy, T. R., Suhasini, M., Rappaport, J., Looney, D. J., Kraus, G., and Wong-Staal, F. (1995). Molecular cloning and characterization of a TAR-binding nuclear factor from T cells. *AIDS Res. Hum. Retroviruses* **11**, 663–669.

Reinberg, D., Orphanides, G., Ebright, R., Akoulitchev, S., Carcamo, J., Cho, H., Cortes, P., Drapkin, R., Flores, O., Ha, I., Inostroza, J. A., Kim, S., Kim, T. K., Kumar, P., Lagrange, T., LeRoy, G., Lu, H., Ma, D. M., Maldonado, E., Merino, A., Mermelstein, F., Olave, I., Sheldon, M., Shiekhattar, R., Stone, N., Sun, X., Weis, L., Yeung, K., and Zawel, L. (1998). The RNA polymerase II general transcription factors: past, present, and future. *Cold Spring Harb. Symp. Quant. Biol.* **63**, 83–103.

Reineke, U., Sabat, R., Misselwitz, R., Welfle, H., Volk, H., and Schneider-Mergener, J. (1999). A synthetic mimic of a discontinuous binding site on interleukin-10. *Nat. Biotechnol.* **17**, 271–275.

Rounseville, M. P., Lin, H. C., Agbottah, E., Shukla, R. R., Rabson, A. B., and Kumar, A. (1996). Inhibition of HIV-1 replication in viral mutants with altered TAR RNA stem structures. *Virology* **216**, 411–417.

Roy, S., Parkin, N., Rosen, C., Itovitch, J., and Sonenberg, N. (1990). Structural requirements for trans-activation of human immunodeficiency virus type 1 long terminal repeat-directed gene expression by tat: Importance of base pairing, loop sequence, and bulges in the tat-responsive sequence. *J. Virol.* **64**, 1402–1406.

Seigel, L. J., Ratner, L., Josephs, S. F., Derse, D., Feinberg, M. B., Reyes, G. R., O'Brien, S. J., and Wong-Staal, F. (1986). Transactivation induced by human T-lymphotropic virus type III (HTLV III) maps to a viral sequence encoding 58 amino acids and lacks tissue specificity. *Virology* **148**, 226–231.

Selby, M., Bain, E., Luciw, P., and Peterlin, B. (1989). Structure, sequence, and position of the stem-loop in tar determine transcriptional elongation by tat through the HIV-1 long terminal repeat. *Genes Dev.* **3**, 547–558.

Selby, M., and Peterlin, B. (1990). Trans-activation by HIV-1 Tat via a heterologous RNA binding protein. *Cell* **62**, 769–776.

Sheldon, M., Ratnasabapathy, R., and Hernandez, N. (1993). Characterization of the inducer of short transcripts, a human immunodeficiency virus type 1 transcriptional element that activates the synthesis of short RNAs. *Mol. Cell. Biol.* **13**, 1251–1263.

Sheridan, P., Mayall, T., Verdin, E., and Jones, K. (1997). Histone acetyltransferases regulate HIV-1 enhancer activity *in vitro*. *Genes Dev.* **11**, 3327–3340.

Southgate, C., Zapp, M. L., and Green, M. R. (1990). Activation of transcription by HIV-1 Tat protein tethered to nascent RNA through another protein. *Nature* **345**, 640–642.

Southgate, C. D., and Green, M. R. (1995). Delineating minimal protein domains and promoter elements for transcriptional activation by lentivirus Tat proteins. *J. Virol.* **69**, 2605–2610.

Southgate, C. D., and Green, M. R. (1991). The HIV-1 Tat protein activates transcription from an upstream DNA-binding site: Implications for Tat function. *Genes Dev.* **5**, 2496–2507.

St Johnston, D., Brown, N. H., Gall, J. G., and Jantsch, M. (1992). A conserved double-stranded RNA-binding domain. *Proc. Natl. Acad. Sci. USA* **89**, 10979–10983.

Taylor, J., Pomerantz, R., Bagasra, O., Chowdhury, M., Rappaport, J., Khalili, K., and Amini, S. (1992). TAR-independent transactivation by Tat in cells derived from the CNS: A novel mechanism of HIV-1 gene regulation. *EMBO J.* **11**, 3395–3403.

Thatikunta, P., Sawaya, B. E., Denisova, L., Cole, C., Yusibova, G., Johnson, E. M., Khalili, K., and Amini, S. (1997). Identification of a cellular protein that binds to Tat-responsive element of TGF beta-1 promoter in glial cells. *J. Cell. Biochem.* **67**, 466–477.

Toyama, R., Bende, S. M., and Dhar, R. (1992). Transcriptional activity of the human immunodeficiency virus-1 LTR promoter in fission yeast *Schizosaccharomyces pombe*. *Nucleic Acids Res.* **20**, 2591–2596.

Van Lint, C., Emiliani, S., Ott, M., and Verdin, E. (1996). Transcriptional activation and chromatin remodeling of the HIV-1 promoter in response to histone acetylation. *EMBO J.* **15**, 1112–1120.

Verdin, E., Paras, J., and Van Lint, C. (1993). Chromatin disruption in the promoter of human immunodeficiency virus type 1 during transcriptional activation. *EMBO J.* **12**, 3249–3259.

Verhoef, K., and Berkhout, B. (1999). A second-site mutation that restores replication of a Tat-defective human immunodeficiency virus. *J. Virol.* **73**, 2781–2789.

Verhoef, K., Tijms, M., and Berkhout, B. (1997). Optimal Tat-mediated activation of the HIV-1 LTR promoter requires a full-length TAR RNA hairpin. *Nucleic Acids Res.* **25**, 496–502.

Veschambre, P., Simard, P., and Jalinot, P. (1995). Evidence for functional interaction between the HIV-1 Tat transactivator and the TATA box binding protein *in vivo*. *J. Mol. Biol.* **250**, 169–180.

Wei, P., Garber, M. E., Fang, S. M., Fischer, W. H., and Jones, K. A. (1998). A novel CDK9-associated C-type cyclin interacts directly with HIV-1 Tat and mediates its high-affinity, loop-specific binding to TAR RNA. *Cell* **92**, 451–462.

Weissman, J. D., Brown, J. A., Howcroft, T. K., Hwang, J., Chawla, A., Roche, P. A., Schiltz, L., Nakatani, Y., and Singer, D. S. (1998). HIV-1 tat binds TAFII250 and represses TAFII250-dependent transcription of major histocompatibility class I genes. *Proc. Natl. Acad. Sci. USA* **95**, 11601–11606.

Wilson, D., and Finlay, B. (1998). Phage display: applications, innovations, and issues in phage and host biology. *Can. J. Microbiol.* **44**, 313–329.

Wu-Baer, F., Lane, W. S., and Gaynor, R. B. (1995). The cellular factor TRP-185 regulates RNA polymerase II binding to HIV-1 TAR RNA. *EMBO J.* **14,** 5995–6009.

Wu-Baer, F., Sigman, D., and Gaynor, R. B. (1995). Specific binding of RNA polymerase II to the human immunodeficiency virus trans-activating region RNA is regulated by cellular cofactors and Tat. *Proc. Natl. Acad. Sci. USA* **92,** 7153–7157.

Xiao, H., Lis, J. T., and Jeang, K.-T. (1997). Promoter activity of Tat at steps subsequent to TATA-binding protein recruitment. *Mol. Cell. Biol.* **17,** 6898–6905.

Yang, X., Gold, M. O., Tang, D. N., Lewis, D. E., Aguilar-Cordova, E., Rice, A. P., and Herrmann, C. H. (1997). TAK, an HIV Tat-associated kinase, is a member of the cyclin-dependent family of protein kinases and is induced by activation of peripheral blood lymphocytes and differentiation of promonocytic cell lines. *Proc. Natl. Acad. Sci. USA* **94,** 12331–12336.

Yang, X., Herrmann, C. H., and Rice, A. P. (1996). The human immunodeficiency virus Tat proteins specifically associate with TAK *in vivo* and require the carboxyl-terminal domain of RNA polymerase II for function. *J. Virol.* **70,** 4576–4584.

Yankulov, K., and Bentley, D. (1998). Transcriptional control: Tat cofactors and transcriptional elongation. *Curr. Biol.* **8,** R447–449.

Yu, L., Loewenstein, P. M., Zhang, Z., and Green, M. (1995). In vitro interaction of the human immunodeficiency virus type 1 Tat transactivator and the general transcription factor TFIIB with the cellular protein TAP. *J. Virol.* **69,** 3017–3023.

Yu, L., Zhang, Z., Loewenstein, P. M., Desai, K., Tang, Q., Mao, D., Symington, J. S., and Green, M. (1995). Molecular cloning and characterization of a cellular protein that interacts with the human immunodeficiency virus type 1 Tat transactivator and encodes a strong transcriptional activation domain. *J. Virol.* **69,** 3007–3016.

Zhou, Q., and Sharp, P. A. (1995). Novel mechanism and factor for regulation by HIV-1 Tat. *EMBO J.* **14,** 321–328.

Zhou, Q., and Sharp, P. A. (1996). Tat-SF1: cofactor for stimulation of transcriptional elongation by HIV-1 Tat. *Science* **274,** 605–610.

Zhu, Y., Pe'ery, T., Peng, J., Ramanathan, Y., Marshall, N., Marshall, T., Amendt, B., Mathews, M. B., and Price, D. H. (1997). Transcription elongation factor P-TEFb is required for HIV-1 tat transactivation in vitro. *Genes Dev.* **11,** 2622–2632.

[illegible] and [illegible] regulation [illegible] *Science* [illegible]

[illegible] and [illegible] (1991). [illegible]

[illegible] acetylation and [illegible] regulate mating-type switching and [illegible] the homeodomain [illegible] with [illegible] RNA [illegible] *Proc. Natl. Acad. Sci.* [illegible]

[illegible] and [illegible] (1991). Polypyrimidine tract [illegible]

[illegible]

[illegible] family of proteins [illegible]

[illegible] and [illegible]

[illegible]

Douglas Noonan and Adriana Albini

Istituto Nazionale per la Ricerca sul Cancro
16132 Genova, Italy

From the Outside In: Extracellular Activities of HIV Tat

I. Introduction

The HIV transactivator (Tat) protein is an accessory protein whose principal function appears to be the trans-activation of the HIV LTR. Tat also transactivates several cellular genes, and it has been demonstrated that Tat plays a critical function in cytopathogenicity that is independent of HIV-LTR transactivation (Huang *et al.*, 1994).

In addition to the nuclear localization and function of the HIV-1 Tat protein, it has been shown that Tat is released into the extracellular environment. A wide range of activities have been attributed to the Tat protein found extracellularly. The most studied Tat protein is based on the HIV-1 LAI/HB10 group of laboratory HIV strains, an 86-amino-acid protein. However, while amino acids 1–86 of this Tat are representative of the Tat sequence of many HIV isolates, most Tat proteins from primary isolates contain an additional C-terminal 15–16 amino acids whose function has

Advances in Pharmacology, Volume 48

1054-3589/00 $35.00

not yet been sufficiently investigated (Jeang, 1994). These additional amino acids bring the sequence of HIV-1 Tat closer to that of HIV-2 Tat (Fig. 1). In addition, like most HIV proteins Tat does show sequence variability from isolate to isolate. In an alignment of numerous Tat proteins, the most conserved domains are the Cysteine-rich and core domains, followed by the basic domain, the N-terminus, and the C-terminus (Jeang, 1994).

A. Tat: On the Way Out

Several studies have shown that the HIV-1 Tat protein can exit from cells which produce it (Fig. 2). While this has been particularly well documented in transfected cells (Ensoli, 1990; 1993; Chang, 1997; Rubartelli and Sitia, 1997; Albini *et al.*, 1998; Milani *et al.*, 1993; Zauli *et al.*, 1993, 1995) it has also been shown that Tat is released from HIV-infected cells at significant levels (Westendorp *et al.*, 1995). As the Tat gene does not encode a signal

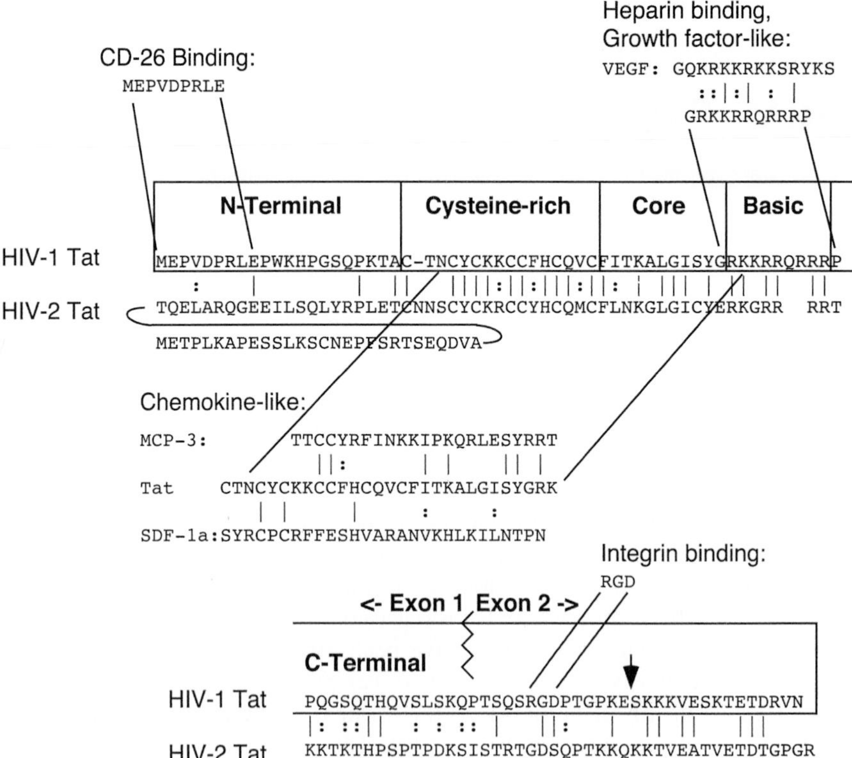

FIGURE 1 Domain structure, extracellular functional domains, and alignment of HIV-1 and HIV-2 Tat sequences.

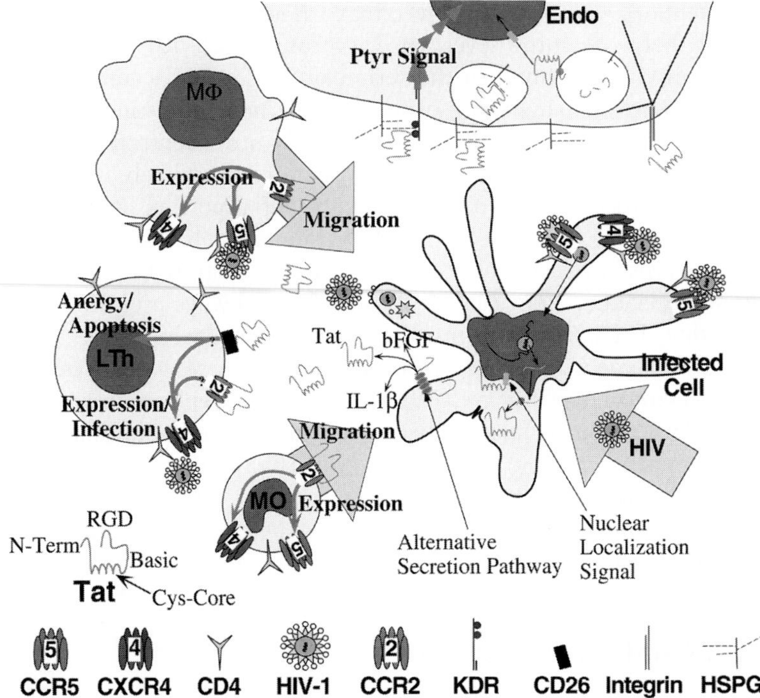

FIGURE 2 Extracellular Tat: putative cell surface receptors for Tat and their possible roles in Tat-associated pathogenesis.

peptide, the release of HIV-Tat has been suggested to occur via an alternative secretion pathway (Chang *et al.*, 1997; Rubartelli and Sitia, 1997). Such alternative secretion pathways have been demonstrated for the cytokines IL-1β and bFGF as well as for others (Rubartelli and Sitia, 1997). These pathways appear to be important for the secretion of proteins without first exposing them to to the redox conditions in the endoplasmic reticulum. This may be particularly crucial for proteins containing unpaired cysteines, which is typical of the Tat protein.

Tat appears to be able to exit from transfected cells in the absence of apoptosis (Chang *et al.*, 1997); the Tat found extracellularly appears to be intact (Chang *et al.*, 1997; Albini *et al.*, 1998). Most critical is that the released Tat has been found to be functional in numerous different assays, which are discussed here. A formal proof for the release of Tat, like that shown for bFGF and IL-1β, is that specific anti-Tat antibodies disrupt autocrine growth loops in Tat-producing cells (Milani *et al.*,1993; Zauli *et al.*, 1996; Ramazzotti *et al.*, 1996).

Westendorp *et al.* (1995) showed that Tat is released from HIV-infected cells and that substantial levels of Tat protein (0.1–1.0 ng/ml, approx. 0.01–0.1 n*M*) were found in the serum of 40% of HIV patients. These serum levels are similar to that of many cytokines/chemokines (McKenzie *et al.*, 1996), which are usually released locally in concentration gradients. The levels of the Tat protein found in AIDS patients most likely corresponded to the viral burden in the individual. A substantial portion of extracellular Tat may also come from cell death of HIV-infected cells. The turnover rates of infected cells in patients has recently been demonstrated to be quite high (Finzi and Siliciano, 1998). Regardless of the pathways utilized *in vivo,* it is clear that Tat is released into the extracellular environment, where it appears to assume a variety of functions (Fig. 2). There is evidence that the effects of extracellular Tat may directly affect HIV replication *in vivo* and *in vitro.* An inverse correlation between anti-Tat antibodies and survival has been reported in some studies (Re *et al.,* 1995, 1996; van Baalen *et al.,* 1997), and anti-Tat antibodies have been found to inhibit viral replication in culture (Steinaa *et al.,* 1994). These data suggest that extracellular Tat may favor HIV replication.

B. Tat: On the Way Back In

Several studies have demonstrated that the HIV Tat protein, or peptides based on Tat, are capable of entering cells cultured *in vitro* (Frankel and Pabo, 1988; Mann and Frankel, 1991; Green and Loewenstein, 1988; Bonifaci *et al.,* 1995; Viscidi *et al.,* 1989). The Tat which enters cells is capable of transactivating the HIV LTR, and the second exon appears to be dispensable for both cellular entry and transactivation. Tat and Tat peptides have even been used to deliver other proteins into cells (Fawell *et al.,* 1994; Vives *et al.,* 1997). Tat has been observed to activate the HIV LTR in cells neighboring the Tat-producing cells (Helland *et al.,* 1991).

However, the potential role of Tat entry into cells and in enhancement of HIV infection *in vivo* should again be approached with caution. In general, there are two prerequisites for observation of cellular entry by the Tat protein: either Tat present in (a) very high concentrations (micromolar or greater) or (b) in the presence of chloroquine or similar agents which perturb lysosomal activity. The vast majority of studies investigating the transactivation activity of exogenous Tat have been done in the presence of chloroquine. Only a few studies have observed that chloroquine does not have a substantial effect on the transactivational activity of Tat (Viscidi *et al.,* 1989). Basic peptides have been found to enhance Tat protein internalization (Green and Loewenstein, 1988; Vives *et al.,* 1997); the fact that Tat is a basic protein which contains strongly basic domains could possibly explain entry into cells with high concentrations of Tat.

It is our opinion that the vast majority of Tat effects *in vivo* are not likely to be due to Tat internalization and transactivation of either the HIV-LTR or cellular genes, but to the interaction of Tat with cellular receptors present on cell surfaces which induce, or interfere with, a signal cascade.

II. Tat and Angiogenesis–The Kaposi Connection

Early studies with transgenic mice had linked expression of the Tat protein to Kaposi's sarcoma (Vogel *et al.*, 1988; Corallini *et al.*, 1993), as the male transgenic mice tended to develop lesions which resembled the angiogenic Kaposi lesion. Kaposi's sarcoma (KS) prior to the outbreak of AIDS was a relatively rare disease, found in elderly men from the mediterranean region (sporadic KS), as an endemic form in some regions of Africa (endemic KS), and in some transplant patients (iatrogenic KS). In the early phases of the AIDS epidemic, KS was found in astonishing frequency in homosexual males with AIDS (epidemic KS). It was often the reason for the first entry of the HIV patient into the clinic and was almost diagnostic (Friedman-Kien *et al.*, 1982). Kaposi's Sarcoma is a highly angiogenic lesion, characterized by new blood vessel formation, an inflammatory infiltrate, and proliferation of a spindle-shaped cell population which is considered to be the "tumor" cell population of the lesion. In spite of the fact that epidemic KS was often an aggressive disease, KS has a number of characteristics which separated it from most tumors, including multifocal origin, frequent regression of several forms, and a normal karyotype of the cells in the lesion. Most cells cultured from KS lesions also have a normal karyotype and undergo senescence in culture. There are only three immortalized KS cell lines to date, and these all show substantial chromosomal rearrangements not normally found in KS cells.

The reason for the high prevalence of KS in homosexual AIDS patients was unknown. Even though HIV was not found in AIDS-KS cell cultures, it was thought that HIV or HIV products may be involved. The reports of KS-like lesions and the frequent occurrance of other tumor types in the transgenic mouse studies strongly implicated HIV Tat. The studies on Tat transgenic animals were closely followed by studies *in vitro*. Tat was found to transactivate several cytokine genes which could be involved in KS; however, it was the discovery that Tat could exit cells, as well as enter others, which prompted extensive studies on the effects of exogenous Tat on many cell types.

Tat was shown to be a growth factor for KS cells (Ensoli *et al.*, 1990) and for endothelial cells (Albini *et al.*, 1995, 1996). It was also shown to induce the migration of both KS and endothelial cells (Albini *et al.*, 1995, 1996). Tat was shown to induce angiogenesis *in vivo* (Albini *et al.*, 1994,

1996a,b; Barbanti-Brodano *et al.*, 1994; Ensoli *et al.*, 1994), giving rise to lesions which resembled KS.

Based on these observations, it was proposed that the Tat protein, in combination with a cytokine imbalance, could be a causative agent for AIDS-KS (Ensoli *et al.*, 1990). Several groups began to investigate the mechanism for the angiogenic and tumorigenic effects of Tat.

A. Tat and Integrins

The sequence of the LAI/IIIB Tat protein, as well as a portion (approximately 60%) of primary isolates, contains an RGD motif in the C-terminal domain. The RGD motif has been identified as the key sequence of a number of extracellular matrix proteins (in particular fibronectin and vitronectin) that is recognized by their receptors. These receptors are members of the integrin family, a large group of heterodimeric cell surface receptors involved in cell–substrate or cell–cell adhesion. Brake *et al.* first demonstrated that the RGD sequence of Tat could mediate cell adhesion (Brake *et al.*, 1990). Later studies showed that the integrins $\alpha 5\beta 1$, $\alpha v\beta 3$, and $\alpha v\beta 5$ could recognize the Tat protein as a substrate (Barillari *et al.*, 1993; Vogel *et al.*, 1993). In addition to the RGD sequence, Vogel *et al.* (1993) suggested that the basic domain of Tat could also be involved in integrin-mediated adhesion (Vogel *et al.*, 1993). Tat–integrin binding has been shown to trigger events typical of integrin–extracellular matrix ligand interactions, including activation of p125FAK (Milani *et al.*, 1998).

The binding of Tat by integrin receptors has been proposed to mediate a wide variety of Tat-induced biological responses *in vitro*. However, there are several considerations which suggest that ascribing functions to Tat RGD motif–integrin binding *in vivo* should be approached with caution. A major consideration is receptor affinity and competition with host ligands. The $\alpha 5\beta 1$ and $\alpha v\beta 3$ integrin receptors have intermediate affinities for their ligands, with dissociation constants in the micromolar range. *In vivo,* the lower affinity of these receptors for their ligands is compensated by ligand concentration. Serum levels of fibronectin average 450 μg/ml (2 μM) and that of vitronectin 250 μg/ml (1.7 μM), more than 10,000-fold higher than the highest serum concentrations reported for Tat (0.1 nM) (Westendorp *et al.*, 1995). In addition, serum levels of fibronectin and vitronectin vary substantially between donors, suggesting that these ligands are generally in excess. These observations would suggest that most cellular integrins would be occupied by host matrix factors rather than HIV Tat.

Although the affinity of integrins for their ligands appears to modulated by specific cell signals, the affinity of Tat for integrin receptors has never been accurately measured. Studies on the adhesion of cells to Tat shows similar levels of activity at comparable concentrations of extracellular matrix substrates such as fibronectin (Barillari *et al.*, 1993; Zauli *et al.*, 1996), suggesting analogous ligand affinity. Function-blocking antibodies are frequently used to interfere with Tat–integrin interactions; however, these studies require proper controls for the general effect of the integrin receptors run in parallel. Integrins are critical for cell–substrate interactions and disruption of integrin interactions with the substrate can have effects ranging from inhibition of movement to induction of apoptosis of a wide variety of cell types. For example, function-blocking anti-$\alpha v\beta 3$ integrins prevented Tat-induced migration of human dendritic cells (Benelli *et al.*, 1998); however, these same antibodies also blocked the migration of the same cells to f-MLP, whose well-defined ligand–receptor pathway does not involve RGD or integrins. The use of peptides may be more informative; however, the best controls are mutant Tat proteins (Brake *et al.*, 1990) or Tat proteins from HIV isolates lacking the RGD sequence.

Cell migration to the RGD peptide of Tat does occur at lower concentrations than that observed for similar peptides from fibronectin (Benelli *et al.*, 1998). In addition, the basic domain of Tat may also play a role in multiple receptor interactions, which could increase affinity (see below). Local concentrations of Tat may be higher in certain tissues where, if endogenous extracellular matrix ligands are limited, Tat might produce integrin-mediated biological effects.

B. Tat–Heparin Interactions—A Biological Role?

Mann and Frankel (1991) observed that the effects of extracellular Tat could be blocked by high doses of heparin or other polysaccharides (dextran sulfate). Several groups have shown that Tat tightly binds to heparin (Albini *et al.*, 1996; Chang *et al.*, 1997; Rusnati *et al.*, 1997). Interactions between Tat and heparin are to be expected, given the transcriptional activity of Tat. Heparin affinity is a common feature of many transcription factors, and this has been extensively exploited for their isolation and purification. A key observation was that heparin or heparan sulfate modified the biological activity of extracellular Tat (Albini *et al.*, 1996). Previous studies had shown that Tat was not chemotactic for endothelial cells unless these cells had been activated by cytokines (Albini *et al.*, 1995). Heparin was shown to overcome the need for cytokine stimulation in a stoichiometrically dose-dependent manner (Albini *et al.*, 1996). As observed with many heparin-dependent growth factors, lower doses stimulated, while a molar excess of heparin inhibited, the activity of Tat.

Tat has been shown to preferentially bind areas of heparin containing 2-*O*-sulfate, 6-*O*-sulfate, and *N*-sulfate groups (Rusnati *et al.*, 1997). These structual requirements overlap those of bFGF, which binds to pentamers containing *N*-sulfates and a single 2-*O*-sulfate groups (see Rusnati *et al.*, 1997). This may partially explain the observation of bFGF displacement from extracellular matrix by Tat (Chang *et al.*, 1997) and the cooperative effects of these two factors (Ensoli *et al.*, 1994).

C. Tat and Tyrosine Kinase Receptors

The ability of heparin and heparan sulfate to modify the biological activity of Tat on endothelial cells suggested that Tat was acting as a heparin-binding angiogenic growth factor. Most heparin-binding angiogenic growth factors have tyrosine kinase receptors on the target cell surface. The role of heparan sulfate has been shown to be critical in ligand–receptor interactions for most heparin-binding angiogenic growth factors. Heparin or heparan sulfate appears to be required for growth factor–receptor interactions to occur and receptor-signaling function (Yayon *et al.*, 1991; Rapraeger *et al.*, 1991). The interaction of bFGF with its receptors has been shown to also depend preferentially on a single proteoglycan, perlecan (Aviezer *et al.*, 1994, 1997) for receptor binding, while other proteoglycans inhibit (Mali *et al.*, 1993).

The observation that Tat-induced endothelial cell growth and migration *in vitro* (Albini *et al.*, 1996), and angiogenesis *in vivo* (Albini *et al.*, 1994, 1996), depended on the presence of heparin or heparan sulfate suggested that Tat interactions with endothelial cells could be mediated by tyrosine kinase receptors belonging to one or more heparin-binding angiogenic growth factors. It was then demonstrated that extracellular Tat could bind to and induce tyrosine phosphorylation and signaling through the VEGF receptor KDR/flk-1 on endothelial cells (Albini *et al.*, 1996). This interaction was shown to be specific and to mediate the migratory response to Tat *in vitro* and the angiogenic response to Tat *in vivo*. Interestingly, the binding of Tat to the VEGF tyrosine kinase receptor KDR/flk-1 was specific, no interaction was observed with the VEGF tyrosine kinase receptor flt-1 or to a series of other tyrosine kinase receptors (Albini *et al.*, 1996). This study showed that the affinity of Tat for KDR/flk-1 was in the picomolar range, similar to that of VEGF for the same receptor. Finally, the RGD peptide of Tat did not induce angiogenesis *in vivo*, whereas the basic peptide was active, and concentrations of Tat close to that reported in human sera (Westendorp *et al.*, 1995) were able to induce an angiogenic response *in vivo* (Albini *et al.*, 1996).

The specificity of Tat for the KDR/flk-1 receptor suggests that this ligand could have an even more potent angiogenic activity than that of the endogenous ligand, VEGF. VEGF binds to KDR/flk-1 with lower affinity than to flt-1, a VEGF receptor more closely associated with vascular differentiation also present on endothelial cells. Thus Tat would not show the competitive binding to another receptor found for VEGF. In addition, the RGD sequence of Tat may further enhance signaling through tyrosine kinase pathways. Recently the association of receptor tyrosine kinases and integrins has been demonstrated (Falcioni *et al.*, 1997), in particular for VEGFR-2 and the $\alpha v\beta 3$ integrin receptors (Soldi *et al.*, 1999), which appear to synergize in signaling. Given the presence of both $\alpha v\beta 3$ and VEGFR-2 binding activities on the Tat protein, this suggests that Tat may be a particularly potent signaling factor.

Kaposi's sarcoma cells are closely related to endothelial cells and also show expression of the KDR/flk-1 VEGF receptors (Masood *et al.*, 1997; Ganju *et al.*, 1998). The ability of Tat to bind to and activate the KDR/flk-1 in Kaposi's sarcoma cells has also been documented (Ganju *et al.*, 1998).

D. Tat and Kaposi's Sarcoma—Cause or Complication?

These data suggested that the Tat protein itself has strong angiogenic activity, explaining the observations made in transgenic animals. However, it did not explain the number of unusual features of KS, particularly for the nonepidemic KS forms. The peculiar features of KS; multifocal origin, frequent regression, and a normal karyotype, suggested that a secondary infectious agent may be involved (Siegal *et al.*, 1990). A very strong candidate for this infectious agent was identified by Chang and Moore (1996) as KSHV, a γ-herpesvirus also named HHV8. HHV8 has been found in essentially every KS lesion tested (Chang and Moore, 1996). The serology closely fits that of those who are likely to develop KS with immunosuppression (Nocera *et al.*, 1998; Parravicini *et al.*, 1997; Moore *et al.*, 1996). Although there is a school of thought that suggests that HHV8 may be a secondary opportunistic infection arising after the KS lesion has formed (Gallo, 1998; Sirianni *et al.*, 1998), HHV8 has been shown to fill almost all the Koch's postulates for being a causative agent (Foreman *et al.*, 1997; Flore *et al.*, 1998). HHV8 infection *in vitro* has been shown to permit perpetual growth of endothelial cells (Flore *et al.*, 1998). HHV8 encodes a number of angiogenic proteins, although the actual timing of the expression of these proteins is a point of controversy. Although the mechanisms of exactly how HHV8 may cause a KS lesion are not yet completely clear, as is the initial infection route and primary symptoms of infection, replication of this agent in the immune-suppressed host is most likely the initial cause of KS. The role of the HIV-Tat protein is probably one as a tumor progressor and angiogenic factor contributing to the aggressive nature of AIDS-KS. Finally, a brief report indicated that Tat was able to activate HHV8 (Harrington *et al.*, 1997), a potential direct contribution to HHV8 replication in KS.

III. Tat and Immunosuppression

The immune suppression seen with AIDS appears to affect cells that are not infected with HIV, aside from those harboring the virus. Several studies have shown that there is immunosuppression of non-HIV-infected cells from AIDS patients and that the number of immunosuppressed cells substantially exceeds that of the potentially HIV-infected cells. Proteins released from HIV-infected cells are clearly potential candidates for mediating such immune suppression, and the HIV *env* and Tat proteins have been among the most extensively studied. Tat has been linked to induction of T-cell anergy, T-cell apoptosis, and to a T-cell hyperactivation which appears to prime cells for infection by HIV. These events are probably all closely linked to the same phenomenon. The potential receptor system(s) involved in this activity include some novel candidate cellular receptors.

A. T-Cell Anergy in AIDS

Several studies have shown that HIV-1 Tat reduces the T-cell response to tetanus toxin and candida antigens (Viscidi *et al.*, 1989; Chirmule *et al.*, 1995; Gutheil *et al.*, 1994; Subramanyam *et al.*, 1993), while activation with PHA is not inhibited by Tat. The T-cell responses to immobilized CD3 (Chirmule *et al.*, 1995) and OKT3 (Subramanyam *et al.*, 1993) also have been reported to be inhibited by Tat. Both CD4+ and CD8+ cells appear to be inhibited equally well (Chirmule *et al.*, 1995). Finally, Tat also seems to reduce the production of chemokines by activated T cells (Zagury *et al.*, 1998), several of which interact with the HIV coreceptor CCR5 and have been proposed to be responsible for apparent resistance to HIV infection by some patients (Zagury *et al.*, 1998). T-cell anergy has been suggested to involve CD26, a dipeptidyl peptidase expressed on T-cell surfaces. Tat binds to CD26 with high affinity (20 p*M* to 1.3 n*M*) (Gutheil *et al.*, 1994); this binding appears to be mediated by the first nine amino acids of Tat (Wrenger *et al.*, 1997), although it is not clear if this is sufficient to produce the Tat inhibitory effect. Antibodies to CD26 (Gutheil *et al.*, 1994) and soluble CD26 (Subramanyam *et al.*, 1993) block the inhibitory effect of Tat on the T-cell response to antigen stimulation. Exogenous IL-2 or costimulation via CD28 appear to override the Tat-induced anergy (Subramanyam *et al.*, 1993). The mechanism of Tat-CD26 effects is not completely clear; however, it is interesting to note that CD26 has recently been found to cleave several chemokines which can dramatically change their activity. This includes cleavage of MDC from a relatively inactive form to a form which inhibits HIV replication (Proost *et al.*, 1999), apparently via acquisition of novel receptor binding.

B. T-Cell Apoptosis Induced by Tat

The Tat protein has been reported to act as a growth factor and protect transfected cell lines from apoptosis (Milani *et al.*, 1993; Zauli *et al.*, 1993, 1995a,b; Gibellini *et al.*, 1995), including the Jurkat lymphocyte cell line.

In contrast, several studies have shown that Tat, in addition to the inhibition of T-cell responses to antigens, also increases the apoptotic rate of T-cells. Notwithstanding the very different systems studied, Tat has been consistently found to up-regulate the expression of CD95-fas (Westendorp *et al.,* 1995; Zauli *et al.,* 1996; Li *et al.,* 1997). Tat-transfected Jurkat cells cultured in low serum showed an increased level of apoptosis. Human PBMCs cultured in normal levels of serum showed increased levels of apoptosis when exposed to 30–60 n*M* exogenous Tat (Li *et al.,* 1995). Again, both CD4+ and CD8+ T cells were affected similarly (Westendorp *et al.,* 1995; Li *et al.,* 1995), whereas monocytes did not show an increase in apoptosis. Interestingly, the level of apoptosis of low-level HIV-infected H9 cells was lowered by the addition of anti-Tat antibodies (Westendorp *et al.,* 1995). The mode of presentation of Tat seems to affect its ability to induce or inhibit antigen-stimulated apoptosis (Zauli *et al.,* 1996). Tat as a substrate has similar effect at similar concentrations as fibronectin, resulting in a relative sparing of cells, whereas Tat in a soluble form increased apoptosis (Zauli *et al.,* 1996).

C. Tat and HIV Infection

An increase in apoptosis is typical for partially activated T cells, as is entry into anergy resulting from an incomplete stimulation through the T-cell receptor. These data suggest that Tat is capable of partial, but incomplete, T-cell activation. HIV does not readily infect resting T cells—T-cell activation is a key requisite for HIV infection of these cells. At the same time, a pan-systemic complete T-cell activation would likely lead to rapid generation of HIV-specific CTL and potential early elimination of the virus. A partial T-cell activation may be sufficient for HIV infection yet detrimental to the host immune response, a potential role which Tat may fulfill.

HIV-infected cells appear to be hypersensitive to CD3/CD28 costimulation, with an increased production of IL-2. Transfection with Tat has been shown to increase IL-2 production (Ott *et al.,* 1997; Westendorp *et al.,* 1994), as has exogenous Tat (Ott *et al.,* 1997; Westendorp *et al.,* 1994). Although Ott *et al.* utilized 5–10 μg/ml of recombinant Tat 101, a similar effect was seen from HeLa cells producing Tat (Westendorp *et al.,* 1994), suggesting that Tat produced by cell lines may be more active than recombinant material in stimulation of IL-2 production or that a cofactor is released by these cells which synergizes with Tat. Exposure of T cells to low concentrations of extracellular Tat (12–24 n*M*) increased not only CD95 fas expression but also the expression of the IL-2 receptor CD25, a marker that correlates with the ability of HIV to infect T cells (Li *et al.,* 1997). The observations of T-cell stimulation by Tat were made not only *in vitro* but also in an *in vivo* model of T cells in nu/nu mice. The stimulation by Tat was not sufficient to induce T-cell proliferation, consistent with the incomplete stimulation provided by Tat (Li *et* *al.*, 1997). These authors provided evidence that function-blocking antibodies to the $\alpha v\beta 3$ and the $\alpha 3\beta 1$ integrins interfered with the Tat stimulation, while anti $\alpha 5\beta 1$ integrin had no effect. These data are surprising, as Tat has not been reported to be a ligand for the $\alpha 3\beta 1$ integrin. However the effects of anti-integrins on T-cell responses to other stimuli or to Tat proteins lacking an RGD sequence were not tested, so nonspecific effects of the anti-integrin antibodies cannot be ruled out. No stimulation of the *lck* tyrosine kinase were observed, although there was stimulation of the MAP kinase pathway (Li *et al.*, 1997).

The observation of partial T-cell activation by Tat suggested that stimulation by extracellular Tat may result in improved HIV infection. Li *et al.* demonstrated that stimulation with 12 n*M* of Tat significantly and dramatically increased infection of primary T cells by NL4-3 (Li *et al.*, 1997). These effects were blocked by anti-Tat antibodies. Exogenous Tat has recently been shown to significantly increase the expression of CXCR4 by monocytes and T-lymphocytes and also of CCR5 on monocytes (Huang *et al.*, 1998). Monocytes were much more sensitive to Tat stimulation, giving responses at 10-fold lower doses. The increase in these HIV coreceptors corresponded to an increase in infection in a model system by R5 and x4 HIV strains (Huang *et al.*, 1998). Most interesting was the observation that Tat increased the infection of monocyte/macrophage cells by T-tropic viruses, which are supposedly not able to infect these cells. Secchiero *et al.* reported a similar Tat-mediated induction of CXCR4 expression on lymphocytes with a corresponding increased rate of infection (Secchiero *et al.*, 1999).

IV. Tat as a Cytokine

A. Pleiotropic Effects on Accessory Cells

Extracellular HIV Tat has been shown to have wide-ranging effects on monocytes, macrophages, dendritic cells, and even NK cells. A cytokine like activity resulting in modulation of specific cytokine/growth factor production has been reported for monocytes and/or macrophages. These include increased production of TGFβ (Gibellini *et al.*, 1994; Zauli *et al.*, 1992), TNFα (Chen *et al.*, 1997), and MCP-1 (Conant *et al.*, 1998) and decreased IL-12 production (Ito *et al.*, 1998). Tat has been reported by several groups to be a strong chemoattractant for monocytes (Albini *et al.*, 1998a; Benelli *et al.*, 1998; Lafrenie *et al.*, 1996a,b; Mitola *et al.*, 1997). This activity could contribute directly to the recruitment of potentially "infectable" cells toward an HIV-infected cell producing and releasing Tat protein, an activity which may have a direct affect on establishment and spread of HIV infection in the host. Enhanced production of TNFα and MCP-1 may also contribute to monocyte recruitment. The chemotactic activity TNFα and MCP-1, as well as of Tat, appear to contribute to the monocyte/macrophage infiltration associated with neurological complications of HIV (Chen *et al.*, 1997; Conant *et al.*, 1998) and most likely enhance the direct recruitment activity of Tat on these cell types.

The mechanism of monocyte chemotaxis induced by Tat was suggested to be due to Tat-integrin interactions (Lafrenie *et al.*, 1996b) or to Tat–Flt-1 interactions (Mitola *et al.*, 1997). However, the Tat RGD and basic peptides, which mediate interaction with these receptors, were relatively poor chemoattractants for monocytes as well as dendritic cells (Benelli *et al.*, 1998). In addition, there was no evidence of cooperation between these peptides. The chemotaxis of monocytes toward Tat was blocked by pertussis toxin but not cholera toxin (Mitola *et al.*, 1997; Albini *et al.*, 1998b), indicating involvement of G_i proteins. Peptide mapping of the entire Tat protein showed that the monocyte chemotactic activity was concentrated in the cysteine-rich and core domains of Tat (Albini *et al.*, 1998a). These domains are the most highly conserved domains of the Tat protein and contain both CC and CXC motifs (Albini *et al.*, 1998b), which show a limited sequence similarity with chemokines, strong monocyte chemoattractants. Tat was shown to signal through G_i proteins, like chemokines, in monocytes and macrophages (Albini *et al.*, 1998b). Receptor desensitization and ligand binding assays indicated that Tat interacted with the β-chemokine receptors CCR2 and CCR3 but not with CCR1, CCR4, and CCR5. However, the lack of a complete desensitization by chemokines suggested that Tat may interact with novel chemokine receptors as well. Chemokines which are ligands for HIV coreceptors have been shown to block infection by viral strains using those same viral receptors. In contrast, chemokines stimulating other receptors have been found to increase HIV infection (Kinter *et al.*, 1998; Cinque *et al.*, 1998). The role of the Tat chemokine like activity in HIV infection still remains to be investigated, although it is possible that interaction with chemokine receptors could be responsible for the up-regulation of CCR5 and CXCR4 (Huang *et al.*, 1998; Secchiero *et al.*, 1999) particularly on monocytes/macrophages.

Unlike many chemokines, Tat appeared to exhibit a strict cell type specificity, showing activity on monocytes, macrophages, and dendritic cells (Albini *et al.*, 1998a,b; Benelli *et al.*, 1998), but not on T cells. The reasons for this specificity are not yet clear; it is possible that Tat interacts with a receptor that is expressed on these cell types but not on T cells or that additional receptors on T cells interfere with Tat-chemokine signaling in these cells. Tat has also been reported to indirectly affect T cells by alteration of T-cell–accessory cell interactions (Wu and Schlossman, 1997).

Exogenous HIV Tat protein appears to inhibit dendritic cell phagocytosis (Zocchi *et al.*, 1997). This inhibitory activity was linked to the Tat blockage of L-type (ligand-dependent) calcium channel activity. Interference with L-type calcium channels also appears to lead to impairment of natural killer (NK) cell function (Zocchi *et al.*, 1998), a phenomenon observed in AIDS patients (Fauci, 1996). Although there is evidence that Tat directly interacts with these calcium channels (Zocchi *et al.*, 1997, 1998), activity of these channels appears to be repressed by activation of G_i proteins (Rubartelli *et al.*, 1998). The activation of G_i proteins via interaction with chemokine receptors (Albini *et al.*, 1998b) may be responsible for the observed block in calcium channel activity. This is strongly supported by the observation that pertussis toxin prevents (1) the Tat-mediated block of calcium channel activity (Rubartelli *et al.*, 1998), (2) the chemokine-receptor-mediated Tat-induced signaling (Albini *et al.*, 1998b), and (3) monocyte chemotaxis (Mitola *et al.*, 1997; Albini *et al.*, 1998b). Tat interactions with chemokine receptors may facilitate infection by (a) recruiting cells toward sites of virus production and (b) partially activating these cells, while it may interfere with the immune response to HIV by blocking (c) dendritic cell function, (d) NK cell function, and possibly even (e) B-cell function (Rubartelli *et al.*, 1998). These activities coupled with the partial activation leading to functional impairment and increased infectivity of T cells sets the stage for HIV infection and destruction of the host immune system.

B. Tat, Dementia, and the Central Nervous System

The dementia associated with AIDS led to the investigation of potential toxicity of HIV proteins on cells of the neural system. Tat has been shown to induce excitation in neurons (Sabatier *et al.*, 1991; Cheng *et al.*, 1998), which is associated with neurotoxicity, although the mechanism of these effects is not yet fully elucidated. Tat transfection of PC12 cells increased cellular proliferation and stimulated differentiation toward sympathetic neurons (Milani *et al.*, 1993). Interestingly, anti-Tat antibodies blocked the cellular proliferation effect, but not the differentiative effect, suggesting that proliferation was mediated by extracellular Tat. A 90-kDa cell surface receptor mediating attachment has been isolated from PC12 cells (Weeks *et al.*, 1993), which interacted with Tat through its basic domain. The basic domain also appears to be mediate Tat neurotoxicity (Sabatier *et al.*, 1991), suggesting that these two observations may be linked. The molecular identity of this receptor has never been reported; however, the observation that the neuroexcitory properties of Tat were blocked by lowering extracellular calcium (Cheng *et al.*, 1998), suggests that interference with calcium channel function may be involved. This may be due to direct effects on L-type calcium channels (Rubartelli *et al.*, 1998) or perhaps even chemokine receptors, which have been recently reported to be on neural cells (Meucci *et al.*, 1998). Alterations in calcium flux may also be involved in induction of neuronal apoptosis (Kruman *et al.,* 1998), although this appeared to be secondary to a state of oxidative stress induced by Tat (Kruman *et al.,* 1998; Shi *et al.,* 1998). Oxidative stress as a result of exposure to exogenous Tat has been reported for other cell types as well (Westendorp *et al.,* 1995).

The injection of Tat into the brain of rats resulted in a rapid influx of neutrophils, followed by monocyte/macrophages and in turn by lymphocytes (Jones *et al.,* 1998). This activity may not be surprising given the chemokine-like activity of Tat. The recruitment of monocytes by Tat both directly (Albini *et al.,* 1998a; Benelli *et al.,* 1998; Lafrenie *et al.,* 1996a,b; Mitola *et al.,* 1997) and indirectly by induction of TNFα (Chen *et al.,* 1997) and MCP-1 (Conant *et al.,* 1998) may be a key factor in the increased monocyte/macrophage presence associated with AIDS dementia (Glass *et al.,* 1993).

C. Tat Induction of Signal Cascades

In addition to receptor-mediated Tat effects on signaling though calcium fluxes, extracellular Tat appears to stimulate specific signal transduction cascades as a result of receptor activation. The adhesion-associated kinases p 125Fak and RAFTK have been reported to be Tat activated (Milani *et al.,* 1998; Ganju *et al.,* 1998). A variety of secondary messengers have been observed to be activated in different cell types, including paxillin, p 130cas, src, plI3kinase, and Phospholipase C (Milani *et al.,* 1998; Ganju *et al.,* 1998; Chen *et al.,* 1997). While pathways involving PKA and PKC were shown not to be involved in macrophages (Chen *et al.,* 1997), these have been reported to be activated in Tat-induced events in endothelial cells (Zidovetzki *et al.,* 1998). Involvement of the MAP kinase pathway has been observed in Tat signaling in monocytes-macrophages (Gibellini *et al.,* 1998), T-cells (Li *et al.,* 1997; Gibellini *et al.,* 1998) and KS cells (Ganju *et al.,* 1998). Downstream activation of NF-kB has been reported in Tat-treated macrophages (Chen *et al.,* 1997) and in TNFα/Tat-treated Jurkat cells (Ramazzotti *et al.,* 1996). Given the involvement of NF-κB in HIV transcription, this suggests that Tat could enhance HIV replication though cell-surface-mediated events. Activation of the nuclear factor CREB (Ramazzotti *et al.,* 1996) and alteration of cyclins (Li *et al.,* 1997) in responses to Tat have also been observed.

The studies reviewed here support a major role for Tat stimulation of—or interference with—cell surface receptors and the resultant signal cascades in mediating the pleiotropic effects of extracellular Tat.

References

Albini, A., Barillari, G., Benelli, R., Gallo, R. C., and Ensoli, B. (1995). Angiogenic properties of human immunodeficiency virus type 1 Tat protein. *Proc. Natl. Acad. Sci. USA* **92,** 4838–4842.

Albini, A., Benelli, R., Giunciuglio, D., Cai, T., Mariani, G., Ferrini, S., and Noonan, D. (1998a). Identification of a novel domain of HIV Tat involved in monocyte chemotaxis. *J. Biol. Chem.* **273,** 15895–15900.

Albini, A., Benelli, R., Presta, M., Rusnati, M., Ziche, M., Rubartelli, A., Paglialunga, G., Bussolino, F., and Noonan, D. (1996). HIV-tat protein is a heparin-binding angiogenic growth factor. *Oncogene* **12,** 289–297.

Albini, A., Ferrini, S., Benelli, R., Sforzini, S., Giunciuglio, D., Aluigi, M. G., Proudfoot, A., Alouani, S., Wells, T., Mariani, G., Rabin, R. L., Farber, J. M., and Noonan, D. M. (1998b). HIV-1 Tat protein mimicry of chemokines. *Proc. Natl. Acad. Sci. USA* **95,** 13153–13158.

Albini, A., Fontanini, G., Masiello, L., Tacchetti, C., Bigini, D., Luzzi, P., Noonan, D. M., and Stetler-Stevenson, W. G. (1994). Angiogenic potential *in vivo* by Kaposi sarcoma cell-free supernatants and HIV1-tat product: Inhibition of KS-like lesions by TIMP-2. *AIDS* **8,** 1237–1244.

Albini, A., Soldi, R., Giunciuglio, D., Giraudo, E., Benelli, R., Primo, L., Noonan, D., Salio, M., Camussi, G., Rockl, W., and Bussolino, F. (1996). HIV-1 Tat induced angiogenesis is mediated by activation of the flk-1/KDR tyrosine kinase receptor on vascular endothelial cells. *Nature Med.* **2,** 1371–1375.

Aviezer, D., Hecht, D., Safran, M., Eisinger, M., David, G., and Yayon, A. (1994). Perlecan, basal lamina proteoglycan, promotes basic fibroblast growth factor-receptor binding, mitogenesis, and angiogenesis. *Cell* **79,** 1005–1013.

Aviezer, D., Iozzo, R. V., Noonan, D. M., and Yayon, A. (1997). Suppression of autocrine and paracrine functions of basic fibroblast growth factor by stable expression of perlecan antisense cDNA. *Mol. Cell Biol.* **17,** 1938–1946.

Barbanti-Brodano, G., Sampaolesi, R., Campioni, D., Lazzarin, L., Altavilla, G., Possati, L., Masiello, L., Benelli, R., Albini, A., and Corallini, A. (1994). HIV-1 tat acts as a growth factor and induces angiogenic activity in BK virus/tat transgenic mice. *Antibiot. Chemother.* **46,** 88–101.

Barillari, G., Gendelman, R., Gallo, R. C., and Ensoli, B. (1993). The Tat protein of human immunodeficiency virus type 1, a growth factor for AIDS Kaposi sarcoma and cytokine-activated vascular cells, induces adhesion of the same cell types by using integrin receptors recognizing the RGD amino acid sequence. *Proc. Natl. Acad. Sci. USA* **90,** 7941–7945.

Benelli, R., Mortarini, R., Anichini, A., Giunciuglio, D., Noonan, D. M., Montalti, S., Tacchetti, C., and Albini, A. (1998). Monocyte-derived dendritic cells and monocytes migrate to HIV-Tat RGD and basic peptides. *AIDS* **12,** 261–268.

Brake, D., Debouck, C., and Biesecker, G. (1990). Identification of an Arg-Gly-Asp (RGD) cell adhesion site in human immunodeficiency virus type 1 transactivation protein, tat. *J. Cell Biol.* **111,** 1275–1281.

Bonifaci, N., Sitia, R., and Rubartelli, A. (1995). Nuclear translocation of an exogenous protein containing tat requires unfolding. *AIDS* **9,** 995–1000.

Campioni, D., Corallini, A., Zauli, G., Possati, L., Altavilla, G., and Barbanti, B. G. (1995). HIV type 1 extracellular Tat protein stimulates growth and protects cells of BK virus/tat transgenic mice from apoptosis. *AIDS Res. Hum. Retroviruses* **11,** 1039–1048.

Chang, H. C., Samaniego, F., Nair, B. C., Buonaguro, L., and Ensoli, B. (1997). HIV-1 Tat protein exits from cells via a leaderless secretory pathway and binds to extracellular matrix-associated heparan sulfate proteoglycans through its basic region. *AIDS* **11,** 1421–1431.

Chang, Y., and Moore, P. S. (1996). Kaposi's Sarcoma (KS)-associated herpesvirus and its role in KS. *Infect. Agents Dis.* **5,** 215–222.

Chen, P., Mayne, M., Power, C., and Nath, A. (1997). The Tat protein of HIV-1 induces tumor necrosis factor-alpha production: Implications for HIV-1-associated neurological diseases. *J. Biol. Chem.* **272,** 22385–22388.

Cheng, J., Nath, A., Knudsen, B., Hochman, S., Geiger, J. D., Ma, M., and Magnuson, D. S. (1998). Neuronal excitatory properties of human immunodeficiency virus type 1 Tat protein. *Neuroscience* **82,** 97–106.

Chirmule, N., Than, S., Khan, S. A., and Pahwa, S. (1995). Human immunodeficiency virus Tat induces functional unresponsiveness in T cells. *J. Virol.* **69**, 492–498.

Cinque, P., Vago, L., Mengozzi, M., Torri, V., Ceresa, D., Vicenzi, E., Transidico, P., Vagani, A., Sozzani, S., Mantovani, A., Lazzarin, A., and Poli, G. (1998). Elevated cerebrospinal fluid levels of monocyte chemotactic protein-1 correlate with HIV-1 encephalitis and local viral replication. *AIDS* **12**, 1327–1332.

Conant, K., Garzino, D. A., Nath, A., McArthur, J. C., Halliday, W., Power, C., Gallo, R. C., and Major, E. O. (1998). Induction of monocyte chemoattractant protein-1 in HIV-1 Tat-stimulated astrocytes and elevation in AIDS dementia. *Proc. Natl. Acad. Sci. USA* **95**, 3117–3121.

Corallini, A., Altavilla, G., Pozzi, L., Bignozzi, F., Negrini, M., Rimessi, P., Gualandi, F., and Barbanti-Brodano, G. (1993). Systemic expression of HIV1 *tat* gene in transgenic mice induces endothelial proliferation and tumors of different histotypes. *Cancer Res.* **53**, 5569–5575.

Ensoli, B., Barillari, G., Salahuddin, S. Z., Gallo, R., and Wong-Staal, F. (1990). Tat protein of HIV-1 stimulates growth of cells derived from Kaposi's sarcoma lesions of AIDS patients. *Nature* **345**, 84–86.

Ensoli, B., Buonaguro, L., Barillari, G., Fiorelli, V., Gendelman, R., Morgan, R. A., Wingfield, P., and Gallo, R. C. (1993). Release, uptake, and effects of extracellular human immunodeficiency virus type 1 Tat protein on cell growth and vital transactivation. *J. Virol.* **67**, 277–287.

Ensoli, B., Gendelman, R., Markham, P., Fiorelli, V., Colombini, S., Raffeld, M., Cafaro, A., Chang, H., Brady, J. N., and Gallo, R. C. (1994). Synergy between basic fibroblast growth factor and HIV-1 tat protein in induction of Kaposi's sarcoma. *Nature* **371**, 674–680.

Falcioni, R., Antonini, A., Nistico, P., Di Stefano, S., Crescenzi, M., Natali, P. G., and Sacchi, A. (1997). Alpha 6 beta 4 and alpha 6 beta 1 integrins associate with erbB-2 in human carcinoma cell lines. *Exp. Cell. Res.* **236**, 76–85.

Fauci, A. S. (1996). Host factors and the pathogenesis of HIV-induced disease. *Nature* **384**, 529–534.

Fawell, S., Seery, J., Daikh, Y., Moore, C., Chen, L. L., Pepinsky, B., and Barsoum, J. (1994). Tat-mediated delivery of heterologous proteins into cells. *Proc. Natl. Acad. Sci USA* **91**, 664–668.

Finzi, D., and Siliciano, R. F. (1998). Viral dynamics in HIV infection. *Cell* **93**, 665–671.

Flore, O., Rafii, S., Ely, S., O'Leary, J. J., Hyjek, E. M., and Cesarman, E. (1998). Transformation of primary human endothelial cells by Kaposi's sarcoma-associated herpesvirus. *Nature* **394**, 588–592.

Foreman, K. E., Bacon, P. E., Hsi, E. D., and Nickoloff, B. J. (1997). *In situ* polymerase chain reaction-based localization studies support role of human herpesvirus-8 as the cause of two AIDS-related neoplasms: Kaposi's sarcoma and body cavity lymphoma. *J. Clin. Invest.* **99**, 2971–2978.

Frankel, A. D., and Pabo, C. O. (1988). Cellular uptake of the tat protein from human immunodeficiency virus. *Cell* **55**, 1189–1193.

Friedman-Kien, A. E., Laubenstein, L. J., Rubenstein, P., Buimovici-Klein, M., Marmor, M., Stahl, R., Spigland, I., Kim, K., and Zolla-Pazner, S. (1982). Disseminated Kaposi's sarcoma in homosexual men. *Ann. Intern. Med.* **96**, 693–700.

Gallo, R. C. (1998). Some aspects of the pathogenesis of HIV-1-associated Kaposi's sarcoma. *J. Natl. Cancer Inst. Monogr.* **1998**, 55–57.

Ganju, R. K., Munshi, N., Nair, B. C., Liu, Z. Y., Gill, P., and Groopman, J. E. (1998). Human immunodeficiency virus tat modulates the Flk-1/KDR receptor, mitogen-activated protein kinases, and components of focal adhesion in Kaposi's sarcoma cells. *J. Virol.* **72**, 6131–6137.

Gibellini, D., Bassini, A., Pierpaoli, S., Bertolaso, L., Milani, D., Capitani, S., La Placa, M., and Zauli, G. (1998). Extracellular HIV-1 Tat protein induces the rapid Ser133 phosphorylation and activation of CREB transcription factor in both Jurkat lymphoblastoid T cells and primary peripheral blood mononuclear cells. *J. Immunol.* **160,** 3891–3898.

Gibellini, D., Caputo, A., Celeghini, C., Bassini, A., La, P. M., Capitani, S., and Zauli, G. (1995). Tat-expressing Jurkat cells show an increased resistance to different apoptotic stimuli, including acute human immunodeficiency virus-type 1 (HIV-1) infection. *Br. J. Haematol.* **89,** 24–33.

Gibellini, D., Zauli, G., Re, M. C., Milani, D., Furlini, G., Caramelli, E., Capitani, S., and La Placa, M. (1994). Recombinant human immunodeficiency virus type-1 (HIV-1) Tat protein sequentially up-regulates IL-6 and TGF-beta 1 mRNA expression and protein synthesis in peripheral blood monocytes. *Br. J. Haematol.* **88,** 261–267.

Glass, J. D., Wesselingh, S. L., Selnes, O. A., and McArthur, J. C. (1993). Clinical-neuropathologic correlation in HIV-associated dementia. *Neurology* **43,** 2230–2237.

Green, M., and Loewenstein, P. M. (1988). Autonomous functional domains of chemically synthesized human immunodeficiency virus tat trans-activator protein. *Cell* **55,** 1179–1188.

Gutheil, W. G., Subramanyam, M., Flentke, G. R., Sanford, D. G., Munoz, E., Huber, B. T., and Bachovchin, W. W. (1994). Human immunodeficiency virus 1 Tat binds to dipeptidyl aminopeptidase IV (CD26): A possible mechanism for Tat's immunosuppressive activity. *Proc. Natl. Acad. Sci. USA* **91,** 6594–6598.

Harrington, W. J., Sieczkowski, L., Sosa, C., Chan, A., Sue, S., Cai, J. P., Cabral, L., and Wood, C. (1997). Activation of HHV-8 by HIV-1 tat. *Lancet* **349,** 774–775.

Helland, D. E., Welles, J. L., Caputo, A., and Haseltine, W. A. (1991). Transcellular transactivation by the human immunodeficiency virus type 1 tat protein. *J. Virol.* **65,** 4547–4547.

Huang, L., Bosch, I., Hofmann, W., Sodroski, J., and Pardee, A. B. (1998). Tat protein induces human immunodeficiency virus type 1 (HIV-1) coreceptors and promotes infection with both macrophage-tropic and T-lymphotropic HIV-1 strains. *J. Virol.* **72,** 8952–8960.

Huang, L. M., Joshi, A., Willey, R., Orenstein, J., and Jeang, K. T. (1994). Human immunodeficiency viruses regulated by alternative trans-activators: Genetic evidence for a novel non-transcriptional function of Tat in virion infectivity. *EMBO J.* **13,** 2886–2896.

Ito, M., Ishida, T., He, L., Tanabe, F., Rongge, Y., Miyakawa, Y., and Terunuma, H. (1998). HIV type 1 Tat protein inhibits interleukin 12 production by human peripheral blood mononuclear cells. *AIDS Res. Hum. Retroviruses* **14,** 845–849.

Jeang, K.-T. (1994). HIV-1 Tat: Structure and function. "The Human Retroviruses and AIDS Compendium On Line." [**http://hiv-web.lanl.gov**]

Jones, M., Olafson, K., DelBigio, M., Peeling, J., and Nath, A. (1998). Intraventricular injection of Human Immunodeficiency Virus type 1 (HIV-1) Tat protein causes inflammation, gliosis, apoptosis and ventricular enlargement. *J. Neuropathol. Exp. Neurol.* **57,** 563–570.

Kinter, A., Catanzaro, A., Monaco, J., Ruiz, M., Justement, J., Moir, S., Arthos, J., Oliva, A., Ehler, L., Mizell, S., Jackson, R., Ostrowski, M., Hoxie, J., Offord, R., and Fauci, A. S. (1998). CC-chemokines enhance the replication of T-tropic strains of HIV-1 in CD4(+) T cells: Role of signal transduction. *Proc. Natl. Acad. Sci. USA* **95,** 11880–11885.

Kruman, I. I., Nath, A., and Mattson, M. P. (1998). HIV-1 protein Tat induces apoptosis of hippocampal neurons by a mechanism involving caspase activation, calcium overload, and oxidative stress. *Exp. Neurol.* **154,** 276–288.

Lafrenie, R. M., Wahl, L. M., Epstein, J. S., Hewlett, I. K., Yamada, K. M., and Dhawan, S. (1996a). HIV-1-Tat modulates the function of monocytes and alters their interactions with microvessel endothelial cells: A mechanism of HIV pathogenesis. *J. Immunol.* **156,** 1638–1645.

Lafrenie, R. M., Wahl, L. M., Epstein, J. S., Hewlett, I. K., Yamada, K. M., and Dhawan, S. (1996b). HIV-1-Tat protein promotes chemotaxis and invasive behavior by monocytes. *J. Immunol.* **157,** 974–977.

Li, C. J., Friedman, D. J., Wang, C., Metelev, V., and Pardee, A. B. (1995). Induction of apoptosis in uninfected lymphocytes by HIV-1 Tat protein. *Science* **268,** 429–431.

Li, C. J., Ueda, Y., Shi, B., Borodyansky, L., Huang, L., Li, Y. Z., and Pardee, A. B. (1997). Tat protein induces self-perpetuating permissivity for productive HIV-1 infection. *Proc. Natl. Acad. Sci, USA* **94,** 8116–8120.

Mali, M., Elenius, K., Miettinen, H. M., and Jalkanen, M. (1993), Inhibition of basic fibroblast growth factor-induced growth promotion by overexpression of syndecan-1. *J. Biol. Chem.* **268,** 24215–24222.

Mann, D. A., and Frankel, A. D. (1991). Endocytosis and targeting of exogenous HIV-1 tat protein. *EMBO J.* **10,** 1733–1739.

Masood, R., Cai, J., Zheng, T., Smith, D. L., Naidu, Y., and Gill, P. S. (1997). Vascular endothelial growth factor/vascular permeability factor is an autocrine growth factor for AIDS–Kaposi sarcoma. *Proc. Natl. Acad. Sci. USA* **94,** 979–984.

McKenzie, S. W., Dallalio, G., North, M., Frame, P., and Means, R. J. (1996). Serum chemokine levels in patients with non-progressing HIV infection. *AIDS* **10,** F29–F33. 1996.

Meucci, O., Fatatis, A., Simen, A. A., Bushell, T. J., Gray, P. W., and Miller, R. J. (1998). Chemokines regulate hippocampal neuronal signaling and gp120 neurotoxicity. *Proc. Natl. Acad. Sci. USA* **95,** 14500–14505.

Milani, D., Mazzoni, M., Zauli, G., Mischiati, C., Gibellini, D., Giacca, M., and Capitani, S. (1998). HIV-1 Tat induces tyrosine phosphorylation of p125FAK and its association with phosphoinositide 3-kinase in PC12 cells. *AIDS* **12,** 1275–1284.

Milani, D., Zauli, G., Neri, L. M., Marchisio, M., Previati, M., and Capitani, S. (1993). Influence of the human immunodeficiency virus type 1 Tat protein on the proliferation and differentiation of PC12 rat pheochromocytoma cells. *J. Gen. Virol.* **74,** 2587–2594.

Mitola, S., Sozzani, S., Luini, W., Primo, L., Bosatti, A., Weich, H., and Bussolino, F. (1997). Tat-human immunodeficiency virus-1 induces human monocyte chemotaxis by activation of vascular endothelial growth factor receptor-1. *Blood* **90,** 1365–1372.

Moore, P. S., Kingsley, L. A., Holmberg, S. D., Spira, T., Gupta, P., Hoover, D. R., Parry, J. P., Conley, L. J., Jaffe, H. W., and Chang, Y. (1996). Kaposi's sarcoma-associated herpesvirus infection prior to onset of Kaposi's sarcoma. *AIDS* **10,** 175–180.

Nocera, A., Corbellino, M., Valente, U., Barocci, S., Torre, F., De, P. R., Sementa, A., Traverso, G. B., Icardi, A., Fontana, I., Arcuri, V., Poli, F., Cagetti, P., Moore, P., and Parravicini, C. (1998). Posttransplant human herpes virus 8 infection and seroconversion in a Kaposi's sarcoma affected kidney recipient transplanted from a human herpes virus 8 positive living related donor. *Transplant Proc.* **30,** 2095–2096.

Ott, M., Emiliani, S., Van, L. C., Herbein, G., Lovett, J., Chirmule, N., McCloskey, T., Pahwa, S., and Verdin, E. (1997). Immune hyperactivation of HIV-1 infected T cells mediated by Tat and the CD28 pathway. *Science* **275,** 1481–1485.

Parravicini, C., Olsen, S. J., Capra, M., Poli, F., Sirchia, G., Gao, S. J., Berti, E., Nocera, A., Rossi, E., Bestetti, G., Pizzuto, M., Galli, M., Moroni, M., Moore, P. S., and Corbellino, M. (1997). Risk of Kaposi's sarcoma-associated herpes virus transmission from donor allografts among Italian posttransplant Kaposi's sarcoma patients. *Blood* **90,** 2826–2829.

Proost, P., Struyf, S., Schols, D., Opdenakker, G., Sozzani, S., Allavena, P., Mantovani, A., Augustyns, K., Bal, G., Haemers, A., Lambeir, A. M., Scharpe, S., Van, D. J., and De, M. I. (1999). Truncation of macrophage-derived chemokine by CD26/ dipeptidyl-peptidase IV beyond its predicted cleavage site affects chemotactic activity and CC chemokine receptor 4 interaction. *J. Biol. Chem.* **274,** 3988–3993.

Ramazzotti, E., Vignoli, M., Re, M. C., Furlini, G., and La Placa, M. (1996). Enhanced nuclear factor-kappa B activation induced by tumour necrosis factor-alpha in stably tat-transfected cells is associated with the presence of cell-surface-bound Tat protein. *AIDS* **10,** 455–461.

Rapraeger, A. C., Krufka, A., and Olwin, B. B. (1991). Requirement of heparan sulfate for bFGF-mediated fibroblast growth and myoblast differentiation. *Science* **252,** 1705–1708.

Re, M. C., Furlini, G., Vignoli, M., Ramazzotti, E., Roderigo, G., DeRosa, V., Zauli, G., Lolli, S., Capitani S., and LaPlaca, M. (1995). Effect of antibody to HIV-1 Tat protein on viral replication *in vitro* and progression of HIV-1 disease *in vivo*. *J. Acquir. Immune Defic. Syndr. Hum. Retrovirol.* **10**, 408–416.

Re, M. C., Furlini, G., Vignoli, M., Ramazzotti, E., Zauli, G., and LaPlaca, M. (1996). Antibody against human immunodeficiency virus type 1 (HIV-1) Tat protein may have influenced the progression of AIDS in HIV-1-infected hemophiliac patients. *Clin. Diagn. Lab. Immunol.* **3**, 230–232.

Rubartelli, A., and Sitia, R. (1997). Secretion of mammalian proteins that lack a signal sequence. *In "Unusual Secretory Pathways: From Bacteria to Man"* (K. Kuchler, A. Rubartelli, and B. Holland, Eds.). Springer-Verlag, Heidelberg.

Rubartelli, A., Poggi, A., Sitia, R., and Zocchi, M. R. (1998). HIV-I Tat: A polypeptide for all seasons. *Immunol. Today* **19**, 543–545.

Rusnati, M., Coltrini, D., Oreste, P., Zoppetti, G., Albini, A., Noonan, D., Giacca, M., D'Adda, F., and Presta, M. (1997). Interaction of HIV-1 tat protein with heparin: Role of the backbone structure, sulfation, and size. *J. Biol. Chem.* **272**, 11313–11320.

Secchiero, P., Zella, D., Capitani, S., Gallo, R. C., and Zauli, G. (1999). Extracellular HIV-1 Tat protein up-regulates the expression of surface CXC-chemokine receptor 4 in resting CD4+ T cells. *J. Immunol.* **162**, 2427–2431.

Sabatier, J., Vives, E., Mabrouk, K., Benjouad, A., Rochat, H., Duval, A., Hue, B., and Bahrahoui, E. (1991). Evidence for neurotoxic activity of tat from Human Immunodeficiency Virus type 1. *J. Virol.* **65**, 961–967.

Shi, B., Raina, J., Lorenzo, A., Busciglio, J., and Gabuzda, D. (1998). Neuronal apoptosis induced by HIV-1 Tat protein and TNF-alpha: Potentiation of neurotoxicity mediated by oxidative stress and implications for HIV-1 dementia. *J. Neurovirol.* **4**, 281–290.

Siegal, B., Levington-Kriss, S., Schiffer, A., Sayar, J., Engelberg, I., Vonsover, A., Ramon, Y., and Rubinstein, E. (1990). Kaposi's Sarcoma in immunosuppression: Possibly the result of a dual viral infection. *Cancer* **65**, 492–498.

Sirianni, M. C., Vincenzi, L., Fiorelli, V., Topino, S., Scala, E., Uccini, S., Angeloni, A., Faggioni, A., Cerimele, D., Cottoni, F., Aiuti, F., and Ensoli, B. (1998). Gamma-Interferon production in peripheral blood mononuclear cells and tumor infiltrating lymphocytes from Kaposi's sarcoma patients: Correlation with the presence of human herpesvirus-8 in peripheral blood mononuclear cells and lesional macrophages. *Blood* **91**, 968–976.

Soldi, R., Mitola, S., Strasly, M., Defilippi, P., Tarone, G., and Bussolino, F. (1999). Role of $\alpha v\beta 3$ integrin in the activation of vascular endothelial growth factor receptor-2. *EMBO J.* **18**, 882–892.

Steinaa, L., Sorensen, A. M., Nielsen, J. O., and Hansen, J. E. (1994). Antibody to HIV-1 Tat protein inhibits the replication of virus in culture. *Arch. Virol.* **139**, 263–271.

Subramanyam, M., Gutheil, W. G., Bachovchin, W. W., and Huber, B. T., (1993). Mechanism of HIV-1 Tat induced inhibition of antigen-specific T cell responsiveness. *J. Immunol.* **150**, 2544–2553.

van Baalen, C., Pontesilli, O., Huisman, R. C., Geretti, A. M., Klein, M. R., De, W. F., Miedema, F., Gruters, R. A., and Osterhaus, A. D. (1997). Human immunodeficiency virus type 1 Rev- and Tat-specific cytotoxic T lymphocyte frequencies inversely correlate with rapid progression to AIDS. *J. Gen. Virol.* **78**, 1913–1918.

Viscidi, R. P., Mayur, K., Lederman, H. M., and Frankel, A. D. (1989). Inhibition of antigen-induced lymphocyte proliferation by Tat protein from HIV-1. *Science* **246**, 1606–1608.

Vives, E., Brodin, P., and Lebleu, B. (1997). A truncated HIV-1 Tat protein basic domain rapidly translocates through the plasma membrane and accumulates in the cell nucleus. *J. Biol. Chem.* **272**, 16010–16017.

Vogel, J., Hinrichs, S. H., Reynolds, R. K., Luciw, P. A., and Jay, G. (1988). The HIV tat gene induces dermal lesions resembling Kaposi's sarcoma in transgenic mice. *Nature* **335**, 606–611.

Vogel, B. E., Lee, S.-J., Hildebrand, A., Craig, W., Pierschbacher, M. D., Wong-Staal, F., and Ruoslahti, E. (1993). A novel integrin specificity exemplified by binding of the $\alpha v\beta 5$

integrin to the basic domain of the HIV tat protein and vitronectin. *J. Cell Biol.* **121,** 461–468.

Weeks, B. S., Desai, K., Loewenstein, P. M., Klotman, M. E., Klotman, P. E., Green, M., and Kleinman, H. K. (1993). Identification of a novel cell attachment domain in the HIV-1 Tat protein and its 90-kDa cell surface binding protein. *J. Biol. Chem.* **268,** 5279–5284.

Westendorp, M. O., Frank, R., Ochsenbauer, C., Stricker, K., Dhein, J., Walczak, H., Debatin, K. M., and Krammer, P. H. (1995). Sensitization of T cells to CD95-mediated apoptosis by HIV-1 Tat and gp120. *Nature* **375,** 497–500.

Westendorp, M. O., Li, W. M., Frank, R. W., and Krammer, P. H. (1994). Human immunodeficiency virus type 1 Tat upregulates interleukin-2 secretion in activated T cells. *J. Virol.* **68,** 4177–4185.

Westendorp, M., Shatrov, V., Schulze-Osthoff, K., Frank, R., Kraft, M., Los, M., Krammer, P., Dröge, W., and Lehmann, V. (1995) HIV-1 tat potentiates TNF-induced NF-κB activation and cytotoxicity by altering the cellular redox state. *EMBO J.* **14,** 546–554.

Wu, M. X., and Schlossman, S. F. (1997). Decreased ability of HIV-1 tat protein-treated accessory cells to organize cellular clusters is associated with partial activation of T cells. *Proc. Natl. Acad. Sci. USA* **94,** 13832–13837.

Wrenger, S., Hoffmann, T., Faust, J., Mrestani, K. C., Brandt, W., Neubert, K., Kraft, M., Olek, S., Frank, R., Ansorge, S., and Reinhold, D. (1997). The N-terminal structure of HIV-1 Tat is required for suppression of CD26-dependent T cell growth. *J. Biol. Chem.* **272,** 30283–30288.

Yayon, A., Klagsbrun, M., Esko, J. D., Lader, P., and Ornitz, D. M. (1991). Cell surface, heparin-like molecules are required for binding of basic fibroblast growth factor to its high affinty receptor. *Cell* **64,** 841–848.

Zagury, D., Lachgar, A., Chams, V., Fall, L. S., Bernard, J., Zagury, J. F., Bizzini, B., Gringeri, A., Santagostino, E., Rappaport, J., Feldman, M., Bumy, A., and Gallo, R. C. (1998a). Interferon alpha and Tat involvement in the immunosuppression of uninfected T cells and C-C chemokine decline in AIDS. *Proc. Natl. Acad. Sci. USA* **95,** 3851–3856.

Zagury, D., Lachgar, A., Chams, V., Fall, L. S., Bernard, J., Zagury, J. F., Bizzini, B., Gringeri, A., Santagostino, E., Rappaport, J., Feldman, M., O'Brien, S. J., Burny, A., and Gallo, R. C. (1998b). C-C chemokines, pivotal in protection against HIV type 1 infection. *Proc. Natl. Acad. Sci. USA* **95,** 3857–3861.

Zauli, G., Davis, B. R., Re, M. C., Visani, G., Furlini, G., and La, P. M. (1992). tat protein stimulates production of transforming growth factor-beta 1 by marrow macrophages: a potential mechanism for human immunodeficincy virus-1-induced hematopoietic suppression. *Blood* **80,** 3036–3043.

Zauli, G., Gibellini, D., Caputo, A., Bassini, A., Negrini, M., Monne, M., Mazzoni, M., and Capitani, S. (1995b). The human immunodeficiency virus type-1 Tat protein upregulates Bcl-2 gene expression in Jurkat T-cell lines and primary peripheral blood mononuclear cells. *Blood* **86,** 3823–3834.

Zauli, G., Gibellini, D., Celeghini, C., Mischiati, C., Bassini, A., La, P. M., and Capitani, S. (1996). Pleiotropic effects of immobilized versus soluble recombinant HIV-1 Tat protein on CD3-mediated activation, induction of apoptosis, and HIV-1 long terminal repeat transactivation in purified CD4+ T lymphocytes. *J. Immunol.* **157,** 2216–2224.

Zauli, G., Gibellini, D., Milani, D., Mazzoni, M., Borgatti, P., La Placa, M., and Capitani, S. (1993). Human immunodeficiency virus type 1 tat protein protects lymphoid epithelial and neuronal cell lines from death and apoptosis. *Cancer Res.* **53,** 4481–4485.

Zauli, G., La, P. M., Vignoli, M., Re, M. C., Gibellini, D., Furlini, G., Milani, D., Marchisio, M., Mazzoni, M., and Capitani, S. (1995a). An autocrine loop of HIV type-1 Tat protein responsible for the improved survival/proliferation capacity of permanently Tat-transfected cells and required for optimal HIV-1 LTR transactivating activity. *J. Acquir. Immune Defic. Syndr. Hum. Retrovirol.* **10,** 306–316.

Zidovetzki, R., Wang, J. L., Chen, P., Jeyaseelan, R., and Hofman, F. (1998). Human immunodeficiency virus Tat protein induces interleukin 6 mRNA expression in human brain endothelial cells via protein kinase C- and cAMP-dependent protein kinase pathways. *AIDS Res. Hum. Retroviruses* **14**, 825–833.

Zocchi, M. R., Poggi, A., and Rubartelli, A. (1997). The RGD-containing domain of exogenous HIV-1 Tat inhibits the engulfment of apoptotic bodies by Dendritic cells. *AIDS* **11**, 1227–1235.

Zocchi, M. R., Rubartelli, A., Morgavi, P., and Poggi, A. (1998). HIV-1 Tat inhibits human natural killer cell function by blocking L- type calcium channels. *J. Immunol.* **161**, 2938–2943.

Jørgen Kjems and Peter Askjaer
Department of Molecular and Structural Biology
University of Aarhus
DK-8000 Aarhus C, Denmark

Rev Protein and Its Cellular Partners

I. Introduction

Retroviruses have evolved a complex gene expression strategy in order to produce multiple proteins from a single precursor mRNA (pre-mRNA) transcript. The main contributor to this diversity originates from alternative splicing. In simple retroviruses that only encode three protein products, Gag, Gag-Pol, and Env, a fraction of the nascent transcripts is spliced once to produce an mRNA for the Env protein, whereas the unspliced transcript is exported to the cytoplasm in a constitutive fashion, where it serves both as mRNA for Gag and Gag-Pol precursor proteins and as genomic RNA in viral progeny. The gene expression pattern is more sophisticated in the group of complex retroviruses which include the lentiviruses [prototypic example is the human immunodeficiency virus type 1 (HIV-1)], the oncoretroviruses [e.g., human T-cell leukemia virus type 1 (HTLV-I)], and the spumavirus [e.g., human foamy virus (HFV)]. In HIV-1, a single primary transcript is

Advances in Pharmacology, Volume 48

1054-3589/00 $35.00

used as a precursor for more than 30 different mRNAs encoding at least nine different proteins or protein precursors: Gag, Gag-Pol, Env, Tat, Rev, Nef, Vif, Vpr, and Vpu (Fig. 1; reviewed by Frankel and Young, 1998). The various mRNAs are produced from the alternative use of at least four different 5′ splice sites and eight different 3′ splice sites (Schwartz *et al.*, 1990; Purcell and Martin, 1993). The mRNA species fall into three main groups expressed at different stages of the viral life cycle. The full-length (~9 kb) unspliced mRNA, which encodes the Gag and Gag-Pol fusion proteins and serves as genomic RNA in the virus progeny; a set of intermediate sized (~4 kb), singly spliced mRNAs encoding Env, Vpr, Vif, and Vpu; and a group of fully spliced (~2 kb) mRNAs encoding Tat, Rev, and Nef (Fig. 1).

All retroviruses are faced with the problem that incompletely spliced pre-mRNA generally is retained in the nucleus until RNA splicing is completed. Simple retroviruses bypass this retention by the action of a constitutive transport element (CTE) that, together with cellular host factors, activates the transport of unspliced viral mRNA to the cytoplasm. The CTE was initially characterized in Mason-Pfizer monkey virus (MPMV) (Bray *et al.*, 1994), and similar elements have subsequently been found in other retroviruses (Zolotukhin *et al.*, 1994; Ogert *et al.*, 1996; Tabernero *et al.*, 1997). Candidate host proteins, that bind the CTE in simian retrovirus type 1 (SRV-1) and mediate nuclear export, have recently been identified (Tang

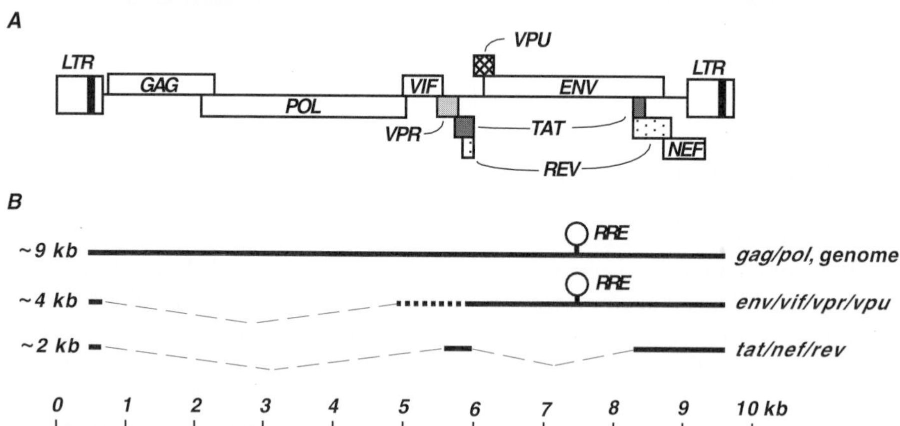

FIGURE 1 Structure and organization of the HIV-1 genome. (A) The HIV-1 provirus contains several overlapping open reading frames (indicated by boxes) flanked by two long terminal repeats (LTRs). (B) Transcription of the HIV-1 provirus initiates in the 5′ LTR and terminates in the 3′ LTR, giving rise to an ~9-kb transcript. Through alternative splicing two classes of mRNA (~4 and ~2 kb) are produced. The proteins encoded by the different mRNA classes are indicated to the right. Also indicated is the position of the Rev response element (RRE).

et al., 1997; Li *et al.*, 1999), whereas no viral proteins appear to be involved in this process which occurs in a constitutive mode. HIV-1, and probably all other complex retroviruses, have evolved an active transport system for incompletely spliced mRNAs that requires a viral trans-acting factor, which in HIV-1 is the Rev protein (Fig. 2; Feinberg *et al.*, 1986; Sodroski *et al.*, 1986; Knight *et al.*, 1987; Malim *et al.*, 1988; Emerman *et al.*, 1989; Felber *et al.*, 1989; Hadzopoulou Cladaras *et al.*, 1989; Hammarskjöld *et al.*, 1989; Pomerantz *et al.*, 1990, 1992; recent reviews include Hammarskjöld, 1997; Cullen, 1998; Emerman and Malim, 1998; Pollard and Malim, 1998). The Rev-regulated expression of incompletely spliced mRNAs has a significant effect on the temporal expression pattern of the individual HIV-1 genes. At the initial stage of HIV-1 expression, the concentration of Rev in the cell is low, and incompletely spliced viral mRNAs accumulate in the nucleus, where they are either spliced to completion or subjected to degradation. Only fully spliced mRNA species are exported to the cytoplasm, where some of them are translated into Rev protein (Fig. 2A). Rev enters the nucleus and binds in an oligomeric manner directly to a viral RNA element, the Rev response element (RRE), present in the *env* gene of all incompletely spliced viral mRNAs. At a certain threshold concentration of Rev protein in the nucleus, functional Rev–RRE complexes are formed, and through interactions with cellular proteins cytoplasmic expression of the singly and nonspliced mRNA species is achieved (Fig. 2B). The temporal delay of expression of mRNAs encoding the structural proteins Gag and Env is probably important for maintenance of viral latency and to prevent elimination of the infected cells by the immune system. Rev-like proteins with a similar mechanism have been characterized in many other complex retroviruses including simian immunodeficiency virus (SIV) (Sakai *et al.*, 1991), feline immunodeficiency virus (FIV) (Kiyomasu *et al.*, 1991), caprine arthritis encephalitis virus (CAEV) (Saltarelli *et al.*, 1994), bovine immunodeficiency virus (BIV) (Oberste *et al.*, 1993), equine infectious anemia virus (EIAV) (Stephens *et al.*, 1990), visna maedi virus (VMV) (Tiley *et al.*, 1990), and HTLV-I and -II (Hidaka *et al.*, 1988; Hanly *et al.*, 1989). The consensus model, based on the large body of HIV-1 Rev evidence, implies that Rev-mediated expression of the RRE-containing mRNAs is mainly caused by a significant increase in their nuclear export (Emerman *et al.*, 1989; Felber *et al.*, 1989; Malim *et al.*, 1989a), but other steps of the RNA processing pathway may contribute to the overall effect. For instance, Rev has been shown to specifically enhance the stability of viral RRE-containing RNA (Felber *et al.*, 1989; Malim and Cullen, 1993), to overcome the retention exerted by the RNA splicing machinery (Chang and Sharp, 1989; Lu *et al.*, 1990), and to increase translational efficiency of the viral mRNAs (Arrigo and Chen, 1991; D'Agostino *et al.*, 1992). Since pre-mRNA splicing, mRNA transport and translation are functionally related processes *in vivo*, it remains a difficult task to ascertain whether the effect of Rev on gene expression is a result of a unique targeting

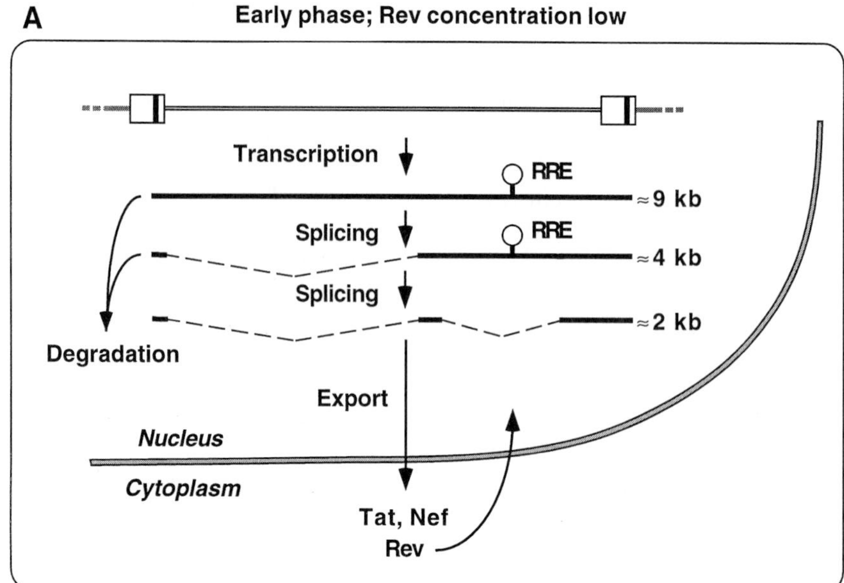

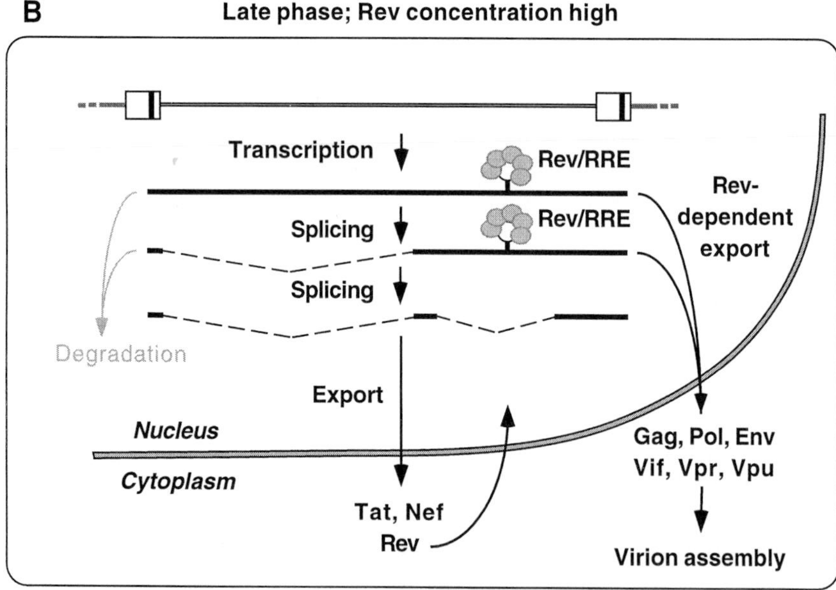

FIGURE 2 HIV-1 gene expression is divided into two temporally separated phases. (A) In the initial phase of HIV-1 gene expression, when no or only little Rev protein has been produced, intron-containing viral RNA is either spliced to completion or degraded in the nucleus. Fully spliced viral RNA is exported to the cytoplasm and translated into the regulatory Tat, Nef, and Rev proteins. Rev is actively imported into the nucleus. (B) In the nucleus Rev binds the RRE present in unspliced and singly spliced viral RNA. When a threshold concentration of Rev is reached, Rev induces nuclear export of these incompletely spliced RNA species, encoding primarily the viral structural proteins. Thus, Rev promotes a shift from an early latent phase to a late virion-producing phase. In the nucleus, Rev also stabilizes RRE-containing RNA.

event or a combined effect of distinct functions at different levels of posttranscriptional processing.

At least six cellular proteins have been reported to interact with Rev and different mechanistic models have been proposed to account for Rev function (Table I). The implications of most of the reported interactions are still speculative, but recently we have obtained a much better understanding of which cellular proteins are responsible for Rev-mediated export of HIV-1 RNA. Moreover, a plausible model for reimport of Rev into the nucleus has also been suggested. The mechanism for transport of Rev across the nuclear membrane will therefore be a major focus of this review.

II. Characteristics of the Rev Protein

A. Functional Domains

HIV-1 Rev is a 116-amino-acid, 18-kDa phosphoprotein predominantly located in the nucleus. Rev has been delineated by extensive mutagenesis into two major functional domains: an N-terminal domain, responsible for RRE binding, oligomerization, and nuclear localization, and a C-terminal

TABLE I Cellular Proteins Proposed to Interact Directly with Rev

Cellular protein	*Rev binding region*[a]	*Cellular localization*	*Suggested function in viral replication*	*References*[b]
CRM-1/ Exportin-1	NES	Nuclear (shuttling)	Nuclear export	1
Rip1/Rab[c] (nucleoporins)	NES	Nuclear	Nuclear export	2
eIF-5A	NES	Nuclear and cytoplasmic	Nuclear export translation	3
B23	NLS/ARM	Nucleolar (shuttling)	Chaperon Nucleolar import	4
p32	NLS/ARM	Primarily mitochondria	Splicing	5
Importin-β	NLS/ARM	Cytoplasmic (shuttling)	Nuclear import	6

Abbreviations:

[a] NES; nuclear export signal, NLS/ARM; nuclear localization signal/arginine-rich motif.

[b] Only selected references are included. See text for further details. 1, Fornerod *et al.*, 1997a; 2, Bogerd *et al.*, 1995; Fritz *et al.*, 1995; 3, Ruhl *et al.*, 1993; 4, Fankhauser *et al.*, 1991; Szebeni and Olson, 1999; 5, Luo *et al.*, 1994; Tange *et al.*, 1996; 6, Henderson and Percipalle, 1997; Truant and Cullen, 1999.

[c] The binding of nucleoporins is probably indirect.

domain, that is important for nuclear export (Fig. 3). The nuclear localization signal (NLS) of Rev has been mapped to a highly basic region (aa 38–46: RRNRRRRWRER) located within the core of the RNA binding domain (Kubota *et al.*, 1989; Perkins *et al.*, 1989; Cochrane *et al.*, 1990;

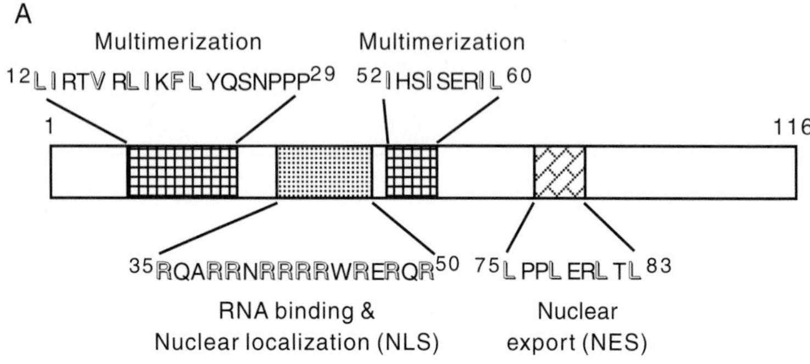

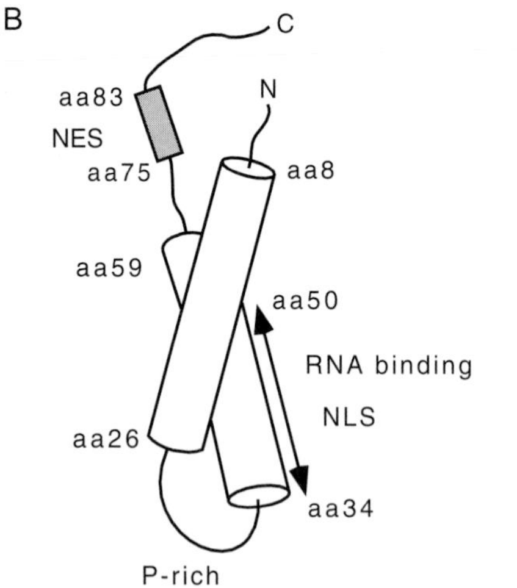

FIGURE 3 Structure of the 116-amino-acid Rev protein. (A) Rev is composed of several domains harboring distinct functions. A leucine-rich stretch in the C-terminus of Rev constitutes a nuclear export signal (NES). The basic region spanning residues 35–50 serves dual functions, namely import of Rev into the nucleus (NLS) and binding of RRE RNA. The basic domain of Rev is flanked by two regions essential for multimerization of Rev. Hydrophobic residues in the NES and multimerization domains, and basic residues in the RNA-binding domain, are written in shadowed type. (B) The N-terminal ~60 residues of Rev is proposed to form a helix-loop-helix structure stabilized by hydrophobic interactions between the multimerization domains. The structure of the C-terminal half of Rev including the NES is drawn arbitrarily.

Venkatesh *et al.*, 1990; Bohnlein *et al.*, 1991; Hammerschmid *et al.*, 1994). An overlapping, but slightly larger region (aa 35–50), constitutes the primary RRE binding site (Hope *et al.*, 1990a; Malim *et al.*, 1990; Bohnlein *et al.*, 1991; Zapp *et al.*, 1991; Kjems *et al.*, 1992; Tan *et al.*, 1993; Hammerschmid *et al.*, 1994; Tan and Frankel, 1995) and has become a prototypic example of the so-called arginine-rich motif (ARM) RNA binding domain (Lazinski *et al.*, 1989). Sequences flanking the basic domain are responsible for multimerization of Rev (Fig. 3) (Olsen *et al.*, 1990; Malim and Cullen, 1991; Zapp *et al.*, 1991; Bogerd and Greene, 1993; Madore *et al.*, 1994; Thomas *et al.*, 1998). The leucine-rich C-terminal domain (aa 75–83: LPPLERLTL) constitutes another functional domain (Malim *et al.*, 1989b; Mermer *et al.*, 1990; Venkatesh and Chinnadurai, 1990; Hope *et al.*, 1992; Weichselbraun *et al.*, 1992). This region was originally termed the activation domain and suggested as a binding site for cellular proteins based on the observed trans-dominant negative phenotype of Rev mutated in this region (Malim *et al.*, 1989b). It has subsequently been shown to function as a nuclear export signal (NES) for Rev (Meyer and Malim, 1994; Fischer *et al.*, 1995).

B. Rev Structure

The elucidation of the three-dimensional structure of Rev by NMR and X-ray crystallography has been impeded by an tendency of Rev to aggregate in solution (Heaphy *et al.*, 1990; Nalin *et al.*, 1990; Heaphy *et al.*, 1991; Karn *et al.*, 1991; Wingfield *et al.*, 1991; Cole *et al.*, 1993). However, the structure of the RNA binding domain of Rev (aa 35–50) in complex with a single high-affinity RNA binding site has been solved at atomic resolution by NMR (Battiste *et al.*, 1996). In agreement with biochemical data using similar constituents (Kjems *et al.*, 1992; Tan *et al.*, 1993; Tan and Frankel, 1994, 1995), the structure revealed that the basic region of Rev forms a regular α-helix, which can interact specifically with an internal loop of the RRE element through major groove interactions (Battiste *et al.*, 1996). For the intact protein only low-resolution structural information exists mainly from circular dichroism and proteolytic mapping experiments. Circular dichroism spectra indicated that Rev contains approximately 50% α-helical structure, mainly positioned within the N-terminal part of the protein (aa 8–66), and that the C-terminal part is more flexible (Auer *et al.*, 1994). The occurrence of a proline-rich region (aa 27–31: PPPNP), which is unfavorable for α-helix formation, suggests that at least two α-helical segments are present in Rev. Moreover, the observation that both predicted α-helices displayed a characteristic patch of hydrophobic amino acids (Fig. 3A) led to the prediction that Rev folds into a helix-loop-helix structure where intramolecular contacts between the two helices are facilitated by hydrophobic interactions (Fig. 3B; Auer *et al.*, 1994). These considerations were later reinforced experimentally by cooperative amino acid substitutions in the two helical regions (Thomas *et al.,* 1997). The helix-loop-helix model is consistent with data obtained from proteolytic digestion of Rev under native conditions in the absence and presence of RRE (Jensen *et al.,* 1995). These studies indicated that only one side of the predicted α-helical basic domain was accessible to proteolytic attack and that RRE binding led to protection of this region. Moreover, the C-terminal region was most prone to proteolytic cleavage, suggesting that this region is less structured. Mapping the epitopes of various Rev-specific monoclonal antibodies has also yielded insight into the overall folding of Rev (Jensen *et al.,* 1997). A discontinuous epitope of a Rev-specific monoclonal antibody was mapped to regions 10-20 and 95-105 of Rev by protein footprinting, predicting that these regions are in close vicinity of each other. This conclusion is consistent with the helix-loop-helix model shown in Fig. 3B.

C. Rev Multimerization

Rev forms multimers both *in vitro* and *in vivo* independent of the RRE (Olsen *et al.,* 1990; Malim and Cullen, 1991; Zapp *et al.,* 1991; Hope *et al.,* 1992; Kubota *et al.,* 1992; Bogerd and Greene, 1993; Madore *et al.,* 1994; Szilvay *et al.,* 1997; Stauber *et al.,* 1998; Thomas *et al.,* 1998). At high concentrations of Rev, it polymerizes into regular hollow fibers which can extend more than one micrometer, but the biological significance of this *in vitro* observation is uncertain (Heaphy *et al.,* 1991; Karn *et al.,* 1991). A 21 Å resolution X-ray structure revealed that the Rev fiber has an outer and inner diameter of approximately 14.8 and 10.4 nm, respectively, formed by a helical arrangement of Rev dimers (Watts *et al.,* 1998).

In the presence of RRE RNA, Rev forms a multimeric complex with an estimated number of up to 12 Rev molecules per RRE (Daly *et al.,* 1989; 1993; Kjems *et al.,* 1991a; Zapp *et al.,* 1991; Mann *et al.,* 1994; Zemmel *et al.,* 1996). However, it is not known whether the same protein–protein interactions are responsible for both RNA-dependent and RNA-independent multimerization. Rev can bind as a monomer to a single high-affinity site within the RRE (Cook *et al.,* 1991; Kjems *et al.,* 1992; Tiley *et al.,* 1992), suggesting that this is the initial event for Rev–RRE complex formation. It is generally believed that additional Rev molecules bind lower affinity sites on the RRE, stabilized by protein–protein interactions (Kjems *et al.,* 1991a; Daly *et al.,* 1993; Mann *et al.,* 1994; Zemmel *et al.,* 1996), and this ability of Rev to multimerize is critical for Rev activity *in vivo* (Olsen *et al.,* 1990; Malim and Cullen, 1991; Zapp *et al.,* 1991). The Rev sequences involved in the multimerization process have been mapped using different assays often yielding conflicting results. It is generally agreed that sequences flanking the basic domain are responsible for multimerization of Rev (Fig. 3; Hope *et al.,* 1990b; Olsen *et al.,* 1990; Malim and Cullen, 1991; Zapp *et al.,* 1991; Bogerd and Greene, 1993; Szilvay *et al.,* 1997; Stauber *et al.,* 1998;

Thomas *et al.,* 1998). However, some reports find that mutations in the NES also influence multimerization of Rev *in vivo* (Bogerd and Greene, 1993; Madore *et al.,* 1994). In a systematic analysis of a large number of Rev mutants a set of hydrophobic residues in two regions of Rev (aa 12–22 and 52–60) was identified as functionally important for multimerization (Fig. 3A; Thomas *et al.,* 1998). Interestingly, these regions coincide with the hydrophobic patches that form the interhelical interaction in the suggested helix-loop-helix model (Fig. 3B; Thomas *et al.,* 1997). It is therefore possible that mutations in these regions perturb the overall structure of Rev, which in turn may be unfavorable for multimerization. Alternatively, the hydrophobic regions may be directly involved in intermolecular contacts, implicating that the proposed structure in Fig. 3B accounts only for monomeric Rev.

D. Rev Phosphorylation

Rev is phosphorylated *in vivo* at two serine residues, positions 54 and 56 (Fig. 3A; Hauber *et al.,* 1988; Cochrane *et al.,* 1989a). The extent of phosphorylation is modulated by phorbol esters, suggesting that a serine/threonine protein kinase from a signal transduction pathway is responsible (Hauber *et al.,* 1988). However, mutating the serines to other amino acids has no measurable effect on Rev function in transfection assays, implying that phosphorylation is not essential for Rev function (Cochrane *et al.,* 1989b; Malim *et al.,* 1989b). More recently, *in vitro* data have shown that phosphorylation of the serine residues in Rev induces a conformational change in the protein structure to a state exhibiting a higher affinity for RRE (Fouts *et al.,* 1997). This observation contrasts other examples of RNA binding phosphoproteins that generally bind RNA less efficiently in the phosphorylated state. Although Ser54 and Ser56 are located in a region implicated in Rev multimerization no effect was observed on the ability of Rev to multimerize upon phosphorylation *in vitro* (Fouts *et al.,* 1997). It is possible that Rev phosphorylation is important for HIV-1 replication in infected cells, where the level of Rev is considerable lower than in most transfection assays.

It is not known which kinase is responsible for Rev phosphorylation *in vivo*. Casein kinase II (CK-II) and mitogen-activated protein kinase (MAPK) have been suggested as potential candidates based on their capacity to phosphorylate Rev *in vitro* (Critchfield *et al.,* 1997; Ohtsuki *et al.,* 1998; Yang and Gabuzda, 1999). Moreover, it has been shown that the basic region of Rev stimulates the general activity of CK-II, including its ability to phosphorylate other HIV-1 proteins (Ohtsuki *et al.,* 1998). The potential role of CK-II in HIV-1 protein phosphorylation may explain the inhibitory effect by CK-II inhibitors like crysin, benzothiophene, and 5,6-dichloro-1-β-D-ribofuranosyl-benzimidazole (DRB) on HIV-1 replication (Critchfield *et al.,*

1996). None of these studies have, however, demonstrated a direct interaction between Rev and the kinase.

E. Rev Localization

Rev accumulates primarily in the nucleolus of the infected cell at steady state (Cullen *et al.,* 1988; Hauber *et al.,* 1988; Cochrane *et al.,* 1989) although a fraction of the Rev molecules are constantly shuttling between the nucleus and cytoplasm (Kalland *et al.,* 1994; Meyer and Malim, 1994; Richard *et al.,* 1994; Wolff *et al.,* 1995). Nucleolar localization is not absolutely required for Rev function (McDonald *et al.,* 1992), suggesting that the nucleolus is used as a storage site for Rev. Since the nucleolar localization requires ongoing pre-ribosomal RNA transcription, it has been suggested that Rev accumulation is caused by interaction with ribosomal RNA (D'Agostino *et al.,* 1995; Dundr *et al.,* 1995) and a potential binding site has been characterized in the 5S rRNA (Lam *et al.,* 1998). Rev localization at different stages of cell division has also been studied. During anaphase and early telophase Rev colocalizes with various nonribosomal nucleolar proteins, including fibrillarin, nucleolin, and B23, within the perichromosomal regions and nucleoli-derived foci. However, in contrast to these proteins, Rev does not enter the nucleolus again before early G1 phase when a nuclear import system and active nucleoli are fully established (Dundr *et al.,* 1996; 1997). This observation is consistent with the idea that Rev accumulation in nucleoli is a retention phenomena caused by association with ribosomal RNA.

III. Rev and Nuclear Export Factors

A. Characterization of the Rev Nuclear Export Signal

The activation domain of Rev has been mapped to amino acids 75-83 on the basis of mutational analyses (Fig. 3A; Malim *et al.,* 1989b, 1991; Mermer *et al.,* 1990; Venkatesh and Chinnadurai, 1990; Hope *et al.,* 1991; Weichselbraun *et al.,* 1992). Mutation of any of three leucines (Leu78, Leu81, or Leu83) in this element not only abolishes Rev function, but also inhibits the activity of coexpressed wild-type Rev protein (WT Rev), a phenomenon known as trans-dominant inhibition. Since these Rev mutants have normal RRE binding activity, it was speculated that the trans-dominant negative phenotype was due to an inability to interact with essential cellular cofactors (see below). In this model, inactive trans-dominant negative Rev (TD Rev) might compete with WT Rev for binding to the target RNA in the nucleus and thereby prevent cytoplasmic appearance of these incompletely spliced RNA species (Malim *et al.,* 1989b). Another possibility is that TD

Rev makes stable but nonfunctional oligomers with WT Rev, thereby sequestering WT Rev (Hope *et al.*, 1992).

As noted earlier, Rev localizes mainly to the nucleolus. This reflects, however, only the steady-state distribution of a more dynamic behavior of Rev. Treatment of cells expressing Rev with either actinomycin D or DRB to block transcription leads to a change in the cellular distribution of Rev from the nucleus to the cytoplasm (Kalland *et al.*, 1994; Meyer and Malim, 1994). In these studies the cells were additionally treated with the translational inhibitor cycloheximide to prevent *de novo* protein synthesis, demonstrating that Rev is exported from the nucleus. Moreover, the export process was shown to be active, as it ceased at 4°C. The actinomycin D-mediated cytoplasmic distribution of Rev was shown to be reversible, suggesting that nuclear entry of Rev requires active transcription. To further demonstrate the shuttling behavior of Rev, without addition of drugs to block transcription, Meyer and Malim (1994) fused murine cells with human cells expressing Rev. After cell fusion, Rev could be detected in the nuclei derived from both species, showing that Rev can be exported from the human nuclei and imported into the nuclei of mouse origin. Importantly, Rev protein mutated in the activation domain (TD Rev) does not shuttle, but is restricted to the nucleus and even impedes the export of WT Rev when the two Rev forms are coexpressed (Meyer and Malim, 1994; Stauber *et al.*, 1995; Szilvay *et al.*, 1995). Microinjection of RRE-containing RNA along with Rev protein into the nuclei of *Xenopus laevis* oocytes extended the view to include a direct and active participation of Rev in export of viral RNA (Fischer *et al.*, 1994). In this system, Rev stimulated the nuclear export of unspliced pre-mRNA and excised intron-lariat RNA, which both contain the RRE. Export of the spliced mRNA, from which the RRE had been removed, was, as it would be predicted, Rev, independent. In agreement with the ability of TD Rev to inhibit WT Rev export in mammalian cells, coinjection of a Rev activation domain mutant reduced the Rev-dependent export of intron-lariat RNA in these injection experiments (Fischer *et al.*, 1994). Furthermore, the study by Fischer and co-workers showed that Rev can also mediate export of RNA that does not contain any splice sites. Thus, an RNA consisting of an exon only as well as an unusual cap structure at the 5′ end to delay export is rendered export-competent by the RRE in the presence of Rev (Fischer *et al.*, 1994). Likewise, a chimera of U3 snRNA, which is normally retained in the nucleus, and the RRE is exported from *X. laevis* nuclei upon coinjection of Rev protein (Pasquinelli *et al.*, 1997a). Together, the above results strongly suggest that Rev associates with its cognate RNA in the nucleus and promotes export of the Rev/RNA complex by interaction with cellular factors via the activation domain. Since Rev activity can be reproduced in amphibian cells as well as in yeast (Stutz and Rosbash, 1994), Rev cofactors were expected to be widely conserved during evolution. To test whether the activation domain of Rev is sufficient for nuclear export, Wen and co-workers (1995) fused the sequence encoding amino acids 73-84 of Rev (LQLPPLERLTLD) to glutathione-*S*-transferase (GST) and microinjected the fluorescently labeled protein into the nuclei of fibroblast cells. After 30 min the fusion protein was exported to the cytoplasm, while, in contrast, a fusion protein where Rev amino acids Leu78 and Glu79 were replaced with Asp78 and Leu79 remained in the nucleus. The same mutation, named M10, has in full-length Rev protein been shown to cause a TD Rev phenotype (Malim *et al.*, 1989b, 1991; Hope *et al.*, 1990b; Bevec *et al.*, 1992) and abolish nucleocytoplasmic shuttling (Meyer and Malim, 1994; Szilvay *et al.*, 1995). Using *X. laevis* oocytes as a biological model, Fischer and co-workers could also demonstrate that the activation domain has an intrinsic export capacity in that it mediates export of a nuclear injected BSA-Rev activation domain conjugate (Fischer *et al.*, 1995). The activation domain of Rev has accordingly been renamed Nuclear Export Signal, or NES.

B. Nuclear Export Receptor CRM1/Exportin 1

The molecular mechanism for Rev-mediated nuclear export has been a subject of intense study in recent years and a great deal of our knowledge on general nucleocytoplasmic transport relies on analysis of Rev as a prototypic shuttling protein. From these investigations, which are described in more detail below, the following scheme can be deduced: In the nucleoplasm, Rev bound to RRE-containing RNA associates with at least two soluble proteins, CRM1 and Ran, to form an export complex (Fig. 4, step 1). This ternary complex interacts with proteins associated with adjustable pores in the nuclear membrane, followed by translocation of the export complex through the pore and disassembly in the cytoplasm (Fig. 4, step 2).

Several independent observations recently led to the discovery of the 112-kDa protein CRM1 (originally described in *Schizosaccharomyces pombe* for its chromosome region maintenance phenotype) as a nuclear export receptor for Rev. In accordance with its newly assigned function, CRM1 has also been renamed Exportin 1 or XPO1 (Fornerod *et al.*, 1997a; Stade *et al.*, 1997). In a screen for inhibitors of Rev export, Wolff and co-workers found four antibiotics of the *Streptomyces* leptomycin/kazusamycin family that blocked this process (Wolff *et al.*, 1997). Of the four drugs, leptomycin B (LMB) proved to be the most potent inhibitor, and concentrations in the low nanomolar range efficiently inhibited Rev export in HeLa cells as well as HIV-1 replication in primary human monocytes. Unfortunately, due to its cytotoxicity, LMB cannot be used therapeutically (Wolff *et al.*, 1997). From studies in *S. pombe* it was known that LMB targets the CRM1 protein since point mutations in CRM1 can confer LMB resistance (Nishi *et al.*, 1994). Thus, together these observations suggested a connection between Rev and CRM1.

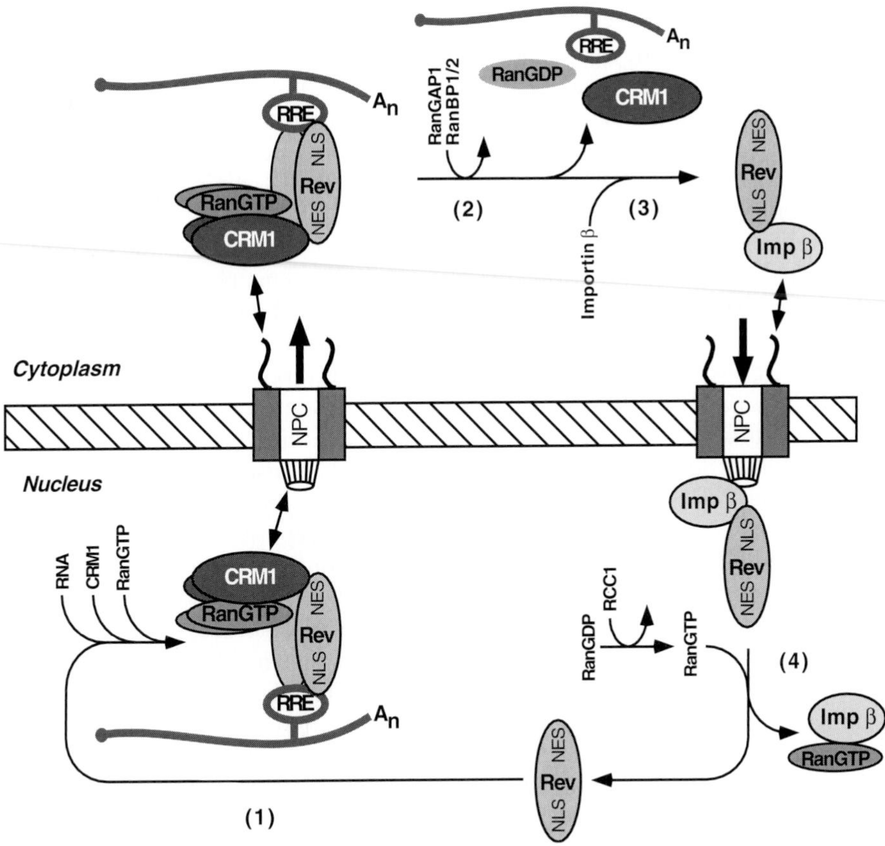

FIGURE 4 Nucleocytoplasmic shuttling of Rev can be divided into four steps. (1) Within the nucleus Rev multimerizes on the RRE via its overlapping RNA-binding domain/nuclear localization signal (NLS). In addition, Rev binds to CRM1 through the nuclear export signal (NES) together with Ran in its GTP bound form (RanGTP). Consecutive interactions of CRM1 with nucleoporins lead to docking and translocation of the export complex through the nuclear pore complex (NPC). (2) At the cytoplasmic face of the NPC, RanGTPase activating protein (RanGAP1) and Ran binding proteins 1 and/or 2 (RanBP1/2) trigger hydrolysis of Ran-bound GTP resulting in complex disassembly. (3) In the cytoplasm the NLS of Rev is recognized by importin-β, which interacts with the NPC and mediates nuclear entry. (4) The import complex docks on the nuclear side of the NPC until nuclear RanGTP generated by Ran guanine nucleotide exchange factor (RCC1) binds importin-β. This leads to release of Rev in the nucleoplasm and a single transport cycle is complete.

Sequence database searches had placed CRM1 in a family of ~20 importin β-like proteins based on homology to the N-terminal Ran-binding site of the nuclear import factor, importin-β (see below) (Fornerod *et al.*, 1997b; Görlich *et al.*, 1997), possibly reflecting a role of CRM1 in nucleocytoplasmic trafficking. Soon after, several groups published results demonstrating a direct role of CRM1, not only in Rev export, but also in export of other proteins carrying a Rev-like NES (Fornerod *et al.,* 1997a; Fukuda *et al.,* 1997; Ossareh-Nazari *et al.,* 1997; Stade *et al.,* 1997). Overexpression of human CRM1 in *X. laevis* oocytes accelerated nuclear export of Rev protein as well as U snRNA, whereas the transport kinetics of tRNA, mRNA and a control protein were unaffected (Fornerod *et al.,* 1997a). CRM1 made by *in vitro* translation was furthermore found to bind a Rev NES peptide, but not a mutant RevM10 peptide, and only in the presence of the GTPase Ran in its GTP-bound form (see Section III,C). Finally, the connection between Rev export and CRM1 was substantiated by demonstrating that overexpression of CRM1 reverses the inhibitory effect of LMB on Rev export seen *in vivo* and that LMB inhibits the observed interaction between Rev and CRM1 *in vitro* (Fornerod *et al.,* 1997a). Interestingly, LMB binds directly to CRM1, but without changing the shuttling behavior of CRM1 itself (Fornerod *et al.,* 1997a). Yeast genetic data also revealed a role of CRM1 on Rev function. In viable *Saccharomyces cerevisiae crm1* mutant strains, the activity of Rev on a RRE-containing reporter RNA is dramatically reduced (Neville *et al.,* 1997). In a more direct assay for the effect of yeast Crm1p on protein localization, Stade and co-workers (1997) demonstrated that a fusion protein, carrying a NES from protein kinase inhibitor (PKI, see Table II) as well as a nuclear localization signal, accumulated in the nucleus as early as 5 min after shifting a temperature-sensitive *crm1* mutant strain to the nonpermissive temperature. Finally, in the yeast two-hybrid assay, Crm1p has been shown to interact with WT Rev and the PKI NES, but not RevM10 nor a PKI NES mutant (Neville *et al.,* 1997; Stade *et al.,* 1997). Contemporary studies using Rev-like NESs from the mitogen-activated protein kinase kinase (MAPKK) and IκBα substantiated the picture of CRM1 being responsible for signal-mediated nuclear protein export (Fukuda *et al.,* 1997; Ossareh-Nazari *et al.,* 1997).

The list of proteins carrying functional leucine-rich NESs is rapidly growing (Table II) and it seems obvious that CRM1-mediated nuclear export is a potential regulatory step in many cellular processes, such as signal transduction, transcriptional regulation, and cell growth (see references in Table II). Interestingly, the Rev proteins of two HIV-1 related lentiviruses, equine infectious anemia virus (EIAV) and feline immunodeficiency virus (FIV), contain a domain which is functional interchangeable with the HIV-1 Rev NES in spite of a quite divergent amino acid composition (Mancuso *et al.,* 1994). Competition experiments suggest, moreover, that the NESs of HIV-1 Rev and EIAV Rev work through the same saturable pathway, sensitive to LMB (Meyer *et al.,* 1996; Otero *et al.,* 1998). It will be important to get more detailed information on how CRM1 recognizes the apparently different NES structures. One possibility is that an adapter is involved in recognition of those NESs that deviate substantially from the somewhat loose consensus sequence shown in Table II.

TABLE II List of Proteins Containing Rev-like NESs

	Protein[a]	Amino acid sequence[b]	NES activity[b]		Reference[c]
			Natural	Heterol.	
Viral	Rev (HIV-1)	L-PPLERLTL	+	+	1,2
	Rex (HTLV-I)	LSAQLYSSLSL	+	+	3,4
	ICP27 (HSV-1)	LIDLGLDLDL	+	+	5
	NS2 (MVM)	MTKKFGTLTI	ND	+	6
	NS1 (Flu)	F-DRLETLIL	+	+	7
	NS2 (Flu)	ILLRMSKMQL	ND	+	8
	Rev (VMV)	M-VGMEMLTL	ND	+	9
	E4-34kD (Ad)	MVLTREELVI	+	+	10
Cellular	PKI	LALKLAGLDI	+	+	2
	MAPKK	LQKKLEELEL	+	+	11
	MK 2	MTSALATMRV	+	−	12
	Dsk-1p	LEGAVSEISL	ND	+	13
	c-Abl	LESNLRELQI	+	+	14
	Cyclin B1	LCQAFSDVIL	+	+	15,16
	RanBP1	VAEKLEALSV	+	+	17,18
	Actin NES1	LPHAIMRLDL	+	+	19
	Actin NES2	IKEKLCYVAL	+	+	19
	An3	LDQQFAGLDL	+	ND	20
	IκBα	IQQQLGQLTL	+	+	21
	IRF-3	LDELLGNMVL	+	+	22
	p53	MFRELNEALEL	+	+	23
	Mex67p	L-ELLNKLHL	ND	+	24
	TFIIIA	L-PVLENLTL	ND	+	25
	FMRP	LKEVDRQLRL	+	+	26
	HDM2	L-SFDESLAL	+	+	27
	Gle1p	L--PLGKLTL	ND	+	28
	Yap1p	I--DVDGLCS	+	+	29
	Pap1p	IDDLCSKLKN	+	+	30
Consensus[d]		$LX_{2-3}LX_{2-3}L$			4,31,32
	Rev (EIAV)	PLESDQWCRVLRQSLPEEKIP			9,33
	Rev (FIV)	KKMMTDLEDRFRKLFGSPSKDEYT			34,35

[a] The origins of viral proteins are shown in brackets (HTLV-I, human T-cell lymphotropic virus type I; HVS-1, herpes simplex virus 1; MVM, minute virus of mice; Flu, influenza virus; VMV, visna maedi virus; Ad, adenovirus type 5; EIAV, equine infectious anemia virus; FIV, feline immunodeficiency virus).

[b] Peptides shown explicitly to constitute functional nuclear export signals in their natural context and/or when fused to a heterologous protein (indicated by + signs; ND denotes not determined).

[c] References: 1, Meyer and Malim, 1994; 2, Wen *et al.,* 1995; 3, Palmeri and Malim, 1996; 4, Kim *et al.,* 1996; 5, Sandri-Goldin, 1998; 6, Ohshima *et al.,* 1999; 7, Li *et al.,* 1998; 8, O'Neill *et al.,* 1998; 9, Meyer *et al.,* 1996; 10, Dobbelstein *et al.,* 1997; 11, Fukuda *et al.,* 1996; 12, Engel *et al.,* 1998; 13, (Fukuda *et al.,* 1997); 14, (Taagepera *et al.,* 1998); 15, (Toyoshima *et al.,* 1998); 16, Yang *et al.,* 1998; 17, Richards *et al.,* 1996; 18, Zolotukhin and Felber, 1997; 19, Wada *et al.,* 1998; 20, Askjaer *et al.,* 1999; 21, Arenzana-Seisdedos *et al.,* 1997; 22, Stommel *et al.,* 1999; 24, Segref *et al.,* 1997; 25, Fridell *et al.,* 1996a;

(*continues*)

TABLE II *Continued*

26, Fridell *et al.*, 1996b; 27, Roth *et al.*, 1998; 28, Murphy and Wente, 1996; 29, Yan *et al.*, 1998; 30, Kudo *et al.*, 1999; 31, Bogerd *et al.*, 1996; 32, Zhang and Dayton, 1998; 33, Harris *et al.*, 1998; 34, Mancuso *et al.*, 1994; 35, Otero *et al.*, 1998.

[d] From sequence alignment and *in vivo* selection studies a consensus sequence can be generated, containing four characteristically spaced hydrophobic residues (shown in bold; X denotes any residue). Although leucine residues are most common, other hydrophobic residues are found at each position as well. Thus, since the consensus is nonstringent, NES activity cannot be predicted solely on sequence homology. Two such peptides which functionally can substitute the HIV-1 Rev NES are included below the consensus sequence.

C. RanGTPase Cycle

Since Rev function is clearly linked to nucleocytoplasmic transport, a brief overview is given of the essential components involved in this process (yeast homologs given in brackets). The 25-kDa GTPase Ran (sometimes referred to as Gsp1p in *S. cerevisiae*) has, through numerous studies, been shown to play a pivotal role in translocation across the nuclear envelope (reviewed recently in Dahlberg and Lund, 1998; Mattaj and Englmeier, 1998). Like other GTPases, Ran cycles between two forms, RanGTP and RanGDP, and requires effectors to modulate its nucleotide-bound state. The low intrinsic GTPase activity of Ran is stimulated by the Ran GTPase-activating protein RanGAP1 (Rna1p), an effect increased by the Ran-binding proteins 1 and 2, RanBP1 (Yrb1p), and RanBP2/Nup358 (Bischoff *et al.*, 1994, 1995a,b; Richards *et al.*, 1995; Mahajan *et al.*, 1997). Conversely, the replacement of Ran-bound guanine nucleotides is greatly stimulated by the exchange factor RCC1 (Prp20p/Pim1p) (Bischoff and Ponstingl, 1991; Klebe *et al.*, 1995). Since the majority of RanGAP1 in the cell is cytoplasmic, whereas RCC1 is associated with chromatin and thus nuclear, it is predicted that cytoplasmic Ran exists primarily in the GDP-bound state, and nuclear Ran is mainly in the GTP form. This asymmetric distribution of the two forms of Ran has been shown to be essential for movement of several classes of RNA and proteins in and out of the nucleus (Görlich *et al.*, 1996a; Izaurralde *et al.*, 1997a; Richards *et al.*, 1997). In contrast to the nuclear import receptors (e.g., importin-β; see Section IV) import factors, which only bind their substrate in the absence of RanGTP (i.e., in the cytoplasm), export receptors like CRM1 form ternary complexes with their substrate(-s) and RanGTP but not RanGDP (i.e., in the nucleus; Fig. 4, step 1). Indeed, in several cases a highly cooperative binding between the export receptor, substrate, and RanGTP has been reported (Fornerod *et al.*, 1997a; Kutay *et al.*, 1997, 1998), and in the study from Mattaj's group, CRM1 was found to crosslink to a Rev NES peptide only in the presence of RanGTP (Fornerod *et al.*, 1997a). In contrast, we have recently reported that full-length Rev protein can interact directly with CRM1 in the absence of Ran-

GTP (Askjaer *et al.,* 1998). The RanGTP-independent interaction is not stringently dependent on the NES core sequence and is resistant to LMB, while formation of the ternary Rev/CRM1/Ran complex is sensitive to LMB, NES mutations, and Ran being charged with GTP. The most likely explanation of the discrepancy in Ran requirement for Rev/CRM1 interaction is in the use of either full-length protein or an isolated NES peptide. Indeed, our experiments indicate that residues outside the NES core (and therefore not included in the NES peptide) play a role in the binding (Askjaer *et al.,* 1998). A direct comparison of multiple NES peptides and NES-containing proteins with respect to heterotrimeric substrate/CRM1/RanGTP complex formation has furthermore shown that the level of cooperativity is significantly lower when Rev is used as substrate as compared to the NES from PKI, An3, and minute virus of mice NS2 (Askjaer *et al.,* 1999). The relatively low affinity of Rev toward CRM1 might suggest that the steady-state localization of Rev in the nucleus is caused by a slower export than import rate. Furthermore, it might also be part of the explanation why Rev multimerization on the RRE is required for Rev to promote export of viral RNA. As the local concentration of Rev increases upon binding of several Rev molecules to the RNA, association with CRM1 and RanGTP becomes perhaps more efficient, thereby increasing the export rate (Askjaer *et al.,* 1999).

When the ternary substrate/CRM1/RanGTP complex has crossed the nuclear envelope (see next section) the concerted action of RanGAP1 and either RanBP1 or RanBP2 has been proposed to disassemble the heterotrimer by triggering GTP hydrolysis (Fig. 4, step 2; Bischoff and Görlich, 1997; Fornerod *et al.,* 1997a). Based on the potent capacity of RanBP1 to stimulate RanGAP1 activity on NES/CRM1/RanGTP heterotrimers *in vitro* (Askjaer *et al.,* 1999), one can imagine that RanBP1 (and perhaps RanBP2) acts *in vivo* by releasing RanGTP from the protein complex, thereby making RanGTP accessible to RanGAP1. This event probably dissociates Rev from CRM1, but whether the exported RRE-containing RNA is released from Rev concomitantly is unknown. In fact, the precise role(-s) that hydrolysis of Ran-bound GTP plays in nuclear export is currently quite uncertain. Active transport in and out of the nucleus has previously been considered to be driven exclusively by Ran-mediated GTP hydrolysis (see Mattaj and Englmeier, 1998 for discussion). However, it has later been shown that a Ran mutant locked in the GTP conformation (RanQ69L) can replace wild-type Ran in export of Rev-like NESs from microinjected mammalian and *X. laevis* nuclei (Izaurralde *et al.,* 1997a; Richards *et al.,* 1997). Likewise, in a study using permeabilized human cells, nuclear NES export was unaffected by the substitution of GTP and ATP with nonhydrolyzable analogous (Englmeier *et al.,* 1998). Thus, at least in shuttling of free NES proteins, hydrolysis of Ran-bound GTP is perhaps only necessary to keep the concentration of RanGTP in the cytoplasm low, which would otherwise disturb binding of import receptors to their substrate (e.g., reimport of export factors).

D. Rev and Nucleoporins

Trafficking across the nuclear envelope takes place through nuclear pore complexes (NPCs), which are large protein complexes with a molecular mass of approximately 125 MDa and estimated to consist of ~50 different components, so-called nucleoporins (see Fabre and Hurt, 1997; Gant *et al.*, 1998; Ohno *et al.*, 1998 for recent reviews on NPC structure and function). In principle, proteins and RNA molecules smaller than about 40–60 kDa can diffuse freely through the NPC, but even these molecules are normally transported by an active process. Due to the complexity of the NPC it has proven very difficult to assign specific functions to the individual nucleoporins, although yeast genetic data are accumulating rapidly. Future investigations of the dynamic interactions between the Rev-containing export complex and the NPC during translocation will therefore elucidate not only Rev-mediated gene expression but also help understand geneal nucleocytoplasmic transport. Most nucleoporins are characterized by the presence of multiple FXFG, GLFG, and FG repeated sequences (single-letter amino acid code), and both entire nucleoporins and individual repeats show a certain degree of redundancy *in vivo* (Fabre and Hurt, 1997). Nucleoporins have traditionally been cloned and subsequently identified as static components of the NPC by immunolocalization. However, more recent data suggest a rather dynamic association of certain nucleoporins with the NPC, and detection of soluble nucleoporins in both the nucleoplasm and cytoplasm points towards an active role of nucleoporins in the transport processes (Zolotukhin and Felber, 1999).

Using the yeast two-hybrid screen to find cellular cofactors for Rev, two groups independently identified a human protein containing numerous FG dipeptide repeats (Bogerd *et al.*, 1995; Fritz *et al.*, 1995). The 59-kDa protein, named hRip (human Rev-interacting protein) or Rab (Rev/Rex activation domain-binding protein), was shown to interact *in vivo* with Rev and other retroviral Rev-like proteins dependent on a functional NES. Furthermore, overexpression of hRip/Rab in mammalian cells increased the activity of Rev if the level of Rev expression was suboptimal or if the intrinsic activity of Rev was impaired by a single amino acid substitution (Bogerd *et al.*, 1995; Fritz *et al.*, 1995). Concurrently with the discovery of hRip/Rab, Stuz and colleagues isolated Rip1p from a yeast genomic library on the basis of interaction with Rev, but not RevM10, in the two-hybrid system (Stutz *et al.*, 1995). Rip1p shares several characteristics with other yeast nucleoporins, including FG repeats and a punctuate perinuclear staining revealed by immunofluorescence microscopy. Both a *RIP1* gene disruption and overexpression of Rip1p reduced specifically the activity of Rev in yeast, implying that Rip1p is required for optimal Rev function and that excess Rip1p may compete with endogenous NPC-localized Rip1p for Rev interaction (Stutz *et al.,* 1995). It is unclear whether Rip1p is the yeast homolog of hRip/Rab, while, in all three reports, the homology of hRip/Rab and Rip1p to the carboxy-terminal FG repeat domain of another mammalian nucleoporin, CAN/Nup214, is discussed (Bogered *et al.,* 1995; Fritz *et al.,* 1995; Stutz *et al.,* 1995). Interestingly, this region of CAN/Nup214 interacts with CRM1 and it causes CRM1 to redistribute away from the nuclear envelope and into the nucleoplasm when overexpressed (Fornerod *et al.,* 1997b). This observation suggests that CRM1 contacts the NPC through CAN/Nup214. However, since CAN/Nup214 is localized on the cytoplasmic face of the NPC it was speculated that other nucleoporins must be involved in docking of the CRM1-containing export complex at the nuclear face of the NPC. Consistent with this notion, Neville and co-workers showed that Crm1p interacts with several yeast and human nucleoporins in the yeast two-hybrid system (Neville *et al.,* 1997). Among these, Nup98, which also belongs to the class of FG repeat-containing nucleoporins, has been reported to be present at the nuclear side of the NPC, but possibly only transiently (Radu *et al.,* 1995). Expression of Rev in HeLa cells resulted in the redistribution of Nup98 from the nuclear rim and nucleoplasm to the nucleolus, while a Rev mutant lacking the NES had no effect (Zolotukhin and Felber, 1999). Similarly, also CRM1 and CAN/Nup214 could be recruited to the nucleolus by Rev, while another nucleoporin, Nup153, remained localized at the nuclear rim and in the nucleoplasm. Treatment of the cells with LMB reversed the effects of Rev on localization of CRM1 and the nucleoporins (Zolotukhin and Felber, 1999). In agreement with the ability of the CAN/Nup214 repeat domain to cause CRM1 redistribution, overexpression of the same domain inhibits Rev nuclear export in human cells and hence also Rev function (Bogerd *et al.,* 1998). Recently, Rev export, as well as export of Rev-dependent RNA, was shown to involve Nup153 in that injection of Nup153 antibodies into *X. laevis* nuclei blocks these transport processes (Ullman *et al.,* 1999). While it seems clear that nucleoporins play a role in Rev function the precise mechanisms are less obvious. In the yeast two-hybrid system Stutz and colleagues observed interactions between Rev and the FG repeats of multiple yeast and mammalian nucleoporins, while three nucleoporins belonging to the FXFG class scored negative (Stutz *et al.,* 1996). However, coinjections of Rev and individual repeat domains into nuclei of *X. laevis* oocytes failed to show any clear correlation between interference with Rev export in *X. laevis* oocytes and the strength of interaction in the yeast two-hybrid study (Stutz *et al.,* 1996). A limitation of most *in vivo* interaction assays is the possibility that an observed association between two molecules is not direct, but bridged by an cellular factor. This appears indeed to be the case for the interaction between Rev and Rip1p. When assayed in a *crm1* mutant strain, Neville and co-workers did not detect any Rev/Rip1p association, demonstrating that wild-type CRM1 activity is required for the interaction (Neville *et al.*, 1997). Indeed, attempts to show direct interaction between purified preparations of Rev and hRip/Rab have been without success (Henderson and Percipalle, 1997).

With the identification of CRM1 as an export receptor for Rev, is it reasonable to suggest that the *in vivo* interactions between Rev and nucleoporins reported above are in general mediated by CRM1, which leads to the following summarized model for Rev export: Either prior to or after binding of Rev to its target RNA inside the nucleus, CRM1 binds cooperatively to Rev and RanGTP and targets the RNP complex for export by binding to soluble and NPC-anchored nucleoporins (Fig. 4, step 1). After translocation through the NPC channel, RanGAP1 together with either RanBP1 or RanBP2 promote disassembly of the exported RNP complex and trigger hydrolysis of the Ran-bound GTP (Fig. 4, step 2). In the cytoplasm the RRE-containing RNA is released from Rev by an unknown mechanism, perhaps involving nuclear import factors binding to Rev.

E. Common Pathways in Rev-Mediated and Cellular Export

The increasing number of shuttling proteins harboring Rev-like NESs makes it clear that Rev, and hence the RRE-containing HIV-1 RNA, is exported via a cellular protein export pathway (Table II and references herein). However, cellular factors utilizing the Rev export pathway could potentially also be involved in RNA transport. Competition experiments in *X. laevis* oocytes have revealed that the different RNA classes are actively exported by distinct saturable pathways, indicative of multiple RNA-type specific export factors (Jarmolowski *et al.*, 1994). Of the vertebrate proteins listed in Table II only transcription factor IIIA (TFIIIA) has been reported to participate in RNA export. TFIIIA regulates RNA polymerase III transcription of the 5S rRNA gene family but also binds the 5S rRNA itself to form a 7S complex in the nucleus, which then migrates to the cytoplasm (Guddat *et al.*, 1990). An NES-dependent shuttling phenotype has not been demonstrated for the 7S RNP complex, but two reports indicate a link to Rev-mediated RNA export: (1) TFIIIA contains a sequence homologous to the Rev NES. These residues can functionally replace the NES of Rev in HIV-1 RNA export and mediate export of a nuclear injected GST fusion protein (Fridell *et al.*, 1996a). (2) Saturation of Rev-dependent RNA export in *X. laevis* oocytes by injection of BSA-Rev NES conjugates leads to inhibition of the export of 5S rRNA and spliceosomal U snRNA, but not mRNA, tRNA, or ribosomal subunits (Fischer *et al.*, 1995). The prospect of TFIIIA being responsible for 5S rRNA export is, however, not clear, since 5S rRNA mutants unable to bind TFIIIA still translocate to the cytoplasm, presumably through interaction with the L5 protein (Guddat *et al.*, 1990). L5, on the other hand, contains no recognizable Rev-like NES, suggestive of yet another transport factor involved in export of 5S rRNA.

The connection between Rev and U snRNA transport mentioned above was further drawn by the observation that overexpression of CRM1 in *X. laevis* oocytes stimulates export of U snRNA while the kinetics of mRNA and tRNA export are unaffected (Fornerod *et al.*, 1997a). Likewise, injection of free NES peptides into oocyte nuclei specifically interferes with U snRNA export, but not processing and export of mRNA and tRNA (Askjaer *et al.*, 1999). All spliceosomal U snRNAs, with the exception of U6, are transported to the cytoplasm for maturation and then reimported to the nucleus (Nakielny *et al.*, 1997; Mattaj and Englmeier, 1998 and references herein). Binding of the newly synthesized transcripts by the cap binding complex (CBC) is known to be critical for U snRNA export, but the exact function of CBC in the transport process remains to be discovered.

Since proteins containing Rev-like NESs are either known or believed to be exported by CRM1, one would expect that these proteins might compete for export of Rev. One of the proteins included in Table II, IκBα, has in fact been suggested to compete with Rev at the level of transport, but the data are ambiguous. IκBα accumulates transiently in the nucleus, where it promotes export of the transcription factor NFκB, leading to reduced NFκB-dependent transcription (e.g., transcription from the HIV-1 promoter) (Arenzana-Seisdedos *et al.*, 1997). In addition, IκBα has also been reported to interfere posttranscriptionally with Rev-regulated gene expression. However, while one report argues that the inhibition of Rev activity is due to NES-dependent competition for CRM1-mediated nuclear export (Bachelerie *et al.*, 1997), another report claims that an alternative, unknown mechanism is responsible (Wu *et al.*, 1997). The reason for this discrepancy is uncertain at the moment since both groups reached their conclusions through similar mutations within the putative IκBα NES.

F. Cellular mRNA Processing and Export

Although RRE-containing HIV-1 RNA seems to be exported by a pathway different from the export of cellular mRNAs, we briefly focus on this important export route, since some common factors might be involved. Moreover, fully spliced HIV-1 RNA (i.e., Rev-independent RNA) is most probably recognized and exported like cellular mRNAs. The primary HIV-1 transcript itself contains cis-elements which could potentially be export signals. Similar to most other RNA polymerase II transcripts, the 5′ end of the viral RNA is cotranscriptionally modified with a "cap," while a poly(A) tail is added to the 3′ end. Both the cap and the poly(A) tail have been suggested to play a role in mRNA export, but neither structure is essential (Nakielny *et al.*, 1997). Except for alternatively spliced messenger RNA, pre-mRNA molecules containing splice sites are retained in the nucleus. It can be imagined that binding of splicing factors to the pre-mRNA prevents export mediators from recognizing the RNA or that splicing factors contain retention signals acting dominantly over the export factors (Legrain and Rosbash, 1989; Nakielny *et al.,* 1997). During transcription, the nascent pre-mRNA is bound by a variety of abundant heterogeneous nuclear proteins to form hnRNP particles (reviewed by Dreyfuss *et al.,* 1993; Mattaj and Englmeier, 1998). One of the well-characterized hnRNP proteins is hnRNP A1, which, based on several observations, is a strong candidate as a mRNA export factor (Michael *et al.,* 1995; Visa *et al.,* 1996; Izaurralde *et al.,* 1997b).

In addition to a potential role in nucleocytoplasmic transport, hnRNP A1 has also been shown to regulate splicing, perhaps creating a link between the two processes (Mayeda and Krainer, 1992; Caceres *et al.,* 1994). Recently, hnRNP A1 was found to bind two distinct HIV-1 RNA elements, a splicing silencer element from the second exon of the Tat gene and an instability element (INS) in the p17gag gene (Del Gatto-Konczak *et al.,* 1999; Najera *et al.,* 1999). While the former element has been shown to repress splicing *in vitro* (Amendt *et al.,* 1995), the mechanism underlying INS function is less clear. Based on experiments showing that the p17gag INS confers Rev responsiveness on a β-globin reporter gene cooperatively with the RRE, it has been proposed that binding of hnRNP A1 to the p17gag INS rescues unspliced RNA from a nuclear splicing compartment into a Rev-accessible export pathway (Mikaelian *et al.,* 1996; Najera *et al.,* 1999). A more direct participation of the shuttling hnRNP A1 protein in HIV-1 RNA export is, however, also a possibility.

In contrast to the situation in higher eukaryotes, analyses of nuclear export in yeast have indicated a similarity between the export of Rev and mRNA via CRM1, but the results are ambiguous. The finding that the *S. cerevisiae* proteins Gle1p and Mex67p both contain an essential NES-like sequence (Table II) and appear to be involved in mRNA transport (Murphy and Wente, 1996; Segref *et al.,* 1997) suggests that CRM1 in yeast also mediates this export. However, it remains to be demonstrated if Gle1p and Mex67p actively shuttle (i.e., contain a functional NES), as expected for mRNA export factors, and in an CRM1-dependent manner. Stade and coworkers have reported that shifting a *crm1* temperature-sensitive mutant strain to the nonpermissive temperature leads to a rapid and pronounced accumulation of poly(A)+ RNA in the nucleus, while importin β-mediated protein import is unaffected (Stade *et al.,* 1997). In contrast, Neville and colleagues did not detect any defect in mRNA export in three other *crm1* mutant strains, although Rev activity was greatly reduced (Neville *et al.,* 1997). A general obstacle to analysis of transport processes in living cells is the problem of distinguishing between direct and indirect effects. A block in nuclear export might be a secondary effect of an import defect and vice versa, in particular when considering the cycling behavior of transport receptors. Due to the existence of multiple import and export pathways, it seems therefore difficult to rule out the possibility that the nuclear poly(A)+ RNA accumulation observed by Stade and co-workers is indirect, although the rapid onset argues against this.

Recently, Rev was implicated in yet another specialized export pathway, involving heat-shock mRNA. When yeast is heat-shocked at 42°C, bulk poly(A)+ RNA accumulates in the nucleus, whereas mRNAs encoding heat-shock proteins are efficiently exported. However, deletion of *RIP1,* which is dispensable for growth under normal conditions, prevents the export of these heat-shock mRNAs upon stress (Saavedra *et al.,* 1997a; Stutz *et al.,* 1997). Based on the original observations of Rip1p being a Rev NES-interacting nucleoporin, Saavedra and co-workers furthermore looked for a correlation between Rev activity and expression of heat-shock proteins. Overexpression of Rev, but not RevM10, leads to a partial inhibition of heat-shock mRNA export, suggesting that Rev titrates a cellular protein(-s) involved in a common export pathway (Saavedra *et al.,* 1997a). However, further investigations are required to elucidate the pathway utilized by heat-shock mRNA and perhaps Rev.

G. Eukaryotic Initiation Factor 5A

The search for cellular Rev cofactors has also identified the 19-kDa eukaryotic initiation factor 5A (eIF5A) as a specific Rev NES binding protein (Ruhl *et al.,* 1993). Transfection experiments with either eIF5A mutants or an antisense eIF5A construct, which are supposed to block expression of endogenous eIF5A, revealed an inhibition of Rev nuclear export and HIV-1 replication (Ruhl *et al.,* 1993; Bevec *et al.,* 1996). More direct evidence was provided by studies in *X. laevis* oocytes, where Ruhl and co-workers found that injection of a Rev plasmid alone could not mediate expression of an RRE-containing reporter gene, but required coexpression of eIF5A (Ruhl *et al.,* 1993). However, this apparent eIF5A requirement in *X. laevis* oocytes is in clear conflict with data from several groups mentioned above, demonstrating a direct export activity of Rev on RRE-containing RNA in the absence of exogenous added eIF5A (Fischer *et al.,* 1994; Pasquinelli *et al.,* 1997a; Saavedra *et al.,* 1997b). A direct role of eIF5A in Rev export is therefore not obvious. The cellular function of eIF5A is unknown, and depletion of eIF5A in yeast does not cause dramatic effects on protein synthesis, implying that eIF5A is a misleading name (Kang and Hershey, 1994). Recent results suggest a role for eIF5A in mRNA turnover, which might account for some of the effects seen on Rev function (Zuk and Jacobson, 1998). Interestingly, eIF5A is the only protein known to contain a hypusine residue, arising from a posttranslational modification of a lysine residue at the N-terminus. The function of this modification is unknown, but seems nevertheless essential since treatment with inhibitors of hypusine formation leads to cell growth arrest and apoptosis in tissue culture (Park *et al.,* 1993). In a recent study Andrus *et al.* (1998) reported that two of these drugs, mimosine and deferiprone, specifically inhibit expression of Rev-dependent HIV-1 transcripts resulting in suppressed viral replication, further suggesting a link between eIF5A and Rev activity.

Using eIF5A as bait in a yeast two-hybrid screen led to the discovery of ribosomal protein L5 as an eIF5A-interacting protein, supported by *in vivo* and *in vitro* coprecipitation assays (Schatz *et al.,* 1998). As described earlier, L5 is probably involved in nuclear export of 5S rRNA, which in turn utilizes the same export pathway as Rev (Guddat *et al.,* 1990; Fischer *et al.,* 1995). These data combined suggest a putative connection between eIF5A and Rev via L5, and, indeed, overexpression of L5 stimulates Rev-dependent gene expression in cell culture (Schatz *et al.,* 1998). However, as long as clear evidence for direct effects on Rev nuclear export is lacking, the involvement of eIF5A and L5 remains speculative.

IV. Rev and Nuclear Import Factors

The Rev NLS region (aa 38–48: RRNRRRRWRER) contains a high proportion of basic amino acids, a feature that bears some resembles to the "classical" NLS elements represented by the SV40 large T antigen (PKKK-RKV; Kalderon *et al.,* 1984) and the bipartite signal in nucleoplasmin (KR-PAATKKAGQAKKKK; Robbins *et al.,* 1991). However, the Rev NLS is strikingly more arginine rich. The mechanism for import of lysine-rich NLS proteins has been studied intensively and is generally viewed as a multiple-step process involving docking of the NLS-containing substrate on the cytoplasmic side of the NPC and translocation through the NPC, followed by cargo release in the nucleus (reviewed by Mattaj and Englmeier, 1998). Although proteins carrying arginine-rich NLSs probably follow the same overall pathway, interesting differences at the docking step exist for the two types of NLS.

A. Docking Process

The classic lysine-rich NLS, which is found in the majority of nuclear proteins, is recognized by the NLS receptor, importin-α (also known as karyopherin-α), which is encoded by a multigene family in humans. Importin-α interacts directly with the NLS and with a second import factor, importin-β (also known as karyopherin-β), which appears to be encoded by a single gene in humans (Adam and Gerace, 1991; Görlich *et al.,* 1994; Imamoto *et al.,* 1995; Weis *et al.,* 1995). Importin-β mediates docking of the complex to nucleoporins associated with the cytoplasmic fibers protruding from the NPC (Görlich *et al.,* 1995; Moroianu *et al.,* 1995a,b). Importin-

α thereby functions as an adapter that links the lysine-rich NLS to the actual import factor, importin-β. The import mechanism of Rev, which contains an arginine-rich NLS, diverges from the classic pathway at this step. Rev has been shown to interact directly with importin-β (Fig. 4, step 3; Henderson and Percipalle, 1997), whereas importin-α is nonessential for Rev import in an *in vitro* nuclear import assay (Truant and Cullen, 1999). The binding sites of the Rev NLS and importin-α on importin-β probably overlap based on the observation that a peptide corresponding to the importin-β-binding domain of importin-α (IBB) competes specifically with the binding of Rev to importin-β and with Rev import in a permeabilized cell assay (Truant and Cullen, 1999). Also, the observation that the IBB domain contains an arginine-rich sequence motif that resembles the NLS of Rev and that IBB can function as a NLS when attached to a heterologous protein supports the idea that Rev NLS structurally mimics the IBB domain (Görlich *et al.*, 1996b). A very similar import mechanism appears to be operating for the HIV-1 Tat protein (Truant and Cullen, 1999) and the HTLV-1 Rex protein (Palmeri and Malim, 1999).

B. Translocation and Release

The subsequent steps have not been studied for the Rev protein in particular, but they probably resemble the import pathway utilized by the lysine-rich NLS. After docking, the complex is translocated through the NPC channel by an active but largely unknown mechanism. The energy may be provided by hydrolysis of GTP by the RanGTPase (Moore and Blobel, 1993; Weis *et al.*, 1996; Paschal *et al.*, 1997), but recent experiments suggest that translocation is independent of RanGTP hydrolysis (Englmeier *et al.*, 1998; Ribbeck *et al.*, 1999). Once the import complex reaches the nuclear face of the NPC, nuclear RanGTP binds importin-β, which then dissociates from importin-α (or Rev; Fig. 4, step 4). This model is supported by the observation that RanGTP is able to dissociate a complex consisting of NLS, importin-α, and importin-β *in vitro* (Rexach and Blobel, 1995; Moroianu *et al.*, 1996). Moreover, using an importin-β mutant which is unable to interact with Ran does not seem to interfere with translocation but blocks release of the imported protein at the nuclear side of the NPC (Görlich *et al.*, 1996a). After the NLS protein has been released importin-α and -β are exported back to the cytoplasm.

C. B23

A second protein, shown to interact with the NLS of Rev, is the B23 protein (also known as NO38, nucleophosmin, or numatrin) (Fankhauser *et al.*, 1991). B23 is a phosphoprotein that shuttles between the nucleolus and the cytoplasm (Borer *et al.*, 1989), but at steady-state predominantly localizes in the nucleolus (Fankhauser *et al.*, 1991). B23 has also been shown to interact with a variety of other NLS sequences including the SV40 large T antigen NLS in a 1:1 stoichiometry (Szebeni *et al.*, 1995), and two highly acidic segments of B23 are implicated in the interaction (Adachi *et al.*, 1993). The functional role of B23 remains elusive but the nucleolar localization and NLS binding characteristics of B23 are consistent with a role of escorting ribosomal proteins into the nucleolus (Szebeni and Olson, 1999). Both the nuclear import of Rev and albumin conjugated to the SV40 large T antigen NLS is stimulated by B23, an effect which is further enhanced if B23 is phosphorylated by casein kinase II (Szebeni *et al.*, 1997). However, even in the presence of optimized concentrations of B23, the import is strongly inhibited by antibodies specific for importin-β, suggesting that importin-β is the actual import factor and that B23 merely facilitates the process. This is in agreement with the observation that B23 is unable to induce import of Rev in the absence of importin-β (Truant and Cullen, 1999). Recent evidence suggests that B23 may act as a molecular chaperone by preventing aggregation of proteins (Szebeni and Olson, 1999). Such a function may be crucial for ribosome biogenesis within the densely packed nucleolus of a cell. In particular, B23 was shown to inhibit the aggregation of Rev protein, suggesting that B23 may also function as a chaperone for Rev *in vivo*. In contrast to other well-characterized chaperones, which function by repeated cycles of substrate association and an ATPase-coupled dissociation, B23 seems to remain constantly associated with Rev under *in vitro* conditions (Szebeni and Olson, 1999). However, it cannot be excluded that necessary release factors are absent in the cell-free system and that the B23-Rev association is only short-lived *in vivo*.

D. Coordination of Factors Interacting with the Basic Region

Since the binding sites for the RRE RNA, importin-β, and B23 all map to the basic region of Rev (aa 35–50), it remains an open question how association of Rev with the various components is temporally and structurally programmed within the cell. Whereas it is possible that the release of Rev from the RRE and binding of importin-β may occur simultaneously in the cytoplasm, it is unlikely that the transfer of Rev from importin-β to RRE RNA is a coupled process in the nucleus, considering that Rev probably is released from importin-β immediately upon RanGTP binding. It is possible that the function of B23 is to bind the basic domain of Rev after Rev is released from importin-β and protect it from aggregation until it encounters the RRE RNA target. The temporary storage of Rev/B23 complexes may occur in the nucleolus, since this is the natural environment of B23.

E. Cytoplasmic Retention of Rev

Ions, metabolites, and small proteins with masses below 40–60 kDa can generally enter the nucleus by passive diffusion. However, in spite of Rev being significantly smaller than this limit, import of Rev appears to be actively controlled. Rev import is efficiently blocked if transcription in the nucleus is ceased by drug treatment with, e.g., actinomycin D or by lowering the temperature. This is also the phenotype of Rev mutants which contain either deletions or single amino acid substitutions in the NLS element (Berger *et al.,* 1991; Kubota *et al.,* 1992; Furuta *et al.,* 1995). These observations have led to the suggestion that in addition to active transport, the Rev protein contains a cytoplasmic retention signal. The region responsible for cytoplasmic accumulation has been mapped to 15 amino acids near the N-terminus (aa 11–25) and it was shown that this element alone coupled either to HTLV-1 p21 or ribosomal protein S25 inhibits nuclear entry; hence it was named "nuclear entry inhibitory signal" (NIS) (Kubota and Pomerantz, 1998). The existence of an NIS domain in Rev is supported by Szilvay *et al.* (1997), who reported an exclusive cytoplasmic distribution of a Rev mutant dysfunctional in both the NLS and NES signals. The aa 11–25 region of Rev is also implicated in Rev multimerization (Fig. 3A; Thomas *et al.,* 1998) and in formation of the interhelical hydrophobic interaction in the putative helix-loop-helix structure of Rev (Fig. 3B; Thomas *et al.,* 1997). It is conceivable that Rev multimerization, via this domain, decreases the level of nuclear diffusion and thereby accounts for the observed cytoplasmic retention of Rev NLS mutants. However, the aa 52–60 region is also essential to multimerization (Thomas *et al.,* 1998), making it less likely that the action of NIS alone can be explained by oligomerization. Hence, an alternative theory may be that unknown cytoplasmic protein(s) is associated with the NIS element.

V. Rev and RNA Splicing Factors

A possible regulatory role for Rev in RNA splicing has been debated a great deal. Several reports have demonstrated that Rev only has a marginal effect *in vivo* on the ratio of nuclear located spliced and unspliced RRE-containing pre-mRNA derived from HIV-1 or heterologous systems (e.g., Chang and Sharp, 1989; Malim and Cullen, 1993), whereas others find that Rev expression causes a modest decrease in fully spliced HIV-1 mRNA in the nucleus (Favaro *et al.,* 1998). Differences in constructs and cell lines may explain these discrepancies. In general, the effect of Rev expression on RNA splicing is measured according to the actual level of spliced product, ignoring the possibility that Rev may affect the composition of hnRNP and prespliceosome complexes assembled on the nascent transcript in the nucleus. Individual splicing factors are likely to play an important role in nuclear retention and may therefore indirectly affect nuclear export. The interplay between Rev function and assembly of pretransport complexes has been studied *in vivo* and *in vitro*. It has been demonstrated that Rev-activated cytoplasmic appearance of an unspliced β-globin pre-mRNA construct containing a single intron with an RRE element is dependent on the combination of a suboptimal and an optimal splice site (Chang and Sharp, 1989). A simple interpretation of this result is that the kinetics of spliceosome assembly *in vivo* must be sufficiently slow to allow Rev to promote active transport of intron-containing RNA species. Moreover, the necessity of at least one optimal splice site has been interpreted as the requirement for being recognized as an intron, which in turn may cause nuclear retention. This model is supported by other reports showing that an intron-less, RRE-containing HIV-1 env mRNA is exported to the cytoplasm in a Rev-independent manner and that a Rev response can be restored by inserting a 5′ splice site upstream from the env gene (Lu *et al.*, 1990; Hammarskjöld *et al.*, 1994). By introducing a genetic suppressor U1 snRNP in the cell, matching a mutated 5′ splice site, Hammarskjöld and co-workers were able to demonstrate that U1 snRNP recognition of the splice site was essential for Rev response (Lu *et al.*, 1990). The importance of splice site recognition has also been demonstrated in a yeast study, where it was shown that mutations in either the 5′ splice site or the branch point eliminate Rev responsiveness (Stutz and Rosbash, 1994). Experiments *in vitro* have demonstrated that unspliced RNA corresponding to the Tat gene binds strongly to the U1 snRNP but only to a limited extent to the other spliceosomal snRNPs, U2, U4, U5, and U6 (Dyhr-Mikkelsen and Kjems, 1995). Moreover, when using a heterologous RNA construct, containing optimal 5′ and 3′ splice sites, Rev is able to block *in vitro* splicing in an RRE-dependent fashion (Kjems *et al.*, 1991b). Analysis of the arrested spliceosome complex revealed a normal level of U1 snRNP binding, whereas the entry of the U5/U4/U6 tri-snRNP is severely retarded by Rev (Kjems and Sharp, 1993). The cellular target for this inhibitory effect by the basic domain of Rev has not been characterized, but circumstantial evidence suggests that a 32-kDa cellular protein (p32) is involved (Tange *et al.*, 1996). p32 was originally identified as a protein associated with the essential splicing factor SF2/ASF (Krainer *et al.*, 1991) and has also been isolated using Rev as bait in a yeast two-hybrid screen (Luo *et al.*, 1994). The functional significance of this interaction was reinforced by the observed stimulatory effect on Rev activity *in vivo* upon transfection with p32 expression plasmids (Luo *et al.*, 1994). p32 interacts strongly with Rev *in vitro* and modulates the inhibitory effect of Rev on *in vitro* splicing (Tange, 1996).

The splicing factor SF2/ASF has been shown to bind RRE in a Rev-dependent manner (Powell *et al.*, 1997), raising the possibility that the p32/SF2/ASF complex may bind cooperatively to the Rev/RRE complex. SF2/

ASF belongs to the group of so-called SR-proteins, which play an important role in recruiting snRNPs and other splicing factors to the splicing signals in pre-mRNA (Manley and Tacke, 1996). A role of Rev/RRE-associated SF2/ASF in the recruitment of U1 snRNP to the 5′ splice site is consistent with the above data, including the obligatory binding of U1 snRNP for Rev response (Lu *et al.*, 1990).

The importance of U1 snRNP binding raises the question how and when the U1 snRNP is removed from the transcript. Generally, snRNPs are released from the mRNA as a consequence of splicing. It is therefore likely that a specific mechanism exists for removal of snRNPs associated with incompletely spliced HIV-1 transcripts before transport. A potential cellular factor engaged in this process is RNA helicase A, based on the observation that injection of antibodies directed toward this protein inhibits Rev-dependent gene expression in human cells (Li *et al.*, 1999). Moreover, RNA helicase A binds weakly to the RRE in a Rev-independent manner (Li *et al.*, 1999) as well as to the constitutive transport element (CTE) from a simian retrovirus (Tang *et al.*, 1997).

In some instances it has been shown that Rev can function independently of functional splice sites. These constructs contain cis-acting elements termed instability sequences (INS) or cis-acting repressive sequences (CRS) (Rosen *et al.*, 1988; Cochrane *et al.*, 1991; Maldarelli *et al.*, 1991; Schwartz *et al.*, 1992; Nasioulas *et al.*, 1994; Mikaelian *et al.*, 1996; Schneider *et al.*, 1997). The negative effects of these elements on cytoplasmic expression, which include increased degradation of the RNA in the nucleus, nuclear retention, and inhibition of translation, can be overcome by the Rev/RRE system. Interaction of host factors with these RNA elements has been reported (Olsen *et al.*, 1992; Mikaelian *et al.*, 1996), but there is no evidence for a mechanism involving a physical link to the Rev/RRE complex.

VI. Rev Targeted Therapy

Several properties of Rev make it an ideal target for therapeutic intervention of HIV-1 replication. Rev is a relatively small protein and the amino acid sequence of the functional domains is highly conserved among different HIV isolates. Also the nucleic acid sequence is highly constrained by 1-2 overlapping reading frames for the Tat and Env proteins, respectively. The therapy need not target all Rev molecules since Rev exhibits favorable threshold kinetics in that Rev concentrations must rise above a critical level to exert an effect. Moreover, a therapeutic intervention would occur at a relatively early step in gene expression, eliminating the expression of cytotoxic Env protein. The disadvantage is that Rev uses the same transport pathways as cellular proteins, which narrows the therapeutic target possibilities. Moreover, Rev functions in the nucleus of the cell which calls for more sophisticated drug delivery systems.

Several anti-Rev strategies have been proposed and tested (reviewed by Gilboa and Smith, 1994; Pomerantz and Trono, 1995). When grouped according to the nature of the effector molecule they include: (1) Protein-based effector molecules. This group includes various transdominant Rev mutants, intracellular expressed antibodies, and short peptide-derived drugs designed to interfere with the binding of Rev to various ligands. (2) Nucleic acid-based effector molecules. This group includes RNA antisense and ribozymes directed toward the Rev gene (mRNA) or the RRE, and RNA sense molecules acting as decoys. (3) Nonprotein, nonnucleic acid-based small molecules with inhibitory effects on Rev function.

A. Trans-Dominant Rev Proteins

Various Rev mutants have been shown to posses trans-dominant negative phenotypes in respect to the function of wild-type Rev protein (WT Rev). The best characterized group of mutants contain amino acid substitutions/deletions within the NES region exemplified by the RevM10 mutant (see above; Malim *et al.*, 1989b). Expression of RevM10 strongly inhibits HIV-1 replication in cell lines of lymphoid lineage (Bevec *et al.*, 1992; Malim *et al.*, 1992; Bahner *et al.*, 1993; Escaich *et al.*, 1995; Ragheb *et al.*, 1995; Vandendriessche *et al.*, 1995; Plavec *et al.*, 1997) and CD34-enriched haematopoietic progenitor stem cells (HPSC) (Bauer *et al.*, 1997; Bonyhadi *et al.*, 1997; Su *et al.*, 1997). RevM10 was also the first HIV-1-derived construct to be evaluated in a clinical trial where it was shown to prolong survival of T cells in HIV-1 infected patients (Woffendin *et al.*, 1996). Using an improved clinical protocol based on retroviral gene transfer of RevM10 to CD4+ T cells, the same group was able to detect RevM10-producing cells in HIV-1-infected individuals for an average of 6 months posttransduction (Ranga *et al.*, 1998). The toxicity of RevM10 expression in human cells is generally reported to be low (Fox *et al.*, 1995). However, nuclear export of certain cellular mRNAs has been reported to be inhibited upon injection of a BSA-M10 NES peptide conjugate into *X. laevis* oocytes (Pasquinelli *et al.*, 1997b), and constitutive RevM10 expression is toxic in CMT3/COS cells (M.-L. Hammerskjöld, personal communication).

Two multimerization deficient Rev mutants, RevSLT26 and RevSLT40 (described above), have also been shown to inhibit WT Rev function in a trans-dominant manner (Thomas *et al.*, 1998), but to a much lower extent than RevM10. Moreover, RevSLT40 was also shown to inhibit HIV-1 replication in a transient expression assay, but again less potently than RevM10 (Thomas *et al.*, 1998). A Rev mutant in which the basic amino acids 38–44 (RRNRRRR) were deleted (Rev38) has also been reported to exhibit a trans-dominant phenotype (Kubota *et al.*, 1992; Furuta *et al.*, 1995). This mutant was originally reported to accumulate in the cytoplasm and it was proposed that it may function by sequestering WT Rev protein in the cytoplasm (Furuta *et al.*, 1995). However, in a more recent study it was found that, in comparison to RevM10, the trans-dominant phenotype of Rev38 was negligible (Stauber *et al.*, 1998). Moreover, in contrast to previous reports it was demonstrated in a dual-color autofluorescent study of living cells that WT Rev expression led to colocalization of both Rev38 and WT Rev in the nucleolus (Stauber *et al.*, 1998).

The trans-dominant potential has also been studied using Rev fusion proteins. A strong effect was obtained when a fusion protein (Trev) was constructed by joining trans-dominant mutants of both Tat and Rev (Chinen *et al.*, 1997). In another example, WT Rev protein, when fused to the 78-amino-acid N-terminal fragment of NS1 protein of influenza A virus, was shown to inhibit Rev function. Since both the RNA binding and oligomerization activity of the fusion protein were retained, it was suggested that the fusion protein and WT Rev protein form mixed oligomers and that the nuclear retention activity of NS1 is dominant over the Rev-mediated nuclear export (Chen *et al.*, 1998).

Expression of mutant cellular cofactors for Rev may also lead to inhibition of viral replication. This principle was utilized by Bevec *et al.*, who demonstrated that expression of a mutant form of eIF-5A, which was competent in Rev binding, inhibited Rev function (Bevec *et al.*, 1996). The inhibitory effect is probably achieved by sequestering of Rev protein by the eIF5A mutant. However, since trans-dominant eIF-5A may also target other cellular proteins containing a NES, this approach can potentially effect important cellular processes.

B. Intracellular Antibodies (Intrabodies)

Intracellular single-chain antibodies (sFv intrabodies), which can be expressed from a single gene, are effective tools in neutralizing intracellular pathogenic molecules (reviewed by Rondon and Marasco, 1997). Expression of anti-Rev sFv intrabodies in human PBMC cells from a retroviral vector has been shown to have a significant negative effect on HIV-1 replication, and it was proposed that sFv interferes with Rev nuclear import by sequestering Rev in the cytoplasm (Duan *et al.*, 1994, 1995).

In an attempt to further disturb the Rev–RRE interaction, Inouye *et al.* (1997) expressed a combination of Anti-Rev sFv intrabodies and an RRE decoy in T cells and macrophages and found that infection with laboratory and clinical virus isolates was effectively inhibited. In a comparative study the effect of monoclonal sFv directed toward either the NES region or the C-terminal region of Rev was compared (Wu *et al.*, 1996). The NES-specific sFv was the most potent inhibitor of HIV-1 replication in PBMC and T-cell lines, even though the binding affinity of the C-terminal sFv for Rev was significantly higher. Simple affinity studies may therefore not be sufficient for prediction of the therapeutic effectiveness of a particular sFv.

C. Aptamers and Peptides

In vitro selection methods, employing an iterative binding-selection procedure, have been used to isolate Rev-binding RNA or peptide molecules from random pools of sequences. RNA-based aptamers with comparable or higher Rev affinity than the natural RNA substrate (RRE) have been isolated this way (Xu and Ellington, 1996). RNA aptamers are superficially similar to peptide antibodies and may recognize epitopes other than the natural RNA binding site. Some of the selected aptamers were subsequently shown to be functional as Rev response elements *in vivo* (Symensma *et al.*, 1996). Increased Rev affinity for the aptamer correlated with increased level of Rev response, suggesting that binding of the RNA, more than the binding of cellular factors, is the limiting step in the Rev-mediated response. In a modified *in vitro* selection protocol Jensen *et al.* selected RNA aptamers, which, upon binding, become covalently attached to the Rev molecule by a spontaneous crosslinking reaction (Jensen *et al.*, 1995). Short peptides, which specifically interact with Rev, have been isolated from a 15-amino-acid randomized library by the M13 phage display method (Jensen *et al.*, 1998). Strikingly, nearly all the isolated peptides appear to interact with the NES of Rev, suggesting that this region is particularly exposed to protein–protein interactions. The therapeutic possibilities of RNA and peptide-based aptamers still remain to be explored.

D. Small Molecules

A number of organic compounds, unrelated to nucleic acids and peptides, have been demonstrated to target Rev function, either by interference with proper function of the host proteins or by binding to the RRE. Leptomycin B, which binds CRM1 and blocks the interaction to NESs, is a very potent inhibitor of Rev function and HIV-1 replication, but is unsuitable as a therapeutic drug, owing to its cytotoxic properties (Wolff *et al.*, 1997). α-Hydroxypyridones, especially mimosine and deferiprone, target the post-translational hypusine-modification of eIF5A, and it has been shown to suppress HIV-1 replication (Andrus *et al.*, 1998). However, changes to the eIF5A-modification may be the cause of the general toxicity observed also with this drug. Aminoglycoside antibiotics have been shown to bind the RRE and inhibit the interaction with Rev at micromolar concentrations. Moreover, their inhibitory effect was also observed *in vivo* in Rev transfection and HIV-1 replication assays, but only when applied in millimolar concentrations (Zapp *et al.*, 1993). The diminished potency *in vivo* probably reflects the polycationic nature of aminoglycosides antibiotics, which is unfavorable for cellular uptake. More recently, the same group have demonstrated an inhibitory effect on the Rev/RRE interaction by several aromatic heterocyclic compounds (Zapp *et al.,* 1997).

VII. Concluding Remarks

The ability of HIV-1 to persist, even when attacked with multiple drugs, has called for continuing the search for novel strategies in HIV therapy. The recent advances in our understanding of HIV-1 posttranscriptional regulation and Rev function will most likely improve the likelihood of success. One important finding has been that Rev interacts with the transport machinery in a fashion that is very similar to other cellular proteins. This may imply that targeting of Rev at the level of nuclear export may cause severe side effects on vital cellular processes. The most promising approach may therefore be to develop effector molecules that will interfere with the Rev–RRE interaction and/or Rev multimerization. The first generation of molecules targeting the Rev–RRE interaction, including negative trans-dominant protein RevM10 and antisense/decoy RRE have proven their effects—even in clinical trails. However, this may only be the starting point for the design of second-generation therapeutical molecules that will interact more strongly with Rev than the natural ligands and thereby inhibit HIV-1 replication more efficiently.

Acknowledgment

We thank Torben Heick Jensen, Jakob Nilsson, Finn Skou Pedersen, and Ray Brown for critically reading of this manuscript. The work was supported in part by the Danish Natural Research Council, the Danish Biotechnology Program, the Danish Medical Research Council, the Danish AIDS Foundation, the EU Biomed Program, and the Karen Elise Jensen Foundation.

References

Adachi, Y., Copeland, T. D., Hatanaka, M., and Oroszlan, S. (1993). Nucleolar targeting signal of Rex protein of human T-cell leukemia virus type I specifically binds to nucleolar shuttle protein B-23. *J. Biol. Chem.* **268**, 13930–13934.

Adam, S. A., and Gerace, L. (1991). Cytosolic proteins that specifically bind nuclear location signals are receptors for nuclear import. *Cell* **66**, 837–847.

Amendt, B. A., Si, Z. H., and Stoltzfus, C. M. (1995). Presence of exon splicing silencers within human immunodeficiency virus type 1 tat exon 2 and tat-rev exon 3: Evidence for inhibition mediated by cellular factors. *Mol. Cell. Biol.* **15**, 4606–4615.

Andrus, L., Szabo, P., Grady, R. W., Hanauske, A. R., Huima-Byron, T., Slowinska, B., Zagulska, S., and Hanauske-Abel, H. M. (1998). Antiretroviral effects of deoxyhypusyl hydroxylase inhibitors: A hypusine-dependent host cell mechanism for replication of human immunodeficiency virus type 1 (HIV-1). *Biochem, Pharmacol.* **55,** 1807–1818.

Arenzana-Seisdedos, F., Turpin, P., Rodriguez, M., Thomas, D., Hay, R. T., Virelizier, J. L., and Dargemont, C. (1997). Nuclear localization of I kappa B alpha promotes active transport of NF-kappaB from the nucleus to the cytoplasm. *J. Cell Sci.* **110,** 369–378.

Arrigo, S. J., and Chen, I. S. (1991). Rev is necessary for translation but not cytoplasmic accumulation of HIV-1 vif, vpr, and env/vpu 2 RNAs. *Genes Dev.* **5,** 808–819.

Askjaer, P., Jensen, T. H., Nilsson, J., Englmeier, L., and Kjems, J. (1998). The specificity of the CRM1-Rev nuclear export signal interaction is mediated by RanGTP. *J. Biol. Chem.* **273,** 33414–33422.

Askjaer, P., Bachi, A., Wilm, M., Bischoff, R., Weeks, D. L., Ogniewski, V., Ohno, M., Kjems, J., Mattaj, I. W., and Fornerod, M. (1999). RanGTP-regulated interactions of CRM1 with nucleoporins and a shuttling DEAD-box helicase. *Mol. Cell Biol.* **19,** 6276–6285.

Auer, M., Gremlich, H. U., Seifert, J. M., Daly, T. J., Parslow, T. G., Casari, G., and Gstach, H. (1994). Helix-loop-helix motif in HIV-1 Rev. *Biochemistry* **33,** 2988–2996.

Bachelerie, F., Rodriguez, M. S., Dargemont, C., Rousset, D., Thomas, D., Virelizier, J. L., and Arenzana-Seisdedos, F. (1997). Nuclear export signal of IkappaBalpha interferes with the Rev-dependent posttranscriptional regulation of human immunodeficiency virus type I. *J. Cell. Sci.* **110,** 2883–2893.

Bahner, I., Zhou, C., Yu, X. J., Hao, Q. L., Guatelli, J. C., and Kohn, D. B. (1993). Comparison of trans-dominant inhibitory mutant human immunodeficiency virus type 1 genes expressed by retroviral vectors in human T lymphocytes. *J. Virol.* **67,** 3199–3207.

Battiste, J. L., Mao, H., Rao, S., Tan, R., Muhandiram, D. R., Kay, L. E., Frankel, A. D., and Williamson, J. R. (1996). α helix-RNA major groove recognition in an HIV-1 Rev peptide-RRE RNA complex. *Science* **273,** 1547–1551.

Bauer, G., Valdez, P., Kearns, K., Bahner, I., Wen, S. F., Zaia, J. A., and Kohn, D. B. (1997). Inhibition of human immunodeficiency virus-1 (HIV-1) replication after transduction of granulocyte colony-stimulating factor-mobilized CD34+ cells from HIV-1-infected donors using retroviral vectors containing anti-HIV-1 genes. *Blood* **89,** 2259–2267.

Berger, J., Aepinus, C., Dobrovnik, M., Fleckenstein, B., Hauber, J., and Bohnlein, E. (1991). Mutational analysis of functional domains in the HIV-1 Rev trans-regulatory protein. *Virology* **183,** 630–635.

Bevec, D., Dobrovnik, M., Hauber, J., and Bohnlein, E. (1992). Inhibition of human immunodeficiency virus type 1 replication in human T cells by retroviral-mediated gene transfer of a dominant-negative Rev trans-activator. *Proc. Natl. Acad. Sci. USA* **89,** 9870–9874.

Bevec, D., Jaksche, H., Oft, M., Wohl, T., Himmelspach, M., Pacher, A., Schebesta, M., Koettnitz, K., Dobrovnik, M., Csonga, R., Lottspeich, F., and Hauber, J. (1996). Inhibition of HIV-1 replication in lymphocytes by mutants of the Rev cofactor eIF-5A. *Science* **271,** 1858–1860.

Bischoff, F. R., and Ponstingl, H. (1991). Catalysis of guanine nucleotide exchange on Ran by the mitotic regulator RCC1. *Nature* **354,** 80–82.

Bischoff, F. R., Krebber, H., Kretschmer, J., Wittinghofer, A., and Ponstingl, H. (1994). RanGAP1 induces GTPase activity of nuclear Ras-related Ran. *Proc. Natl. Acad. Sci. USA* **91,** 2587–2591.

Bischoff, F. R., Krebber, H., Kempf, T., Hermes, I., and Ponstingl, H. (1995a). Human RanGTPase-activating protein RanGAP1 is a homologue of yeast Rna1p involved in mRNA processing and transport. *Proc. Natl. Acad. Sci. USA* **92,** 1749–1753.

Bischoff, F. R., Krebber, H., Smirnova, E., Dong, W., and Ponstingl, H. (1995b). Coactivation of RanGTPase and inhibition of GTP dissociation by Ran-GTP binding protein RanBP1. *EMBO J.* **14,** 705–715.

Bischoff, F. R., and Görlich, D. (1997). RanBP1 is crucial for the release of RanGTP from importin beta-related nuclear transport factors. *FEBS Lett.* **419,** 249–254.

Bogerd, H. P., and Greene, W. C. (1993). Dominant negative mutants of human T-cell leukemia virus type I Rex and human immunodeficiency virus type 1 Rev fail to multimerize *in vivo*. *J. Virol.* **67,** 2496–2502.

Bogerd, H. P., Fridell, R. A., Madore, S., and Cullen, B. R. (1995). Identification of a novel cellular confactor for the Rev/Rex class of retroviral regulatory proteins. *Cell* **82,** 485–494.

Bogerd, H. P., Fridell, R. A., Benson, R. E., Hua, J., and Cullen, B. R. (1996). Protein sequence requirements for function of the human T-cell leukemia virus type 1 Rex nuclear export signal delineated by a novel in vivo randomization-selection assay. *Mol. Cell. Biol.* **16,** 4207–4214.

Bogerd, H. P., Echarri, A., Ross, T. M., and Cullen, B. R. (1998). Inhibition of human immunodeficiency virus Rev and human T-cell leukemia virus Rex function, but not Mason-Pfizer monkey virus constitutive transport element activity, by a mutant human nucleoporin targeted to Crm1. *J. Virol.* **72,** 8627–8635.

Bohnlein, E., Berger, J., and Hauber, J. (1991). Functional mapping of the human immunodeficiency rus type 1 Rev RNA binding domain: New insights into the domain structure of Rev and Rex. *J. Virol.* **65,** 7051–7055.

Bonyhadi, M. L., Moss, K., Voytovich, A., Auten, J., Kalfoglou, C., Plavec, I., Forestell, S., Su, L., Bohnlein, E., and Kaneshima, H. (1997). RevM10-expressing T cells derived *in vivo* from transduced human hematopoietic stem-progenitor cells inhibit human immunodeficiency virus replication. *J. Virol.* **71,** 4707–4716.

Borer, R. A., Lehner, C. F., Eppenberger, H. M., and Nigg, E. A. (1989). Major nucleolar proteins shuttle between nucleus and cytoplasm. *Cell* **56,** 379–390.

Bray, M., Prasad, S., Dubay, J. W., Hunter, E., Jeang, K. T., Rekosh, D., and Hammarskjöld, M. L. (1994). A small element from the Mason-Pfizer monkey virus genome makes human immunodeficiency virus type 1 expression and replication Rev-independent. *Proc. Natl. Acad. Sci. USA* **91,** 1256–1260.

Caceres, J. F., Stamm, S., Helfman, D. M., and Krainer, A. R. (1994). Regulation of alternative splicing *in vivo* by overexpression of antagonistic splicing factors. *Science* **265,** 1706–1709.

Chang, D. D., and Sharp, P. A. (1989). Regulation by HIV Rev depends upon recognition of splice sites. *Cell* **59,** 789–795.

Chen, Z., Li, Y., and Krug, R. M. (1998). Chimeras containing influenza NS1 and HIV-1 Rev protein sequences: Mechanism of their inhibition of nuclear export of Rev protein–RNA complexes. *Virology* **241,** 234–250.

Chinen, J., Aguilar-Cordova, E., Ng-Tang, D., Lewis, D. E., and Belmont, J. W. (1997). Protection of primary human T cells from HIV infection by Trev: A transdominant fusion gene. *Hum. Gene. Ther.* **8,** 861–868.

Cochrane, A. W., Kramer, R., Ruben, S., Levine, J., and Rosen, C. A. (1989a). The human immunodeficiency virus rev protein is a nuclear phosphoprotein. *Virology* **171,** 264–266.

Cochrane, A. W., Golub, E., Volsky, D., Ruben, S., and Rosen, C. A. (1989b). Functional significance of phosphorylation to the human immunodeficiency virus Rev protein. *J. Virol.* **63,** 4438–4440.

Cochrane, A. W., Perkins, A., and Rosen, C. A. (1990). Identification of sequences important in the nucleolar localization of human immunodeficiency virus Rev: Relevance of nucleolar localization to function. *J. Virol.* **64,** 881–885.

Cochrane, A. W., Jones, K. S., Beidas, S., Dillon, P. J., Skalka, A. M., and Rosen, C. A. (1991). Identification and characterization of intragenic sequences which repress human immunodeficiency virus structural gene expression. *J. Virol.* **65,** 5305–5313.

Cole, J. L., Gehman, J. D., Shafer, J. A., and Kuo, L. C. (1993). Solution oligomerization of the rev protein of HIV-1; Implications for function. *Biochemistry* **32,** 11769–11775.

Cook, K. S., Fisk, G. J., Hauber, J., Usman, N., Daly, T. J., and Rusche, J. R. (1991). Characterization of HIV-1 REV protein: binding stoichiometry and minimal RNA substrate. *Nucleic Acids Res.* **19,** 1577–1583.

Critchfield, J. W., Butera, S. T., and Folks, T. M. (1996). Inhibition of HIV activation in latently infected cells by flavonoid compounds. *AIDS Res. Hum. Retroviruses* **12,** 39–46.

Critchfield, J. W., Coligan, J. E., Folks, T. M., and Butera, S. T. (1997). Casein kinase II is a selective target of HIV-1 transcriptional inhibitors, *Proc. Natl. Acad. Sci. USA* **94,** 6110–6115.

Cullen, B. R., Hauber, J., Campbell, K., Sodroski, J. G., Haseltine, W. A., and Rosen, C. A. (1988). Subcellular localization of the human immunodeficiency virus trans-acting art gene product. *J. Virol.* **62,** 2498–2501.

Cullen, B. R. (1998). Retroviruses as model systems for the study of nuclear RNA export pathways. *Virology* **249,** 203–210.

D'Agostino, D. M., Felber, B. K., Harrison, J. E., and Pavlakis, G. N. (1992). The Rev protein of human immunodeficiency virus type 1 promotes polysomal association and translation of gag/pol and vpu/env mRNAs. *Mol. Cell. Biol.* **12,** 1375–1386.

D'Agostino, D. M., Ciminale, V., Pavlakis, G. N., and Chieco-Bianchi, L. (1995). Intracellular trafficking of the human immunodeficiency virus type 1 Rev protein: Involvement of continued rRNA synthesis in nuclear retention. *AIDS Res. Hum. Retroviruses* **11,** 1063–1071.

Dahlberg, J. E., and Lund, E. (1998). Functions of the GTPase Ran in RNA export from the nucleus. *Curr. Opin. Cell. Biol.* **10,** 400–408.

Daly, T. J., Cook, K. S., Gray, G. S., Maione, T. E., and Rusche, J. R. (1989). Specific binding of HIV-1 recombinant Rev protein to the Rev-responsive element *in vitro*. *Nature* **342,** 816–819.

Daly, T. J., Doten, R. C., Rennert, P., Auer, M., Jaksche, H., Donner, A., Fisk, G., and Rusche, J. R. (1993). Biochemical characterization of binding of multiple HIV-1 Rev monomeric proteins to the Rev responsive element. *Biochemistry* **32,** 10497–10505.

Del Gatto-Konczak, F., Olive, M., Gesnel, M. C., and Breathnach, R. (1999). hnRNP A1 recruited to an exon *in vivo* can function as an exon splicing silencer. *Mol. Cell. Biol.* **19,** 251–260.

Dobbelstein, M., Roth, J., Kimberley, W. T., Levine, A. J., and Shenk, T. (1997). Nuclear export of the EIB 55-kDa and E4 34-kDa adenoviral oncoproteins mediated by a rev-like signal sequence. *EMBO J.* **16,** 4276–4284.

Dreyfuss, G., Matunis, M. J., Pinol-Roma, S., and Burd, C. G. (1993). hnRNP proteins and the biogenesis of mRNA. *Annu. Rev. Biochem.* **62,** 289–321.

Duan, L., Bagasra, O., Laughlin, M. A., Oakes, J. W., and Pomerantz, R. J. (1994). Potent inhibition of human immunodeficiency virus type 1 replication by an intracellular anti-Rev single-chain antibody. *Proc. Natl. Acad. Sci. USA* **91,** 5075–5079.

Duan, L., Zhu, M., Bagasra, O., and Pomerantz, R. J. (1995). Intracellular immunization against HIV-1 infection of human T lymphocytes: Utility of anti-rev single-chain variable fragments. *Hum. Gene Ther.* **6,** 1561–1573.

Dundr, M., Leno, G., Hammarskjöld, M., Rekosh, D., Helga-Maria, C., and Olson, M. (1995). The roles of nucleolar structure and function in the subcellular location of the HIV-1 Rev protein. *J. Cell. Sci.* **108,** 2811–2823.

Dundr, M., Leno, G., Lewis, N., Rekosh, D., Hammarskjoid, M., and Olson, M. (1996). Location of the HIV-1 Rev protein during mitosis: Inactivation of the nuclear export signal alters the pathway for postmitotic reentry into nucleoli. *J. Cell. Sci.* **109,** 2239–2251.

Dundr, M., Meier, U., Lewis, N., Rekosh, D., Hammarskjöld, M., and Olson, M. (1997). A class of nonribosomal nucleolar components is located in chromosome periphery and in nucleolus-derived foci during anaphase and telophase. *Chromosoma* **105,** 407–417.

Dyhr-Mikkelsen, H., and Kjems, J. (1995). Inefficient spliceosome assembly and abnormal branch site selection in splicing of an HIV-1 transcript in vitro. *J. Biol. Chem.* **270,** 24060–24066.

Emerman, M., Vazeux, R., and Peden, K. (1989). The rev gene product of the human immunodeficiency virus affects envelope-specific RNA localization. *Cell* **57,** 1155–1165.

Emerman, M., and Malim, M. H. (1998). HIV-1 regulatory/accessory genes: Keys to unraveling viral and host cell biology. *Science* **280,** 1880–1884.

Engel, K., Kotlyarov, A., and Gaestel, M. (1998). Leptomycin B-sensitive nuclear export of MAPKAP kinase 2 is regulated by phosphorylation. *EMBO J.* **17,** 3363–3371.

Englmeier, L., Olivo, J.-C., and Mattaj, I. W. (1998). Receptor-mediated substrate translocation through the nuclear pore complex without nucleotide triphosphate hydrolysis. *Curr. Biol.* **9,** 30–41.

Escaich, S., Kalfoglou, C., Plavec, I., Kaushal, S., Mosca, J. D., and Bohnlein, E. (1995). RevM10-mediated inhibition of HIV-1 replication in chronically infected T cells. *Hum. Gene. Ther.* **6,** 625–634.

Fabre, E., and Hurt, E. (1997). Yeast genetics to dissect the nuclear pore complex and nucleocytoplasmic trafficking. *Annu. Rev. Genet.* **31,** 277–313.

Fankhauser, C., Izaurralde, E., Adachi, Y., Wingfield, P., and Laemmli, U. (1991). Specific complex of human immunodeficiency virus type 1 rev and nucleolar B23 proteins: Dissociation by the Rev response element. *Mol. Cell. Biol.* **11,** 2567–2575.

Favaro, J. P., Borg, K. T., Arrigo, S. J., and Schmidt, M. G. (1998). Effect of Rev on the intranuclear localization of HIV-1 unspliced RNA. *Virology* **249,** 286–296.

Feinberg, M. B., Jarrett, R. F., Aldovini, A., Gallo, R. C., and Wong Staal, F. (1986). HTLV-III expression and production involve complex regulation at the levels of splicing and translation of viral RNA. *Cell* **46,** 807–817.

Felber, B. K., Hadzopoulou-Cladaras, M., Cladaras, C., Copeland, T., and Pavlakis, G. N. (1989). rev protein of human immunodeficiency virus type 1 affects the stability and transport of the viral mRNA. *Proc. Natl. Acad. Sci. USA* **86,** 1495–1499.

Fischer, U., Meyer, S., Teufel, M., Heckel, C., Lührmann, R., and Rautmann, G. (1994). Evidence that HIV-1 Rev directly promotes the nuclear export of unspliced RNA. *EMBO J.* **13,** 4105–4112.

Fischer, U., Huber, J., Boelens, W. C., Mattaj, I. W., and Lührmann, R. (1995). The HIV-1 Rev activation domain is a nuclear export signal that accesses an export pathway used by specific cellular RNAs. *Cell* **82,** 475–483.

Fornerod, M., Ohno, M., Yoshida, M., and Mattaj, I. W. (1997a). CRM1 is an export receptor for leucine-rich nuclear export signals. *Cell* **90,** 1051–1060.

Fornerod, M., van Deursen, J., van Baal, S., Reynolds, A., Davis, D., Murti, K. G., Fransen, J., and Grosveld, G. (1997b). The human homologue of yeast CRM1 is in a dynamic subcomplex with CAN/Nup214 and a novel nuclear pore component Nup88. *EMBO J.* **16,** 807–816.

Fouts, D. E., True, H. L., Cengel, K. A., and Celander, D. W. (1997). Site-specific phosphorylation of the human immunodeficiency virus type-1 Rev protein accelerates formation of an efficient RNA-binding conformation. *Biochemistry* **36,** 13256–13262.

Fox, B. A., Woffendin, C., Yang, Z. Y., San, H., Ranga, U., Gordon, D., Osterholzer, J., and Nabel, G. J. (1995). Genetic modification of human peripheral blood lymphocytes with a transdominant negative form of Rev: Safety and toxicity. *Hum. Gene. Ther.* **6,** 997–1004.

Frankel, A. D., and Young, J. A. (1998). HIV-1: Fifteen proteins and an RNA. *Annu. Rev. Biochem.* **67,** 1–25.

Fridell, R. A., Fischer, U., Luhrmann, R., Meyer, B. E., Meinkoth, J. L., Malim, M. H., and Cullen, B. R. (1996a). Amphibian transcription factor IIIA proteins contain a sequence element functionally equivalent to the nuclear export signal of human immunodeficiency virus type 1 Rev. *Proc. Natl. Acad. Sci. USA* **93,** 2936–2940.

Fridell, R. A., Benson, R. E., Hua, J., Bogerd, H. P., and Cullen, B. R. (1996b). A nuclear role for the Fragile X mental retardation protein. *EMBO J.* **15,** 5408–5414.

Fritz, C. C., Zapp, M. L., and Green, M. R. (1995). A human nucleoporin-like protein that specifically interacts with HIV Rev. *Nature* **376,** 530–533.

Fukuda, M., Gotoh, I., Gotoh, Y., and Nishida, E. (1996). Cytoplasmic localization of mitogen-activated protein kinase kinase directed by its NH2-terminal, leucine-rich short amino acid sequence, which acts as a nuclear export signal. *J. Biol. Chem.* **271,** 20024–20028.

Fukuda, M., Asano, S., Nakamura, T., Makoto, A., Yoshida, M., Yanagida, M., and Nishida, E. (1997). CRM1 is responsible for intracellular transport mediated by the nuclear export signal. *Nature* **390**, 308–311.

Furuta, R. A., Kubota, S., Maki, M., Miyazaki, Y., Hattori, T., and Hatanaka, M. (1995). Use of a human immunodeficiency virus type 1 Rev mutant without nucleolar dysfunction as a candidate for potential AIDS therapy. *J. Virol.* **69**, 1591–1599.

Gant, T. M., Goldberg, M. W., and Allen, T. D. (1998). Nuclear envelope and nuclear pore assembly: Analysis of assembly intermediates by electron microscopy. *Curr. Opin. Cell. Biol.* **10**, 409–415.

Gilboa, E., and Smith, C. (1994). Gene therapy for infectious diesases: The AIDS model. *Trends Genet.* **10**, 139–144.

Görlich, D., Prehn, S., Laskey, R. A., and Hartmann, E. (1994). Isolation of a protein that is essential for the first step of nuclear protein import. *Cell* **79**, 767–778.

Görlich, D., Vogel, F., Mills, A. D., Hartmann, E., and Laskey, R. A. (1995). Distinct functions for the two importin subunits in nuclear protein import. *Nature* **377**, 246–488.

Görlich, D., Pante, N., Kutay, U., Aebi, U., and Bischoff, F. R. (1996a). Identification of different roles for RanGDP and RanGTP in nuclear protein import. *EMBO J.* **15**, 5584–5594.

Görlich, D., Henklein, P., Laskey, R. A., and Hartmann, E. (1996b). A 41 amino acid motif in importin-alpha confers binding to importin-beta and hence transit into the nucleus. *EMBO J.* **15**, 1810–1917.

Görlich, D., Dabrowski, M., Bischoff, F. R., Kutay, U., Bork, P., Hartmann, E., Prehn, S., and Izaurralde, E. (1997). A novel class of RanGTP binding proteins. *J. Cell. Biol.* **138**, 65–80.

Guddat, U., Bakken, A. H., and Pieler, T. (1990). Protein-mediated nuclear export of RNA: 5S rRNA containing small RNPs in *Xenopus oocytes*. *Cell* **60**, 619–628.

Hadzopoulou-Cladaras, M., Felber, B. K., Cladaras, C., Athanassopoulos, A., Tse, A., and Pavlakis, G. N. (1989). The rev (trs/art) protein of human immunodeficiency virus type 1 affects viral mRNA and protein expression via a cis-acting sequence in the env region. *J. Virol.* **63**, 1265–1274.

Hammarskjöld, M. L. (1997). Regulation of Retroviral RNA Export. *Semin Cell Dev. Biol.* **8**, 83–90.

Hammarskjöld, M. L., Heimer, J., Hammarskjöld, B., Sangwan, I., Albert, L., and Rekosh, D. (1989). Regulation of human immunodeficiency virus env expression by the rev gene product. *J. Virol.* **63**, 1959–1966.

Hammarskjöld, M. L., Li, H., Rekosh, D., and Prasad, S. (1994). Human immunodeficiency virus env expression becomes Rev-independent if the env region is not defined as an intron. *J. Virol.* **68**, 951–958.

Hammerschmid, M., Palmeri, D., Ruhl, M., Jaksche, H., Weichselbraun, I., Böhnlein, E., Malim, M. H., and Hauber, J. (1994). Scanning mutagenesis of the arginine-rich region of the human immunodeficiency virus type 1 Rev *trans* activator. *J. Virol.* **68**, 7329–7335.

Hanly, S. M., Rimsky, L. T., Malim, M. H., Kim, J. H., Hauber, J., Duc, D. M., Le, S. Y., Maizel, J. V., Cullen, B. R., and Greene, W. C. (1989). Comparative analysis of the HTLV-I Rex and HIV-1 Rev trans-regulatory proteins and their RNA response elements. *Genes Dev.* **3**, 1534–1544.

Harris, M. E., Gontarek, R. R., Derse, D., and Hope, T. J. (1998). Differential requirements for alternative splicing and nuclear export functions of equine infectious anemia virus Rev protein. *Mol. Cell. Biol.* **18**, 3889–3899.

Hauber, J., Bouvier, M., Malim, M. H., and Cullen, B. R. (1988). Phosphorylation of the rev gene product of human immunodeficiency virus type 1. *J. Virol.* **62**, 4801–4804.

Heaphy, S., Dingwall, C., Ernberg, I., Gait, M. J., Green, S. M., Karn, J., Lowe, A. D., Singh, M., and Skinner, M. A. (1990). HIV-1 regulator of virion expression (Rev) protein binds to an RNA stem-loop structure located within the Rev response element region. *Cell* **60**, 685–693.

Heaphy, S., Finch, J. T., Gait, M. J., Karn, J., and Singh, M. (1991). Human immunodeficiency virus type 1 regulator of virion expression, rev, forms nucleoprotein filaments after binding to a purine-rich "bubble" located within the rev-responsive region of viral mRNAs. *Proc. Natl. Acad. Sci. USA* **88,** 7366–7370.

Henderson, B. R., and Percipalle, P. (1997). Interactions between HIV Rev and nuclear import and export factors: The Rev nuclear localisation signal mediates specific binding to human importin-beta. *J. Mol. Biol.* **274,** 693–707.

Hidaka, M., Inoue, J., Yoshida, M., and Seiki, M. (1988). Post-transcriptional regulator (rex) of HTLV-1 initiates expression of viral structural proteins but suppresses expression of regulatory proteins. *EMBO J.* **7,** 519–523.

Hope, T. J., Huang, X. J., McDonald, D., and Parslow, T. G. (1990a). Steroid-receptor fusion of the human immunodeficiency virus type 1 Rev transactivator: Mapping cryptic functions of the arginine-rich motif. *Proc. Natl. Acad. Sci. USA* **87,** 7787–7791.

Hope, T. J., McDonald, D., Huang, X. J., Low, J., and Parslow, T. G. (1990b). Mutational analysis of the human immunodeficiency virus type 1 Rev transactivator: Essential residues near the amino terminus. *J. Virol.* **64,** 5360–5366.

Hope, T. J., Bond, B. L., McDonald, D., Klein, N. P., and Parslow, T. G. (1991). Effector domains of human immunodeficiency virus type 1 Rev and human T-cell leukemia virus type I Rex are functionally interchangeable and share an essential peptide motif. *J. Virol.* **65,** 6001–6007.

Hope, T. J., Klein, N. P., Elder, M. E., and Parslow, T. G. (1992). *trans*-dominant inhibition of human immunodeficiency virus type 1 Rev occurs through formation of inactive protein complexes. *J. Virol.* **66,** 1849–1855.

Imamoto, N., Tachibana, T., Matsubae, M., and Yoneda, Y. (1995). A karyophilic protein forms a stable complex with cytoplasmic components prior to nuclear pore binding. *J. Biol. Chem.* **270,** 8559–8565.

Inouye, R. T., Du, B., Boldt-Houle, D., Ferrante, A., Park, I. W., Hammer, S. M., Duan, L., Groopman, J. E., Pomerantz, R. J., and Terwilliger, E. F. (1997). Potent inhibition of human immunodeficiency virus type 1 in primary T cells and alveolar macrophages by a combination anti-Rev strategy delivered in an adeno-associated virus vector. *J. Virol.* **71,** 4071–4078.

Izaurralde, E., Kutay, U., von Kobbe, C., Mattaj, I. W., and Görlich, D. (1997a). The asymmetric distribution of the constituents of the Ran system is essential for transport into and out of the nucleus. *EMBO J.* **16,** 6535–6547.

Izaurralde, E., Jarmolowski, A., Beisel, C., Mattaj, I. W., Dreyfuss, G., and Fischer, U. (1997b). A role for the M9 transport signal of hnRNP A1 in mRNA nuclear export. *J. Cell. Biol.* **137,** 27–35.

Jarmolowski, A., Boelens, W. C., Izaurralde, E., and Mattaj, I. W. (1994). Nuclear export of different classes of RNA is mediated by specific factors. *J. Cell. Biol.* **124,** 627–635.

Jensen, A., Jensen, T. H., and Kjems, J. (1998). HIV-1 Rev nuclear export signal binding peptides isolated from a random phage display peptide library. *J. Mol. Biol.* **283,** 245–254.

Jensen, K. B., Atkinson, B. L., Willis, M. C., Koch, T. H., and Gold, L. (1995). Using *in vitro* selection to direct the covalent attachment of human immunodeficiency virus type 1 Rev protein to high-affinity RNA ligands. *Proc. Natl. Acad. Sci. USA* **92,** 12220–12224.

Jensen, T. H., Leffers, H., and Kjems, J. (1995). Intermolecular binding sites of HIV-1 Rev protein determined by protein footprinting. *J. Biol. Chem.* **270,** 13777–13784.

Jensen, T. H., Jensen, A., Szilvay, A. M., and Kjems, J. (1997). Probing the structure of HIV-1 Rev by protein footprinting of multiple monoclonal antibody binding sites. *FEBS Lett.* **414,** 50–54.

Kalderon, D., Roberts, B. L., Richardson, W. D., and Smith, A. E. (1984). A short amino acid sequence able to specify nuclear location. *Cell* **39,** 499–509.

Kalland, K. H., Szilvay, A. M., Brokstad, K. A., Sætrevik, W., and Haukenes, G. (1994). The human immunodeficiency virus type 1 Rev protein shuttles between the cytoplasm and nuclear compartments. *Mol. Cell. Biol.* **14,** 7436–7444.

Kang, H. A., and Hershey, J. W. (1994). Effect of initiation factor eIF-5A depletion on protein synthesis and proliferation of *Saccharomyces cerevisiae*. *J. Biol. Chem.* **269**, 3934–3940.

Karn, J., Dingwall, C., Finch, J. T., Heaphy, S., and Gait, M. J. (1991). RNA binding by the tat and rev proteins of HIV-1. *Biochimie* **73**, 9–16.

Kim, F. J., Beeche, A. A., Hunter, J. J., Chin, D. J., and Hope, T. J. (1996). Characterization of the nuclear export signal of human T-cell lymphotropic virus type 1 Rex reveals that nuclear export is mediated by position-variable hydrophobic interactions. *Mol. Cell. Biol.* **16**, 5147–5155.

Kiyomasu, T., Miyazawa, T., Furuya, T., Shibata, R., Sakai, H., Sakuragi, J., Fukasawa, M., Maki, N., Hasegawa, A., Mikami, T., and *et al.* (1991). Identification of feline immunodeficiency virus rev gene activity. *J. Virol.* **65**, 4539–4542.

Kjems, J., Brown, M., Chang, D. D., and Sharp, P. A. (1991a). Structural analysis of the interaction between the human immunodeficiency virus Rev protein and the Rev response element. *Proc. Natl. Acad. Sci. USA* **88**, 683–687.

Kjems, J., Frankel, A. D., and Sharp, P. A. (1991b). Specific regulation of mRNA splicing *in vitro* by a peptide from HIV-1 Rev. *Cell* **67**, 169–178.

Kjems, J., Calnan, B. J., Frankel, A. D., and Sharp, P. A. (1992). Specific binding of a basic peptide from HIV-1 Rev. *EMBO J.* **11**, 1119–1129.

Kjems, J., and Sharp, P. A. (1993). The basic domain of Rev from human immunodeficiency virus type 1 specifically blocks the entry of U4/U6.U5 small nuclear ribonucleoprotein in spliceosome assembly. *J. Virol.* **67**, 4769–4776.

Klebe, C., Bischoff, F. R., Ponstingl, H., and Wittinghofer, A. (1995). Interactions of the nuclear GTP-binding protein Ran with its regulatory proteins RCC1 and RanGAP1. *Biochemistry* **34**, 639–647.

Knight, D. M., Flomerfelt, F. A., and Ghrayeb, J. (1987). Expression of the art/trs protein of HIV and study of its role in viral envelope synthesis. *Science* **236**, 837–840.

Krainer, A. R., Mayeda, A., Kozak, D., and Binns, G. (1991). Functional expression of cloned human splicing factor SF2: Homology to RNA-binding proteins, U1 70K, and *Drosophila* splicing regulators. *Cell* **66**, 383–394.

Kubota, S., Siomi, H., Satoh, T., Endo, S., Maki, M., and Hatanaka, M. (1989). Functional similarity of HIV-I rev and HTLV-I rex proteins: Identification of a new nucleolar-targeting signal in rev protein. *Biochem. Biophys. Res. Commun.* **162**, 963–970.

Kubota, S., Furuta, R., Maki, M., and Hatanaka, M. (1992). Inhibition of human immunodeficiency virus type 1 Rev function by a Rev mutant which interferes with nuclear/nucleolar localization of Rev. *J. Virol.* **66**, 2510–2513.

Kubota, S., and Pomerantz, R. J. (1998). A cis-acting peptide signal in human immunodeficiency virus type I Rev which inhibits nuclear entry of small proteins. *Oncogene* **16**, 1851–1861.

Kudo, N., Taoka, H., Toda, T., Yoshida, M., and Horinouchi, S. (1999). A novel nuclear export signal sensitive to oxidative stress in the fission yeast transcription factor Pap1. *J. Biol. Chem.* **274**, 15151–15158.

Kutay, U., Bischoff, F. R., Kostka, S., Kraft, R., and Gorlich, D. (1997). Export of importin α from the nucleus is mediated by a specific nuclear transport factor. *Cell* **90**, 1061–1071.

Kutay, U., Lipowsky, G., Izaurralde, E., Bischoff, F. R., Schwarzmaier, P., Hartmann, E., and Gorlich, D. (1998). Identification of a tRNA-specific nuclear export receptor. *Mol. Cell.* **1**, 359–369.

Lam, W., Seifert, J. M., Amberger, F., Graf, C., Aur, M., and Millar, D. P. (1998). Structural dynamics of HIV-1 Rev and its complexes with RRE and 5S RNA. *Biochemistry* **37**, 1800–1809.

Lazinski, D., Grzadzielska, E., and Das, A. (1989). Sequence-specific recognition of RNA hairpins by bacteriophage antiterminators requires a conserved arginine-rich motif. *Cell* **59**, 207–218.

Legrain, P., and Rosbash, M. (1989). Some cis- and trans-acting mutants for splicing target pre-mRNA to the cytoplasm. *Cell* **57**, 573–583.

Li, J., Tang, H., Mullen, T. M., Westberg, C., Reddy, T. R., Rose, D. W., and Wong-Staal, F. (1999). A role for RNA helicase A in post-transcriptional regulation of HIV type 1. *Proc. Natl. Acad. Sci. USA* **96**, 709–714.

Li, Y., Yamakita, Y., and Krug, R. M. (1998). Regulation of a nuclear export signal by an adjacent inhibitory sequence: The effector domain of the influenza virus NS1 protein. *Proc. Natl. Acad. Sci. USA* **95**, 4864–4869.

Lu, X. B., Heimer, J., Rekosh, D., and Hammarskjöld, M. L. (1990). U1 small nuclear RNA plays a direct role in the formation of a rev-regulated human immunodeficiency virus env mRNA that remains unspliced. *Proc. Natl. Acad. Sci. USA* **87**, 7598–7602.

Luo, Y., Yu, H., and Peterlin, B. M. (1994). Cellular protein modulates effects of human immunodeficiency virus type 1 Rev. *J. Virol.* **68**, 3850–3856.

Madore, S. J., Tiley, L. S., Malim, M. H., and Cullen, B. R. (1994). Sequence requirements for Rev multimerization *in vivo*. *Virology* **202**, 186–194.

Mahajan, R., Delphin, C., Guan, T., Gerace, L., and Melchior, F. (1997). A small ubiquitin-related polypeptide involved in targeting RanGAP1 to nuclear pore complex protein RanBP2. *Cell* **88**, 97–107.

Maldarelli, F., Martin, M. A., and Strebel, K. (1991). Identification of posttranscriptionally active inhibitory sequences in human immunodeficiency virus type 1 RNA: Novel level of gene regulation. *J. Virol.* **65**, 5732–5743.

Malim, M. H., Hauber, J., Fenrick, R., and Cullen, B. R. (1988). Immunodeficiency virus rev trans-activator modulates the expression of the viral regulatory genes. *Nature* **335**, 181–183.

Malim, M. H., Hauber, J., Le, S. Y., Maizel, J. V., and Cullen, B. R. (1989a). The HIV-1 rev trans-activator acts through a structured target sequence to activate nuclear export of unspliced viral mRNA. *Nature* **338**, 254–257.

Malim, M. H., Bohnlein, S., Hauber, J., and Cullen, B. R. (1989b). Functional dissection of the HIV-1 Rev trans-activator—derivation of a trans-dominant repressor of Rev function. *Cell* **58**, 205–214.

Malim, M. H., Tiley, L. S., McCarn, D. F., Rusche, J. R., Hauber, J., and Cullen, B. R. (1990). HIV-1 structural gene expression requires binding of the Rev trans-activator to its RNA target sequence. *Cell* **60**, 675–683.

Malim, M. H., and Cullen, B. R. (1991). HIV-1 structural gene expression requires the binding of multiple Rev monomers to the viral RRE: Implications for HIV-1 latency. *Cell* **65**, 241–248.

Malim, M. H., McCarn, D. F., Tiley, L. S., and Cullen, B. R. (1991). Mutational definition of the human immunodeficiency virus type 1 Rev activation domain. *J. Virol.* **65**, 4248–4254.

Malim, M. H., Freimuth, W. W., Liu, J., Boyle, T. J., Lyerly, H. K., Cullen, B. R., and Nabel, G. J. (1992). Stable expression of transdominant Rev protein in human T cells inhibits human immunodeficiency virus replication. *J. Exp. Med.* **176**, 1197–1201.

Malim, M. H., and Cullen, B. R. (1993). Rev and the fate of pre-mRNA in the nucleus: Implications for the regulation of RNA processing in eukaryotes. *Mol. Cell. Biol.* **13**, 6180–6189.

Mancuso, V. A., Hope, T. J., Zhu, L., Derse, D., Phillips, T., and Parslow, T. G. (1994). Posttranscriptional effector domains in the Rev proteins of feline immunodeficiency virus and equine infectious anemia virus. *J. Virol.* **68**, 1998–2001.

Manley, J. L., and Tacke, R. (1996). SR proteins and splicing control. *Genes Dev.* **10**, 1569–1579.

Mann, D. A., Mikaelian, I., Zemmel, R. W., Green, S. M., Lowe, A. D., Kimura, T., Singh, M., Butler, P. J., Gait, M. J., and Karn, J. (1994). A molecular rheostat: Cooperative rev binding to stem I of the rev-response element modulates human immunodeficiency virus type-1 late gene expression. *J. Mol. Biol.* **241**, 193–207.

Mattaj, I. W., and Englmeier, L. (1998). Nucleocytoplasmic transport: The soluble phase. *Ann. Rev. Biochem.* **67**, 265–306.

Mayeda, A., and Krainer, A. R. (1992). Regulation of alternative pre-mRNA splicing by hnRNP A1 and splicing factor SF2. *Cell* **68**, 365–375.

McDonald, D., Hope, T. J., and Parslow, T. G. (1992). Posttranscriptional regulation by the human immunodeficiency virus type 1 Rev and human T-cell leukemia virus type I Rex proteins through a heterologous RNA binding site. *J. Virol.* **66**, 7232–7238.

Mermer, B., Felber, B. K., Campbell, M., and Pavlakis, G. N. (1990). Identification of trans-dominant HIV-1 rev protein mutants by direct transfer of bacterially produced proteins into human cells. *Nucleic Acids Res.* **18**, 2037–2044.

Meyer, B. E., and Malim, M. H. (1994). The HIV-1 Rev trans-activator shuttles between the nucleus and the cytoplasm. *Genes Dev.* **8**, 1538–1547.

Meyer, B. E., Meinkoth, J. L., and Malim, M. H. (1996). Nuclear transport of human immunodeficiency virus type 1, visna virus, and equine infectious anemia virus Rev proteins: Identification of a family of transferable nuclear export signals. *J. Virol.* **70**, 2350–2359.

Michael, W. M., Choi, M., and Dreyfuss, G. (1995). A nuclear export signal in hnRNP A1: A signal-mediated, temperature-dependent nuclear protein export pathway. *Cell* **83**, 415–422.

Mikaelian, I., Krieg, M., Gait, M. J., and Karn, J. (1996). Interactions of INS (CRS) elements and the splicing machinery regulate the production of Rev-responsive mRNAs. *J. Mol. Biol.* **257**, 246–264.

Moore, M. S., and Blobel, G. (1993). The GTP-binding protein Ran/TC4 is required for protein import into the nucleus. *Nature* **365**, 661–663.

Moroianu, J., Blobel, G., and Radu, A. (1995a). Previously identified protein of uncertain function is karyopherin alpha and together with karyopherin beta docks import substrate at nuclear pore complexes. *Proc. Natl. Acad. Sci. USA* **92**, 2008–2011.

Moroianu, J., Hijikata, M., Blobel, G., and Radu, A. (1995b). Mammalian karyopherin alpha 1 beta and alpha 2 beta heterodimers: alpha 1 or alpha 2 subunit binds nuclear localization signal and beta subunit interacts with peptide repeat-containing nucleoporins. *Proc. Natl. Acad. Sci. USA* **92**, 6532–6536.

Moroianu, J., Blobel, G., and Radu, A. (1996). Nuclear protein import: Ran-GTP dissociates the karyopherin alphabeta heterodimer by displacing alpha from an overlapping binding site on beta. *Proc. Natl. Acad. Sci. USA* **93**, 7059–7062.

Murphy, R., and Wente, S. R. (1996). An RNA-export mediator with an essential nuclear export signal. *Nature* **383**, 357–360.

Najera, I., Krieg, M., and Karn, J. (1999). Synergistic stimulation of HIV-1 rev-dependent export of unspliced mRNA to the cytoplasm by hnRNP A1. *J. Mol. Biol.* **285**, 1951–1964.

Nakielny, S., Fischer, U., Michael, M. W., and Dreyfuss, G. (1997). RNA transport. *Annu. Rev. Neurosci.* **20**, 269–301.

Nalin, C. M., Purcell, R. D., Antelman, D., Mueller, D., Tomchak, L., Wegrzynski, B., McCarney, E., Toome, V., Kramer, R., and Hsu, M. C. (1990). Purification and characterization of recombinant Rev protein of human immunodeficiency virus type 1. *Proc. Natl. Acad. Sci. USA* **87**, 7593–7597.

Nasioulas, G., Zolotukhin, A. S., Tabernero, C., Solomin, L., Cunningham, C. P., Pavlakis, G. N., and Felber, B. K. (1994). Elements distinct from human immunodeficiency virus type 1 splice sites are responsible for the Rev dependence of env mRNA. *J. Virol.* **68**, 2986–2993.

Neville, M., Stutz, F., Lee, L., Davis, L. I., and Rosbash, M. (1997). The importin-beta family member Crm1p bridges the interaction between Rev and the nuclear pore complex during nuclear export. *Curr. Biol.* **7**, 767–775.

Nishi, K., Yoshida, M., Fujiwara, D., Nishikawa, M., Horinouchi, S., and Beppu, T. (1994). Leptomycin B targets a regulatory cascade of crm1, a fission yeast nuclear protein, involved in control of higher order chromosome structure and gene expression. *J. Biol. Chem.* **269**, 6320–6324.

O'Neill, R. E., Talon, J., and Palese, P. (1998). The influenza virus NEP (NS2 protein) mediates the nuclear export of viral ribonucleoproteins. *EMBO J.* **17,** 288–296.

Oberste, M. S., Williamson, J. C., Greenwood, J. D., Nagashima, K., Copeland, T. D., and Gonda, M. A. (1993). Characterization of bovine immunodeficiency virus rev cDNAs and identification and subcellular localization of the Rev protein. *J. Virol.* **67,** 6395–6405.

Ogert, R. A., Lee, L. H., and Beemon, K. L. (1996). Avian retroviral RNA element promotes unspliced RNA accumulation in the cytoplasm. *J. Virol.* **70,** 3834–3843.

Ohno, M., Fornerod, M., and Mattaj, I. W. (1998). Nucleocytoplasmic transport: The last 200 nanometers. *Cell* **92,** 327–336.

Ohshima, T., Nakajima, T., Oishi, T., Imamoto, N., Yoneda, Y., Fukamizu, A., and Yagami, F. (1999). CRM1 mediates nuclear export of nonstructural protein 2 from parvovirus minute virus of mice. *Biochem. Biophys. Res. Commun.* **264,** 144–150.

Ohtsuki, K., Maekawa, T., Harada, S., Karino, A., Morikawa, Y., and Ito, M. (1998). Biochemical characterization of HIV-1 Rev as a potent activator of casein kinase II *in vitro. FEBS Lett.* **428,** 235–240.

Olsen, H. S., Cochrane, A. W., Dillon, P. J., Nalin, C. M., and Rosen, C. A. (1990). Interaction of the human immunodeficiency virus type 1 Rev protein with a structured region in env mRNA is dependent on multimer formation mediated through a basic stretch of amino acids. *Genes Dev.* **4,** 1357–1364.

Olsen, H. S., Cochrane, A. W., and Rosen, C. (1992). Interaction of cellular factors with intragenic cis-acting repressive sequences within the HIV genome. *Virology* **191,** 709–715.

Ossareh-Nazari, B., Bachelerie, F., and Dargemont, C. (1997). Evidence for a role of CRM1 in signal-mediated nuclear protein export. *Science* **278,** 141–144.

Otero, G. C., Harris, M. E., Donello, J. E., and Hope, T. J. (1998). Leptomycin B inhibits equine infectious anemia virus Rev and feline immunodeficiency virus rev function but not the function of the hepatitis B virus posttranscriptional regulatory element. *J. Virol.* **72,** 7593–7597.

Palmeri, D., and Malim, M. H. (1996). The human T-cell leukemia virus type 1 posttranscriptional trans-activator Rex contains a nuclear export signal. *J. Virol.* **70,** 6442–6445.

Palmeri, D., and Malim, M. H. (1999). Importin beta can mediate the nuclear import of an arginine-rich nuclear localization signal in the absence of importin alpha. *Mol. Cell. Biol.* **19,** 1218–1225.

Park, M. H., Wolff, E. C., and Folk, J. E. (1993). Is hypusine essential for eukaryotic cell proliferation? *Trends Biochem. Sci.* **18,** 475–479.

Paschal, B. M., Fritze, C., Guan, T., and Gerace, L. (1997). High levels of the GTPase Ran/TC4 relieve the requirement for nuclear protein transport factor 2. *J. Biol. Chem.* **272,** 21534–21539.

Pasquinelli, A. E., Ernst, R. K., Lund, E., Grimm, C., Zapp, M. L., Rekosh, D., Hammarskjöld, M. L., and Dahlberg, J. E. (1997a). The constitutive transport element (CTE) of Mason-Pfizer monkey virus (MPMV) accesses a cellular mRNA export pathway. *EMBO J.* **16,** 7500–7510.

Pasquinelli, A. E., Powers, M. A., Lund, E., Forbes, D., and Dahlberg, J. E. (1997b). Inhibition of mRNA export in vertebrate cells by nuclear export signal conjugates. *Proc. Natl. Acad. Sci. USA* **94,** 14394–14399.

Perkins, A., Cochrane, A. W., Ruben, S. M., and Rosen, C. A. (1989). Structural and functional characterization of the human immunodeficiency virus rev protein. *J. Acquir. Immune. Defic. Syndr.* **2,** 256–263.

Plavec, I., Agarwal, M., Ho, K. E., Pineda, M., Auten, J., Baker, J., Matsuzaki, H., Escaich, S., Bonyhadi, M., and Bohnlein, E. (1997). High transdominant RevM10 protein levels are required to inhibit HIV-1 replication in cell lines and primary T cells: Implication for gene therapy of AIDS. *Gene Ther.* **4,** 128–139.

Pollard, V. W., and Malim, M. H. (1998). The HIV-1 Rev protein. *Annu. Rev. Microbiol.* **52,** 491–532.

Pomerantz, R. J., Trono, D., Feinberg, M. B., and Baltimore, D. (1990). Cells nonproductively infected with HIV-1 exhibit an aberrant pattern of viral RNA expression: A molecular model for latency. *Cell* **61**, 1271–1276.

Pomerantz, R. J., Seshamma, T., and Trono, D. (1992). Efficient replication of human immunodeficiency virus type 1 requires a threshold level of Rev: Potential implications for latency. *J. Virol.* **66**, 1809–1813.

Pomerantz, R. J., and Trono, D. (1995). Genetic therapies for HIV infections: Promise for the future. *AIDS* **9**, 985–993.

Powell, D. M., Amaral, M. C., Wu, J. Y., Maniatis, T., and Greene, W. C. (1997). HIV Rev-dependent binding of SF2/ASF to the Rev response element: Possible role in Rev-mediated inhibition of HIV RNA splicing. *Proc. Natl. Acad. Sci. USA* **94**, 973–978.

Purcell, D. F., and Martin, M. A. (1993). Alternative splicing of human immunodeficiency virus type 1 mRNA modulates viral protein expression, replication, and infectivity. *J. Virol.* **67**, 6365–6378.

Radu, A., Moore, M. S., and Blobel, G. (1995). The peptide repeat domain of nucleoporin Nup98 functions as a docking site in transport across the nuclear pore complex. *Cell* **81**, 215–222.

Ragheb, J. A., Bressler, P., Daucher, M., Chiang, L., Chuah, M. K., Vandendriessche, T., and Morgan, R. A. (1995). Analysis of trans-dominant mutants of the HIV type 1 Rev protein for their ability to inhibit Rev function, HIV type 1 replication, and their use as anti-HIV gene therapeutics. *AIDS Res. Hum. Retroviruses* **11**, 1343–1353.

Ranga, U., Woffendin, C., Verma, S., Xu, L., June, C. H., Bishop, D. K., and Nabel, G. J. (1998). Enhanced T cell engraftment after retroviral delivery of an antiviral gene in HIV-infected individuals. *Proc. Natl. Acad. Sci. USA* **95**, 1201–1206.

Rexach, M., and Blobel, G. (1995). Protein import into nuclei: Association and dissociation reactions involving transport substrate, transport factors, and nucleoporins. *Cell* **83**, 683–692.

Ribbeck, K., Kutay, U., Paraskeva, E., and Gorlich, D. (1999). The translocation of transportin-cargo complexes through nuclear pores is independent of both ran and energy. *Curr. Biol.* **9**, 47–50.

Richard, N., Iacampo, S., and Cochrane, A. W. (1994). HIV-1 Rev is capable of shuttling between the nucleus and cytoplasm. *Virology* **204**, 123–131.

Richards, S. A., Lounsbury, K. M., and Macara, I. G. (1995). The C terminus of the nuclear RAN/TC4 GTPase stabilizes the GDP-bound state and mediates interactions with RCC1, RAN-GAP, and HTF9A/RANBP1. *J. Biol. Chem.* **270**, 14405–14411.

Richards, S. A., Lounsbury, K. M., Carey, K. L., and Macara, I. G. (1996). A nuclear export signal is essential for the cytosolic localization of the Ran binding protein, RanBP1. *J. Cell. Biol.* **134**, 1157–1168.

Richards, S. A., Carey, K. L., and Macara, I. G. (1997). Requirement of guanosine triphosphate-bound ran for signal-mediated nuclear protein export. *Science* **276**, 1842–1844.

Robbins, J., Dilworth, S. M., Laskey, R. A., and Dingwall, C. (1991). Two interdependent basic domains in nucleoplasmin nuclear targeting sequence: Identification of a class of bipartite nuclear targeting sequence. *Cell* **64**, 615–623.

Rondon, I. J., and Marasco, W. A. (1997). Intracellular antibodies (intrabodies) for gene therapy of infectious diseases. *Annu. Rev. Microbiol.* **51**, 257–283.

Rosen, C. A., Terwillinger, E., Dayton, A., Sodroski, J. G., and Haseltine, W. A. (1988). Intragenic cis-acting art gene-responsive sequences of the human immunodeficiency virus. *Proc. Natl. Acad. Sci. USA* **85**, 2071–2075.

Roth, J., Dobbelstein, M., Freedman, D. A., Shenk, T., and Levine, A. J. (1998). Nucleo-cytoplasmic shuttling of the hdm2 oncoprotein regulates the levels of the p53 protein via a pathway used by the human immunodeficiency virus rev protein. *EMBO J.* **17**, 554–564.

Ruhl, M., Himmelspach, M., Bahr, G., Hammerschmid, F., Jaksche, H., Wolff, B., Aschauer, H., Farrington, G., Probst, H., Bevec, D., and et, a. 993). Eukaryotic initiation factor 5A is a cellular target of the human immunodeficiency virus type 1 Rev activation domain mediating trans-activation. *J. Cell. Biol.* **123,** 1309–1320.

Saavedra, C. A., Hammell, C. M., Health, C. V., and Cole, C. N. (1997a). Yeast heat shock mRNAs are exported through a distinct pathway defined by Rip1p. *Genes Dev.* **11,** 2845–2856.

Saavedra, C., Felber, B., and Izaurralde, E. (1997b). The simian retrovirus-1 constitutive transport element, unlike the HIV-1 RRE, uses factors required for cellular mRNA export. *Curr. Biol.* **7,** 619–628.

Sakai, H., Shibata, R., Sakuragi, J., Kiyomasu, T., Kawamura, M., Hayami, M., Ishimoto, A., and Adachi, A. (1991). Compatibility of rev gene activity in the four groups of primate lentiviruses. *Virology* **184,** 513–520.

Saltarelli, M. J., Schoborg, R., Pavlakis, G. N., and Clements, J. E. (1994). Identification of the caprine arthritis encephalitis virus Rev protein and its cis-acting Rev-responsive element. *Virology* **199,** 47–55.

Sandri-Goldin, R. M. (1998). ICP27 mediates HSV RNA export by shuttling through a leucine-rich nuclear export signal and binding viral intronless RNAs through an RGG motif. *Genes Dev.* **12,** 868–879.

Schatz, O., Oft, M., Dascher, C., Schebesta, M., Rosorius, O., Jaksche, H., Dobrovnik, M., Bevec, D., and Hauber, J. (1998). Interaction of the HIV-1 rev cofactor eukaryotic initiation factor 5A with ribosomal protein L5. *Proc. Natl. Acad. Sci. USA* **95,** 1607–1612.

Schneider, R., Campell, M., Nasioulas, G., Felber, B. K., and Pavlakis, G. N. (1997). Inactivation of the human immunodeficiency virus type 1 inhibitory elements allows Rev-independent expression of Gag and Gag/protease and particle formation. *J. Virol.* **71,** 4892–4903.

Schwartz, S., Felber, B. K., Benko, D. M., Fenyo, E. M., and Pavlakis, G. N. (1990). Cloning and functional analysis of multiply spliced mRNA species of human immunodeficiency virus type 1. *J. Virol.* **64,** 2519–2529.

Schwartz, S., Felber, B. K., and Pavlakis, G. N. (1992). Distinct RNA sequences in the gag region of human immunodeficiency virus type 1 decrease RNA stability and inhibit expression in the absence of Rev protein. *J. Virol.* **66,** 150–159.

Segref, A., Sharma, K., Doye, V., Hellwig, A., Huber, J., Luhrmann, R., and Hurt, E. (1997). Mex67p, a novel factor for nuclear mRNA export binds to both poly(A)$^+$ RNA and nuclear pores. *EMBO J.* **16,** 3256–3271.

Sodroski, J., Goh, W. C., Rosen, C., Dayton, A., Terwilliger, E., and Haseltine, W. (1986). A second post-transcriptional trans-activator gene required for HTLV-III replication. *Nature* **321,** 412–417.

Stade, K., Ford, C., Guthrie, C., and Weis, K. (1997). Exportin 1 (Crm1p) is an essential nuclear export factor. *Cell* **90,** 1041–1050.

Stauber, R., Gaitanaris, G. A., and Pavlakis, G. N. (1995). Analysis of trafficking of Rev and transdominant Rev proteins in living cells using green fluorescent protein fusions: Transdominant Rev blocks the export of Rev from the nucleus to the cytoplasm. *Virology* **213,** 439–449.

Stauber, R. H., Afonina, E., Gulnik, S., Erickson, J., and Pavlakis, G. N. (1998). Analysis of intracellular trafficking and interactions of cytoplasmic HIV-1 Rev mutants in living cells. *Virology* **251,** 38–48.

Stephens, R. M., Derse, D., and Rice, N. R. (1990). Cloning and characterization of cDNAs encoding equine infectious anemia virus tat and putative Rev proteins. *J. Virol.* **64,** 3716–3725.

Stommel, J. M., Marchenko, N. D., Jimenez, G. S., Moll, U. M., Hope, T. J., and Wahl, G. M. (1999). A leucine-rich nuclear export signal in the p53 tetramerization domain: regulation of subcellular localization and p53 activity by NES masking. *EMBO J.* **18,** 1660–1672.

Stutz, F., and Rosbash, M. (1994). A functional interaction between Rev and yeast pre-mRNA is related to splicing complex formation. *EMBO J.* **13,** 4096–4104.

Stutz, F., Neville, M., and Rosbash, M. (1995). Identification of a novel nuclear pore-associated protein as a functional target of the HIV-1 Rev protein in yeast. *Cell* **82**, 495–506.

Stutz, F., Izaurralde, E., Mattaj, I., and Rosbash, M. (1996). A role for nucleoporin FG repeat domains in export of human immunodeficiency virus type 1 Rev protein and RNA from the nucleus. *Mol. Cell. Biol.* **16**, 7144–7150.

Stutz, F., Kantor, J., Zhang, D., McCarthy, T., Neville, M., and Rosbash, M. (1997). The yeast nucleoporin rip1p contributes to multiple export pathways with no essential role for its FG-repeat region. *Genes Dev.* **11**, 2857–2868.

Su, L., Lee, R., Bonyhadi, M., Matsuzaki, H., Forestell, S., Escaich, S., Bohnlein, E., and Kaneshima, H. (1997). Hematopoietic stem cell-based gene therapy for acquired immunodeficiency syndrome: Efficient transduction and expression of RevM10 in myeloid cells *in vivo* and *in vitro*. *Blood* **89**, 2283–2290.

Symensma, T. L., Giver, L., Zapp, M., Takle, G. B., and Ellington, A. D. (1996). RNA aptamers selected to bind human immunodeficiency virus type 1 Rev *in vitro* are Rev responsive *in vivo*. *J. Virol.* **70**, 179–187.

Szebeni, A., Herrera, J. E., and Olson, M. O. (1995). Interaction of nucleolar protein B23 with peptides related to nuclear localization signals. *Biochemistry* **34**, 8037–8042.

Szebeni, A., Mehrotra, B., Baumann, A., Adam, S. A., Wingfield, P. T., and Olson, M. O. (1997). Nucleolar protein B23 stimulates nuclear import of the HIV-1 Rev protein and NLS-conjugated albumin. *Biochemistry* **36**, 3941–3949.

Szebeni, A., and Olson, M. O. (1999). Nucleolar protein B23 has molecular chaparone activities. *Prot. Sci.* **8**, 905–912.

Szilvay, A. M., Brokstad, K. A., Kopperud, R., Haukenes, G., and Kalland, K. H. (1995). Nuclear export of the human immunodeficiency virus type 1 nucleocytoplasmic shuttle protein Rev is mediated by its activation domain and is blocked by transdominant negative mutants. *J. Virol.* **69**, 3315–3323.

Szilvay, A. M., Brokstad, K. A., Boe, S. O., Haukenes, G., and Kalland, K. H. (1997). Oligomerization of HIV-1 Rev mutants in the cytoplasm and during nuclear import. *Virology* **235**, 73–81.

Taagepera, S., McDonald, D., Loeb, J. E., Whitaker, L. L., McElroy, A. K., Wang, J. Y., and Hope, T. J. (1998). Nuclear-cytoplasmic shuttling of C-ABL tyrosine kinase. *Proc. Natl. Acad. Sci. USA* **95**, 7457–7462.

Tabernero, C., Zolotukhin, A. S., Bear, J., Schneider, R., Karsenty, G., and Felber, B. K. (1997). Identification of an RNA sequence within an intracisternal-A particle element able to replace Rev-mediated posttranscriptional regulation of human immunodeficiency virus type 1. *J. Virol.* **71**, 95–101.

Tan, R., Chen, L., Buettner, J. A., Hudson, D., and Frankel, A. D. (1993). RNA recognition by an isolated alpha helix. *Cell* **73**, 1031–1040.

Tan, R., and Frankel, A. D. (1994). Costabilization of peptide and RNA structure in an HIV Rev peptide-RRE complex. *Biochemistry* **33**, 14579–14585.

Tan, R., and Frankel, A. D. (1995). Structural variety of arginine-rich RNA-binding peptides. *Proc. Natl. Acad. Sci. USA* **92**, 5582–5586.

Tang, H., Gaietta, G. M., Fischer, W. H., Ellisman, M. H., and Wong-Staal, F. (1997). A cellular cofactor for the constitutive transport element of type D retrovirus. *Science* **276**, 1412–1415.

Tange, T. Ø., Jensen, T. H., and Kjems, J. (1996). In vitro interaction between human immunodeficiency virus type 1 Rev protein and splicing factor ASF/SF2-associated protein, p32. *J. Biol. Chem.* **271**, 10066–10072.

Thomas, S. L., Hauber, J., and Casari, G. (1997). Probing the structure of the HIV-1 transactivator protein by functional analysis. *Protein Engineering* **10**, 103–107.

Thomas, S. L., Oft, M., Jaksche, H., Casari, G., Heger, P., Dobrovnik, M., Bevec, D., and Hauber, J. (1998). Functional analysis of the human immunodeficiency virus type 1 Rev protein oligomerization interface. *J. Virol.* **72**, 2935–2944.

Tiley, L. S., Brown, P. H., Le, S. Y., Maizel, J. V., Clements, J. E., and Cullen, B. R. (1990). Visna virus encodes a post-transcriptional regulator of viral structural gene expression. *Proc. Natl. Acad. Sci. USA* **87,** 7497–7501.

Tiley, L. S., Malim, M. H., Tewary, H. K., Stockley, P. G., and Cullen, B. R. (1992). Identification of a high-affinity RNA-binding site for the human immunodeficiency virus type 1 Rev protein. *Proc. Natl. Acad. Sci. USA* **89,** 758–762.

Toyoshima, F., Moriguchi, T., Wada, A., Fukuda, M., and Nishida, E. (1998). Nuclear export of cyclin B1 and its possible role in the DNA damage-induced G2 checkpoint. *EMBO J.* **17,** 2728–2735.

Truant, R., and Cullen, B. R. (1999). The arginine-rich domains present in human immunodeficiency virus type 1 Tat and Rev function as direct importin beta-dependent nuclear localization signals. *Mol. Cell. Biol.* **19,** 1210–1217.

Ullman, K. S., Shah, S., Powers, M. A., and Forbes, D. J. (1999). The nucleoporin Nup 153 plays a critical role in multiple types of nuclear export. *Mol. Biol. Cell.* **10,** 649–664.

Vandendriessche, T., Chuah, M. K., Chiang, L., Chang, H. K., Ensoli, B., and Morgan, R. A. (1995). Inhibition of clinical human immunodeficiency virus (HIV) type 1 isolates in primary CD4+ T lymphocytes by retroviral vectors expressing anti-HIV genes. *J. Virol.* **69,** 4045–4052.

Venkatesh, L. K., and Chinnadurai, G. (1990). Mutants in a conserved region near the carboxy-terminus of HIV-1 Rev identify functionally important residues and exhibit a dominant negative phenotype. *Virology* **178,** 327–330.

Venkatesh, L. K., Mohammed, S., and Chinnadurai, G. (1990). Functional domains of the HIV-1 rev gene required for trans-regulation and subcellular localization. *Virology* **176,** 39–47.

Visa, N., Izaurralde, E., Ferreira, J., Daneholt, B., and Mattaj, I. W. (1996). A nuclear cap-binding complex binds Balbiani ring pre-mRNA cotranscriptionally and accompanies the ribonucleoprotein particle during nuclear export. *J. Cell. Biol.* **133,** 5–14.

Wada, A., Fukuda, M., Mishima, M., and Nishida, E. (1998). Nuclear export of actin: A novel mechanism regulating the subcellular localization of a major cytoskeletal protein. *EMBO J.* **17,** 1635–1641.

Watts, N. R., Misra, M., Wingfield, P. T., Stahl, S. J., Cheng, N., Trus, B. L., Steven, A. C., and Williams, R. W. (1998). Three-dimensional structure of HIV-1 Rev protein filaments. *J. Struct. Biol.* **121,** 41–52.

Weichselbraun, I., Farrington, G. K., Rusche, J. R., Bohnlein, E., and Hauber, J. (1992). Definition of the human immunodeficiency virus type 1 Rev and human T-cell leukemia virus type I Rex protein activation domain by functional exchange. *J. Virol.* **66,** 2583–2587.

Weis, K., Mattaj, I. W., and Lamond, A. I. (1995). Identification of hSRP1 alpha as a functional receptor for nuclear localization sequences. *Science* **268,** 1049–1053.

Weis, K., Dingwall, C., and Lamond, A. I. (1996). Characterization of the nuclear protein import mechanism using Ran mutants with altered nucleotide binding specificities. *EMBO J.* **15,** 7120–7128.

Wen, W., Meinkoth, J. L., Tsien, R. Y., and Taylor, S. S. (1995). Identification of a signal for rapid export of proteins from the nucleus. *Cell* **82,** 463–473.

Wingfield, P. T., Stahl, S. J., Payton, M. A., Venkatesan, S., Misra, M., and Steven, A. C. (1991). HIV-1 Rev expressed in recombinant Escherichia coli: Purification, polymerization, and conformational properties. *Biochemistry* **30,** 7527–7534.

Woffendin, C., Ranga, U., Yang, Z., Xu, L., and Nabel, G. J. (1996). Expression of a protective gene-prolongs survival of T cells in human immunodeficiency virus-infected patients. *Proc. Natl. Acad. Sci. USA* **93,** 2889–2894.

Wolff, B., Cohen, G., Hauber, J., Meshcheryakova, D., and Rabeck, C. (1995). Nucleocytoplasmic transport of the Rev protein of human immunodeficiency virus type 1 is dependent on the activation domain of the protein. *Exp. Cell Res.* **217,** 31–41.

Wolff, B., Sanglier, J., and Wang, Y. (1997). Leptomycin B is an inhibitor of nuclear export: Inhibition of nucleo-cytoplasmic translocation of the human immunodeficiency virus type 1 (HIV-1) Rev protein and Rev-dependent mRNA. *Chem. Biol.* **4**, 139–147.

Wu, B. Y., Woffendin, C., MacLachlan, I., and Nabel, G. J. (1997). Distinct domains of IkappaB-alpha inhibit human immunodeficiency virus type 1 replication through NF-kappaB and Rev. *J. Virol.* **71**, 3161–3167.

Wu, Y., Duan, L., Zhu, M., Hu, B., Kubota, S., Bagasra, O., and Pomerantz, R. J. (1996). Binding of intracellular anti-Rev single chain variable fragments to different epitopes of human immunodeficiency virus type 1 rev: Variations in viral inhibition. *J. Virol.* **70**, 3290–3297.

Xu, W., and Ellington, A. D. (1996). Anti-peptide aptamers recognize amino acid sequence and bind a protein epitope. *Proc. Natl. Acad. Sci. USA* **93**, 7475–7480.

Yan, C., Lee, L. H., and Davis, L. I. (1998). Crm1p mediates regulated nuclear export of a yeast AP-1-like transcription factor. *EMBO J.* **17**, 7416–7429.

Yang, J., Bardes, E. S., Moore, J. D., Brennan, J., Powers, M. A., and Kornbluth, S. (1998). Control of cyclin B1 localization through regulated binding of the nuclear export factor CRM1. *Genes Dev.* **12**, 2131–2143.

Yang, X., and Gabuzda, D. (1999). Regulation of human immunodeficiency virus type 1 infectivity by the ERK mitogen-activated protein kinase signaling pathway. *J. Virol.* **73**, 3460–3466.

Zapp, M. L., Hope, T. J., Parslow, T. G., and Green, M. R. (1991). Oligomerization and RNA binding domains of the type 1 human immunodeficiency virus Rev protein: A dual function for an arginine-rich binding motif. *Proc. Natl. Acad. Sci. USA* **88**, 7734–7738.

Zapp, M. L., Stern, S., and Green, M. R. (1993). Small molecules that selectively block RNA binding of HIV-1 Rev protein inhibit Rev function and viral production. *Cell* **74**, 969–978.

Zapp, M. L., Young, D. W., Kumar, A., Singh, R., Boykin, D. W., Wilson, W. D., and Green, M. R. (1997). Modulation of the Rev-RRE interaction by aromatic heterocyclic compounds. *Bioorg. Med. Chem.* **5**, 1149–1155.

Zemmel, R. W., Kelley, A. C., Karn, J., and Butler, P. J. (1996). Flexible regions of RNA structure facilitate co-operative Rev assembly on the Rev-response element. *J. Mol. Biol.* **258**, 763–777.

Zhang, M. J., and Dayton, A. I. (1998). Tolerance of diverse amino acid substitutions at conserved positions in the nuclear export signal (NES) of HIV-1 Rev. *Biochem. Biophys. Res. Commun.* **243**, 113–116.

Zolotukhin, A. S., Valentin, A., Pavlakis, G. N., and Felber, B. K. (1994). Continuous propagation of RRE(-) and Rev(-)RRE(-) human immunodeficiency virus type 1 molecular clones containing a cis-acting element of simian retrovirus type 1 in human peripheral blood lymphocytes. *J. Virol.* **68**, 7944–7952.

Zolotukhin, A. S., and Felber, B. K. (1997). Mutations in the nuclear export signal of human ran-binding protein RanBP1 block the Rev-mediated posttranscriptional regulation of human immunodeficiency virus type 1. *J. Biol. Chem.* **272**, 11356–11360.

Zolotukhin, A. S., and Felber, B. K. (1999). Nucleoporins nup98 and nup214 participate in nuclear export of human immunodeficiency virus type 1 Rev. *J. Virol.* **73**, 120–127.

Zuk, D., and Jacobson, A. (1998). A single amino acid substitution in yeast eIF-5A results in mRNA stabilization. *EMBO J.* **17**, 2914–2925.

A. L. Greenway, G. Holloway, and D. A. McPhee

AIDS Cellular Biology Unit
Macfarlane Burnet Centre for Medical Research
Fairfield, Victoria, Australia, 3078

HIV-1 Nef: A Critical Factor in Viral-Induced Pathogenesis

I. Introduction

The HIV-1 *nef* gene and protein product have been the focus of many researchers in understanding disease pathogenesis induced by the primate lentiviruses HIV and SIV. These cytopathic lentiviruses have evolved into more complex retroviruses with multiple regulatory and accessory genes. These additional genes play essential/critical roles in the infection process and disease development. Of these and the respective protein products the most fascinating is *nef*. This gene encodes a protein that has a plethora of perceived functions designed to control the host cell to the advantage of virus infection.

The genomic complexity of the retroviruses, in particular the lentiviruses, indicates that HIV and SIV have evolved by acquiring several gene products not encoded by the nonprimate lentiviruses and the less complex

Advances in Pharmacology, Volume 48

1054-3589/00 $35.00

retroviruses (Fig. 1). Interestingly, when the simple murine retroviruses are compared with the more complex immunodeficiency viruses SIV and HIV the expanded tropism for the hemopoietic system is coincident with the acquisition of the six "regulatory/accessory" genes (Cereseto *et al.,* 1996; Datta *et al.,* 1975; Perry *et al.,* 1992; Ugolini *et al.,* 1999; Zink *et al.,* 1990). Even within the lentivirus genus there is a further expansion in tropism from restricted infection of macrophages by the ungulate lentiviruses to infection of both lymphocytes and macrophages by the primate lentiviruses (Clements and Zink, 1996). The nonprimate or ungulate lentiviruses most closely related to HIV and SIV certainly have many common elements in pathogenesis including prolonged incubation periods, persistent viremia, a weak neutralizing antibody response, neuropathology, rapid genetic changes, and a lytic infection of selected blood-cell populations; however, there is little evidence indicating the condition is associated with immunosuppression. Coincident with this is the generation of an increased number of accessory genes. The marked change from the nonprimate to the primate lentiviruses with the successive acquisition of the *tat, vpr, vpu, vif,* and *nef* genes suggest derivation through gene capture, particularly for *nef* (Myers *et al.,* 1992; Tristem *et al.,* 1998). This gene acquisition and altered tropism correlates with altered disease pathogenesis. The primate lentiviruses have developed to cause a complex disease of the immune system and it is this last point we wish to explore in the context of the *nef* gene for HIV and SIV.

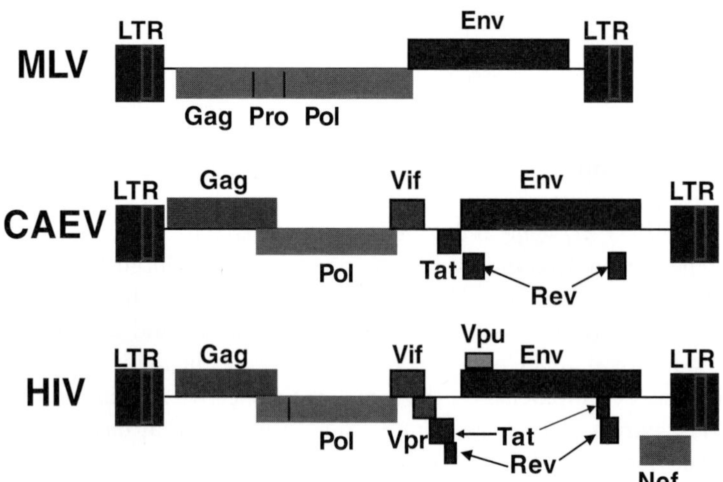

FIGURE 1 Coding regions of the retroviruses moloney murine leukemia virus (MLV), caprine arthritis encephalitits virus (CAEV), and human immunodeficiency virus (HIV) (Coffin, 1996; Myers *et al.,* 1997). The genes are listed according to current nomenclature. The complexity increases as to HIV with the most gene products (9).

II. Historical Background of Nef Involvement in Pathogenesis

A. Animal Models

A landmark publication describing infection of macaques with SIV provided the first insight that the *nef* gene product was required for maintaining high viral loads and the full pathogenic potential of the cytolytic lentivirus SIV (Kestler *et al.*, 1991).

Both point mutations and deletion mutants within SIV *nef* showed there was intense selection pressure for reversion to wild-type virus, implicating the gene as being very important in disease development. This was confirmed and extended in vaccine-related studies with SIV_{MAC239} where animals infected with a *nef*-deleted virus and subsequently challenged with wild-type virus were completely protected (Daniel *et al.*, 1992). Targeting the *nef* gene product showed the promise of the development of a live-attenuated vaccine for SIV. Subsequent detailed studies with *nef*-attenuated SIV strains confirmed this early result but identified that there was a very fine line between nonpathogenic and pathogenic attenuated viruses (Baba *et al.*, 1995, 1999; Carl *et al.*, 1999; Johnson *et al.*, 1999; Kirchhoff *et al.*, 1994; Novembre *et al.*, 1996; Saucier *et al.*, 1998; Schwiebert *et al.*, 1997; Whatmore *et al.*, 1995; Wyand *et al.*, 1996).

In initial experiments SIV *nef* was deleted in the N-terminal region postenvelope sequence [SIV_{MAC239} nt 9250–9433; (Kestler *et al.*, 1991)]. Unlike virus which had a point mutation introduced into *nef*, this virus was unable to revert to wild type and infected animals did not develop disease over the course of the experiments. The importance of the *nef* gene to the virus was demonstrated by the quick reversion of the virus with a point mutation of wild type. Furthermore, in other experiments with a natural mutant of SIV_{MAC32H}, termed SIV_{MACC8}, the introduction of a 12-bp deletion in *nef* (nt 9501–9512) was also rapidly repaired upon inoculation into rhesus macaques (Whatmore *et al.*, 1995). The initial 12-bp deletion caused a marked reduction in protein stability. The restoration of Nef protein stability was gradual, but even prior to full recovery, it still contributed to increased viral replication (Carl *et al.*, 1999). In contrast to restoration of the *nef* gene, virus that had a major deletion introduced further deletions upstream in the U3 region (Kirchhoff *et al.*, 1994).

This attenuated pathogenicity due to a *nef* deletion was further confirmed with the highly virulent SIV_{SMMPBj} virus infection of pig-tailed macaques, which normally causes an acute disease and death 5 to 14 days postinfection (Novembre *et al.*, 1996). When the *nef* gene was deleted through a frameshift mutation at amino acid 57 a marked attenuation of pathogenesis occurred despite a resultant infection. A more subtle mutation of the *nef* gene at amino acids 17 Y to R of this highly virulent SIV strain also attenuated virus upon infection of macaques (Saucier *et al.*, 1998).

Mutation or alteration within the HIV-1 *nef* gene has been assessed using SCID-hu mouse model systems repopulated with thymocytes (Aldrovandi *et al.*, 1993, 1998; Aldrovandi and Zack, 1996) or leukocytes (Gulizia *et al.*, 1997). The results were quite concordant despite the different target cells, thymocytes, and leukocytes. Deletion of HIV-1 *nef* in either a T-tropic or M-tropic background significantly delayed but did not abrogate the depletion of the CD4-positive cell population. Further, a region in the protein from amino acids 41 to 49 was defined as being the most important in this attenuation (Aldrovandi *et al.*, 1998). Both of these *in vivo* experiments point to HIV *nef* being important in virulence, as do the infections of SIV in macaques.

B. Natural Infection

Long-term survival or long-term nonprogression in HIV infection is a well-recognized phenomenon. With the definition of several cohorts of such individuals the genetic and biological markers responsible for this lack of disease progression have been sought (Easterbrook and Schrager, 1998). These include coreceptor modification, HLA alleles and viral makeup. One obvious means of attenuated infection from the animal model experiments is modification of the virus in the *nef* gene. Investigation of several cohorts has defined that indeed there is a rare but significant group of HIV-infected individuals where the *nef* gene has been modified or deleted during natural infection (L. J. Ashton and D. I. Rhodes, personal communication; Blaak *et al.*, 1998; Deacon *et al.*, 1995; Easterbrook and Schrager, 1998; Huang *et al.*, 1995; Kirchhoff *et al.*, 1995, 1999; Mariani *et al.*, 1996; Michael *et al.*, 1995a,b; Premkumar *et al.*, 1996; Salvi *et al.*, 1998; Switzer *et al.*, 1998). Review of several cohorts has revealed one or more members that have alterations in the *nef* gene where the patients are classified as long-term nonprogressors at the time of assessment.

The most complete cohort implicating the *nef* gene in disease progression was the study of the Sydney Blood Bank Cohort (SBBC), where eight recipients were infected with the blood from a single donor (Deacon *et al.*, 1995; Learmont *et al.*, 1999). All of the viral strains infecting these patients contain defective *nef*/LTR sequences. A hemophiliac long-term nonprogressor from a cohort of five individuals was found to have a viral genotype with only gross deletions in the *nef* gene alone (Kirchhoff *et al.*, 1995). Interestingly, this patient subsequently developed further deletions in the U3 region upstream; a very similar finding to that observed in experimentally infected macaques with SIV (Kirchhoff *et al.*, 1994). This pattern of increasing deletions in the *nef*/LTR over time has also been observed for the SBBC (Deacon *et al.*, 1995) (N. Deacon and D. Rhodes, personal communication). In a more recent long-term nonprogressor cohort 1 of 11 members has a replication-incompetent virus that has a deletion in the *nef*/LTR (Salvi *et al.*, 1998). A fourth cohort of 38 long-term nonprogressors with viral loads less than 200 copies/ml has revealed 3 to have gross deletions in the *nef*/LTR region (L. J. Ashton and D. I. Rhodes, personal communication).

In addition to identification of long-term nonprogressors with gross deletions in the *nef* gene there have also been several individuals seen with more subtle deletions/mutations (Blaak *et al.,* 1998; Mariani *et al.,* 1996; Premkumar *et al.,* 1996). In one of seven individuals a *nef* deletion of four amino acids (156 to 159) has been consistently seen, although wild-type sequences were observed earlier during infection (Blaak *et al.,* 1998). A long-term nonprogressor with a grossly normal *nef* gene sequence consistently showed a cysteine at position 138, a rare occurrence (Premkumar *et al.,* 1996). In a group of four long-term survivors one individual yielded an unusually high frequency of disrupted *nef* open reading frames and Nef protein defective in CD4 down-regulation, one of the perceived functions of Nef (Mariani *et al.,* 1996).

As for HIV-1, infection with HIV-2 results in disease progression but the virus is quite distinct with a 5- to 10-fold lower rate of heterosexual transmission and time to disease progression is much longer (Marlink *et al.,* 1994). In the only detailed study of the *nef* gene product in HIV-2 (Switzer *et al.,* 1998) the results are interesting in that 10% of HIV-2-infected individuals studied (both asymptomatic and symptomatic) had truncated Nef proteins. The truncations occurred predominantly in the asymptomatic group. These results raise the question of involvement of *nef* in the slow progression to disease.

Despite the importance of the *nef* gene in disease progression from the naturally infected individuals the attenuated virus is still able to replicate and, from longer term follow-up, can cause disease. This has been recognized for the SBBC of the infected donor and recipients where long-term follow-up has revealed that three of the six living members have declining CD4 cell counts and increasing viral loads despite the gross *nef*/LTR deletion (Learmont *et al.,* 1999). Progression to disease has also been observed for a long-term nonprogressor hemophiliac with a gross *nef* deletion (Greenough *et al.,* 1999). In both of these studies the virus has changed over the course of infection, showing more extensive deletions with time of infection; for the SBBC these viruses have been now demonstrated to be more fit (D. Rhodes, personal communication).

III. Nef Structure

A. Gene Loci

HIV-1, HIV-2, and SIV *nef* genes were originally designated as 3′ ORF and also as F, ORF B O or E, and finally as *nef*. The *nef* gene relative to the molecular clone HXB2 of HIV-1 spans nucleotide residues 8797 to 9417 and partially overlaps the nucleotide sequence for the 3′ LTR, which begins at nucleotide residue 9085. The *nef* gene from HIV-2 (HIV-2_{ROD}) and SIV(SIV_{MAC239}) spans nucleotide residues 8557–9327 and 8789–9580 respectively and also overlaps with the respective 3′ LTR and *env* sequences.

B. Nucleotide and Amino Acid Sequence Conservation

A high degree of polymorphism of the *nef* gene exists among different HIV-1 isolates and between HIV-1, HIV-2, and SIV (Myers *et al.*, 1997). Vast numbers of *nef* sequences from HIV-1 isolates have been sequenced and deposited in sequence data banks. Similarly, nucleotide sequences from large numbers of *nef* genes have been determined from PBMC isolated from unrelated infected individuals (Myers *et al.*, 1997). Pairwise comparisons of nucleotide and predicted amino acid sequences reveal significant diversity within an individual. In general, and as expected, a larger difference between those sequences derived from different individuals was found (Shugars *et al.*, 1993). At the nucleotide level, the intraindividual variation ranged from 0.6 to 2.3%, while between individuals the variation ranged from 1.2 to 11.2%. At the amino acid level, variation between Nef proteins, within the same individual, was estimated to range from 0.5 to 20.2%, while the interindividual variation was 0.5 to 22.7% (Shugars *et al.*, 1993). A smaller number of *nef* sequences from HIV-2 and SIV isolates have been sequenced. Alignment of a consensus HIV-2 amino acid sequence derived from the HIV-2 strains ROD, GH1, D194, BEN, ISY, and ST with a consensus HIV-1 sequence based on the alignment of 54 *nef* sequences isolated from 12 individuals showed strong sequence homology. Twenty-six percent of amino acid residues were identical and a further 16% were represented by conservative changes (Shugars *et al.*, 1993). A similar degree of amino acid sequence conservation was demonstrated between the HIV-1 consensus sequence and some isolates of SIV (Shugars *et al.*, 1993).

C. Characteristics of Nef Sequences

The amino acid sequence variation among the Nef proteins of HIV-1 are not distributed evenly. Alignment of the different sequences showed several regions and sequence motifs which are highly conserved among HIV-1 and to varying degrees among their HIV-2 and SIV counterparts (Shugars *et al.*, 1993). The most highly conserved feature displayed by HIV-1 Nef proteins is an N-terminal myristylation signal present between amino acid residues 2 to 7. The myristylation signal is virtually invariant among the HIV-1 Nef proteins analyzed and is highly conserved among HIV-2 and SIV Nef (Delassus *et al.*, 1991; Shugars *et al.*, 1993). The myristylation sequence is followed by a run of 7 or 8 amino acids which exhibit high sequence diversity as compared to other regions of Nef and has been proposed to act as a flexible spacer region for the molecule (Shugars *et al.*, 1993). An internal translation initiation site exists at the methionine residue at position 20, resulting in a truncated nonmyristylated 25-kDa protein relative to its full-length 27-kDa counterpart. A variably duplicated region between positions 22 and 23 is present in a large number of Nef protein sequences. This region represents an imperfect duplication of adjacent amino acid sequences from amino acids 23 to 29 and ranges in length from 3 to 10 amino acid residues (Shugars *et al.*, 1993).

Nef also contains highly conserved domains which may be targeted by cellular protein tyrosine or serine/threonine kinases, resulting in its biochemical modification. A highly conserved sequence between amino acid residues 72 and 82 (RXPXMTXYXK) resembles a phosphorylation site for protein kinase C (PKC) (Shugars *et al.*, 1993). Other potential sites for PKC phosphorylation as well as sites for phosphorylation by casein kinase II (S/TXXD/E, positions 15 to 18) and a tyrosine kinase recognition site (R/KXXXD/EXXXY at positions 94 to 102) also exist in Nef but are not highly conserved (Shugars *et al.*, 1993).

Other regions within Nef which are highly conserved and which may respresent functional domains include a proline-rich repeat motif present between amino acid residues 69 to 78 (Shugars *et al.*, 1993). This domain bears strong resemblance to a *src* homology 3 domain binding motif and most likely represents an interactive domain for multiple *src* family kinase members (Collette *et al.*, 1996a; Greenway *et al.*, 1996; Ren *et al.*, 1993; Saksela *et al.*, 1995; Shugars *et al.*, 1993). The SH3 binding capacity of Nef is discussed at length below. A highly conserved, acidic domain region, composed of mainly glutamic acid residues between amino acid residues 62 to 65 is most likely structurally important (Shugars *et al.*, 1993). The polypurine tract, which is an essential element for reverse transcription and is the site for second-strand DNA synthesis, is located within the *nef* coding sequence. The amino acid sequences 91 to 96, which make up the polypurine tract, are highly conserved throughout Nef proteins. The high degree of conservation most likely reflects constraint at the level of the RNA sequence for virus replication rather than conservation of amino acid sequence for protein function.

Comparison of HIV-1 and HIV-2 Nef for amino acid sequences which may be critical for functional aspects has identified four highly preserved or invariant regions. These have been denoted as blocks A through to D. Block A, which spans amino-acid residues 64 to 90, includes the acidic/charged region, the proline repeat motif, and the potential PKC phosphorylation site (Shugars *et al.*, 1993). The polypurine tract was not included in block A, as it was thought that its conservation was at the nucleotide level for replication rather than for protein function. Block B, at positions 106 to 114, block C, at positions 130 to 148, and block D, at positions 179 to 190, were also identified to contain highly conserved residues (Shugars *et al.*, 1993). These blocks contain most of the highly conserved stretches of amino acid sequences. Block B contains the diarginine repeat, which has been implicated as essential for interaction of Nef with serine/threonine kinases and may modulate virus replication (Sawai *et al.*, 1995, 1996; Shugars *et al.*, 1993). Block C contains residues which have not yet been implicated in any known function, while block D residues contain a site for Raf-1 kinase interaction, although the consequence of this interaction with Nef is unknown (Hodge *et al.*, 1998; Shugars *et al.*, 1993).

The generation of a consensus sequence for SIV Nef is difficult because of the diverse sequence variation between viral isolates and the different species used for their generation. However, several features conserved among HIV-1 and HIV-2 Nef proteins are also present in SIV Nef. SIV Nef contains a highly conserved myristylation sequence as well as the highly invariant potential PKC phosphorylation site present on block A (Shugars *et al.*, 1993). Interestingly, SIV Nef also contains the proline repeat motif; however, in this case the Pxx sequence is repeated two or three times rather than four as observed in HIV-1 and HIV-2 Nef proteins (Shugars *et al.*, 1993). The absence of an intact polyproline repeat motif and its potential consequences are discussed below. Other features conserved in HIV-1 Nef, such as the diarginine residues, are also present in SIV Nef. However, none of the reported SIV Nef sequences contain a second initiating methionine, as observed in HIV-1 Nef sequences, suggesting that at least in SIV infection a second truncated form of Nef is not produced and not required for efficient virus replication.

D. Expression and Subcellular Localization of Nef

The HIV-1 proviral DNA genome is 9.8 kb in length and encodes at least 15 known proteins. A number of these proteins have overlapping reading frames and are expressed from unspliced as well as alternatively spliced mRNAs (Arrigo and Chen, 1991; Bruggeman *et al.*, 1994; Furtado *et al.*, 1991; Klotman *et al.*, 1991; Purcell and Martin, 1993). HIV-1 generates an extremely complex pattern of spliced RNA to encode the essential regulatory gene products Tat and Rev as well as several other proteins including Nef (Arrigo and Chen, 1991; Bruggeman *et al.*, 1994; Furtado *et al.*, 1991; Klotman *et al.*, 1991; Purcell and Martin, 1993). The splicing of HIV-1 is complex because of the presence of competing constitutive and alternatively used splice acceptor and donor sites. Their alternate selection usually determines the protein encoded by the mature RNA. Nef is transcribed early during infection from a number of variant transcripts (Klotman *et al.*, 1991; Ranki *et al.*, 1994). While the predominant *rev, tat, vpr,* and *env* RNAs produced during HIV-1 infection of PBMC *in vitro* contain a minimum of noncoding sequences the predominant *nef* transcripts are incompletely spliced and invariably include noncoding exons. The role that the noncoding exon plays in the translation of Nef is unclear but may contribute to the stability of the mRNA species or regulate the efficiency of its translation (Klotman *et al.*, 1991; Purcell and Martin, 1993).

A combination of immunocytochemistry and *in situ* hybridization techniques have been used to study the expression of Nef relative to other HIV-1 regulatory proteins and its subcellular localization. In acutely infected MT-4 cells four distinct phases of infection were defined (Ranki *et al.*, 1994). Unlike other viral proteins the expression of Nef was consistent throughout the phases of infection when viral proteins were produced. For example, Nef was shown to be produced abundantly during the second regulatory phase (6–9 h postinfection) as was Rev and Tat. mRNA-encoding Nef was still abundant during this phase. In contrast, *in situ* hybridization studies showed that mRNA-encoding gp160 was absent at this period as was its corresponding mature protein. Nef was again abundantly expressed during the productive phase of infection (12–48 h), at which time expression of full-length mRNA and gp160 was also abundant. During the fourth cytopathic phase the expression of mRNA and viral proteins decreased (Ranki *et al.*, 1994).

Numerous studies describing the cellular distribution of HIV-1 Nef protein have been reported (Kaminchik *et al.*, 1991; Kienzle *et al.*, 1992; Kohleisen *et al.*, 1992; Murti *et al.*, 1993; Ranki *et al.*, 1994; Yu and Felsted, 1992). These studies have analyzed the subcellular distribution of Nef when expressed by transfection in mammalian cells or during *in vitro* HIV-1 infection. Cell extracts fractionated by low- and high-speed centrifugation and by nonionic detergents have shown that two Nef-related proteins can be expressed, a 27-kDa form (Nef 27) and a 25-kDa form (Nef 25). Nef 27, an N-myristylated form of Nef, was found in the cytosol and in association with a particulate fraction of the cytoplasm (Kienzle *et al.*, 1992). Treatment of the particulate cytoplasmic fraction with nonionic detergents, using three different protocols designed to isolate the cytoskeletal matrix, indicated that part of Nef was sensitive and part was resistant to detergent solubilization. The isolation of the two different cellular fractions represent membrane- and cytoskeleton-associated Nef. The 25-kDa form of Nef, initiated from an in-frame AUG codon, was not modified with myristic acid at the amino terminus. Consequently, this protein was present in a soluble form in the cytosol. Furthermore, a mutant of the full-length 27-kDa form of Nef, in which the myristylation signal is deleted, appeared as a cytoplasmic soluble protein highlighting the role of the N-terminal myristylation signal in localizing Nef to the plasma membrane and the cytoskeleton (Kienzle *et al.*, 1992).

Interestingly, the subcellular localization of Nef appeared to change according to the different phases of the infection cycle (Ranki *et al.*, 1994). Up until 24 h postinfection Nef was expressed almost exclusively in the cytoplasm with a characteristic polar pattern, indicating localization in the Golgi complex and endoplasmic reticulum (Kienzle *et al.*, 1992; Mangasarian *et al.*, 1997). A substantial number of cells showed Nef localization in the nucleus 24 h after infection after which time, however, Nef was mainly localized in the cytoplasm (Ranki *et al.*, 1994). These findings are congruent with those which identified significant proportions of Nef in the cytosolic (15–50%), membrane (32–48%), and cytoskeletal (16–42%) fractions of Jurkat cells stably expressing Nef (Niederman *et al.*, 1993b). As mentioned above, the cytoskeletal association of Nef is significantly enhanced by myristylation, as a nonmyristylated version of Nef was located predominantly in the cytosolic fraction. The result that 50–85% of Nef associates with the plasma membrane or cytoskeleton may reflect that only a proportion of Nef is myristylated at a given time and that a dynamic state may exist such that Nef may localize and translocate within the cell depending upon posttranslational modification(s) (Niederman *et al.*, 1993b).

Indeed, numerous studies have identified the nuclear localization of Nef (Kienzle *et al.*, 1992; Kohleisen *et al.*, 1992; Macreadie *et al.*, 1993; Murti *et al.*, 1993; Ranki *et al.*, 1994; Yu and Felsted, 1992). High-resolution immunogold labeling and electron microscopic studies of cells expressing Nef derived from the isolate HIV-1 SF2 have extended the original observations of Nef colocalization within the nucleus (Murti *et al.*, 1993). Specific staining using antibodies directed toward Nef revealed that a small fraction of Nef is in the nucleus and it is localized in specific curvilinear tracks that extend between the nuclear envelope and the nucleoplasm. The presence of Nef in distinct nuclear tracks suggests that Nef is transported along a specific pathway that extends from the nuclear envelope into the nucleoplasm. The appearance of Nef in the nucleus will, like other proteins which can be shuttled into the nucleus, be influenced by the cell cycle. Indeed, in nondividing cells Nef is observed both in the cytoplasm as well as the nucleus, while in dividing cells the viral protein is present in the cytoplasm and at the nuclear membrane but not in the nucleus. The localization of Nef within the nucleus suggests a potential nuclear function for this protein, which may be relevant to its association with p53 (A. L. Greenway, unpublished). Furthermore, the relationship between nuclear localization of Nef and cell division may be relevant to the function of Nef in CD4+ T cells versus nondividing HIV-1 susceptible cells such as monocyte-derived macrophages (Murti *et al.*, 1993).

The functional characteristics of Nef will obviously be dictated by multiple factors, including its cellular localization, which can be determined, at least in part, by its biochemical modification by myristylation. Other biochemical and biophysical properties of Nef may also influence its biological role during HIV-1 infection. Nef can form multimers with itself to form structures including dimers, trimers, and tetramers, some of which are reliant on covalent disulphide bonding but others which form independent of this type of linkage (Kienzle *et al.*, 1992). Under nonreducing conditions eukaryotically and prokaryotically expressed Nef forms oligomers which are covalently linked by disulfide bonds. Nuclear magnetic resonance studies show also that even under strong reducing conditions Nef is able to form noncovalently linked oligomers that are stabilized mainly by polar and electrostatic interactions. Since T cells of HIV-1-infected patients display very low levels of reduced glutathione and since Nef molecules are not found exclusively in the cytoplasm it is possible that oligomeric complexes of Nef may be biologically functional (Davis *et al.*, 1997; Kienzle *et al.*, 1993).

IV. Nef Function(s)

A. Nef-Induced Down-Regulation of Cell-Surface Molecules

Given that the *nef* gene products of both HIV-1 and SIV have been shown during *in vivo* studies and analysis to be critical determinants of disease outcome in adult infection of humans and macaques, it is of utmost importance to obtain a clear understanding of the role of Nef during infection and the molecular basis of its action(s). Following its identification, the *nef* gene was the center of unprecedented controversy surrounding its function in the virus life cycle. However, following at least 15 years of intensive research an integrated picture of Nef function is slowly emerging. A plethora of information regarding the role of Nef during HIV-1 infection exists with numerous mechanisms of action proposed. These studies have adopted various strategies to determine the role of Nef during the virus life cycle. These include investigations of the Nef protein for its effect on host-cell functions, which may relate to its ability to determine the outcome of infection and the use of molecular clones of HIV-1 deleted within *nef* to recreate *in vitro* the effect of Nef on virus replication. Useful information has also been generated using artificial *in vivo* models for HIV-1 infection such as the scid-hu system and comparison with SIV infection of rhesus macaques (Daniel *et al.*, 1992; Jamieson *et al.*, 1994; Kestler *et al.*, 1991). Proposed mechanisms of Nef function have largely been based upon information derived from coprecipitation or yeast-2 hybrid studies, which have determined the potential cellular binding proteins of Nef (Baur *et al.*, 1997; Collette *et al.*, 1996a; Dutartre *et al.*, 1998; Greenway *et al.*, 1995, 1996, 1999; Rossi *et al.*, 1997; Saksela *et al.*, 1995; Sawai *et al.*, 1994, 1995). The careful integration of all the studies performed clearly highlights major functions of Nef. It is likely that several if not all of the observed *in vitro* effects of Nef contribute to the overall importance of this protein *in vivo*.

The contribution of the protein to the induction of AIDS can be explained by at least three properties which have been ascribed to Nef: down-modulation of cell surface receptors including CD4 (Garcia and Miller,

1991; Aiken *et al.*, 1994; Garcia *et al.*, 1993; Greenway *et al.*, 1994), the α-chain of the interleukin 2 receptor (IL-2R) (Greenway *et al.*, 1994), and, most recently, major histocompatibility complex class 1 (MHC class 1) (Collins *et al.*, 1998; Le Gall *et al.*, 1997; Schwartz *et al.*, 1996); enhancement of virus production and the alteration of T-cell activation pathways (Bandres and Ratner, 1994; Baur *et al.*, 1994; Collette *et al.*, 1996a,b 1997; Greenway *et al.*, 1995, 1996, 1999; Iafrate *et al.*, 1997; Luria *et al.*, 1991; Niederman *et al.*, 1989, 1992, 1993a; Sawai *et al.*, 1994, 1995; Sawai *et al.*, 1996; Skowronski *et al.*, 1993). This last activity most likely has a direct bearing on the contribution by Nef to virion production and regulation of key T-cell receptors. We now discuss each of these aspects of Nef function and its proposed mechanisms of action and attempt to portray the magnificent orchestration of multiple systems by this complex molecule which undoubtedly determines the fate of the HIV-1-infected individual.

1. Down-Regulation of CD4

Although the role of Nef *in vivo* is unclear, one well-documented effect of the viral protein *in vitro* is the down-regulation of cell surface CD4. This effect is well conserved and has been demonstrated using a number of naturally occurring and reference isolates of both HIV-1 and SIV in a variety of lymphoid and nonlymphoid cells. Nef-induced CD4 down-regulation occurs early after viral infection, thus preceding the effects of Env- and Vpu-induced retention and degradation of newly synthesized molecules within the ER (Chen *et al.*, 1996; Crise *et al.*, 1990; Willey *et al.*, 1992). Nef does not affect levels of CD4 mRNA (Garcia and Miller, 1991), protein synthesis, or transport from the ER to the golgi (Aiken *et al.*, 1994; Rhee and Marsh, 1994b), but acts chiefly by promoting an acceleration of CD4 endocytosis in clathrin-coated pits followed by degradation of CD4 in lysosomes (Aiken *et al.*, 1994).

A likely mechanism of Nef-induced CD4 down-regulation has been proposed whereby Nef acts as a connector protein between CD4 and the endocytic machinery. In fact, chimeric CD4 or CD8 proteins which have their cytoplasmic tails replaced by Nef, and therefore lack receptor-based internalization signals, are rapidly internalized in clathrin-coated pits (Mangasarian *et al.*, 1997). The binding of Nef to CD4 has been demonstrated in a number of experimental systems including transfection of insect cells (Harris and Neil, 1994), *in vitro* using recombinant proteins (Grzesiek *et al.*, 1996b), in the yeast two-hybrid system (Rossi *et al.*, 1996), and by precipitation of CD4 from MT-2 cell extracts by a GST–Nef fusion protein (Greenway *et al.*, 1995). The membrane proximal region of CD4 that contains a predicted α-helix is important for the Nef-induced effect, as mutations significantly affecting this predicted structure, at least partially, disrupt Nef-induced down-regulation of CD4 (Gratton *et al.*, 1996; Yao *et al.*, 1995). Included in this region is a dileucine motif that is critical for binding to Nef and for Nef-induced down-regulation (Aiken *et al.*, 1994; Grzesiek *et al.*, 1996b; Salghetti *et al.*, 1995). CD4 down-regulation is induced by phorbol esters such as phorbol 12-myristate 13-acetate (PMA) through a process that also relies on the dileucine motif within CD4 (Shin *et al.*, 1991). However, the mechanisms of PMA and Nef action appear to differ as the PMA effect requires serine phosphorylation of the CD4 cytoplasmic tail by PKC (Acres *et al.*, 1986; Garcia and Miller, 1991). Substitution of certain amino acids around the dileucine motif blocks CD4 down-regulation induced by PMA but not by Nef (Aiken *et al.*, 1994; Gratton *et al.*, 1996). It is of note that the α-helical structure in the CD4 cytoplasmic domain is absent from the dileucine containing CD3γ and -δ chains which are not down-regulated by Nef (Aiken *et al.*, 1994; Gratton *et al.*, 1996).

The region of CD4 necessary for its down-regulation by Nef overlaps the region required for binding of the *src* family protein tyrosine kinase Lck (Gratton *et al.*, 1996; Salghetti *et al.*, 1995). However, binding of Lck to CD4 is not essential for Nef-mediated CD4 endocytosis as demonstrated in Lck-deficient cell lines (Anderson *et al.*, 1994; Garcia *et al.*, 1993; Garcia and Miller, 1991). Nef actually appears to displace Lck from its association with CD4, which may facilitate CD4 down-modulation (Anderson *et al.*, 1994; Salghetti *et al.*, 1995). N-terminal myristylation is necessary for Nef localization at the plasma membrane (Kaminchik *et al.*, 1991; Yu and Felsted, 1992). This localization is also important for Nef-induced CD4 down-regulation as mutation of the myristylation site greatly reduces the ability of Nef to decrease cell-surface CD4 levels and also abolishes detectable Nef binding to CD4 in insect cells (Aiken *et al.*, 1994; Harris and Neil, 1994; Mariani and Skowronski, 1993). This is most likely due to the disruption of the plasma membrane localization of nonmyristylated Nef. Other regions within Nef necessary for CD4 endocytosis have been mapped to two disordered loops within the protein, particularly residues 36 and 56–59 in the N-terminal loop and residues 174–179 (Iafrate *et al.*, 1997; Mariani and Skowronski, 1993). Mutational studies have revealed that these regions, in both HIV-1 and SIV Nef, seem to be involved in the interaction with the cytoplasmic tail of CD4 (Greenberg *et al.*, 1997; Grzesiek *et al.*, 1996b; Mangasarian and Trono, 1997; Piguet *et al.*, 1998). However, they are not highly conserved between the two forms of HIV-1 and SIV Nef. Therefore it is likely that the two proteins make slightly different connections with the CD4 molecule.

Recent evidence suggests that the main link between Nef and the endocytic machinery occurs through a direct interaction with the μ-chain of adaptor complexes (Greenberg *et al.*, 1997; Piguet *et al.*, 1998). Adaptor protein (AP) complexes interact with cytosolic clathrin, allowing the formation of clathrin-coated pits and mediating sorting of proteins at the plasma membrane and the trans-Golgi (reviewed in Robinson, 1994). Adaptor protein complexes recruit proteins for internalization in clathrin-coated pits through internalization signals, usually containing tyrosine or dileucine motifs (Letourneur and Klausner, 1992). In the yeast two-hybrid system HIV-1 Nef binds to the μ-chain of AP-2, which is specifically involved in clathrin-coated pit formation at the plasma membrane (Piguet *et al.,* 1998). HIV-1 Nef also colocalizes with a component of AP-2 at the cell margin (Greenberg *et al.,* 1997). SIV and HIV-2 Nef also bind to AP-2 in the yeast system, apparently through N-terminal tyrosine-based motifs (Piguet *et al.,* 1998). However HIV-1 Nef does not contain these tyrosine residues and may bind AP-2 through a region of Nef containing a dileucine motif at position 165 (Mangasarian *et al.,* 1999). Nef also binds the medium chain of AP-1 (μ1) in the yeast system and in cell-free binding assays and this interaction possibly mediates routing of CD4 from the golgi to the endosome, bypassing the cell surface (Le Gall *et al.,* 1998; Piguet *et al.,* 1998). This mechanism, however, is thought to play only a minor role in the down-regulation of cell-surface CD4 by Nef (Mangasarian and Trono, 1997). Sequences within the disordered loops of Nef may also play a role in interaction with AP complexes, as mutations in these regions abolish Nef colocalization with AP components at the cell margin (Greenberg *et al.,* 1997).

CD4, a type-1 integral cell surface glycoprotein required for T-lymphocyte ontogeny and activation of mature helper T-lymphocytes, is also the main primate lentivirus receptor. Therefore possible advantages of Nef-induced CD4 down-regulation could be to prevent detrimental superinfection events or promotion of virus particle release (Benson *et al.,* 1993). CD4 can enhance antigen-driven TCR signaling events through interactions between its extracellular domain and the MHC class II molecule, with the association of the Lck tyrosine kinase with the CD4 cytoplasmic domain also being important. The effect of Nef on surface CD4 levels may interfere with TcR signaling to the advantage of the virus, possibly by increasing levels of free Lck within the infected cell.

2. Down-Modulation IL-2 Receptor

Nef also modulates T-cell activation by modulating IL-2R (CD25). Electroporation of PHA-activated PBMC and CD4+ T-cell lines with mature Nef protein markedly reduced the expression of IL-2 R (Greenway *et al.,* 1994). Other cell-surface antigens such as CD2, CD7, and transferrin receptor were not affected by the introduction of Nef into cells, confirming the relative specificity of Nef on modulation of CD4 and CD25. The mechanism of Nef action on CD25 may not represent a direct effect on the receptor itself, as levels of CD25 were not affected when Nef and CD25 were coexpressed in cells by transfection (Goldsmith *et al.,* 1995). However, these experiments do not consider involvement of the β- and γ-chains of the IL-2 R complex in maintenance of CD25 expression during activation. Other studies arguing against an effect of Nef on CD25 expression have looked only at its effect on induction rather than down-modulation of CD25 (Luria *et al.,* 1991) or have used cell systems which usually express only low levels of CD25 (Schwartz *et al.*, 1995). As both CD4 and the IL-2 R play crucial roles in antigen-driven helper T-cell signaling and T-cell proliferation, respectively, one critical role of Nef in the viral life cycle may be to perturb signaling pathways emanating from these receptors.

3. Nef Down-Regulation of MHC Class I

Nef has been shown to down-regulate cell-surface major histocompatibility complex class 1 (MHC I), protecting infected cells from lysis by cytotoxic T-lymphocytes (CTL) which recognize their targets through viral peptide epitopes attached to MHC I molecules (Collins *et al.*, 1998). MHC I down-regulation is a process also used by other pathogenic viruses to evade CTL killing (Fruh *et al.*, 1997). As with CD4 down-regulation, Nef does not affect MHC I sythesis but promotes its internalization from the plasma membrane in clathrin-coated pits (Schwartz *et al.*, 1996) and, to some extent, its redirection from the exocytic pathway, leading to retention in the trans-golgi (Le Gall *et al.*, 1998). Spontaneous internalization relies on determinants within MHC I encoded by exon 7 (Vega and Strominger, 1989), while Nef-induced internalization requires sequences encoded by exon 6, particularly a tyrosine residue at position 320 within the HLA-A and HLA-B heavy chains (Le Gall *et al.*, 1998). This tyrosine is not present in HLA-C, molecules that are not down-regulated by Nef. As HLA-C is an inhibitory ligand against lysis by natural killer (NK) cells (Colonna *et al.*, 1993) the selective effect of Nef on MHC I down-regulation may protect infected cells from both CTL and NK (Le Gall *et al.*, 1998).

Direct interaction between Nef and MHC I has not been reported. MHC I colocalizes with clathrin and AP-1 in the trans-golgi network in the presence of Nef, and Nef also partially localizes with these complexes, suggesting an active role for Nef in targeting of MHC I to these regions (Greenberg *et al.*, 1998). Although Nef binds $\mu1$ and $\mu2$ subunits of adaptor protein complexes in yeast, the dileucine motif in HIV-1 Nef which mediates this binding is dispensable for MHC I down-regulation (Mangasarian *et al.*, 1999). This suggests Nef does not act as an adaptor between MHC I and the endocytic machinery, but is somehow involved in exposing the tyrosine containing motif in HLA-A and B molecules which acts as an endocytosis signal.

Regions of Nef necessary for down-regulation of MHC I include the N-terminal myristylation signal, an acidic sequence (position 65 in HIV-1 Nef; Greenberg *et al.*, 1998), the two C-terminal prolines in the conserved core (Greenberg *et al.*, 1998; Mangasarian *et al.*, 1999), and the N-terminal α-helix (Mangasarian *et al.*, 1999). It has been speculated that phosphotyrosine-based signaling events are involved in the Nef-induced down-regulation of MHC I. This hypothesis is based on the known interactions between the proline-repeat motif and the N-terminal α-helix of Nef with members of the *src* family kinases and the presence of a tyrosine-based motif within the MHC I cytoplasmic region. However, the tyrosine residue within the HLA-A and -B molecules is not phosphorylated in response to Nef (Le Gall *et al.*, 1998). Furthermore, Nef-mediated MHC I down-regulation is not blocked by the tyrosine kinase-inhibiting drug herbimycin A (Mangasarian *et al.*, 1999), decreasing the likelihood of tyrosine kinase signaling being involved. A clear mechanism for MHC I down-modulation by Nef remains to be delineated.

B. Nef Control of Signal Transduction Events in HIV-1 Susceptible Cell Types

One of the most rigorously debated issues surrounding the role of Nef in the HIV-1 life cycle has been its effect on signal transduction events which occur in cells susceptible to HIV-1 infection. This most controversial argument was compounded by early reports which described Nef, produced in bacteria, to possess GTPase, autophosphorylation, and GTP-binding activities reported for the *ras* gene product (Guy *et al.*, 1987). Intense follow-up of this study showed that a bacterial contaminant was responsible for the GTP-ase activity associated with Nef. Despite this first report, considerable evidence now supports a major role for Nef in disease progression by controlling T-cell activation (Bandres and Ratner, 1994; Luria *et al.*, 1991; Niederman *et al.*, 1992, 1993a, 1989; Baur *et al.*, 1994; Collette *et al.*, 1997, 1996a,b; Greenway *et al.*, 1995, 1996, 1999; Iafrate *et al.*, 1997; Sawai *et al.*, 1994, 1995, 1996; Skowronski *et al.*, 1993).

As discussed above, down-modulation of CD4 and IL-2 R by Nef has the potential to affect T-cell activation. Indeed, down-regulation of CD4 by Nef results in the dissociation of CD4 from the signaling molecule Lck (Aiken *et al.*, 1994; Goldsmith *et al.*, 1995). Liberation of Lck from CD4 could impair the ability of this kinase to participate in the phosphorylation of antigen recognition activation motif (ARAM) sites in the ζ-chain of the CD3 complex. This event is thought to be required for the recruitment and binding of ZAP-70 to CD3 and the eventual signaling events that lead to activation and nuclear translocation of transcription factors and ultimately IL-2 production (Weiss and Littman, 1994). Indeed a body of literature exists reporting that Nef affects the T-cell signal transduction pathway originating from the T-cell receptor (TcR) (Bandres and Ratner, 1994; Baur *et al.*, 1994; Brady *et al.*, 1993; Iafrate *et al.*, 1997; Luria *et al.*, 1991; Niederman *et al.*, 1992, 1993a; Rhee and Marsh, 1994a; Skowronski *et al.*, 1993).

Normally, when the CD4+ T-cell line Jurkat is treated with PMA plus either phytohemagglutinin (PHA) or antibodies against the TcR/CD3 complex there is a prompt increase in interleukin 2 (IL-2) mRNA, intracellular calcium, and in the IL-2 receptor α-chain on the cell surface (Weiss and

Littman, 1994). In sharp contrast to this, stable transformants of the Jurkat T-cell line which express Nef fail to upregulate the expression of IL-2 mRNA in response to these stimuli. This suggests that Nef interferes with a signal emanating from receptors such as the TcR complex that induces IL-2 gene transcription (Luria *et al.*, 1991). These findings concur with further reports of Nef modulation of cytokines involved in the development of immune responsiveness. For example, Jurkat cells, which contain an inducible stably integrated *nef* gene, also failed to express significant levels of IL-2 and interferon-γ upon TcR triggering (Collette *et al.*, 1996b). The effect of Nef on IL-2 production appears to be specific to the TcR pathway as the cosignals provided by CD28 to up-regulate IL-2 induction were unaffected by Nef (Collette *et al.*, 1996a,b). These data suggest a selective immunosuppression induced by Nef in human T cells by altering TcR signaling without detectable impact on CD28 coreceptor function. Utilization of systems other than T-cell lines transiently or stably transfected for *nef* expression also report Nef-effect on T-cell signaling pathways. Nef-transgenic thymocytes show a decrease in total activation in response to cross-linking of CD3 (Brady *et al.*, 1993). Further, introduction, by square-wave electroporation, of highly purified recombinant Nef protein into primary peripheral blood mononuclear cells or Jurkat cells also substantiated the negative effect of Nef on signals generated from the TcR (Greenway *et al.*, 1995). The use of primary cells rather then transformed cell lines in determining the effect of Nef on signaling is of paramount importance if one is to relate these effects to signaling events and virus replication during HIV-1 infection *in vivo*.

Negative regulation of TcR signaling by Nef is congruent to reports describing Nef-mediated down-regulation of the transcriptional factors NFκB and AP-1 (Bandres and Ratner, 1994; Niederman *et al.*, 1992, 1993a). Stimulation of T cells by mitogens or antibodies to the TcR–CD3 complex resulted in the down-regulation of transcriptional factors NF-κB and AP-1 in cells expressing the *nef* gene compared with control cells. As Nef did not affect the surface expression of the TcR–CD3 complex in these studies it was concluded that the Nef protein down-regulates the transcriptional factors NF-κB and AP-1 through an effect on the TcR-dependent signal transduction pathway (Niederman *et al.*, 1992, 1993a; Bandres and Ratner, 1994). While Nef inhibits the induction of NF-κB DNA-binding activity by T-cell mitogens, it does not affect the DNA-binding activity of other transcription factors implicated in HIV-1 regulation, including SP-1, USF, URS, and NF-AT (Bandres and Ratner, 1994; Niederman *et al.*, 1992, 1993a). The defective recruitment of NF-κB may underlie the negative effect of Nef on transcriptional activation of the interleukin-2 promoters.

However, the reports documented above describing Nef inhibition of TcR-mediated signaling are in direct contrast with others which describe augmentation of T-cell signaling as a result of Nef expression (Baur *et al.*, 1994, 1997; Fackler *et al.*, 1999; Rhee and Marsh, 1994a; Schrager and

Marsh, 1999). For example, Nef expression in an Ag-specific murine T-cell hybridoma results in both the down-modulation of CD4, as seen in primary cells and human T-cell lines, and a positive enhancement of the TcR response to stimuli (Rhee and Marsh, 1994a). Investigation demonstrated that the positive enhancement of the TcR response was independent of CD4 expression or modulation (Rhee and Marsh, 1994a). These results constrast with those of Brady *et al.* (1993) and Skowronski *et al.* (1993). The complexity of Nef action on signal transduction is further illustrated by the finding that a CD8-Nef chimera can inhibit or potentiate activation signals, particularly relevant to the TcR, depending on its localization within the cell and depending upon the nature of the signal used to activate the T cell (Baur *et al.*, 1994; Schrager and Marsh, 1999).

Modulation of TcR-mediated signaling is seemingly also a property of Nef encoded by a particularly unusual strain of SIV. Its action, in this regard, appears to align with those HIV-1 Nef sequences which lead to cellular activation (Du *et al.*, 1995, 1996). This Nef is from a highly virulent strain of SIV, SIVpbj14, which induces an acute lethal disease in monkeys. This virus displays the unusual property of replicating well in resting peripheral blood mononuclear cell cultures, unlike all other naturally occurring SIV or HIV-1 isolates examined to date, causing extensive T-lymphocyte activation (Du *et al.*, 1995, 1996). Amino acid sequence comparison of SIV pbj14 Nef with other SIV Nef sequences highlights a unique YXXL motif, present within its N-terminus from residues 10–43 (SRPSGDLYERLLRARGETYGRLLGEVEDGYSQSP), which matches very well with consensus sequences for SH2 binding domains present in the cytoplasmic tail of T- and B-cell antigen receptors (Weiss and Littman, 1994). Creation of this YXXL-containing Nef sequence by changing amino acid residues 17 to 18 in Nef of SIVmac239 from RQ to YE recreated the phenotype of the aggressive SIVpbj14 isolate (Du *et al.*, 1995, 1996). The YE variant of SIVmac239, unlike SIVmac239 but like SIVpbj14, replicated well in resting peripheral blood mononuclear cell cultures without prior lymphocyte activation and without the addition of exogenous interleukin-2. It caused extensive lymphocyte activation in these cultures and produced an acute disease in rhesus and pigtailed monkeys characterized by severe diarrhea (Du *et al.*, 1995, 1996).

Further investigation of the YE-Nef variant shows that the YXXL motif within YE-Nef may perform the same function as those present in the T-cell and B-cell receptors and may even act in place of the T-cell receptor itself. This conclusion is based upon findings that YE-Nef alone increases the activity of the transcription factor NFAT, which is one of the downstream targets of T-cell activation (Du *et al.*, 1996). Second, the YXXL motif from YE-Nef can be phosphorylated on tyrosine residues by the *src* family kinase Lck and associates with ZAP-70, a T-cell-specific tyrosine kinase. The phosphorylation of both conserved tyrosine residues on the YXXL motif appears to be required for the recruitment of ZAP-70 and correlates with the unusual pathogenicity associated with infection with SIVpbj14. Conversely, the lymphocyte-activating properties of SIVpbj14 are lost by the single change of Y to R at position 17 of Nef (Du *et al.,* 1995). These results provide compelling evidence that SIV Nef may activate T-cell receptor signaling, providing a mechanism by which it enhances virus replication *in vivo.* By extrapolation HIV-1 *nef* may also act to positively regulate T-cell activation events. However, SIVpbj is a unique variant of SIV and is the only naturally occurring virus isolate (HIV-1 or SIV) which can replicate in quiescent T cells, suggesting that not all SIV Nef sequences function congruently. SIVmac239 does not possess the capability of replicating in quiescent T cells and any effect SIV Nef has on T-cell activation is reduced in comparison with SIVmac239YE Nef (Du *et al.,* 1995). Similarly, differences have been observed between HIV-1 and SIV Nef and indeed between different HIV-1 *nef* isolates with regard to function, indicating that some differences in Nef function may be attributed to the diversity of *nef* sequences. Further investigation of multiple Nef sequences from HIV-1 and SIV Nef in systems which utilize primary T cells are required for a truly accurate assessment of the effect of Nef on T-cell signaling and how it relates to the HIV-1 infection process.

Regulation of the TcR signaling pathway is not the only signaling cascade modulated by Nef. Introduction of highly purified Nef protein into peripheral blood mononuclear cells (PBMC) caused reduced proliferative responsiveness to IL-2 in this cell population. This may be a direct effect on signaling by the TcR but may be a consequence of the effect of Nef on the IL-2 R (Greenway *et al.,* 1994). Normally, stimulation of T cells by IL-2 or PMA provokes both augmentation of Lck activity and corresponding posttranslational modification of the kinase (Horak *et al.,* 1991). These changes were also inhibited by treatment of PBMC with Nef. Further evidence for Nef interfering with cell activation and proliferation is its negative effect on the production of the proto-oncogene *c-myb,* which is required for cell-cycle progression, in Nef-treated cells (Greenway *et al.,* 1995). Further, transduction of murine NIH-3T3 cells with a retroviral HIV-1 *nef* expression system resulted in significantly decreased proliferative responsiveness to bombesin and platelet-derived growth factor (PDGF), suggesting that multiple pathways may be affected at common intersection points by Nef.

Although studies are limited, signal transduction pathways in other cell types susceptible to HIV-1 infection are also modulated by Nef which may also result in regulation of virus replication (Ambrosini *et al.,* 1999; Biggs *et al.,* 1997, 1999; Murphy *et al.,* 1994; Greenway *et al.,* submitted; Romero *et al.,* 1998). Introduction of Nef by electroporation into monocyte-derived macrophages caused a dramatic increase in the expression of IL-6, TNF-α, and IL-1β (A. L. Greenway, unpublished). The upregulation of these cell factors, or monokines, may be as a result of Nef modulation of several *src* family tyrosine kinases, which is discussed at length below (Biggs *et al.*, 1997, 1999; Collette *et al.*, 1996a,b; Greenway *et al.*, 1995, 1996, 1999; Lee *et al.*, 1995; Moarefi *et al.*, 1997; Saksela *et al.*, 1995). Similarly, Nef has been shown to induce the phosphorylation of cellular proteins that may impact on their function or the function of their target molecules. Upregulation of monokine production by Nef may augment HIV-1 replication and contribute to production of disease by enhancing viral gene transcription. Activation pathways of cells of neuronal origin are also modulated by the expression of Nef. For example, intracellular signaling pathways associated with the growth factor activity of ET-1 are impaired in Nef-expressing and HIV-1-infected astrocytes (He *et al.*, 1997). Furthermore, Nef may act to promote the expression of c-kit and the eventual induction of apoptosis in this cell population, suggesting that infection of astrocytes and Nef in particular may play a significant role in the neuropathogenesis of HIV-1 encephalopathy (He *et al.*, 1997).

The conserved ability of Nef derived from HIV-1 and SIV to modulate cell signaling suggests that this function may play a prominent role in regulating virus production during infection. The research described in this section provides compelling evidence that Nef acts to modulate cell signaling. As mentioned above the differing effects of Nef on T-cell signaling may be related to different *nef* alleles or assay systems used in the studies. However, another likely possibility is that Nef may be able to modulate signaling in both a positive and a negative sense depending upon the context under which it is expressed and on the signals which the Nef-expressing cells receive. The initial stage of the infectious process in T cells may require Nef to promote or initiate signaling events from the TcR for the competent activation of the cell. As HIV-1 replication is dependent upon T-cell activation it is easy to imagine how this would be beneficial to the virus. However, as repeated stimulation of the T cell through the TcR can result in activation-induced apoptosis, it may be essential for the virus to dampen the signaling cascades through this receptor at later time points of the life cycle to maintain the viability of the infected cell so as to maximize virus production. Indeed, introduction of Nef into Jurkat cells specifically inhibited apoptosis of the cells in response to anti-CD3 stimulation (A. L. Greenway, unpublished). Thus, Nef may alter the balance of signals in T cells during the course of the HIV-1 life cycle. The up-regulation of *src* kinases expressed in monocyte-derived macrophages may also enhance the production of virus in this cell population by directly increasing the production of factors which stimulate HIV-1 production.

I. Nef-Associated Host-Cell Proteins Involved in Signal Transduction

The molecular mechanisms underlying the effect of Nef on cell-signaling pathways or other functions requires the identification of the cellular mediators of Nef. Extensive investigation of the cellular proteins which interact with Nef has been undertaken and many of the proteins identified are protein tyrosine or serine/threonine kinases which play key roles in the generation of signals in multiple signaling cascades (Dutartre *et al.*, 1998; Fackler *et al.*, 1999; Greenway *et al.*, 1999; Lock *et al.*, 1999; Manninen *et al.*, 1998; Wiskerchen and Cheng-Mayer, 1996; Xu *et al.*, 1997, 1999; Benichou *et al.*, 1994, 1997; Bodeus *et al.*, 1995; Collette *et al.*, 1996a,b; Greenway *et al.*, 1996; Harris and Neil, 1994; Lee *et al.*, 1995; Liu *et al.*, 1997; Smith *et al.*, 1996). Indeed, the pleiotropic functions of Nef are revealed by the identification of numerous Nef-interacting proteins. Nevertheless this complex network of interactions is likely to be tightly regulated at different levels. The effect of Nef on at least cellular activation and virus replication events may be explained in part by its interaction with specific cellular proteins involved in signal transduction.

Many reports now document the binding of HIV-1 Nef to multiple members of the *src* family kinases (Benichou *et al.*, 1994, 1997; Bodeus *et al.*, 1995; Collette *et al.*, 1996a; Dutartre *et al.*, 1998; Fackler *et al.*, 1999; Greenway *et al.*, 1995, 1996, 1999; Harris and Coates, 1993; Harris and Neil, 1994; Lee *et al.*, 1995; Liu *et al.*, 1997; Lock *et al.*, 1999; Manninen *et al.*, 1998; Smith *et al.*, 1996; Wiskerchen and Cheng-Mayer, 1996; Xu *et al.*, 1999). At least nine members of the *src* family are expressed in various hematopoietic cell types and each most likely play integral roles in signaling (Weiss, 1993). Several different systems have shown that Nef can interact with at least Lck, Fyn, Lyn, and Hck (Greenway *et al.*, 1995, 1996, 1999; Collette *et al.*, 1996, Dutartre *et al.*, 1998, Arold *et al.*, 1998. Karn, *et al.*, 1998; Briggs *et al.*, 1997; Biggs *et al.*, 1999; Saksela *et al.*, 1995; Moarefi *et al.*, 1997). Lck and Fyn, expressed in T-lymphocytes, are intricately involved in mediating signals derived from the coreceptors CD4 and CD8, IL-2R (Lck), and the TcR (Lck and Fyn) (Weiss, 1993; Weiss and Littman, 1994). Hck and Lyn, which are expressed in multiple cell types including monocyte-derived macrophages, may be involved in the expression of monokine factors expressed in this cell type (Beaty *et al.*, 1994; Bohuslav *et al.*, 1995; Schmid-Alliana *et al.*, 1998). Although it is apparent that Nef can bind to *src* kinases, the precise members which are targeted by Nef and the regions of each of the molecules involved in the interactions are still contentious issues.

The first observation that Nef could coprecipitate *src* family kinases was based on studies using a recombinant glutathione-*S*-transferase (GST)–Nef fusion protein to probe cytoplasmic extracts prepared from T-cell lines and activated PBMC (Greenway *et al.*, 1995). Specific immunoblotting showed the presence of Lck among the cellular proteins within the coprecipitate. Other researchers also pursued the hypothesis that Nef may contribute to virus replication by targeting the *src* family since Nef contains a highly conserved proline-repeat motif which bears strong resemblance to the binding motif for SH3 domains which are a feature of *src* kinases (Saksela *et al.*, 1995). GST-fusion proteins corresponding to the SH3 domain of Hck or Lyn in particular were used with Nef in direct protein-binding studies to show that Nef could interact efficiently with the SH3 domain of each kinase (Saksela *et al.*, 1995). In the case of Hck, Nef is described as a high-affinity ligand (Lee *et al.*, 1995). Similar binding and coprecipitation assays with short synthetic peptides corresponding to the proline-rich repeat sequence [$(Pxx)_4$] of Nef, and the SH2, SH3 or SH2, and SH3 domains of Lck revealed that the interaction between these two proteins can also be mediated by the proline-repeat sequence of Nef and the SH3 domain of Lck (Collette *et al.*, 1996a; Greenway *et al.*, 1996, 1999). Alteration of the proline residues within the proline-repeat motif of Nef has verified that an intact proline repeat motif is essential for direct interaction of Nef with each of these kinases (Dutartre *et al.*, 1998; Greenway *et al.*, 1996, 1999; Saksela *et al.*, 1995). An additional mechanism of Nef interaction with Lck has also been proposed whereby the first N-terminal 22 amino acid residues of Nef can support indirect binding of Lck (Baur *et al.*, 1997). A third intermediate partner, yet to be identified, is thought to facilitate this interaction. Further work is necessary to fully elucidate the relationship between each binding domain to the Nef–Lck interaction event. Whether the two domains of Nef cooperate to bind to Lck or whether they operate independently under different circumstances is being investigated. The N-terminus of Nef may allow a stable interaction between Nef and Lck since a GST–Nef fusion protein corresponding to amino acid residues 20 to 206 of Nef binds poorly to Lck compared to its full-length counterpart, despite containing the proline-repeat motif (Greenway *et al.*, 1995). Hence, while the proline-repeat motif may afford Nef direct binding to Lck, the N-terminal region may stabilize this interaction or offer a further site for indirect interaction. Further analysis using Fyn suggests it, too, can bind to Nef via its SH3 domain and the proline motif within Nef. In fact both Fyn and Hck have been useful in the derivation of the crystal structure of a truncated form of Nef (Arold *et al.*, 1997; Grzesiek *et al.*, 1996b). All of these interactions have now been verified during HIV-1 infection of CD4+ T cells or monocyte-derived macrophages *in vitro*, suggesting that the interactions may be relevant to the infection process (Baur *et al.*, 1997) (A. L., Greenway, unpublished).

Nef association with each of the *src* family kinases mentioned above results in the modulation of their catalytic activities. Several reports describe the dramatic up-regulation of Hck catalytic activity when its SH3 domain is bound by Nef (Briggs *et al.*, 1997; Greenway *et al.*, 1999; Moarefi *et al.*, 1997). Normally, *src* kinases are inhibited by tyrosine-phosphorylation at a carboxy-terminal site (Hunter, 1987). The SH2 domains of these enzymes play an essential role in this regulation by binding to the tyrosine-phosphorylated tail. The crystal structure of the down-regulated form of

Hck has been determined and reveals that the SH2 domain regulates enzymatic activity indirectly; intramolecular interactions between the SH3 and catalytic domains appear to stabilize an inactive form of the kinase (Moarefi *et al.*, 1997; Sicheri *et al.*, 1997). The addition of the HIV-1 Nef protein to either the down-regulated or activated form of Hck causes a large increase in Hck catalytic activity. The intact proline-rich motif in Nef is crucial for Hck activation. This effect is considered a consequence of Nef displacement of the SH3 domain of the kinase from a polyproline type II helix chain linking the SH2 and the catalytic domains in an inactive form, causing a conformational change in the amino terminal lobe of the catalytic domain which enhances phosphotransfer (Moarefi *et al.*, 1997). Such displacement is proposed as a mechanism by which the catalytic activity of all *src* family kinase members may be regulated. Lyn kinase activity has also been observed to be up-regulated by Nef–SH3 binding, presumably in a similar manner (A. L. Greenway, unpublished).

In contrast to these findings, however, direct binding of Nef to either Lck or Fyn has been shown to result in the inhibition of Lck and Fyn catalytic activities (Collette *et al.*, 1996a; Greenway *et al.*, 1996, 1999) (A. L. Greenway, unpublished). Addition of purified recombinant Nef protein to purified Lck or Fyn kinases significantly impaired their abilities to phosphorylate target peptide substrates. The inhibition of Lck and Fyn kinase activities is dependent upon the proline repeat motif within Nef and correlates with binding directly to each of the kinases. The autophosphorylation of each of these kinases was also inhibited by Nef (Greenway *et al.*, 1996) (A. L. Greenway, unpublished).

The direct interaction *in vitro* of *src* kinases with Nef and the differential regulation of their activities highlights the proposed role for these kinases in Nef function in a variety of cell types. Members of the *src* kinase family are cell type specific (Bolen *et al.*, 1992; Frank *et al.*, 1990; Horak *et al.*, 1991; Minami *et al.*, 1993; Rudd *et al.*, 1988; Veillette *et al.*, 1988; Weiss and Littman, 1994). The differential regulation of *src* kinase activity by Nef suggests it has adopted different strategies in the CD4+ T cells and monocyte-derived macrophages to augment virus replication. Both Lck and Fyn are expressed in T lymphocytes and are intimately involved in mediating signals derived from the coreceptors CD4 and CD8 (Lck), IL-2R (Lck), and the TcR (Lck and Fyn) (Bolen *et al.*, 1992; Frank *et al.*, 1990; Horak *et al.*, 1991; Minami *et al.*, 1993; Rudd *et al.*, 1988; Veillette *et al.*, 1988).

The modulation of Lck and Fyn activities by Nef could explain the inhibition of CD3-mediated signaling. The inhibition of Lck and Fyn kinases by Nef may also result in the observed inhibition of IL-2 mRNA and protein production in Nef-expressing/-containing cells (Collette *et al.*, 1996a; Greenway *et al.*, 1995; Luria *et al.*, 1991). As Lck signaling plays an essential role in the induction of apoptosis in Jurkat cells following TcR stimulation (Oyaizu *et al.*, 1990), it is also likely that the inhibition by Nef of TcR/anti-

CD3-induced apoptosis described above is most likely due to the ability of Nef to inhibit Lck and Fyn activities (A. L. Greenway, unpublished). This hypothesis is supported by the findings that Nef protein-containing mutations within its proline-repeat motif to abrogate binding to Lck and Fyn did not protect Jurkat cells from activation-induced death when stimulated through the TcR (A. L. Greenway, unpublished). Interception of the TcR pathway at an early point by Nef may prolong the life span of an HIV-1-infected cell, allowing increased virus production. Prevention of premature killing of HIV-1-infected cells by Nef may represent one mechanism, in addition to others proposed, by which Nef augments virus production (Goldsmith *et al.,* 1995; Miller *et al.,* 1994; Spina *et al.,* 1994). It is interesting to note that the cell populations infected with HIV-$I_{\text{NL43-nef-stop}}$ or HIV-1 containing the gene for green fluorescent protein in place of Nef undergo high levels of apoptosis at or before peak virus replication, suggesting the expression of Nef during HIV-1 infection may be protective for the cell (Gandhi *et al.,* 1998; Herbein *et al.,* 1998a,b).

Of course there have been reports indicating that the expression of Nef may actually facilitate apoptosis of the HIV-1-infected T cell and neighboring cells (Baur *et al.,* 1994; Xu *et al.,* 1997; 1999). Recently, cells selected for stable expression of Nef were shown to up-regulate Fas-ligand expression, presumably in an Lck-dependent manner, while cells expressing a CD8–Nef chimeric protein underwent apoptosis when the protein was localized at the plasma membrane (Xu *et al.,* 1999). Obviously, further investigation is necessary to clarify this effect of Nef. Again, it may be possible that Nef can inhibit or promote apoptosis and T-cell activation depending on the context under which it is expressed. It should be pointed out, however, that the stable expression of Nef may result in the deletion or alteration of *nef* sequence during the cell-selection procedure such that its function is altered. Further, the use of a CD8–Nef chimeric molecule should be viewed with caution until the same results are obtained with wild-type Nef by multiple research groups. Further investigation will clarify these issues.

The role of the *src* kinase family in SIV Nef function is less clear. Although HIV-1 and SIV Nef share many conserved functional and structural features some subtle structural differences do exist. These differences may translate into different modes of action by the proteins. The proline-repeat motif within HIV-1 Nef, as described above, plays an integral role in interaction with members of the *src* kinase family. The proline-repeat motif within HIV-1 Nef is characterized by the sequence $(Pxx)_4$. Although the proline motif is also present in SIV Nef it is repeated only twice $(Pxx)_2$, at least with SIV mac239 (Shugars *et al.,* 1993). While SIV Nef does interact directly with *src* kinase family members, including Lck, the interaction is not mediated by the proline-repeat motif, nor via the SH3 domain of the kinase (Greenway *et al.,* 1999). Instead the N-terminal region of SIV Nef appears to be a principal determinant of binding (Greenway *et al.,* 1999).

The different regions of Nef involved in *src* kinase interaction also results in the differential regulation of catalytic activities. Unlike HIV-1 Nef, interaction of SIV Nef with Lck results in the up-regulation of its catalytic activity (Collette *et al.*, 1996a; Greenway *et al.*, 1995, 1996, 1999). Thus, SIVmac239 Nef certainly appears to have similar if not diluted characteristics of the Nef protein from the highly virulent strain of SIV, SIVpbj14 (Du *et al.*, 1995, 1997). As mentioned above infection of T cells with this strain results in the activation of the T cell. This activating phenotype is dependent upon the ARAM motif present in the N-terminus of Nef, which can be phosphorylated by Lck and which has been shown to recruit and associate with the ζ-chain of the T-cell receptor (Luo and Peterlin, 1997). Normally, Lck is involved in the phosphorylation of ARAM residues present in the ζ-chain of the CD3 complex, allowing the recruitment of ZAP-70 to CD3 and eventual activation of the T cell. Nef from SIVpbj14 may act in place of the TcR and promote activation in the absence of normal stimulation. While SIV Nef does not possess the same number of YxxL motifs present in SIVpbj14 Nef it can be phosphorylated by Lck and does activate this kinase, suggesting that it, too, may lead to the recruitment of ZAP 70 and T-cell activation. The differential effect of HIV-1 and SIV Nef on T-cell activation occurring particularly through the TcR pathway may be indicative of a different mode of action to augment virus replication during virus infection.

Interaction of HIV-1 Nef with Hck, which is expressed predominantly in B-cells and cells of the macrophage/monocyte lineage, pinpoints a principal role for Nef in perturbing cell activation pathways in a number of cell types susceptible to HIV infection. Nef causes a dramatic increase in the kinase activities of Hck and Lyn and induces high-level constitutive expression of IL-6, TNF-α, and IL-1β in monocyte-derived macrophages. As Hck and Lyn are involved in the transduction of signals leading to the production of monokines such IL-6 and TNF-α it appears likely that Nef interaction with the *src* kinases up-regulates the expression of these factors (Beaty *et al.*, 1994; Ernst *et al.*, 1994; Gupta *et al.*, 1995). TNF-α, IL-1β, and IL-6 have been shown to be potent activators of HIV replication and the production of each of these monokines is increased in patients with HIV-1 infection (Weissman *et al.*, 1996a, b). Up-regulation of monokine production by Nef may augment HIV-1 replication and contribute to production of disease by enhancing viral gene transcription or virus replication or by contributing to HIV-induced macrophage dysfunction (Biggs *et al.*, 1999).

Numerous other cellular kinases or proteins involved in signal transduction have also been shown to bind to HIV-1 Nef. These include a 21-kDa PAK familylike kinase termed NAK (Nef-associated kinase), Raf-1, MAPK (Erk-1), CD4, the tumor suppressor protein p53, and protein kinase C theta (Greenway *et al.*, 1995, 1996, 1999; Hodge *et al.*, 1998; Lu *et al.*, 1996; Manninen *et al.*, 1998; Nunn and Marsh, 1996; Sawai *et al.*, 1994, 1995,

1996; Smith *et al.*, 1996). Some of these interactions have been characterized further in relation to the domains of each protein involved and the possible consequence of interaction. The interaction of Nef with p53 has been localized to the N-terminal region of Nef (A. L. Greenway, unpublished). As p53 is located in both the nucleus and the cytoplasm of the cell, it is possible that the localization of Nef in the nucleus, which was described earlier, is relevant to this interaction. Preliminary studies suggest that this interaction may relate to the antiapoptotic effect of Nef described above (A. L. Greenway, unpublished). Interestingly, Nef shares striking homology with an acidic sequence at the c-Raf 1-binding site within the Ras effector region. Deletion and site-specific mutagenesis of Nef proteins mapped the specific interaction between the HIV-1LAI Nef and c-Raf1 to a conserved acidic sequence motif containing the core sequence Asp-Asp-X-X-X-Glu (positions 174–179) (Hodge *et al.*, 1998). Additionally, lysates from a permanent CEM T-cell line constitutively expressing the native HIV-1 Nef protein were used to coimmunoprecipitate a stable Nef-c-Raf1 complex, suggesting that molecular interactions between Nef and c-Raf1, an important downstream transducer of cell signaling through the c-Raf1-MAP kinase pathway, occurs *in vivo*.

The proline repeat motif of Nef has also been implicated in further interactions with signaling molecules other than *src* kinase members, including MAPK and NAK (Greenway *et al.*, 1996; Manninen *et al.*, 1998). In the case of MAPK the interaction with Nef is direct, suggesting that it may compete with *src* kinases for this binding site or that the interaction occurs in a cellular compartment where *src* kinases are not located. Nef protein significantly decreased the *in vitro* kinase activity MAPK. This interaction has been shown to occur during HIV-1 infection *in vitro*, highlighting the potential relevance of the association (A. L. Greenway, unpublished). Two additional serine/threonine kinases, NAK and PKC-θ associate with Nef (Hodge *et al.*, 1998; Lu *et al.*, 1996; Manninen *et al.*, 1998; Nunn and Marsh, 1996; Sawai *et al.*, 1995, 1996; Smith *et al.*, 1996). The interaction of NAK with Nef occurs not only with HIV-1 Nef but also with Nef derived from SIVmac239. Two principal regions of Nef appear to be involved in binding to NAK including the proline-repeat motif and also a diarginine motif located at amino acid residues 109/110 of HIV-1 Nef (Sawai *et al.*, 1995). This association may be mediated by another protein(s). Using two approaches for detecting interactions between Nef and PKC isozymes in Jurkat cells shows that Nef interacts preferentially with PKC-θ (Smith *et al.*, 1996). The interaction of Nef with PKC-θ MAPK, and/or NAK could contribute to the various impairments of T-cell function associated with HIV infection and Nef expression. Immunoprecipitation of Nef with multiple cellular serine/threonine kinases in addition to tyrosine kinases highlights the involvement of Nef in manipulating signal transduction events through a multiplicity of pathways.

Nef binding cellular proteins have also been implicated in the down-modulation of CD4. As described above, Nef expression results in the down-modulation of the cell-surface marker CD4 (Aiken *et al.*, 1994; Garcia *et al.*, 1993; Greenberg *et al.*, 1998; Liu *et al.*, 1997; Mangasarian *et al.*, 1997; Mariani *et al.*, 1996; Rhee and Marsh, 1994a,b). Down-modulation requires amino acid sequences within the cytoplasmic domain of CD4 (Aiken *et al.*, 1994; Anderson *et al.*, 1994) but occurs by a mechanism distinct from the normal serine phosphorylation-dependent pathway (Garcia and Miller, 1991). As CD4 is a transmembrane glycoprotein and Nef is a myristylated protein targeted to the cytoplasmic face of the plasma membrane, it has been considered that a direct interaction between Nef and CD4 might play a role in down-modulation. Several groups have investigated this hypothesis and have shown that Nef associates with CD4 (Greenway *et al.*, 1995; Grzesiek *et al.*, 1996a; Harris and Neil, 1994). This interaction appears to be dependent on Nef myristylation (Harris and Neil, 1994). The site of Nef interaction maps to the cytoplasmic domain of CD4, as a deletion mutant lacking this domain fails to interact with Nef. These observations shed new light on the biochemical function of Nef and offer a further mechanism, additional to the binding of Nef to $\mu 2$ and thioesterase, in Nef-down-regulation of CD4 expression (Aiken *et al.*, 1994; Le Gall *et al.*, 1998; Liu *et al.*, 1997; Watanabe *et al.*, 1997).

C. Nef and Virus Replication

Studies in SIV-infected rhesus macaques have provided evidence for the requirement of Nef in viral infection. This was elegantly demonstrated when macaques infected with SIV deleted within *nef* exhibited low viral loads and failed to progress to disease (Kestler *et al.*, 1991). Further demonstrating the high requirement for an intact open *nef* reading frame within the virus was the fact that within 2 weeks of inoculation only wild-type virus could be isolated from macaques infected with SIV containing a point mutation within *nef*. These animals progressed to disease and showed high virus load. These studies were the first to demonstrate the critical importance of *nef* during infection and revolutionized our thinking regarding the function of Nef. The importance of Nef in HIV-1 infection and disease has also been strengthened by the finding that humans infected with viral strains containing deletions within the *nef* gene also have low viral loads and have not, after 18 years of infection, progressed to disease (Deacon *et al.*, 1995; Learmont *et al.*, 1999).

Despite the obvious relationship between the expression of Nef and disease progression during HIV-1 and SIV infection *in vivo* the reproduction of this phenotype during *in vitro* infection has been somewhat confusing. Initially, Nef, which is an acronym for negative effect factor, was reported to have a negative effect on viral replication in T-cell lines (Ahmad and

Venkatesan, 1988; Cheng-Mayer *et al.*, 1989; Maitra *et al.*, 1991; Niederman *et al.*, 1989; Terwilliger *et al.*, 1991; Tsunetsugu-Yokota *et al.*, 1992). Proviruses with mutations in the *nef* gene were shown to replicate to higher levels than their parental counterparts during transient expression. The mutant virus maintained its enhanced replication even after serial passages in T lymphocytes. The negative effect of Nef on virus replication was demonstrated in a number of T-cell and non-T-cell lines (Ahmad and Venkatesan, 1988; Maitra *et al.*, 1991; Niederman *et al.*, 1989) and was associated with inhibition of LTR-mediated gene transcription (Ahmad and Venkatesan, 1988; Maitra *et al.*, 1991; Niederman *et al.*, 1989). Nef trans-suppressed, in a dose-dependent manner, the expression of reporter genes linked to the HIV-1 long terminal repeat (LTR). Comparison of basal transcription from the HIV-1 LTR in the presence of Nef or a mutated counterpart showed a greater than 10-fold repression of LTR transcription (Ahmad and Venkatesan, 1988; Maitra *et al.*, 1991). It was proposed that the repression of transcription from the HIV-1 LTR induced by Nef was linked to the upstream *cis* element within the LTR, which is recognized to be a negative regulatory element) between 340 and 156 nucleotides upstream of the RNA initiation site (Ahmad and Venkatesan, 1988). These findings were extended to analysis of *nef*-defective virus replication in monocyte cell lines (Tsunetsugu-Yokota *et al.*, 1992). Viral replication in Nef-expressing monocytic cells was inhibited; however, variation in the extent of the inhibition dependent upon the virus strain and the cell was observed.

The inhibitory effect of *nef* on virus replication was also described as a conserved feature of the SIV *nef* gene (Binninger *et al.*, 1991; Niederman and Ratner, 1992). Construction of a series of simian immunodeficiency virus SIV mac *nef* mutants by partial deletion and insertions in the *nef* gene established that *nef* insertion mutants replicated faster than wild-type SIV mac, suggesting that the *nef* gene product acts as a negative factor for replication. Surface phenotyping revealed that cultures permanently infected with *nef* mutants exhibited an enhanced expression of viral proteins on the outer cell surface (Binninger *et al.*, 1991).

However, although many laboratories reproduced the findings that *nef* inhibited virus replication, other investigations demonstrated that *nef* had no effect on virus replication (Bachelerie *et al.*, 1990; Hammes *et al.*, 1989; Kim *et al.*, 1989). These results are obviously disparate with those mentioned above and do not fit well with the SIV/rhesus macaque and HIV-1/human *in vivo* studies which clearly show that Nef plays an important role in facilitating virus replication (Deacon *et al.*, 1995; Kestler *et al.*, 1991; Kirchhoff *et al.*, 1995). The differences are most likely attributable to the use of immortalized cell lines for analysis of virus replication and allelic variation among *nef* sequences. Indeed, very elegant studies by Spina *et al.* (1994) and Miller *et al.* (1994) which defined a tissue culture system that would approximate the *in vivo* setting for virus infection have demonstrated that *nef* confers a positive growth advantage to HIV-1. This effect by *Nef* becomes readily discernible only when the target populations are primary lymphocytes that are quiescent at the time of initial infection but subsequently activated at a specific time after infection. For example, infection of mitogen-activated peripheral blood mononuclear cells (PBMC) with Nef+ HIV resulted in enhanced replication as evidenced by earlier gag p24 expression when compared with infections performed with *nef* mutant viruses. Furthermore, the positive effect of *nef* on virus replication was exaggerated when unstimulated freshly isolated PBMC were infected with Nef+ and Nef- viruses and then subsequently activated with mitogen. The Nef- viruses required a significantly greater time in culture to show appreciable growth. These results show the critical importance of the timing in T-cell activation in the effect of *nef* on virus replication. Indeed, *nef* shows a reduced ability to augment virus replication when the T cell is already activated at the time of infection. The multiplicity of infection used was also an important determinant of *nef* effect on virus replication. A positive effect of Nef on viral replication was also observed in primary macrophages infected with the macrophage-tropic molecular clone HIV-1 YU-2, showing that the effect of *nef* on virus replication is not restricted to certain cell types (Miller *et al.*, 1994). These results confirm previous reports describing the effect of Nef on virus replication when CD4+ T-cell lines were used for assessment of virus replication. As these cell lines are immortal and in a constant state of activation the positive effect of Nef may not have been so readily observed. Furthermore aberrant signaling pathways in T-cell lines may promote functions of Nef which are not strictly relevant to primary cells.

The positive effect of Nef on virus replication is now supported by a number of studies (Aiken and Trono, 1995; Blumberg *et al.*, 1992; Chowers *et al.*, 1995; Chowers *et al.*, 1994; de Ronde *et al.*, 1992; Goldsmith *et al.*, 1995; Jamieson *et al.*, 1994; Schwartz *et al.*, 1995). However, as observed with virus replication in CD4+ T-cell lines different *nef* alleles may influence virus replication to differing extents. Luo and Garcia (1996), noted that deletion of *nef* from HIV-1 SF2 impaired virus replication to a greater extent than deletion of *nef* from HIV-1 NL4-3.

Several mechanisms have been proposed to account for the positive effect Nef has on virus replication in primary CD4+ T cells and monocyte-derived macrophages. These include an ability of Nef to increase the efficiency of reverse transcription in virus-infected cells (Schwartz *et al.*, 1995), to down-regulate the expression of cell-surface CD4 to avoid superinfection (Benson *et al.*, 1993) or increase virus release from cells (Schwartz *et al.*, 1993, 1995), to enhance virion infectivity (Chowers *et al.*, 1994; Miller *et al.*, 1994; Pandori *et al.*, 1996), and to regulate apoptosis of the HIV-1-infected cell for maximum virus production (A. L. Greenway, unpublished). Some of these issues have already been discussed. All of these mechanisms may be relevant to the positive effect of Nef on virus replication depending upon the HIV-1 isolate.

Down-modulation of CD4 by Nef also correlates with its ability to augment virus replication and may do so in a number of ways. First, the down-modulation of cell-surface CD4 results in the increased release of viral progeny from cells, suggesting that the ability of Nef to promote down-regulation of CD4 acts to facilitate the efficient release of infectious progeny virions and, hence, viral replication (Ross *et al.*, 1999). Moreover, the expression of Nef is reported to decrease the cell-surface levels of gp120 in a CD4-dependent manner, as surface levels of mutant envelope glycoproteins unable to bind CD4 were not altered in Nef-expressing cells. The reduction in surface levels of gp120 correlated with a dramatic reduction of fusion-mediated cell death. It was postulated that the intracellular accumulation of fully processed envelope glycoproteins could significantly delay the cytopathic effect associated with envelope surface expression in HIV-infected cells, suggesting that Nef may provide a selective advantage during the infectious process, thereby promoting virus replication by inhibiting cell death (Schwartz *et al.*, 1993). The down-regulation of cell-surface CD4 receptor for HIV-1 by Nef also correlates with the acquisition of resistance to superinfection by HIV-1, which may result in the maintenance of cell viability during HIV-1 infection.

Analysis of reverse transcription within cells infected with HIV-1 or HIV-1 deleted within *nef* showed that the expression of Nef in virus-producing cells resulted in larger amounts of viral DNA accumulating in target cells (Schwartz *et al.*, 1995). The overall amounts of viral DNA synthesized during reverse transcription were 5- to 10-fold higher for HIV than for its *nef* deleted counterpart. The reduced amounts of viral DNA during infection of cell with HIV-1 deleted within *nef* were not a result of inhibited reverse transcriptase activity per se, as virion-associated reverse transcriptase derived from HIV-1 or HIV-1 deleted within *nef* virions demonstrated similar activity. These results indicate HIV-1 deleted within *nef* performed reverse transcription in a suboptimal cellular environment and that the expression of Nef in virus-producing cells is required for efficient processing of the early stages of virus replication in target cells (Schwartz *et al.*, 1995).

Other studies have shown that the positive influence of *nef* on viral growth rate is due, at least in part, to an infectivity advantage of virus produced with an intact *nef* gene. Comparison of wild-type and *nef*-deleted virus production during single-cycle replication, initiated by infection with high-titer virus stocks or by transfection with viral DNA, showed that wild-type virus yielded a 5-fold increase in p24 production relative to its *nef*-deleted counterpart (Miller *et al.*, 1994). In contrast single-cycle transfection yielded equal amounts of p24 production. These results imply that Nef does not affect replication after the provirus is established, supporting the findings of Schwartz *et al.* (1995), but does suggest prior events are *nef*-sensitive. End-point titrations of isogenic wild-type and *nef*-deleted viruses have determined that virus containing an intact *nef* gene have a greater infectivity per picogram of HIV p24 antigen than *nef*-deleted virus. HIV-1 encoding mutated nef reading frames are 10- to 30-fold less infectious than are isogenic viruses in which the *nef* gene is intact (Pandori *et al.,* 1996).

The infectivity of HIV-1 defective or deleted within *nef* can be restored to near wild-type levels by coexpression of Nef in trans in the cell line producing the virus. This observation implies that the HIV-1 virions produced in the presence or absence of Nef are intrinsically different. Studies utilizing a CEM derivative cell line (designated CLN) that expresses Nef under the control of the viral long terminal repeat for the growth of *nef*-deleted virus show that its growth is restored to near wild-type levels. However, this effect was not immediate, as the output of *nef*-deleted virus during the first 72 h after infection was not restored, suggesting that the expression of Nef in *trans* does not enhance the productivity of *nef*-deleted virus. In contrast the *nef*-deleted virus derived from the Nef-expressing cells restored the infectivity of the mutant virus. The restoration of infectivity was dependent upon myristylation and subsequently the plasma membrane localization of Nef (Pandori *et al.,* 1996). Hence, Nef results in viral particles that are more infectious, and this increased infectivity is manifested at a stage after viral entry but prior to or coincident with HIV-1 gene expression.

Biochemical analysis of HIV-1 virus particles produced in the presence or absence of *nef* shows that the expression of Nef does not lead to any obvious alteration of virion composition (Chowers *et al.,* 1994; Miller *et al.,* 1995). This finding led to the obvious question: *Is Nef incorporated into the virion*? Numerous researchers have now shown that Nef is present within the virus particle (Pandori *et al.,* 1996, 1998; Welker *et al.,* 1996, 1998). Initial quantitative analysis revealed Nef to be incorporated in the order of 10% of reverse transcriptase incorporation, which corresponds to 5 to 10 molecules of Nef per virion (Welker *et al.,* 1996). While it should still be considered a possibility that the detection of Nef in the virion is as a consequence of membrane-vesicle contamination of purified virion preparation, it is thought that this figure is most likely an underestimate as further quantitation suggests that a 10-fold higher level may be present (Welker *et al.,* 1996).

Interestingly, while Nef is detected within HIV-1-infected cells as a full-length 27-kDa protein, approximately 50% of particle-associated Nef corresponds to an 18- to 20-kDa species (Welker *et al.,* 1996). This species comigrates with purified Nef protein which has been cleaved between amino acids 57 and 58 by purified HIV-1 protease, suggesting that this enzyme is responsible for the cleavage of Nef (Freund *et al.,* 1994; Schorr *et al.,* 1996; Welker *et al.,* 1996). Indeed, Nef cleavage in particle preparations can be completely abolished by a specific inhibitor of HIV-1 protease (Bukovsky *et al.*, 1997). N-terminal truncations of Nef abolish its incorporation into HIV particles. Mutational analysis revealed that both myristylation and an N-terminal cluster of basic amino acids are required for virion incorporation and for plasma membrane targeting of Nef (Bukovsky *et al.*, 1997).

The functional significance of the virion incorporation and subsequent cleavage of Nef is unclear (Chen *et al.*, 1998; Miller *et al.*, 1997). Whether the inclusion of Nef into the virion occurs only as a bystander event—a consequence of its association with the plasma membrane at the time of virus budding—or whether this event is related to increased virus production is a hotly debated issue. Indeed, the incorporation of Nef into the virus particle occurs independent of other HIV-1-specific proteins and can be incorporated into other budding viruses (Bukovsky *et al.*, 1997). However, several hypotheses have been put forward to explain its role in virus replication, including an involvement in enhancement of virus replication. However, to date, this hypothesis has not been substantiated. Substitution of alanine for tryptophan 57 and leucine 58 within Nef almost completely abrogates Nef proteolytic cleavage. Comparison of virus containing this mutation within *nef* with wild-type virus in viral infectivity assays showed no correlation between the levels of cleavage and the ability to stimulate virion infectivity (Chen *et al.*, 1998; Miller *et al.*, 1997). Furthermore, SIV Nef, which lacks the sequence recognized by the protease and as a consequence is not cleaved despite its incorporation into virions, could stimulate the infectivity of a *nef*-defective HIV-1 variant as efficiently as HIV-1 Nef (Flaherty *et al.*, 1998). Similarly, while virion incorporation of Nef correlated with enhanced infectivity of the respective viruses in a single-round replication assay the phenotypes of HIV mutants with reduced Nef incorporation only partly correlated with their ability to replicate in primary lymphocytes (Welker *et al.*, 1998). This indicates that additional or different mechanisms may be involved in this system (Welker *et al.*, 1998). On the basis of this data it was concluded that the proteolytic processing of Nef is not required absolutely for the ability of this protein to enhance virion infectivity.

However, the effect of Nef incorporation and cleavage on virion infectivity may be dependent upon the cell type which is used to generate the virus. For example, when viruses were produced using T-cell lines or primary lymphoblasts, amino acid residues 57 and 58 were shown to be essential for optimal viral infectivity. In contrast, virus containing a mutation within amino acid residues 57 and 58 is only minimally impaired when produced from 293 or HeLa cells (Pandori *et al.*, 1998). This mutant is resistant to protease cleavage, indicating that proteolytic processing of Nef is dispensable for infectivity enhancement when virions are assembled in certain non-T-cells.

In addition to the virion incorporation of Nef, kinases such as the cellular serine/threonine kinase designated NAK and the serine/threonine MAPK, which are postulated to associate with Nef, may also be present in the virion (Flaherty *et al.*, 1998; Jacque *et al.*, 1998). This has led to the hypothesis that Nef may assist in the virion incorporation of these kinases and that the kinases may be involved in modification of virion-associated proteins leading to increased virion infectivity. Investigations regarding these issues are preliminary; however, there are reports of virion inclusion of a Nef-associated kinase (Flaherty *et al.*, 1998; Greenway *et al.*, 1996; Jacque *et al.*, 1998; Sawai *et al.*, 1995). Substantiating the hypothesis, a Nef-associated kinase has also been shown to increase the phosphorylation of matrix (Swingler *et al.*, 1997). Nevertheless, these findings must be tempered by those reports that suggest that binding of certain cellular kinases including the one identified in the virion by Nef is isolate dependent, as is the effect of Nef on virion infectivity (Luo *et al.*, 1997). These findings warrant further careful investigation into the role of Nef incorporation and cleavage in virion infectivity. A possible role for Nef-associated kinases in the modification of virion associated proteins as a mechanism for the augmentation by Nef of virion infectivity complements the additional, proposed regulation by Nef of kinases for the promotion of virus replication.

V. Conclusions and Discussion

Viral pathogenesis can be defined in terms of a series of successive interactions between the virus and its target host (Tyler and McPhee, 1987). For HIV, infection is initially established in the hemopoietic system and replication occurs to a high level within cells expressing the CD4 and either an α- or β-chemokine receptor (Pantaleo *et al.*, 1998). Continued virus replication in these cells and virion production beyond the acute phase of infection, termed the viral set point, then determines the time to development of clinical disease (Ho, 1996; Mellors *et al.*, 1996). One of the important parameters determining this viral load is the *nef* gene product. The acquisition or development of the *nef* gene, which extensively overlaps the LTR, is intimately linked to the complexity of this retrovirus. From experimental infection of macaques with SIV and from individuals naturally infected with a *nef* deleted virus showing marked attenuation of virus replication and disease progression it is obvious this gene product plays a key role in the time to development of AIDS. The function of the protein is complex in that there are many protein interactions and effects within the virus-infected cell, all apparently augmenting virus infectivity.

Thus HIV-1 Nef protein manipulates the infected host cell to advantage the virus through down-modulation of the cell surface receptors CD4, IL2R, and MHC I, controls the signaling of events through regulation of numerous cellular kinases, and enhances virion infectivity both directly and via the former mechanisms. It appears Nef has entrusted its activities to a number of key cellular kinases and adaptor molecules to regulate cell activation events and virion and host-cell modification for the purpose of augmenting virus production. The possible manipulation by Nef of cellular kinases for more than one purpose as well as adaptor molecules such as the AP complexes, which prevent the association of some of these kinases with CD4, represents a finely tuned and economically shrewd way for the virus to achieve its ultimate goal—survival.

References

Acres, R. B., Conlon, P. J., Mochizuki, D. Y., and Gallis, B. (1986). Rapid phosphorylation and modulation of the T4 antigen on cloned helper T cells induced by phorbol myristate acetate or antigen. *J. Biol. Chem.* **261,** 16210–16214.

Ahmad, N., and Venkatesan, S. (1988). Nef protein of HIV-1 is a transcriptional repressor of HIV-1 LTR. *Science* **241,** 1481–1485.

Aiken, C., and Trono, D. (1995). Nef stimulates human immunodeficiency virus type 1 proviral DNA synthesis. *J. Virol.* **69,** 5048–5056.

Aiken, C., Konner, J., Landau, N. R., Lenburg, M. E., and Trono, D. (1994). Nef induces CD4 endocytosis: Requirement for a critical dileucine motif in the membrane-proximal CD4 cytoplasmic domain. *Cell* **76,** 853–864.

Aldrovandi, G. M., and Zack, J. A. (1996). Replication and pathogenicity of human immunodeficiency virus type 1 accessory gene mutants in SCID-hu mice. *J. Virol.* **70,** 1505–1511.

Aldrovandi, G. M., Feuer, G., Gao, L., Jamieson, B., Kristeva, M., Chen, I. S., and Zack, J. A. (1993). The SCID-hu mouse as a model for HIV-1 infection. *Nature* **363,** 732–736.

Aldrovandi, G. M., Gao, L., Bristol, G., and Zack, J.A. (1998). Regions of human immunodeficiency virus type 1 nef required for function *in vivo*. *J. Virol.* **72,** 7032–7039.

Ambrosini, E., Slepko, N., Kohleisen, B., Shumay, E., Erfle, V., Aloisi, F., and Levi, G. (1999). HIV-1 Nef alters the expression of betaII and epsilon isoforms of protein kinase c and the activation of the long terminal repeat promoter in human astrocytoma cells. *Glia* **27,** 143–151.

Anderson, S. J., Lenburg, M., Landau, N. R., and Garcia, J. V. (1994). The cytoplasmic domain of CD4 is sufficient for its down-regulation from the cell surface by human immunodeficiency virus type 1 Nef. *J. Virol.* **68,** 3092–3101.

Arold, S., Franken, P., Strub, M. P., Hoh, F., Benichou, S., Benarous, R., and Dumas, C. (1997). The crystal structure of HIV-1 Nef protein bound to the Fyn kinase SH3 domain suggests a role for this complex in altered T cell receptor signaling. *Structure* **5,** 1361–1372.

Arrigo, S. J., and Chen, I. S. (1991). Rev is necessary for translation but not cytoplasmic accumulation of HIV-1 vif, vpr, and env/vpu 2 RNAs. *Genes Dev.* **5,** 808–819.

Baba, T. W., Jeong, Y. S., Pennick, D., Bronson, R., Greene, M. F., and Ruprecht, R. M. (1995). Pathogenicity of live, attenuated SIV after mucosal infection of neonatal macaques [see comments]. *Science,* **267,** 1820–1825.

Baba, T. W., Liska, V., Khimani, A. H., Ray, N. B., Dailey, P. J., Penninck, D., Bronson, R., Greene, M. F., McClure, H. M., Martin, L. N., and Ruprecht, R. M. (1999). Live attenuated, multiply deleted simian immunodeficiency virus causes AIDS in infant and adult macaques. *Nat. Med.* **5,** 194–203.

Bachelerie, F., Alcami, J., Hazan, U., Israel, N., Goud, B., Arenzana-Seisdedos, F., and Virelizier, J. L. (1990). Constitutive expression of human immunodeficiency virus (HIV) nef protein in human astrocytes does not influence basal or induced HIV long terminal repeat activity. *J. Virol.* **64,** 3059–3062.

Bandres, J. C., and Ratner, L. (1994). Human immunodeficiency virus type 1 Nef protein down-regulates transcription factors NF-kappa B and AP-1 in human T cells *in vitro* after T-cell receptor stimulation. *J. Virol.* **68**, 3243–3249.

Baur, A. S., Sass, G., Laffert, B., Willbold, D., Cheng-Mayer, C., and Peterlin, B. M. (1997). The N-terminus of Nef from HIV-1/SIV associates with a protein complex containing Lck and a serine kinase. *Immunity* **6**, 283–291.

Baur, A. S., Sawai, E. T., Dazin, P., Fantl, W. J., Cheng-Mayer, C., and Peterlin, B. M. (1994). HIV-1 Nef leads to inhibition or activation of T cells depending on its intracellular localization. *Immunity* **1**, 373–384.

Beaty, C. D., Franklin, T. L., Uehara, Y., and Wilson, C. B. (1994). Lipopolysaccharide-induced cytokine production in human monocytes: role of tyrosine phosphorylation in transmembrane signal transduction. *Eur. J. Immunol.* **24**, 1278–1284.

Benichou, S., Bomsel, M., Bodeus, M., Durand, H., Doute, M., Letourneur, F., Camonis, J., and Benarous, R. (1994). Physical interaction of the HIV-1 Nef protein with beta-COP, a component of non-clathrin-coated vesicles essential for membrane traffic. *J. Biol. Chem.* **269**, 30073–30076.

Benichou, S., Liu, L. X., Erdtmann, L., Selig, L., and Benarous, R. (1997). Use of the two-hybrid system to identify cellular partners of the HIV1 Nef protein. *Res. Virol.* **148**, 71–73.

Benson, R. E., Sanfridson, A., Ottinger, J. S., Doyle, C., and Cullen, B. R. (1993). Downregulation of cell-surface CD4 expression by simian immunodeficiency virus Nef prevents viral super infection. *J. Exp. Med.* **177**, 1561–1566.

Biggs, T. E., Cooke, S. J., Barton, C. H., Harris, M. P., Saksela, K., and Mann, D. A. (1999). Induction of activator protein 1 (AP-1) in macrophages by human immunodeficiency virus type-1 NEF is a cell-type-specific response that requires both hck and MAPK signaling events. *J. Mol. Biol.* **290**, 21–35.

Binninger, D., Ennen, J., Bonn, D., Norley, S. G., and Kurth, R. (1991). Mutational analysis of the simian immunodeficiency virus SIVmac nef gene. *J. Virol.* **65**, 5237–5243.

Blaak, H., Brouwer, M., Ran, L. J., de Wolf, F., and Schuitemaker, H. (1998). *In vitro* replication kinetics of human immunodeficiency virus type 1 (HIV-1) variants in relation to virus load in long-term survivors of HIV-1 infection. *J. Infect. Dis.* **177**, 600–610.

Blumberg, B. M., Epstein, L. G., Saito, Y., Chen, D., Sharer, L. R., and Anand, R. (1992). Human immunodeficiency virus type 1 nef quasispecies in pathological tissue. *J. Virol.* **66**, 5256–5264.

Bodeus, M., Marie-Cardine, A., Bougeret, C., Ramos-Morales, F., and Benarous, R. (1995). *In vitro* binding and phosphorylation of human immunodeficiency virus type 1 Nef protein by serine/threonine protein kinase. *J. Gen. Virol.* **76**, 1337–1344.

Bohuslav, J., Horejsi, V., Hansmann, C., Stockl, J., Weidle, U. H., Majdic, O., Bartke, I., Knapp, W., and Stockinger, H. (1995). Urokinase plasminogen activator receptor, beta 2-integrins, and Src-kinases within a single receptor complex of human monocytes. *J. Exp. Med.* **181**, 1381–1390.

Bolen, J. B., Rowley, R. B., Spana, C., and Tsygankov, A.Y. (1992). The Src family of tyrosine protein kinases in hemopoietic signal transduction. *FASEB J.* **6**, 3403–3409.

Brady, H. J., Pennington, D. J., Miles, C. G., and Dzierzak, E. A. (1993). CD4 cell surface downregulation in HIV-1 Nef transgenic mice is a consequence of intracellular sequestration. *EMBO J.* **12**, 4923–4932.

Briggs, S. D., Sharkey, M., Stevenson, M., and Smithgall, T. E. (1997). SH3-mediated Hck tyrosine kinase activation and fibroblast transformation by the Nef protein of HIV-1. *J. Biol. Chem.* **272**, 17899–17902.

Bruggeman, L. A., Thomson, M. M., Nelson, P. J., Kopp, J. B., Rappaport, J., Klotman, P. E., and Klotman, M. E. (1994). Patterns of HIV-1 mRNA expression in transgenic mice are tissue-dependent. *Virology* **202**, 940–948.

Bukovsky, A. A., Dorfman, T., Weimann, A., and Gottlinger, H. G. (1997). Nef association with human immunodeficiency virus type 1 virions and cleavage by the viral protease. *J. Virol.* **71**, 1013–1018.

Carl, S., Iafrate, A. J., Skowronski, J., Stahl-Hennig, C., and Kirchhoff, F. (1999). Effect of the attenuating deletion and of sequence alterations evolving *in vivo* on simian immunodeficiency virus C8-Nef function. *J. Virol.* **73,** 2790–2797.

Cereseto, A., Mulloy, J. C., and Franchini, G. (1996). Insights on the pathogenicity of human T-lymphotropic/leukemia virus types I and II. *J. Acquir, Immune Defic. Syndr. Hum. Retrovirol.* **13,** S69–75.

Chen, B. K., Gandhi, R. T., and Baltimore, D. (1996). CD4 down-modulation during infection of human T cells with human immunodeficiency virus type 1 involves independent activities of vpu, env, and nef. *J. Virol.* **70,** 6044–6053.

Chen, Y. L., Trono, D., and Camaur, D. (1998). The proteolytic cleavage of human immunodeficiency virus type 1 Nef does not correlate with its ability to stimulate virion infectivity. *J. Virol.* **72,** 3178–3184.

Cheng-Mayer, C., Iannello, P., Shaw, K., Luciw, P. A., and Levy, J. A. (1989). Differential effects of nef on HIV replication: Implications for viral pathogenesis in the host. *Science* **246,** 1629–1632.

Chowers, M. Y., Pandori, M. W., Spina, C. A., Richman, D. D., and Guatelli, J. C. (1995). The growth advantage conferred by HIV-1 nef is determined at the level of viral DNA formation and is independent of CD4 downregulation. *Virology* **212,** 451–457.

Chowers, M. Y., Spina, C. A., Kwoh, T. J., Fitch, N. J., Richman, D. D., and Guatelli, J. C. (1994). Optimal infectivity *in vitro* of human immunodeficiency virus type 1 requires an intact nef gene. *J. Virol.* **68,** 2906–2914.

Clements, J. E., and Zink, M. C. (1996). Molecular biology and pathogenesis of animal lentivirus infections. *Clin. Microbiol. Rev.* **9,** 100–117.

Coffin, J. (1996). Retroviridae: The viruses and their replication. *In* "Fields Virology" (B. N. Fields, D. M. Kripe, and P. M. Howley, Eds.). Vol. 2, pp. 1767–1848. Lippincott–Raven, Philadelphia.

Collette, Y., Dutartre, H., Benziane, A., and Olive, D. (1997). The role of HIV1 Nef in T-cell activation: Nef impairs induction of Th1 cytokines and interacts with the Src family tyrosine kinase Lck. *Res. Virol.* **148,** 52–58.

Collette, Y., Dutartre, H., Benziane, A., Ramos, M., Benarous, R., Harris, M., and Olive, D. (1996a). Physical and functional interaction of Nef with Lck: HIV-1 Nef-induced T-cell signaling defects. *J. Biol. Chem.* **271,** 6333–6341.

Collette, Y., Mawas, C., and Olive, D. (1996b). Evidence for intact CD28 signaling in T cell hyporesponsiveness induced by the HIV-1 nef gene. *Eur. J. Immunol.* **26,** 1788–1793.

Collins, K. L., Chen, B. K., Kalams, S. A., Walker, B. D., and Baltimore, D. (1998). HIV-1 Nef protein protects infected primary cells against killing by cytotoxic T lymphocytes. *Nature* **391,** 397–401.

Colonna, M., Borsellino, G., Falco, M., Ferrara, G. B., and Strominger, J. L. (1993). HLA-C is the inhibitory ligand that determines dominant resistance to lysis by NK1- and NK2-specific natural killer cells. *Proc. Natl. Acad. Sci. USA* **90,** 12000–12004.

Crise, B., Buonocore, L., and Rose, J. K. (1990). CD4 is retained in the endoplasmic reticulum by the human immunodeficiency virus type 1 glycoprotein precursor. *J. Virol.* **64,** 5585–5593.

Daniel, M. D., Kirchhoff, F., Czajak, S. C., Sehgal, P. K., and Desrosiers, R. C. (1992). Protective effects of a live attenuated SIV vaccine with a deletion in the nef gene. *Science* **258,** 1938–1941.

Datta, S. K., Melief, C. J., and Schwartz, R. S. (1975). Lymphocytes and leukemia viruses: Tropism and transtropism of murine leukemia virus. *J. Natl. Cancer Inst.* **55,** 425–432.

Davis, D. A., Newcomb, F. M., Starke, D. W., Ott, D. E., Mieyal, J. J., and Yarchoan, R. (1997). Thioltransferase (glutaredoxin) is detected within HIV-1 and can regulate the activity of glutathionylated HIV-1 protease *in vitro*. *J. Biol. Chem.* **272,** 25935–25940.

de Ronde, A., Klaver, B., Keulen, W., Smit, L., and Goudsmit, J. (1992). Natural HIV-1 NEF accelerates virus replication in primary human lymphocytes. *Virology* **188,** 391–395.

Deacon, N. J., Tsykin, A., Solomon, A., Smith, K., Ludford-Menting, M., Hooker, D. J., McPhee, D. A., Greenway, A. L., Ellett, A., Chatfield, C., *et al.* (1995). Genomic structure of an attenuated quasi species of HIV-1 from a blood transfusion donor and recipients. *Science* **270**, 988–991.

Delassus, S., Cheynier, R., and Wain-Hobson, S. (1991). Evolution of human immunodeficiency virus type 1 nef and long terminal repeat sequences over 4 years *in vivo* and *in vitro*. *J. Virol.* **65**, 225–231.

Du, Z., Ilyinskii, P. O., Sasseville, V. G., Newstein, M., Lackner, A. A., and Desrosiers, R. C. (1996). Requirements for lymphocyte activation by unusual strains of simian immunodeficiency virus. *J. Virol.* **70**, 4157–4161.

Du, Z., Lang, S. M., Sasseville, V. G., Lackner, A. A., Ilyinskii, P. O., Daniel, M. D., Jung, J. U., and Desrosiers, R. C. (1995). Identification of a nef allele that causes lymphocyte activation and acute disease in macaque monkeys. *Cell* **82**, 665–674.

Dutartre, H., Harris, M., Olive, D., and Collette, Y. (1998) The human immunodeficiency virus type 1 Nef protein binds the Src- related tyrosine kinase Lck SH2 domain through a novel phosphotyrosine independent mechanism. *Virology* **247**, 200–211.

Easterbrook, P. J., and Schrager, L. K. (1998). Long-term nonprogression in HIV infection: Methodological issues and scientific priorities [Report of an international European community—National Institutes of Health Workshop, The Royal Society, London, England, November 27–29, 1995]. *AIDS Res. Hum. Retroviruses* **14**, 1211–1228.

Ernst, C. A., Zhang, Y. J., Hancock, P. R., Rutledge, B. J., Corless, C. L., and Rollins, B. J. (1994). Biochemical and biologic characterization of murine monocyte chemoattractant protein-1: Identification of two functional domains. *J. Immunol.* **152**, 3541–3549.

Fackler, O. T., Luo, W., Geyer, M., Alberts, A. S., and Peterlin, B. M. (1999). Activation of Vav by Nef induces cytoskeletal rearrangements and downstream effector functions. *Mol. Cell.* **3**, 729–739.

Flaherty, M. T., Barber, S. A., and Clements, J. E. (1998). Neurovirulent simian immunodeficiency virus incorporates a Nef-associated kinase activity into virions. *AIDS Res. Hum. Retroviruses* **14**, 163–170.

Frank, S. J., Samelson, L. E., and Klausner, R. D. (1990). The structure and signalling functions of the invariant T cell receptor components. *Semin. Immunol.* **2**, 89–97.

Freund, J., Kellner, R., Houthaeve, T., and Kalbitzer, H. R. (1994). Stability and proteolytic domains of Nef protein from human immunodeficiency virus (HIV) type 1. Eur. *J. Biochem.* **221**, 811–819.

Fruh, K., Ahn, K., and Peterson, P. A. (1997). Inhibition of MHC class I antigen presentation by viral proteins. *J. Mol. Med.* **75**, 18–27.

Furtado, M. R., Balachandran, R., Gupta, P., and Wolinsky, S. M. (1991). Analysis of alternatively spliced human immunodeficiency virus type-1 mRNA species, one of which encodes a novel tat-env fusion protein. *Virology* **185**, 258–270.

Gandhi, R. T., Chen, B. K., Straus, S. E., Dale, J. K., Lenardo, M. J., and Baltimore, D. (1998). HIV-1 directly kills CD4+ T cells by a Fas-independent mechanism. *J. Exp. Med.* **187**, 1113–1122.

Garcia, J. V., and Miller, A. D. (1991). Serine phosphorylation-independent downregulation of cell-surface CD4 by nef. *Nature* **350**, 508–511.

Garcia, J. V., Alfano, J., and Miller, A. D. (1993). The negative effect of human immunodeficiency virus type 1 Nef on cell surface CD4 expression is not species specific and requires the cytoplasmic domain of CD4. *J. Virol.* **67**, 1511–1516.

Goldsmith, M. A., Warmerdam, M. T., Atchison, R. E., Miller, M. D., and Greene, W. C. (1995). Dissociation of the CD4 downregulation and viral infectivity enhancement functions of human immunodeficiency virus type 1 Nef. *J. Virol.* **69**, 4112–4121.

Gratton, S., Yao, X. J., Venkatesan, S., Cohen, E. A., and Sekaly, R. P. (1996). Molecular analysis of the cytoplasmic domain of CD4: Overlapping but noncompetitive requirement for lck association and down-regulation by Nef. *J. Immunol.* **157**, 3305–3311.

Greenberg, M. E., Bronson, S., Lock, M., Neumann, M., Pavlakis, G. N., and Skowronski, J. (1997). Co-localization of HIV-1 Nef with the AP-2 adaptor protein complex correlates with Nef-induced CD4 down-regulation. *EMBO J.* **16**, 6964–6976.

Greenberg, M. E., Iafrate, A. J., and Skowronski, J. (1998). The SH3 domain-binding surface and an acidic motif in HIV-1 Nef regulate trafficking of class I MHC complexes. *EMBO J.* **17**, 2777–2789.

Greenough, T. C., Sullivan, J. L., and Desrosiers, R. C. (1999). Declining CD4 T-cell counts in a person infected with nef-deleted HIV-1. N. *Engl. J. Med.* **340**, 236–237.

Greenway, A., Azad, A., and McPhee, D. (1995). Human immunodeficiency virus type 1 Nef protein inhibits activation pathways in peripheral blood mononuclear cells and T-cell lines. *J. Virol.* **69**, 1842–1850.

Greenway, A., Azad, A., Mills, J., and McPhee, D. (1996). Human immunodeficiency virus type 1 Nef binds directly to Lck and mitogen-activated protein kinase, inhibiting kinase activity. *J. Virol.* **70**, 6701–6708.

Greenway, A. L., Allen, K., Crowe, S., Mills, J., and McPhee, D. A. (submitted). HIV-1 Nef associates with multiple src family kinases and MAPK during infection of CD4+ T-lymphocytes and macrophages causing differential effects on their kinase activities.

Greenway, A. L., Dutartre, H., Allen, K., McPhee, D. A., Olive, D., and Collette, Y. (1999). Simian immunodeficiency virus and human immunodeficiency virus type 1 nef proteins show distinct patterns and mechanisms of Src kinase activation. *J. Virol.* **73**, 6152–6158.

Greenway, A. L., McPhee, D. A., Grgacic, E., Hewish, D., Lucantoni, A., Macreadie, I., and Azad, A. (1994). Nef 27, but not the Nef 25 isoform of human immunodeficiency virus-type 1 pNL4.3 down-regulates surface CD4 and IL-2R expression in peripheral blood mononuclear cells and transformed T cells. *Virology* **198**, 245–256.

Grzesiek, S., Bax, A., Clore, G. M., Gronenborn, A. M., Hu, J. S., Kaufman, J., Palmer, I., Stahl, S. J., and Wingfield, P. T. (1996a). The solution structure of HIV-1 Nef reveals an unexpected fold and permits delineation of the binding surface for the SH3 domain of Hck tyrosine protein kinase. *Nat. Struct. Biol.* **3**, 340–345.

Grzesiek, S., Stahl, S. J., Wingfield, P. T., and Bax, A. (1996b). The CD4 determinant for downregulation by HIV-1 Nef directly binds to Nef. Mapping of the Nef binding surface by NMR. *Biochemistry* **35**, 10256–10261.

Gulizia, R. J., Collman, R. G., Levy, J. A., Trono, D., and Mosier, D. E. (1997). Deletion of nef slows but does not prevent CD4-positive T-cell depletion in human immunodeficiency virus type 1-infected human-PBL-SCID mice. *J. Virol.* **71**, 4161–4164.

Gupta, D., Jin, Y. P., and Dziarski, R. (1995). Peptidoglycan induces transcription and secretion of TNF-alpha and activation of lyn, extracellular signal-regulated kinase, and rsk signal transduction proteins in mouse macrophages. *J. Immunol.* **155**, 2620–2630.

Guy, B., Kieny, M. P., Riviere, Y., Le Peuch, C., Dott, K., Girard, M., Montagnier, L., and Lecocq, J. P. (1987). HIV F/3′ orf encodes a phosphorylated GTP-binding protein resembling an oncogene product. *Nature* **330**, 266–269.

Hammes, S. R., Dixon, E. P., Malim, M. H., Cullen, B. R., and Greene, W. C. (1989). Nef protein of human immunodeficiency virus type 1: Evidence against its role as a transcriptional inhibitor. *Proc. Natl. Acad. Sci. USA* **86**, 9549–9553.

Harris, M., and Coates, K. (1993). Identification of cellular proteins that bind to the human immunodeficiency virus type 1 nef gene product *in vitro*: A role for myristylation. *J. Gen. Virol.* **74**, 1581–1589.

Harris, M. P., and Neil, J. C. (1994). Myristoylation-dependent binding of HIV-1 Nef to CD4. *J. Mol. Biol.* **241**, 136–142.

He, J., deCastro, C. M., Vandenbark, G. R., Busciglio, J., and Gabuzda, D. (1997). Astrocyte apoptosis induced by HIV-1 transactivation of the c-kit protooncogene. *Proc. Natl. Acad. Sci. USA* **94**, 3954–3959.

Herbein, G., Mahlknecht, U., Batliwalla, F., Gregersen, P., Pappas, T., Butler, J., O'Brien, W. A., and Verdin, E. (1998a). Apoptosis of CD8+ T cells is mediated by macrophages through interaction of HIV gp120 with chemokine receptor CXCR4. *Nature* **395,** 189–194.

Herbein, G., Van Lint, C., Lovett, J. L., and Verdin, E. (1998b). Distinct mechanisms trigger apoptosis in human immunodeficiency virus type 1-infected and in uninfected bystander T lymphocytes. *J. Virol.* **72,** 660–670.

Ho, D. D. (1996). Viral counts count in HIV infection. *Science* **272,** 1124–1125.

Hodge, D. R., Dunn, K. J., Pei, G. K., Chakrabarty, M. K., Heidecker, G., Lautenberger, J. A., and Samuel, K. P. (1998). Binding of c-Raf1 kinase to a conserved acidic sequence within the carboxyl-terminal region of the HIV-1 Nef protein. *J. Biol. Chem.* **273,** 15727–15733.

Horak, I. D., Gress, R. E., Lucas, P. J., Horak, E. M., Waldmann, T. A., and Bolen, J. B. (1991). T-lymphocyte interleukin 2-dependent tyrosine protein kinase signal transduction involves the activation of p56lck. *Proc. Natl. Acad. Sci. USA* **88,** 1996–2000.

Huang, Y., Zhang, L., and Ho, D. D. (1995). Characterization of nef sequences in long-term survivors of human immunodeficiency virus type 1 infection. *J. Virol.* **69,** 93–100.

Hunter, T. (1987). A tail of two src's: Mutatis mutandis. *Cell* **49,** 1–4.

Iafrate, A. J., Bronson, S., and Skowronski, J. (1997). Separable functions of Nef disrupt two aspects of T cell receptor machinery: CD4 expression and CD3 signaling. *EMBO J.* **16,** 673–684.

Jacque, J. M., Mann, A., Enslen, H., Sharova, N., Brichacek, B., Davis, R. J., and Stevenson, M. (1998). Modulation of HIV-1 infectivity by MAPK, a virion-associated kinase. *EMBO J.* **17,** 2607–2618.

Jamieson, B. D., Aldrovandi, G. M., Planelles, V., Jowett, J. B., Gao, L., Bloch, L. M., Chen, I. S., and Zack, J. A. (1994). Requirement of human immunodeficiency virus type 1 nef for *in vivo* replication and pathogenicity. *J. Virol.* **68,** 3478–3485.

Johnson, R. P., Lifson, J. D., Czajak, S. C., Cole, K. S., Manson, K. H., Glickman, R., Yang, J., Montefiori, D. C., Montelaro, R., Wyand, M. S., and Desrosiers, R. C. (1999). Highly attenuated vaccine strains of simian immunodeficiency virus protect against vaginal challenge: Inverse relationship of degree of protection with level of attenuation. *J. Virol.* **73,** 4952–4961.

Kaminchik, J., Bashan, N., Itach, A., Sarver, N., Gorecki, M., and Panet, A. (1991). Genetic characterization of human immunodeficiency virus type 1 nef gene products translated *in vitro* and expressed in mammalian cells. *J. Virol.* **65,** 583–588.

Kestler, H. W. D., Ringler, D. J., Mori, K., Panicali, D. L., Sehgal, P. K., Daniel, M. D., and Desrosiers, R. C. (1991). Importance of the nef gene for maintenance of high virus loads and for development of AIDS. *Cell* **65,** 651–662.

Kienzle, N., Bachmann, M., Muller, W. E., and Muller-Lantzsch, N. (1992). Expression and cellular localization of the Nef protein from human immunodeficiency virus-1 in stably transfected B-cells. *Arch. Virol.* **124,** 123–132.

Kienzle, N., Freund, J., Kalbitzer, H. R., and Mueller-Lantzsch, N. (1993). Oligomerization of the Nef protein from human immunodeficiency virus (HIV) type 1. *Eur. J. Biochem.* **214,** 451–457.

Kim, S., Ikeuchi, K., Byrn, R., Groopman, J., and Baltimore, D. (1989). Lack of a negative influence on viral growth by the nef gene of human immunodeficiency virus type 1. *Proc. Natl. Acad. Sci. USA* **86,** 9544–9548.

Kirchhoff, F., Easterbrook, P. J., Douglas, N., Troop, M., Greenough, T. C., Weber, J., Carl, S., Sullivan, J. L., and Daniels, R. S. (1999). Sequence variations in human immunodeficiency virus type 1 Nef are associated with different stages of disease. *J. Virol.* **73,** 5497–5508.

Kirchhoff, F., Greenough, T. C., Brettler, D. B., Sullivan, J. L., and Desrosiers, R. C. (1995). Brief report: Absence of intact nef sequences in a long-term survivor with nonprogressive HIV-1 infection. *N. Engl. J. Med.* **332,** 228–232.

Kirchhoff, F., Kestler, H. W., III and Desrosiers, R. C. (1994). Upstream U3 sequences in simian immunodeficiency virus are selectively deleted *in vivo* in the absence of an intact nef gene. *J. Virol.* **68**, 2031–2037.

Klotman, M. E., Kim, S., Buchbinder, A., DeRossi, A., Baltimore, D., and Wong-Staal, F. (1991). Kinetics of expression of multiply spliced RNA in early human immunodeficiency virus type 1 infection of lymphocytes and monocytes. *Proc. Natl. Acad. Sci. USA* **88**, 5011–5015.

Kohleisen, B., Neumann, M., Herrmann, R., Brack-Werner, R., Krohn, K. J., Ovod, V., Ranki, A., and Erfle, V. (1992). Cellular localization of Nef expressed in persistently HIV-1-infected low-producer astrocytes. *AIDS* **6**, 1427–1436.

Le Gall, S., Erdtmann, L., Benichou, S., Berlioz-Torrent, C., Liu, L., Benarous, R., Heard, J. M., and Schwartz, O. (1998). Nef interacts with the mu subunit of clathrin adaptor complexes and reveals a cryptic sorting signal in MHC I molecules. *Immunity* **8**, 483–495.

Le Gall, S., Heard, J. M., and Schwartz, O. (1997). Analysis of Nef-induced MHC-I endocytosis. *Res. Virol.* **148**, 43–47.

Learmont, J. C., Geczy, A. F., Mills, J., Ashton, L. J., Raynes-Greenow, C. H., Garsia, R. J., Dyer, W. B., McIntyre, L., Oelrichs, R. B., Rhodes, D. I., Deacon, N. J., and Sullivan, J. S. (1999). Immunologic and virologic status after 14 to 18 years of infection with an attenuated strain of HIV-1: A report from the Sydney Blood Bank Cohort. *N. Engl. J. Med.* **340**, 1715–1722.

Lee, C. H., Leung, B., Lemmon, M. A., Zheng, J., Cowburn, D., Kuriyan, J., and Saksela, K. (1995). A single amino acid in the SH3 domain of Hck determines its high affinity and specificity in binding to HIV-1 Nef protein. *EMBO J.* **14**, 5006–5015.

Letourneur, F., and Klausner, R. D. (1992). A novel di-leucine motif and a tyrosine-based motif independently mediate lysosomal targeting and endocytosis of CD3 chains. *Cell* **69**, 1143–1157.

Liu, L. X., Margottin, F., Le Gall, S., Schwartz, O., Selig, L., Benarous, R., and Benichou, S. (1997). Binding of HIV-1 Nef to a novel thioesterase enzyme correlates with Nef-mediated CD4 down-regulation. *J. Biol. Chem.* **272**, 13779–13785.

Lock, M., Greenberg, M. E., Iafrate, A. J., Swigut, T., Muench, J., Kirchhoff, F., Shohdy, N., and Skowronski, J. (1999). Two elements target SIV Nef to the AP-2 clathrin adaptor complex, but only one is required for the induction of CD4 endocytosis. *EMBO J.* **18**, 2722–2733.

Lu, X., Wu, X., Plemenitas, A., Yu, H., Sawai, E. T., Abo, A., and Peterlin, B. M. (1996). CDC42 and Rac1 are implicated in the activation of the Nef-associated kinase and replication of HIV-1. *Curr. Biol.* **6**, 1677–1684.

Luo, T., and Garcia, J. V. (1996). The association of Nef with a cellular serine/threonine kinase and its enhancement of infectivity are viral isolate dependent. *J. Virol.* **70**, 6493–6496.

Luo, T., Livingston, R. A., and Garcia, J. V. (1997). Infectivity enhancement by human immunodeficiency virus type 1 Nef is independent of its association with a cellular serine/threonine kinase. *J. Virol.* **71**, 9524–9530.

Luo, W., and Peterlin, B. M. (1997). Activation of the T-cell receptor signaling pathway by Nef from an aggressive strain of simian immunodeficiency virus. *J. Virol.* **71**, 9531–9537.

Luria, S., Chambers, I., and Berg, P. (1991). Expression of the type 1 human immunodeficiency virus Nef protein in T cells prevents antigen receptor-mediated induction of interleukin 2 mRNA. *Proc. Natl. Acad. Sci. USA* **88**, 5326–5330.

Macreadie, I. G., Ward, A. C., Failla, P., Grgacic, E., McPhee, D., and Azad, A. A. (1993). Expression of HIV-1 nef in yeast: The 27 kDa Nef protein is myristylated and fractionates with the nucleus. *Yeast* **9**, 565–573.

Maitra, R. K., Ahmad, N., Holland, S. M., and Venkatesan, S. (1991). Human immunodeficiency virus type 1 (HIV-1) provirus expression and LTR transcription are repressed in NEF-expressing cell lines. *Virology* **182**, 522–533.

Mangasarian, A., and Trono, D. (1997). The multifaceted role of HIV Nef. *Res. Virol.* **148,** 30–33.

Mangasarian, A., Foti, M., Aiken, C., Chin, D., Carpentier, J. L., and Trono, D. (1997). The HIV-1 Nef protein acts as a connector with sorting pathways in the Golgi and at the plasma membrane. *Immunity* **6,** 67–77.

Mangasarian, A., Piguet, V., Wang, J. K., Chen, Y. L., and Trono, D. (1999). Nef-induced CD4 and major histocompatibility complex class I (MHC-I) down-regulation are governed by distinct determinants: N-terminal alpha helix and proline repeat of Nef selectively regulate MHC-I trafficking. *J. Virol.* **73,** 1964–1973.

Manninen, A., Hiipakka, M., Vihinen, M., Lu, W., Mayer, B. J., and Saksela, K. (1998). SH3-Domain binding function of HIV-1 Nef is required for association with a PAK-related kinase. *Virology* **250,** 273–282.

Mariani, R., and Skowronski, J. (1993). CD4 down-regulation by nef alleles isolated from human immunodeficiency virus type 1-infected individuals. *Proc. Natl. Acad. Sci. USA* **90,** 5549–5553.

Mariani, R., Kirchhoff, F., Greenough, T. C., Sullivan, J. L., Desrosiers, R. C., and Skowronski, J. (1996). High frequency of defective nef alleles in a long-term survivor with nonprogressive human immunodeficiency virus type 1 infection. *J. Virol.* **70,** 7752–7764.

Marlink, R., Kanki, P., Thior, I., Travers, K., Eisen, G., Siby, T., Traore, I., Hsieh, C. C., Dia, M. C., Gueye, E. H., *et al.* (1994). Reduced rate of disease development after HIV-2 infection as compared to HIV-1. *Science* **265,** 1587–1590.

Mellors, J. W., Rinaldo, C. R., Jr., Gupta, P., White, R. M., Todd, J. A., and Kingsley, L. A. (1996). Prognosis in HIV-1 infection predicted by the quantity of virus in plasma. *Science* **272,** 1167–1170.

Michael, N. L., Chang, G., d'Arcy, L. A., Ehrenberg, P. K., Mariani, R., Busch, M. P., Birx, D. L., and Schwartz, D. H. (1995a). Defective accessory genes in a human immunodeficiency virus type 1-infected long-term survivor lacking recoverable virus. *J. Virol.* **69,** 4228–4236.

Michael, N. L., Chang, G., d'Arcy, L. A., Tseng, C. J., Birx, D. L., and Sheppard, H. W. (1995b). Functional characterization of human immunodeficiency virus type 1 nef genes in patients with divergent rates of disease progression. *J. Virol.* **69,** 6758–6769.

Miller, M. D., Warmerdam, M. T., Ferrell, S. S., Benitez, R., and Greene, W. C. (1997). Intravirion generation of the C-terminal core domain of HIV-1 Nef by the HIV-1 protease is insufficient to enhance viral infectivity. *Virology* **234,** 215–225.

Miller, M. D., Warmerdam, M. T., Gaston, I., Greene, W. C., and Feinberg, M. B. (1994). The human immunodeficiency virus-1 nef gene product: A positive factor for viral infection and replication in primary lymphocytes and macrophages. *J. Exp. Med.* **179,** 101–113.

Miller, M. D., Warmerdam, M. T., Page, K. A., Feinberg, M. B., and Greene, W. C. (1995). Expression of the human immunodeficiency virus type 1 (HIV-1) nef gene during HIV-1 production increases progeny particle infectivity independently of gp160 or viral entry. *J. Virol.* **69,** 579–584.

Minami, Y., Kono, T., Yamada, K., Kobayashi, N., Kawahara, A., Perlmutter, R. M., and Taniguchi, T. (1993). Association of p561ck with IL-2 receptor beta chain is critical for the IL-2-induced activation of p561ck. *EMBO J.* **12,** 759–768.

Moarefi, I., LaFevre-Bernt, M., Sicheri, F., Huse, M., Lee, C. H., Kuriyan, J., and Miller, W. T. (1997). Activation of the Src-family tyrosine kinase Hck by SH3 domain displacement. *Nature* **385,** 650–653.

Murphy, K. M., Sweet, M. J., and Hume, D. A. (1994). The HIV-1 regulatory protein Nef has a specific function in viral expression in a murine macrophage cell line. *J. Leukoc. Biol.* **56,** 294–303.

Murti, K. G., Brown, P. S., Ratner, L., and Garcia, J. V. (1993). Highly localized tracks of human immunodeficiency virus type 1 Nef in the nucleus of cells of a human CD4+ T-cell line. *Proc. Natl. Acad. Sci. USA* **90,** 11895–11899.

Myers, G., Hahn, B., Mellors, J., Henderson, L., Korber, B., Jeang, F., McCutchan, F., and Pavlakis, G. (Eds.) (1997). "Human Retroviruses and AIDS Database Compendium." Theoretical Biology and Biophysics Group, Los Alamos National Laboratory, Los Alamos, New Mexico. [http://hiv-web.lanl.gov]

Myers, G., MacInnes, K., and Korber, B. (1992). The emergence of simian/human immunodeficiency viruses. *AIDS Res. Hum. Retroviruses* **8**, 373–386.

Niederman, T. M., Garcia, J. V., Hastings, W. R., Luria, S., and Ratner, L. (1992). Human immunodeficiency virus type 1 Nef protein inhibits NF-kappa B induction in human T cells. *J. Virol.* **66**, 6213–6219.

Niederman, T. M., and Ratner, L. (1992). Functional analysis of HIV1 and SIV Nef proteins. *Res. Virol.* **143**, 43–46.

Niederman, T. M., Hastings, W. R., Luria, S., Bandres, J. C., and Ratner, L. (1993a). HIV-1 Nef protein inhibits the recruitment of AP-1 DNA-binding activity in human T-cells. *Virology* **194**, 338–344.

Niederman, T. M., Hastings, W. R., and Ratner, L. (1993b). Myristoylation-enhanced binding of the HIV-1 Nef protein to T cell skeletal matrix. *Virology* **197**, 420–425.

Niederman, T. M., Thielan, B. J., and Ratner, L. (1989). Human immunodeficiency virus type 1 negative factor is a transcriptional silencer. *Proc. Natl. Acad. Sci. USA* **86**, 1128–1132.

Novembre, F. J., Lewis, M. G., Saucier, M. M., Yalley-Ogunro, J., Brennan, T., McKinnon, K., Bellah, S., and McClure, H. M. (1996). Deletion of the nef gene abrogates the ability of SIV smmPBj to induce acutely lethal disease in pigtail macaques. *AIDS Res. Hum. Retroviruses* **12**, 727–736.

Nunn, M. F., and Marsh, J. W. (1996). Human immunodeficiency virus type 1 Nef associates with a member of the p21-activated kinase family. *J. Virol.* **70**, 6157–6161.

Oyaizu, N., Chirmule, N., Kalyanaraman, V. S., Hall, W. W., Pahwa, R., Shuster, M., and Pahwa, S. (1990). Human immunodeficiency virus type 1 envelope glycoprotein gp120 produces immune defects in CD4+ T lymphocytes by inhibiting interleukin 2 mRNA. *Proc. Natl. Acad. Sci. USA* **87**, 2379–2383.

Pandori, M., Craig, H., Moutouh, L., Corbeil, J., and Guatelli, J. (1998). Virological importance of the protease-cleavage site in human immunodeficiency virus type 1 Nef is independent of both intravirion processing and CD4 down-regulation. *Virology* **251**, 302–316.

Pandori, M. W., Fitch, N. J., Craig, H. M., Richman, D. D., Spina, C. A., and Guatelli, J. C. (1996). Producer-cell modification of human immunodeficiency virus type 1: Nef is a virion protein. *J. Virol.* **70**, 4283–4290.

Pantaleo, G., Cohen, O. J., Schacker, T., Vaccarezza, M., Graziosi, C., Rizzardi, G. P., Kahn, J., Fox, C. H., Schnittman, S. M., Schwartz, D. H., Corey, L., and Fauci, A. S. (1998). Evolutionary pattern of human immunodeficiency virus (HIV) replication and distribution in lymph nodes following primary infection: Implications for antiviral therapy. *Nat. Med.* **4**, 341–345.

Perry, S. T., Flaherty, M. T., Kelley, M. J., Clabough, D. L., Tronick, S. R., Coggins, L., Whetter, L., Lengel, C. R., and Fuller, F. (1992). The surface envelope protein gene region of equine infections anemia virus is not an important determinant of tropism in vitro. *J. Virol.* **66**, 4085–4097.

Piguet, V., Chen, Y. L., Mangasarian, A., Foti, M., Carpentier, J. L., and Trono, D. (1998). Mechanism of Nef-induced CD4 endocytosis: Nef connects CD4 with the mu chain of adaptor complexes. *EMBO J.* **17**, 2472–2481.

Premkumar, D. R., Ma, X. Z., Maitra, R. K., Chakrabarti, B. K., Salkowitz, J., Yen-Lieberman, B., Hirsch, M. S., and Kestler, H. W. (1996). The nef gene from a long-term HIV type 1 nonprogressor. *AIDS Res. Hum. Retroviruses* **12**, 337–345.

Purcell, D. F., and Martin, M. A. (1993). Alternative splicing of human immunodeficiency virus type 1 mRNA modulates viral protein expression, replication, and infectivity. *J. Virol.* **67**, 6365–6378.

Ranki, A., Lagerstedt, A., Ovod, V., Aavik, E., and Krohn, K. J. (1994). Expression kinetics and subcellular localization of HIV-1 regulatory proteins Nef, Tat and Rev in acutely and chronically infected lymphoid cell lines. *Arch. Virol.* **139,** 365–378.

Ren, R., Mayer, B. J., Cicchetti, P., and Baltimore, D. (1993). Identification of a tenamino acid proline-rich SH3 binding site. *Science* **259,** 1157–1161.

Rhee, S. S., and Marsh, J. W. (1994a). HIV-1 Nef activity in murine T cells. CD4 modulation and positive enhancement. *J. Immunol.* **152,** 5128–5134.

Rhee, S. S., and Marsh, J. W. (1994b). Human immunodeficiency virus type 1 Nef-induced down-modulation of CD4 is due to rapid internalization and degradation of surface CD4. *J. Virol.* **68,** 5156–5163.

Robinson, M. S. (1994). The role of clathrin, adaptors and dynamin in endocytosis. *Curr. Opin. Cell Biol.* **6,** 538–544.

Romero, I. A., Teixeira, A., Strosberg, A. D., Cazaubon, S., and Couraud, P. O. (1998). The HIV-1 nef protein inhibits extracellular signal-regulated kinase- dependent DNA synthesis in a human astrocytic cell line. *J. Neurochem.* **70,** 778–785.

Ross, T. M., Oran, A. E., and Cullen, B. R. (1999). Inhibition of HIV-1 progeny virion release by cell-surface CD4 is relieved by expression of the viral nef protein. *Curr. Biol.* **9,** 613–621.

Rossi, F., Evstafieva, A., Pedrali-Noy, G., Gallina, A., and Milanesi, G. (1997). HsN3 proteasomal subunit as a target for human immunodeficiency virus type 1 Nef protein. *Virology* **237,** 33–45.

Rossi, F., Gallina, A., and Milanesi, G. (1996). Nef-CD4 physical interaction sensed with the yeast two-hybrid system. *Virology* **217,** 397–403.

Rudd, C. E., Trevillyan, J. M., Dasgupta, J. D., Wong, L. L., and Schlossman, S. F. (1988). The CD4 receptor is complexed in detergent lysates to a protein-tyrosine kinase (pp58) from human T lymphocytes. *Proc. Natl. Acad. Sci. USA* **85,** 5190–5194.

Saksela, K., Cheng, G., and Baltimore, D. (1995). Proline-rich (PxxP) motifs in HIV-1 Nef bind to SH3 domains of a subset of Src kinases and are required for the enhanced growth of Nef+ viruses but not for down-regulation of CD4. *EMBO J.* **14,** 484–491.

Salghetti, S., Mariani, R., and Skowronski, J. (1995). Human immunodeficiency virus type 1 Nef and p56lck protein-tyrosine kinase interact with a common element in CD4 cytoplasmic tail. *Proc. Natl. Acad. Sci. USA* **92,** 349–353.

Salvi, R., Garbuglia, A. R., Di Caro, A., Pulciani, S., Montella, F., and Benedetto, A. (1998). Grossly defective nef gene sequences in a human immunodeficiency virus type 1-seropositive long-term nonprogressor. *J. Virol.* **72,** 3646–3657.

Saucier, M., Hodge, S., Dewhurst, S., Gibson, T., Gibson, J. P., McClure, H. M., and Novembre, F. J. (1998). The tyrosine-17 residue of Nef in SIV smmPBj14 is required for acute pathogenesis and contributes to replication in macrophages. *Virology* **244,** 261–272.

Sawai, E. T., Baur, A., Struble, H., Peterlin, B. M., Levy, J. A., and Cheng-Mayer, C. (1994). Human immunodeficiency virus type 1 Nef associates with a cellular serine kinase in T lymphocytes. *Proc. Natl. Acad. Sci. USA* **91,** 1539–1543.

Sawai, E. T., Baur, A. S., Peterlin, B. M., Levy, J. A., and Cheng-Mayer, C. (1995). A conserved domain and membrane targeting of Nef from HIV and SIV are required for association with a cellular serine kinase activity. *J. Biol. Chem.* **270,** 15307–15314.

Sawai, E. T., Khan, I. H., Montbriand, P. M., Peterlin, B. M., Cheng-Mayer, C., and Luciw, P. A. (1996). Activation of PAK by HIV and SIV Nef: Importance for AIDS in rhesus macaques. *Curr. Biol.* **6,** 1519–1527.

Schmid-Alliana, A., Menou, L., Manie, S., Schmid-Antomarchi, H., Millet, M. A., Giuriato, S., Ferrua, B., and Rossi, B. (1998). Microtubule integrity regulates src-like and extracellular signal-regulated kinase activities in human pro-monocytic cells. Importance for interleukin-1 production. *J. Biol. Chem.* **273,** 3394–3400.

Schorr, J., Kellner, R., Fackler, O., Freund, J., Konvalinka, J., Kienzle, N., Krausslich, H. G., Mueller-Lantzsch, N., and Kalbitzer, H. R. (1996). Specific cleavage sites of Nef proteins from human immunodeficiency virus types 1 and 2 for the viral proteases. *J. Virol.* **70,** 9051–9054.

Schrager, J. A., and Marsh, J. W. (1999). HIV-1 nef increases T cell activation in a stimulus-dependent manner. *Proc. Natl. Acad. Sci. USA* **96,** 8167–8172.

Schwartz, O., Marechal, V., Danos, O., and Heard, J. M. (1995). Human immunodeficiency virus type 1 Nef increases the efficiency of reverse transcription in the infected cell. *J. Virol.* **69,** 4053–4059.

Schwartz, O., Marechal, V., Le Gall, S., Lemonnier, F., and Heard, J. M. (1996). Endocytosis of major histocompatibility complex class I molecules is induced by the HIV-1 Nef protein. *Nat. Med.* **2,** 338–342.

Schwartz, O., Riviere, Y., Heard, J. M., and Danos, O. (1993). Reduced cell surface expression of processed human immunodeficiency virus type 1 envelope glycoprotein in the presence of Nef. *J. Virol.* **67,** 3274–3280.

Schwiebert, R. S., Tao, B., and Fultz, P. N. (1997). Loss of the SIV smmPBj14 phenotype and nef genotype during long-term survival of macaques infected by mucosal routes. *Virology* **230,** 82–92.

Shin, J., Dunbrack, R. L., Jr., Lee, S., and Strominger, J. L. (1991). Phosphorylation-dependent down-modulation of CD4 requires a specific structure within the cytoplasmic domain of CD4. *J. Biol. Chem.* **266,** 10658–10665.

Shugars, D. C., Smith, M. S., Glueck, D. H., Nanterment, P. V., Seillier-Moiseiwitsch, F., and Swanstrom, R. (1993). Analysis of human immunodeficiency virus type 1 nef gene sequences present *in vivo*. *J. Virol.* **67,** 4639–4650.

Sicheri, F., Moarefi, I., and Kuriyan, J. (1997). Crystal structure of the Src family tyrosine kinase Hck. *Nature* **385,** 602–609.

Skowronski, J., Parks, D., and Mariani, R. (1993). Altered T cell activation and development in transgenic mice expressing the HIV-1 nef gene. *EMBO J.* **12,** 703–713.

Smith, B. L., Krushelnycky, B. W., Mochly-Rosen, D., and Berg, P. (1996). The HIV nef protein associates with protein kinase C theta. *J. Biol. Chem.* **271,** 16753–16757.

Spina, C. A., Kwoh, T. J., Chowers, M. Y., Guatelli, J. C., and Richman, D. D. (1994). The importance of nef in the induction of human immunodeficiency virus type 1 replication from primary quiescent CD4 lymphocytes. *J. Exp. Med.* **179,** 115–123.

Swingler, S., Gallay, P., Camaur, D., Song, J., Abo, A., and Trono, D. (1997). The Nef protein of human immunodeficiency virus type 1 enhances serine phosphorylation of the viral matrix. *J. Virol.* **71,** 4372–4377.

Switzer, W. M., Wiktor, S., Soriano, V., Silva-Graca, A., Mansinho, K., Coulibaly, I. M., Ekpini, E., Greenberg, A. E., Folks, T. M., and Heneine, W. (1998). Evidence of Nef truncation in human immunodeficiency virus type 2 infection. *J. Infect. Dis.* **177,** 65–71.

Terwilliger, E. F., Langhoff, E., Gabuzda, D., Zazopoulos, E., and Haseltine, W. A. (1991). Allelic variation in the effects of the nef gene on replication of human immunodeficiency virus type 1. *Proc. Natl. Acad. Sci. USA* **88,** 10971–10975.

Tristem, M., Purvis, A., and Quicke, D. L. (1998). Complex evolutionary history of primate lentiviral vpr genes. *Virology* **240,** 232–237.

Tsunetsugu-Yokota, Y., Matsuda, S., Maekawa, M., Saito, T., Takemori, T., and Takebe, Y. (1992). Constitutive expression of the nef gene suppresses human immunodeficiency virus type 1 (HIV-1) replication in monocytic cell lines. *Virology* **191,** 960–963.

Tyler, K. L., and McPhee, D. A. (1987). Molecular and genetic aspects of the pathogenesis of viral infections of the central nervous system. *Crit. Rev. Neurobiol.* **3,** 221–243.

Ugolini, S., Mondor, I., and Sattentau, Q. J. (1999). HIV-1 attachment: another look. *Trends Microbiol.* **7,** 144–149.

Vega, M. A., and Strominger, J. L. (1989). Constitutive endocytosis of HLA class I antigens requires a specific portion of the intracytoplasmic tail that shares structural features with other endocytosed molecules. *Proc. Natl. Acad. Sci. USA* **86,** 2688–2692.

Veillette, A., Horak, I. D., Horak, E. M., Bookman, M. A., and Bolen, J. B. (1988). Alterations of the lymphocyte-specific protein tyrosine kinase (p56lck) during T-cell activation. *Mol. Cell. Biol.* **8**, 4353–4361.

Watanabe, H., Shiratori, T., Shoji, H., Miyatake, S., Okazaki, Y., Ikuta, K., Sato, T., and Saito, T. (1997). A novel acyl-CoA thioesterase enhances its enzymatic activity by direct binding with HIV Nef. *Biochem. Biophys. Res. Commun.* **238**, 234–239.

Weiss. A. (1993). T cell antigen receptor signal transduction: A tale of tails and cytoplasmic protein-tyrosine kinases. *Cell* **73**, 209–212.

Weiss, A., and Littman, D. R. (1994). Signal transduction by lymphocyte antigen receptors. *Cell* **76**, 263–274.

Weissman, D., Barker, T. D., and Fauci, A. S. (1996a). The efficiency of acute infection of CD4+ T cells is markedly enhanced in the setting of antigen-specific immune activation. *J. Exp. Med.* **183**, 687–692.

Weissman, D., Daucher, J., Barker, T., Adelsberger, J., Baseler, M., and Fauci, A. S. (1996b). Cytokine regulation of HIV replication induced by dendritic cell-CD4-positive T cell interactions. *AIDS Res. Hum. Retroviruses* **12**, 759–767.

Welker, R., Harris, M., Cardel, B., and Krausslich, H. G. (1998). Virion incorporation of human immunodeficiency virus type 1 Nef is mediated by a bipartite membrane-targeting signal: Analysis of its role in enhancement of viral infectivity. *J. Virol.* **72**, 8833–8840.

Welker, R., Kottler, H., Kalbitzer, H. R., and Krausslich, H. G. (1996). Human immunodeficiency virus type 1 Nef protein is incorporated into virus particles and specifically cleaved by the viral proteinase. *Virology* **219**, 228–236.

Whatmore, A. M., Cook, N., Hall, G. A., Sharpe, S., Rud, E. W., and Cranage, M. P. (1995). Repair and evolution of nef in vivo modulates simian immunodeficiency virus virulence. *J. Virol.* **69**, 5117–5123.

Willey, R. L., Maldarelli, F., Martin, M. A., and Strebel, K. (1992). Human immunodeficiency virus type 1 Vpu protein induces rapid degradation of CD4. *J. Virol.* **66**, 7193–7200.

Wiskerchen, M., and Cheng-Mayer, C. (1996). HIV-1 Nef association with cellular serine kinase correlates with enhanced virion infectivity and efficient proviral DNA synthesis. *Virology* **224**, 292–301.

Wyand, M. S., Manson, K. H., Garcia-Moll, M., Montefiori, D., and Desrosiers, R. C. (1996). Vaccine protection by a triple deletion mutant of simian immunodeficiency virus. *J. Virol.* **70**, 3724–3733.

Xu, X. N., Laffert, B., Screaton, G. R., Kraft, M., Wolf, D., Kolanus, W., Mongkolsapay, J., McMichael, A. J., and Baur, A. S. (1999). Induction of Fas ligand expression by HIV involves the interaction of Nef with the T cell receptor zeta chain. *J. Exp. Med.* **189**, 1489–1496.

Xu, X. N., Screaton, G. R., Gotch, F. M., Dong, T., Tan, R., Almond, N., Walker, B., Stebbings, R., Kent, K., Nagata, S., Stott, J. E., and McMichael, A. J. (1997). Evasion of cytotoxic T lymphocyte (CTL) responses by nef-dependent induction of Fas ligand (CD95L) expression on simian immunodeficiency virus-infected cells. *J. Exp. Med.* **186**, 7–16.

Yao, X. J., Friborg, J., Checroune, F., Gratton, S., Boisvert, F., Sekaly, R. P., and Cohen, E. A. (1995). Degradation of CD4 induced by human immunodeficiency virus type 1 Vpu protein: A predicted alpha-helix structure in the proximal cytoplasmic region of CD4 contributes to Vpu sensitivity. *Virology* **209**, 615–623.

Yu, G., and Felsted, R. L. (1992). Effect of myristoylation on p27 nef subcellular distribution and suppression of HIV-LTR transcription. *Virology* **187**, 46–55.

Zink, M. C., Yager, J. A., and Myers, J. D. (1990). Pathogenesis of caprine arthritis encephalitis virus: Cellular localization of viral transcripts in tissues of infected goats. *Am. J. Pathol.* **136**, 843–854.

Jean-Luc Darlix
Gaël Cristofari
Michael Rau
Christine Péchoux
Lionel Berthoux
Bernard Roques

LaboRetro
Unité de Virologie Humaine INSERM 412
Ecole Normale Supérieure de Lyon
69364 Lyon, France

Pharmacochimie moléculaire et structurale INSERM 266
75270 Paris, France

Nucleocapsid Protein of Human Immunodeficiency Virus as a Model Protein with Chaperoning Functions and as a Target for Antiviral Drugs

I. The Nucleocore

Our understanding of HIV-1 has largely benefited from our knowledge of the structure and replication of murine leukemia viruses (MLV), avian sarcoma leukemia viruses (ASLV), and other oncoviruses. We therefore begin by introducing basic notions of virion structure and replication resulting from investigations on these oncoviruses.

The nucleocore was first described in the 1970s and shown to be the most stable substructure of the viral particle (Davis and Rueckert, 1972), exhibiting a condensed helical conformation (Sarkar *et al.*, 1971). Upon partial denaturation the nucleocore shows extended circular and nucleosomelike structures (Chen *et al.*, 1980; Pager *et al.*, 1994). At present the nucleocore can be viewed as a highly organized nucleoprotein complex composed of a genomic RNA dimer coated by 2000–2500 molecules of nucleocapsid protein (NCp) as well as 20 to 50 molecules of reverse tran-

Advances in Pharmacology, Volume 48

1054-3589/00 $35.00

scriptase (RT) and integrase (IN) (Chen *et al.,* 1980; Dickson *et al.,* 1985).

In the mature virus, NC protein results from proteolytic processing of the GAG polyprotein by the viral protease (PR) (see also below) (Dickson *et al.,* 1985). Two categories of NC protein binding sites on genomic RNA have been described, namely high-affinity sites (of which there are a small number) and low-affinity sites (of which there are a large number) (Darlix *et al.,* 1995). In addition, biochemical and genetic studies show that NC protein molecules can interact with one another to form oligomers along the genomic RNA (Darlix and Spahr, 1982; Méric *et al.,* 1984; Tanchou *et al.,* 1995b). The genomic RNA monomers, present in dimeric form, are held together by a number of intermolecular interactions, contributing to the stability of the individual RNA monomers despite nicks that can be present in each monomer (reviewed in Coffin, 1985; Darlix *et al.,* 1995). Major interactions between the RNA monomers have been characterized close to the 5′ end of the genome and are known as the DLS and the DIS (dimer linkage structure and dimerization initiation sequence, respectively) (Darlix *et al.,* 1990; Prats *et al.,* 1990; Bieth *et al.,* 1990; Paillart *et al.,* 1994). Both the DLS and DIS represent intermolecular interactions, but only the DLS appears unique to the unspliced genomic RNA and critical for packaging (Lever *et al.,* 1989; reviewed in Berkhout, 1996). Last, interactions between the 5′ and 3′ ends have been documented for ASLV genomic RNA (Darlix, 1986), in agreement with electron microscopy data (Chen *et al.,* 1980).

Taken together the above data suggest a possible helical conformation for the nucleocore as shown in Fig. 1. This highly schematicized condensed conformation takes into account RNA–NCp as well as RNA–RNA and NCp–NCp interactions. The top of the proposed structure contains the DLS–DIS domain as well as the 5′ and 3′ end regions of the genomic RNA. This polar zone, also containing the packaging (E or Psi) and reverse transcription initiation (PBS) signals, may well guide formation of the nucleocore since it contains high-affinity sites for NC protein (Darlix and Spahr, 1982; Darlix *et al.,* 1990; Dannull *et al.,* 1994; reviewed in Darlix *et al.,* 1995). Furthermore, the top segment of the structure should not only favor recognition of the primer tRNA by RT and initiation of reverse transcription but also the two strand transfers that occur during proviral DNA synthesis and that are required to generate the LTRs (Gilboa *et al.,* 1979). In agreement with this, mutations in the DIS were found to impair proviral DNA synthesis (Paillart *et al.,* 1996). The parallel orientation of the two RNA monomers in this helical conformation is thought to facilitate strand transfers during reverse transcription and therefore copy choice recombination (Coffin, 1985; Linial and Blair, 1985), giving rise to a high level of genetic variability (Temin, 1991, 1993).

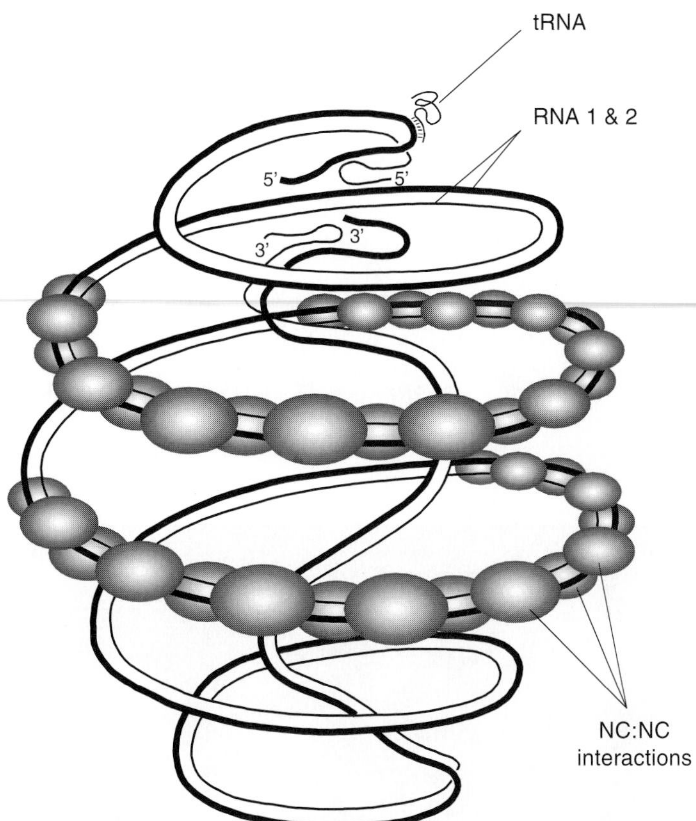

FIGURE 1 Scheme of a possible nucleocore structure. The helicoidal conformation shows the two genomic RNA monomers in parallel orientation. The dimeric RNA structure is maintained by virtue of major interactions close to the 5′ end (DLS and DIS sequences) and a number of secondary interactions between sequences dispersed along the genome, including 5′–3′ end interactions. Nucleocapsid protein molecules coat the dimeric RNA and form NC protein oligomers, ensuring partial protection of the RNA against degradation by nucleases. Binding of primer tRNA to the PBS of one RNA monomer is also shown.

Much attention has been paid to NCp7 as a critical component of the nucleocore. Interestingly, mutations known to unfold, or partly delete, the central globular domain of NCp7 (see Fig. 2 for a three-dimensional representation of HIV-1 NCp7), causing defaults in the condensation of the nucleocore such as those summarized in Fig. 3 (Berthoux *et al.,* 1997; Tanchou *et al.,* 1998) and resulting in the production of noninfectious viral particles (Méric *et al.,* 1988; Gorelick *et al.,* 1991, 1993; Dorfman *et al.,* 1993; Dupraz and Spahr, 1992; Deméné *et al.,* 1994; Ottmann *et al.,* 1995; see also review by Darlix *et al.,* 1995). Moreover, proviral DNA synthesis is either incomplete or DNA is unstable in the newly infected cells, which

A

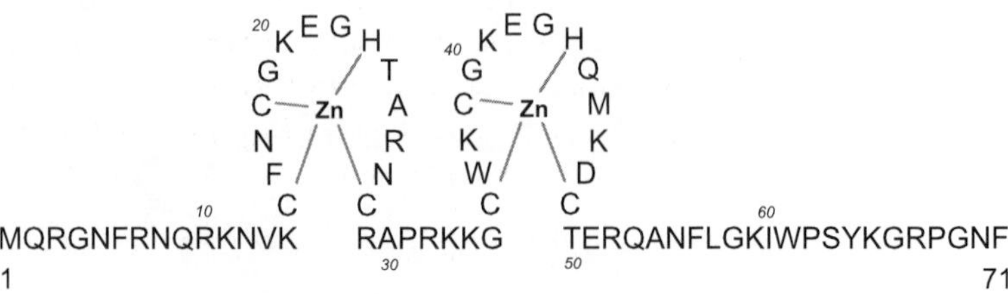

B

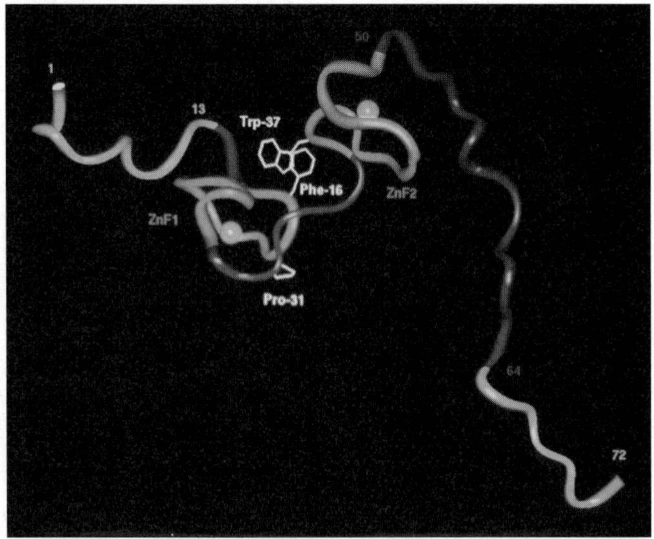

FIGURE 2 Sequence and conformation of HIV-1 nucleocapsid protein NCp7. The primary sequence of HIV-1 NCp7 (mal isolate) is shown using the one-letter code (A) and its three-dimensional structure (B) determined by 1H-NMR (Morellet *et al.*, 1994). Numbers indicate positions of amino acids. Zinc fingers 1 and 2 are labeled ZnF1 and ZnF2 with the Zn^{2+} ion as a dot. Phe-16, Pro-31, and Trp-37 appear to be structurally important for zinc finger proximity. With respect to the affinity of the "CCHC" motif for Zn^{2+} ions see Mély *et al.* (1991) and Green and Berg (1990). The N- and C-terminal domains of NCp7 (1–13 and 64–72, respectively) are flexible in the free protein.

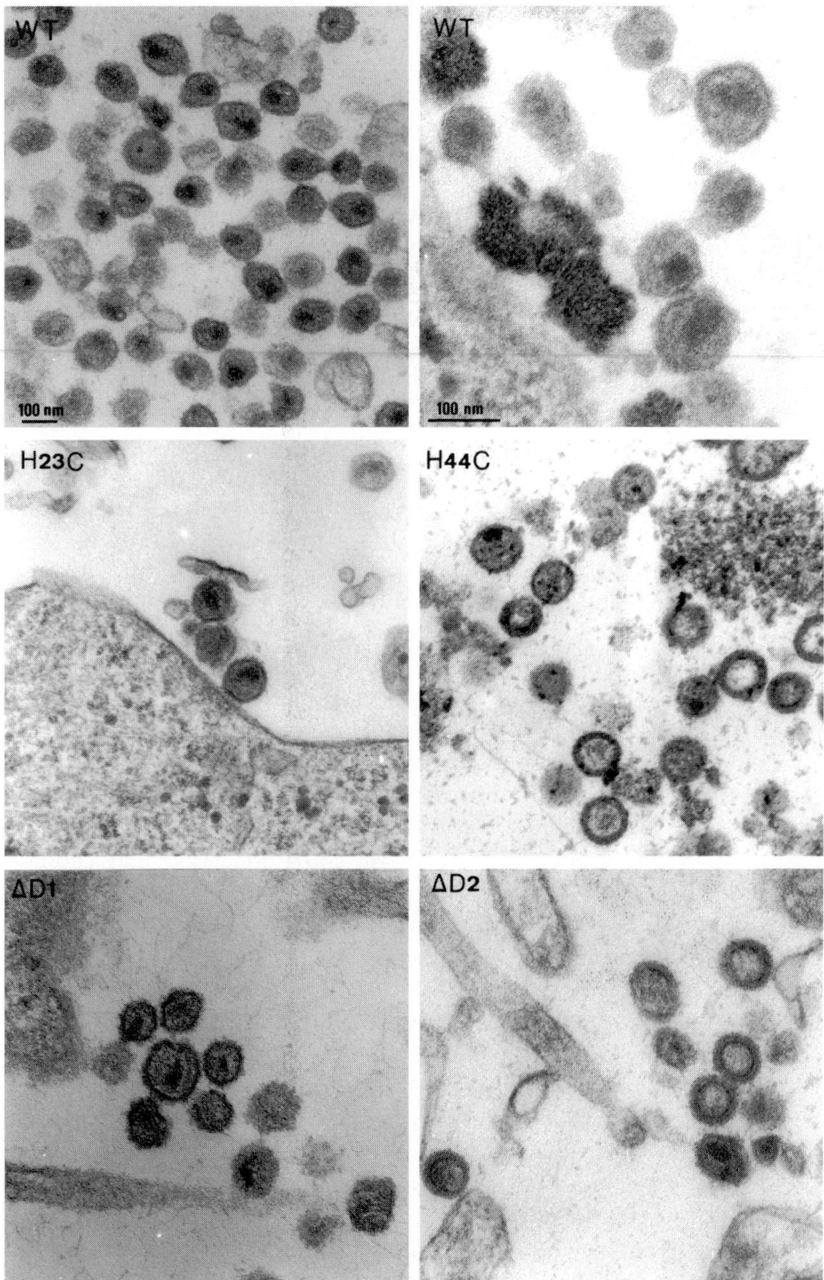

FIGURE 3 Visualization of HIV-1 NC mutant viruses by electron microscopy. HIV-1-infected wild-type and NC-mutant HIV virions were produced and prepared for electron microscopy as described in Berthoux *et al.* (1997) and observed under the electron microscope (bar, 100 nm). Wild-type viral particles exhibit the usual conical shape of the core, while the zinc finger mutants (Gelderblom *et al.,* 1987) (H23C contained His23 substituted by a Cys residue and H44C contained His44 substituted by a Cys residue; ΔD1 and ΔD2 correspond to a deletion of the first, or the second finger, respectively) show an immature core morphology.

explains, at least in part, why mutant viruses are completely noninfectious (Berthoux *et al.*, 1997; Tanchou *et al.*, 1998).

II. Nucleic Acid Chaperoning Activities of Nucleocapsid Protein during Proviral DNA Synthesis

Nucleocapsid protein binds to both RNA and DNA with the following order of affinity: retroviral RNA > ss DNA > ds DNA > oligonucleotides (Berkowit and Goff, 1994; Lapadat-Tapolsky *et al.*, 1993). The binding of NC protein to retroviral RNA and DNA appears to be cooperative, probably due to NCp–NCp interactions (Tanchou *et al.*, 1995b). At saturating levels NC protein binds to any nucleic acid with an occluded site size of 5 to 7 nt, resulting in its partial protection from nuclease attack (Tanchou *et al.*, 1995b).

The nucleic acid annealing activity of NC protein was first described for MuLV NCp10 and ASLV NCp12 (Prats *et al.*, 1988) and later extended to HIV-1 NCp7 (Barat *et al.*, 1989) and NCp9 of the Ty3 retrotransposon (Gabus *et al.*, 1998). For example, NC protein chaperones a rapid hybridization of small DNA oligonucleotides to structured RNA or DNA sequences (Lapadat-Tapolsky *et al.*, 1993, 1995). However, addition of excess NC protein to small (<20 nt) double-stranded nucleic acid complexes results in their destabilization (Khan and Giedroc, 1992; Li *et al.*, 1996). Additionally, NC protein has strand-transfer and strand-exchange activity, favoring the most stable double-stranded complex as the end product (Lapadat-Tapolsky *et al.*, 1995; You and McHenry, 1994; Tsuchihasi *et al.*, 1993). This chaperoning activity of NC protein is probably accomplished by an initial destabilization of nucleic acid secondary structures, as shown for tRNA (Khan and Giedroc, 1992; Gregoire *et al.*, 1997). Subsequently, it is thought that recruitment of nucleic acids with complementary sequences into high-molecular-weight complexes leads to their high local concentration and therefore to their hybridization (Dib-Hajj *et al.*, 1993; Tsuchihashi *et al.*, 1993; Tsuchihashi and Brown, 1994; Lapadat-Tapolsky *et al.*, 1995).

Annealing of primer tRNA to the PBS (primer binding site) of genomic RNA and the first strand transfer during reverse transcription are two highly demonstrative examples of the chaperoning functions of NC protein during the initial steps of proviral DNA synthesis (Prats *et al.*, 1988; Barat *et al.*, 1989; Darlix *et al.*, 1993, 1995; Allain *et al.*, 1994). Under these conditions NC protein can be viewed as an enhancer of intermolecular interactions between nucleic acids. Nucleocapsid protein also appears to be a silencer of intramolecular interactions since it can inhibit self-initiation of reverse transcription or DNA replication (Li *et al.*, 1996; Lapadat *et al.*, 1997). Both enhancement and silencing of inter- and intramolecular interactions, respectively, are illustrated in Fig. 4, showing the initial steps of proviral DNA synthesis. To be optimal *in vitro* the chaperoning function of HIV-1 NCp7 was found to require both zinc fingers and the flanking basic residues (de Rocquigny *et al.*, 1992; Darlix, Gabus, and Roques, unpublished data).

Interestingly, the viral RNA template is largely resistant to small nucleases such as RNaseA when present in these nucleoprotein complexes, mimicking the nucleocore formed *in vitro* (Tanchou *et al.*, 1995b). Nevertheless primer tRNALys3 is accessible to reverse transcriptase, which then transcribes the RNA template and transfers the newly made minus-strand DNA onto the 3′ end of the RNA if it is present in the complex (Darlix *et al.*, 1993; Tanchou *et al.*, 1995b). Recent data show that this is achieved through the recruitment of RT by NCp7 (Lener *et al.*, 1998).

During proviral DNA synthesis the error-prone RT incorporates mutations into the newly made double-stranded DNA, and this appears to be a major source of the high genetic variability of HIV in particular and of intensively replicating retroviruses in general (Hu and Temin, 1990a,b; Temin, 1991, 1993). NCp7 was found to facilitate cDNA elongation (Tanchou *et al.*, 1995b; Rodriguez-Rodriguez *et al.*, 1995; Guo *et al.*, 1997) and to promote extension of mutated minus-strand cDNA by RT (Lapadat-Tapolsky *et al.*, 1997; Rascle *et al.*, 1998), probably due to its active recruitment of RT (Lener *et al.*, 1998). As indicated above, NC protein chaperones minus-strand and plus-strand DNA transfers (Darlix *et al.*, 1993; Allain *et al.*, 1994; Auxilien *et al.*, 1999) that are essential to generate the provirus LTRs (Gilboa *et al.*, 1979; Hu and Temin, 1990a,b). Generally, NCp7 has strong strand-transfer activity, which appears critical for genetic recombination leading to the reassortment of mutations in replicating populations of HIV and is thus a major contributor to HIV variability (Linial and Blair, 1985; Temin, 1991, 1993; Coffin, 1995).

Very similar chaperoning effects have been obtained with NCp10 during MoMuLV DNA synthesis (Prats *et al.*, 1988, 1991; Allain *et al.*, 1994; Rascle *et al.*, 1998). These data favor the notion that NC protein can be viewed as a critical cofactor of RT, chaperoning proviral DNA synthesis and contributing to its genetic diversity.

At the end of the reverse transcription process, NC protein molecules are found associated with the viral DNA in nucleoprotein complexes partially resistant to nuclease degradation (Lapadat-Tapolsky *et al.*, 1993). This is illustrated in Fig. 5 where NCp7–LTR DNA complexes formed *in vitro* were examined by electron microscopy and found to exhibit extended oligomeric conformations where DNA can be locally denatured. Highly compact nucleoprotein structures can also be seen, possibly ensuring protection of the DNA (Fig. 5). Upon HIV-1 infection of human T cells proviral DNA is made and a fraction of NCp7 is found in the nucleus 18 h postinfection (Gallay *et al.*, 1995). In Fig. 6 NCp7 is shown to be detectable in the nucleus by immunoelectron microscopy (Fig. 6C, see arrowheads at 18 h

RNA template + tRNA *HEAT* → RNA-tRNA complex

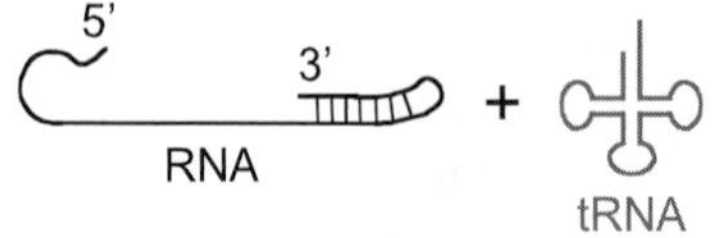

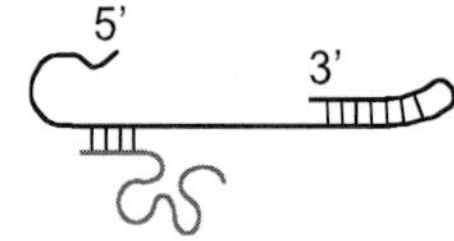

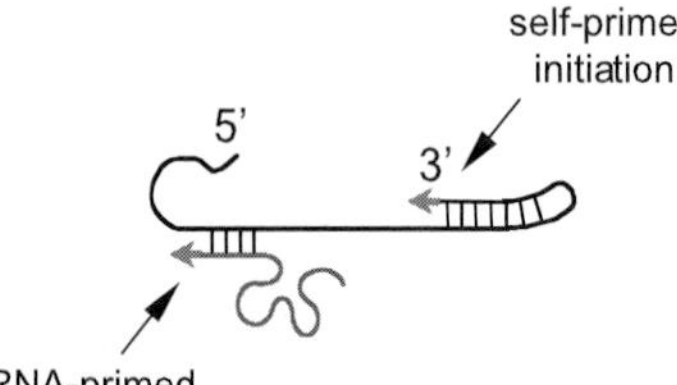

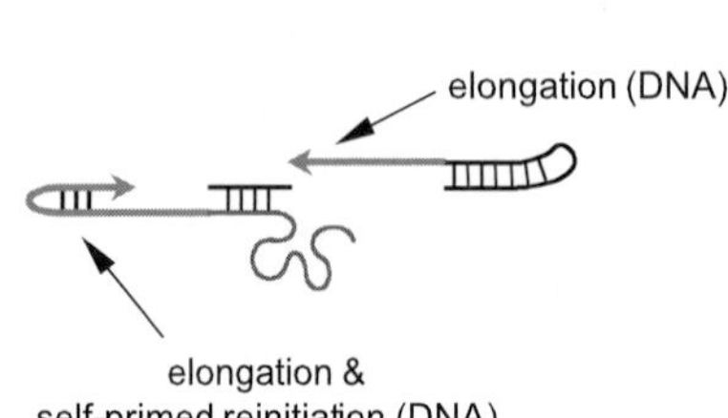

B

RNA template + tRNA + NCp → NC nucleoprotein complex

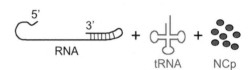

5' 3' RNA + tRNA + NCp

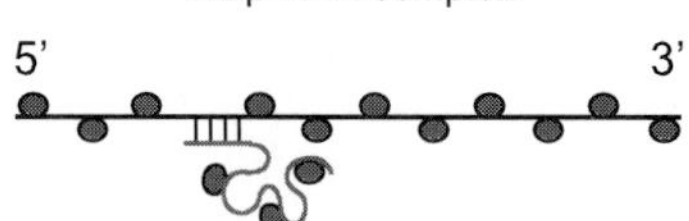

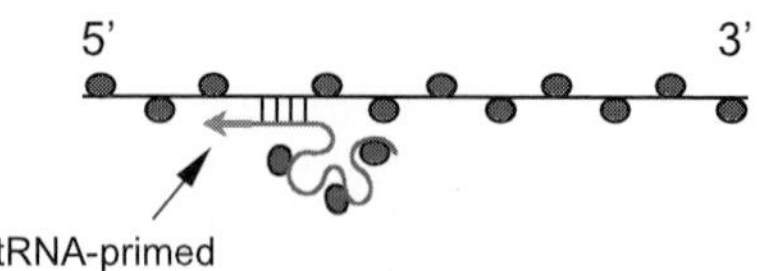

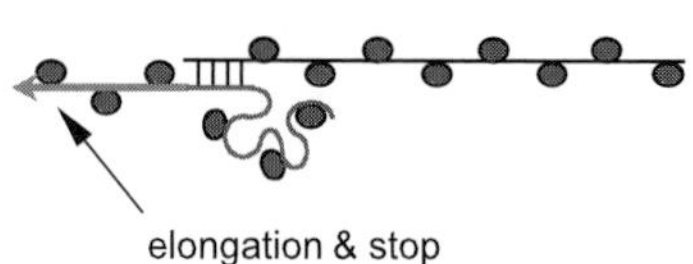

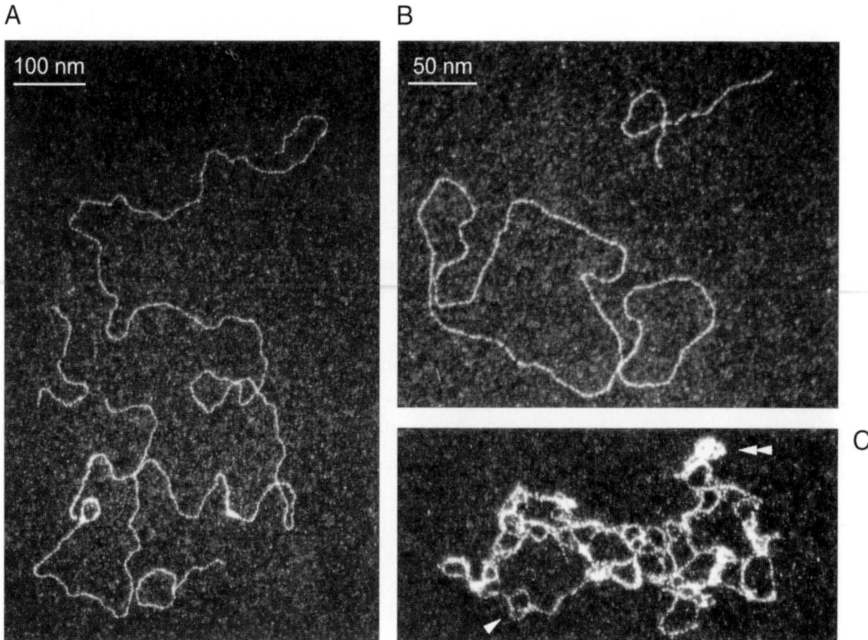

FIGURE 5 Visualization of NCp7–DNA complexes by electron microscopy. Double-stranded DNA (HIV-1 LTR, 1012 bp in length) fragments were incubated 10 min at 20°C in 10 m*M* Tris–Cl, pH 7.5; 50 m*M* NaCl; 0.1 m*M* ZnCl2 at an NCp7: nucleotide ratio of 1:20. Electron microscopic observation of nucleoprotein complexes was performed as previously described (Le Cam and Delain, 1995). (A and B) Electron micrograph shows naked DNA (top) and odd structures (bottom) where NC protein molecules have locally denatured the DNA strands. Linear DNA molecules appear to be joined to one another probably due to NCp7–NCp7 interactions at DNA ends. Circlelike structures can also be seen (B). (C) High molecular weight complexes are shown with local denaturation of the DNA strand (arrowhead) and highly compact nucleoprotein structures (double arrow head). (A) Bar, 100 nm; (B) bar, 50 nm. Electron micrographs kindly provided by Eric Le Cam and Etienne Delain (IGR, Villejuif).

FIGURE 4 A scheme of the nucleic acid chaperoning function of NC protein in the specificity of cDNA synthesis. (A) Viral RNA and primer tRNA are incubated together at 65°C to promote tRNA annealing to the PBS. cDNA synthesis takes place at 37°C after addition of RT and dNTPs and, as well documented (Li *et al.*, 1996; Lapadat-Tapolsky *et al.*, 1997; Rascle *et al.*, 1998), corresponds not only to primer tRNA extension but also to a 3′ extension upon self-initiation. When the RT complex reaches the 5′ end of template RNA, newly made cDNA can eventually fold back and self initiate as indicated. (B) Viral RNA, primer tRNA, and NC protein are incubated at 37°C under physiological conditions, which leads to the formation of NC complexes with tRNA annealed to the PBS. cDNA synthesis takes place only by extension of primer tRNA. RNA template and the newly made cDNA cannot fold back because NCp inhibits intramolecular interactions, as shown.

postinfection). In agreement with these data, NC protein, in the form of NCp15 or NCp7, was reported to interact with the regulatory protein VPR (de Rocquigny *et al.*, 1997; Kondo *et al.*, 1995), believed to be important for the nuclear import of HIV-1 DNA (Nie *et al.*, 1998).

The central globular domain of NCp7 containing the two zinc fingers appears to be critical for proviral DNA synthesis and the stability of the proviral DNA ends since mutating one Zn^{2+} coordinating residue or deleting one zinc finger causes instability of the LTRs (Tanchou *et al.*, 1998; Gorelick *et al.*, 1999). In an attempt to understand the role of the central globular domain in contacting a nucleic acid, the 3D structure of NCp7 complexed with a small oligonucleotide has recently been investigated by NMR. Data shows that the nucleic acid is almost perpendicular to the amino acid sequence linking the two zinc fingers and the 1:1 NCp7–oligonucleotide complex is stabilized by hydrophobic interactions and hydrogen bonds with the side chain of amino acids Val13, Phe16, Thr24, Ala25, Arg26, Arg32, Trp37, Gln45, Met46, and Lys47 (Fig. 7) (De Guzman *et al.*, 1998; Morellet *et al.*, 1998; see also review by Turner and Summers, 1999). Interestingly, in the first finger mutant H23C, residues Val13, Phe16, Thr24, and Ala25 have a different spatial orientation, thus probably preventing their interaction with the nucleic acid (Morellet *et al.*, 1998). These data favor the notion that mutations in the zinc finger might destabilize complexes between NCp7 and proviral DNA explaining, at least in part, why NC zinc finger mutant viruses are replication defective (Lapadat-Tapolsky *et al.*, 1993; Tanchou *et al.*, 1998; Gorelick *et al.*, 1999). Finally, interactions between NCp7 and proviral DNA are probably functionally implicated in provirus integration since NCp7 was reported to strongly enhance integration of HIV-1 LTR DNA into plasmid DNA using integrase *in vitro* (Carteau *et al.*, 1997).

III. Nucleocapsid Protein and Virion Core Assembly

In the infected cells the viral core proteins (MA, CA, and NC) and enzymes (PR, RT, and IN) are synthesized as myristylated GAG and GAG–POL polyprotein precursors (Dickson *et al.*, 1985). Virus assembly is most probably a highly dynamic molecular process (Hansen *et al.*, 1990; Wills and Craven, 1991), starting with the targeting of GAG and GAG–POL precursors to the inner face of the plasma membrane by virtue of the myristate group and the basic residues present at the N-terminus of MA protein and their interaction with acidic phospholipids (Zhou *et al.*, 1994). A further critical interaction takes place, that of GAG and GAG–POL with RNA (Aldovini and Young, 1990; Gorelick *et al.*, 1991; Berkowitz *et al.*, 1993; Darlix *et al.*, 1995; Berkhout, 1996; Berkowitz *et al.*, 1996). In fact, the NC domain of GAG as well as the RT and IN domains of POL can bind to RNA and therefore should be able to recruit the viral genomic RNA and

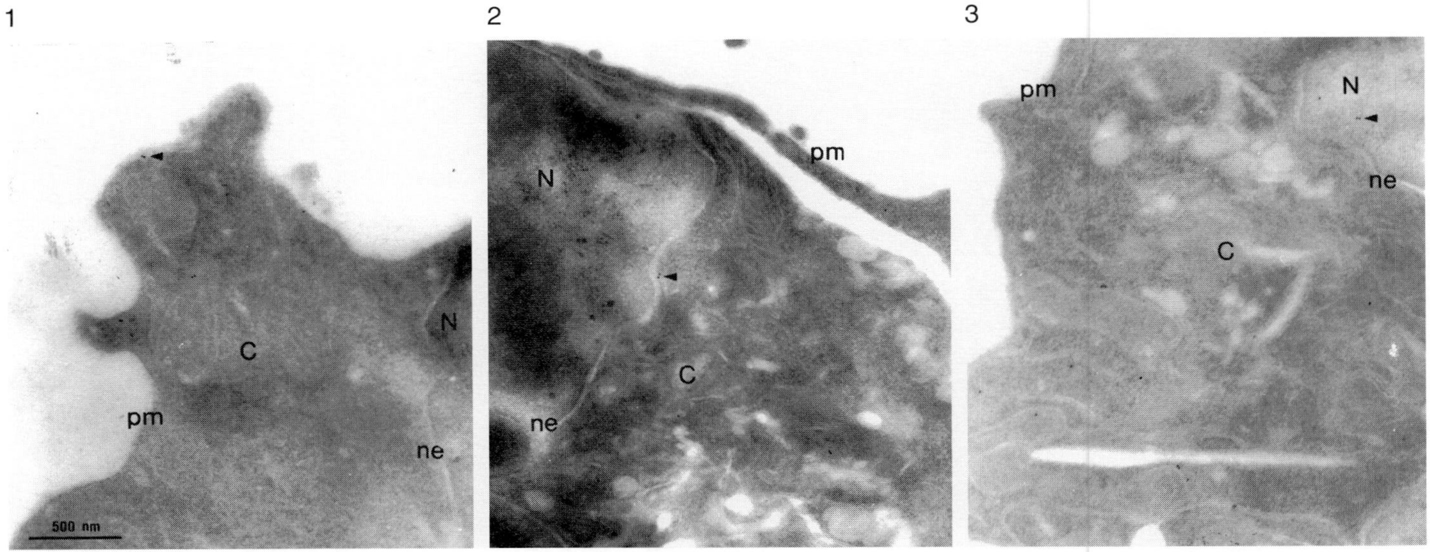

FIGURE 6 Nuclear import of NC protein as observed by immunoelectron microscopy. Human cells were infected with HIV-1 and cells processed for immunodetection using anti-NCp7 monoclonal antibodies 4, 12, and 18 h postinfection (Tanchou *et al.*, 1995). Arrowheads point to grains revealing NC protein at the plasma membrane (**pm** in 1), at the nuclear envelope (**ne** in 2) and in the nucleus (**N** in 3). Bar, 500 nm.

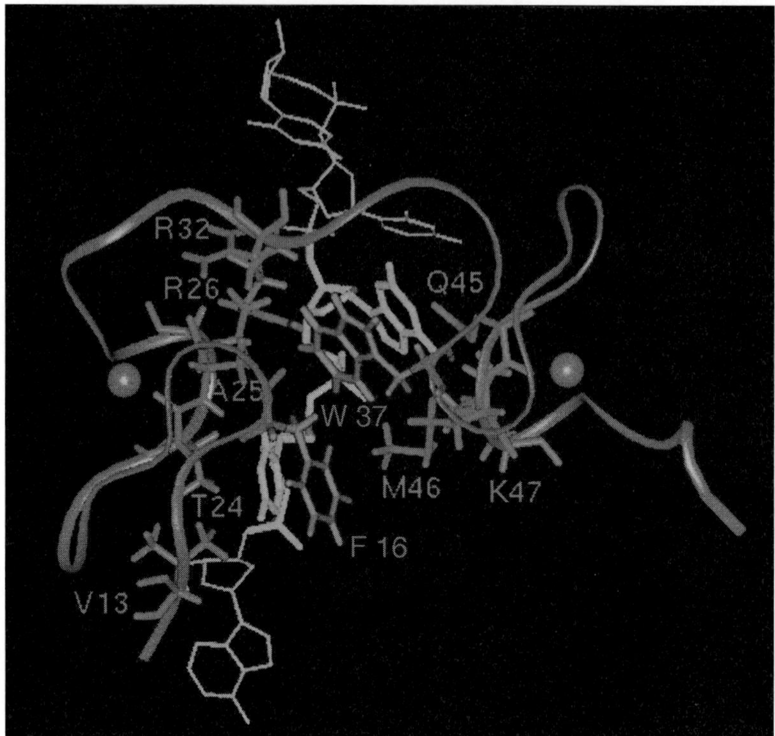

FIGURE 7 Three-dimensional structure of HIV-1 NCp7 complexed with a small oligonucleotide. The complex between (12–53) NCp7 and d (ACGCC) at a ratio of 1:1 was investigated by 1H 2D NMR. As shown in the structure, the nucleic acid is perpendicular to the amino acid sequence linking the two zinc fingers and in bottom-to-top orientation. Residues Vall3, Phe16, Thr24, Ala25, Arg26, Arg32, Trp37, Gln45, Met46, and Lys47 interacting with d(ACGCC) are indicated using the one-letter code (Morellet *et al.*, 1998). Note the stacking of Phe16 onto C2 and Trp37 onto G3 and this appears to be of critical importance *in vivo* (reviewed in Darlix *et al.*, 1995).

cellular tRNAs necessary for virus replication. More specifically, the NC domain can recognize, with some degree of preference, structural determinants of the packaging signal on the genomic RNA (Yang and Temin, 1994; Richardson *et al.*, 1993; McBride and Panganiban, 1997; Harrison *et al.*, 1998), while the RT domain preferentially selects replication primer tRNA (Barat *et al.*, 1989; Huang *et al.*, 1997; Mak and Kleiman, 1997).

GAG and GAG–POL then accumulate on the genomic RNA, causing a drastic increase of their local concentration, which is likely to promote numerous homologous interactions such as CA–CA, NC–NC, PR–PR, RT–RT, and IN–IN (Mammano *et al.*, 1994; Tanchou *et al.*, 1995b). Coating of the genomic RNA by NC is expected to cause RNA dimerization (Prats *et al.*, 1990; Darlix *et al.*, 1990; Harrison *et al.*, 1998) and hence further local accumulation of GAG and GAG–POL precursors. An expected consequence is the dimerization of PR resulting in its activation and the start of GAG and GAG–POL processing (Tracktman and Baltimore, 1982; Kaplan *et al.*, 1993). In agreement with this, partially and completely processed GAG products such as MAp17, CAp25-p24, and NCp15 are found in virus-producing cells (Kaplan *et al.*, 1993, 1994; Pettit *et al.*, 1994). Activation of PR is probably an important factor influencing the rate of virion formation, since inhibition of HIV-1 PR by drugs causes a strong reduction of virion production (Kaplan *et al.*, 1994) as do mutations preventing CA–NC cleavage (Göttlinger *et al.*, 1989; Housset and Darlix, 1996). Cleavage of the MA–CA junction causes the N-terminus of CA to refold into a B-hairpin/helix structure which creates a CA–CA interface essential for core assembly (von Schwedler *et al.*, 1998).

Figure 8 illustrates the sequence of events leading to the formation of a viral core and ultimately to production of a new viral particle by budding (Bolognesi *et al.*, 1978). The genomic RNA appears to be in an open dimeric conformation within the newly formed virion core and then to undergo condensation (Canaani *et al.*, 1973; Stoltzfus and Snyder, 1975; Fu and Rein, 1993), probably necessitating a high concentration of mature NC protein. The central globular structure of NCp7 was found to be important for the condensation step since Zinc finger mutations unfolding the globular structure prevent, at least in part, condensation of the core (see Fig. 2). It was recently found that the same mutations also impair NCp–NCp interactions, which thus may provide an explanation for their impact on core condensation. The start of reverse transcription appears to shortly follow core formation and processing since cDNA transcripts are found in newly made virions (Darlix *et al.*, 1977; Lori *et al.*, 1992; Zhang *et al.*, 1994), and this is thought to necessitate mature RT (Katz and Skalka, 1994; Stewart *et al.*, 1990).

In conclusion, NCp7 of HIV-1, the major structural protein of the nucleocore, has a strong affinity for single- and double-stranded nucleic acids and exhibits nucleic acid chaperoning activities that are critical for core formation due to the ability of NC protein to recruit and package RNA and to form oligomers. At later stages of viral replication, NCp7 chaperones proviral DNA synthesis by reverse transcriptase and possibly provirus integration by integrase.

IV. Phylogenic Relationships between HIV-1 NCp7 and Nucleocapsid Protein of Other Retroviruses

Retroviruses are an abundant family of viruses widespread in vertebrates which, as exogenous agents, cause malignancies, immunodeficiencies, and neurological diseases (for HIV see Ho *et al.*, 1995; Fauci, 1993). Retroviruses have highly divergent nucleotide sequences due to high rates of mutation

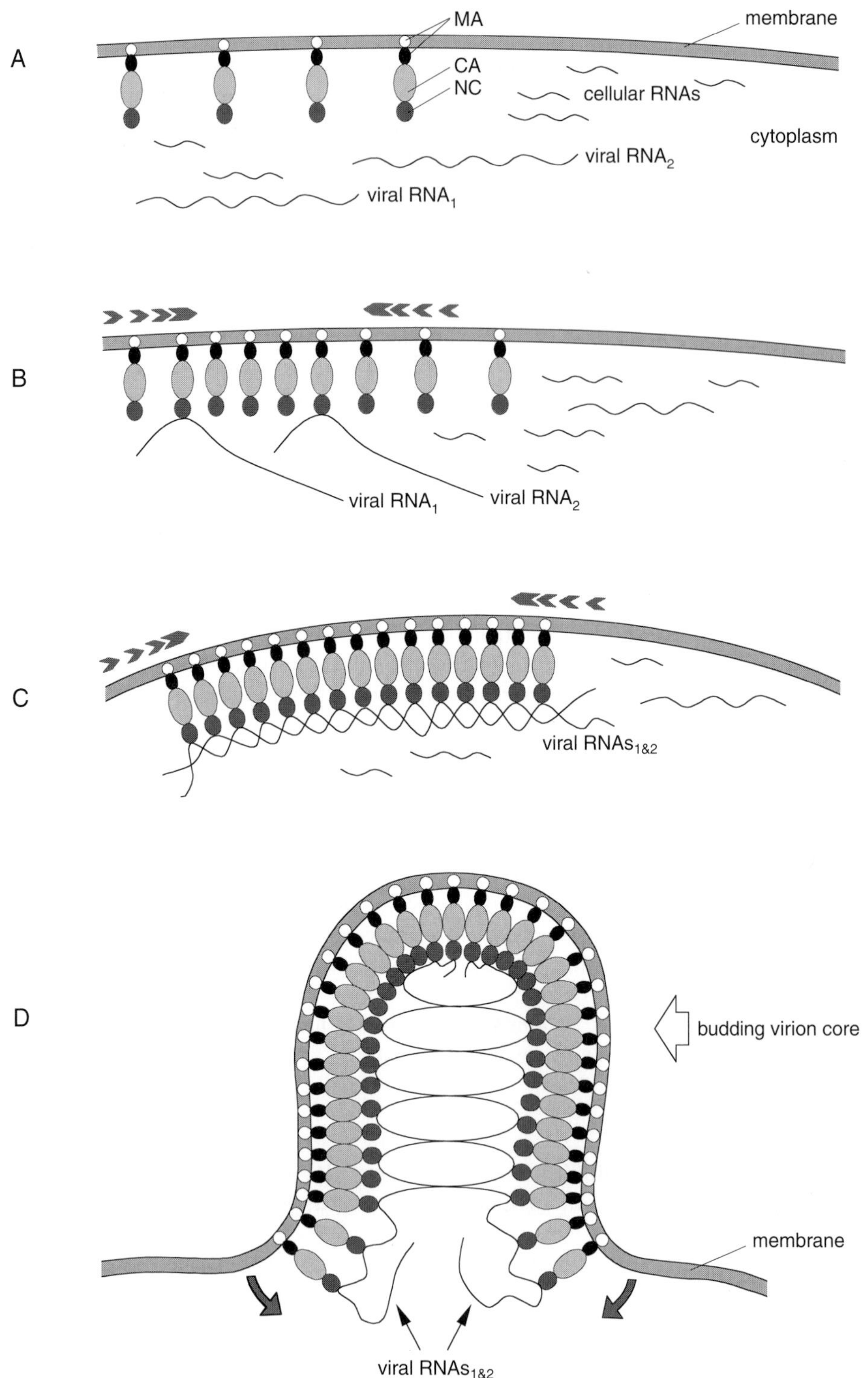

A
MA
membrane
CA
NC
cellular RNAs
cytoplasm
viral RNA2
viral RNA1
B
viral RNA1
viral RNA2
C
viral RNAs1&2
D
budding virion core
membrane
viral RNAs1&2

and recombination during conversion of the single-stranded genomic RNA into a replication-competent viral DNA (Varmus and Swanstrom, 1985; Linial and Blair, 1985; Hu and Temin, 1990a,b; Coffin, 1995). As pointed out above, both the RT enzyme and the NC structural protein are essential to the process of proviral DNA synthesis. Interestingly RT and NC protein are ubiquitous in retroviruses, and gene amplification (PCR) techniques as well as sequencing of RT motifs have been used to identify new members of the retrovirus family (Herniou *et al.,* 1998). These RT motif sequences have also been exploited to study retroviral taxonomy and phylogeny (Xiong and Eickbush, 1990).

The central zinc finger domain of NC protein is ubiquitous among all retroviruses, except for spumaviruses (Darlix *et al.,* 1995), and NC-like proteins have been identified in a family of plant viruses called caulimoviruses (Chapdelaine and Hohn, 1998). Retroviral NC protein deserves much attention from a phylogenic point of view since it possesses both highly conserved amino acids (in the form of the zinc finger), allowing comparison of distantly related sequences, and more variable domains, permitting comparison of closely related sequences.

Amino acid sequence alignements of NC proteins of the major retrovirus genera are shown in Fig. 9 together with that of cauliflower mosaic virus (CaMV) belonging to the plant caulimovirus family (Fig. 8A). Clearly, the Zn^{2+} coordinating residues "CCHC" are entirely conserved. In addition, a glycine before the histidine as well as aromatic and basic residues are conserved in the zinc finger and basic regions flanking the finger motif (Fig. 9A). Interestingly, both the "CCHC" finger motif and the basic residues are critically important in HIV, MLV, and ASLV replication (i.e., Méric *et al.,* 1988; Aldovini and Young, 1990; Dorfman *et al.,* 1993; Deméné *et al.,* 1994; Dupraz and Spahr, 1992; Ottmann *et al.,* 1995; Housset *et al.,* 1993). Figure 9B shows a bootstrapped neighbor-joining tree of the isolates shown in Fig. 9A (from spleen necrosis virus, SNV, to equine infectious anemia virus, EIAV; see legend to Fig. 9). The phylogenic tree was rooted to NC protein of CaMV. It is apparent from this analysis that retroviral NC se-

FIGURE 8 Scheme of virus assembly. The sequence of events schematically illustrating viral core assembly is shown (only GAG precursors). (A) GAG molecules are targeted to the inner face of the plasma membrane due to interaction of the N-terminus of MA protein with acidic phospholipids. (B) The NC domain of GAG recognizes the packaging signal on the genomic RNA (viral RNA 1 and 2). (C) GAG molecules accumulate on the genomic RNA, which increases their local concentration and thus promotes homologous interactions like CA–CA and NC–NC. In addition, coating of viral RNA by NC causes RNA dimerization contributing to the local accumulation of GAG (arrows). (D) High local accumulation of GAG and GAG-POL (not shown here) anchored to the plasma membrane on the one hand, and bound to the RNA on the other results in the formation of the budding virion, processing of GAG and GAG-POL with concomitant initiation of cDNA synthesis (not shown).

A

```
                 10         20         30         40         50         60
SNV        ....QESRAERGSKKTPPGKGRPPLGKN.........QCAYCKEEGHWKKNCPKLVSGAAP 48
GaLV       STGRQTGNLSNQAKKTPR.DGRPPLDKD.........QCAYCKEKGHWARECPRKKHVREA 51
BaEV       AAVVTEKR.AGKSG...ETRRRPKVDKD.........QCAYCKERGHWIKDCPKRPRDQKK 48
MoMuLV     ATVVSGQK.QDRQG...GERRRSQLDRD.........QCAYCKEKGHWAKDCPKKPRGPRG 48
FeLV       ATVVAQNRDKDREENKLGDQRKIPLGKD.........QCAYCKEKGHWVRDCPKRPRKKPA 52
JSRV       ..GKSGNSGCFVCGQPGHRAAVCPQKHQTSVNT..PNLCPRCKKGKHWARDCRSKTDVQGN 57
MMTV       ..QGAEGPVCFSCGKTGHIRKDCKDEKGSKRAP..PGLCPRCKKGYHWKSECKSKFDKDGN 57
MPMV       ..NKEKGGCCFKCGKKGHFAKNCHEHAHNNAEPKVPGLCPRCKRGKHWANECKSKTDNQGN 59
BLV        ..VHTPG.PCYRCLKEGHWARDCPTKTT..GPP..PGPCPICKDPSHWKRDCPTLKSKN.. 52
HTLV-I     ..KPPPNQPCFRCGKAGHWSRDCTQPR....PP..PGPCPLCQDPTHWKRDCPRLKPTIPE 53
HTLV-II    ..RPPPTQPCFRCGKVGHWSRDCTQPR....PP..PGPCPLCQDPSHWKRDCPQLKPPQEE 53
RSV        ..GGRARGLCYTCGSPGHYQAQCPKKRKSGNSR...ERCQLCNGMGHNAKQCRKRDGNQGQ 56
ALV        ..GGRARGLCYTCGSPGHYQAQCPKKRKSGNSR...ERCQLCDGMGHNARQCRKRDGNQGQ 56
FIV        ..SKGSGPVCFNCKKPGHLARQCRDV..........KKCNKCGKPGHLAAKCWQGGKRNSG 49
HIV-1      ..NQRKIVKCFNCGKEGHIARNCRAPR.....K...KGCWKCGKEGHQMKDCTERQANFLG 51
HIV-2      ..QQRKAFKCWNCGKEGHSARQCRAPR.....R...QGCWKCGKPGHIMTNCPDRQAGFLG 51
SIVmac     ..QGRRTVKCWNCGKEGHTAKQCKAPR.....R...QGCWKCGKPGHQMAKCPERQVGFLG 51
CAEV       ..GNGQPQRCYNCGKPGHQARQCRQG..........IICHNCGKRGHMQKECRGKRDIRGK 49
Visna      ..HKGVNQKCYNCGKPGHLARQCRQG..........IICHHCGKRGHMQKDCRQKKQQGNN 49
OMVV       ..KQGV.QKCYYCGKPGHLARQCRQG. ........IICHHCGKRGHMQKDCRQKKGNPTS 48
EIAV       ..PLKAAQTCYNCGKPGHLSSQCRAP..........KVCFKCKQPGHFSKQCRSVPKNGKQ 49

CaMV       ..KFRSGKYFKPKEKKGSKRKYCPKGKK.......DCRCWICNIEGHYANECPNRQSSEKA 52
consensus  ---------**-***-**--****--*------------!--!---*!*-**!*-*------ 61
```

B

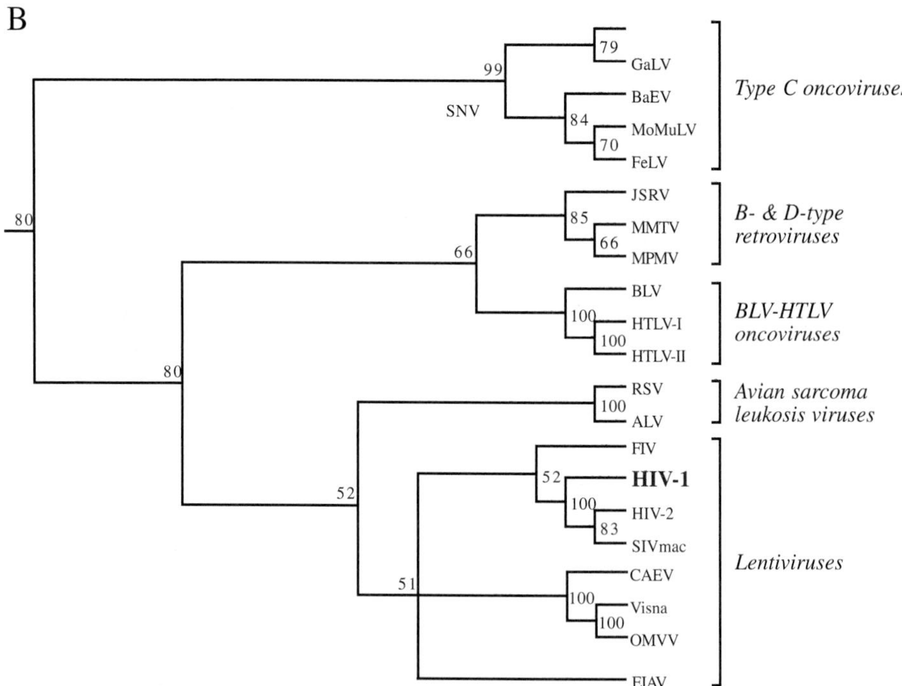

FIGURE 9 Phylogenic relationships between retroviral NC proteins. (A) Amino acid sequence alignment of NC protein of different retrovirus genera as indicated on the left: ALV, avian leukosis virus; BaEV, baboon endogenous virus; BLV, bovine leukemia virus; CAEV, caprine arthritis–encephalitis virus; CaMV, cauliflower mosaic virus; FeLV, feline leukemia virus; FIV, feline immunodeficiency virus; GaLV, gibbon ape leukemia virus; HIV-1, human immunodeficiency virus1; HIV-2, human immunodeficiency virus 2; HTLV-1, human T-cell leukemia virus1; HTLV-2, human T-cell leukemia virus 2; EIAV, equine infectious anemia virus; JSRV, Jaagsiekte retrovirus; MMTV, mouse mammary tumor virus; MoMuLV, Moloney murine leukemia virus; MPMV, Mason-Pfizer monkey virus; OMVV, ovine maedi visna virus;

quences cluster into several main viral groups in a manner very similar to that described by Xiong and Eickbush (1990), whose analysis was based on RT sequences. Such close topological similarities between the NC tree and the RT tree supports the notion that NC protein is a vital cofactor of RT as indicated by biochemical and genetic data (described above). It is interesting to note that single- and double-zinc-finger virus isolates fall into two distinct groups (Fig. 9B), suggesting that the viruses possessing two zinc fingers have probably been generated by duplication of the finger motif.

V. NCp7 as a Target for Anti-HIV Inhibitors

Ideally an anti-HIV drug needs be highly potent, i.e., be able to completely block replication of all HIV subtypes (Chun *et al.,* 1997), should not cause drug resistant viruses to emerge and spread (Coffin, 1995; Wainberg, 1997; Quan *et al.,* 1996), and should be easily accessible and inexpensive. Since 1995/1996 highly active antiretroviral therapies (HAART) (Chun *et al.,* 1997; Li *et al.,* 1998; Autran *et al.,* 1997) have been in use to treat HIV-infected persons. HAARTs impose a triple antiviral drug treatment using two anti-RTs and one anti-PR, known to inhibit both the early and late stages of HIV replication. Most anti-RTs are nucleoside analogs and therefore should block proviral DNA synthesis as soon as one of these analogs is incorporated into cDNA by the error-prone RT enzyme. Anti-PRs are targeted to the active site of the protease and as competitive inhibitors of the proteolytic activity of this viral enzyme they can prevent processing of GAG and GAG–POL precursors and thus formation of mature infectious viruses (Kaplan *et al.,* 1994). HAARTs have been a success as judged by their ability to decrease the viral load by between 50- to more than 100-fold, with substantial recovery of immune functions (Autran *et*

RSV, Rous sarcoma virus; SIVmac, simian immunodeficiency of macacus rhesus; SNV, spleen necrosis virus; Visna, visna virus see also Covey, 1986; Henderson *et al.,* 1992). The one-letter code for amino acids has been used. For *similar* or *conserved* residues the baseline has been set up at 35% similarity or conservation, respectively. For residues under the 35% baseline (or no consensus) for similar residues (* for consensus) and for conserved residues (* for consensus), and for identical residues (! for consensus). Note that all sequences are rich in basic residues and that at least one "CCHC" motif coordinating Zn^{2+} is entirely conserved. (B) This phylogenic tree of vertebrates retroviruses is based on NC protein sequence alignement shown in A, rooted to CaMV NC protein and drawn using the Neighbor-Joining method. The figure on each branch represents percentage of bootstrap support (from 1000 replicates) and the unsolved branches (<50% bootstrap support) have been collapsed. The retrovirus groups are indicated on the right with the single zinc finger group at the top (see also Xiong and Eickbush, 1990; Herniou *et al.,* 1998).

al., 1997) and improvements in the life expectancy and health of HIV-infected persons.

However, HAART is a permanent daily treatment, with a large number of tablets to be taken at very regular intervals, and HIV cannot be eradicated from the individual. A growing number of HIV-infected persons under HAART experience either drug intolerance or resistance and side effects. In fact, resistance to anti-RT drugs like AZT, DDI, DDC, D4T, and 3TC have been described and correspond to mutations mapping in the RT coding sequence, more precisely in the active and nucleotide site domains (reviewed in Johnson, 1995). Similarly, treatments with anti-PR generate a large panel of anti-PR resistant viruses with mutations mapping in the PR coding sequence and/or in the GAG processing sites (Doyon *et al.*, 1998; Zhang *et al.*, 1994; Dulioust *et al.*, 1999).

Since HAART is not yet able to cure HIV infection, mutant viruses resistant to such therapies probably emerge for a number of reasons; for example, (1) nucleoside analogs must be converted into their triphosphonucleotide counterpart by cellular enzymes and then be constantly available at the site of proviral DNA synthesis; (2) HIV is a highly replicating virus, formed of a population of quasispecies (Pezo and Wain-Hobson, 1997), with potentially varying sensitivity to a given drug (Wainberg, 1997); and (3) the viral population facing the immune response and drug treatments remains highly dynamic due to the RT–NC protein replicating complex being able to incorporate a large number of errors in newly synthesized viral DNAs and to combine mutations present in two different genotypes during recombination events (Hu and Temin, 1990; Darlix *et al.*, 1995; Lapadat-Tapolsky *et al.*, 1997). Finally, due to the very nature of HAART, the level of compliance and thus the effective pressure on viral replication in persons receiving treatment invariably wavers with time.

Despite the fact that some of the functions of NCp7 have been known for a decade (Barat *et al.*, 1989), little attention has been paid to the development of anti-NCp inhibitors. As is clear from the above, NCp7 chaperones proviral DNA synthesis and viral genetic variability when part of a complex with RT by promoting mutations and recombination events. NCp7 also acts as one of the assembly domains of GAG. The central domain of NCp7 appears to be the natural target of choice for antiviral drugs since this domain is required early as well as late in HIV replication, is entirely conserved in all HIV subtypes (Myers *et al.*, 1995), and possesses a well-defined globular structure (see above).

Two groups have recently started to develop anti-NCp7 inhibitors capable of actively interacting with the central globular domain of NC protein. Rice and collaborators have shown that compounds such as the DIBAs can eject Zn^{2+} ions from free HIV-1 NCp7 or MoMuLV NCp10 (Tummino *et al.*, 1996; Huang *et al.*, 1998). When added during virion formation DIBA-1 can generate dimeric NC proteins in which NC protein monomers are linked by a disulfide bridge, resulting in the production of noninfectious viral particles (Turpin *et al.,* 1996). Interestingly, DIBAs do not affect the biological activity of other zinc finger enzymes such as poly-ADP-ribozylase and a number of transcription factors like Sp1 (Huang *et al.,* 1998). Interestingly, DIBA-2 does not inhibit replication of human spumaretrovirus (HSRV), which lacks an NC–zinc finger protein (Rein *et al.,* 1996).

The impact of several NC protein inhibitors on HIV-1 replication has been extensively investigated in the laboratory. The most potent inhibitor was found to be DIBA-1, while ADA (azodicarbonamide) did not exhibit any inhibitory effect. Interestingly, DIBA-1 appears to modify the assembly process by impairing virion core maturation in a manner similar to that of an antiprotease, and the virus produced possesses an "immature" morphology (Berthoux, 1998). In addition, proviral DNA synthesis is partially inhibited both at the levels of minus- and plus-strand synthesis, in agreement with the functions described for NCp7 (see above).

However, DIBA-1 is poorly water soluble and rather unstable *in vivo.* This is why new derivatives of DIBA, called PATE, are being developed (Turpin *et al.,* 1999). Using molecular modeling and peptide synthesis techniques (Morellet *et al.,* 1994, 1998), Roques and colleagues have synthesized cyclic peptides capable of binding to NCp7 zinc fingers, causing an inhibition of RT/NCp7 interactions and therefore impairing proviral DNA synthesis (Le Druillenec *et al.,* 1999).

VI. Future Prospects

The findings on retroviral NC protein, and in particular on HIV-1 NCp7, summarized above, result from joint efforts in laboratories in both Europe and North America. Molecular, biochemical, biological, and virological data all clearly show that NC protein is a key component of the virus and chaperones its replication. In the future it is hoped that the effort aimed at the development and discovery of new anti-NCp7 drugs will continue and be reinforced so that compounds aimed at blocking the exceedingly conserved central globular domain of NCp7 may be employed in clinical trials.

It is surprising that so relatively little is known about the immunobiology of NCp7. Analyses carried out in the laboratory indicate that less than 10% of HIV-1(+) persons in France raise antibodies against NCp7 (Rogemond and Darlix, unpublished data), and yet the reason for this is not known. On the other hand, it has been possible to generate anti-NCp7 monoclonal antibodies (mAbs) directed against either a linear NC epitope or certain NC protein conformations (Tanchou *et al.,* 1995a). Interestingly, the conformational anti-NCp7 mAbs can recognize free NCp7 but not when it is part of the Pr55gag precursor (Tanchou *et al.,* 1995a), suggesting that NC protein undergoes conformational and functional changes during GAG processing as does capsid protein (Bacharach and Goff, 1998; von Schwedler *et al.*, 1998). This area of research is presently receiving further attention.

The molecular structure of free NCp7, or NCp7 complexed with a very small nucleic acid, has been elucidated by NMR (South *et al.*, 1990; South and Summers, 1993; Morellet *et al.*, 1992, 1994, 1998; De Guzman *et al.*, 1998). The 3D structure of MoMuLV NCp10 has also been investigated by NMR, revealing a central globular domain with a single zinc finger (Deméné *et al.*, 1994a). In an attempt to understand the molecular basis of NC protein-packaging RNA sequence interactions and specificity, it will be of the utmost interest to compare the structure of NCp7 to that of NC protein of other retroviruses such as HTLV and RSV, either in the free form or in a complex with the homologous or a heterologous packaging sequence (for the second structure of HIV-1 leader RNA see Baudin *et al.*, 1993; Berkhout, 1996). Investigations of the organization of NC protein molecules along the genomic RNA and the dynamics of these functional interactions are underway (Tanchou *et al.*, 1995b; Lener *et al.*, 1998; Le Cam, 1998); however, much remains to be learned of NC protein–nucleic acid interactions in virus formation, structure, stability, and proviral DNA synthesis.

A further very interesting avenue of research on NC protein is to understand how and what contacts are made with cellular components during entry of the viral nucleocore into the cytoplasma of the newly infected cell. HIV-1 NC protein appears to associate with actin (Wilk *et al.*, 1999) and this might be linked to proviral DNA synthesis in the cytoplasm as well as its nuclear import (Jacque *et al.*, 1998). Recently, we have been able to identify SmD1 as a cellular protein partner of HIV-1 NCp7. SmD1 is a protein ubiquitous from yeast to humans, a component of the splicing machinery, and found to be important for HIV-1 virus formation. Interestingly SmD1 is present in HIV-1 viral particles; its role as a virion component is presently under investigation.

Acknowledgments

We acknowledge Eric Le Cam and Etienne Delain (IGR Villejuif) for examination of NCp7-LTR DNA nucleoprotein complexes under the electron microscope. Thanks are due to the "Agence Nationale de Recherches sur le SIDA" (ANRS), the "Mutuelle Générale de l'Education Nationale" (MGEN) and SIDACTION for their continuous support.

References

Aldovini, A., and Young, R. (1990). Mutations of RNA and protein sequences involved in human immunodeficiency virus type 1 packaging result in production of noninfectious virus. *J. Virol.* **64**, 1920–1926.

Allain, B., Lapadat-Tapolsky, M., Berlioz, C., and Darlix, J.-L. (1994). Trans-activation of the minus-strand DNA transfer by nucleocapsid protein during reverse transcription of the retroviral genome. *EMBO J.* **13**, 973–981.

Autran, B., Carcelain, G., Li, T., Blanc, C., Mathez, D., Tubiana, R., Katlama, C., Debre, P., and Leibowitch, J. (1997). Positive effects of combined antiretroviral therapy on CD4+ T cell homeostasis and function in advanced HIV disease. *Science* **277**, 112–116.

Auxilien, S., Keith, G., Le Grice, S., and Darlix, J. L. (1999). Role of post-transcriptional modifications of primer tRNAlys, 3 in the fidelity and efficacy of plus strand DNA transfer during HIV-1 reverse transcription. *J. Biol. Chem.* **274**, 4412–4420.

Bacharach, E., and Goff, S. (1998). Binding of the human immunodeficiency virus type 1 Gag protein to the viral RNA encapsidation signal in the yeast three-hybrid system. *J. Virol.* **72**, 6944–6949.

Barat, C., Lullien, V., Schatz, O., Grüninger-Leitch, M., Le Grice, S., Nugeyre, M.-T., Barré-Sinoussi, F., and Darlix, J.-L. (1989). HIV-1 reverse transcriptase specifically interacts with the anti-codon domain of its cognate primer tRNA. *EMBO J.* **8**, 3279–3285.

Baudin, F., Marquet, R., Isel, C., Darlix, J.-L., Ehresmann, B., and Ehresmann, C. (1993). Functional sites in the 5′ region of HIV1 RNA form defined structural domains. *J. Mol. Biol.* **229**, 382–397.

Berkhout, B. (1996). Structure and function of the human immunodeficiency virus leader RNA. *Prog. Nucleic Acid Res. Mol. Biol.* **54**, 1–34.

Berkowitz, R., and Goff, S. (1994). Analysis of binding elements in HIV-1 genomic RNA and nucleocapsid protein. *Virology* **202**, 233–246.

Berkowitz, R., Fisher, J., and Goff, S. (1996). RNA packaging. *Curr. Top. Microbiol. Immunol.*, **214**, 177–218.

Berkowitz, R., Luban, J., and Goff, S. (1993). Specific binding of HIV type 1 gag polyprotein and nucleocapsid protein to viral RNAs detected by RNA mobility shift assays. *J. Virol.* **67**, 7190–7200.

Berthoux, L. (1998). "Fonctions de la NCp7 du HIV-1 dans la réplication virale, et le mécanisme d'action d'inhibiteurs de type DIBA." Thesis Ecole Normale Supérieure de Lyon, France.

Berthoux, L., Péchoux, C., Ottmann, M., Morel, G., and Darlix, J.-L. (1997). Mutations in the N-terminal domain of HIV-1 nucleocapsid protein affect virion core structure and proviral DNA synthesis. *J. Virol.* **71**, 6973–6981.

Bieth, E., Gabus, C., and Darlix, J.-L. (1990). A study of the dimer formation of Rous sarcoma virus RNA and of its effect on viral protein synthesis *in vitro*. *Nucleic Acids Res.* **18**, 119–126.

Bolognesi, D. P., Montelaro, R. C., Franck, H., and Shäfer, W. (1978). Assembly of type C oncorna virus: A model. *Science* **199**, 183–186.

Canaani, E., von Der Helm, K., and Duesberg, P. (1973). Evidence for 30-40S RNA as precursor of the 60-70S RNA of Rous sarcoma virus. *Proc. Natl. Acad. Sci. USA* **72**, 401–405.

Carteau, S., Batson, S. C., Poljak, L., Mouscadet, J. F., de Rocquigny, H., Darlix, J. L., Roques, B. P., Kas, E., and Auclair, C. (1997). Human immunodeficiency virus type 1 nucleocapsid protein specifically stimulates Mg2+-dependent DNA integration *in vitro*. *J. Virol.* **71**, 6225–6229.

Chapdelaine, Y., and Hohn, T. (1998). The cauliflower mosaic virus capsid protein: Assembly and nucleic acid binding *in vitro*. *Virus Genes* **17**, 139–150.

Chen, M., Garon, C., and Papas, T. (1980). Native ribonucleoprotein is an efficient transcriptional complex of avian myeloblastosis virus. *Proc. Natl. Acad. Sci. USA* **77**, 1296–1300.

Chun, T., Stuyver, L., Mizell, S., Ehler, L., Mican, J., Baseler, M., Lloyd, A., Nowak, M., and Fauci, A. (1997). Presence of an inducible HIV-1 latent reservoir during highly active antiretroviral therapy. *Proc. Natl. Acad. Sci. USA* **94**, 13193–13197.

Coffin, J. (1985). Genome structure, p. 17–74. *In* "RNA Tumor Viruses." (R. Weiss, N. Teich, H. Varmus, and J. Coffin, Eds.), Part 2, 2nd ed. Cold Spring Harbor Laboratory, Cold Spring Harbor, New York.

Coffin, J. (1995). HIV population dynamics *in vivo:* Implications for genetic variation, pathogenesis, and therapy. *Science* **267**, 483–489.

Covey, S. N. (1986). Amino acid sequence homology in *GAG* region of reverse transcribing elements and the coat protein gene of cawliflower mosaic virus. *Nucleic Acids Res.* **14**, 623–633.

Dannull, J., Surovoy, A., Jung, G., and Moelling, K. (1994). Specific binding of HIV-1 nucleocapsid protein to PSI RNA *in vitro* requires N-terminal zinc finger and flanking basic amino acid residues. *EMBO J.* **13**, 1525–1533.

Darlix, J.-L. (1986). Control of RSV genome translation and packaging by the 5′ and 3′ untranslated sequences. *J. Mol. Biol.* **189**, 421–434.

Darlix, J.-L., and Spahr, P.-F. (1982). Binding sites of viral protein p19 onto RSV RNA and possible control of viral functions. *J. Mol. Biol.* **160**, 147–161.

Darlix, J.-L., Bromley, P. A., and Spahr, P. F. (1977). New procedure for the direct analysis of *in vitro* reverse transcription of Rous sarcoma virus RNA. *J. Virol.* **22**, 118–129.

Darlix, J.-L., Gabus, C., Nugeyre, M.-T., Clavel, F., and Barré-Sinoussi, F. (1990). Cis elements and trans acting factors involved in the RNA dimerization of HIV-1. *J. Mol. Biol.* **216**, 689–699.

Darlix, J.-L., Lapadat-Tapolsky, M., de Rocquigny, H., and Roques, B. (1995). First glimpses at structure-function relationships of the Nucleocapsid Protein of Retroviruses. Review article. *J. Mol. Biol.* **254**, 523–537.

Darlix, J.-L., Vincent, A., Gabus, C., de Rocquigny, H., and Roques, B. (1993). Trans-activation of the 5′ to 3′ viral DNA strand transfer by nucleocapsid protein during reverse transcription of HIV1 RNA. *Compte Rendus Acad. Sci. Life Sci.* **316**, 763–771.

Davis, N., and Rueckert, R. (1972). Properties of a ribonucleoprotein particle isolated from NP 40 treated RSV. *J. Virol.* **10**, 1010–1020.

De Guzman, R., Wu, Z. R., Stalling, C., Pappalardo, L., Borer, P., and Summers, M. (1998). Structure of the HIV-1 nucleocapsid protein bound to the SL3 psi-RNA recognition element. *Science* **279**, 384–388.

Deméné, H., Dong, C., Ottmann, M., Rouyez, M., Jullian, N., Morellet, N., Mely, Y., Darlix, J.-L., Fournié-Zaluski, M.-C., Saragosti, S., and Roques, B. (1994b). 1H NMR structure and biological studies of the His23 >Cys mutant nucleocapsid protein of HIV-1 indicate that the conformation of the first zinc finger is critical for virus infectivity. *Biochemistry* **33**, 11707–11716.

Deméné, H., Jullian, N., Morellet, N., de Rocquigny, H., Cornille, F., Maigret, B., and Roques, B. (1994a). Three-dimensional 1H NMR structure of the nucleocapsid protein NCp10 of MoMuLV. *J. Biomol. NMR* **4**, 153–170.

De Rocquigny, H., Gabus, C., Vincent, A., Fournié-Zaluski, M.-C., Roques, B., and Darlix, J.-L. (1992). Viral RNA annealing activities of HIV-1 nucleocapsid protein require only peptide domains outside the zinc fingers. *Proc. Nat. Acad. Sci. USA* **89**, 6472–6476.

De Rocquigny, H., Petitjean, P., Tanchou, V., Decimo, D., Drouot, L., Delaunay, T., Darlix, J. L., Roques, B. (1997). The zinc fingers of HIV nucleocapsid protein NCp7 direct interactions with the viral regulatory protein Vpr. *J. Biol. Chem.* **272**, 30753–30759.

Dib-Hajj, F., Khan, R., and Giedroc, D. P. (1993). Retroviral nucleocapsid proteins possess potent nucleic acid strand renaturation activity. *Prot. Sci.* **2**, 231–243.

Dickson, C., Eisenman, R., Fan, H., Hunter, E., and Reich, N. (1985). Protein biosynthesis and assembly. *In* "RNA Tumor Viruses" (R. Weiss, N. Teich, H. Varmus, and J. Coffin, Eds.), Part 2, 2nd ed, pp. 513–648. Cold Spring Harbor Laboratory, Cold Spring Harbor, New York.

Dorfman, T., Luban, J., Goff, S., Haseltine, W., and G. Göttlinger, H. (1993). Mapping of functionally important residues of a cysteine-histidine box in the human immunodeficiency virus type-1 nucleocapsid protein. *J. Virol.* **67**, 6159–6169.

Doyon, L., Payant, C., Brakier-Gingras, L., and Lamarr, E. D. (1998). Novel Gag-Pol frameshift site in human immunodeficiency virus type 1 variants resistant to protease inhibitors. *J. Virol.* **72**, 6146–6150.

Dupraz, P., and Spahr, P.-F. (1992). Specificity of Rous sarcoma virus nucleocapsid protein in genomic RNA pacaking. *J. Virol.* **66,** 4662–4670.

Dulioust, A., Paulous, S., Guillemot, L., Delavalle, A. M., Boue, F., and Clavel, F. (1999). Constrained evolution of human immunodeficiency virus type 1 protease during sequential therapy with two distinct protease inhibitors. *J. Virol.* **73,** 850–854.

Fauci, A. S. (1993). Immunopathogenesis of HIV Infection. *J. Acq. Immun. Defic. Syndr.* **6,** 655–662.

Fu, W., and Rein, A. (1993). Maturation of dimeric viral RNA of MoMuLV. *J. Virol.* **67,** 5443–5449.

Gabus, C., Ficheux, D., Rau, M., Keith, G., Sandmeyer, S., and Darlix, J.-L. (1998). The yeast Ty3 Retrotransposon contains a bipartite primer binding site and encodes nucleocapsid protein NCp9 functionally homologous to HIV-1 NCp7. *EMBO J.* **17,** 4873–4880.

Gallay, P., Swingler, S., Song, J., Bushman, F., and Trono, D. (1995). HIV nuclear import is governed by the phosphotyrosine-mediated binding of matrix to the core domain of integrase. *Cell* **83,** 569–576.

Gelderblom, H., Hausmann, E., Ozel, M., Pauli, G., and Koch, M. (1987). Fine structure of Human Immunodeficiency Virus (HIV) and immunolocalization of structural proteins. *Virology* **156,** 171–176.

Gilboa, E., Goff, S., Shields, A., Yoshimura, F., Mitra, S., and Baltimore, D. (1979). *In vitro* synthesis of a 9 Kbp terminally redundant DNA carrying the infectivity of MoMuLV. *Cell* **16,** 863–874.

Gorelick, R., Gagliardi, T., Bosche, W., Wiltrout, T., Coren, L., Chabot, D., Lifson, J., Henderson, L., and Arthur, L. (1999). Strict conservation of the retroviral nucleocapsid (NC) protein zinc finger is strongly influenced by its role in viral infection processes: Characterization of HIV-1 particles containing mutant NC zinc-coordinating sequences. *Virology* (in press).

Gorelick, R. J., Chabot, D. J., Rein, A., Henderson, L. E., and Arthur, L. O. (1993). The two zinc fingers in the immunodeficiency virus type 1 nucleocapsid protein are not functionally equivalent. *J. Virol.* **67,** 4027–4036.

Gorelick, R. J., Henderson, L. E., Hanser, J. P., and Rein, A. (1991). "Roles of Nucleocapsid Cysteine Arrays in Retroviral Assembly and Replication: Possible Mechanisms in RNA Replication Advances in Molecular Biology and Targeted Treatment for AIDS" (A. Kumar, Ed.). Plenum, New York.

Göttlinger, H. G., Sodroski, J. G., and Haseltine, W. (1989). Role of capsid precursor processing and myristoylation in morphogenesis and infectivity of human immunodeficiency virus type-1. *Proc. Natl. Acad. Sci. USA* **86,** 5781–5785.

Green, L., and Berg, J. (1990). Retroviral nucleocapsid protein–ion interactions: Folding and sequence variants. *Proc. Natl. Acad. Sci.* (Wash.) **87,** 6403–6407.

Gregoire, C. J., Gautheret, D., and Loret, E. (1997). No tRNA3Lys unwinding in a complex with HIV NCp7. *J. Biol. Chem.* **272,** 25143–25148.

Guo, J., Henderson, L. E., Bess, J., Kane, B., and Levin, J. G. (1997). Human immunodeficiency virus type 1 nucleocapsid protein promotes efficient strand transfer and specific viral DNA synthesis by inhibiting TAR-dependent self-priming from minus-strand strong-stop DNA. *J. Virol.* **71,** 5178–5188.

Hansen, M., Jelinek, L., Whiting, S., and Barklis, E. (1990). Transport and assembly of gag proteins into MoMuLV. *J. Virol.* **64,** 5306–5316.

Harrison, G., and Lever, A. (1992). The HIV-1 packaging signal and major splice donor site region have a conserved stable secondary structure. *J. Virol.* **66,** 4144–4153.

Harrison, G., Miele, G., Hunter, E., and Lever, A. (1998). Functional analysis of the core human immunodeficiency virus type 1 packaging signal in a permissive cell line. *J. Virol.* **72,** 5886–5896.

Henderson, L., Bowers, M., Sowder, R., *et al.* (1992). Gag proteins of the highly replicative MN strain of HIV-1: Posttranslational modifications, proteolytic processings and complete amino acid sequences. *J. Virol.* **66,** 1856–1865.

Herniou, E., Martin, J., Miller, K., Cook, J., Wilkinson, M., and Tristem, M. (1998). Retroviral diversity and distribution in vertebrates. *J. Virol.* **72**, 5955–5966.

Ho, D., Neumann, A., Perelson, A., Chen, W., Leonard, J., and Markowitz, M. (1995). Rapid turnover of plasma virions and CD4 lymphocytes in HIV-1 infection. *Nature* **373**, 123–126.

Housset, V., and Darlix, J. L. (1996). Mutations at the capsid-nucleocapsid cleavage site of gag polyprotein of Moloney murine leukemia virus abolish virus infectivity. *C. R. Acad. Sci. III* **319**, 81–89.

Housset, V., de Rocquigny, H., Roques, B., and Darlix, J.-L. (1993). Basic amino acids flanking the zinc finger of Moloney Murine Leukaemia Virus are critical for virus infectivity. *J. Virol.* **67**, 2537–2545.

Hu, W., and Temin, H. (1990a). Genetic consequence of packaging two RNA genomes in one retroviral particle: Pseudodiploïdy and high rate of genetic recombination. *Proc. Natl. Acad. Sci. USA* **87**, 1556–1560.

Hu, W., and Temin, H. (1990b). Retroviral recombination and reverse transcription. *Science* **250**, 1227–1233.

Huang, Y., Khorchid, A., Wang, J., Parniak, M. A., Darlix, J. L., Wainberg, M., and Kleiman, L. (1997). Effect of mutations in the nucleocapsid protein (NCp7) upon Pr160 (gag-pol) and tRNA (Lys) incorporation into human immunodeficiency virus type 1. *J. Virol.* **71**, 4378–4384.

Huang, M., Maynard, A., Turpin, J. A., Graham, L., Janini, G. M., Covell, D. G., and Rice, W. (1998). Anti-HIV agents that selectively target retroviral nucleocapsid protein zinc fingers without affecting cellular zinc finger proteins. *J. Med. Chem.* **41**, 1371–1381.

Jacque, J., Mann, A., Enslen, H., Sharova, N., Brichacek, B., Davis, R. J., and Stevenson, M. (1998). Modulation of HIV-1 infectivity by MAPK, a virion-associated kinase. *EMBO J.* **17**, 2607–2618.

Johnson, V. (1995). Nucleoside reverse transcriptase inhibitors and resistance of HIV-1. *J. Infect. Dis.* **171**, S140–S149.

Kaplan, A., Manchester, M., and Swanstrom, R. (1994). The activity of the protease of HIV-1 is initiated at the membrane of infected cells before the release of viral proteins and is required for the release to occur with maximum efficiency. *J. Virol.* **68**, 6782–6786.

Kaplan, A. H., Zack, J. A., Knigge, M., Paul, D. A., Kempf, D. J., Norbeck, D., W., and Swanstrom, R. (1993). Partial inhibition of the human immunodeficiency virus type 1 protease results in aberrant assembly and the formation of noninfectious particles. *J. Virol.* **67**, 4050–4055.

Katz, R., and Skalka, A. M. (1994). The retroviral enzymes. *Annu. Rev. Biochem.* **63**, 133–173.

Khan, R., and Giedroc, D. P. (1992). Recombinant human immunodeficiency virus type 1 nucleocapsid (NCp7) protein unwinds tRNA. *J. Biol. Chem.* **267**, 6689–6695.

Kondo, E., Mammano, F., Cohen, E., and Gottlinger, H. (1995). The p6gag domain of human immunodeficiency virus type 1 is sufficient for the incorporation of Vpr into heterologous viral particles. *J. Virol.* **69**, 2759–2764.

Lapadat-Tapolsky, M., de Rocquigny, H., van Gent, D., Roques, B., Plasterk, R., and Darlix, J.-L. (1993). Interactions between HIV-1 nucleocapsid protein and viral DNA may have important functions in the viral life cycle. *Nucleic Acids Res.* **21**, 831–839.

Lapadat-Tapolsky, M., Gabus, C., Rau, M., and Darlix, J-L. (1997). Possible roles of HIV-1 Nucleocapsid protein in the specificity of proviral DNA synthesis and in its variability. *J. Mol. Biol.* **268**, 250–260.

Lapadat-Tapolsky, M., Pernelle, C., Borie, C., and Darlix, J.-L. (1995). Analysis of the nucleic acid annealing activities of nucleocapsid protein from HIV-1. *Nucleic Acids Res.* **23**, 2434–2441.

Le Cam, E., and Delain, E. (1995). Nucleic acids–ligand interactions. In "Visualization of Nucleic Acids" (G. Morel, ed.), pp. 333–358. CRC Press, Boca Raton, FL.

Le Cam, E., Coulaud, D., Delain, E., Petitjean, P., Roques, B., Gerard, D., Stoylova, E., Vuilleumier, C., Stoylov, S. P., and Mely, Y. (1998). Properties and growth mechanism of the ordered aggregation of a model RNA by the HIV-1 nucleocapsid protein: An electron microscopy investigation. *Biopolymers* **45**, 217–229.

Lener, D., Tanchou, V., Roques, B. P., Le Grice S., and Darlix, J. L. (1998). Involvement of HIV-1 NC protein in the recruitment of RT into nucleoprotein complexes formed *in vitro*. *J. Biol. Chem.* **273**, 33781–33786.

Lever, A., Göttlinger, H., Haseltine, W., and Sodroski, J. (1989). Identification of a sequence required for efficient packaging of HIV-1 RNA into virions. *J. Virol.* **63**, 4085–4087.

Li, T. S., Tubiana, R., Katlama, C., Calvez, V., Ait Mohand, H., and Autran, B. (1998). Long-lasting recovery in CD4 T-cell function and viral-load reduction after highly active antiretroviral therapy in advanced HIV-1 disease. *Lancet* **351**, 1682–1686.

Li, X., Quan, Y., Arts, E., Li, Z., Preston, B., de Rocquigny, H., Roques, B., Darlix, J.-L., Kleiman, L., Parniak, M., and Wainberg, M. (1996). HIV type 1 nucleocapsid protein (NCp7) directs specific initiation of minus strand DNA synthesis primed by human tRNALys,3 *in vitro:* Studies of viral RNA molecules mutated in regions that flank the primer binding site. *J. Virol.* **70**, 4996–5004.

Linial, M., and Blair, D. (1985). Genetic recombination. *In* "RNA Tumor Viruses" (R. Weiss, N. Teich, H. Varmus, and J. Coffin, Eds.), 2nd ed., pp. 719–734. Cold Spring Harbor Laboratory, Cold Spring Harbor, New York.

Lori, F., Veronese, F., Devico, A., Lusso, P., Reitz, M., and Gallo, R. (1992). Viral DNA carried by HIV-1 virions. *J. Virol.* **66**, 5067–5074.

Mak, J., and Kleiman, L. (1997). Primer tRNAs for reverse transcription. *J. Virol.* **71**, 8087–8095.

Mammano, F., Öhagen, A., Höglung, S., and Göttlinger, H. (1994). Role of the major homology region of HIV-1 in virion morphogenesis. *J. Virol.* **68**, 4927–4936.

McBride, M., and Panganiban, A. (1997). Position dependence of functional hairpins important for human immunodeficiency virus type 1 RNA encapsidation *in vivo*. *J. Virol.* **71**, 2050–2058.

Mély, Y., Cornille, F., Fournié-Zaluski, M.-C., Darlix, J.-L., Roques, B., and Gérard, D. (1991). Investigation of Zinc binding affinities of MoMuLV nucleocapsid protein and its related Zinc finger and modified peptides. *Biopolymers* **31**, 899–906.

Méric, C., Darlix, J. L., and Spahr, P. F. (1984). It is Rous sarcoma virus protein p12 and not p19 that binds tightly to Rous sarcoma virus RNA. *J. Mol. Biol.* **173**, 531–538.

Méric, C., Gouilloud, E., and Spahr, P. F. (1988). Mutations in Rous sarcoma virus nucleocapsid protein p12 (NC): Deletions of Cys-His boxes. *J. Virol.* **62**, 3328–3333.

Morellet, N., Déméné, H., Teilleux, V., Huynh-Dinh, T., de Rocquigny, H., Fournie-Zaluski, M. C., and Roques, B. (1998). Structure of the complex between the HIV-1 nucleocapsid protein NCp7 and the single-stranded pentanucleotide d(ACGCC). *J. Mol. Biol.* **283**, 419–434.

Morellet, N., de Rocquigny, H., Mély, Y., Jullian, N., Deméné, H., Ottmann, M., Gérard, D., Darlix, J.-L., Fournié-Zaluski, M.-C., and Roques, B. (1994). Conformational behaviour of the active and inactive forms of the nucleocapsid NCp7 of HIV-1 studied by ^{1}H-NMR. *J. Mol. Biol.* **235**, 287–301.

Myers, G., Wain-Hobson, S., Korber, B., and Smith, R. (1995). "Human Retroviruses and AIDS," Part II. Los Alamos National Laboratory, Los Alamos, NM.

Nie, Z., Bergeron, D., Subbramanian, R. A., Yao, X. J., Checroune, F., Rougeau, N., and Cohen, E. (1998). The putative alpha helix 2 of human immunodeficiency virus type 1 Vpr contains a determinant which is responsible for the nuclear translocation of proviral DNA in growth-arrested cells. *J. Virol.* **72**, 4104–4115.

Ottmann, M., Gabus, C., and Darlix, J.-L. (1995). The central globular domain of HIV-1 NC protein is critical for virion formation and infectivity. *J. Virol.* **69**, 1778–1784.

Pager, J., Coulaud, D., and Delain, E. (1994). Electron microscopy of the nucleocapsid from disrupted MoMuL V and of associated type VI collagen-like filaments. *J. Virol.* **68**, 223–232.

Paillart, J. C., Berthoux, L., Ottmann, M., Darlix, J. L., Marquet, R., Ehresmann, B., and Ehresmann, C. (1996). A dual role of the putative RNA dimerization initiation site of human immunodeficiency virus type 1 in genomic RNA packaging and proviral DNA synthesis. *J. Virol.* **70**, 8348–8354.

Paillart, J.-C., Marquet, R., Skripkin, E., Ehresmann, B., and Ehresmann, C. (1994). Mutational analysis of the bipartite dimer linkage structure of HIV-1 genomic RNA. *J. Biol. Chem.* **269**, 27486–27493.

Pettit, S. C., Moody, M. D., Wehbie, R. S., Kaplan, A., Nantermet, P. V., Klein, C. A., and Swanstrom, R. (1994). The p2 domain of human immunodeficiency virus type 1 Gag regulates sequential proteolytic processing and is required to produce fully infectious virions. *J. Virol.* **68**, 8017–8027.

Pezo, V., and Wain-Hobson, S. (1997). HIV genetic variation: Life at the edge. *J. Infect.* **34**, 201–203.

Prats, A., Roy, C., Wang, P., Roy, C., Paoletti, C., and Darlix, J.-L. (1990). Cis elements and trans-acting factors involved in dimer formation of MuLV RNA. *J. Virol.* **64**, 774–783.

Prats, A.-C., Housset, V., de Billy, G., Cornille, F., Prats, H., Roques, B., and Darlix, J.-L. (1991). Viral annealing activity of the nucleocapsid protein of MoMuLV is zinc independent. *Nucleic Acids Res.* **13**, 3533–3541.

Prats, A.-C., Sarih, L., Gabus, C., Litvak, S., Keith, G., and Darlix, J.-L. (1988). Small finger protein of avian and murine retroviruses has nucleic acid annealing activity and positions the replication primer tRNA onto genomic RNA. *EMBO J.* **7**, 1777-1783.

Quan, Y., Gu, Z., Li, X., Li, Z., Morrow, C., and Wainberg, M. (1996). Endogenous reverse transcription assays reveal high-level resistance to the triphosphate of (−)2′-dideoxy-3′-thiacytidine by mutated M184V human immunodeficiency virus type 1. *J. Virol.* **70**, 5642–5645.

Rascle, J. B., Ficheux, D., and Darlix, J. L. (1998) Possible roles of MuLV Nucleocapsid proetin in proviral DNA synthesis and in its variability. *J. Mol. Biol.* **280**, 215–225.

Rein, A., Ott, D. E., Mirro, J., Arthur, L., Rice, W., and Henderson, L. (1996). Inactivation of murine leukemia virus by compounds that react with the zinc finger in the viral nucleocapsid protein. *J. Virol.* **70**, 4966–4972.

Rice, W., Schaeffer, C., Graham, L., *et al.*, (1993). The site of antiviral action of 3-nitrosobenzamide on the infectivity process of HIV in human lymphocytes. *Proc. Natl. Acad. Sci. USA* **90**, 9721–9724.

Richardson, J., Child, L., and Lever, A. (1993). Packaging of HIV-1 RNA requires cis acting sequences outside the 5′ leader region. *J. Virol.* **67**, 3997–4005.

Rodriguez-Rodriguez, L., Tsuchihashi, Z., Fuentes, G., Bambara, R., and Fay, P. (1995). Influence of human immunodeficiency virus nucleocapsid protein on synthesis and strand transfer by the reverse transcriptase *in vitro*. *J. Biol. Chem.* **270**, 15005–15011.

Sarkar, N., Nowinski, R., and Moore, D. (1971). Helical nucleocapsid structure of the oncogenic ribonucleic acid viruses (oncornaviruses). *J. Virol.* **8**, 564–572.

South, T., and Summers, M. (1993). Zinc and sequence dependent binding to nucleic acids by the N-terminal zinc finger of the HIV-1 nucleocapsid protein: NMR structure of the complex with the Psi site analog, dACGCC. *Prot. Sci.* **2**, 3–19.

South, T., Blake, P., Sowder, R. *et al.*, (1990). The nucleocapsid protein isolated from HIV-l particle binds zinc and forms retroviral-type zinc fingers. *Biochemistry* **29**, 7786–7789.

Stoltzfus, M., and Snyder, P. (1975). Structure of B77 sarcoma virus RNA: stabilization of RNA after packaging. *J. Virol.* **16**, 1161–1170.

Summers, M., Henderson, L., Chance, M. *et al.* (1992). Nucleocapsid zinc fingers detected in retroviruses: EXAFS studies of intact viruses and the solution-state structure of the nucleocapsid protein from HIV1. *Prot. Sci.* **1**, 563–574.

Stewart, L., Schatz, G., and Vogt, V. (1990). Properties of avian particles defective in viral protease. *J. Virol.* **64**, 5076–5092.

Tanchou, V., Decimo, D., Péchoux, C., Lener, D., Rogemond, V., Berthoux, L., Ottmann, M., and Darlix, J.-L. (1998) Role of the N-terminal Zinc finger of HIV-1 NCp7 in virus structure and replication. *J. Virol.* **72**, 4442–4447.

Tanchou, V., Delaunay, T., Bodeus, M., Roques, B., and Darlix, J. L., and Benarous, R. (1995a). Conformational changes between human immunodeficiency virus type 1 nucleocapsid protein NCp7 and its precursor NCp15 as detected by anti-NCp7 monoclonal antibodies. *J. Gen. Virol.* **76**, 2457–2466.

Tanchou, V., Gabus, C., Rogemond, V., and Darlix, J.-L. (1995b). Formation of stable and functional HIV-1 Nucleoprotein complexes *in vitro*. *J. Mol. Biol.* **252**, 563–571.

Temin, H. (1991). Sex and recombination in retroviruses. *Trends Genet* **7**, 71–74.

Temin, H. (1993). Retrovirus variation and reverse transcription: abnormal strand transfers result in retrovirus genetic variation. *Proc. Natl. Acad. Sci. USA* **90**, 6900–6903.

Tracktman, P., and Baltimore, D. (1982). Protease bypass of temperature-sensitive murine leukemia virus maturation mutants. *J. Virol.* **44**, 1039–1046.

Tsuchihashi, Z., and Brown, P. O. (1994). DNA strand exchange and selective DNA annealing promoted by the human immunodeficiency virus type 1 nucleocapsid protein. *J. Virol.* **68**, 5863–5870.

Tsuchihashi, Z., Khosla, M., and Hershlag, D. (1993). Protein enhancement of hammerhead ribozyme catalysis. *Science* **262**, 99–102.

Tummino, P. J., Scholten, J. D., Harvey, P. J., Holler, T. P., Maloney, L., Gogliotti, R., Domagala, J., and Hupe, D. (1996). The *in vitro* ejection of zinc from human immunodeficiency virus (HIV) type 1 nucleocapsid protein by disulfide benzamides with cellular anti-HIV activity. *Proc. Natl. Acad. Sci. USA* **93**, 969–973.

Turner, B., and Summers, M. (1999). Structural biology of HIV. *J. Mol. Biol.* **285**, 1–32.

Turpin, J. A., Song, Y., Inman, J. K., Huang, M., Wallqvist, A., Maynard, A., Covell, D. G., Rice, W., and Appella, E. (1999). Synthesis and biological properties of novel pyridinioalkanoyl thiolesters (PATE) as anti-HIV-1 agents that target the viral nucleocapsid protein zinc fingers. *J. Med. Chem.* **14**, (42), 67–86.

Turpin, J. A., Terpening, S. J., Scheaffer, C. A., Yu, G., Glover, C. J., Felsted, R. L., Sausville, E. A., Rice and W. (1996). Inhibitors of human immunodeficiency virus type 1 zinc fingers prevent normal processing of gag precursors and result in the release of noninfectious virus particles. *J. Virol.* **70**, 6180–6189.

Varmus, H., and Swanstrom, R. (1985). Replication of retroviruses. *In* "RNA Tumor Viruses" (R. Weiss, N. Teich, H. Varmus, and J. Coffin, Eds.), 2nd ed, pp. 369–512. Cold Spring Harbor Laboratory Press, Cold Spring Harbor, New York.

von Schwedler, U., Stemmler, L., Klishk, o. V., Li, S., Albertine, K., Davi, S. D., and Sundquist, W. (1998). Proteolytic refolding of the HIV-1 capsid protein amino-terminus facilitates viral core assembly. *EMBO J.* **17**, 1555–1568.

Wainberg, M. (1997). Increased fidelity of drug-selected M184V mutated HIV-1 reverse transcriptase as the basis for the effectiveness of 3TC in HIV clinical trials. *Leukemia* 3(Suppl.), 85–88.

Wilk, T., Gowen, B., and Fuller, S. (1999). Actin associates with the nucleocapsid domain of HIV Gag polyprotein. *J. Virol.* **73**, 1931–1940.

Wills, J., and Craven, R. (1991). Form, function, and use of retroviral gag proteins. *AIDS* **5**, 639–654.

Xiong, Y., and Eickbush, T. (1990). Origin and evolution of retroelements based upon their reverse transcriptase sequences. *EMBO J.* **9**, 3353–3362.

Yang, S., and Temin, H. (1994). A double hairpin structure is necessary for the efficient encapsidation of spleen necrosis virus retroviral RNA. *EMBO J.* **13**, 713–726.

You, J.-C., and McHenry, C. S. (1994). HIV nucleocapsid protein accelerates strand transfer of the terminally redundant sequences involved in reverse transcription. *J. Biol. Chem.* **269**, 31491–31495.

Zhang, H., Basgara, O., Niikura, M., Poiesz, B., and Pomerantz, R. (1994). Intravirion reverse transcripts in the peripheral blood plasma of HIV-1 infected individuals. *J. Virol.* 68, 7591–7597.

Zhou, W., Parent, L., Wills, J., and Resh, M. (1994). Identification of a membrane binding domain within the amino-terminal region of HIV-1 gag protein which interacts with acidic phospholipids. *J. Virol.* 68, 2556–2569.

Laurence Briant and Christian Devaux

Laboratoire Infections Rétrovirales et Signalisation Cellulaire
CNRS EP 2104
Institut de Biologie
34060 Montpellier, France

Bioactive CD4 Ligands as Pre- and/or Postbinding Inhibitors of HIV-1

I. Introducing the CD4 Molecule

Thymic differentiation results in the generation of two mutually exclusive subsets of peripheral T lymphocytes that can be distinguished by the expression of CD4 and CD8 cell-surface molecules. Thereby, the human CD4 antigen was identified for the first time in 1979 with an anti-CD4 monoclonal antibody and was originally described as a marker defining a subset of mature T lymphocytes (for review see Litman, 1987). Yet, CD4 is also found on distinct populations of thymocytes, macrophages, monocytes, microglial cells, and langerhans cells. The earliest *in vitro* studies of the peripheral T-cell subsets revealed that expression of CD4 and CD8 correlated both with function and with specificity for human leukocyte antigen (HLA) class II and class I molecules, respectively. Cloning of the CD4 gene in the mid-1980s (Maddon *et al.*, 1985) made it possible to explore the structure

Advances in Pharmacology, Volume 48

1054-3589/00 $35.00

and function of CD4 in depth (for reviews see Litman 1987; Bour *et al.*, 1995a; Brady and Barclay, 1996).

CD4 is an integral membrane glycoprotein of about 125 Å long (Kwong *et al.*, 1990) that shows homology to members of the immunoglobulin superfamily (IgSF) (Maddon *et al.*, 1986). The CD4 molecule consists of an extracellular region of 370 amino acid residues organized in four domains (D1–D4), of which the three-dimensional structure has been resolved (Ryu *et al.*, 1990; Wang *et al.*, 1990; Wu *et al.*, 1997) and includes a hydrophobic membrane-spanning region of 25 amino acids and a highly charged cytoplasmic tail of 38 amino acids (Fig. 1). Domain 1 shows clear similarities to IgSF variable domains. Domains 2 and 4 also resemble IgSF domains, whereas domain 3 lacks the disulfide bridge between the β-sheets that is normally conserved among IgSF domains. CD4 contains two carbohydrates located in D3D4.

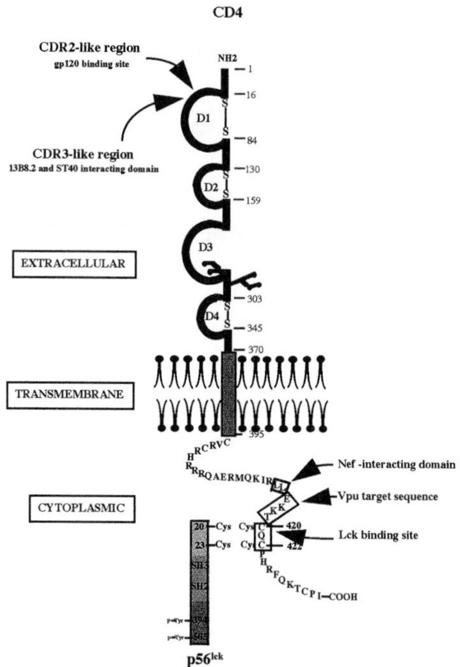

FIGURE 1 Schematic diagram of a human CD4 glycoprotein as expressed at the surface of a T cell. Numbering of CD4 amino acid residues is based on the structure of the mature processed protein. The CDR2- and CDR3-like loop regions in D1 are indicated by arrows. CD4 amino acid residues in the intracytoplasmic domain that are important for interaction with the cellular protein tyrosine kinase p56^{lck}, and viral proteins Nef and Vpu, are shown by boxes.

CD4 acts as a signal transduction molecule. Signals transduced through CD4 deal with physiological functions such as T-cell thymic ontogeny (Kruisbeek *et al.,* 1985), T-cell receptor (TCR)-dependent antigen recognition (Doyle and Stromiger, 1987; Lamarre *et al.,* 1989), IL-16 response (Cruikshank *et al.,* 1996), or with pathological events, such as T-cell activation associated with autoimmune diseases (Jameson *et al.,* 1994; Marini *et al.,* 1996) or interaction with HIV-1 envelope glycoprotein (Mizukami *et al.,* 1988; Arthos *et al.,* 1989; Clayton *et al.,* 1989). A common mechanism of extracellular signal transduction through integral cell membrane receptors associated with cytoplasmic tyrosine kinases involves dimerization of the receptors following ligand binding. Given the association of CD4 with the protein tyrosine kinase $p56^{lck}$, it is likely that dimerization of CD4 represents a key element in activation of CD4 T cells. It is believed that the regions of CD4 required for dimerization are located in different domains of CD4 but there is no definitive consensus regarding the nature of domains involved (this point is discussed later in this chapter).

Most of the results reviewed in this article are related to ligands of domain 1 of CD4. In the past few years, a large number of studies have been performed to investigate the functional role of the three complementarity determining region (CDR) loops in the aminoterminal domain (D1) during physiological and pathological CD4-dependent cell-signaling processes. The consequence of anti-CD4 antibodies binding to CD4 T cells is described here in the context of HIV infection. The function of synthetic peptides that bind CD4 are also reviewed in details. Several authors previously argued that the conformational properties of CDR loops or reverse turns determine the biological activity of members of IgSF (Jameson *et al.,* 1994; Dougall *et al.,* 1994). As a result, biological antagonists were designed after information gained on the sequence and/or structure of CDRs. Special attention was devoted to the CDR3 loop, which plays a key role in regulating CD4 signal transduction in T cells. However, the importance of ligands of domains 2, 3, and 4 should by no means be ignored, although much less information is presently available about the function and mechanism of action of these ligands.

During the past 2 decades, application of hybridoma technology, gene technology, and synthetic peptide technology led to an explosion in the discovery of CD4 biological function. Many CD4 binding ligands are currently available to manipulate the immune response. However, we are certainly at the very beginning of a new era for applying CD4 ligands to human diseases. A comprehensive understanding of the interactions of CD4 with its ligands should contribute to better selecting the appropriate CD4 ligand aimed at controlling human diseases in which CD4-positive cells play a major role. Moreover, new approaches based on structural biology concepts are likely to permit the development of a generation of new drugs more suitable for treatment of patients.

II. Ligands of CD4

A. Natural Ligands of CD4 Extracellular Domains

1. HLA Class II Molecules

Over the past 20 years, a clear correlation was established between CD4 expression at the surface of T cells and antigens recognition by $\alpha\beta$-T-cell receptors in the context of HLA class II molecules. The CD4 molecule acts as an adhesion molecule recognizing conserved monomorphic determinants on a class II molecule. Competition experiments with synthetic peptides encompassing amino acid residues 35–45 and 41–55 of HLA-DR1 or DR2 (Mazerolles *et al.*, 1990; Brogdon *et al.*, 1998) and mutagenesis studies of HLA-DR β1 domain (Brogdon *et al.*, 1998) indicated that the β1 domain binds CD4. Competition experiments with synthetic peptides encompassing amino acid residues 121–135 and 141–155 of HLA-DR1 β2 domain (Brogdon *et al.*, 1998) and mutagenesis studies of the β2 domain also provided evidence that several amino acid residues in the β2 121–155 segment are involved in the interactions with CD4 (König *et al.*, 1992; Cammarota *et al.*, 1992). Mutations of CD4 domains 1, 2, and 3 (Clayton *et al.*, 1989; Moebius *et al.*, 1993) suggested that an extended surface area not confined to a single face of CD4 that includes domain 1 and a region of domain 2 close to the interface with domain 1 interacts with HLA class II. The dissociation constant ($K_d = 3.2 \times 10^{-6}\,M$) for the interaction between HLA-DR4 and sepharose-immobilized human CD4 was determined by Scatchard analysis (Cammarota *et al.*, 1992).

Moreover, CD4 contributes to T-cell activation by acting as a signal-transducing molecule tightly associated with the $\alpha\beta$-T-cell receptor (TCR)/CD3 molecule complex. Transduction of extracellular signal through CD4 usually requires its association with the protein tyrosine kinase $p56^{lck}$ (Rudd *et al.*, 1988; Veillette *et al.*, 1988). The mechanisms by which CD4 contributes to TCR/CD3 complex signaling are discussed in Section II,C,1.

2. A Soluble CD4 Ligand: Interleukin-16

Interleukin (IL)-16, originally named lymphocyte chemoattractant factor (LCF), is a 14- to 17-kDa protein secreted from activated $CD8^+$ and described as a $CD4^+$ T-cell-specific chemoattractant molecule (Center and Cruikshank, 1982; Center *et al.*, 1996). It has been demonstrated that recombinant IL-16 physically binds soluble CD4 (Cruikshank *et al.*, 1994). The IL-16 binding site on CD4 is probably not the CDR2-like loop in domain 1, as anti-CD4 mAb OKT4a still binds CD4 in the presence of IL-

16 (Center *et al.,* 1996). Although IL-16 appears to interact with the CD4 region in the vicinity of the OKT4 mAb binding site in domain 4, further investigations will be required to precisely determine the location of IL-16 binding site on CD4. Moreover, the region of IL-16 required for CD4 binding probably involves the C-terminal hydrophilic domain, as a synthetic peptide (RRKSLQSKETTAAGDS) from this region partially inhibited OKT4 mAb binding to CD4 (Keane *et al.,* 1998).

Interleukin-16 is synthesized as a precursor protein (pro-IL-16) and processed into a bioactive form after cleavage by caspase-3 (Zhang *et al.,* 1998). The bioactive form of IL-16 is composed of several polypeptidic chains that multimerize into homotetramers (56 kDa) (Cruikshank *et al.,* 1994). Interleukin-16 is thought to trigger CD4 aggregation and CD4-dependent signal transduction in CD4-positive T cells, thereby inducing the migration of cells and their accumulation at sites of inflammation (Cruikshank *et al.,* 1994). The chemoattractant activity of IL-16 is proportional to the amount of CD4 molecules expressed at the surface of target cells (Center *et al.,* 1996).

The fact that IL-16 induced motile response of cells bearing chimeric CD4-p56^{lck} molecules which lack the *lck* kinase domain has brought Center and co-workers (Center *et al.,* 1996) to hypothesize that the migratory signal is not associated with the kinase domain but rather with the SH2 and SH3 domains of *lck;* the SH2/SH3 domains of *lck* (see Section II,C,1 for details) might be required for the recruitment of other intracellular proteins, such as phosphoinositide 3-kinase (PI3-k) and phospholipase C_γ (PLC_γ). In agreement with this hypothesis, it was found that wortmannin, a PI-3K specific inhibitor, and selective inhibitors of protein kinase C (PKC) prevent IL-16-induced cell migration.

3. *Glycoprotein-17 and Its Function as a Natural CD4 Ligand*

A glycoprotein found at high concentrations in human seminal plasma (Autiero *et al.,* 1991) was recently found to bind recombinant soluble CD4 (Autiero *et al.,* 1997). This glycoprotein, named gp17, which apparently exists as tetramers, is similar to three other proteins, namely secretory actin-binding protein (SABP), derived from seminal plasma (Schaller *et al.,* 1991), gross-cyctic-disease fluid protein-15 (GCDFP-15) (Haagensen *et al.,* 1990), and prolactin-inducible protein PIP (Murphy *et al.,* 1987), derived from breast tumors. The biological function of these molecules is presently unknown.

Autiero and co-workers (1997) demonstrated that gp17–CD4 binding affinity is high ($K_d = 9 \times 10^{-9}\,M$). Glygoprotein-17 binding to CD4 does not require the CD4 domains 3 and 4 and is expected to bind a site distinct from but close to the gp120 binding site in domain 1. Indeed, gp17 binding to CD4 was inhibited by B66.1.6 (an anti-CD4 mAb specific for CD4 domain 1) and by HIV-1 gp120^{env}. Conversely, engagement of gp17 with CD4 decreases both the binding of recombinant soluble gp120 to immobilized recombinant CD4 and the formation of syncytia (Autiero *et al.*, 1997).

Very recently, Gaubin and co-workers (1999) demonstrated that gp17 inhibits T lymphocyte apoptosis induced by CD4 cross-linking and subsequent TCR activation.

B. Viral Components That Bind CD4 Extracellular Domains

1. HIV-1 gp120^{env}

Human immunodeficiency virus type 1 (HIV-1) has been originally described as a $CD4^+$ T-lymphotropic retrovirus. The external envelope glycoprotein gp120 plays essential roles in the initial phases of the infectious cycle and binds CD4. The sequence of the *env* gene greatly varies among different isolates. This variation, however, is not randomly distributed throughout the gene and consists of a succession of variable (V1 to V5) and constant (C1 to C5) regions. The CD4-binding site, which is located within the carboxyl portion of gp120, is composed of discontinuous sequences with major contributions for constant regions. Soon after the identification of CD4 as primary cellular receptor for HIV-1, mapping of the gp120 binding site on CD4, using anti-CD4 mAbs, synthetic peptides, and mutant forms of CD4, revealed that the principal site is located on and around the CDR2-like loop in domain 1. The first reports indicating that HIV-1 binds to CD4 were published by Dalgleish and co-workers (1984) and Klatzmann and co-workers (1984), who found that incubation of $CD4^+$ cells with anti-CD4 mAbs (including OKT4a) completely inhibited virus replication in these cells. Two additional studies performed with a large series of anti-CD4 mAbs and published in 1986 (Sattentau *et al.*, 1986; McDougal *et al.*, 1986) contributed to the emergence of a new important concept about the mechanisms of interactions between HIV envelope glycoproteins and cell-surface CD4 since they demonstrated that only some epitopes of CD4 are important for virus binding. Of particular importance for the binding of HIV-1 gp120^{env} are positively charged amino acid residues at positions 46 and 59 surrounding the phenylalanine residue at position 43. The role of sites outside CDR2-like loop, particularly the CDR3-like loop, has been the source of conflicting reports (for reviews see James *et al.*, 1996; Devaux, 1996) and this aspect is described later in this chapter. The interaction between HIV-1 gp120^{env} and the primary CD4 receptor was reported to occur with a high affinity, the dissociation constant for gp120-binding to CD4 being $K_d = 3 \times 10^{-9}\,M$.

It is worth noting that the binding site for HIV-1 on CD4 is also utilized by HIV-2 and SIV (Sattentau *et al.*, 1988). However, a few variant isolates of the HIV-1 ($HIV\text{-}1_{NDK}$) and HIV-2 ($HIV\text{-}2_{ROD/B}$) laboratory strains, as well as some primary isolates, have been shown to infect cells independent of CD4 expression.

2. HHV7 Virion Molecule Binds to CD4

Human herpesvirus 7 (HHV-7) is a recently identified T-lymphotropic herpesvirus. Reciprocal interference between HHV-7 and HIV-1 for infection of $CD4^+$ cells suggested that CD4 is a critical component of the receptor for HHV-7. Indeed, preexposure to HHV-7 rendered $CD4^+$ T cells and monocytes/macrophages resistant to HIV-1 infection (Lusso *et al.*, 1994; Crowley *et al.*, 1996). It is worth noting that, in contrast to HIV-1, HHV-7 was unable to productively infect HeLa–CD4 cells. It is not clear at present which virion molecule specifically binds to CD4; herpesvirus possess multiple glycoproteins at their virion surface which are likely capable of interacting with different molecules expressed at the cell surface.

HHV-7 infection of CD4 cells was found to be inhibited by OKT4a and Leu3a that bind domain 1 of CD4 (Lusso *et al.*, 1994). Unexpectedly, HHV-7 infection and syncytia formation were described as blocked by anti-CD4 mAb OKT4, which binds to domain 3 and/or 4 of CD4 (Lusso *et al.*, 1994). However, another study performed using cells from individuals with OKT4 epitope deficiency, indicates that the interaction of HHV-7 with $CD4^+$ T cells does not require participation of the CD4 region recognized by OKT4 mAb (Yasukawa *et al.*, 1996). Blocking experiments performed with anti-CD4 mAbs, recombinant soluble CD4, or recombinant HIV-1 $gp120^{env}$ support the hypothesis that CD4 plays a role in attachment and/or penetration of HHV-7 into $CD4^+$ T cells (Secchiero *et al.*, 1997).

C. The Cellular Partner and Viral Molecules That Bind the CD4 Intracytoplasmic Domain

Although no attempt will be made in this review to describe bioactive CD4 ligands that might act through interactions with the CD4 cytoplasmic tail, a very brief description of the cellular and viral ligands of CD4 intracytoplasmic domain is hereafter provided in order to summarize the most important knowledge required to understand CD4 function in normal and pathological situations. Obviously, there are several reviews providing a recent overview of this field (see references in Sections II,C,1; II,C,2; and II,C,3), and therefore the reader is strongly encouraged to refer to these reference reviews for more details.

1. The Cellular Tyrosine Kinase $p56^{lck}$

Lck is a 56-kDa protein of the Src family of tyrosine kinases, composed of a unique NH2-terminal domain, a src-homology 3 (SH3) domain, a SH2 domain, and a kinase domain. Amino acid residues 20 and 23 of Lck interact with the cysteine residues 420 and 422 of CD4. Cross-linking of CD4 results in rapid phosphorylation of Lck at tyrosine residue 394, the major site of autophosphorylation, and at several serine residues. This phosphorylation process increases the Lck catalytic function. It appears likely that CD4 potentiates TCR/CD3 complex signaling both by activating Lck and by recruiting the active forms of Lck in the vicinity of the TCR/CD3 complex. Another mechanism has been proposed that involves the SH2 domain of Lck, which may contribute to stabilize the association between CD4 and the TCR–CD3 complex by acting as a bridge between CD4 and phosphorylated CD3 (for reviews see Ravichandran *et al.*, 1996; Weil and Veillette, 1996).

The natural ligands of CD4 extracellular domains trigger activation of $p56^{lck}$. As a ligand capable of binding CD4, the HIV-1 envelope glycoprotein has also been considered for its capacity to activate T-cell signals. Although there is some controversy as to the nature of the signals delivered to $CD4^+$ cells, it is now generally accepted that HIV-1 envelope glycoproteins modulate T-cell activation (for review see Devaux *et al.*, 1999). Over the past few years, it was demonstrated that HIV-1 envelope glycoproteins mediated cross-linking of CD4 triggers signals in $CD4^+$ T cells which can be monitored by the nuclear translocation of NF-κB DNA-binding complex. It is worth noting that we have recently demonstrated the involvement of $p56^{lck}$ in the induction of NF-κB nuclear translocation after virus envelope binding to CD4 (Briant *et al.*, 1998a).

2. *Viral Nef Protein*

Although the Nef protein of HIV is thought to play an important role *in vivo* for HIV replication and pathogenesis, such as induction of CD4 down-regulation, alteration of T-cell activation pathways, and enhancement of virus infectivity, we are still far from fully understanding the different aspects of Nef function (for a review see Luo *et al.*, 1997). Initial observations by Guy and co-workers (1987), suggested that Nef, a 27-kDa protein which is myristylated at its amino terminus, triggers down-modulation of cell-surface-expressed CD4. The cytoplasmic domain of CD4 was required for Nef-induced internalization. However, neither the cysteine residues involved in interaction with $p56^{lck}$ nor the serine residues involved in phorbol ester-mediated down-regulation of CD4 appear to be critical for Nef-mediated CD4 down-regulation. Indeed, the first 20 membrane-proximal amino acid residues of the CD4 intracytoplasmic domain containing a functionally important dileucine motif at position 413/414 are sufficient to permit Nef-mediated CD4 down-regulation (Aiken *et al.*, 1994). CD4-associated Lck activity was found to be significantly decreased in *nef*-transfected cells. The fact that Nef binds directly to CD4 has been controversial. Some observations suggest that Nef binds the CD4 intracytoplasmic tail, but experiments performed in the yeast two-hybrid system indicate a weak interaction at best (Harris *et al.*, 1994; Rossi *et al.*, 1996). Other reports suggest that the mechanism of Nef-mediated CD4 down-regulation would involve a cellular Nef partner controlling CD4 trafficking (Mangasarian *et al.,* 1997). Indeed, Nef probably acts as a connector between CD4 and the cellular endocytic machinery by the targeting of CD4 into clathrin-coated pits due to direct interaction with both CD4 and the μ2 component of adaptor protein (AP)-2 (for a review see Cullen, 1998). Interestingly, Nef was also found able to bind p56^{lck}. Nef contains a prolin-rich region involved in the interactions with the SH3 domain of p56^{lck} (Lee *et al.,* 1996). In addition, Nef also interacts with the SH2 domain of Lck (Dutartre *et al.,* 1998).

3. *Viral Vpu Protein*

HIV-1 Vpu is a 16-kDa transmembrane protein having an N-terminal hydrophobic domain of 27 amino acids and a 54-amino-acid-residue cytoplasmic domain (Strebel *et al.,* 1988). The cytoplasmic domain contains two phosphoacceptor sites at serine residues 52 and 56, which are targets for the casein kinase 2 (Schubert *et al.,* 1994). Vpu plays a role in augmenting viral particle release at the plasma membrane and is also considered as having ion channel activity (for reviews see Lamb and Pinto, 1997; Cullen, 1998). Another known function of Vpu is to induce the proteolysis of CD4. Vpu-mediated CD4 degradation occurs in the endoplasmic reticulum and requires the phosphorylation of the Vpu cytoplasmic domain (Margottin *et al.,* 1998). There is evidence that Vpu-mediated CD4 degradation requires the presence of sequences in both the CD4 transmembrane and cytoplasmic domains. Coimmunoprecipitation experiments showed that Vpu can specifically bind to the cytoplasmic tail of CD4 (Bour *et al.,* 1995b). The use of CD4 deletion mutants lacking residues in the cytoplasmic domain indicated that removal of the EKKTCQCP region extending from residue 416 to residue 423 suppressed CD4 sensitivity to Vpu-mediated degradation (Lenburg *et al.,* 1993). Moreover, other experiments performed with chimeric molecules defined the amino acid sequence LSEKKT (residues 414 to 419) as a minimal element required for to Vpu-mediated degradation of CD4 (Vincent *et al.,* 1993).

III. Dissection of CD4 Structure–Function Relationship by Mean of Antibodies and Synthetic Peptides

A. Anti-CD4 mAbs

During the past 20 years anti-CD4 mAbs have allowed precise epitope mapping of the CD4 molecule and structure–function correlation in conjunction with mutational analysis of CD4. One of the major informations gained from these studies is that most anti-CD4 mAbs can influence T-cell activation (for review see Olive and Mawas, 1993). Initially, anti-CD4 mAbs were classified according to their capacity at inhibiting *in vitro* assays such as mixed-lymphocyte reaction or antigen-induced proliferation. Since the identification of CD4 as a major receptor for HIV, these mAbs have also been classified in terms of their capacity to inhibit syncitium formation, gp120 binding, and HIV replication.

It has been thought for a long time that interactions between virion-bound gp120 (HIV-entry) or infected cell-surface gp120 (syncytium formation) and CD4 are equivalent phenomena which involve a region in the first domain (D1) of CD4 localized around amino acid positions 38–55, including the CDR2-like region. However, several observations remained enigmatic with respect to a unique CD4–gp120 interaction model common to viral entry and syncytium formation. Further investigations supported a model that considers the CD4–gp120 interactions in syncytium formation as distinct from the CD4–gp120 interactions involved in viral entry (for review see Corbeau *et al.*, 1993). In line with these studies, two groups of anti-CD4 mAbs able of interfering with gp 120–CD4 interactions have been identified; mAbs of the first group, the main representant of which are Leu3a and OKT4a, recognize epitopes linked to the CDR2-like region of D1. They block $_{s}$gp120^{env} binding to CD4, syncytia formation and HIV production (Sattentau *et al.*, 1986, 1988, 1989; Mizukami *et al.*, 1988; Peterson and Seed, 1988; Wilks *et al.*, 1990). mAbs of the second group (e.g., OKT4e, MT321, ViT4, and 13B8.2) are specific for determinants close to the CDR3-like region of D1. They have been mainly studied for their antisyncytial capacity (Dalgleish *et al.*, 1994; Sattentau *et al.*, 1986, 1988).

In 1990, a few reports suggested that there are epitopes on CD4, unrelated to the binding site of gp120, which can affect postbinding events that usually lead to infection (Celada *et al.*, 1990; Healey *et al.*, 1990). For example, Healey and colleagues reported that mAb Q425 and Q428, which map to D3–D4 domains of CD4, block syncytium formation and HIV infection but do not inhibit HIV/CD4 interaction. Later, another study (Moore *et al.*, 1992) indicated that the 5A8 mAb, which is reactive with domain 2 of CD4, also can affect postbinding events. These authors speculate that 5A8 mAb acts by blocking the conformational changes in Env or in CD4 required for virus–cell fusion. To explore whether domains of CD4 not involved in HIV-1 binding may be important in postbinding events of HIV infection, a number of studies have investigated the ability of cells expressing mutant CD4 molecules to support HIV infection and/or replication and the antiviral properties of anti-CD4 mAbs. Although the kinetics of HIV replication has been shown altered in cells expressing different mutant forms of CD4, most human T-cell lines expressing mutant CD4 molecules efficiently support HIV infection and replication (Bedinger *et al.*, 1988; Tremblay *et al.*, 1994; Benkirane *et al.*, 1994; Briant *et al.*, 1997). The only exception concerns cells expressing CD4 molecules mutated for the HIV binding site.

We have focused our attention on one anti-CD4 mAb named 13B8.2 (or IOT4a) and found that this mAb specific for the CDR3 loop of CD4 domain 1 (see Fig. 2 and Table I) inhibits HIV-1 replication in addition to inhibiting syncytium formation (Corbeau *et al.*, 1993). To investigate HIV-1 replication in CEM cells cultured in the presence of 13B8.2 mAb, cells were exposed to HIV-1 for 30 min at 4°C, extensively washed, and cultured at 37°C in medium alone or medium supplemented with 10 μg/ml anti-CD4 mAb. The 13B8.2 mAb as well as AZT inhibited virus replication during the first 2 weeks of culture, whereas BL4 mAb, which is reactive with D1D2

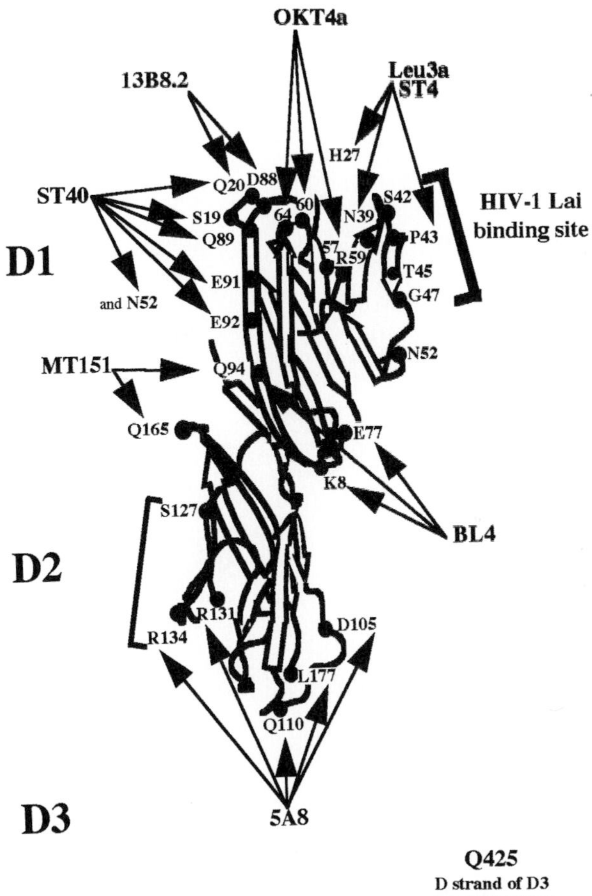

FIGURE 2 Schematic diagram illustrating the folding of CD4 D1D2 domains (adapted from Ryu *et al.*, 1990 and Wang *et al.*, 1990). Location of amino acid residues involved in interactions with anti-CD4 mAbs OKT4a, Leu3a, ST4, 13B8.2, ST40, MT151, 5A8 and BL4 is indicated. mAb Q425 binds to the D strand of D3 domain. The region in D1 involved in interactions with HIV-1 gp120^{env} is also indicated (upper right).

TABLE I Mechanisms of HIV-1 Inhibition Ascribed to Representative Anti-CD4 mAbs That Interfere with Different Steps of the Virus Replication Cycle

	mAb[a]/CD4 domain				
	Leu3a/D1(CDR2)	*13B8.2/D1(CDR3)*	*BL4/D1/D2*	*5A8/D2*	*Q425/D3*
Inhibition of					
Virions/CD4 binding	Yes	No	No	No	No
rgp120/CD4 binding	Yes	Yes	No	No	No
HIV-1-induced					
Syncytium formation	Yes	Yes	No	Yes	Yes
CD4 conformational changes	No	No	No	Yes	ND[b]
HIV-1-activated					
ERK pathway	NT[c]	Yes	No	NT	NT
HIV-1-induced					
NF-κB activation	NT	Yes	No	NT	NT
Tat dependent					
HIV transcription	NT	Yes	No	NT	NT
HIV-1-induced apoptosis	Yes	Yes	No	NT	NT
Mechanisms of inhibition	Binding entry	Postentry (T cell activation an HIV transcription)	No effect	Postbinding entry	Postbinding

[a] Main references: Mizukami *et al.* (1988); Sattentau *et al.*, (1989); Healey *et al.*(1990); Moore *et al.* (1992); Burkly *et al.* (1992); Corbeau *et al.* (1993); Benkirane *et al.* (1993); Houlgatte *et al.* (1994); Benkirane *et al.* (1995a); Corbeil and Richman (1995); Jabado *et al.* (1997b); Huang *et al.* (1997); Guillerm *et al.* (1998); Coudronnière *et al.* (1998).

[b] ND, not defined.

[c] NT, not tested.

of CD4, was devoid of antiviral properties (Corbeau *et al.*, 1993; Benkirane *et al.*, 1993). The effect of 13B8.2 mAb was reversible, as evidenced by the fact that virus production was found after cells had been extensively washed and grown in antibody-free culture medium (Benkirane *et al.*, 1993). 13B8.2 mAb was found capable of preventing viral particle production by cells infected with HIV-1_{Lai}, HIV-1_{Eli}, HIV-1_{SF2}, HIV-1_{Ger}, and HIV-2_{Rod} and delayed production of HIV-$1_{Lai13EM}$, but failed to block HIV-1_{NDK} and the unrelated CD4-independent retrovirus HTLV-I (Rey *et al.*, 1991; Benkirane *et al.*, 1993; Emiliani *et al.*, 1996; Lemasson *et al.*, 1996; Coudronnière and Devaux, 1998). Although less efficient than 13B8.2 mAb, another CDR3-like loop-specific anti-CD4 mAb (ST40) was found to inhibit virus replication (Benkirane *et al.*, 1995a).

To further investigate the consequences of 13B8.2 mAb treatment on cell phenotype, we compared the expression of CD4 on infected cells treated or not with 13B8.2 mAb. As expected, HIV-1 infection induced down-regulation of cell-surface CD4. In contrast, most cells treated with 13B8.2 mAb remained positive for CD4 expression. In search for a mechanism of action, we observed that the cytoplasmic tail of CD4, which is known to act as a signal transduction region through its association with the $p56^{lck}$, is required for 13B8.2 mAb-mediated inhibition of HIV replication (Benkirane *et al.*, 1995b). This result suggested that 13B8.2 acts, at least in part, by regulating a signaling pathway involved in the inactivation process. However, expression of a functional $p56^{lck}$ was not required for 13B8.2 mAb inhibitory effect (Lemasson *et al.*, 1996; Coudronnière *et al.*, 1998), suggesting the existence of another signaling partner for CD4. In agreement with this hypothesis, we have recently found that there is no apparent difference in 13B8.2 responsiveness in CD45-deficient versus CD45-positive Jurkat cells, CD45 being known to modulate the activity of $p56^{lck}$ (Péléraux *et al.*, 1998). Finally, the 13B8.2 anti-HIV effect was also observed in the presence of pertussis toxin (PTX), a known inhibitor of Gi proteins involved in the transduction of signal that may originate from the HIV coreceptor CXCR4 (Coudronnière *et al.*, 1998).

B. CD4-Binding Peptides

In 1986, a report described a peptide that apparently inhibited gp120 binding to CD4 and HIV infection of human T cells at a concentration of 1×10^{-10} *M*; this peptide (ASTTTNYT), termed "peptide T", showed a similarity with the $gp120^{env}$ region extending from residues 280 to 287 (Pert *et al.*, 1986). Two years later, the same group reported on a vasoactive intestinal peptide (VIP)-derived peptide (HSDAVFTDNYTR) showing a high similarity with $gp120^{env}$ that *in vitro* antagonized neuronal death induced by $gp120^{env}$ (Brenneman *et al.*, 1988). The anti-HIV property of a peptide (VVVRSLTFKTNKKT) derived from jacalin (a lectin from jackfruit that binds CD4) and presenting a high similarity with amino acid sequence 273–288 of the C2 domain of gp120^{env} has also been reported (Favero *et al.,* 1993).

Multibranched peptide constructs (MBPCs) MBPC1 ([GPGRAF]8-[K]4-[K]2-K-βA-OH) and MBPC2 ([RKSIHIGPGRAFYT]4-[K]2-K-βA-OH) derived from the HIV-1 gp120^{env} V3-loop (Yahi *et al.,* 1994; Benjouad *et al.,* 1995) have been reported to inhibit HIV-1 infection at a concentration of 5×10^{-6} *M*. It was claimed that this effect was obtained by interaction with the CDR3-like region in D1 of CD4 or with a region in its vicinity (Benjouad *et al.,* 1995). However, it was also reported that the effect of these MBPCs is independent of CD4 binding but occurs through recognition of galactosylceramide (GalCer) (Fantini *et al.,* 1997). More recently, a V3-loop-derived peptide named V3Cs (CTRPNNNTRKIHIGPGRAFYTTGEIIGDIRQAHC) was found to bind a cell-surface protein identified as being the chemokine receptor CCR5 (Rabehi *et al.,* 1998) and V3 peptides derived from X4 and R5X4 strains were shown to bind CXCR4 (Sakaida *et al.,* 1998). Finally, a 17-amino-acid cyclized peptide (CDLIYYDYEEDYYFDYC) derived from the third CDR of the heavy chain of mAb F58 that recognizes the V3-loop, named MicroAb/MAb F58, was found to inhibit fusion without inhibiting attachment of virus to CD4 (Jackson *et al.,* 1999). Relevant observations indicate that changes in coreceptor usage (evolution from a CCR5 to a CXCR4 specificity) are largely due to changes in HIV-1 gp120^{env} protein sequence characterized by an increase in V3-loop basic charge. It is worth noting that the extracellular domains 1 and 2 of CXCR4 are much more negatively charged than the corresponding domains of CCR5, supporting the hypothesis that the V3-loop is involved in CXCR4 binding. Moreover, several reports in the literature have shown that antibodies to the V3-loop inhibit HIV-1 infection of cells without interfering with the binding of HIV-1 gp120^{env} to the CD4 molecule. Therefore, it is much more likely that V3-loop derived peptides inhibit interaction with chemokine receptors than the interaction between HIV-1 gp120^{env} V3-loop and CD4.

A 32-mer HIV-1 gp120^{env}-derived peptide (KSSGGDPEIVTHSFNCVC-SSNITGLLLTRDGG), named GC-1, designed to mimic a discontinuous region representing a part of C3 and C4 regions of HIV-1 gp120^{env} thought to be critically important for CD4 binding, was found to bind CD4 and to inhibit the interaction between recombinant HIV-1 gp120^{env} and soluble CD4 (Cotton *et al.,* 1996).

C. CD4-Derived Peptides

The rationale for the synthesis of short (10–20 residues) peptides comprising different regions of CD4 was originally to identify peptides that inhibit HIV infection in order to precisely determine a gp120 binding site(s) of CD4. The start point in the development of CD4-derived peptides with anti-HIV properties was the publication by Lifson and co-workers (1988).

Among other CD4-derived peptides, peptides containing all or most of the CDR3-like region of domain 1, such as CD4[76-94], CD4[74-92], or CD4[81-92], were capable of inhibiting HIV-induced cell fusion and HIV replication (for a review see Rausch *et al.*, 1990). For example, the CD4[74-92] peptide inhibited HIV replication at a concentration of 12.5×10^{-5} *M*. In contrast, peptides corresponding to CD4[1-25], CD4[26-50], CD4[51-75], CD4[95-118], CD4[261-285], and CD4[331-355] were completely inactive in inhibiting HIV-induced cell fusion. The next studies confirmed the observations, indicating that CDR3-like loop-derived peptides showed antiviral properties, whereas CDR2-like loop-derived peptides most often did not (reviewed in James *et al.*, 1996).

It is interesting to remember that, surprisingly, the active form of CD4-derived peptides containing the CDR3-like region of domain 1, provisionally named CD4[CDR3], was not the pure peptide but a benzylated peptide found in the crude extract. Removal of the benzyl group to glutamic acid afforded a completely inactive compound. The prototype compound CD4 [81-92] ($T_1C_4E_5$-tribenzyl-K_{10}-acetyl-TYICEVEDQKEE, named GLH328) inhibited HIV-induced cell fusion at 3.2×10^{-5} *M* and HIV replication at 1×10^{-5} *M* and did not inhibit HTLV-induced cell fusion (Rausch *et al.*, 1990). Increasing attention was devoted to CD4[CDR3] peptides and many research groups synthesized different forms of CD4[CDR3] peptides to improve their function. Simultaneously, these results were controversed; Repke and co-workers (1992) reported that the CD4[CDR3] peptides inhibit the binding of CDR2-directed mAbs and syncytium formation by HTLV-I. Moreover, as already mentioned above, the fact that CD4[CDR3] peptides activity were found to depend crucially on their derivatization with aromatic groups, was a source of criticisms concerning their specificity. Indeed, in CD4 the cystein at position 84 is involved in a disulfide bridge; it is therefore possible that the presence of a benzyl on the cystein at position 4 of the GLH328 peptide changes the structure into one closer to that found in the wild-type CD4 molecule. It is very likely that the discrepancies between the results reported by different research groups about anti-HIV properties of CD4[CDR3] peptides are related to the three-dimensional structure of the CD4[CDR3] peptides tested and their solubility, global charge, and stability in cell cultures (reviewed in Murali and Greene, 1998). Kumagai and co-workers (1993) reported that the S1 peptide, a sequence-scrambled form of CD4[CDR3] peptide, has more potent inhibitory activities on HIV-1 than the original peptide; moreover, a homodimer of S1 which formed an α-helix was at least 10-fold more potent than S1.

D. CD4[CDR3] Peptides Have Potential Therapeutic Utility

A large body of data indicates that bioactive derivatized CD4[CDR3] peptides interfere with HIV-1 replication (Lifson *et al.*, 1988; Nara *et al.*,

1989; Rausch *et al.*, 1990; Ohki *et al.*, 1990, 1992; Repke *et al.*, 1992; Batinic and Robey, 1992; Lasarte *et al.*, 1994; Chang *et al.*, 1996; Monnet *et al.*, 1999). In *in vitro* cultures of human cells and in animal models CD4[CDR3] peptides have also been shown to exhibit specific inhibition of CD4-dependent T-cell responses, such as antigen-induced T lymphocyte activation, mixed-lymphocyte reaction (MLR), graft rejection and IL-2 production, and to inhibit HLA-Class II interaction with CD4 (McDonnel *et al.*, 1992; Chang *et al.*, 1996; Koch and Korngold, 1997). Indeed, the CDR3 loop plays a key role in regulating CD4 signal transduction in T cells and is involved in postbinding stages of HIV-1 infection (Camerini and Seed, 1990; Corbeau *et al.*, 1993; Benkirane *et al.*, 1993; Broder *et al.*, 1993; Briant *et al.*, 1997). This fact may explain why different authors raised conflicting conclusions; for example, if CD4[CDR3] peptides do inhibit HTLV-I-mediated syncytium formation, as reported by Repke and colleagues (1992), it remains possible that this effect does not involve the receptor function of CD4 but the signal transduction function of CD4 (this point is discussed later in this chapter). However, it is worth remembering that Rausch and co-workers claimed that CD4[CDR3] peptides do not inhibit HTLV-I-mediated syncytium formation (Rausch *et al.*, 1990).

More recently, a functionally interesting synthetic molecule, the aromatically modified exocyclic (AME) peptide analog CDR3.AME(82-89) (FCYICEVEDQCY), was designed by Greene and collaborators on the basis of the crystal structure of CD4-D1 and computer-assisted modeling; this molecule interacts with CD4 (Fig. 3) and was found to inhibit antigen-induced T-

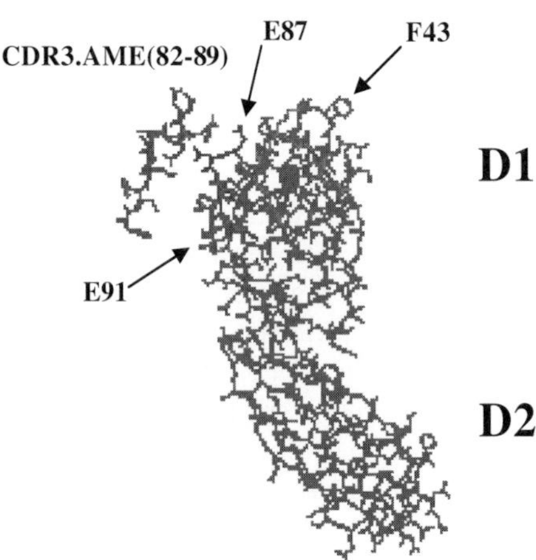

FIGURE 3 Figure showing the atomic structure of D1D2 domains of CD4 and CDR3.AME(82-89) analog adapted from Briant *et al.*, 1997. Interaction of CDR3.AME(82-89) analog with CDR3 loop of CD4 is shown as bond. Amino acid residues E_{87} and and E_{91} in the CDR3 loop and F_{43} in the CDR2-loop are indicated.

lymphocyte activation and HIV-1 replication (Zhang *et al.*, 1996, 1997). Resolution of the molecular mechanism by which the CDR3.AME(82-89) analog prevents HIV-1 replication in cultures of CD4-positive cells exposed to the virus has contributed to understand the unique antiviral properties of CD4[CDR3] peptides (see below). We found that the CDR3.AME(82-89) binds to the CDR3 loop of CD4 domain 1 and inhibits HIV-1 replication (Zhang *et al.*, 1997, Briant *et al.*, 1997). A strong inhibition of HIV-1 replication was found in the presence of CDR3.AME(82-89) synthetic peptides at 40 μg/ml, whereas CDR3.LIN(82-89) (Zhang *et al.*, 1997), a linear peptide with primary sequence identical to that of CDR3.AME(82-89), or the C2-LAI control peptide (Lemasson *et al.*, 1995), were devoid of antiviral properties (Zhang *et al.*, 1997; Roland *et al.*, 1999). To further investigate the consequences of CDR3.AME(82-89) analog treatment on cell phenotype, we compared the expression of CD4 and HLA class II molecules on infected cells treated or not with CDR3.AME(82-89). HIV-1 infection induced down-regulation of cell-surface CD4 and an up-regulation of HLA class II. In contrast, most cells treated with CDR3.AME(82-89) analog remained positive for CD4 expression and kept low surface expression of HLA class II activation marker (Roland *et al.*, 1999).

Very recently, a new strategy, using what are called paratope-derived peptides (PDPs), has been developed to design peptides that bind the CDR3-like loop. The systematic exploration of the antigen-binding capacity of short peptides derived from the primary sequence of variable heavy (V_H) and light (V_L) chains of an antibody (ST40) that bind the CDR3-like loop has lead to the identification of numerous PDPs, such as the cyclized peptide CM9 (KCDSYMNWYQQKPGCK), that display significant binding for the CD4 antigen ($K_d = 1 \times 10^{-8}\,M$) and inhibit HIV-1 promoter activation (Monnet *et al.*, 1999). Exploration of the functional properties of PDPs derived from 13B8.2 mAb is currently underway in our laboratory, and preliminary results suggest that some peptides are much more potent inhibitors of HIV-1 promoter activation than CM9.

Another strategy which should be further explored to isolate novel peptides able to bind CD4 consists of using phage-displayed peptide libaries. This approach has already been successful for the identification of peptides with antiviral properties that bind HIV-1 gp120^{env} (Ferrer and Harrison, 1999).

E. Mechanism(s) of HIV Inhibition by CDR3-Loop Ligands

1. Anti-CD4 mAb 13B8.2 and CDR3.AME(82-89) Do Not Inhibit Virus Entry but Down-Regulate HIV-1 Gene Expression

To determine if the lack of virus production in infected cultures treated with 13B8.2 mAb and CDR3.AME(82-89) may be ascribed to a defect in virus entry or, rather, is due to the inhibition of a postinfection stage of HIV-1 life cycle, the presence of HIV-1 DNA 24 h postinfection was studied. Direct PCR were performed to probe the presence of HIV-1 DNA in infected cells using an oligonucleotide primer pair which detects intermediate molecules formed during HIV-1 reverse transcription. Under these experimental conditions, HIV-1 DNA was found in cells exposed to virus and treated with either 13B8.2 mAb or the CDR3.AME(82-89) analog. These results indicated that 13B8.2 mAb and CDR3.AME(82-89) do not block virus entry. Similar conclusions could be raised from different types of assays (Corbeau *et al.*, 1993; Benkirane *et al.*, 1995a; Roland *et al.*, 1999). Further investigations of the early steps of virus replication in cells treated by 13B8.2 mAb suggested that retrotranscription was complete and that integration, studied by "Alu"-PCR (Benkirane *et al.*, 1993), occured.

To determine whether the lack of virus production from infected cells treated with 13B8.2 mAb or the CDR3.AME(82-89) analog was due to a down-regulation of viral RNA expression, RT-PCR experiments were performed 72 h postinfection with oligonucleotide primers designed to flank the common splice donor and acceptor sites of the Env, Tat, and Rev genes. Spliced HIV-1 mRNAs were found in cells exposed to virus and cultured in the absence of additive as well as in infected cells treated with anti-CD4 mAbs BL4 or OKT4 or with C2-LAI peptide. In contrast, HIV-1 spliced mRNAs were absent or very poorly detected in infected cells treated with 13B8.2 mAb or the CDR3.AME(82-89) analog, indicating that these antiviral molecules inhibit a stage of the virus life cycle that takes place in between integration of the provirus and the provirus transcription, likely HIV-1 promoter activation (Benkirane *et al.*, 1993, 1995a; Roland *et al.*, 1999).

2. *Anti-CD4 mAb 13B8.2 and CDR3.AME(82-89) Inhibit the Induction of HIV-1 LTR Activation by NF-κB*

To further investigate the ability of 13B8.2 mAb and CDR3.AME(82-89) analog to down-regulate HIV-1 gene expression, different assays were performed. CD4-positive Hela P4 indicator cells (that contain Lac Z gene under control of the HIV-1 promoter) were infected by HIV-1 and treated postinfection with different molecules. Strong β-galactosidase activity was measured in cell lysates prepared 72 h after infection by HIV-1 and in lysates of infected cells treated with the C2-LAI peptide. In contrast, no β-galactosidase activity was found in infected cells treated with 13B8.2 mAb or CDR3.AME(82-89). The inhibition of β-galactosidase activity was strictly dependent on the concentration of 13B8.2 mAb or CDR3.AME(82-89) analog added to the cell culture. A complete inhibition was observed at a CDR3.AME(82-89) concentration down to 10 μg/ml, whereas a partial inhibition was already induced by a 5 μg/ml concentration. Inhibition of HIV-1 promoter activation was also analyzed by CAT assay; CEM cells were transfected with a vector containing the CAT gene under control of the HIV-1 promoter and a Tat expression vector and CAT gene transcription induced by Tat was evaluated. Treatment of cells with 13B8.2 mAb resulted in a strong inhibition of CAT synthesis compared with that in untreated cells (Benkirane *et al.*, 1993, 1995a). In contrast, 13B8.2 mAb did not inhibit CAT gene expression driven by the HTLV-I promoter and induced by the Tax protein of HLTV-I (Lemasson *et al.*, 1996).

Although we failed for technical reasons in our first attempt to demonstrate that 13B8.2 mAb inhibits nuclear factor (NF)-κB translocation (Benkirane *et al.*, 1993), we were more successful later at demonstrating that both 13B8.2 mAb and CDR3.AME(82-89) analog can achieve this function (Lemasson *et al.*, 1996; Roland *et al.*, 1999). Electrophoretic mobility shift assay (EMSA) was used to investigate the ability of 13B8.2 mAb and CDR3.AME(82-89) to inhibit NF-κB translocation upon CEM cells stimulation by heat-inactivated HIV-1 (iHIV-1). NF-κB nuclear translocation was strongly reduced under 13B8.2 mAb and CDR3.AME(82-89) treatment, whereas it remained unaffected by several control molecules. These results indicated that 13B8.2 mAb and the CDR3.AME(82-89) analog inhibit the activation of HIV-1 promoter probably by preventing NF-κB translocation.

IV. Which Conclusions Can Be Drawn about the Function of CDR3-Loop Ligands?

A. CDR3-Loop Ligands Inhibit T-Cell Signaling and HIV-1 Promoter Activation

During the past few years we have analyzed the mechanism(s) responsible for the inhibitory effect of 13B8.2 mAb and CDR3.AME(82-89) on viral replication in HIV-1-infected cells. The comparison of the effects of these CD4 ligands, provisionally designated hereafter as CDR3-loop ligands, on HIV-1 replication emphasizes several common features for both molecules, suggesting that they act on HIV by identical or similar ways. The recent determination of 13B8.2 mAb V_H and V_L sequences (Chardès *et al.*, 1999; also see Table II) demonstrates the absence of sequence homology with CDR3.AME(82-89), thereby supporting the hypothesis that these two molecules, although different in their amino acid sequences, share biological properties because they bind to the same or, more likely, to two closely related sites on CD4 which control important functions. Our results indicate that viral spliced mRNAs cannot be found in cultures of CEM cells treated with CDR3-loop ligands despite evidence for HIV-1 infection. Our multiple attempts to understand the mechanism(s) by which CDR3-loop ligands inhibit HIV provide sufficient answers to speculate on the function(s) of CDR3-loop ligands.

The CDR3-loop ligands inhibit the activation of HIV-1 LTR (Benkirane *et al.*, 1993, 1995b; Roland *et al.*, 1999). As previously mentioned, bioactive

TABLE II Characteristics of Some Murine Anti-CD4 mAbs: Heavy and Light-Chain Variable Regions Composition

mAb[a]	*$V_HD_HJ_H$ rearrangement*	*V_lJ_L rearrangement*
OKT4a	V_H5(7183)-DFL16.1-J_H2	V_κ31/38C-J_κ1/2/4
Leu3a	V_H1(J558)-DSP2.5-J_H3	V_κ21-J_κ4[b]
13B8.2	V_H2-DQ52-J_H3	V_κ12/13-J_κ2
ST40	V_H9(Gam3.8)-DSP2.4-J_H2	V_κ21-J_κ1
MT151	V_H9(Gam3.8)-DSP2.2/4/6-J_H4	V_κ1-J_κ2
MT310	V_H1(J558)-DSP2.9-J_H3	V_κ21-J_κ1
L34	V_H1(J558)-DSP2.5-J_H2	V_κ21-J_κ4
L69	V_H9(Gam3.8)-DSP2.7-J_H4	V_κ21-J_κ4
L71	V_H1(J558)-DSP2.3/4-J_H3	V_κ21-J_κ4
L77	V_H1(J558)-DFL16.1-J_H4	V_κ21-J_κ4
L93	V_H1(J558)-DQ52-J_H3	V_κ21-J_κ4
L202	V_H1(J558)-DSP2.3/4-J_H3	V_κ21-J_κ4

[a] Main references: Lohman *et al.* (1992); Weissenhorn *et al.* (1992); Attanasio *et al.* (1993); Pulito *et al.* (1996); Monnet *et al.* (1999); Chardès *et al.* (1999).

[b] The fact that most anti-CD4 mAbs exhibit an apparent preferential expression for V_κ21 should not be considered as a characteristic of this group of mAbs (Devaux *et al.*, 1985, 1986). The same consideration is true for the apparent preferential expression of V_H1.

CD4[CDR3] peptides exhibit specific inhibition of CD4-dependent T-cell responses (McDonnel *et al.*, 1992; Chang *et al.*, 1996; Koch and Korngold, 1997). It has been suggested (Jameson *et al.*, 1994) that the CDR3 analog may act in EAE by uncoupling CD4 from the signal transduction machinery, thereby preventing helper T-cell activation. Moreover, the CDR3.AME(82-89) peptide and 13B8.2 mAb behave as immunosuppresive agents capable of inhibiting antigen-induced proliferative responses of $CD4^+$ T cells (Jabado *et al.*, 1994; Zhang *et al.*, 1996). We have reported that 13B8.2 mAb inhibits MAP kinase (MAPK/ERK) activation triggered by heat-inactivated HIV-1 (iHIV-1) binding to CD4 (Benkirane *et al.*, 1995b; Schmid-Antomarchi *et al.*, 1996) and that MAPK is able to regulate NF-κB activation (Briant *et al.*, 1998b). These observations have been recently confirmed by Jabado and co-workers (1997a,b), who found that 13B8.2 mAb decreases Raf-1, JNK, and ERK-2 activation induced by PMA or PMA + anti-CD3 mAb in $CD4^+$ T cells. This suggests that 13B8.2 mAb-induced inhibition of NF-κB translocation generated by iHIV-1 binding to CD4 may be related to its effect on ERK activation. Relevant observations were reported (Jabado *et al.*, 1994), indicating that the 13B8.2 mAb significantly decreases the binding activity of NF-AT, NF-κB, and AP-1 to IL-2 gene promoter in CD4-positive T cells after triggering with an anti-CD3 antibody plus a protein kinase C activator. We actually know that inhibition of HIV-1 replication is partly due to signals that prevent NF-κB nuclear translocation, which is involved in HIV-1 promoter stimulation through binding to the two NF-κB sites found in the LTR (Benkirane *et al.*, 1994; Lemasson *et al.*, 1996). Recent results suggest that 13B8.2 mAb negative effects occur at the level of Iκ-B degradation (Briant and Devaux, manuscript in preparation), a protein which acts as a major regulatory element in NF-κB activation. However, we cannot exclude the involvement of nuclear factors other than NF-κB in the hyporesponsiveness of HIV-1 promoter since we have also recently demonstrated that 13B8.2 mAb inhibits HIV-1-induced nuclear translocation of several DNA-binding proteins (Coudronnière and Devaux, 1998). Finally, we still have no satisfactory explanation to the lack of effect of 13B8.2 mAb treatment on virus production in cells infected by HIV-1_{NDK} and HIV-$1_{Lai13EM}$, two viruses which should be sensitive to NF-κB since they contain intact NF-κB sites in their LTR. It is, however, likely that the down-regulation of the Raf-1/ERK-2/I-κB/NF-κB activation pathway significantly contributes to the lack of virus transcription in infected cells treated by CDR3-loop ligands.

An important question that remains to be addressed concerns the mechanism by which CDR3-loop ligands influence the activation of the ERK signaling pathway. CD4 dimerization is required for signal transduction triggering $p56^{lck}$ autophosphorylation (Adam *et al.*, 1993). Crystallographic analysis suggests that CD4 oligomerizes through D3–D4 interactions (Sakihama *et al.*, 1995; Wu *et al.*, 1997). Yet, we and others (Langedijk *et al.*, 1993; Briant *et al.*, 1997) have reported evidence that CD4[CDR3] peptides bind to CD4 itself, suggesting that a putative dimerization site of CD4 involves the CDR3-loop of CD4 domain 1. Interestingly, the folding of the CDR3-loop region is relatively independent from the rest of the structure of the D1 domain of CD4, and the tip of the CDR3-loop represents the only prominent patch of negative potential (Ryu *et al.*, 1990). It is possible that the D1 domain makes a primary contact followed by D3–D4 contacts during CD4 dimer formation. In agreement with this model, we have shown that the E91K,E92K substitution in the CDR3-loop results in the inability of this mutant molecule to transduce signals triggering NF-$_{\kappa}$B activation through engagement of CD4 and also leads to impaired replication of HIV-1 (Briant *et al.*, 1997). Alternatively, Li *et al.*, (1998) recently proposed a hypothetical model of CD4 dimerization and oligomerization. In this model CD4 dimerizes at the D4 interface. Once CD4 dimers are formed, they further associate through the D1 to form an oligomer. Whatever the correct model, CD4[CDR3]-loop ligands may act by uncoupling CD4 from the signal transduction machinery or by inducing immunosuppresive signal in $CD4^+$ T cells or both.

In addition to the effects on T-cell activation, it is very likely that both the 13B8.2 mAb and CDR3.AME(82-89) analog are able to inhibit cell-to-cell transmission of the virus without inhibiting virus propagation in cultures by infectious cell-free virions. Indeed, we have found that the CDR3-loop ligands inhibit recombinant HIV-1 gp120^{env} binding to cell-surface CD4 and cell-surface-expressed gp120-mediated syncytium formation (Corbeau *et al.*, 1993; Zhang *et al.*, 1997). Interestingly, we and others have recently demonstrated that the 13B8.2 mAb inhibits HIV-1-induced apoptosis (Guillerm *et al.*, 1998; Moutouh *et al.*, 1998).

B. Functional Similarities between CDR3-Loop Ligands and IL-16

Interleukin-16 has been described as a suppressor of HIV-1 and SIV infectivity (Baier *et al.*, 1995; Truong *et al.*, 1999). While IL-16 and HIV-1 share a common receptor, the mechanism of IL-16 supression of HIV is probably not a steric hindrance preventing viral binding. Indeed, it was found that the CDR2-like loop in domain 1 of CD4 remains accessible after IL-16 binding to CD4 (Center *et al.*, 1996). Rather, IL-16 inhibits HIV-1 replication by other mechanisms including CD4-dependent signal-transduction events mediating repression of HIV-1 promoter activity. In transient transfections of CD4^{+} lymphoblastoid T cells with HIV-1 LTR-reporter gene constructs, IL-16 was found to repress HIV-1 promoter activity up to 60-fold preventing both PMA activation and Tat transactivation (Maciaszek *et al.*, 1997). Similarly, Zhou and co-workers (1997) reported data suggesting that IL-16-mediated inhibition of HIV-1 occurs at the level of mRNA expression.

Interleukin-16 binding to CD4 influences T-cell signaling and demonstrates immunosuppressive effects, as it inhibits CD3-dependent lymphocytes activation and proliferation (Cruikshank *et al.*, 1996). Recently, IL-16 was found to induce phosphorylation of SEK-1 in CD4^{+} macrophages and activation of p38^{MAPK} and JNK but not ERK-1 and ERK-2 (Krautwald, 1998). Although IL-16 is not considered as able to bind the CDR3-loop in D1 of CD4 (Center *et al.*, 1996), Kurth's group, at the 2nd ECEAR meeting held in Stockholm in 1997, reported preliminary data suggesting cross-competition of CD4 binding between 13B8.2 mAb and IL-16. It may be interesting to reinvestigate this possibility. Finally, IL-16 protected cells against activation-induced apoptosis (Idziorek *et al.*, 1998).

It was recently reported (Ross *et al.*, 1999) that down-regulation of cell-surface CD4 by the Nef protein of HIV is important for the spread of virus from infected cells. Indeed, high levels of surface CD4 reduce the overall amount of virus released from the cell, but the absolute infectivity of released virus is unaffected. According to Harris' model (Harris, 1999), HIV presumably use Nef-mediated CD4 downregulation to prevent the transduction of signals that negatively regulate HIV transcription and concomitant virus production. Such signals might be induced *in vivo* by IL-16 binding to the CDR3-like loop in D1 of CD4. Clearly, this interesting model needs to be further investigated.

V. Trials of Anti-CD4 mAbs for Treatment of HIV-Infected Patients

A. Trials of Anti-CD4 mAbs to Autoimmunity and Transplantation

The first human trials of anti-CD4 mAbs for treatment of autoimmune diseases performed in 1987/1988 concerned arthritis and multiple sclerosis. Since this period, chimeric or humanized mAbs with various isotypes have been developed for most CD4 epitopes and assayed in autoimmune disease and transplantation (for a review see Olive and Mawas, 1993). These anti-CD4 mAbs retain the properties of native parental mAbs. In most reports, with few exceptions, trials that have been performed in autoimmune disease (rheumatoid arthritis, multiple sclerosis, systemic vasculitis, relapsing polychondritis, systemic lupus erythematosus, and psoriasis) emphasize that anti-CD4 regimens are remarkably well tolerated and induce remission of the autoimmune disease symptoms for prolonged periods. Moreover, anti-CD4 trials in human allotransplantation (kidney and heart transplantation) indicate that anti-CD4 mAbs may behave as potent inhibitors for organ rejection. However, it is currently important to explore precisely the efficiency of anti-CD4 mAbs specific for different epitopes in true phase I–II trials; one should expect some anti-CD4 mAbs to be much more potent than others for treatment of diseases.

Very recently, Schedel *et al.* (1999) performed a randomized double-blind, placebo-controlled clinical phase II trial with 13B8.2 mAb in a group of 158 HIV-positive volunteers with 350–500 CD4+ cells/μl. The mAb was well tolerated and the course of $p24^{gag}$ antigen levels was found to be significantly in favor of the 13B8.2 mAb-treated group. The authors concluded that this treatment has a positive impact on the course of HIV disease.

B. Trials of Anti-CD4 mAbs for Treatment of HIV Infected Patients

The use of anti-CD4 mAbs in therapeutic strategies against HIV-1 propagation and AIDS has already been considered by a few medical teams. Rieber and colleagues (1990) have used anti-CD4 mAbs to prevent contamination after there were many accidental injuries of healthcare workers with HIV-contaminated needles. Dhiver and colleagues (1989) have started anti-CD4 treatment as an adjuvant in full-blown AIDS patients; this phase I trial, performed on seven AIDS patients with CD4 levels <80 CD4 T cells/mm^3, combined AZT treatment and a 10-day treatment with 10 mg $\times$ 1 to 40 mg $\times$ 2 of anti-CD4 mAb (13B8.2). The same team (Olive and Mawas, 1993) also reported a phase I–II trial with 13B8.2 mAb in a second series of seven AIDS patients with CD4 levels >100 CD4 T cells/mm^3, who received

AZT and several intravenous infusions of anti-CD4 mAbs (2 × 20 mg/day). These regimens were relatively well tolerated. Interestingly, HIV-1 $p24^{gag}$ antigenemia was lowered to indetectable levels for several patients and viremia became negative for some of them. Finally, because most patients develop anti-idiotypic antibodies to the foreign mAb such immune responses could contribute to prolonging the anti-CD4 mAb therapy through the idiotypic network; according to this speculation, Deckert and colleagues (1996) in a phase I trial inoculated 10 HIV-infected volunteers with about 1–2 mg of 13B8.2 mAb on days 1, 3, 7, 21, 35, and 63 and found the emergence of internal image antibodies that inhibit gp120 binding to CD4.

VI. Trials of CD4-Interacting Peptides in Human Therapy

Although CD4[CDR3] peptides have shown potent therapeutical effect *in vivo* to enhance tissue engraftment (Koch and Korngold, 1997; Koch *et al.*, 1998) and to prevent experimental autoimmune disease such as experimental allergic encephalomyelitis (Jameson *et al.*, 1994; Marini *et al.*, 1996) and mouse colitis, a murine model for the human Crohn's disease (Okamoto *et al.*, 1999), trials in humans have not been documented. Preliminary assessment of potential therapeutical efficency CD4[CDR3] peptides (GLH328 compound) against retrovirus was performed in macaques inoculated with a dose of SIV_{SM} that causes a fatal disease within 7 months (Rausch *et al.*, 1990). Rausch and co-workers reported that infusion of peptide before inoculation of monkeys with SIV_{SM} followed by a weekly injection of 200 mg of GLH328 does not block virus propagation but apparently attenuates the lethal course of infection. No trial has been reported to date in HIV-infected patients.

In contrast to CD4[CDR3] peptides, phase I–II trials using 2–20 mg Peptide T/day/patient have been reported by several medical teams. The effect of treatment with peptide T was documented for patients with psoriasis (Delfino *et al.*, 1992; Talme *et al.*, 1995). For example, Delfino and colleagues (1992) reported that a 2-months treatment with peptide T led to complete remission or improvement of lesions in four of five patients suffering from severe psoriasis. Interestingly, peptide T has also been used for the treatment of neuropathy associated with AIDS and several reports describe improvement in AIDS patients on peptide T (Bridge *et al.*, 1989; Simpson *et al.*, 1996; Kosten *et al.*, 1997; Heseltine *et al.*, 1998). The multibranched peptide construct MBPC1 ([GPGRAF]8-[K]4-[K]2-K-βA-OH), also named SPC3, derived from the HIV-1 $gp120^{env}$ V3 loop, is currently under evaluation in phase I–II trials. It is worth noting that other peptides which do not bind CD4 are currently tested in HIV-1-infected patients; for example, short (14 days) administration of a synthetic peptide (T-20, corresponding to a region of the transmembrane subunit of the HIV-1 gp41^{env}) to 16 HIV-infected adults in four dose groups (3, 10, 30, and 100 mg twice daily) has been recently reported (Kilby *et al.,* 1998). This trial indicated dose-related declines in plasma HIV RNA in all subjects who received the highest dose levels.

VII. Future Developments of CDR3-Loop Ligands

New perspectives are actually open for designing CD4-binding compounds (peptidomimetics, recombinant chimeric molecules, and small organic molecules) having optimal pharmacological properties in terms of potency and stability. Recently, Li and colleagues (1997) reported on a computer-based screening approach that allows identification of nonpeptidic organic ligands in a critical binding pocket consisting of the FG (CDR3-like region) and CC' loops in CD4 D1 that showed therapeutical potential. This strategy is currently being developed in several laboratories, and preliminary *in vitro* data are encouraging. These recent breakthroughs together with the results summarized in this chapter indicate that therapies based on CDR3-loop ligands are useful for treatment of autoimmune diseases, graft rejection, and/or AIDS and that the therapeutical applications of CDR3-loop ligands could be increased in the future.

Acknowledgments

We thank all our colleagues who contributed to the success of the experiments summarized in this chapter. This work was supported by institutional funds from the Centre National de la Recherche Scientifique (CNRS) and the Ministère de l'Enseignement Supérieur de la Recherche et de la Technologie (MESRT 97CO107).

References

Adam, D., Klages, S., Bishop, P., Mahajan, S., Escobedo, J. A., and Bolen, J. B. (1993). Signal transduction through a bimolecular receptor tyrosine protein kinase composed of a platelet-derived growth factor receptor-CD4 chimera and the nonreceptor tyrosine kinase lck. *J. Biol. Chem.* **268,** 19882–19888.

Aiken, C., Konner, J., Landau, N. R., Lenburg, M. E., and Trono, D. (1994). Nef induces CD4 endocytosis: Requirement for a critical dileucine motif in the membrane-proximal CD4 cytoplasmic domain. *Cell* **76,** 853–864.

Arthos, J., Deen, K. C., Chaikin, M. A., Fornwald, J. A., Sathe, G., Sattentau, Q. J., Clapham, P. R., Weiss, R. A., Mc Dougal, J. S., Pietropaolo, C., Axel, R., Truneh, A., Maddon, P. J., and Sweet, R. W. (1989). Identification of the residues in human CD4 critical for the binding of HIV. *Cell* **57,** 469–481.

Attanasio, R., Kanda, P., Stunz, G. W., Buck, D. W., and Kennedy, R. C. (1993). Anti-peptide reagent identifies a primary-structure-dependent, cross-reactive idiotype expressed on heavy and light chains from a murine monoclonal anti-CD4. *Mol. Immunol.* **30,** 9–17.

Autiero, M., Abrescia, P., and Guardiola, J. (1991). Interaction of seminal plasma proteins with cell surface antigens: Presence of a CD4-binding glycoprotein in human seminal plasma. *Exp. Cell Res.* **197**, 268–271.

Autiero, M., Gaubin, M., Mani, J.-C., Castejon, C., Martin, M., El Marhomy, S., Guardiola, J., and Piatier-Tonneau, D. (1997). Surface plasmon resonance analysis of gp17, a natural CD4 ligand from human seminal plasma inhibiting human immunodeficiency virus type-1 gp120-mediated syncytium formation. *Eur. J. Biochem.* **245**, 208–213.

Baier, M., Werner, A., Bannert, N., Metzner, K., and Kurth, R. (1995). HIV suppression by interleukin-16. *Nature* **378**, 563.

Batinic, D., and Robey, F. A. (1992). The V3 region of the envelope glycoprotein of human immunodeficiency virus type 1 binds sulfated polysaccharides and CD4-derived synthetic peptides. *J. Biol. Chem.* **267**, 6664–6671.

Bedinger, P., Moriarty, A., von Borstel II, R. C., Donovan, N. J., Steimer, K. S., and Littman, D. R. (1988). Internalization of the human immunodeficiency virus does not require the cytoplasmic domain of CD4. *Nature* **334**, 162–165.

Benjouad, A., Chapuis, F., Fenouillet, E., and Gluckman, J.-C. (1995). Multibranched peptide constructs derived from the V3 loop of envelope glycoprotein gp120 inhibit human immunodeficiency virus type 1 infection through interaction with CD4. *Virology* **206**, 457–464.

Benkirane, M., Corbeau, P., Housset, V., and Devaux, C. (1993). An antibody that binds the immunoglobulin CDR3-like region of the CD4 molecule inhibits provirus transcription in HIV-infected T cells. *EMBO J.* **12**, 4909–4921.

Benkirane, M., Jeang, K.-T., and Devaux, C. (1994). The cytoplasmic domain of CD4 plays a critical role during the early stages of HIV infection in T-cells. *EMBO J.* **13**, 5559–5569.

Benkirane, M., Hirn, M., Carrière, D., and Devaux, C. (1995a). Functional epitope analysis of the human CD4 molecule: Antibodies that inhibit human immunodeficiency virus type 1 gene expression bind to the immunoglobulin CDR3-like region of CD4. *J. Virol,* **69**, 6898–6903.

Benkirane, M., Schmid-Antomarchi, H., Littman, D. R., Hirn, M., Rossi, B., and Devaux, C. (1995b). The cytoplasmic tail of CD4 is required for inhibition of human immunodeficiency virus type 1 replication by antibodies that bind to the immunoglobulin CDR3-like region in domain 1 of CD4. *J. Virol.* **69**, 6904–6910.

Bour, S., Geleziunas, R., and Wainberg, M. A. (1995a). The human immunodeficiency virus type 1 (HIV-1) CD4 receptor and its central role in promotion of HIV-1 infection. *Microbiol. Rev.* **59**, 63–93.

Bour, S., Schubert, U., and Strebel, K. (1995b). The human immunodeficiency virus type 1 Vpu protein specifically binds to the cytoplasmic domain of CD4: Implication for the mechanism of degradation. *J. Virol.* **69**, 1510–1520.

Brady, R. L., and Barclay, A. N. (1996). The structure of CD4. *In* "Current Topics in Microbiology and Immunology: The CD4 molecule: Roles in T lymphocytes and in HIV Disease" (D. R. Littman, Ed.), Vol. 205, pp. 1–18. Springer-Verlag, New York.

Brenneman, D. E., Westbrook, G. L., Fitzgerald, S. P., Ennist, D. L., Elkins, K. L., Ruff, M. R., and Pert, C. B. (1988). Neuronal cell killing by the envelope protein of HIV and its prevention by vasoactive intestinal peptide. *Nature* **335**, 639–642.

Briant, L., Signoret, N., Gaubin, M., Robert–Hebmann, V., Zhang, X., Murali, R., Greene, M. I., Piatier-Tonneau, D., and Devaux, C. (1997). Transduction of activation signal that follows HIV-1 binding to CD4 and CD4 dimerization involves the immunoglobulin CDR3-like region in domain 1 of CD4. *J. Biol. Chem.* **272**, 19441–19450.

Briant, L., Robert-Hebmann, V., Acquaviva, C., Pelchen-Matthews, A., Marsh, M., and Devaux, C. (1998a). The protein tyrosine kinase $p56^{lck}$ is required for triggering NF-κB activation upon interaction of human immunodeficiency virus type 1 envelope glycoprotein gp120 with cell surface CD4. *J. Virol.* **72**, 6207–6214.

Briant, L., Robert-Hebmann, V., Sivan, V., Brunet, A., Pouysségur, J., and Devaux, C. (1998b). Involvement of extracellular signal-regulated kinase module in HIV-mediated CD4 signals controlling activation of nuclear factor-$_{\kappa}$B and AP-1 transcription. *J. Immunol.* **160,** 1875–1885.

Bridge, T. P., Heseltine, P. N., Parker, E. S., Eaton, E., Ingraham, I. J., Gill, M., Ruff, M., Pert, C. B., and Goodwin, F. K. (1989). Improvement in AIDS patients on peptide T. *Lancet* **2,** 226–227.

Broder, C. C., and Berger, E. A. (1993). CD4 molecules with a diversity of mutations encompassing the CDR3 region efficiently support human immunodeficiency virus type 1 envelope glycoprotein-mediated cell fusion. *J. Virol.* **67,** 913–926.

Brogdon, J., Eckels, D. D., Davies, C., White, S., and Doyle, C. (1998). A site for CD4 binding in the β1 domain of the MHC class II protein HLA–DR1. *J. Immunol.* **161,** 5472–5480.

Burkly, L. C., Olson, D., Shapiro, R., Winkler, G., Rosa, J. J., Thomas, D. W., Williams, C., and Chisholm, P. (1992). Inhibition of HIV infection by a novel CD4 domain 2-specific monoclonal antibody: Dissecting the basis for its inhibitory effect on HIV-induced cell fusion. *J. Immunol.* **149,** 1779–1787.

Camerini, D., and Seed, B. (1990). A CD4 domain important for HIV-mediated syncytium lies outside the virus binding site. *Cell* **60,** 747–754.

Cammarota, G., Scheirle, A., Takacs, B., Doran, D. M., Knorr, R., Bannwarth, W., Guardiola, J., and Sinigaglia, F. (1992). Identification of a CD4 binding site on the β2 domain of HLA-DR molecules. *Nature* **356,** 799–801.

Celada, F., Cambiaggi, C., Maccari, J., Burastero, S., Gregory, T., Patzer, E., Porter, J., McDanal, C., and Matthews, T. (1990). Antibody raised against soluble CD4-rgp120 complex recognizes the CD4 moiety and blocks membrane fusion without inhibiting CD4-gp120 binding. *J. Exp. Med.* **172,** 1143–1150.

Center, D. M., and Cruikshank, W. W. (1982). Modulation of lymphocyte migration by human lymphokines.I. Identification and characterization of chemoattractant activity for lymphocytes from mitogen-stimulated mononuclear cells. *J. Immunol.* **128,** 2563–2568.

Center, D. M., Kornfeld, H., and Cruikshank, W. W. (1996). Interleukin 16 and its function as a CD4 ligand. *Immunol. Today* **17,** 476–481.

Chang, D. K., Chien, W.-J., and Cheng, S.-F. (1996). Characterization of conformation and dynamics of CD4 fragment (81-92) TYICEVEDQKEE and its benzylated derivative by^{1}H NMR spectroscopy and molecular modeling: relevance of conformation to biological function. *J. AIDS Hum. Retrovirology* **11,** 222–232.

Chardès, T., Villard, S., Ferrières, G., Piechaczyk, M., Cerruti, M., Devauchelle, G., and Pau, B. (1999). Efficient amplification and direct sequencing of mouse variable regions from any immunoglobulin gene family. *FEBS Lett.* **452,** 386–394.

Clayton, L. K., Sieh, M., Pious, D. A., and Reinherz, E. L. (1989). Identification of human CD4 residues affecting class II MHC-versus HIV-1 gp120 binding. *Nature* **339,** 548–551.

Corbeau, P., Benkirane, M., Weil, R., David, C., Emiliani, S., Olive, D., Mawas, C., Serre, A., and Devaux, C. (1993). Ig CDR3-like region of the CD4 molecule is involved in HIV-induced syncytia formation but not viral entry. *J. Immunol.* **150,** 290–301.

Corbeil, J., and Richman, D. D. (1995). Productive infection and subsequent interaction of CD4-gp120 at the cellular membrane is required for HIV-induced apoptosis of CD4+ T cells. *J. Gen. Virol.* **76,** 681–690.

Cotton, G., Howie, S. E. M., Heslop, I., Ross, J. A., Harrison, D. J., and Ramage, R. (1996). Design and synthesis of a highly immunogenic discontinuous epitope of HIV-1 gp120 which binds to CD4+ve transfected cells. *Mol. Immunol.* **33,** 171–178.

Coudronnière, N., and Devaux, C. (1998). A novel complex of proteins binds the HIV-1 promoter upon virus interaction with CD4. *J. Biomed. Sci.* **5,** 281–289.

Coudronnière, N., Corbeil, J., Robert-Hebmann, V., Mesnard, J.-M., and Devaux, C. (1998). The lck protein tyrosine kinase is not involved in antibody-mediated CD4(CDR3-loop) signal transduction that inhibits HIV-1 transcription. *Eur. J. Immunol.* **28,** 1445–1457.

Crowley, R. W., Secchiero, P., Zella, D., Cara, A., Gallo, R. C., and Lusso, P. (1996). Interference between human herpesvirus 7 and HIV-1 in mononuclear phagocytes. *J. Immunol.* **156,** 2004–2008.

Cruikshank, W. W., and Center, D. M. (1982). Modulation of lymphocyte migration by human lymphokines.II. Purification of a lymphotactic factor (LCF). *J. Immunol.* **128,** 2569–2571.

Cruikshank, W. W., Center, D. M., Nisar, N., Wu, M., Natke, B., Theodore, A. C., and Kornfeld, H. (1994). Molecular and functional analysis of a lymphocyte chemoattractant factor: Association of biologic function with CD4 expression. *Proc. Natl. Acad. Sci. USA* **91,** 5109–5113.

Cruikshank, W. W., Lim, K., Theodore, A. C., Cook, J., Fine, G., Weller, P. F., and Center, D. M. (1996). IL-16 inhibition of CD3-dependent lymphocyte activation and proliferation. *J. Immunol.* **157,** 5240–5248.

Cullen, B. R. (1998). HIV-1 auxiliary proteins: Making connections in a dying cell. *Cell* **93,** 685–692.

Dalgleish, A. G., Beverley, P. C. L., Clapham, P. R., Crawford, D. H., Greaves, M. R., and Weiss, R. A. (1984). The CD4 (T4) antigen is an essential component of the receptor for the AIDS retrovirus. *Nature* **312,** 763–767.

Deckert, P. M., Ballmaier, M., Lang, S., Deicher, H., and Schedel, I. (1996). CD4-imitating human antibodies in HIV infection and anti-idiotypic vaccination. *J. Immunol.* **156,** 826–833.

Delfino, M., Fabbrocini, G., Brunetti, B., Procaccini, E. M., and Santoianni, P. (1992). Peptide T in the treatment of severe psoriasis. *Acta Derm. Venereol.* **72,** 68–69.

Devaux, C. (1996). Multiple roles played by a unique actor. *Trends Microbiol.* **4,** 411–412.

Devaux, C., Briant, L., and Biard-Piechaczyk, M. (1999). Involvement of CD4 and CXCR4 molecules in HIV-1 gp120^{env}-induced T cell activation and apoptosis. *In* "Recent Research Developments in Virology" (S. Pandälai, Ed.), Vol 1, pp. 139–155. Transworld Research Network, Trivandrum.

Devaux, C., Moinier, D., Mazza, G., Guo, X., Marchetto, S., Fougereau, M., and Pierres, M. (1984). Preferential expression of Vκ21E light chains on IdX Ia.7 positive monoclonal anti-I-E antibodies. *J. Immunol.* **134,** 4024–3766.

Devaux, C., Pierres, M., Epstein, S., and Sachs, D. H. (1986). Xenogeneic antibodies with apparent public idiotypic specificity for anti-Ia.7 antibodies are directed in part against Vκ21D and E subgroup marker. *J. Immunol.* **136,** 3760–3766.

Dhiver, C., Olive, D., Rousseau, S., Tamalet, C., Lopez, M., Galindo, J. R., Mourens, M., Hirn, M., Gastaut, J. A., and Mawas, C. (1989). Pilot phase I study using zidovudine in association with a 10-day course of anti-CD4 monoclonal antibody in seven AIDS patients. *AIDS* **3,** 835–842.

Dougall, W. C., Peterson, N. C., and Greene, M. I. (1994). Design of pharmacologic agents based on antibody structure. *Trends Biotechnol.* **12,** 372–379.

Doyle, C., and Strominger, J. L. (1987). Interaction between œ4 and class II MHC molecules mediates cell adhesion. *Nature* **330,** 256–259.

Dutartre, H., Harris, M., Olive, D., and Colette, Y. (1998). The human immunodeficiency virus type 1 Nef protein binds the Src-related tyrosine kinase Lck SH2 domain through a novel phosphotyrosine independent mechanism. *Virology* **247,** 200–211.

Emiliani, S., Coudronnière, N., Delsert, C., and Devaux, C. (1996). Structural and functional properties of HIV-1GER TAR sequences. *J. Biomed. Sci.* **3,** 31–40.

Fantini, J., Hammache, D., Delézay, O., Yahi, N, André-Barrès, C., Rico-Lattes, I., and Lattes, A. (1997). Synthetic soluble analogs of galactosylceramide (GalCer) bind to the V3 domain of HIV-1 gp120 and inhibit HIV-induced fusion and entry. *J. Biol. Chem.* **272,** 7245–7252.

Favero, J., Corbeau, P., Nicolas, M., Benkirane, M., Travé, G., Dixon, J. F. P., Aucouturier, P., Rasheed, S., Parker, J. W., Liautard, J.-P., Devaux, C., and Dornand, J. (1993). Inhibition of human immunodeficiency virus infection by the lectin jacalin and by a derived peptide showing a sequence similarity with gp120. *Eur. J. Immunol.* **23,** 179–185.

Ferrer, M., and Harrison, S. C., (1999). Peptide ligands to human immunodeficiency virus type 1 gp120 identified from phage display libraries. *J. Virol.* **23**, 5795–5802.

Friedman, T. M., Reddy, A. P., Wassell, R., Jameson, B. A., and Korngold, R. (1996). Identification of a human CD4-CDR3-like surface involved in CD4+ T cell function. *J. Biol. Chem.* **271**, 22635–22640.

Gaubin, M., Autiero, M., Basmaciogullari, S., Métivier, D., Misëhal, Z., Culerrier, R., Oudin, A., Guardiola, J., and Piatier-Tonneau, D. (1999). Potent inhibition of CD4/TCR-mediated T cell apoptosis by a CD4-binding glycoprotein secreted from breast tumor and seminal vesicle cells. *J. Immunol.* **162**, 2631–2638.

Guillerm, C., Robert-Hebmann, V., Hibner, U., Hirn, M., and Devaux, C. (1998). An anti-CD4 (CDR3-loop) monoclonal antibody inhibits human immunodeficiency virus type 1 envelope glycoprotein-induced apoptosis. *Virology* **248**, 254–263.

Guy, B., Kieny, M. P., Rivière, Y., Le Peuch, C., Dott, K., Girard, M., Montagnier, L., and Lecocq, J. P. (1987). HIV F/3′ orf encodes a phosphorylated GTP-binding protein resembling an oncogene product. *Nature* **330**, 266–269.

Haagensen, D. E. Jr, Deilley, W. G., Mazoujian, G., and Wells, S. A. (1990). Review of GCDFP-15, an aprocrine marker protein. *Ann. N.Y. Acad. Sci.* **586**, 161–173.

Harris, M. (1999). HIV: A new role for Nef in the spread of HIV. *Curr. Biol.* **9**, 459–461.

Harris, M. P., and Neil, J. C. (1994). Myristoylation-dependent binding of HIV-1 Nef to CD4. *J. Mol. Biol.* **241**, 136–142.

Healey, D., Dianda, L., Moore, J. P., McDougal, J. S., Moore, M. J., Estess, P., Buck, D., Kwong, P. D., Beverley, P. C. L., and Sattentau, Q. J. (1990). Novel anti-CD4 monoclonal antibodies separate human immunodeficency virus infection and fusion of CD4 cells from virus binding. *J. Exp. Med.* **172**, 1233–1242.

Heseltine, P. N., Goodkin, K., Atkinson, J. H., Vitiello, B., Rochon, J., Heaton, R. K., Eaton, E. M., Wilkie, F. L., Sobel, E., Brown, S. J., Feaster, D., Schneider, L., Goldschmidts, W. L., and Stover, E. S. (1998). Randomized double-blind placebo-controlled trial of peptide T for HIV-associated cognitive impairment. *Arch. Neurol.* **55**, 41–51.

Hivroz, C., Mazerolles, F., Soula, M., Fagard, R., Graton, S., Meloche, S., Sekaly, R. P., and Fischer, A. (1992). Human immunodeficiency virus gp120 and derived peptides activate protein tyrosine kinase p56lck in human CD4 T lymphocytes. *Eur. J. Immunol.* **23**, 600–607.

Houlgatte, R., Scarmato, P., El Marhomy, S., Martin, M., Ostankovitch, M., Lafosse, S., Vervisch, A., Auffray, C., and Piatier-Tonneau, D. (1994). HLA Class II antigens and the HIV envelope glycoprotein gp120 bind to the same face of CD4. *J. Immunol.* **152**, 4475–4488.

Huang, B., Yachou, A., Fleury, S., Hendrickson, W. A., and Sekaly, R.-P. (1997). Analysis of the contact sites on the CD4 molecule with class II MHC molecule: Co-ligand versus co-receptor function. *J. Immunol.* **158**, 216–225.

Idziorek, T., Khalife, J., Billaut-Mulot, O., Hermann, E., Aumercier, M., Mouton, Y., Capron, A., and Bahr, G. (1998). Recombinant human IL-16 inhibits HIV-1 replication and protects against activation-induced cell death (AICD). *Clin. Exp. Immunol.* **112**, 84–91.

Jabado, N., Le Deist, F., Fisher, A., and Hivroz, C. (1994). Interaction of HIV gp120 and anti-CD4 antibodies with the CD4 molecule on human CD4+ T cells inhibits the binding activity of NF-AT, NF-κB and AP-1, three nuclear factors regulating interleukin-2 gene enhancer activity. *Eur. J. Immunol.* **24**, 2646–2652.

Jabado, N., Pallier, A., Jauliac, S., Fisher, A., and Hivroz, C., (1997a). gp160 of HIV or anti-CD4 monoclonal antibody ligation of CD4 induces inhibition of JNK and ERK-2 activities in human peripheral CD4+ T lymphocytes. *Eur. J. Immunol.* **27**, 397–404.

Jabado, N., Pallier, A., Le Deist, F., Bernard, F., Fisher, A., and Hivroz, C. (1997b). CD4 ligands inhibit the formation of multifunctional transduction complexes involved in T cell activation. *J. Immunol.* **158**, 94–103.

Jackson, N. A. C., Levi, M., Wahren, B., and Dimmock, N. J. (1999). Properties and mechanism of action of a 17 amino acid, V3 loop-specific microantibody that binds to and neutralizes human immunodeficiency virus type 1 virions. *J. Gen. Virol.* **80**, 225–236.

James, W., Weiss, R. A., and Simon, J. H. M. (1996). The receptor for HIV: dissection of CD4 and studies on putative accessory factors. *In* "Current Topics in Microbiology and Immunology: The CD4 Molecule: Roles in T Lymphocytes and in HIV Disease" (D. R. Littman, Ed.), Vol. 205, pp. 47–62. Springer-Verlag, New York.

Jameson, B. A., McDonnell, J. M., Marini, J. C., and Korngold, R. (1994). A rationally designed CD4 analogue inhibits experimental allergic encephalomyelitis. *Nature* **368**, 744–746.

Keane, J., Nicoll, J., Kim, S., Wu, D. M. H., Cruikshank, W. W., Brazer, W., Natke, B., Zhang, Y., Center, D. M., and Kornfeld, H. (1998). Conservation of structure and function between human and murine IL-16. *J. Immunol.* **160**, 5945–5954.

Kilby, J. M., Hopkins, S., Venetta, T. M., Di Massimo, B., Cloud, G. A., Lee, J. Y., Alldredge, L., Hunter, E., Lambert, D., Bolognesi, D., Matthews, T., Johnson, M. R., Nowak, M. A., Shaw, G. M., and Saag, M. S. (1998). Potent suppression of HIV-1 replication in humans by T-20, a peptide inhibitor of gp41-mediated virus entry. *Nature Med.* **4**, 1302–1307.

Klatzmann, D., Champagne, E., Chamaret, S., Gruest, J., Guetard, D., Hercend, T., Gluckman, J.-C., and Montagnier, L. (1984). T-lymphocyte T4 molecule behaves as the receptor for human retrovirus LAV. *Nature* **312**, 767–768.

Koch, U., and Korngold, R. (1997). A synthetic CD4-CDR3 peptide analog enhance bone marrow engraftment across major histocompatibility barriers. *Blood* **89**, 2880–2890.

Koch, U., Choksi, S., Marcucci, L., and Korngold, R. (1998). A synthetic CD4-CDR3 peptide analog enhances skin allograft survival across a MHC class II barrier. *J. Immunol.* **161**, 421–429.

König, R., Huang, L. Y., and Germain, R. N. (1992). MHC class II interaction with CD4 mediated by a region analogous to the MHC class I binding site for CD8. *Nature* **356**, 796–798.

Kosten, T. R., Rosen, M. I., McMahon, T. L., Bridge, T. P., O'Malley, S. S., Pearsall, R., and O'Connor, P. G. (1997). Treatment of early AIDS dementia in intravenous drug users: High versus low dose peptide T. *Am. J. Drug Alcohol Abuse* **23**, 543–553.

Krautwald, S. (1998). IL-16 activates the SAPK signaling pathway in $CD4^+$ macrophages. *J. Immunol.* **160**, 5874–5879.

Kruisbeek, A. M., Mond, J. J., Fowlkes, B. J., Carmen, J. A., Bridges, S., and Longo, D. L. (1985). Absence of the Lyt-2^-, L3T4^+ lineage of T cells in mice treated neonatally with anti-I-A correlates with absence of intrathymic I-A-bearing antigen-presenting cell function. *J. Exp. Med.* **161**, 1029–1047.

Kumagai, K., Tokunaga, K., Tsutsumi, M., and Ikuta, K. (1993). Increased anti-HIV-1 activity of CD4 CDR3-related synthetic peptides by scrambling and further structural modifications, including D-isomerization and dimerization. *FEBS Lett.* **330**, 117–121.

Kwong, P. D., Ryu, S. E., Hendrickson, W. A., Axel, R., Sweet, R. M., Folena, W. G., Hensley, P., and Sweet, R. W. (1990). Molecular characteristics of recombinant human CD4 as deduced from polymorphic crystals. *Proc. Natl. Acad. Sci. USA* **87**, 6423–6427.

Lamarre, D., Capon, D. J., Karp, D. R., Gregory, T., Long, E. O., and Sekaly, R. P. (1989). Class II MHC molecules and the HIV gp120 envelope protein interact with functionally distinct regions of the CD4 molecule. *EMBO J.* **8**, 3271–3277.

Lamb, R. A., and Pinto, L. H. (1997). Do Vpu and Vpr of human immunodeficiency virus type 1 and NB of influenza B virus have ion channel activities in the viral life cycles? *Virology* **229**, 1–11.

Langedijk, J. P. M., Puijk, W. C., van Hoorn, W. P., and Meloen, R. H. (1993). Location of CD4 dimerization site explains critical role of CDR3-like region in HIV-1 infection and T-cell activation and implies a model for complex of coreceptor-MHC. *J. Biol. Chem.* **268**, 16875–16878.

Lasarte, J. J., Sarobe, P., Golvano, J., Prieto, I., Civeira, M. P., Gullon, A., Sarin, P. S., Prieto, J., and Borras-Cuesta, F. (1994). CD4-modified synthetic peptides containing phenylalanine inhibit HIV-1 infection *in vitro*. *J. AIDS* 7, 129–134.

Lee, C.-H., Saksela, K., Mirza, U. A., Chait, B. T., and Kuriyan, J. (1996). Crystal structure of the conserved core of HIV-1 Nef complexed with a Src family SH3 domain. *Cell* **85**, 931–942.

Lemasson, I., Briant, L., Hague, B., Coudronnière, N., Heron, L., David, C., Rebouissou, C., Kindt, T., and Devaux, C. (1996). An antibody that binds domain 1 of CD4 inhibits replication of HIV-1, but not HTLV-I, in a CD4-positive/p561ck-negative HTLV-I-transformed cell line. *J. Immunol.* **156**, 859–865.

Lemasson, I., Housset, V., Calas, B., and Devaux, C. (1995). Antigenic analysis of HIV type 1 external envelope (Env) glycoprotein C2 region: Implication for the structure of Env. *AIDS Res. Hum. Retroviruses* **11**, 1177–1186.

Lenburg, M. E., and Landau, N. R. (1993). Vpu-induced degradation of CD4: Requirement for specific amino acid residues in the cytoplasmic domain of CD4. *J. Virol.* **67**, 7238–7245.

Li, S., Gao, J., Satoh, T., Friedman, T. M., Edling, A. E., Koch, U., Choksi, S., Han, X., Korngold, R., and Huang, Z. (1997). A computer screening approach to immunoglobulin superfamily structures and interactions: discovery of small non-peptidic CD4 inhibitors as novel immunotherapeutics. *Proc. Natl. Acad. Sci. USA* **94**, 73–78.

Li. S., Gao, J., Satoh, T., Korngold, R., and Huang, Z., (1998). CD4 dimerization and oligomerization: Implications for T-cell function and structure-based drug design. *Immunol. Today* **19**, 455–462.

Lifson, J. D., Hwang, K. M., Nara, P. L., Fraser, B., Padgett, M., Dunlop, N. M., and Eiden, L. E. (1988). Synthetic CD4 peptide derivatives that inhibit HIV infection and cytopathicity. *Science* **241**, 712–716.

Littman, D. R. (1987). The structure of the CD4 and CD8 genes. *Ann. Rev. Immunol.* **5**, 561–584.

Lohman, K. L., Attanasio, R., Buck, D., Carrillo, M. A., Allan, J. S., and Kennedy, R. C. (1992). Characteristics of murine monoclonal anti-CD4: Epitope recognition, idiotype expression, and variable region gene sequence. *J. Immunol.* **149**, 3247–3353.

Luo, T., Foster, J. L., and Garcia, J. V. (1997). Molecular determinants of Nef function. *J. Biomed. Sci.* **4**, 132–138.

Lusso, P., Secchiero, P., Crowley, R. W., Garzino-Demo, A., Berneman, Z. N., and Gallo, R. C. (1994). CD4 is a critical component of the receptor for human herpesvirus 7: Interference with human immunodeficiency virus. *Proc. Natl. Acad. Sci. USA* **91**, 3872–3876.

Maciaszek, J. W., Parada, N. A., Cruikshank, W. W., Center, D. M., Kornfeld, H., and Viglianti, G. A. (1997). IL-16 represses HIV-1 promoter activity. *J. Immunol.* **158**, 5–8.

Maddon, P. J., Dalgleish, A. G., Mc Dougal, J. S., Clapham, P. R., Weiss, R. A., and Axel, R. (1986). The T4 gene encodes the AIDS virus receptor and is expressed in the immune system and the brain. *Cell* **47**, 333–348.

Maddon, P. J., Littman, D. R., Godfrey, M., Maddon, D. E., Chess, L., and Axel, R. (1985). The isolation and nucleotide sequence of a cDNA encoding the T cell surface protein T4: A new member of the immunoglobulin gene family. *Cell* **42**, 93–104.

Mangasarian, A., Foti, M., Aiken, C., Chin, D., Carpentier, J.-L., and Trono, D. (1997). The HIV-1 Nef protein act as a connector with sorting pathways in the Golgi and at the plasma membrane. *Immunity* **6**, 67–77.

Margottin, F., Bour, S. P., Durand, H., Selig, L., Benichou, S., Richard, V., Thomas, D., Strebel, K., and Benarous, R. (1998). A novel human WD protein, h-βTrCP, that interacts with HIV-1 Vpu connects CD4 to the ER degradation pathway through an F-box motif. *Mol. Cell.* **1**, 565–574.

Marini, J. C., Jameson, B. A., Lublin, F. D., and Korngold, R. (1996). A CD4-CDR3 peptide analog inhibits both primary and secondary autoreactive CD4+ T cell responses in experimental allergic encephalomyelitis. *J. Immunol.* **157**, 3706–3715.

Mazerolles, F., Amblard, F., Lumbroso, C., Lecomte, O., Van-de-Moortele, P. F., Barbat, C., Piatier-Tonneau, D., Auffray, C., and Fischer, A. (1990). Regulation of T helper-B lymphocyte adhesion through CD4-HLA class II interaction. *Eur. J. Immunol,* **20,** 637–644.

McDonnel, J. M., Blank, K. J., Rao, P. E., and Jameson, B. A. (1992). Direct involvement of the CDR3-like domain of CD4 in T helper cell activation. *J. Immunol.* **149,** 1626–1635.

McDougal, J. S., Nicholson, J. K. A., Cross, G. D., Cort, S. P., Kennedy, M. S., and Mawle, A. C. (1986). Binding of the human retrovirus HTLV-III/LAV/ARV/HIV to the CD4 (T4) molecule: Conformation dependence, epitope mapping, antibody inhibition, and potential for idiotypic mimicry. *J. Immunol.* **137,** 2937–2944.

Mizukami, T., Fuerst, T. R., Berger, E. A., and Moss, B. (1988). Binding region for human immunodeficiency virus (HIV) and epitopes for HIV-blocking monoclonal antibodies of the CD4 molecule defined by site-directed mutagenesis. *Proc. Natl. Acad. Sci. USA* **85,** 9273–9277.

Moebius, U., Pallai, P., Harrison, S. C., and Reinherz, E. L. (1993). Delineation of an extended surface contact area on human CD4 involved in class II major histocompatibility complex binding. *Proc. Natl. Acad. Sci. USA* **90,** 8259–8263.

Monnet, C., Laune, D., Laroche-Traineau, J., Piechaczyk, M., Briant, L., Bès, C., Pugnière, M., Mani, J.-C., Pau, B., Cerutti, M., Devauchelle, G., Devaux, C., Granier, C., and Chardès, T. (1999). Synthetic peptides derived from the variable regions of an anti-CD4 monoclonal antibody bind to CD4 and inhibit HIV-1 promoter activation in virus infected cells. *J. Biol. Chem.* **274,** 3789–3796.

Moore, J. P., Sattentau, Q. J., Klasse, P. J., and Burkly, L. C. (1992). A monoclonal antibody to CD4 domain 2 blocks soluble CD4-induced conformational changes in the envelope glycoproteins of human immunodeficiency virus type 1 (HIV-1) and HIV-1 infection of $CD4^+$ cells. *J. Virol.* **66,** 4784–4793.

Moutouh, L., Estaquier, J., Richman, D. D., and Corbeil, J. (1998). Molecular and cellular analysis of human immunodeficiency virus-induced apoptosis in lymphoblastoid T-cell-line expressing wild-type and mutated CD4 receptors. *J. Virol.* **72,** 8061–8072.

Murali, R., and Greene, M. L. (1998). Structure-based design of immunologically active therapeutic peptides. *Immunol. Res.* **17,** 163–169.

Murphy, L. C., Tsuyuki, D., Myal, Y., and Shiu, R. P. C. (1987). Isolation and sequencing of a cDNA clone for a prolactin-inducible protein (PIP). *J. Biol. Chem.* **262,** 15236–15241.

Nara, P. L., Hwang, K. M., Rausch, D. M., Lifson, J. D., and Eiden, L. E. (1989). CD4 antigen-based antireceptor peptides inhibit infectivity of human immunodeficiency virus *in vitro* at multiple stages of the viral life cycle. *Proc. Natl. Acad. Sci. USA* **86,** 7139–7143.

Ohki, K., Kimura, T., Ohmura, K., Kato, S., and Ikuta, K. (1990). Blocking of HIV-1 infection, but not HIV-1-induced syncytium formation, by a CD4 peptide derivative partly corresponding to an immunoglobulin CDR3. *AIDS* **4,** 1160–1161.

Ohki, K., Kimura, T., Ohmura, K., Morikawa, Y., Jones, I. M., Azuma, I., and Ikuta, K. (1992). Monoclonal antibodies to a CD4 peptide derivative which includes the region corresponding to an immunoglobulin CDR3: Evidence of the involvement of pre-CDR3 related region in HIV-1 and host cell interaction. *Mol. Immunol.* **29,** 1391–1400.

Okamoto, S., Watanabe, M., Yamazaki, M., Yajima, T., Hayashi, T., Ishii, H., Mukai, M., Yamada, T., Watanabe, N., Jameson, B. A., and Hibi, T. (1999). A synthetic mimetic of CD4 is able to suppress disease in a rodent model of immune colitis. *Eur. J. Immunol.* **29,** 355–366.

Olive, D., and Mawas, C. (1993). Therapeutic applications of anti-CD4 antibodies. *Crit. Rev. Therapeut. Drug Carrier Syst.* **10,** 29–63.

Péléraux, A., Peyron, J.-F., and Devaux, C. (1998). Inhibition of HIV-1 replication by a monoclonal antibody directed toward the complementarity determining region 3-like domain of CD4 in CD45 expressing and CD45-deficient cells. *Virology* **242,** 233–237.

Pert, C. B., Hill, J. M., Ruff, M. R., Berman, R. M., Robey, W. G., Arthur, L. O., Ruscetti, F. W., and Farrar, W. L. (1986). Octapeptides deduced from the neuropeptide receptor-like pattern of antigen T4 in brain potently inhibit human immunodeficiency virus receptor binding and T-cell infectivity. *Proc. Natl. Acad. Sci. USA* **83**, 9254–9258.

Peterson, A., and Seed, B. (1988). Genetic analysis of monoclonal antibody and HIV binding sites of the human lymphocyte antigen CD4. *Cell* **54**, 65–72.

Pulito, V. L., Roberts, V. A., Adair, J. R., Rothermel, A. L., Collins, A. M., Varga, S. S., Martocello, C., Bodmer, M., Jolliffo, L. K., and Zivin, R. A. (1996). Humanization and molecular modeling of the anti-CD4 monoclonal antibody, OKT4A. *J. Immunol.* **156**, 2840–2850.

Rabehi, L., Seddiki, N., Benjouad, A., Gluckman, J.-C., and Gattegno, L. (1998). Interaction of human immunodeficiency virus type 1 envelope glycoprotein V3 loop with CCR5 and CD4 at the membrane of human primary macrophages. *AIDS Res. Hum. Retroviruses* **14**, 1605–1615.

Rausch, D. M., Hwang, K. M., Padgett, M., Voltz, A. H., Rivas, A., Engleman, E., Gaston, I., Mc Grath, M., Fraser, B., Kalyanamaran, V. S., Nara, P. L., Dunlop, N., Martin, L., Murphey-Corb, M., Kibort, T., Lifson, J. D., and Eiden, L. D. (1990). Peptides derived from the CDR3-homologous domain of the CD4 molecule are specific inhibitors of HIV1 and SIV infection, virus-induced cell fusion, and post-infection viral transmission *in vitro:* Implication for the design of small-peptide anti-HIV therapeutic agents. *Ann. N.Y. Acad. Sci.* **61**, 125–148.

Ravichandran, K. S., Collins, T. L., and Burakoff, S. J. (1996). CD4 and signal transduction. *In* "Current Topics in Microbiology and Immunology: The CD4 Molecule: Roles in T Lymphocytes and in HIV Disease" (D. R. Littman, Ed.), Vol. 205, pp. 47–62. Springer-Verlag, New York.

Repke, H., Gabuzda, D., Palù, G., Emmrich, F., and Sodroski, J. (1992). Effect of CD4 synthetic peptides on HIV type I envelope glycoprotein function. *J. Immunol.* **149**, 1809–1816.

Rey, F., Donker, G., Hirsch, I., and Chermann, J. C. (1991). Productive infection of CD4+ cells by selected HIV strains is not inhibited by anti-CD4 monoclonal antibodies. *Virology* **181**, 165–171.

Rey, M.-A., Spire, B., Dormont, D., Barré-Sinoussi, F., Montagnier, L., and Chermann, J. C. (1984). Characterization of the RNA-dependent DNA polymerase of a new human T-lymphotropic retrovirus (LAV) *Biochem. Biophys. Res. Commun.* **121**, 126–133.

Rieber, E. P., Reiter, C., Gurtler, L., Deinhardt, F., and Riethmuller, G. (1990). Monoclonal CD4 antibodies after accidental HIV infection. *Lancet* **336**, 1007–1008.

Roland, J., Berezov, A., Greene, M. I., Murali, R., Piatier-Tonneau, D., Devaux, C., and Briant, L. (1999). The synthetic CD4 exocyclic CDR3.AME(82-89) inhibits NF-$_{\kappa}$B nuclear translocation, HIV-1 promoter activation, and viral gene expression. *DNA Cell. Biol.* **18**, 819–829.

Ross. T. M., Oran. A. E., and Cullen, B. R. (1999). Inhibition of HIV-1 progeny virion release by cell-surface CD4 is relieved by expression of the Nef protein, *Curr. Biol.* **9**, 613–621.

Rossi, F., Gallina, A., and Milanes, G. (1996). Nef-CD4 physical interaction sensed with the yeast two-hybrid system. *Virology* **217**, 397–403.

Rudd, C. E., Trevillyan, J. M., Dasgupta, J. D., Wong, L. L., and Schlossman, S. F. (1988). The CD4 receptor is complexed in detergent lysates to a protein-tyrosine kinase from human T lymphocytes. *Proc. Natl. Acad. Sci. USA* **85**, 5190–5194.

Ryu, S. E., Kwong, P. D., Truneh, A., Porter, T. G., Arthos, J., Rosenberg, M., Dai, X., Xuong, N. H., Axel, R., Sweet, R. W., and Hendrickson, W. A. (1990). Crystal structure of an HIV-binding recombinant fragment of human CD4. *Nature* **348**, 419–426.

Sakaida, H., Hori, T., Yonezawa, A., Sato, A., Isaka, Y., Yoshie, O., Hattori, T., and Uchiyama, T. (1998). T-tropic human immunodeficiency virus type 1 (HIV-1)-derived V3 loop peptides directly bind to CXCR-4 and inhibit T-tropic HIV-1 infection). *J. Virol.* **72**, 9763–9770.

Sakihama, T., Smolyar, A., and Reinherz, E. L. (1995). Oligomerization of CD4 is required for stable binding to class II major histocompatibility complex proteins but not for interaction with human immunodeficiency virus gp120. *Proc. Natl. Acad. Sci. USA* **92,** 6444–6448.

Sattentau, Q. J., Arthos, J., Deen, K., Hanna, N., Healey, D., Beverley, P. C. L., Sweet, R., and Truneh, A. (1989). Structural analysis of the human immunodeficiency virus-binding domain of CD4: Epitope mapping with site-directed mutants and anti-idiotypes. *J. Exp. Med.* **170,** 1319–1334.

Sattentau, Q. J., Clapham, P. R., Weiss, R. A., Beverley, P., Montagnier, L., Alhalabi, M. F., Gluckmann, J.-C., and Klatzmann, D. (1988). The human and simian immunodeficiency viruses HIV-1, HIV-2, and SIV interact with similar epitopes on their cellular receptor, the CD4 molecule. *AIDS* **2,** 101–105.

Sattentau, Q. J., Dalgleish, A. G., Weiss, R. A., and Beverley, P. C. L. (1986). Epitopes of the CD4 antigen and HIV infection. *Science* **234,** 1120–1123.

Schaller, J., Akiyama, K., Hess, D., Affolter, M., and Rickli, E. E. (1991). Primary structure of a new actin-binding protein from human seminal plasma. *Eur. J. Biochem.* **196,** 743–750.

Schedel, I., Sutor, G.-C., Hunsmann, G., and Jurkiewicz, E. (1999). Phase II study of anti-CD4 idiotype vaccination in HIV positive volunteers. *Vaccine* **17,** 1837–1845.

Schmid-Antomarchi, H., Benkirane, M. Breittmayer, V., Husson, H., Ticchioni, M., Devaux, C., and Rossi, B. (1996). HIV induces activation of phosphatidylinositol 4-kinase and mitogen-activated protein kinase by interacting with T cell CD4 surface molecules. *Eur. J. Immunol.* **26,** 717–720.

Schubert, U., Henklein, P., Boldyreff, B., Wingender, E., Strebel, K., and Porstmann, T. (1994). The human immunodeficiency virus type 1 encoded Vpu protein is phosphorylated by casein kinase-2 (CK-2) at positions Ser52 and Ser56 within a predicted alpha-helix-turn-alpha-helix-motif. *J. Mol. Biol.* **236,** 16–25.

Secchiero, P., Gibellini, D., Flamand, L., Robuffo, I., Marchisio, M., Capitani, S., Gallo, R. C., and Giorgio, Z. (1997). Human herpesvirus 7 induces the down-regulation of CD4 antigen in lymphoid T cells without affecting $p56^{lck}$ levels. *J. Immunol.* **159,** 3412–3423.

Simon, J. H. M., Somoza, C., Schockmel, G. A., Collins, M., Davis, S. J., Williams, A. F., and James W. (1993). A rat CD4 mutant containing the gp120-binding site mediates human immunodeficiency virus type 1 infection. *J. Exp. Med.* **177,** 949–954.

Simpson, D. M., Dorfman, D., Olney, R. K., McKinley, G., Dobkin, J., So, Y., Berger, J., Ferdon, M. B., and Friedman, B. (1996). Peptide T in the treatment of painful distal neuropathy associated with AIDS: Results of a placebo-controlled trial. *Neurobiology* **47,** 1254–1259.

Strebel, K., Klimkait, T., and Martin, M. A. (1988). A novel gene of HIV-1, *vpu,* and its 16-kilodalton product. *Science* **241,** 1221–1223.

Talme, T., Rozell, B. L., Sundqvist, K. G., Wetterberg, L., and Marcusson, J. A. (1995). Histopathological and immunohistochemical changes in psoriatic skin during peptide T treatment. *Arch. Dermatol. Res.* **287,** 553–557.

Tremblay, M., Meloche, S., Gratton, S., Wainberg, M., and Sekaly, R. (1994). Association of p56lck with the cytoplasmic domain of CD4 modulates HIV-1 expression. *EMBO J.* **13,** 774–783.

Truong, M.-J., Darcissac, E. C. A., Hermann, E., Dewulf, J., Capron, A., and Bahr, G. M. (1999). Interleukin-16 inhibits human immunodeficiency virus type 1 entry and replication in macrophages and in dendritic cells. *J. Virol.* **73,** 7008–7013.

Veillette, A., Bookman, M. A., Horak, E. M., and Bolen, J. B (1988). The CD4 and CD8 T cell surface antigens are associated with the internal membrane tyrosine-protein kinase $p56^{lck}$. *Cell* **55,** 301–308.

Vincent, M. J., Raja, N. U., and Jabbar, M. A. (1993). Human immunodeficiency virus type 1 Vpu protein induces degradation of chimeric envelope glycoproteins bearing the cytoplasmic and anchor domains of CD4: Role of the cytoplasmic domain in Vpu-induced degradation in the endoplasmic reticulum. *J. Virol.* **67,** 5538–5549.

Wang, J., Yan, Y., Garrett, T. P. J., Liu, J., Rodgers, D. W., Garlick, R. L., Tarr, G. E., Husain, Y., Reinherz, E. L., and Harrison, S. C. (1990). Atomic structure of a fragment of human CD4 containing two immunoglobulin-like domains. *Nature* **348,** 411–418.

Weil, R., and Veillette, A. (1996). Signal transduction by the lymphocyte-specific tyrosine protein kinase $p56^{lck}$. *In* "Current Topics in Microbiology and Immunology: The CD4 Molecule: Role in T Lymphocytes and in HIV Disease" (D. R. Littman, Ed.), Vol. 205, pp. 63–87. Springer-Verlag, New York.

Weissenhorn, W., Scheuer, W., Kaluza, B., Schwirzke, M., Reiter, C., Flieger, D., Lenz, H., Weiss, E. H., Rieber, E. P., Riethmüller, G., and Weidle, U. H. (1992). Combinatorial functions of two chimeric antibodies directed to human CD4 and one directed to the a-chain of the human interleukin-2 receptor. *Gene* **121,** 271–278.

Wilks, D., Walker, L., O'Brien, J., Habeshaw, J., and Dalgleish, A. (1990). Differences in affinity of anti-CD4 monoclonal antibodies predict their effects on syncytium induction by immunodeficiency virus. *Immunology* **71,** 10–15.

Wu, R., Kwong, P. D., and Hendrickson, W. A. (1997). Dimeric association and segmental variability in the structure of human CD4. *Nature* **387,** 527–529.

Yahi, N., Fantini, J., Mabrouk, K., Tamalet, C., De Micco, P., Van Rietschoten, J., Rochat, H., and Sabatier, J.-M. (1994). Multibranched V3 peptides inhibit human immunodeficiency virus infection in human lymphocytes and macrophages. *J. Virol.* **68,** 5714–5720.

Yasukawa, M., Hatta, N., Sada, E., Inoue, Y., Murakami, T., Onji, M., and Fujita, S. (1996). Human herpesvirus 7 infection of $CD4^+$ T cells does not require expression of the OKT4 epitope. *J. Gen. Virol.* **77,** 3103–3106.

Zhang, X., Piatier-Tonneau, D., Auffray, C., Murali, R., Mahapatra, A., Zhang, F., Maier, C. C., Saragovi, H., and Greene, M. I. (1996). Synthetic CD4 exocyclic peptides antagonize CD4 holoreceptor binding and T cell activation. *Nature Biotechnol* **14,** 472–475.

Zhang, X., Gaubin, M., Briant, L., Srikantan, V., Murali, R., Saragovi, U., Weiner, D., Devaux, C., Autiero, M., Piatier-Tonneau, D., and Greene, M. I. (1997). Synthetic CD4 exocyclics inhibit binding of human immunodeficiency virus type 1 envelope to CD4 and virus replication in T lymphocytes. *Nature Biotechnol.* **15,** 150–154.

Zhang, Y., Center, D. M., Wu, D. M. H., Cruikshank, W. W., Yuan, J., Andrews, D. W., and Kornfeld, H. (1998). Processing and activation of pro-interleukin-16 by caspase-3. *J. Biol. Chem.* **273,** 1144–1149.

Zhou, P., Goldstein, S., Devadas, K., Tewari, D., and Notkins, A. L. (1997). Human CD4+ cells transfected with IL-16 cDNA are resistant to HIV-1 infection: Inhibition of mRNA expression. *Nature Med.* **3,** 659–664.

Keith W. C. Peden* and Joshua M. Farber†

*Laboratory of Retrovirus Research
Center for Biologics Evaluation and Research
Food and Drug Administration and

†Laboratory of Clinical Investigation
National Institute of Allergy and Infectious Diseases
National Institutes of Health
Bethesda, Maryland 20892

Coreceptors for Human Immunodeficiency Virus and Simian Immunodeficiency Virus

I. Introduction

The related discoveries that chemokines can suppress HIV infection and that chemokines receptors are obligate coreceptors for HIV entry into cells have provided major insights into the biology of immunodeficiency-causing retroviruses and the pathogenesis of AIDS that may lead to new therapies.

This chapter addresses some aspects of coreceptor use by human immunodeficiency virus (HIV) and simian immunodeficiency virus (SIV), particularly with respect to *in vitro* systems. It is not intended to be an exhaustive review of the field, since there are many recent and excellent publications to which the reader is referred (Berger *et al.*, 1999; Broder and Collman, 1997; Choe *et al.*, 1998b; Doranz *et al.*, 1997; Lee *et al.*, 1998; Moore *et al.*, 1997). Rather, this chapter concentrates mainly on how *in vitro* systems can provide insights into virus–coreceptor interactions and viral-cellular tropism. We review some of the biology of HIV and SIV as it pertains to

1054-3589/00 $35.00

viral tropism and coreceptor use and the history of receptor and coreceptor discovery for HIV and SIV. Because *in vitro* studies have been instrumental in this field, we review some of the practical aspects of HIV biology, such as the assays that have been used both to reveal coreceptor activity as well as to compare coreceptor efficiencies; which coreceptors are used by some of the commonly used HIV-1, HIV-2, and SIV isolates; which coreceptors are expressed on the CD4-positive cell lines that have been widely used for *in vitro* studies on HIV and SIV and how this knowledge can predict the existence of coreceptors; and how coreceptor use may explain viral tropism. Finally, we summarize briefly the possible role of coreceptors in transmission and disease progression and propose possible roles for alternative coreceptors *in vivo* and how they may take on added importance during therapy targeted at the major coreceptors.

II. Cellular Tropism of HIV and SIV

A. HIV Strains Have Different Tropisms

With the discovery of HIV-1 in 1983 (Barré-Sinoussi *et al.*, 1983; Gallo *et al.*, 1984; Levy *et al.*, 1984) and the recognition that it is the causative agent for AIDS, it was soon found that CD4-positive human lymphocytes were a host cell for the virus. Although in the first published report of HIV-1 the virus was isolated on mitogen-stimulated peripheral blood mononuclear cells (PBMC) (Barré-Sinoussi *et al.*, 1983), many of the early isolates of HIV-1 were isolated on CD4-positive lymphocyte cell lines (Adachi *et al.*, 1986; Gallo *et al.*, 1984; Levy *et al.*, 1984). With hindsight, the use of cell lines for the initial isolation may have been unfortunate. As more HIV-1 isolates and strains became available, it became apparent that many primary viruses were unable to infect *any* CD4-positive cell lines despite being able to establish a productive infection on PBMC to a greater or lesser extent. [In this chapter we use the term "strain" as referring to viruses that are genetically related and an "isolate" as a particular virus from an individual; in general, "isolates" from an individual will belong to the same strain. With this definition, HIV-1_{IIIB} and HIV-1_{LAI}, for example, are different isolates of the same strain (Wain-Hobson *et al.*, 1991), as are HIV-$1_{JR\text{-}FL}$ and HIV-$1_{JR\text{-}CSF}$ (Koyanagi *et al.*, 1987).] It was also found that only a subset of viruses that could infect PBMC could establish a productive infection on monocyte-derived macrophages (MDM) (Gartner *et al.*, 1986). These viruses are referred to as macrophage (M)-tropic isolates. Viruses that can infect primary CD4 T lymphocytes but fail to infect MDM are called T-tropic viruses. Because not all isolates can infect T-cell lines, those that can have been termed T-cell-line (TCL)-tropic viruses. Finally, a subset of TCL-tropic viruses are those that have been adapted to replicate in T-cell lines, and we refer to these as T-cell-line-adapted (TCLA) viruses. The HIV-1_{IIIB} isolate is an example of a TCLA virus, whereas the HIV-1_{LAI} isolate is a TCL-tropic virus, since the former was passaged on HuT78/H9 cells prior to cloning (Fisher *et al.,* 1985), while the latter was only propagated on PBMC prior to cloning (Peden *et al.,* 1991; Wain-Hobson *et al.,* 1985). Therefore, HIV can be characterized as M-tropic, T-tropic, and TCL-tropic. That a virus is M-tropic does not necessarily mean that it cannot replicate in T-cell lines. Certain "dual tropic" isolates, such as the 89.6 (Collman *et al.,* 1990, 1992) and DH12 (Shibata *et al.,* 1995) strains of HIV-1 and the sbl/isy strain of HIV-2 (Hattori *et al.,* 1990), can replicate in both macrophages and T-cell lines.

Another way of classifying HIV isolates was by how well they replicate and whether they induce the formation of mutinucleated giant cells called syncytia. Viruses can be syncytium inducing (SI) or nonsyncytium inducing (NSI) (Tersmette *et al.,* 1988). A related classification referred to viruses as slow/low, which roughly correspond to NSI viruses, or as rapid/high, which roughly correspond to SI viruses (Fenyö *et al.,* 1988). Nonsyncytium-inducing viruses are frequently but not always M-tropic; SI viruses are TCL-tropic. The value of the NSI/SI classification of clinical isolates was the recognition that viral phenotype frequently correlated with clinical status, where the SI viruses typically appear late in the course of disease and are associated with rapid progression of AIDS (Connor and Ho, 1994; Connor *et al.,* 1997; Goudsmit, 1995; Jurriaans *et al.,* 1994; Schuitemaker *et al.,* 1992a; Tersmette *et al.,* 1988; Tersmette *et al.,* 1989a).

It should be emphasized that, while we refer to some isolates as being M-tropic and others as being TCLA viruses, these are not absolute phenotype assignments, but rather they are useful working classifications. As assays become more sensitive and culture conditions are modified (as more knowledge is obtained and culture techniques improved), it is likely that a block to infection may be more apparent than real, and cells that were previously refractory to one virus may now become infectable. Of particular note is the controversy as to whether TCL-tropic and TCLA viruses can infect monocyte-derived macrophages (MDM). Certain laboratories have suggested that TCL-tropic and TCLA viruses can infect MDM. While not doubting the veracity of these reports, it also appears to be true that the efficiency of infection of MDM with the classical M-tropic virus is much higher compared with TCL-tropic or TCLA viruses. This subject is addressed in Section IX,A. More recently, a classification has been adopted based on coreceptor use, and this is detailed below.

B. Viral Determinants of Tropism

Once infectious molecular clones of strains of HIV-1 with different phenotypes and tropisms were available, it became possible to map the regions of the viral genome that determined which cells the viruses could infect. As is the case for other retroviruses, the major HIV determinants for cell tropism lie in the viral envelope (*env*) gene (Cann *et al.*, 1992; Cheng-Mayer *et al.*, 1990, 1991; Chesebro *et al.*, 1991, 1992; Cordonnier *et al.*, 1989; Liu *et al.*, 1990; O'Brien *et al.*, 1990; York-Higgins *et al.*, 1990). Additional mapping implicated the V3 region (the third variable region) of the envelope glycoprotein (Env) as providing the main determinant for M-tropism (Cann *et al.*, 1992; Chesebro *et al.*, 1991, 1992, 1996; De Jong *et al.*, 1992; Fouchier *et al.*, 1992; Hwang *et al.*, 1991; O'Brien *et al.*, 1990; Shioda *et al.*, 1991; Westervelt *et al.*, 1991, 1992), although regions outside V3 clearly contribute quantitatively to the ability of the virus to infect MDM (Shioda *et al.*, 1991; Westervelt *et al.*, 1992). Reciprocal exchanges have been less successful with identifying simple determinants for TCL-tropism (Carrillo and Ratner, 1996a,b; Carrillo *et al.*, 1993).

Demonstrating that the *env* gene carries the main viral determinants for cell tropism of HIV clearly defines entry as the major step that governs the cell tropism of the virus.

III. A History of HIV Receptors

Shortly after the discovery of HIV-1, CD4 was identified as its primary receptor (Dalgleish *et al.*, 1984; Klatzmann *et al.*, 1984). However, with the recognition that CD4 was a receptor for HIV-1 also came the realization that CD4 alone was not sufficient to confer HIV permissivity upon a nonhuman cell (Aoki *et al.*, 1991; Ashorn *et al.*, 1990; Clapham *et al.*, 1991; Maddon *et al.*, 1986) and even upon certain human cells (Chesebro *et al.*, 1990; Clapham *et al.*, 1991; Dragic and Alizon, 1993). Human factors provided via heterokaryon formation (Broder *et al.*, 1993; Dragic and Alizon, 1993; Dragic *et al.*, 1992) or from a subset of human chromosomes in interspecies cell hybrids (Ramarli *et al.*, 1993; Tersmette *et al.*, 1989b; Weiner *et al.*, 1990, 1991) were able to confer upon mouse cells the capacity of human CD4 to allow fusion with cells expressing the HIV-1 Env. Thus, cellular entry factors in addition to CD4 were postulated to exist. Proteins such as CD26 (Callebaut *et al.*, 1993) and even nonprotein factors (Dragic *et al.*, 1992) were invoked. Despite considerable effort and the failure of promising candidates such as CD26 to be confirmed as a coreceptor (Alizon and Dragic, 1994; Broder *et al.*, 1994; Camerini *et al.*, 1994; Lazaro *et al.*, 1994; Patience *et al.*, 1994; Watkins *et al.*, 1996), it was only in 1996 that the first genuine fusion cofactor, or coreceptor, for HIV-1 was identified by Berger and colleagues using a functional enrichment assay to isolate a clone from a cDNA library prepared from HeLa cells that, when expressed in target cells with CD4, allowed CD4-dependent fusion of the Env from HIV-1_{NL4-3} (Feng *et al.*, 1996). The authors called this gene/protein "fusin" to indicate its activity with the HIV Env, although this gene had been cloned and described as encoding an orphan receptor termed HUMSTR and LESTR among others (Federsppiel *et al.*, 1993; Herzog *et al.*, 1993; Jazin *et al.*, 1993; Nomura *et al.*, 1993; Loetscher *et al.*, 1994). LESTR/HUMSTR/fusin was a protein that appeared to belong to the chemokine subfamily of seven transmembrane (7TM), G-protein coupled receptors. The ligand for this receptor was subsequently determined to be stromal-derived factor 1 (SDF-1) (Bleul *et al.*, 1996; Oberlin *et al.*, 1996), which was a chemokine belonging to the CXC or α-chemokine subfamily, and LESTR/HUMSTR/fusin was renamed CXCR4.

The identification that the chemokine receptor CXCR4 was a coreceptor for TCL-tropic and TCLA HIV-1 strains had fortuitously followed closely on work reported by Gallo and colleagues, who had found that factors produced by CD8 cells could inhibit certain M-tropic isolates of HIV-1 but not TCL-tropic or TCLA viruses (Cocchi *et al.*, 1995). The factors they identified—RANTES, MIP-1α, and MIP-1β—were also chemokines, but these belonged to the CC or β-chemokine family. The coincidence between CXCR4 as a coreceptor for HIV-1 and the blocking of M-tropic HIV-1 strains by certain β-chemokines rapidly led to the identification of CCR5 as a second coreceptor for HIV-1, this time for M-tropic viruses (Alkhatib *et al.*, 1996; Choe *et al.*, 1996; Deng *et al.*, 1996; Doranz *et al.*, 1996; Dragic *et al.*, 1996).

In an effort to simplify and standardize the various ways of describing viral phenotypes, a revised classification system has been proposed based on the coreceptor use by the virus (Berger *et al.*, 1998). In this system, a virus that uses CCR5 as coreceptor is termed an R5 virus, one that uses CXCR4 is an X4 virus, and viruses that use both are termed R5X4 viruses. Viruses that were classified as slow/low and NSI used CCR5, whereas those that were rapid/high and SI were found to use CXCR4 and sometimes used additional coreceptors such as CCR3, CCR2B, and STRL33 (Björndal *et al.*, 1997). Although such a classification system has the advantage of simplicity and clarity and can be adapted to describe viruses that use other coreceptors, it may be an oversimplification, and this is discussed in Section X.

Now that two chemokine receptors had been shown to be coreceptors, other chemokine receptors as well as a number of presumed orphan chemokine receptors and other 7TM, G-protein coupled receptors were examined for coreceptor activity. Before describing these additional findings, we review the assays used to assess coreceptor activity.

IV. Assays for Coreceptor Activity

Because the identification of coreceptors has been assay-dependent and because there are some disparities in the literature with respect to whether a certain chemokine receptor or 7TM, G-protein coupled protein can function as an HIV coreceptor, it is relevant to discuss the types of assays, since the disparities are the likely consequences of the particular assay used to assess activity.

A. Fusion Assays

1. Syncytium Formation (SF) Assay

The simplest type of fusion assay measures the formation of multinucleated giant cells, or syncytia, by the fusion of cells expressing the HIV Env with cells expressing CD4 and a coreceptor; syncytium formation is monitored by light microscopy. While the SF assay can be used with HIV infection, it is more commonly used with expression systems. For example, vaccinia virus vectors are used to introduce CD4 and a coreceptor into one cell, the target cell, and HIV Env into another, the effector cell. Syncytium formation is measured after mixing the target and effector cells (Choe *et al.,* 1996; Zaitseva *et al.,* 1997). The disadvantages of this assay are (1) that it is not conveniently made quantitative and (2) not all HIV-1 isolates are syncytium inducing (SI) and therefore the SF assay is only applicable with a subset of viruses/Envs.

2. Vaccinia T7 System with the β-Galactosidase Reporter Gene

This assay is notable in that it was the assay used by the group that first identified an HIV coreceptor (Feng *et al.,* 1996). In one version of this assay, one cell, the effector cell, is infected with a vaccinia virus vector that expresses an HIV Env together with a vaccinia virus that expresses the *E. coli* β-galactosidase gene under the control of the phage T7 promoter. A second cell, the target cell, is infected with two vaccinia virus vectors: one that expresses CD4 and one that expresses the phage T7 DNA-dependent RNA polymerase. Into the target cell is also introduced an expression vector for a coreceptor, usually by transfection prior to vaccinia virus infection. Mixing the target and effector cells leads to fusion if the HIV Env functions with the coreceptor; the amount of this fusion is scored by the activity of β-galactosidase (Feng *et al.,* 1996; Nussbaum *et al.,* 1994). The advantage of this assay is that it is easy to perform, highly sensitive, and quantitative. The disadvantages are that all reagents need to be generated in the vaccinia virus vector and the assay may be so sensitive that it reveals activities that may not be biologically meaningful. This latter point is discussed later.

3. Vaccinia T7 System with the Luciferase Reporter Gene

A variation of the above assay is one that uses the firefly luciferase (*luc*) gene from *Photinus pyralis* as the reporter gene (Doranz *et al.,* 1996). In this assay, the target cells, usually quail QT6 cells, are transfected with expression plasmids for human CD4, the particular coreceptor, and the luciferase gene under the control of the phase T7 promoter. The effector cells are usually HeLa cells that have been infected with one vaccinia virus vector that expresses an HIV Env and another that expresses the T7 RNA polymerase. The extent of fusion is assessed by mixing the target and effector cells and scoring luciferase activity. This assay has similar advantages and disadvantages as the β-galactosidase version above.

4. Vaccinia T7 System with the Secreted Alkaline Phosphatase (SEAP) Reporter Gene

Another variation of the assay has recently been reported that uses the SEAP reporter (Lee *et al.*, 1999). This system has the advantage that the activity is measured in the medium, since the reporter protein is secreted.

B. PCR-Based Entry Assays

These assays measure the production of the complementary (c) DNA from the genomic RNA by the action of the viral reverse transcriptase (RT) after infection. Thus, they measure an event subsequent to fusion. By the use of appropriate primers, the production of viral DNA rather than any contaminating DNA present in the virus stock can be specifically measured. As such, these PCR-based entry assays can be selective and sensitive measurements of the early events of viral infection. This type of assay was used to assess whether resting T cells could be infected by HIV *in vitro* (Spina *et al.*, 1995; Stevenson *et al.*, 1990; Zack *et al.*, 1990, 1992). The assay is particularly useful when small quantities of material and the inability to culture infected cells may preclude the use of other assays. For example, Golding and colleagues have used the assay to study coreceptor expression in freshly isolated thymocytes (Zaitseva *et al.*, 1998) and Langerhans cells (Zaitseva *et al.*, 1997).

C. Single-Cycle Infectivity Assays

1. Virus Production from a Single Cycle of Productive Infection

The ability to measure the virus produced from the initial infection cycle is technically challenging with HIV, since stocks of this virus are generally of low titer and to produce stocks capable of infecting at a multiplicity of infection (MOI) of greater than one requires the preparation of significant volumes of virus and the ability to concentrate the stocks. While infection at an MOI of 1 was reported by Kim *et al.* (1989b) in an analysis of the RNA species produced after a single round of HIV infection, this study was possible because the newly expressed RNA species is distinguishable from the input viral genomic RNA by size. Had this not been the case, the assay would not have been successful due to the presence of large amounts of residual input virus, which is difficult to remove and would have obscured a read-out if an HIV virion protein such as p24 had been monitored.

Recently, it has been possible to measure the virus produced from the initial round of infection using much lower MOIs (Shapiro *et al.,* 1999). This has been possible due to the development of highly sensitive reverse transcriptase (RT) assays that incorporate a PCR step to measure RT activity on a heteromorphic RNA template. These PCR-based RT (PBRT) assays are a millionfold more sensitive than conventional RT assays and have the theoretical capacity to detect a single retroviral particle (Heneine *et al.,* 1995; Maudru and Peden, 1997; Pyra *et al.,* 1994; Silver *et al.,* 1993). These types of sensitive RT assays remain to be exploited for the analysis of coreceptor activity.

2. Single-Cycle Infection with a Pseudotyped or "Heterotyped" Virus with a Reporter Gene

The simplest way to obtain a single-cycle infection is to use a virus that is able to infect cells but is unable to produce infectious virus, and thus the infection is biologically limited to the initial round. Such a system usually incorporates a virus defective in Env expression but has the other genes necessary for the production of a particle and uses an expression plasmid in trans to provide the Env. In this way, particles produced from one plasmid are provided with a functional Env from the other plasmid. The particles are infectious because they have a viral RNA genome that can be reverse transcribed by the virion RT to a functional two-LTR double-stranded DNA molecule that, in the presence of the viral integrase, is integrated into the host genome and becomes the provirus. The Env may be from the same HIV-1 strain or from different HIV-1 strains or even from HIV-2, SIV, or nonlentiviruses such as the G protein from vesicular stomatitis virus (VSV) or an Env from an amphotropic murine leukemia virus. Such particles have been termed pseudotyped or xenotyped when the Env is from a different virus type, and while pseudotype has been used for HIV particles with an Env supplied in trans from a different HIV strain, the term "heterotype" may be more apt. The read-out for this type of assay is generally a reporter gene product, the gene for which is incorporated into the *env*-deleted HIV genome, usually in the place of the *nef* gene but sometimes in the deleted *env* region. Since Nef is not absolutely required for HIV replication in CD4-positive cell lines (Ahmad and Venkatesan, 1988; Kim *et al.,* 1989a; Luciw *et al.,* 1987; Ryan-Graham and Peden, 1995; Terwilliger *et al.,* 1986), elimination of Nef function is presumed to have an insignificant effect on the infectivity of these pseudotyped particles. More recently, *nef*$^{+}$ versions of HIV-1 with reporter genes have been developed by using an internal ribosome entry site (IRES) to express Nef after the reporter gene in a bicistronic mRNA (Chen *et al.,* 1996).

Different reporter genes have been used and provide sensitive read-out assays.

1. Cat reporter expression. Sodroski and colleagues (Helseth *et al.*, 1990) pioneered this Env-complementation approach for the analysis of HIV Env activity and more recently used it to measure coreceptor activity with different Envs. In their system, the chloramphenicol acetyl transferase (*cat*) gene is inserted into the *nef* gene and is expressed from the Nef RNA. Infection results in the expression of Cat, and the amount of Cat activity is a reflection of the infection efficiency.

2. Luciferase reporter expression. A similar system was developed by Baltimore and colleagues (Chen *et al.*, 1994) and Landau and colleagues (Connor *et al.*, 1995) but using the firefly *Photinus pyralis* luciferase (*luc*) gene as the reporter. For optimum sensitivity, the luciferase assay requires a luminometer, but even using a scintillation counter, the assay is still about 100-fold more sensitive than the Cat assay (Alam and Cook, 1990). This assay is now probably the most commonly used of the reporter-gene assays for the assessment of coreceptor activity.

3. Green fluorescent protein (GFP) reporter expression. The green fluorescent protein from the jellyfish *Aequorea victoria* has become widely used in cell and molecular biology. As a reporter gene in the single-cycle infection assays, two approaches have been taken. In one approach, the *GFP* gene (or more usually a "humanized" version of the gene) is inserted into the HIV-1 genome in a similar position to the *cat* and *luc* genes described above (He *et al.*, 1997; Herbein *et al.*, 1998; Lee *et al.*, 1997; Page *et al.*, 1997); an analogous virus has been constructed for SIV (Alexander *et al.*, 1999). The infected cells can be identified by flow cytometry or by *in situ* fluorescence microscopy. This type of assay has the power to identify the infected cell, and, with multiple-color flow cytometry and several cell-surface marker-specific antibodies, the precise coreceptor display on the infected cell could be described.

In the second approach, the *GFP* gene was placed under the control of the HIV-1 LTR and cell lines were derived that contain integrated copies of this expression cassette (Dorsky *et al.*, 1996; Gervaix *et al.*, 1997). In these cells, the *GFP* gene remains silent in the absence of HIV infection, as significant expression from the LTR requires the HIV-1 Tat protein. Infection with HIV-1 provides Tat in trans and results in the expression of GFP in the infected cells, which can be detected by fluorescence microscopy or quantified by flow cytometry. This approach is derived from the earlier assays that used the β-galactosidase gene under the control of an HIV LTR as the reporter gene in a CD4-positive HeLa cell line (Akrigg *et al.*, 1991; Kimpton and Emerman, 1992; Rocancourt *et al.*, 1990); infection with HIV leads to production of β-galactosidase, which is measured by *in situ* staining and light microscopy. The first generation of cells with the LTR-*GFP*, CD4, and different coreceptors were termed GHOST4 and were developed by Littman and colleagues (KewalRamani and Littman, unpublished; Zhang *et al.*, 1998b) from a human osteosarcoma (HOS) cell line; these cells have been disseminated widely. The GHOST4 flow cytometry assay was used by Littman and colleagues to characterize some coreceptors (Deng *et al.*, 1997; Michael *et al.*, 1998).

4. The human placental alkaline phosphatase (PLAP) reporter expression. The use of this reporter was introduced by Cepko and colleagues (Cepko *et al.*, 1993; Fields-Berry *et al.*, 1992) and applied to HIV by Landau and colleagues (He *et al.*, 1995; He and Landau, 1995) and Baltimore and colleagues (Chen *et al.*, 1996, 1997). This reporter system has the advantage that the infected cells can be identified by flow cytometry or microscopy after either an *in situ* phosphatase assay or immunohistochemistry staining with antibodies to PLAP.

D. Productive Infection Assays

Historically, these assays are the most common, as they can be used with the virus itself and therefore no additional reagents need be developed. Productive infection assays measure the kinetics of virus infection and have the advantage that subtle differences in replication capacity can often be discerned, since the assay measures the total of all the steps in the viral life cycle. Moreover, the ability of cells expressing a given coreceptor to support productive infection may indicate that the coreceptor has activity *in vivo*. The main disadvantage is that, if a virus fails to establish a productive infection in a given cell type, the stage at which the infection is blocked is not immediately apparent.

Any permissive cell can be used for the productive infection assay, including primary cells and cell lines. Because most and perhaps all of the T-cell lines express more than one coreceptor, the use of such cells to examine coreceptor use can be complicated. For this reason, the cells generated to express single coreceptors are invaluable. Examples of these are the above-mentioned HOS cell derivatives and also the U87.CD4 lines (Björndal *et al.*, 1997; Clapham *et al.*, 1991; Deng *et al.*, 1996). However, as discussed in Section VII, HOS and U87 cells endogenously express at least one coreceptor.

V. Comparison of the Different Assays for Coreceptor Activity

Because different assays have been used to assess coreceptor activity and different results with the same coreceptors have been seen using different assays, we discuss some of the likely reasons for this finding. From our own work, the alternative coreceptor STRL33 supported the fusion of a wide range of Envs from HIV-1 (Liao *et al.*, 1997) as well as from SIVmac (Alkhatib *et al.*, 1997) using the vaccinia virus-based method. However, when assayed in the context of a productive infection, many fewer Envs could be demonstrated to use STRL33 (unpublished results). In fact, we only identified the ELI1 primary-like isolate of HIV-1 (Alizon *et al.*, 1986; Peden *et al.*, 1991) as being able to use STRL33, and this was only in the context of Jurkat cells (Liao *et al.*, 1997), since HOS cells expressing Bonzo (Deng *et al.*, 1997), another name for STRL33, were unable to support the replication of ELI1 (unpublished results). While there are several possible reasons for this, the simplest and most likely explanation is that fusion is an efficient process, particularly when mediated by the high levels of coreceptor and CD4 on the target cells and the HIV Env on the effector cells that were achieved through the vaccinia virus T7 expression system and transient transfection. Such a highly efficient fusion system can detect coreceptor activity of candidate coreceptors that are only weakly able to catalyze fusion and entry. These *in vitro*-defined fusion coreceptors may not be active enough to function where the levels of the participating proteins are not elevated, and they may not act as coreceptors *in vivo*. Therefore, coreceptors defined solely by such "maximal" systems will need to be demonstrated as biologically relevant by other methods. Nevertheless, these maximal assays are useful for the initial demonstration of coreceptor activity, since it is unlikely that candidate coreceptors will be missed by using them.

In conclusion, *in vitro* coreceptor assays can be ranked on the basis of sensitivity and stringency: the vaccinia virus-based fusion assays and the PCR-based entry assays are the most sensitive, the productive infection assay the most stringent, and the single-cycle infection assays fall in between.

VI. Discovery of Additional Coreceptors (1996 to 1999)

While the main coreceptors for HIV are CCR5 and CXCR4, a number of other 7TM proteins have been found to have coreceptor activity. At present, 14 7TM proteins in all have been shown to function as coreceptors for HIV and/or SIV. These are listed in Table I. Most of these were identified by screening known and presumed chemokine receptors as well as orphan receptors using mainly a vaccinia virus-based fusion assay or a single-cycle infection assay. Others were identified using a functional cloning strategy. For example, Littman and colleagues used a selection system to identify two coreceptors for SIV (Deng *et al.*, 1997). A cDNA library prepared from a human T-cell clone in a retrovirus expression vector was used to infect mouse 3T3.CD4 cells, a mouse fibroblast cell line engineered to express human CD4. The transduced cells were infected with an HIV mutant carrying a puromycin-resistance gene marker that was pseudotyped either with

TABLE I Coreceptors for HIV and/or SIV

Coreceptors	*References*
CCR2B	Doranz *et al.* (1996); Frade *et al.* (1997)
CCR3	Alkhatib *et al.* (1997); Choe *et al.* (1996); Doranz *et al.* (1996); He *et al.* (1997)
CCR5	Choe *et al.* (1996); Deng *et al.* (1996); Doranz *et al.* (1996); Dragic *et al.* (1996) Alkhatib *et al.* (1996)
CCR8	Goya *et al.* (1998); Horuk *et al.* (1998)
D6	Choe *et al.* (1998)
CXCR4	Feng *et al.* (1996)
CX_3CR1	Combadière *et al.* (1998); Reeves *et al.* (1997); Rucker *et al.* (1997)
Gpr1	Farzan *et al.* (1997); Marchese *et al.* (1994)
Gpr15/BOB	Deng *et al.* (1997); Farzan *et al.* (1997); Heiber *et al.* (1996)
STRL33/Bonzo	Deng *et al.* (1997); Liao *et al.* (1997)
ChemR23	Samson *et al.* (1998)
Apj	Choe *et al.* (1998); Edinger *et al.* (1998); O'Dowd *et al.* (1993)
US28	Pleskoff *et al.* (1997)
BLTR	Owman *et al.* (1996; 1997a,b; 1998)

an Env from SIVagmTY01 or from SIVmac1A11. Those particular mouse cells in the culture that had been transduced by a virus carrying a cDNA that provides coreceptor activity will be infected by the pseudotyped virus, and these cells can be selected for through the expression of the puromycin-resistance gene. With this system, two coreceptors were found, which were called Bonzo and BOB. It turned out that both genes had been previously described—BOB was Gpr15 (Heiber *et al.*, 1996) and Bonzo was STRL33 (Liao *et al.*, 1997), the latter having been shown to be an HIV-1 coreceptor. Nevertheless, that two coreceptors were isolated demonstrates the power of this functional cloning approach.

The following have been identified as alternative coreceptors either for HIV or for the various types of SIV and sometimes for both.

1. CCR2B. Subsequent to its initial description as an HIV coreceptor (Choe *et al.*, 1996; Doranz *et al.*, 1996), additional studies have not found that CCR2B is used by many HIV-1 isolates (Connor *et al.*, 1997). While the original assays used were both a fusion assay (Doranz *et al.*, 1996) as well as a single-cycle infection assay (Choe *et al.*, 1998a), both with the 89.6 strain of HIV-1, so far CCR2B has not been shown to function in a productive infection assay.

2. CCR3. This chemokine receptor has been postulated to be an important coreceptor for the infection of microglial cells (He *et al.*, 1997) and is used by some primary and laboratory strains of HIV-1 (Choe *et al.*, 1996, 1998a; Connor *et al.*, 1997; Doranz *et al.*, 1996; He *et al.*, 1997) and HIV-

2 (Bron *et al.*, 1997; Sol *et al.*, 1997). Subsequent studies have not revealed such an important role for CCR3 in the brain, as most brain isolates use CCR5 and not CCR3 (Albright *et al.*, 1999; Shieh *et al.*, 1998).

3. CCR8. This chemokine receptor, whose ligand is I309 (Horuk *et al.*, 1998; Tiffany *et al.*, 1997), has been shown to act as a fusion cofactor for several SIVsm, HIV-2, and HIV-1 isolates by the fusion assay (Rucker *et al.*, 1997), although its activity in a single-cycle infection assay is quantitatively low (Choe *et al.*, 1998a). This receptor is not widely used and has not been shown to function in a productive infection assay.

4. D6 (Formally Referred to as CCR9). Whether this is a functional chemokine receptor is controversial, as the ligands originally proposed were not confirmed to signal in subsequent studies. Thus, D6 remains an orphan receptor and will have to be named officially if/once a signaling ligand(s) is (are) identified. D6 has only been shown to function well with one isolate so far, the UG21 primary isolate of HIV-1, and weakly with the ELI1 and 89.6 isolates of HIV-1 and with SIVmac316 in a single-cycle infection assay (Choe *et al.*, 1998a).

5. CX_3CR1. Formally known as V28 or CMKBRL1, this protein was shown to be the receptor for fractalkine (Imai *et al.*, 1997; Combadière *et al.*, 1998), a CX_3C chemokine. It was shown to be a coreceptor for several isolates of HIV-1 and HIV-2 (Combadière *et al.*, 1998; Reeves *et al.*, 1997; Rucker *et al.*, 1997; Zhang *et al.*, 1998b).

6. Gpr1. This orphan 7TM protein has only shown limited coreceptor activity with HIV and SIV isolates. Originally, only SIV isolates were shown to use Gpr1 as a coreceptor (Farzan *et al.*, 1997; Zhang *et al.*, 1998b), but recently Gpr1 has been shown to function for certain HIV-1 and HIV-2 strains (Shimizu *et al.*, 1999).

7. Gpr15/BOB. Before Gpr15 was identified as a coreceptor, the existence of another coreceptor was predicted (Kirchhoff *et al.*, 1997), since many SIVmac strains were able to establish productive infections on the B cell–T cell hybrid cell line CEMx174. Because this cell line expresses CXCR4 (Kirchhoff *et al.*, 1997), which does not function with all the isolates known to infect CEMx174, but not CCR5 (Kirchhoff *et al.*, 1997), Kirchhoff and colleagues suggested that these cells must have a hitherto unidentified coreceptor for SIVsm strains. [Recently, CEMx174 cells have also been shown to express CCR8 (C.-R. Yu and J.M.F., unpublished), but this receptor is not used by many HIV or SIV isolates.] The major SIVmac coreceptor on CEMx174 cells is likely to be Gpr15 (Heiber *et al.*, 1996), which was identified as a coreceptor and named BOB by Littman and colleagues, who showed this receptor to be active with SIVmac239, SIVmac1A11, and SIVagmTY01 (Deng *et al.*, 1997). While the extent of its use with HIV-1 remains to be determined (Edinger *et al.*, 1998a), Gpr15 is certainly used by the majority of SIVsm and SIVmac isolates. (The various types of SIV are described in the Appendix.) Other SIV types do not appear to use Gpr15

as coreceptor (K.P., unpublished). These include SIVagmSAB, SIVagmTAN, and SIVcpzGAB, a virus more closely related to HIV-1 rather than to the other SIV types (Gao *et al.*, 1999; Huet *et al.*, 1990); SIVcpzGAB seems to use CCR5 exclusively (K.P., unpublished). Recently, several primary HIV-2 isolates (Mörner *et al.*, 1999) and HIV-1 isolates (Pöhlmann *et al.*, 1999) have been described that use Gpr15.

8. STRL33/Bonzo. STRL33 was cloned from activated T cells using degenerate primers in PCR, and STRL33 RNA was found to be expressed in all lymphoid tissues but not in monocytes or macrophages (Liao *et al.*, 1997). This orphan receptor was shown to act as a fusion coreceptor for M-tropic, dual-tropic, and TCLA HIV-1 isolates (Liao *et al.*, 1997) as well as for SIVmac (Alkhatib *et al.*, 1997) and other SIV strains and types (Deng *et al.*, 1997). However, despite the fusion activity of STRL33 with a broad range of HIV-1 and SIV Envs, identifying a virus that could use this orphan receptor as a coreceptor for productive infection was difficult, although one virus, the ELI1 strain of HIV-1, could be shown to have an enhanced replication capacity in Jurkat cells that express STRL33 compared with the parent Jurkat cells (Liao *et al.*, 1997). Others have not been able to confirm that the ELI1 Env can use STRL33 efficiently as coreceptor (Choe *et al.*, 1998a). This group also could not confirm that other HIV-1 Envs, such as the ADA isolate, could use STRL33, although they did confirm that STRL33 provided coreceptor activity for the SIVmac239 and SIVmac316 Envs. The reasons for these discrepancies are unknown, although, as discussed above, they are likely due to the use of assays with different sensitivities. STRL33 was also cloned and identified as a coreceptor and called Bonzo by Littman and colleagues (Deng *et al.*, 1997) and called TYMSTR by Moser and colleagues (Loetscher *et al.*, 1997). In addition to SIV strains, recent work has demonstrated that STRL33 can be used by some HIV-2 isolates as well as by certain primary HIV-1 isolates (Björndal *et al.*, 1997; Zhang *et al.*, 1998b). However, most HIV-1 isolates do not use STRL33 well (Edinger *et al.*, 1998a; Pöhlmann *et al.*, 1999; Zhang *et al.*, 1998a,b), and recent work has shown that even a primary isolate that uses STRL33 as well as CCR5, nevertheless uses CCR5 given the choice in PBMC, since this STRL33-using virus fails to replicate in PBMC derived from a homozygous Δ32 CCR5 individual and replication in CCR5 wild-type PBMC is blocked by ligands for CCR5 (Zhang and Moore, 1999). One caveat to this result is that, although T-cell activation can induce the expression of STRL33, there was no demonstration that the PBMC used in the infection assay were expressing this coreceptor.

9. ChemR23. This orphan receptor was cloned by PCR using degenerate primers designed from opioid and somatostatin receptors (Samson *et al.*, 1998). The sequence of ChemR23 was closer to the receptors for the chemoattractants anaphylatoxin C3a and C5a rather than to the CC or CXC receptor families. ChemR23 is only expressed at low levels in T lym-

phocytes but is expressed at high levels in macrophages and dendritic cells. Using a vaccinia-based fusion assay (Doranz *et al.,* 1996), ChemR23 was shown to function with many SIVsm isolates but with only one HIV-1 isolate. No HIV-2 Envs tested could fuse with cells expressing ChemR23.

10. BLTR. An orphan receptor, CMKRL1, was cloned from CD4 lymphocytes (Owman *et al.,* 1996). CMKRL1 was subsequently shown to be the receptor for leukotriene B_4 and renamed BLTR (Owman *et al.,* 1997; Yokomizo *et al.,* 1997). BLTR is expressed in CD4 lymphocytes and is found in lymphoid tissues (Owman *et al.,* 1996; Yokomizo *et al.,* 1997). When tested against 10 primary and two laboratory isolates in an entry assay that measured cDNA synthesis by PCR, BLTR was shown to be active as a coreceptor for 7 of 10 primary HIV-1 isolates but poorly for the TCLA HIV-1_{IIIB} and not at all for the M-tropic HIV-$1_{Ba\text{-}L}$ (Owman *et al.,* 1998). The phenotype of the isolates that used BLTR was of the SI class and used CXCR4 in addition to BLTR. None of the 7 isolates used BLTR alone, and this receptor was not tested in a productive infection assay. Therefore, how broadly active this coreceptor will prove to be awaits additional studies.

11. Apj. This receptor was cloned using PCR and found to be highly expressed in the brain (O'Dowd *et al.,* 1993); its sequence demonstrates a close relationship to the angiotensin receptor and has been reported to be the receptor for apelin, a peptide isolated from bovine stomach (Tatemoto *et al.,* 1998). Apj has been shown to be a coreceptor for several HIV-1 and SIVmac strains in a fusion assay (Edinger *et al.,* 1998b), in a single-cycle infection assay (Choe *et al.,* 1998a), and in a productive infection assay (Choe *et al.,* 1998a). Recently, Apj was shown to be widely used for fusion with isolates from PBMC and alveolar macrophages (Singh *et al.,* 1999).

12. US28. Several herpes viruses carry in their genomes analogues to chemokines and chemokine receptors (for reviews see Ahuja *et al.,* 1994; Murphy, 1994). Because the protein from the US28 open reading frame of human cytomegalovirus (CMV) signals in response to MIP-1α, MIP-1β, and RANTES (Gao and Murphy, 1994; Neote *et al.,* 1993), it was tested by Alizon and colleagues to determine if it functioned as an HIV coreceptor. Several M-tropic (ADA, JR-CSF) and TCL-tropic (LAI, NDK) strains of HIV-1 and the TCLA ROD10 strain of HIV-2 were able to use US28 as coreceptor in a single-cycle infection assay (Pleskoff *et al.,* 1997).

Whether the list in Table I is complete or whether there are other coreceptors to be discovered remains an open question. As discussed below (Section VII), we predict that others await description. An important question is whether any of these alternative coreceptors have a role *in vivo* or whether their activities are limited to *in vitro* systems. Inefficient use of a coreceptor in PBMC *in vitro* may not be predictive of use *in vivo,* particularly under circumstances where selective pressure, such as therapeutic coreceptor blockade, may drive viruses to adapt to use alternative coreceptors. And

TABLE II Ligands for HIV/SIV Coreceptors

CCR2B	MCP-1, MCP-2, MCP-3, MCP-4
CCR3	Eotaxin-1, -2, RANTES, MCP-2, MCP-3, MCP-4
CCR5	RANTES, MIP-1α, MIP-1β, MCP-2
CCR8	I-309
CXCR4	SDF-1α, SDF-1β
CX_3CR1	Fractalkine
BLTR	Leukotriene B_4
Apj	Apelin
D6	?
STRL33/Bonzo	?
Gpr15/BOB	?
Gpr1	?
ChemR23	?
US28	MIP-1α, MIP-1β, RANTES, MCP-1

even the study of receptors that cannot be used by HIV *in vivo* may be informative for defining the mechanisms of coreceptor function (see Table II).

VII. Coreceptor Use for HIV-1, HIV-2, and SIV

From the earliest studies, it was found that the HIV-1 strains and isolates commonly used in research could infect almost all of the CD4-positive cell lines tested. This, we now know, is because these cells express the fusion cofactor CXCR4 and most of the early HIV isolates were obtained by coculture of PBMC from an HIV-infected person with a T-cell line. In fact, finding a human cell line that does not express CXCR4 and could be used to examine the activities of individual coreceptors in the absence of CXCR4 has been difficult. Even the HOS lines produced by Landau and colleagues (Liu *et al.*, 1996) and Littman and colleagues (KewalRamani *et al.*, unpublished; Trkola *et al.*, 1998; Zhang *et al.*, 1998b) have low but functional levels of CXCR4, the former higher than the latter (unpublished results). The human astroglioblastoma cells U87MG (Ponten and Macintyre, 1968) and U373 (Harrington and Geballe, 1993) do not express either CCR5 or CXCR4 and have been modified to express CD4 (Chesebro *et al.*, 1990; Clapham *et al.*, 1991; Harrington and Geballe, 1993) and various coreceptors (Björndal *et al.*, 1997; Deng et al., 1996; Vodicka *et al.*, 1997). U87MG cells, however, express the alternative coreceptor STRL33 (Deng *et al.*, 1997). Therefore, interpretation of results obtained with any cell system must take into account the endogenous expression of known coreceptors and the possible expression of unsuspected ones.

A. Coreceptor Analysis by Infection of GHOST4 Cells

Because many T-cell lines express multiple coreceptors (Section VII,B), a reductionist approach has been taken to avoid the complexity afforded by using these cell lines. A simpler method is to use a cell that expresses a single coreceptor and CD4, although as just mentioned, these cell lines may not be entirely without other coreceptors. To illustrate how these types of cells have been used, we summarize some of our unpublished work with the GHOST4 lines, which were generated from human osteosarcoma cells by Veneet KewalRamani and Dan Littman (NYU) and express human CD4 and individual coreceptors (and also GFP under the control of the HIV-2 LTR). In the experiments summarized here, the ability of a particular virus to use a coreceptor was assessed by the capacity of that cell line to support a productive infection. As discussed above, this is likely to be a stringent assay for coreceptor activity, and coreceptors that are used inefficiently may be missed.

We have only included viruses derived from molecular clones, since this is the only way to be sure of the sequence of the starting virus. Techniques for these studies have been described in detail (Peden *et al.*, 1991; Peden and Martin, 1995), and infectious molecular clones were either produced by us (Fujita *et al.*, 1992; Peden *et al.*, 1991; Ryan-Graham and Peden, 1995; Theodore *et al.*, 1996) or were obtained from others (Adachi *et al.*, 1986; Collman *et al.*, 1992; Dewhurst *et al.*, 1990; Ghosh *et al.*, 1993; Huet *et al.*, 1990; Jin *et al.*, 1994; Kestler *et al.*, 1990; Koyanagi *et al.*, 1987; Luciw *et al.*, 1992; Naidu *et al.*, 1988; Shibata *et al.*, 1995). Virus stocks were obtained by transfection of the molecular clones into cells that are not permissive for HIV infection so as to avoid the possibility of subsequent rounds of infection and replication, which can lead to adaptation; this is the only way to assure that the genotype and thus phenotype of the starting virus is known. The different cell lines were infected with these virus stocks, and the replication kinetics were followed for up to three weeks. The results are presented in Table III.

HIV-1 isolates LAI, NL4-3, and ELI1 infected the GHOST4-CXCR4 line but no other, demonstrating that these viruses use CXCR4. LAI and NL4-3 have *env* genes from different clones of the same strain, LAI, the difference being that LAI was cloned from DNA obtained from infected PBMC (Peden *et al.*, 1991; Wain-Hobson *et al.*, 1985), while the *env* gene from NL4-3 was cloned from the DNA of an infected T-cell line (Adachi *et al.*, 1986). Thus, LAI is a TCL-tropic virus while NL4-3 is a TCLA virus; both are X4 viruses. As shown below (Table VI), both LAI and NL4-3 infect all T-cell lines tested.

ELI1 was cloned from infected PBMC (Alizon *et al.*, 1986), and virus from an infectious molecular clone (Peden *et al.*, 1991) has the properties of a primary virus (Kozak *et al.*, 1997; Willey *et al.*, 1994). This virus could

TABLE III Coreceptor Use of HIV And SIV Strains On GHOST4 Cells

	Replication in GHOST4 cells								
Virus	*Parent*	*CCR1*	*CCR2B*	*CCR3*	*CCR4*	*CCR5*	*CXCR4*	*BOB*	*Bonzo*
HIV-1									
LAI	−	−	−	−	−	−	+++	−	−
NL4-3	−	−	−	−	ND	−	+++	−	−
ELI1	−	−	−	−	−	−	++	−	−
MAL	−	−	−	−	−	++	−	−	+
89.6	−	−	−	−	−	+	++	−	−
JR-CSF	−	−	−	−	ND	++	−	−	−
AD	−	−	−	−	ND	+++	−	−	−
SIVcpzGAB1	−	−	−	−	ND	+++	−	−	−
SIVmac									
SIVmac239	−	−	−	−	−	+++	−	+++	+
SIVmac1A11	−	−	−	−	−	+++	−	+++	−
SIVsmPBj14	−	−	−	−	ND	++	−	−	−
SIVagm									
SIVagmSAB	+	+	+	+	+	+++	+	+	+++

Note. +, virus growth detectable by reverse transcriptase activity; −, no reverse transcriptase activity detected; and ND, not done.

infect some T-cell lines inefficiently (Jurkat, MT-4) and adapted to infect others (CEM, H9) (Peden *et al.,* 1991). Of the coreceptors tested in this system, only CXCR4 was used by ELI1.

MAL was isolated from the PBMC of a patient with AIDS-related complex from central Africa (Alizon *et al.,* 1986). When the virus obtained from a molecular clone was assessed for its host range, only PBMC and SupT1 were initially found to be permissive for infection (Peden *et al.,* 1991). Because MAL cannot infect most T-cell lines, it was unlikely that this virus uses CXCR4, and the lack of infection of the GHOST4-CXCR4 line is consistent with this. MAL infected the GHOST4-CCR5 line, and thus CCR5 in one of its coreceptors. It also infected the Bonzo line less efficiently, and thus has some activity with STRL33.

The dual-tropic 89.6 virus (Collman *et al.,* 1992) infected the lines expressing CXCR4 and CCR5, although no productive infection was established on the CCR3-expressing line despite the fact that this virus has been reported to use CCR3 (Choe *et al.,* 1998a,b) and even CCR2B weakly (Choe *et al.,* 1998a) in other types of assays. Although achieving adequate expression of CCR3 is often an issue in assessing its activity, we were able to detect expression of CCR3 on the GHOST4-CCR3 cells by flow cytometry (F. Liao and J.M.F., unpublished).

Two M-tropic viruses were tested: the strongly M-tropic AD isolate (Theodore *et al.,* 1996) and the weakly M-tropic isolate JR-CSF (Koyanagi *et al.,* 1987). These viruses only used CCR5 in this system.

The SIVcpzGAB1 virus (Huet *et al.,* 1990), a virus closely related to HIV-1, used CCR5 exclusively in this assay system. The various types of SIV are described in the Appendix.

All of the SIV isolates from lower primates tested used CCR5. The sooty mangabey viruses SIVmac239 and SIVmac1A11 used Gpr15 (BOB) in addition to CCR5, whereas SIVsmPBj14 only used CCR5. The African green monkey virus SIVagmSAB used STRL33 (Bonzo) as well as it did CCR5, while SIVmac239 used this coreceptor less well than it did CCR5. Interestingly, SIVagmSAB could establish a productive infection on all GHOST4 lines including the parent, albeit to a lesser extent than with GHOST4 cells expressing CCR5 or STRL33. Because SIVagmSAB does not use CXCR4, which is expressed at low levels on the GHOST4 cells, this would argue for the existence of an additional unrecognized coreceptor for this virus on the GHOST4 cells.

B. Expression and Activity of Coreceptors on CD4-Positive Cell Lines

Although they have been used extensively and continue to be, the use of CD4-positive cell lines for HIV research has been controversial almost from the start. There are two basic points of view. One is that only primary cells are the appropriate *in vitro* cell system to use to study HIV. The other is that, because cell lines are more homogeneous than primary cells and less expensive and cumbersome to culture, they can be used, but results obtained with them would need to be confirmed with primary cells. One of the concerns with cell lines was that they were derived from human tumors and that the process of transformation that led to the establishment of the tumor is unknown and may influence how HIV infects and replicates in those cells. An additional factor is that some T-cell lines, such as MT-2, MT-4, and C8166, contain the genome of HTLV-I, and the possible interaction between the gene products from this virus with HIV-1 could complicate any interpretation of the results obtained with these lines. Perhaps more importantly, the use of cell lines has been criticized because not all viruses are able to establish productive infections in all CD4-positive cell lines despite being able to do so on PBMC. Until the identification of the chemokine receptor as coreceptor, this result was perplexing. Now that the major reason for this differential permissivity can be explained by the differential expression of the various coreceptors on the surface of the various CD4-positive cell lines used in HIV/SIV research, this reservation about the use of cell lines has been diminished. [It has not been eliminated, though, as there are other differences between primary cells and cell lines that can be revealed by the use of Vif or Nef HIV mutants, which show a reduced capacity to infect primary cells (Chowers *et al.*, 1994; Fan and Peden, 1992; Gabuzda *et al.*, 1992, 1994; Miller *et al.*, 1994; Ryan-Graham and Peden, 1995; Spina *et al.*, 1994; Theodore *et al.*, 1996).] In fact, as we have demonstrated in the preceding section for GHOST4 cells, a comparison of HIV/SIV tropism on different cell lines with coreceptor expression on those cells can be useful to predict the existence of novel coreceptors.

As stated above, primary isolates frequently failed to infect CD4-positive cell lines, and this was due to the absence of functional levels of CCR5 on most cell lines. Some of the cell lines commonly used for HIV/SIV studies are shown in Table IV; these include lymphocyte and promonocyte CD4-positive cell lines together with monolayer cell lines that have been engineered to express CD4 and a single coreceptor. Table V compiles the cell lines with their known coreceptors and some inferred from their heritage.

Over the years, data on HIV-1, HIV-2, and SIV tropism for different cell lines have been obtained in many laboratories. As an example of how infection data on CD4-positive cell lines can be predictive of which viruses use which coreceptors, and also how the existence of new coreceptors can be predicted, we compile some results obtained by us (Fan and Peden, 1992; Fujita *et al.*, 1992; Peden *et al.*, 1991; Peden and Martin, 1995; Ryan-Graham and Peden, 1995; Willey *et al.*, 1994; and unpublished results) and by others in Table VI. In the top part of Table VI the cell lines and their coreceptors are presented. In some cases, the existence of a coreceptor is inferred rather than demonstrated, such as the existence of Gpr15 in PM1,

TABLE IV Cell Lines Commonly Used For HIV/SIV

Cell lines	*References*
Lymphocyte-derived	
CEM	Foley *et al.* (1965)
Clone A301	Folks *et al.* (1985)
Clone CEM-SS	Nara *et al.* (1987)
HuT78	Gazdar *et al.* (1980)
Clone H9	Mann *et al.* (1989)
Clone PM1	Lusso *et al.* (1995)
Jurkat E6-1	Weiss *et al.* (1984)
Jurkat-CCR5	Alkhatib *et al.* (1996)
Jurkat-STRL33	Liao *et al.* (1997)
SupT1	Smith *et al.* (1984)
MT2	Harada *et al.* (1985)
MT4	Harada *et al.* (1985)
Molt4 clone 8	Kikukawa *et al.* (1986)
C8166	Salahuddin *et al.* (1983)
CEMx174	Salter *et al.* (1985)
Promonocytic-derived	
U937	Sundstrom and Nilsson (1976)
THP-1	Tsuchiya *et al.* (1980, 1982)
HL60	Collins *et al.* (1977, 1978)
Modified monolayer cells	
HOS.CD4	Liu *et al.* (1996)
GHOST4	KewalRamani and Littman (unpublished)
U87.CD4	Chesebro *et al.* (1990); Clapham *et al.* (1991)
U373.CD4	Harrington and Geballe (1993)
Cf2Th	Nelson-Rees *et al.* (1976)
CCC	Clapham *et al.* (1991); Crandell, Fabricant, and Nelson-Rees (1973)
HeLa	Scherer *et al.* (1953)
HeLa-MAGI	Kimpton and Emerman (1992)
HEK293	Graham *et al.* (1977)
NIH/3T3	Jainchill (1969)

since both its parent, HuT78 (Table V), and another subclone, H9, have been reported to express this protein. In the lower part of Table VI, results from replication kinetic studies with selected viruses are presented. As discussed above, we have only included viruses derived from molecular clones. In addition to the ones already described, the following clones were used: SG3.1 (Ghosh *et al.*, 1993) and DH123 (Shibata *et al.*, 1995) of HIV-1; ROD10 and ROD14 of HIV-2 (Ryan-Graham and Peden, 1995); and SIVagmTAN (Soares *et al.*, 1997).

The different cell lines were infected with these virus stocks (Peden and Martin, 1995), and the replication kinetics were followed for up to 2 months.

TABLE V Cell Line Expression of Known and Potential Coreceptors for HIV/SIV

	Known and potential coreceptors												
Cell line	*CCR2B*	*CCR3*	*CCR5*	*CCR8*	*D6*	*CXCR4*	*CX_3CR1*	*BLTR*	*STRL33*	*Gpr1*	*Gpr15*	*ChemR23*	*Apj*
CEM	–	–	–			++			–		–		–
HuT78			–	–	–	+			+/–	–	++		–
H9			–	–	–	+	–		–	–	++		–
PM1			+++	–	–	+			+/–	–	(+)		–
Jurkat	–	–	–	–	–	+	–		+/–	–	–	–	–
Jurkat-CCR5	–	–	+++	–	–	+	–		+/–	–	–	–	–
Jurkat-STRL33	–	–	–	–	–	+	–		+++	–	–	–	–
SupT1	–	–	–	–	–	+++	–		+/–	–	–		–
MT-4			–	–	–	+	–		–	–	++		–
MT-2			–	–	–	+				–	+		–
C8166						+					+		+
Molt4 clone8			+	–	–	+	–	+	–	–	–		–
CEMx174	–	+	–	+++	–	+	–		–	–	+++		–
U937	–	–	–	–	–	+	–		–	–	–		–
THP-1	+		–	+		+	+		–			–	–
HL60			–	+		+		+				–	
HeLa			–		++		–	–		–			
HEK293			–	–		++			–				
U87 MG	–	–	–	–		–			+	+	–		–
U873 MG	–	–	–			–							
HOS	–	–	–	–		+/–							
GHOST4	–	–	–	–		+/–							
NIH/3T3	–	–	–	–		–							
QT6	–	–	–	–		–							

Note. +/–, Determined by PCR only; +, determined by Northern analysis, antibody staining, or by function; and (+), presence of the coreceptor is inferred

This length of time was used, first, to ensure that a negative result is truly negative and, second, to look for viruses that have adapted in culture. This approach has revealed that certain viruses can adapt in culture (Alkhatib *et al.,* 1997; Fujita *et al.,* 1992; Peden *et al.,* 1991) and has resulted in the derivation of TCLA versions of the ELI1 strain (e.g., ELI4) and viruses (HIV-1_{MAL} and SIVmac239) that can use STRL33 more efficiently (unpublished results).

The HIV-1-like viruses, including SIVcpzGAB1, are presented in Table VI together with HIV-2 and various SIV types. The LAI (TCL-tropic) and NL4-3 (TCLA) viruses infected all CD4-positive cell lines, consistent with their use of CXCR4 as their major coreceptor (see above). SG3.1 (Ghosh *et al.,* 1993), a virus that was isolated from human PBMC by coculture with HuT78 cells, was found to be extremely cytopathic in all cells tested but does not infect macrophages. Gabuzda and colleagues have shown that this virus uses CXCR4 and other coreceptors (e.g., CCR3) but not CCR5 (Ohagen *et al.,* 1999). Interestingly, this virus can efficiently infect chimpanzee PBMC. ELI1 could infect some T-cell lines inefficiently (Jurkat, MT-4) and adapted to infect others (CEM, H9). Adapted variants (e.g., ELI4) could infect all T-cell lines tested as efficiently as LAI.

MAL infected PBMC and T-cell lines that express CCR5 (PM1, Jurkat-CCR5) but not most other lines, consistent with this virus using CCR5 but not CXCR4, as shown above for the GHOST4 cells. However, it did not infect MDM, demonstrating that M-tropism is not necessarily synonymous with CCR5 tropism. MAL infected the SupT1 line, as reported (Peden *et al.,* 1991), and Molt 4 clone 8 cells were also found to be permissive, although the infection was low and slow (unpublished results). Because Molt 4 clone 8 cells express low levels of CCR5 (Lee *et al.,* 1999; Yu, C.-R. and J.M.F., unpublished) and signal with β-chemokines (Yu, C.-R. and J.M.F., unpublished), it is possible that MAL is using CCR5 on these cells. However, because MAL can infect SupT1 cells, which have very low levels of CCR5 (Dejucq *et al.,* 1999), it is possible that SupT1 cells express a coreceptor that has not been identified.

Two dual-tropic HIV-1 isolates were included. DH123 (Shibata *et al.,* 1995) and 89.6 (Collman *et al.,* 1992) infected almost all cells tested, as expected from their reported use of CXCR4 and CCR5. 89.6 also uses CCR3, Apj, and, weakly, CCR2B (Choe *et al.,* 1996, 1998a), and DH123 uses several alternative coreceptors, such as CCR8, CX_3CR1, and Apj (Edinger *et al.,* 1998b; Zhang *et al.,* 1998b).

Two M-tropic viruses were tested: the strongly M-tropic AD isolate (Theodore *et al.,* 1996) and the weakly M-tropic isolate JR-CSF (Koyanagi *et al.,* 1987). AD replicated in PBMC and MDM and also in PM1 cells, which express CCR5, and in Jurkat-CCR5. JR-CSF exhibited a similar spectrum. We have not been able to demonstrate that these two viruses use a coreceptor other than CCR5 for a productive infection.

TABLE VI Phenotypes of HIV-1, HIV-2, and SIV Strains

	Human cell type and virus replication												
Coreceptor	*PBMC*	*CEM*	*H9*	*SupT1*	*Jurkat*	*J-CCR5*	*J-STRL33*	*MT-4*	*PM1*	*Molt4*	*U937*	*CEMx174*	*MDM*
CCR2B	+	–		–	–	–	–				–		+
CCR3	+	–		–		–	–			–			+
CCR5	+	–	–	–	–	+	–	–	+		–	–	+
CCR8	+							+		+		+	
D6	+		–	–	–	–	–	–		–	–	–	–
CXCR4	+	+	+	+	+	+	+	+	+	+	+	+	+?
STRL33	+	–	–	–	–	–	+	–	–	–	–	–	–
Gpr15	+	–	+					+	+[a]	–		+	
CX_3CR1	+							+		–			
Apj	+		–	–	–	–	–	–		–	–	–	–
ChemR23	–												
BLTR	+												

HIV-1													
LA1	+++	+++	+++	++	+++	+	+++	+++	+++	d, ++	+++	+++	−
NL4-3	+++	+++	+++	++	+++	ND	ND	+++	+++	ND	+++	+++	−
SG3.1	+[b]	ND	+[b]	+++	+++	ND	ND	ND	ND	++	ND	+++[b]	−[b]
EL11	+++	d, +++	d, +++	−	+	−	+++	++	−	−	−	−	−
El14	+++	+++	+++	+++	+++	+++	+++	+++	++	ND	+++	ND	−
MAL	++	−	−	++	−	++	d, ++	−	++	+	−	−	−
89.6	+[b]	+[b]	+[b]	−	++	++	+++	ND	+[b]	ND	ND	++	+[b]
DH12	++	++	++	ND	+++	ND	ND	ND	+++	ND	ND	ND	++
JR-CSF	+[b]	−	−	−	−	++	−	ND	+[b]	ND	ND	−	+[b]
AD	++	−	−	−	−	++	−	−	++	−	−	−	+++
SIVcpz	ND	ND	ND	+	−	+	−	ND	ND	ND	ND	−	ND
HIV-2													
ROD10	++	+++	++	++	++	ND	ND	++	ND	ND	+	ND	−
ROD14	++	+	++	++	++	ND	ND	++	ND	ND	+++	ND	−
SIV													
SIVmac239	ND	ND	ND	+	−	+++	d, ++	ND	+[b]	++	ND	+++	−[b]
SIVmac1A11	ND	ND	ND	++	−	+++	−	ND	+[b]	++	ND	+++	+[b]
SIVsmPBj14	ND	ND	ND	−	ND	+	ND	ND	ND	ND	ND	++	+[b]
SIVagmSAB	−[b]	ND	ND	+++	+	+++	+++	ND	ND	+++	ND	−	−[b]
SIVagmTAN	−[b]	ND	ND	+	−	++	++	ND	ND	++	ND	−	−[b]

Note. +, Virus growth detectable by reverse transcriptase activity; for our work, efficiency of replication indicated indicated by number of symbols; −, no reverse transcriptase activity detected; d, delayed appearance of virus; ND, not done; and +?, expression found but activity controversial.

[a] Expression predicted but not determined.

[b] Result reported by others: no attempt to indicate efficiency of replication.

The SIVcpzGAB1 virus (Huet *et al.*, 1990) failed to replicate in Jurkat or CEMx174, replicated to low levels in SupT1, but replicated well in Jurkat-CCR5 cells. This suggests that SIVcpzGAB1 uses CCR5 but not CXCR4 or Gpr15, which is present on CEMx174 cells. In this way, SIVcpzGAB1 also resembles HIV-1, since these viruses rarely use Gpr15, whereas all of the SIVsm-derived viruses tested use this coreceptor.

The HIV-2 TCLA viruses ROD10 and ROD14 (Guyader *et al.*, 1987; Ryan-Graham and Peden, 1995) infected PBMC and all cell lines tested, but could not establish a productive infection in MDM. These results are consistent with both viruses using CXCR4 and not CCR5.

The virus from the sooty mangabey, SIVsm, gave rise to SIVmac in primate centers in the United States (Daniel *et al.*, 1985; Kestler *et al.*, 1988). Various isolates of this virus have been cloned. SIVmac239 (Kestler *et al.*, 1989; Regier and Desrosiers, 1990; K.W.C.P., unpublished) infects macaque T cells but not MDM, whereas SIVmac1A11 (Luciw *et al.*, 1992; Marthas *et al.*, 1989) and SIVsmPBj14 (Dewhurst *et al.*, 1990) can infect both cell types. When tested with different cell lines, all three viruses could infect cells that express CCR5 (Jurkat-CCR5, PM1) but not cells expressing CXCR4 (Jurkat). These viruses could infect CEMx174, consistent with their using Gpr15. SIVmac239 and SIVmac1A11 could infect SupT1, and SIVmac239 could infect Jurkat-STRL33 and adapted to infect it better.

Two viruses derived from two African green monkey species (the sabaeus monkey *Cercopithicus sabaeus* and the tantalus monkey *Cercopithicus tantalus*) were tested. Viruses from molecular clones of SIVagmSAB (Jin *et al.*, 1994) and SIVagmTAN (Soares *et al.*, 1997) were unable to infect human PBMC or MDM, most likely because the Vif and Vpr proteins are not functional in human primary cells (Simon *et al.*, 1998; Stivahtis *et al.*, 1997) but are not required to infect transformed human cell lines (Fan and Peden, 1992; Gabuzda *et al.*, 1992). These viruses could not use CXCR4 or Gpr15, since they were unable to infect CEMx174 (nor most other T-cell lines), but were able to infect Jurkat-CCR5 and Jurkat-STRL33. SIVagmTAN was unable to infect the parental Jurkat cells, whereas SIVagmSAB established slow/low infection kinetics in this line. These results indicate that both viruses can use CCR5 and STRL33. That SIVagmSAB can infect Jurkat cells, albeit poorly, and cannot use CXCR4 may suggest that Jurkat cells express another coreceptor, which may be the same coreceptor that we predict that ELI1 can use. However, some clones of Jurkat cells have been shown to express some CCR5 (Lee *et al.*, 1999), and further work will be required to determine if this is the case with our cells. In addition, both SIVagmSAB and SIVagmTAN were able to infect SupT1 and Molt 4 clone 8 cells, indicating that these cells express another coreceptor(s) for these viruses. This predicted coreceptor may or may not be the same coreceptor used by the MAL strain of HIV-1 and weakly by SIVcpzGAB1.

In summary, the results described in Table VI demonstrate that the use of multiple cell lines with different viruses can be useful to predict the existence of additional coreceptors for HIV/SIV. For example, that MAL can infect SupT1 cells and does not use CXCR4, suggests that this cell line expresses a hitherto unrecognized coreceptor. Because ELI1 can infect Jurkat and MT4 cells inefficiently and does not use CXCR4 well may indicate that these cell lines express another coreceptor. This may also be the one that SIVagmSAB is using on Jurkat cells. Molt 4 clone 8 cells are permissive for many SIV types, and since this line is not permissive for R5 HIV-1 isolates, such as AD, despite expressing low levels of CCR5, the existence of another coreceptor is predicted. However, it is possible that these SIV strains can fuse with cells expressing lower levels of CCR5 than can R5 HIV-1 isolates.

VIII. Summary of Coreceptor Use for HIV-1, HIV-2, and SIV

In addition to the viruses described in Tables III and VI, we have compiled a list of many of the commonly used viruses and which coreceptors they have been shown to use in Table VII. We have selected mainly viruses derived from complete molecular clones (although we have included results using partial molecular clones, such as the JR-FL, SF162, ADA, and Ba-L isolates of HIV-1) but have included commonly used viruses as well. By necessity, the list is incomplete both in the chosen viruses and also in the coreceptors that they have been shown to use. Some estimate of efficiency of coreceptor use is attempted (+ to +++), although this is difficult to assess, since different assays were used by multiple groups. In some cases, discrepant results were obtained, and we have indicated both results by −,+. The data come from published reports and from our own unpublished experiments.

A. Coreceptor Use for HIV-1

Every HIV-1 isolate to date uses either CCR5 or CXCR4. They may use both and additional coreceptors, but an isolate has not yet been found that cannot use CCR5 or CXCR4, thus demonstrating the primacy of these coreceptors.

Table VII shows that discrepant results have sometimes been obtained for the same viruses. For example, ELI1 was reported to be able to use CCR3 in a single-cycle infection by Sodroski and colleagues (Choe *et al.*, 1998a), but we were unable to show that this virus could use CCR3 in a productive infection (Table III; unpublished results). Also, Sodroski and colleagues were unable to demonstrate that ELI1 used STRL33 in a single-cycle infection assay (Choe *et al.*, 1998a), and we could not confirm that

TABLE VII Coreceptor Use of Common Isolates of HIV-1, HIV-2, and SIV Strains

	Coreceptor														
Virus	*CCR2B*	*CCR3*	*CCR5*	*CCR8*	*D6*	*9-6*	*CXCR4*	*CX_3CR1*	*STRL33*	*Gpr15*	*Gpr1*	*APJ*	*ChemR23*	*BLTR*	*US28*
HIV-1															
LAI	–	–	–				+++								
IIIB (HXB2)	–	–	–				+++		f			+			–
IIIB (BH8)		+	–				+++					+, fff	–		
NL4–3	–	–, +	–	–			+++	–	–, f		–	–			–
MN			–	–			+++					+			
SF2							+++								
SF33			–				+++								
ELI1	–	–, +	–	+	–		++		–, +	–		++			–
ELI4		–, +					+++								
NDK							+++								
SG3.1		++	–				+++								
RF			++				+++		–				–		
89.6	+	++	++	fff		–	+++	f	+, f		–	++	–		
DH123	–	–, +	++	+			+++	–	f	–		–			–
GUN-1	–	–	++	–			+++	–	–	–	+	–			–
MAL	–	–	+++	–			–		+	–					
JR-CSF		–	+++	–			–		–	–					
JR-FL	–	+	+++	–	–	–	–	–	–	–, f	–	–	–		–
SF162	–	–	+++	–		–	–	–	f	–, f		–, +			–
AD	–	–	+++	–			–		–	–					
ADA	–	++	+++	fff	–	–	–	–	f	–, f	–	–	–		–
YU-2		+	+++		–		–		–	–, f	–	–			
Ba-L		+	+++	–		–	–		–, f	–			–		
SIVcpz	–	–	+++				–		–	–					

HIV-2													
ROD10			–			+++							
ROD14			–			+++							
EHO						+++							
MIR						+++							
UC2		+	++			+++		++	++				
SBL6669	–	–	f, +++	fff	–	+++	ff					–	
SBL/ISY			(++)			(++)							
ST	–	–	++	–	–	–		f, +	+				
SIVsm													
mac251			+++	–	–	–	–	fff	f	f	–	–	
mac251 (BK28)			fff			–		fff	fff	fff	f		
mac239	–	–	+++	–	–	–		+	+++	+	–, f	f	–
mac316	–	–	+++	fff	–	–		fff	++	+	+, fff	fff	–
mac1A11			+++			–		+	+++		ff		
mac17E-FR			+++	f	–			fff, +	ff	fff	f	ff	
macCP			+++					++			f		
smpbj14			+++			–		fff	ff	fff	f		
smD670-C13			+++	–	–	–		+	++		f	–	
SIVrcm													
rcmGBA1	+++	–	–	–		–		–	–				
SIVagm													
agmSAB			+++			–		++	–, fff	ff	f		
agmTAN			+++			–		++					
agmTY01			+++			–		+++	++				
SIVmnd													
SIVmndGB-1	–	–	–			+++							

Note. +, Coreceptor used by viral Env in several assays; f, coreceptor used by viral Env in fusion assay; –, coreceptor not used by viral Env in any assay tested; –, +, coreceptor reported negative in one assay but positive in another; and (), indicates that use of coreceptor is inferred from the known tropism of the virus.

this virus could use STRL33 in GHOST4-Bonzo cells in a productive infection assay (Table III), whereas STRL33 provided the same virus with accelerated replication kinetics in the context of Jurkat cells (Liao *et al.*, 1997). At present, the explanation for such discrepancies is not known, but in part likely reflects the different systems used to measure coreceptor activity as discussed above.

B. Coreceptor Use for HIV-2

Coreceptor use by HIV-2 strains has been less studied, but examples of viruses that use the major coreceptors CXCR4 and CCR5 (Bron *et al.*, 1997; Deng *et al.*, 1997; Hill *et al.*, 1997; Sol *et al.*, 1997) as well as some that use the alternative coreceptors such as CCR3, Gpr15, and STRL33 (Bron *et al.*, 1997; Deng *et al.*, 1997; McKnight *et al.*, 1998; Mörner *et al.*, 1999; Sol *et al.*, 1997) and the CMV 7TM, receptor-like protein US28 (Pleskoff *et al.*, 1997) have been described (Table VII). And even CXCR2, CCR1, and CCR4 have been reported to be used by HIV-2 (Bron *et al.*, 1997; McKnight *et al.*, 1998). It has been stated that HIV-2 may be more promiscuous than HIV-1 in its coreceptor preferences (Bron *et al.*, 1997; Guillon *et al.*, 1998; McKnight *et al.*, 1998; Mörner *et al.*, 1999), although fewer primary isolates of HIV-2 have been examined to date, so that this statement may reflect the smaller sample size.

C. Coreceptor Use for SIV

The use of the term SIV obscures the diversity of these viruses (see Appendix). SIV refers to lentiviruses that infect different simian species and are a diverse group. In fact, the use of "immunodeficiency" viruses for these viruses is also misleading in that infection of a species with their "cognate" viruses does not lead to immunodeficiency or perceptible disease, suggesting that the virus and host have adapted so as avoid the parasite killing its host. For example, the natural infection of sooty mangabeys with SIVsm, African green monkey subspecies with SIVagm viruses, SIVmnd infection of mandrills, SIVsyk infection of the Sykes monkey, and perhaps the chimpanzee with SIVcpz all lead to clinically inapparent infections and do not lead to obvious disease. It seems likely that only when zoonotic infection occurs does disease eventuate (Baskin *et al.*, 1988; Daniel *et al.*, 1985; Fultz *et al.*, 1989; Gravell *et al.*, 1989; Hirsch *et al.*, 1995; Novembre *et al.*, 1997). Because of the diversity of these viruses, it is perhaps surprising that almost all SIV isolates to date regardless of their origin have been shown to use CCR5 as their principal coreceptor. Some may use certain alternative coreceptors in addition, such as Gpr15, STRL33, and CCR2B, but the use of CXCR4 until recently had not been demonstrated for any SIV, and thus its use is likely rare *in vivo*.

1. CCR5 use by SIV. All isolates of SIVsm, including the SIVmac isolates and SIVagm, have been found to use CCR5 as their main coreceptor (Table VII). The exceptions to the use of CCR5 are SIVrcmGBA1 and SIVmndGB-1. The almost ubiquitous use of CCR5 by the simian immunodeficiency viruses is consistent with a central role of this coreceptor to HIV/SIV biology.

2. CXCR4 use by SIV. The only SIV described to date that uses CXCR4 is the SIVmndGB-1 isolate of the virus obtained from mandrill monkeys (Tsujimoto *et al.,* 1988). This virus used only CXCR4 and not CCR5, CCR1, CCR2B, or CCR3 (Schols and De Clercq. 1998). Whether all SIVmnd isolates use CXCR4 remains to be seen. It could be that, because the original isolation of SIVmndGB-1 was done by the coculture of infected activated mandrill PBMC with Molt 4 clone 8 (Tsujimoto *et al.,* 1988), a virus that uses CXCR4 was inadvertently selected for, since this cell line expresses CXCR4 and BLTR, but low levels of CCR5 (Table V).

3. CCR2B use by SIV. The red cap mangabey *Cercocebus torquatus torquatus* (Georges-Courbot *et al.,* 1998) is a subspecies related to the sooty mangabey. One isolate, SIVrcmGAB1, was found to use CCR2B and not CCR5 or other coreceptors (CCR1, CCR3, CCR4, CCR8, CXCR4, Gpr15, or STRL33) (Chen *et al.,* 1998). This was the first case of an SIV *not* using CCR5 and is likely explained by the high prevalence of a 24-bp deletion in the *CCR5* gene of the red cap mangabey (Chen *et al.,* 1998). This *CCR5del24* mutation was present at an allelic frequency of 86.6% in red cap mangabeys from Africa and in captivity in the United States. These results demonstrate that when CCR5 is unavailable, SIV can adapt to use an alternative co-receptor.

4. CXCR4 use by SHIV. SHIV-4, an SIV/HIV chimera that has the *env* gene of HXBc2 in the infectious molecular clone of SIVmac239, is infectious for cynomolgus macaques but does not cause disease (Li *et al.,* 1992). Since the HXBc2 Env uses CXCR4 and not CCR5, this result demonstrates that the use of CCR5 is not obligatory for viruses to infect monkeys, at least by direct inoculation, and that the restricted use of CXCR4 can nevertheless lead to the establishment of an *in vivo* infection.

Passage of SHIV-4 through multiple animals, both rhesus and pig-tail macaques, generated a pathogenic variant KU-1 (Joag *et al.,* 1996; Stephens *et al.,* 1996). When the phenotype of several isolates of KU-1 was examined, an increase in macrophage tropism was found (Stephens *et al.,* 1997). Therefore, one question was whether disease development depended upon the switch of a T-tropic virus to an M-tropic virus and whether this switch of phenotype was due to a change in coreceptor use. The *env* gene of one isolate of KU-1, PNb5, that was obtained from the cerebrospinal fluid (CSF) of an infected animal was amplified by PCR and the sequence determined. Fourteen amino acid differences were found in PNb5 compared with the parental SHIV-4 (Hoffman *et al.,* 1998; Stephens *et al.,* 1997). A fragment from PNb5 containing these amino acid differences was substituted in the SHIV-4, and the phenotype of this virus, SHIV-PNb5, was determined. SHIV-PNb5 replicated in rhesus MDM, although not as well as did the uncloned KU-1 virus but better than SHIV-4 (Hoffman *et al.,* 1998). When coreceptor use of the PNb5 Env was examined using a single-cycle infection assay in feline CCCS cells with human CD4 as the target cells, this Env was found to use CCR2B, CCR3, STRL33, and Apj in preference to CXCR4. Thus, animal passage had resulted in a broadened coreceptor use from almost exclusive use of CXCR4 to a virus that could infect macrophages. What coreceptor Pnb5 uses to infect macrophages is not clear, since it did not adapt to use CCR5. CCR2B, which is expressed on MDM (Table VIII), is one candidate.

In another study of $\mathrm{SHIV_{KU\text{-}1}}$, the *env* gene was amplified by PCR and the sequence determined (Cayabyab *et al.,* 1999). In one clone, SHIV-HXBc2P 3.2, 12 amino acids were different from the parent virus. Importantly, this pathogenic variant retained the use of CXCR4 and could not use CCR5, CCR2B, CCR3, STRL33, Gpr15, or Gpr1, at least in the Env complementation *in vitro* assay used with canine cells (Cayabyab *et al.,* 1999). In contrast to the uncloned KU-1 virus, SHIV-HXBc2P 3.2 was unable to infect rhesus MDM any better than the parent SHIV-HXBc2, although it had an increased capacity to infect rhesus PBMC over the parent virus (Cayabyab *et al.,* 1999). SHIV-HXBc2P 3.2 was able to induce disease in rhesus macaques, demonstrating that CCR5 use and infection of macrophages are not required for pathogenicity.

The reason for why Hoffman *et al.* (1998) identified a macrophage-tropic *env* gene from KU-1 stock, whereas Cayabyab *et al.* (1999) obtained an *env* clone whose virus only used CXCR4 may be by chance, since only one clone was examined, or, more likely, was be due to the fact that the latter's stocks were prepared on CEMx174 cells, a line that expresses CXCR4, CCR8, Gpr15, and some CCR3, but does not express CCR5 (Table V), whereas PNb5 was obtained from virus obtained directly from the CSF (Hoffman *et al.,* 1998; Stephens *et al.,* 1997). These results point to the importance of *not* preparing virus in cells that select for viruses with a particular phenotype. For this reason and to avoid such selection pressure, perhaps the best cell system to prepare viruses would be PBMC from the appropriate animal, since all viruses tested replicate in activated PBMC, most likely because activated T cells express multiple coreceptors (Section IX,E and Table VIII).

IX. Expression and Activity of Coreceptors in Cells and Tissues

Few data are available on the tissue distribution of many of the coreceptors, although this information is gradually accumulating. What is known

TABLE VIII Cell Expression of HIV/SIV Coreceptors

	Cells							
Coreceptor	*T Lymphocytes*	*Monocytes*	*Macrophages*	*Thymocytes*	*Microglia*	*Dendritic cells*	*Eosinophils*	*Neutrophils*
CCR2B	+	+				+		
CCR3	+	+, −	+		+	+	+	+/−
CCR5	+	+	+	+	+	+		
CCR8	+	+, −		+				
D6	+							
CXCR4	+	+	+[a]	+	+	+	−/+	+
CX_3CR1	+	+						+
BLTR	+		+				+	+
STRL33	+	−	−				−	
Gpr15	+		+					
Gpr1	−		+					
Apj	+							
ChemR23	+/−		+			+		

Note. +, Demonstrated by protein or RNA; −, demonstrated by absence of RNA; +, −, different results between groups; and +/−, low expression, detectable by PCR.

[a] Expression positive, but coreceptor activity controversial.

at present is summarized in Table VIII. The difficulty in interpreting and comparing data from different groups is that various methods were used to ascertain the expression of the coreceptors. These vary from detection of the protein, usually by antibody staining and flow cytometry, to analysis of RNA expression, which can be by PCR, RNase protection, or by Northern analysis. What is not clear is whether expression detected by PCR translates into cell-surface protein expression and activity. In the main, we have called a tissue positive for coreceptor expression when done so by the authors, although in some cases discrepancies between groups may be explained by the use of different assays.

Many of the coreceptors that are widely used by HIV-1 and HIV-2 strains in artificial systems *in vitro* are expressed in relevant cells and tissues. Lymphocytes express the major coreceptors CCR5 and CXCR4 as well as the alternative coreceptors CCR2B, CCR3, CCR8, CX_3CR1, BLTR, STRL33, Gpr15, D6, and Apj. Several are also expressed on monocytes and macrophages (CCR2B, CCR3, CCR5, CXCR4, Gpr15), while for others (Gpr1, ChemR23) expression is higher on monocytic cells and are found at lower levels on lymphocytes. CCR5, CXCR4, CCR2B and ChemR23 are expressed on dendritic cells.

A. Monocytes and Macrophages

There is universal agreement that the predominant functional coreceptor on monocyte-derived macrophages (MDM) is CCR5. In fact, the definition of macrophage (M) tropism has been taken to be synonymous with CCR5 tropism, although this is an oversimplification, as we discuss in Section X. Whether other coreceptors are expressed or are functional on MDM is controversial, and inconsistent results have been obtained. Also not resolved is whether freshly isolated, nonadherent monocytes can be infected *in vitro*, and there is disagreement over the proportion of circulating monocytes that are HIV-infected *in vivo*.

In the case of the *in vitro* studies with monocytes, some of the disagreements are likely due to methodologies in that how soon freshly isolated monocytes are examined prior to culture influences whether monocytes become activated and/or begin to differentiate into macrophages. An early study reported that monocytes rather than MDM could be infected with HIV-1 *in vitro* (Kazazi *et al.*, 1989). Subsequently, it was found that freshly isolated monocytes are resistant to infection and become permissive only after time in culture to allow differentiation to macrophages (Naif *et al.*, 1998; Rich *et al.*, 1992; Schuitemaker *et al.*, 1992b; Sonza *et al.*, 1995; Tuttle *et al.*, 1998). The block to infection in freshly isolated monocytes is at stage prior to reverse transcription (Sonza *et al.*, 1996). The inability to infect monocytes *in vitro* is consistent with the low numbers of HIV-infected circulating monocytes *in vivo* (Innocenti *et al.*, 1992; McElrath *et al.*, 1989,

1991; McIlroy *et al.*, 1996; Mikovits *et al.*, 1992; Schuitemaker *et al.*, 1992b) and may simply be a consequence of the resting state of most circulating monocytes. This would be analogous to the resistance of resting CD4 lymphocytes to infection with HIV (Spina *et al.*, 1994, 1995; Stevenson *et al.*, 1990; Zack *et al.*, 1990, 1992).

Can the expression of the HIV receptors, CD4 and coreceptors, fully explain viral tropism with respect to MDM? Freshly isolated monocytes express high levels of CD4, which gradually decrease over time in culture as the macrophages differentiate and become susceptible to infection with R5 isolates (Sonza *et al.*, 1995). Thus, CD4 levels do not explain the lack of susceptibility of monocytes to HIV infection and the susceptibility of MDM. When coreceptor expression was measured, freshly isolated monocytes were found to have high levels of surface CXCR4 (Di Marzio *et al.*, 1998; McKnight *et al.*, 1997; Naif *et al.*, 1998; Tuttle *et al.*, 1998; Zaitseva *et al.*, 1997) and RNA for CXCR4 (Di Marzio *et al.*, 1998; Tuttle *et al.*, 1998). That CXCR4 is functional on monocytes is shown by the finding that monocytes can fuse with Envs from X4 viruses (Zaitseva *et al.*, 1997). With time in culture as the monocytes differentiate into macrophages, the level of CXCR4 declines (Di Marzio *et al.*, 1998; Tuttle *et al.*, 1998), while the level of CCR5 protein (Di Marzio *et al.*, 1998; Lapham *et al.*, 1999; Tuttle *et al.*, 1998) and CCR5 RNA (Di Marzio *et al.*, 1998; Tuttle *et al.*, 1998) increases. Therefore, the increase in CCR5 expression of macrophages combined with the establishment of the differentiated state are sufficient to explain the permissiveness of these cells to M-tropic isolates.

The lack of infectability of MDM with most X4 viruses remains a puzzle, particularly since CXCR4 is still present on the cell surface (McKnight *et al.*, 1997; Verani *et al.*, 1998). While it is generally found that most TCLA viruses that use CXCR4 as their sole coreceptor are unable to establish a productive infection in MDM, certain primary isolates that use CXCR4 as their main coreceptor have been reported to infect productively MDM (Schmidtmayerova *et al.*, 1998; Simmons *et al.*, 1998; Verani *et al.*, 1998). In one study (Verani *et al.*, 1998), TCLA X4 viruses were only able to use CXCR4 on MDM with variable efficiencies for entry but could not establish productive infections. Importantly, SDF-1, the ligand for CXCR4, was able to reduce significantly infection with these TCLA X4 viruses, whereas RANTES, a ligand for CCR5, had no effect on infection. In another study (Simmons *et al.*, 1998), several primary isolates that use CXCR4 and not CCR5 replicated equally efficiently in MDM derived from CCR5 +/+ wild type and CCR5 Δ32/Δ32 individuals. Importantly, replication was blocked by the CXCR4 inhibitor AMD3100 whether CCR5 was present or not, indicating that these viruses are using CXCR4 and not other coreceptors for the infection of macrophages. In a third study (Schmidtmayerova *et al.*, 1998), the block to efficient productive infection in MDM by TCLA viruses was shown to be not at the level of entry but at the postentry steps of nuclear importation and, to a lesser extent, reverse transcription. Entry was inhibited both by a monoclonal antibody to CXCR4 and by SDF-1.

Why some primary X4 viruses can establish productive infection in MDM when TCLA X4 viruses cannot is not known. Other studies have shown that MDM can become infectable with X4 viruses after certain treatments. For example, MDM exposed to certain activators of macrophages, such as LPS, become permissive for T-cell-tropic, X4 viruses (Moriuchi *et al.,* 1998). And treatment of MDM with IL-4 allows X4 viruses to infect these cells more efficiently (Valentin *et al.,* 1998). Recent results have demonstrated that CXCR4 may exist in different forms in monocytes versus macrophages (Lapham *et al.,* 1999). Whether the forms of CXCR4 that exist in MDM are altered by certain treatments and whether the different forms of CXCR4 play a role in viral tropism remains to be seen.

B. Dendritic Cells and Langerhans Cells

Dendritic cells (DC) and Langerhans cells (LC), which are the immature DC in the epidermis of the skin and mucosa, are antigen-presenting cells in mucosal tissues and may be the first cell that HIV interacts with during mucosal transmission (Spira *et al.,* 1996; Zoeteweij and Blauvelt, 1998). These cells have been shown to be permissive for HIV infection, and in the presence of activated CD4 T cells a strong productive infection ensues (Pope *et al.,* 1994). Whether pure DC support HIV infection has been controversial (Cameron *et al.,* 1992a,b, 1994; Pope *et al.,* 1995) and may depend on the differentiated state of the DC (Granelli-Piperno *et al.,* 1999; Sozzani *et al.,* 1998) and on DC proliferation (Blauvelt *et al.,* 1997).

Recent reports have tried to relate permissivity of DC/LC to infection with the phenotype and coreceptor use of the virus. In one study, mature DC were isolated from blood and from skin and tested for their capacity to be infected with R5 and X4 viruses using a viral PCR-based entry assay (Granelli-Piperno *et al.,* 1996). Both types of virus entered both DC types, but reverse transcription was incomplete, indicating that HIV replication in pure mature DC is inefficient. Entry of X4 viruses was specifically blocked by SDF-1, and entry of R5 viruses was blocked by RANTES, indicating that both CXCR4 and CCR5 are functional coreceptors on mature DC (Granelli-Piperno *et al.,* 1996).

In another study, immature and mature monocyte-derived DC (MDDC) were found to express CCR5 and CCR3 but not CXCR4 when assayed by flow cytometry, but immature MDDC did express CXCR4 RNA by PCR (Rubbert *et al.,* 1998). However, despite not having detectable CXCR4 protein, immature MDDC were found to signal with SDF-1, and entry of X4 virus was blocked by SDF-1. From this result, the authors propose the existence of a non-CXCR4 SDF-1 receptor, although low levels of CXCR4 may have been missed by flow cytometry. In the case of R5 viruses, mature and immature MDDC could be infected, but only infection of mature MDDC could be substantially blocked by β-chemokines. In MDDC derived from a person lacking CCR5—the homozygous delta32 mutation—R5 viruses were still able to infect; surprisingly, this infection was blocked by SDF-1 and not RANTES, implicating the use of CXCR4 as the coreceptor in these cells, although the use of other coreceptors has not been ruled out. With the 89.6 isolate, a virus that can use CCR3 in addition to CXCR4 and CCR5, both eotaxin (the ligand for CCR3) and SDF-1 were found to block entry (Rubbert *et al.,* 1998).

Using a culture system from fetal liver $CD34^+$ cells, Warren and colleagues devised a method to differentiate these cells specifically into DC or macrophages. When these DC or macrophage cultures were then infected with X4 or R5 HIV-1 strains, only the R5 viruses replicated in the macrophage cultures. Surprisingly, only the X4 viruses replicated in the DC cultures (Warren *et al.,* 1997). These results seem to be at odds with those of others. For example, when DC were differentiated from cord blood $CD34^+$ cells or PBMC, both X4 and R5 viruses replicated in both of these cells (Blauvelt *et al.,* 1997). It is not known what accounts for the discrepant results, but the different culture conditions likely play a role.

Freshly isolated LC were shown to express CCR5 and not CXCR4 and could fuse with Envs from R5 isolates of HIV-1 but not from X4 viruses (Zaitseva *et al.,* 1997). As the LC were cultured, fusion with X4 Envs was observed concomitant with the demonstration of surface staining of CXCR4. The authors suggest that the freshly isolated LC may be more like the LC *in vivo* and may provide an explanation for a selective transmission of R5 viruses through mucosal surfaces.

C. Microglia

These monocyte-derived brain cells have been shown to be the major HIV-infected cell in the brain (Gabuzda *et al.,* 1998; Vazeux, 1991). Although microglia express CCR5, CXCR4, and CCR3 (He *et al.,* 1997; Lavi *et al.,* 1997, 1998; Sanders *et al.,* 1998). it was proposed that CCR3 may be a major coreceptor for HIV on microglial cells (He *et al.,* 1997). However, subsequent studies have implicated CCR5 as the major coreceptor (Albright *et al.,* 1999; Shieh *et al.,* 1998) or perhaps other coreceptors (Ghorpade *et al.,* 1998), such as the fractalkine receptor CX_3 CR1 (Nishiyori *et al.,* 1998).

D. Thymus

The thymus is a tissue known to be infected by HIV (Calabro *et al.,* 1995; Davis, 1984; Grody *et al.,* 1985; Joshi and Oleske, 1985; Reichert *et al.,* 1983; Seemayer *et al.,* 1984) and whose infection is thought to be particularly important in pediatric HIV disease (Joshi and Oleske, 1985;

Joshi *et al.*, 1984; Kourtis *et al.*, 1996; Papiernik *et al.*, 1992; Rosenzweig *et al.*, 1993, 1994; Shearer *et al.*, 1997). The level of coreceptor expression was found to vary according to the developmental state of the T cell; the immature double-positive cells—the $CD34^+$, $CD3^-$ $CD4^+$, $CD8^+$ subset—were found to express high levels of CXCR4 and low levels of CCR5 (Berkowitz *et al.*, 1998b; Kitchen and Zack, 1997; Zaitseva *et al.*, 1998). As the T cells mature to single-positive cells—the $CD34^+$, $CD3^+$, $CD4^+$, $CD8^-$ subset—CXCR4 levels decrease while CCR5 remains low. Susceptibility to infection of purified thymocyte subsets *in vitro* (Zaitseva *et al.*, 1998) or of thymus in the SCID-Hu mouse *in vivo* model (Berkowitz *et al.*, 1998a; Jamieson *et al.*, 1995; Pedroza-Martins *et al.*, 1998) with different strains of HIV-1 reflected the coreceptor expression on these cells, viz., viruses that could use CXCR4 infected thymocytes better and caused greater thymus destruction than did viruses that use CCR5. Whether other coreceptors are used in thymocytes remains to be determined. CCR8 (Horuk *et al.*, 1998) and STRL33 (Liao *et al.*, 1997) are expressed in the thymus, although in which cells is not known. As most HIV-1 strains do not use these coreceptors (Pöhlmann *et al.*, 1999; Rucker *et al.*, 1997; Xiao *et al.*, 1998; Zhang *et al.*, 1998a,b), alone they are unlikely to facilitate infection of this tissue. However, for the thymus as well as other tissues, it should be pointed out that the absence of infection *in vitro* with viruses that were isolated in PBMC, MDM, or T-cell lines does not preclude the possibility that tissue-specific variants exist *in vivo* and that these viruses use alternative coreceptors on these cells.

E. Expression and Activity of Coreceptors on T-Cell Subsets

Expression of coreceptors on T-cell subsets has been investigated to some extent for the major coreceptors for which monoclonal antibodies are available, but information on the effects of specific stimuli and of the cells' state of activation/development on coreceptor expression and function is limited. This information is likely to be important to understand the pathophysiology of HIV disease, since expression of different coreceptors or combinations of coreceptors will determine which cells will become infected with what virus and how efficiently.

Although an early report suggested that CXCR4 and CCR5 are expressed on mutually exclusive subsets (Bleul *et al.*, 1997), this is a simplification. While CCR5 is limited to the effector/memory subset of CD4-positive T cells, CXCR4 is expressed on all T-cell subsets (Rabin *et al.*, 1999), albeit with preferential expression on naïve as compared with memory cells (Lee *et al.*, 1999). T-cell activation using PHA (Bleul *et al.*, 1997) or OKT3 (Rabin *et al.*, 1999) has a more rapid effect on upregulating CXCR4 as compared with CCR5. Like CCR5, CCR2 is expressed on CD26-bright effector/memory cells and is upregulated after T-cell activation (Qin *et al.*, 1996). Similarly, T-cell activation upregulates expression of the orphan receptors STRL33/Bonzo (Deng *et al.*, 1997; Liao *et al.*, 1997), GPR15/BOB (Deng *et al.*, 1997), and D6 and Apj (Choe *et al.*, 1998a), and incubation with IL-2 upregulates expression of CX_3CR1 (Imai *et al.*, 1997). The upregulation of coreceptors after T-cell activation may be one of the requirements for cellular stimulation as a prerequisite for efficient HIV replication. Despite the general positive association between T-cell activation and coreceptor expression, the expression of specific coreceptors can be differentially affected depending on the nature of the activating signals. For example, although in one study PHA plus IL-2 led to the induction of mRNA for both CXCR4 and CCR5, stimulation with anti-CD3 plus anti-CD28 led to the induction of CXCR4 without induction of CCR5 (Carroll *et al.*, 1997).

Cytokines have also been found to affect coreceptor expression on both resting lymphocytes and during activation and differentiation. IL-4 has been shown to upregulate expression of CXCR4 (Jourdan *et al.*, 1998; Valentin *et al.*, 1998) in both resting and activated cells and to decrease expression of CCR5, leading to selective enhancement of infection with X4 viruses in *in vitro* cultures (Valentin *et al.*, 1998). CD4-positive T cells differentiated in defined cytokine environments into Th1 or Th2 cells have been reported to express selectively either CCR5 (Bonecchi *et al.*, 1998; Loetscher *et al.*, 1998; Siveke and Hamann, 1998) or CCR3 (Sallusto *et al.*, 1997) and CCR8 (D'Ambrosio *et al.*, 1998), respectively.

The differences in coreceptor expression on CD4-positive T cells, as well as on other CD4-positive cells depending on the stage of differentiation of the cells and their exposures to activating stimuli and cytokines, suggest a basis for selective interactions among viral isolates displaying differing patterns of coreceptor use with particular subsets of target cells. These relationships may help explain some aspects of HIV pathogenesis, such as an accelerated decline in CD4 T cells associated with infection with isolates able to use the widely expressed CXCR4 coreceptor (Goudsmit, 1995). Moreover, regulation of coreceptor expression may be a mechanism whereby inflammatory conditions, such as concurrent infections, could alter the outcome of HIV disease. Finally, cell-state-specific changes in coreceptor expression per se may not be the only means of regulating coreceptor activity, since, as noted above, there is evidence that the differentiation of monocytes to macrophages is associated with structural changes in CXCR4, or changes in CXCR4-containing complexes, that may affect coreceptor activity (Lapham *et al.*, 1999).

X. Can Coreceptor Use Always Explain Tropism?

While almost all M-tropic viruses use CCR5 to enter macrophages, as we have pointed out above and as have others (Cheng-Mayer *et al.*, 1997), there are examples where the use of CCR5 by HIV-1 does not always result in the efficient replication in MDM. This is also true for SIVmac, where SIVmac239 uses CCR5 (Marcon *et al.,* 1997) but fails to establish a productive infection in MDM (Mori *et al.,* 1993). In this case, the virus can enter MDM and reverse transcribe its RNA, but the infection is abortive. There are related reports of X4 HIV-1 isolates that can enter MDM but also cannot establish a productive infection (Verani *et al.,* 1998). Also, viruses can enter dendritic cells without establishing a productive infection (Granelli-Piperno *et al.,* 1996).

At present, why viruses are able to use a given coreceptor to infect some cells but not others is not clear. There could be quantitative as well as qualitative factors. It could be that relatively low-affinity interactions between Env and its coreceptor make entry particularly sensitive to coreceptor levels and that these levels differ among cell types. Alternatively, establishing a productive infection may depend on the quality of Env–coreceptor interactions, such as those leading to receptor signaling (Davis *et al.,* 1997; Weissman *et al.,* 1997). Because coreceptor use does not explain fully virus tropism, it may be sensible to retain aspects of the older phenotype classification for HIV and combine it with the new one based on coreceptor use.

XI. Envelope Interactions with CD4 and Coreceptor

While many studies over the past decade have addressed the interactions between CD4 and HIV Env, information regarding structure/function with the HIV/SIV Env and the coreceptors is at an early stage. What is clear is that the contacts between the Env and both CD4 and coreceptor involve multiple regions of all. In the case of the HIV Env, regions in both gp120 and gp41 appear to be involved with the interactions with CD4 and the coreceptor. Because changes in the V3 loop of gp120 alone can switch coreceptor use from CXCR4 to CCR5, this has been used to implicate this region as the one involved in the direct binding of gp120 to the coreceptors. However, the heterogeneity of V3 sequences in HIV-1 may argue against a direct interaction and for an indirect one. This remains to be determined.

Golding and colleagues showed that CD4 and CXCR4 could be coimmunoprecipitated using an antibody to either protein when soluble gp120 was added to cells (Lapham *et al.,* 1996). Even in the absence of added gp120, small amounts of CD4 and CXCR4 could be found in a complex. Subsequent work has confirmed this result and showed that it is also true for CCR5 (Lapham *et al.,* 1999; Xiao *et al.,* 1999). Thus, CD4 and the coreceptors are likely in close proximity in the cell membrane, an association that HIV seems to exploit for entry. So far, no information is available whether one coreceptor has higher affinity for CD4 than another, although that CCR5 and CD4 complexes are more easily immunoprecipitated than CXCR4 and CD4 may suggest that the former are in a tighter association than the latter.

The generally accepted model for Env interaction with the cell receptors is as follows. The gp120 component of Env interacts first with CD4 on the cell surface and this leads to conformational changes in both molecules. These structural changes expose the binding sites on the coreceptor for gp120 and expose the fusion region of gp41, which allow the latter to induce fusion between the viral envelope and the cell membrane. The reader is referred to several learned reviews for a more detailed discussion (Berson and Doms, 1998; Choe, 1998; Clapham *et al.*, 1999; Dimitrov *et al.*, 1998).

XII. Coreceptors and Pathogenesis

Because all primary isolates of HIV require a seven-transmembrane-domain coreceptor in addition to CD4 in order to enter cells, it is obvious that, just as for CD4, the coreceptors play a central role in HIV disease. What is less obvious is which among the coreceptors are important *in vivo* and at which stages in the development of disease a given coreceptor may participate. These data will be relevant to attempts to treat AIDS by designing coreceptor antagonists to block viral entry. Information on both the distribution of coreceptor expression on cells and in tissues and on the breadth and classes of viral isolates that use a coreceptor provide circumstantial evidence for use of a coreceptor *in vivo*. As stated above, most of the coreceptors can be found on one or more of the cell types that are infected by HIV (Table VIII). The primacy of CCR5 and CXCR4 is strongly supported by the fact that virtually all primary isolates use CCR5 and/or CXCR4. However, the use of other coreceptors *in vivo* and the possibility of their contributing to pathogenesis has not been excluded.

Additional data linking CCR5 and CXCR4 to the course of HIV disease come from the work described above linking disease progression to changes in viral phenotype. Recently, the original observations that M-tropic isolates are present early after infection (Roos *et al.*, 1992; Schuitemaker *et al.*, 1992a; van't Wout *et al.*, 1994; Zhang *et al.*, 1993; Zhu *et al.*, 1993, 1996) and that TCL-tropic isolates emerge in some individuals with progressive disease (Fenyö *et al.*, 1988, 1989; Tersmette *et al.*, 1988, 1989a) have been extended by analyzing the evolution of coreceptor use by sequential isolates from infected individuals (Connor *et al.*, 1997; Scarlatti *et al.*, 1997). These results demonstrate, as anticipated, that R5 viruses predominate early in infection and X4 and R5X4 viruses emerge later. While it is tempting to assume that this expansion in coreceptor use contributes to disease progression, the basis for changes in coreceptor use, and whether it contributes to or is a consequence of disease progression, is unknown.

The definitive evidence linking coreceptors with human disease has come from the analysis of polymorphisms in coreceptor genes. The first and most informative of these to be identified was a 32-bp deletion (Δ32 mutation) in the CCR5 gene (Dean *et al.*, 1996; Liu *et al.*, 1996; Samson *et al.*, 1996a; Zimmerman *et al.*, 1997) that results in the synthesis of a truncated protein that is not expressed at the cell surface (Benkirance *et al.*, 1997; Liu *et al.*, 1996; Rana *et al.*, 1997). With a frequency of 10% for the heterozygous Δ32/+ genotype in the Caucasian population (Dean *et al.*, 1996; Huang *et al.*, 1996; Libert *et al.*, 1998; Liu *et al.*, 1996; Martinson *et al.*, 1997; Samson *et al.*, 1996a; Stephens *et al.*, 1998; Zimmerman *et al.*, 1997) and an absence of any deleterious effects from the mutation, approximately 1% of Caucasians are homozygous for the mutation (Δ32/Δ32). Besides facilitating its recognition, the high prevalence of the homozygous genotype allowed for statistically meaningful analysis of its frequency in cohorts of HIV-infected individuals as well as in the population as a whole. The first report of the Δ32/Δ32 genotype was in two individuals who had remained uninfected despite multiple sexual exposures to HIV-1 (Liu *et al.*, 1996), and this was soon followed by reports showing a highly significant underrepresentation of the Δ32/Δ32 genotype in infected Caucasian individuals (Dean *et al.*, 1996; Samson *et al.*, 1996a; Zimmerman *et al.*, 1997). Importantly, the basis for resistance to infection conferred by the homozygous genotype was clear: PBMC from Δ32/Δ32 individuals could not be infected by R5, M-tropic isolates *in vitro*. Although individuals have been described subsequently who have become infected (with CXCR4-using viruses) despite the Δ32/Δ32 genotype (Biti *et al.*, 1997; Michael *et al.*, 1998; O'Brien *et al.*, 1997; Theodorou *et al.*, 1997), it is clear that this genotype confers significant protection against infection. As an answer to our initial question, these findings provide convincing evidence that CCR5 is central for the establishment of HIV infection, and CCR5 is the only coreceptor where the genetic evidence has demonstrated such a role.

A number of polymorphisms in receptor genes have been associated with a second aspect of HIV disease, namely disease progression. For the Δ32/+ mutation, some reports have noted a delay in progression to AIDS among heterozygous individuals, although the delay reported has been modest, from 0 to 2 years (Dean *et al.*, 1996; Huang *et al.*, 1996; Michael *et al.*, 1997; Zimmerman *et al.*, 1997). Besides the Δ32 mutation in the coding region of the CCR5 gene, searches of the CCR5 promoter have uncovered a variety of polymorphisms (Carrington *et al.*, 1997; Cohen *et al.*, 1998; Kostrikis *et al.*, 1998; Martin *et al.*, 1998; McDermott *et al.*, 1998; Mummidi *et al.*, 1997; 1998), and two polymorphisms/alleles have been described as associated with altered rates of disease progression: a G/A polymorphism at position 59029, where individuals with the G/G genotype showed approximately a 4-year delay in progression to AIDS compared with A/A individuals; and an allele designated *CCR5P1*, which contains a number of polymorphic positions and is associated with more rapid progression from seroconversion to AIDS (Martin *et al.*, 1998). It has not been ruled out that *CCR5P1* contains the A at 59029, so that these two studies may have been looking at the same effects. Of note, however, one group reported that a 59029 A promoter fragment was somewhat more active in a reporter-gene assay than a fragment with 59029 G, consistent with the protective effect of the G/G genotype (McDermott *et al.*, 1998), while the *CCR5P1* allele was reported not to show any differential activity in a reporter assay (Martin *et al.*, 1998).

In addition to the variations in the CCR5 gene, polymorphisms in the genes for CCR2 (CCR2-64I) (Smith *et al.*, 1997) and for the CXCR4 ligand SDF-1 (SDF-1 3'A) (Winkler *et al.*, 1998) have been associated with delayed progression to AIDS. Experimental data suggesting a mechanism for the protective effects of the CCR2-64I polymorphism (Mariani *et al.*, 1999) and others have not been forthcoming. In this regard it is important to note that there is a chemokine receptor gene cluster on 3p21-p24 (Samson *et al.*, 1996b) that includes CCR1, CCR2, CCR3, CCR4, CCR5, CCR8, and CX_3CR1, so that outcomes associated with particular polymorphisms in these genes may in fact reflect the effect of linked sequences.

Taken together, the data establish a role for CCR5 in the acquisition of HIV infection and strongly suggest that CCR5 and CXCR4 are important for the pathogenesis of HIV disease *in vivo*. The roles for the "alternative" coreceptors *in vivo* is discussed below.

XIII. Possible Roles for Alternative Coreceptors

While there are no data that compel us to consider a role for these coreceptors, such as the identification of isolates that use neither CCR5 or CXCR4, there are nonetheless no data that rule out roles for other coreceptors at any stage of HIV disease, since only for CCR5 has a naturally occurring "knockout" been recognized. It is clear that both HIV and SIV can show flexibility in coreceptor use *in vivo,* since individuals of the Δ32/Δ32 genotype are not absolutely resistant to infection and the studies described above in Section VII,C,4 indicate that SHIVs using coreceptors other than CCR5 can still cause disease.

There are a number of ways that alternative coreceptors may be important to HIV infection and pathogenesis. Alternative coreceptors could be involved in certain types of transmission; they could be transitional coreceptors, where a virus alters its major coreceptor use from, say, CCR5 to CXCR4 by first adapting to use another coreceptor as an intermediate; there could be tissue-specific coreceptor use; viral entry of a cell could be conferred by more than one coreceptor acting in concert; finally, it is possible that some of the alternative coreceptors will become major coreceptors once therapies designed to block the interaction of HIV with its coreceptor are in the clinic.

A. Transmission and Coreceptors

At present, little is known about how coreceptor use influences mucosal transmission. Because the viruses first seen following infection all use CCR5 as their coreceptor (Björndal *et al.,* 1997; Connor *et al.,* 1997; Scarlatti *et al.,* 1997), it has been assumed that mucosal transmission, as well as others, may require the use of CCR5. However, viruses that use alternative coreceptors have been described in maternal–infant transmission (Zhang *et al.,* 1998b; and unpublished results), and, thus, it is possible that more subtle coreceptor use may be occurring *in vivo* and this is being missed by the *in vitro* assays employed to date. Animal models using viruses that use selectively individual coreceptors will be necessary to test this.

B. Alternative Coreceptors as Intermediates to Coreceptor Use Change

Alternative coreceptors may have a role as transitional coreceptors as an HIV Env switches coreceptor use from CCR5 to CXCR4 or vice versa. To date, few reports have appeared where an R5 virus switches *in vitro* to an X4 virus, or an X4 virus to an R5 one. In one, passage in the PM1 cell line, which expresses both CCR5 and CXCR4, resulted in the R5 virus HIV-$1_{\text{JR-CSF}}$ adapting to infect cell lines that do not express CCR5, and while it was not shown that these adapted viruses use CXCR4, it seems likely (Bou-Habib *et al.,* 1994). While there could be several reasons why coreceptor switching has not been seen frequently—such as the number of generations employed, the absence of an appropriate selective pressure, the choice of a virus that has been adapted *in vitro* to use CCR5 or CXCR4 to an artificially high efficiency, among others—the absence of such a demonstration *in vitro* does not mean that it does not occur *in vivo.* And while the existence of dual-tropic viruses that use both CCR5 and CXCR4 does not necessarily mean that dual tropism for CCR5 and CXCR4 is the intermediate in an eventual switch from R5 to X4 monotropism, it certainly could be. There may be several routes to dual-tropism or multitropism. One route could be the direct way from R5 through R5X4 to X4, whereas another could be through adaptation to a third (or more) alternative coreceptor. Our example of the MAL isolate of HIV-1, which uses CCR5 and possibly an unknown coreceptor on SupT1 cells well and STRL33 less efficiently but adapts to use STRL33 quite readily (in preparation), could be an example of a virus that is capable of switching coreceptor use.

C. Cooperation between Coreceptors

Viral entry of a cell could be conferred by more than one coreceptor. At present, there is no direct evidence for the idea that different coreceptors act in concert to provide fusion activity for the HIV Env. Nevertheless, there are observations that are consistent with there being some interaction between coreceptors. For example, in unpublished experiments, we have found that engineered expression of individual coreceptors in a particular cell line can have unexpected consequences. Expression of CCR5 in a cell line that expresses CXCR4, while conferring the ability to be infected by R5 viruses, led to a reduction in the ability of an X4 virus to infect these cells compared with the parent line. Expression of alternative coreceptors in several cell lines has also had effects on viral tropism that were not predicted, and expression of the same coreceptor can have variable consequences in different cells. Coreceptor interaction may provide an explanation for some perplexing results. For example, infection of microglia with HIV-1 was inhibited by both eotaxin and MIP-1β, suggesting that both CCR3 and CCR5 were being used, but that the level of inhibition was greater than would be expected from the coreceptors acting independently (He *et al.*, 1997). Also, the finding that SDF-1 inhibited infection of CCR5-null dendritic cells by an R5 virus (Rubbert *et al.*, 1998) may be a consequence of coreceptor interaction.

While less interesting explanations for these observations are possible, such as levels of receptors on the cell surface, the prospect that different combinations of coreceptors on the cell surface may lead to more subtle coreceptor requirements has not been recognized to date from the *in vitro* studies. Coreceptor interaction could provide an explanation for the finding that viruses that are able to use an alternative coreceptor and CCR5 for entry fail to use the alternative coreceptor when CCR5 use is blocked (Zhang and Moore, 1999).

D. Use in Specific Tissues

Although there is no evidence yet to support a role for alternative coreceptors in the infection of CD4-positive cells *in vivo* (or even non-CD4-positive cells) in specific tissues, it remains a possibility. For example, HIV infection in the brain may involve coreceptors other than CCR5 and CCR3, since the receptor Apj is expressed in the brain (O'Dowd *et al.*, 1993). Also, alternative coreceptors such as STRL33 may be involved in transmission *in utero*, since this orphan receptor is expressed in the placenta (Liao *et al.*, 1997).

E. Therapeutics Targeted to HIV—Coreceptor Interaction

The design and *in vitro* testing of several CCR5 and CXCR4 drugs is underway with the goal of adding this modality to the armamentarium of therapies to combat HIV infection. Such compounds as the bicyclam AMD3100 is active in blocking X4 viruses but not R5 viruses (Donzella *et al.*, 1998; Schols *et al.*, 1997a,b, 1998), and AOP-RANTES is active against R5 viruses but not X4 viruses (Simmons *et al.*, 1997). A likely consequence of such therapies targeted to block HIV interacting with its coreceptor is the generation of viral variants that evade the drugs. In fact, a virus was described that evolved to become resistant to AMD3100, but the resistant virus retained use of CXCR4 and did not change its coreceptor preference (Schols *et al.*, 1998). In another study, viruses that evolved to become resistant against the modified RANTES derivatives AOP-RANTES and NNY-RANTES were found to switch their coreceptor use from CCR5 to CXCR4 (Mosier *et al.*, 1999). However, the starting virus for this study was HIV-1_{242}, a derivative of the X4 virus NL4-3 modified *in vitro* in the V3 region to convert the virus to use CCR5 (Chesebro *et al.*, 1996). Thus, reversion to the use of CXCR4 requires the substitution of only one or a few amino acids in a single region of the protein. Whether such rapid reversion will be seen in primary viruses remains to be determined. Nevertheless, these two studies confirm the necessity of using multiple drugs targeted at least the two main coreceptors simultaneously, as suggested by Zhang and Moore (1999).

XIV. Some Remaining Questions Regarding Coreceptor Use

Throughout this chapter several issues regarding coreceptors and HIV/SIV have been presented. We end by stating some of the issues that have not been resolved or need further study.

1. Does coreceptor use by HIV/SIV as determined by *in vitro* assays accurately reflect what is used *in vivo?*
2. Do different coreceptors interact, and does this complex interact directly with CD4?
3. What are the coreceptor determinants of cellular tropism?
4. Are there qualitative consequences of Env–coreceptor interaction and coreceptor signaling that determine the outcome of infection?
5. Do coreceptors compete with themselves for the binding to CD4?
6. Is there a hierarchy among the coreceptors for CD4 association?
7. We have shown that viruses can adapt *in vitro* to use a coreceptor more efficiently. Does this occur *in vivo,* and, if so, does it have relevance to pathogenesis?
8. Can a virus adapt in a single step to switch coreceptor use from CCR5 to CXCR4, or is a third (or more) coreceptor necessary to effect this switch in coreceptor use?

9. Do coreceptors other than CCR5 and CXCR4 have a role *in vivo?* If they have a role, what is it? Do additional coreceptors await discovery?
10. Is CCR5 the main or only coreceptor used on macrophages? Does CXCR4 contribute to HIV entry on macrophages with primary viruses?
11. What is the role of coreceptors in mucosal transmission and in maternal–infant transmission? Is a single coreceptor used or are multiple ones required? What cells mediate transmission and what are their coreceptor repertoire?
12. Why are R5 but not X4 or R5X4 viruses ordinarily those that establish infection?
13. If use of a coreceptor is blocked by a targeted inhibitor, can escape variants be derived *in vitro* and *in vivo?* What is the mechanism of escape? Do these escape variants use other coreceptors or use the same coreceptor differently?

XV. Concluding Comments

It has been postulated that the ancestral strains of immunodeficiency viruses may have used what we now call the coreceptor as their primary receptor and only came to use CD4 later, perhaps as an evolutionary adaptation either to avoid host responses or to target selectively a subset of the cells that express CCR5 or CXCR4. That certain HIV-2 (Endres *et al.*, 1996; Hoxie *et al.*, 1998), SIVsm (Edinger *et al.*, 1997, 1999), and HIV-1 (Dumonceaux *et al.*, 1998; Hoxie *et al.*, 1998) isolates have been adapted or shown to use the coreceptor as their sole receptor may be consistent with this idea, although even these CD4-independent viruses fuse more efficiently if CD4 is available.

Coreceptor use does not always define tropism. Despite using CCR5, several viruses are unable to infect MDM. Whether this reflects a limitation in the *in vitro* systems or a genuine difference in CCR5 use in T cells versus macrophages, or even if there is another coreceptor on macrophages, remains to be determined.

Cell lines can predict coreceptor use. We have demonstrated that consideration of cell-line tropism of a virus can be useful to predict the existence of novel coreceptor use. In the example we present, SupT1 cells possibly express a coreceptor that is used by the MAL isolate of HIV-1. That several primary HIV-1 isolates as well as many SIV isolates replicate in SupT1 cells may suggest that this coreceptor is quite widely used.

Viruses can adapt *in vitro* to use a coreceptor more efficiently. Adaptation in culture is a well-described phenomenon with HIV. We have stated that the MAL isolate of HIV-1 and SIVmac239 adapt to use STRL33 well on passage in Jurkat cells engineered to express this orphan receptor. Both viruses retain use of CCR5. Whether this type of adaptation reflects a property found *in vivo* is unknown but seems probable.

As coreceptor antagonists are being developed and evaluated for clinical use, a detailed understanding of this complex family of HIV/SIV receptors should help us to anticipate potential pitfalls and interpret outcomes so as to speed the introduction of effective coreceptor-based therapies.

Appendix

Types of SIV

The various types of SIV described to date are listed below.

1. SIVsm/SIVmac. This is numerically the largest group, due to its historical role as identifying a simian analogue for HIV and AIDS in an animal model as well as to it being the likely progenitor of HIV-2, as discussed in the chapter. The host is the sooty mangabey monkey *Cercocebus torquatus atys,* which exists in west Africa.

2. SIVrcm. Georges-Courbot and colleagues recently isolated an SIV from the red-cap mangabey *Cercocebus torquatus torquatus* (Georges-Courbot *et al.,* 1998), a subspecies related to the sooty mangabey. This monkey is found in Gabon and its territory is separate from that of the sooty mangabey. Phylogenetically, SIVrcm appears to be a hybrid, as the *pol* gene clusters with the HIV-1/SIVcpz viruses while the *gag* gene is a new lineage related to SIVagm (Chen *et al.,* 1998).

3. SIVagm. There are several subtypes of the virus (Daniel *et al.,* 1988; Hirsch *et al.,* 1993b; Ohta *et al.,* 1988) that infect the different species of African green monkey—vervet (*Chlorocebus aethiops aethiops* formally *Cercopithicus pygerythrus*), grivet (*Chlorocebus aethiops aethiops* formally *Cercopithicus aethiops*), tantalus monkey (*Chlorocebus aethiops tantalus* formally *Cercopithicus tantalus*), and sabaeus monkey (*Chlorocebus aethiops sabaeus* formally *Cercopithicus sabaeus*).

4. SIVmnd. This virus was isolated from the mandrill *Mandrillus sphinx* from Gabon (Tsujimoto *et al.,* 1988, 1989).

5. SIVsyk. This virus was isolated from the Sykes monkey *Cercopithicus mitis albogularis,* an African monkey that resides in east Africa (Emau *et al.,* 1991; Hirsch *et al.,* 1993a).

6. SIVlhoest. A virus was isolated from a l'hoesti monkey (*Cercopithicus l'hoesti*) from a U.S. zoo (Hirsch *et al.,* 1999). The l'hoesti monkey is from east Africa, although the SIVl'hoest sequence clusters with SIVmnd.

7. SIVcpz. Of all the SIV isolates examined to date, the SIVcpz is the closest to HIV-1 in genome organization (Huet *et al.,* 1990), sequence (Huet *et al.,* 1990), and cross-transactivation of their Tat and Rev proteins (Sakur-

agi *et al.*, 1992). In fact, SIVcpz clusters with HIV-1 rather than with other SIVs. The close evolutionary relationship between SIVcpz and HIV-1 may be due to the close relationship of their hosts, it may suggest that the SIVcpz was the virus that crossed over to humans to give rise to HIV-1, or both. Recent data have strongly suggested that SIVcpz from the chimpanzee subspecies *Pan troglodytes troglodytes,* a chimpanzee that lives in a part of west central Africa that overlaps the region where the first HIV-1 cases were found, may well be the progenitor of HIV-1 (Gao *et al.*, 1999; Weiss and Wrangham, 1999).

References

Adachi, A., Gendelman, H. E., Koenig, S., Folks, T., Willey, R., Rabson, A., and Martin, M. A. (1986). Production of acquired immunodeficiency syndrome-associated retrovirus in human and nonhuman cells transfected with an infectious molecular clone. *J. Virol.* **59**, 284–291.

Ahmad, N., and Venkatesan, S. (1988). Nef protein of HIV-1 is a transcriptional repressor of HIV-1 LTR. *Science* **241**(4872), 1481–1485.

Ahuja, S. K., Gao, J. L., and Murphy, P. M. (1994). Chemokine receptors and molecular mimicry. *Immunol. Today* **15**, 281–287.

Akrigg, A., Wilkinson, G. W., Angliss, S., and Greenaway, P. J. (1991). HIV-1 indicator cell lines. *AIDS* **5**, 153–158.

Alam, J., and Cook, J. L. (1990). Reporter genes: Application to the study of mammalian gene transcription. *Anal. Biochem.* **188**, 245–254.

Albright, A. V., Shieh, J. T., Itoh, T., Lee, B., Pleasure, D., O'Connor, M. J., Doms, R. W., and Gonzalez-Scarano, F. (1999). Microglia express CCR5, CXCR4, and CCR3, but of these, CCR5 is the principal coreceptor for human immunodeficiency virus type 1 dementia isolates. *J. Virol.* **73**, 205–2013.

Alexander, L., Veazey, R. S., Czajak, S., DeMaria, M., Rosenzweig, M., Lackner, A. A., Desrosiers, R. C., and Sasseville, V. G. (1999). Recombinant simian immunodeficiency virus expressing green fluorescent protein identifies infected cells in rhesus monkeys [In Process Citation]. *AIDS Res. Hum. Retroviruses* **15**, 11–21.

Alizon, M., and Dragic, T. (1994). CD26 antigen and HIV fusion? [letter; comment]. *Science* **264**(5162), 1161–1162. [discussion, 1162–1165]

Alizon, M., Wain-Hobson, S., Montagnier, L., and Sonigo, P. (1986). Genetic variability of the AIDS virus: Nucleotide sequence analysis of two isolates from African patients. *Cell* **46**, 63–74.

Alkhatib, G., Combadiere, C., Broder, C. C., Feng, Y., Kennedy, P. E., Murphy, P. M., and Berger, E. A. (1996). CC CKR5: A RANTES, MIP-1alpha, MIP-1beta receptor as a fusion cofactor for macrophage-tropic HIV-1. *Science* **272**, 1955–1958.

Alkhatib, G., Liao, F., Berger, E. A., Farber, J. M., and Peden, K. W. C. (1997). A new SIV co-receptor, STRL33. *Nature* **388**, 238.

Aoki, N., Shioda, T., Satoh, H., and Shibuta, H. (1991). Syncytium formation of human and non-human cells by recombinant vaccinia viruses carrying the HIV env gene and human CD4 gene. *AIDS* **5**, 871–875.

Ashorn, P. A., Berger, E. A., and Moss, B. (1990). Human immunodeficiency virus envelope glycoprotein/CD4-mediated fusion of nonprimate cells with human cells. *J. Virol.* **64**, 2149–2156.

Barré-Sinoussi, F., Chermann, J. C., Rey, F., Nugeyre, M. T., Chamaret, S., Gruest, J., Dauguet, C., Axler-Blin, C., Vézinet-Brun, F., Rouzioux, C., Rozenbaum, W., and Montagnier, L. (1983). Isolation of a T-lymphotropic retrovirus from a patient at risk for acquired immune deficiency syndrome (AIDS). *Science* **220**, 868–871.

Baskin, G. B., Murphey-Corb, M., Watson, E. A., and Martin, L. N. (1988). Necropsy findings in rhesus monkeys experimentally infected with cultured simian immunodeficiency virus (SIV)/delta. *Vet. Pathol.* **25**, 456–467.

Benkirane, M., Jin, D. Y., Chun, R. F., Koup, R. A., and Jeang, K. T. (1997). Mechanism of transdominant inhibition of CCR5-mediated HIV-1 infection by ccr5delta32. *J. Biol. Chem.* **272**, 30603–30606.

Berger, E. A., Doms, R. W., Fenyo, E. M., Korber, B. T., Littman, D. R., Moore, J. P., Sattentau, Q. J., Schuitemaker, H., Sodroski, J., and Weiss, R. A. (1998). A new classification for HIV-1. *Nature* **391**, 240.

Berger, E. A., Murphy, P. M., and Farber, J. M. (1999). Chemokine receptors as HIV-1 coreceptors: roles in viral entry, tropism, and disease. *Annu. Rev. Immunol.* **17**, 657–700.

Berkowitz, R. D., Alexander, S., Bare, C., Linquist-Stepps, V., Bogan, M., Moreno, M. E., Gibson, L., Wieder, E. D., Kosek, J., Stoddart, C. A., and McCune, J. M. (1998a). CCR5- and CXCR4-utilizing strains of human immunodeficiency virus type 1 exhibit differential tropism and pathogenesis *in vivo*. *J. Virol.* **72**, 10108–10117.

Berkowitz, R. D., Beckerman, K. P., Schall, T. J., and McCune, J. M. (1998b). CXCR4 and CCR5 expression delineates targets for HIV-1 disruption of T cell differentiation. *J. Immunol.* **161**, 3702–3710.

Berson, J. F., and Doms, R. W. (1998). Structure-function studies of the HIV-1 coreceptors. *Semin. Immunol.* **10**, 237–248.

Biti, R., Ffrench, R., Young, J., Bennetts, B., Stewart, G., and Liang. T. (1997). HIV-1 infection in an individual homozygous for the CCR5 deletion allele. *Nat. Med.* **3**, 252–253.

Björndal, Å., Deng, H., Jansson, M., Fiore, J. R., Colognesi, C., Karlsson, A., Albert, J., Scarlatti, G., Littman, D. R., and Fenyö, E. M. (1997). Coreceptor usage of primary human immunodeficiency virus type 1 isolates varies according to biological phenotype. *J. Virol.* **71**, 7478–7487.

Blauvelt, A., Asada, H., Saville, M. W., Klaus-Kovtun, V., Altman, D. J., Yarchoan, R., and Katz, S. I. (1997). Productive infection of dendritic cells by HIV-1 and their ability to capture virus are mediated through separate pathways. *J. Clin. Invest.* **100**, 2043–2053.

Bleul, C. C., Farzan, M., Choe, H., Parolin, C., Clark-Lewis, I., Sodroski, J., and Springer, T. A. (1996). The lymphocyte chemoattractant SDF-1 is a ligand for LESTR/fusin and blocks HIV-1 entry. *Nature* **382**, 829–833.

Bleul, C. C., Wu, L., Hoxie, J. A., Springer, T. A., and Mackay, C. R. (1997). The HIV coreceptors CXCR4 and CCR5 are differentially expressed and regulated on human T lymphocytes. *Proc. Natl. Acad. Sci. USA* **94**, 1925–1930.

Bonecchi, R., Bianchi, G., Bordignon, P. P., D'Ambrosio, D., Lang, R., Borsatti, A., Sozzani, S., Allavena, P., Gray, P. A., Mantovani, A., and Sinigaglia, F. (1998). Differential expression of chemokine receptors and chemotactic responsiveness of type 1 T helper cells (Th1s) and Th2s. *J. Exp. Med.* **187**, 129–134.

Bou-Habib, D. C., Roderiquez, G., Oravecz, T., Berman, P. W., Lusso, P., and Norcross, M. A. (1994). Cryptic nature of envelope V3 region epitopes protects primary monocytotropic human immunodeficiency virus type 1 from antibody neutralization. *J. Virol.* **68**, 6006–6013.

Broder, C. C., and Collman, R. G. (1997). Chemokine receptors and HIV. *J. Leukoc. Biol.* **62**, 20–29.

Broder, C. C., Dimitrov, D. S., Blumenthal, R., and Berger, E. A. (1993). The block to HIV-1 envelope glycoprotein-mediated membrane fusion in animal cells expressing human CD4 can be overcome by a human cell component(s). *Virology* **193**, 483–491.

Broder, C. C., Nussbaum, O., Gutheil, W. G., Bachovchin, W. W., and Berger, E. A. (1994). CD26 antigen and HIV fusion? *Science* **264,** 1156–1159. [discussion 1162–1165]

Bron, R., Klasse, P. J., Wilkinson, D., Clapham, P. R., Pelchen-Matthews, A., Power, C., Wells, T. N., Kim, J., Peiper, S. C., Hoxie, J. A., and Marsh, M. (1997). Promiscuous use of CC and CXC chemokine receptors in cell-to-cell fusion mediated by a human immunodeficiency virus type 2 envelope protein. *J. Virol.* **71,** 8405–8415.

Calabro, M. L., Zanotto, C., Calderazzo, F., Crivellaro, C., Del Mistro, A., De Rossi, A., and Chieco-Bianchi, L. (1995). HIV-1 infection of the thymus: Evidence for a cytopathic and thymotropic viral variant *in vivo. AIDS Res. Hum. Retroviruses* **11,** 11–19.

Callebaut, C., Krust, B., Jacotot, E., and Hovanessian, A. G. (1993). T cell activation antigen, CD26, as a cofactor for entry of HIV in CD4+ cells. *Science* **262,** 2045–2050.

Camerini, D., Planelles, V., and Chen, I. S. (1994). CD26 antigen and HIV fusion? [letter; comment]. *Science* **264,** 1160–1161. [discussion 1162–1165]

Cameron, P. U., Forsum, U., Teppler, H., Granelli-Piperno, A., and Steinman, R. M. (1992a). During HIV-1 infection most blood dendritic cells are not productively infected and can induce allogeneic CD4+ T cells clonal expansion. *Clin. Exp. Immunol.* **88,** 226–236.

Cameron, P. U., Freudenthal, P. S., Barker, J. M., Gezelter, S., Inaba, K., and Steinman, R. M. (1992b). Dendritic cells exposed to human immunodeficiency virus type-1 transmit a vigorous cytopathic infection to CD4+ T cells. *Science* **257** (5068), 383–387.

Cameron, P. U., Lowe, M. G., Crowe, S. M., O'Doherty, U., Pope, M., Gezelter, S., and Steinman, R. M. (1994). Susceptibility of dendritic cells to HIV-1 infection *in vitro. J. Leukoc. Biol.* **56,** 257–265.

Cann, A. J., Churcher, M. J., Boyd, M., O'Brien, W., Zhao, J. Q., Zack, J., and Chen, I. S. (1992). The region of the envelope gene of human immunodeficiency virus type 1 responsible for determination of cell tropism. *J. Virol.* **66,** 305–309.

Carrillo, A., and Ratner, L. (1996a). Cooperative effects of the human immunodeficiency virus type 1 envelope variable loops V1 and V3 in mediating infectivity for T cells. *J. Virol.* **70,** 1310–1316.

Carrillo, A., and Ratner, L. (1996b). Human immunodeficiency virus type 1 tropism for T-lymphoid cell lines: Role of the V3 loop and C4 envelope determinants. *J. Virol.* **70,** 1301–1309.

Carrillo, A., Trowbridge, D. B., Westervelt, P., and Ratner, L. (1993). Identification of HIV1 determinants for T lymphoid cell line infection. *Virology* **197,** 817–824.

Carrington, M., Kissner, T., Gerrard, B., Ivanov, S., O'Brien, S. J., and Dean, M. (1997). Novel alleles of the chemokine-receptor gene CCR5. *Am. J. Hum. Genet.* **61,** 1261–1267.

Carroll, R. G., Riley, J. L., Levine, B. L., Feng, Y., Kaushal, S., Ritchey, D. W., Bernstein, W., Weislow, O. S., Brown, C. R., Berger, E. A., June, C. H., and St. Louis, D. C. (1997). Differential regulation of HIV-1 fusion cofactor expression by CD28 costimulation of CD4+ T cells. *Science* **276,** 273–276.

Cayabyab, M., Karlsson, G. B., Etemad-Moghadam, B. A., Hofmann, W., Steenbeke, T., Halloran, M., Fanton, J. W., Axthelm, M. K., Letvin, N. L., and Sodroski, J. G. (1999). Changes in human immunodeficiency virus type 1 envelope glycoproteins responsible for the pathogenicity of a multiply passaged simian-human immunodeficiency virus (SHIV-HXBc2). *J. Virol.* **73,** 976–984.

Cepko, C. L., Ryder, E. F., Austin, C. P., Walsh, C., and Fekete, D. M. (1993). Lineage analysis using retrovirus vectors. *Methods Enzymol.* **225,** 933–960.

Chen, B. K., Feinberg, M. B., and Baltimore, D. (1997). The kappaB sites in the human immunodeficiency virus type 1 long terminal repeat enhance virus replication yet are not absolutely required for viral growth. *J. Virol.* **71,** 5495–5504.

Chen, B. K., Gandhi, R. T., and Baltimore, D. (1996). CD4 down-modulation during infection of human T cells with human immunodeficiency virus type 1 involves independent activities of vpu, env, and nef. *J. Virol.* **70,** 6044–6053.

Chen, B. K., Saksela, K., Andino, R., and Baltimore, D. (1994). Distinct modes of human immunodeficiency virus type 1 proviral latency revealed by superinfection of nonproductively infected cell lines with recombinant luciferase-encoding viruses. *J. Virol.* **68**, 654–660.

Chen, Z., Kwon, D., Jin, Z., Monard, S., Telfer, P., Jones, M. S., Lu, C. Y., Aguilar, R. F., Ho, D. D., and Marx, P. A. (1998). Natural infection of a homozygous delta24 CCR5 red-capped mangabey with an R2b-tropic simian immunodeficiency virus. *J. Exp. Med.* **188**, 2057–2065.

Cheng-Mayer, C., Liu, R., Landau, N. R., and Stamatatos, L. (1997). Macrophage tropism of human immunodeficiency virus type 1 and utilization of the CC-CKR5 coreceptor. *J. Virol.* **71**, 1657–1661.

Cheng-Mayer, C., Quiroga, M., Tung, J. W., Dina, D., and Levy, J. A. (1990). Viral determinants of human immunodeficiency virus type 1 T-cell or macrophage tropism, cytopathogenicity, and CD4 antigen modulation. *J. Virol.* **64**, 4390–4398.

Cheng-Mayer, C., Shioda, T., and Levy, J. A. (1991). Host range, replicative, and cytopathic properties of human immunodeficiency virus type 1 are determined by very few amino acid changes in tat and gp120. *J. Virol.* **65**, 6931–6941.

Chesebro, B., Buller, R., Portis, J., and Wehrly, K. (1990). Failure of human immunodeficiency virus entry and infection in CD4-positive human brain and skin cells. *J. Virol.* **64**, 215–221.

Chesebro, B., Nishio, J., Perryman, S., Cann, A., O'Brien, W., Chen, I. S., and Wehrly, K. (1991). Identification of human immunodeficiency virus envelope gene sequences influencing viral entry into CD4-positive HeLa cells, T-leukemia cells, and macrophages. *J. Virol.* **65**, 5782–5789.

Chesebro, B., Wehrly, K., Nishio, J., and Perryman, S. (1992). Macrophage-tropic human immunodeficiency virus isolates from different patients exhibit unusual V3 envelope sequence homogeneity in comparison with T-cell-tropic isolates: Definition of critical amino acids involved in cell tropism. *J. Virol.* **66**, 6547–6554.

Chesebro, B., Wehrly, K., Nishio, J., and Perryman, S. (1996). Mapping of independent V3 envelope determinants of human immunodeficiency virus type 1 macrophage tropism and syncytium formation in lymphocytes. *J. Virol.* **70**, 9055–9059.

Choe, H. (1998). Chemokine receptors in HIV-1 and SIV infection. *Arch. Pharm. Res.* **21**, 634–639.

Choe, H., Farzan, M., Konkel, M., Martin, K., Sun, Y., Marcon, L., Cayabyab, M., Berman, M., Dorf, M. E., Gerard, N., Gerard, C., and Sodroski, J. (1998a). The orphan seven-transmembrane receptor Apj supports the entry of primary T-cell-line-tropic and dual-tropic human immunodeficiency virus type 1. *J. Virol.* **72**, 6113–6118.

Choe, H., Farzan, M., Sun, Y., Sullivan, N., Rollins, B., Ponath, P. D., Wu, L., Mackay, C. R., LaRosa, G., Newman, W., Gerard, N., Gerard, C., and Sodroski, J. (1996). The beta-chemokine receptors CCR3 and CCR5 facilitate infection by primary HIV-1 isolates. *Cell* **85**, 1135–1148.

Choe, H., Martin, K. A., Farzan, M., Sodroski, J., Gerard, N. P., and Gerard, C. (1998b). Structural interactions between chemokine receptors, gp120 Env and CD4. *Semin. Immunol.* **10**, 249–257.

Chowers, M. Y., Spina, C. A., Kwoh, T. J., Fitch, N. J., Richman, D. D., and Guatelli, J. C. (1994). Optimal infectivity *in vitro* of human immunodeficiency virus type 1 requires an intact nef gene. *J. Virol.* **68**, 2906–2914.

Clapham, P. R., Blanc, D., and Weiss, R. A. (1991). Specific cell surface requirements for the infection of CD4-positive cells by human immunodeficiency virus type 1 and 2 and by Simian immunodeficiency virus. *Virology* **181**, 703–715.

Clapham, P. R., Reeves, J. D., Simmons, G., Dejucq, N., Hibbitts, S., and McKnight, A. (1999). HIV coreceptors, cell tropism and inhibition by chemokine receptor ligands. *Mol. Membr. Biol.* **16**, 49–55.

Cocchi, F., DeVico, A. L., Garzino-Demo, A., Arya, S. K., Gallo, R. C., and Lusso, P. (1995). Identification of RANTES, MIP-1 alpha, and MIP-1 beta as the major HIV-suppressive factors produced by CD8+ T cells. *Science* **270**, 1811–1815.

Cohen, O. J., Paolucci, S., Bende, S. M., Daucher, M., Moriuchi, H., Moriuchi, M., Cicala, C., Davey, R. T., Jr., Baird, B., and Fauci, A. S. (1998). CXCR4 and CCR5 genetic polymorphisms in long-term nonprogressive human immunodeficiency virus infection: Lack of association with mutations other than CCR5-Delta32. *J. Virol.* **72**, 6215–6217.

Collman, R., Balliet, J. W., Gregory, S. A., Friedman, H., Kolson, D. L., Nathanson, N., and Srinivasan, A. (1992). An infectious molecular clone of an unusual macrophage-tropic and highly cytopathic strain of human immunodeficiency virus type 1. *J. Virol.* **66**, 7517–7521.

Collman, R., Godfrey, B., Cutilli, J., Rhodes, A., Hassan, N. F., Sweet, R., Douglas, S. D., Friedman, H., Nathanson, N., and Gonzalez-Scarano, F. (1990). Macrophage-tropic strains of human immunodeficiency virus type 1 utilize the CD4 receptor. *J. Virol.* **64**, 4468–4476.

Combadière, C., Salzwedel, K., Smith, E. D., Tiffany, H. L., Berger, E. A., and Murphy, P. M. (1998). Identification of CX3CR1. A chemotactic receptor for the human CX3C chemokine fractalkine and a fusion coreceptor for HIV-1. *J. Biol. Chem.* **273**, 23799–23804.

Connor, R. I., and Ho, D. D. (1994). Human immunodeficiency virus type 1 variants with increased replicative capacity develop during the asymptomatic stage before disease progression. *J. Virol.* **68**, 4400–4408.

Connor, R. I., Chen, B. K., Choe, S., and Landau, N. R. (1995). Vpr is required for efficient replication of human immunodeficiency virus type-1 in mononuclear phagocytes. *Virology* **206**, 935–944.

Connor, R. I., Sheridan. K. E., Ceradini, D., Choe, S., and Landau, N. R. (1997). Change in coreceptor use coreceptor use correlates with disease progression in HIV-1-infected individuals. *J. Exp. Med.* **185**, 621–628.

Cordonnier, A., Montagnier, L., and Emerman, M. (1989). Single amino-acid changes in HIV envelope affect viral tropism and receptor binding. *Nature* **340**, 571–574.

D'Ambrosio, D., Iellem, A., Bonecchi, R., Mazzeo, D., Sozzani, S., Mantovani, A., and Sinigaglia, F. (1998). Selective up-regulation of chemokine receptors CCR4 and CCR8 upon activation of polarized human type 2 Th cells. *J. Immunol.* **161**, 5111–5115.

Dalgleish, A. G., Beverley, P. C., Clapham, P. R., Crawford, D. H., Greaves, M. F., and Weiss, R. A. (1984). The CD4 (T4) antigen is an essential component of the receptor for the AIDS retrovirus. *Nature* **312**, 763–7.

Daniel, M. D., Letvin, N. L., King, N. W., Kannagi, M., Sehgal, P. K., Hunt, R. D., Kanki, P. J., Essex, M., and Desrosiers, R. C. (1985). Isolation of T-cell tropic HTLV-III-like retrovirus from macaques. *Science* **228**, 1201–1204.

Daniel, M. D., Li, Y., Naidu, Y. M., Durda, P. J., Schmidt, D. K., Troup, C. D., Silva, D. P., MacKey, J. J., Kestler, H. W. III., Sehgal, P. K., King, N. W., Ohta, Y., Hayami, M., and Desrosiers, R. C. (1988). Simian immunodeficiency virus from African green monkeys. *J. Virol.* **62**, 4123–4128.

Davis, A. E., Jr. (1984). The histopathological changes in the thymus gland in the acquired immune deficiency syndrome. *Ann. N. Y. Acad. Sci.* **437**, 493–502.

Davis, C. B., Dikic, I., Unutmaz, D., Hill, C. M., Arthos, J., Siani, M. A., Thompson, D. A., Schlessinger, J., and Littman, D. R. (1997). Signal transduction due to HIV-1 envelope interactions with chemokine receptors CXCR4 or CCR5. *J. Exp. Med.* **186**, 1793–1798.

De Jong, J. J., De Ronde, A., Keulen, W., Tersmette, M., and Goudsmit, J. (1992). Minimal requirements for the human immunodeficiency virus type 1 V3 domain to support the syncytium-inducing phenotype: Analysis by single amino acid substitution. *J. Virol.* **66**, 6777–6780.

Dean, M., Carrington, M., Winkler, C., Huttley, G. A., Smith, M. W., Allikmets, R., Goedert, J. J., Buchbinder, S. P., Vittinghoff, E., Gomperts, E., Donfield, S., Vlahov, D., Kaslow,

R., Saah, A., Rinaldo, C., Detels, R., and O'Brien, S. J. (1996). Genetic restriction of HIV-1 infection and progression to AIDS by a deletion allele of the CKR5 structural gene: Hemophilia Growth and Development Study, Multicenter AIDS Cohort Study, Multicenter Hemophilia Cohort Study, San Francisco City Cohort, ALIVE Study. *Science* **273**, 1856–1862.

Dejucq, N., Simmons, G. and Clapham, P. R. (1999). Expanded tropism of primary immunodeficiency virus type 1 R5 strains to $CD4^+$ T-cell lines determined by the capacity to exploit low concentrations of CCR5. *J. Virol.* **73**, 7842–7847.

Deng, H., Liu, R., Ellmeier, W., Choe, S., Unutmaz, D., Burkhart, M., Di Marzio, P., Marmon, S., Sutton, R. E., Hill, C. M., Davis, C. B., Peiper, S. C., Schall, T. J., Littman, D. R., and Landau, N. R. (1996). Identification of a major co-receptor for primary isolates of HIV-1. *Nature* **381**, 661–666.

Deng, H. K., Unutmaz, D., KewalRamani, V. N., and Littman, D. R. (1997). Expression cloning of new receptors used by simian and human immunodeficiency viruses. *Nature* **388**, 296–300.

Dewhurst, S., Embretson, J. E., Anderson, D. C., Mullins, J. I., and Fultz, P. N. (1990). Sequence analysis and acute pathogenicity of molecularly cloned SIVSMM-PBj14. *Nature* **345**, 636–640.

Di Marzio, P., Tse, J., and Landau, N. R. (1998). Chemokine receptor regulation and HIV type 1 tropism in monocyte- macrophages. *AIDS Res. Hum. Retroviruses.* **14**, 129–138.

Dimitrov, D. S., Xiao, X., Chabot, D. J., and Broder, C. C. (1998). HIV coreceptors. *J. Membr. Biol.* **166**, 75–90.

Donzella, G. A., Schols, D., Lin, S. W., Este, J. A., Nagashima, K. A., Maddon, P. J., Allaway, G. P., Sakmar, T. P., Henson, G., De Clercq. E., and Moore, J. P. (1998). AMD3100, a small molecule inhibitor of HIV-1 entry via the CXCR4 co-receptor. *Nat. Med.* **4**, 72–77.

Doranz, B. J., Berson, J. F., Rucker, J., and Doms, R. W. (1997). Chemokine receptors as fusion cofactors for human immunodeficiency virus type 1 (HIV-1). *Immunol. Res.* **16**, 15–28.

Doranz, B. J., Rucker, J., Yi, Y., Smyth, R. J., Samson, M., Peiper, S. C., Parmentier, M., Collman, R. G., and Doms, R. W. (1996). A dual-tropic primary HIV-1 isolate that uses fusin and the beta-chemokine receptors CKR-5, CKR-3, and CKR-2b as fusion cofactors. *Cell* **85**, 1149–1158.

Dorsky, D. I., Wells, M., and Harrington, R. D. (1996). Detection of HIV-1 infection with a green fluorescent protein reporter system. *J. Acquir. Immune Defic. Syndr. Hum. Retrovirol.* **13**, 308–313.

Dragic, T., and Alizon, M. (1993). Different requirements for membrane fusion mediated by the envelopes of human immunodeficiency virus types 1 and 2. *J. Virol.* **67**, 2355–2359.

Dragic, T., Charneau, P., Clavel, F., and Alizon, M. (1992). Complementation of murine cells for human immunodeficiency virus envelope/CD4-mediated fusion in human/murine heterokaryons. *J. Virol.* **66**, 4794–4802.

Dragic, T., Litwin, V., Allaway, G. P., Martin, S. R., Huang, Y., Nagashima, K. A., Cayanan, C., Maddon, P. J., Koup, R. A., Moore, J. P., and Paxton, W. A. (1996). HIV-1 entry into CD4+ cells is mediated by the chemokine receptor CC-CKR-5. *Nature* **381**, 667–673.

Dumonceaux, J., Nisole, S., Chanel, C., Quivet, L., Amara, A., Baleux, F., Briand, P., and Hazan, U. (1998). Spontaneous mutations in the env gene of the human immunodeficiency virus type 1 NDK isolate are associated with a CD4-independent entry phenotype. *J. Virol.* **72**, 512–519.

Edinger, A. L., Blanpain, C., Kunstman, K. J., Wolinsky, S. M., Parmentier, M., and Doms, R. W. (1999). Functional dissection of CCR5 coreceptor function through the use of CD4-independent simian immunodeficiency virus strains. *J. Virol.* **73**, 4062–4073.

Edinger, A. L., Hoffman, T. L., Sharron, M., Lee, B., O'Dowd, B., and Doms, R. W. (1998a). Use of GPR1, GPR15, and STRL33 as coreceptors by diverse human immunodeficiency virus type 1 and simian immunodeficiency virus envelope proteins. *Virology* **249**, 367–378.

Edinger, A. L., Hoffman, T. L., Sharron, M., Lee, B., Yi, Y., Choe, W., Kolson, D. L., Mitrovic, B., Zhou, Y., Faulds, D., Collman, R. G., Hesselgesser, J., Horuk, R., and Doms, R. W. (1998b). An orphan seven-transmembrane domain receptor expressed widely in the brain functions as a coreceptor for human immunodeficiency virus type 1 and simian immunodeficiency virus. *J. Virol.* **72,** 7934–7940.

Edinger, A. L., Mankowski, J. L., Doranz, B. J., Margulies, B. J., Lee, B., Rucker, J., Sharron, M., Hoffman, T. L., Berson, J. F., Zink, M. C., Hirsch, V. M., Clements, J. E., and Doms, R. W. (1997). CD4-independent, CCR5-dependent infection of brain capillary endothelial cells by a neurovirulent simian immunodeficiency virus strain. *Proc. Natl. Acad. Sci. USA* **94,** 14742–14747.

Emau, P., McClure, H. M., Isahakia, M., Else, J. G., and Fultz, P. N. (1991). Isolation from African Sykes' monkeys (*Cercopithecus mitis*) of a lentivirus related to human and simian immunodeficiency viruses. *J. Virol.* **65,** 2135–2140.

Endres, M. J., Clapham, P. R., Marsh, M., Ahuja, M., Turner, J. D., McKnight, A., Thomas, J. F., Stoebenau-Haggarty, B., Choe, S., Vance, P. J., Wells, T. N., Power, C. A., Sutterwala, S. S., Doms, R. W., Landau, N. R., and Hoxie, J. A. (1996). CD4-independent infection by HIV-2 is mediated by fusin/CXCR4. *Cell* **87,** 745–756.

Fan, L., and Peden, K. (1992). Cell-free transmission of Vif mutants of HIV-1. *Virology* **190,** 19–29.

Farzan, M., Choe, H., Martin, K., Marcon, L., Hofmann, W., Karlsson, G., Sun, Y., Barrett, P., Marchand, N., Sullivan, N., Gerard, N., Gerard, C., and Sodroski, J. (1997). Two orphan seven-transmembrane segment receptors which are expressed in CD4-positive cells support simian immunodeficiency virus infection. *J. Exp. Med.* **186,** 405–411.

Federsppiel, B., Melhado, I. G., Duncan, A. M., Delaney, A., Schappert, K., Clark-Lewis, I., and Jirik, F. R. (1993). Molecular cloning of the cDNA and chromosomal localization of the gene for a putative seven-transmembrane segment (7-TMS) receptor isolated from human spleen. *Genomics* **16,** 707–712.

Feng, Y., Broder, C. C., Kennedy, P. E., and Berger, E. A. (1996). HIV-1 entry cofactor: Functional cDNA cloning of a seven-transmembrane, G protein-coupled receptor. *Science* **272,** 872–877.

Fenyö, E. M., Albert, J., and Åsjö, B. (1989). Replicative capacity, cytopathic effect and cell tropism of HIV. *AIDS* **3**(Suppl. 1), S5–S12.

Fenyö, E. M., Morfeldt-Manson, L., Chiodi, F., Lind, B., von Gegerfelt, A., Albert, J., Olausson, E., and Åsjö, B. (1988). Distinct replicative and cytopathic characteristics of human immunodeficiency virus isolates. *J. Virol.* **62,** 4414–4419.

Fields-Berry, S. C., Halliday, A. L., and Cepko, C. L. (1992). A recombinant retrovirus encoding alkaline phosphatase confirms clonal boundary assignment in lineage analysis of murine retina. *Proc. Natl. Acad. Sci. USA* **89,** 693–697.

Fisher, A. G., Collalti, E., Ratner, L., Gallo, R. C., and Wong-Staal, F. (1985). A molecular clone of HTLV-III with biological activity. *Nature* **316,** 262–265.

Fouchier, R. A., Groenink, M., Kootstra, N. A., Tersmette, M., Huisman, H. G., Miedema, F., and Schuitemaker, H. (1992). Phenotype-associated sequence variation in the third variable domain of the human immunodeficiency virus type 1 gp120 molecule. *J. Virol.* **66,** 3183–3187.

Fujita, K., Silver, J., and Peden, K. (1992). Changes in both gp120 and gp41 can account for increased growth potential and expanded host range of human immunodeficiency virus type 1. *J. Virol.* **66,** 4445–4451.

Fultz, P. N., McClure, H. M., Anderson, D. C., and Switzer, W. M. (1989). Identification and biologic characterization of an acutely lethal variant of simian immunodeficiency virus from sooty mangabeys (SIV/SMM). *AIDS Res. Hum. Retroviruses* **5,** 397–409.

Gabuzda, D., He, J., Ohagen, A., and Vallat, A. V. (1998). Chemokine receptors in HIV-1 infection of the central nervous system. *Semin. Immunol.* **10,** 203–213.

Gabuzda, D. H., Lawrence, K., Langhoff, E., Terwilliger, E., Dorfman, T., Haseltine, W. A., and Sodroski, J. (1992). Role of vif in replication of human immunodeficiency virus type 1 in CD4+ T lymphocytes. *J. Virol.* **66**, 6489–6495.

Gabuzda, D. H., Li, H., Lawrence, K., Vasir, B. S., Crawford, K., and Langhoff, E. (1994). Essential role of vif in establishing productive HIV-1 infection in peripheral blood T lymphocytes and monocyte/macrophages. *J. Acquir. Immune Defic. Syndr.* **7**, 908–915.

Gallo, R. C., Salahuddin, S. Z., Popovic, M., Shearer, G. M., Kaplan, M., Haynes, B. F., Palker, T. J., Redfield, R., Oleske, J., Safai, B. *et al.* (1984). Frequent detection and isolation of cytopathic retroviruses (HTLV-III) from patients with AIDS and at risk for AIDS. *Science* **224**, 500–503.

Gao, F., Bailes, E., Robertson, D. L., Chen, Y., Rodenburg, C. M., Michael, S. F., Cummins, L. B., Arthur, L. O., Peeters, M., Shaw, G. M., Sharp, P. M., and Hahn, B. H. (1999). Origin of HIV-1 in the chimpanzee Pan troglodytes troglodytes. *Nature* **397**, 436–441.

Gao, J. L., and Murphy, P. M. (1994). Human cytomegalovirus open reading frame US28 encodes a functional beta chemokine receptor. *J. Biol. Chem.* **269**, 28539–28542.

Gartner, S., Markovits, P., Markovitz, D. M., Kaplan, M. H., Gallo, R. C., and Popovic, M. (1986). The role of mononuclear phagocytes in HTLV-III/LAV infection. *Science* **233**, 215–219.

Georges-Courbot, M. C., Lu, C. Y., Makuwa, M., Telfer, P., Onanga, R., Dubreuil, G., Chen, Z., Smith, S. M., Georges, A., Gao, F., Hahn, B. H., and Marx, P. A. (1998). Natural infection of a household pet red-capped mangabey (*Cercocebus torquatus torquatus*) with a new simian immunodeficiency virus. *J. Virol.* **72**, 600–608.

Gervaix, A., West, D., Leoni, L. M., Richman, D. D., Wong-Staal, F., and Corbeil, J. (1997). A new reporter cell line to monitor HIV infection and drug susceptibility *in vitro*. *Proc. Natl. Acad. Sci. USA* **94**, 4653–4658.

Ghorpade, A., Xia, M. Q., Hyman, B. T., Persidsky, Y., Nukuna, A., Bock, P., Che, M., Limoges, J., Gendelman, H. E., and Mackay, C. R. (1998). Role of the beta-chemokine receptors CCR3 and CCR5 in human immunodeficiency virus type 1 infection of monocytes and microglia. *J. Virol.* **72**, 3351–3361.

Ghosh, S. K., Fultz, P. N., Keddie, E., Saag, M. S., Sharp, P. M., Hahn, B. H., and Shaw, G. M. (1993). A molecular clone of HIV-1 tropic and cytopathic for human and chimpanzee lymphocytes. *Virology* **194**, 858–864.

Goudsmit, J. (1995). The role of viral diversity in HIV pathogenesis. *J. Acquir. Immune Defic. Syndr. Hum. Retrovirol.* **10**(Suppl. 1), S15–S19.

Granelli-Piperno, A., Finkel, V., Delgado, E., and Steinman, R. M. (1999). Virus replication begins in dendritic cells during the transmission of HIV-1 from mature dendritic cells to T cells. *Curr. Biol.* **9**, 21–29.

Granelli-Piperno, A., Moser, B., Pope, M., Chen, D., Wei, Y., Isdell, F., O'Doherty, U., Paxton, W., Koup, R., Mojsov, S., Bhardwaj, N., Clark-Lewis, I., Baggiolini, M., and Steinman, R. M. (1996). Efficient interaction of HIV-1 with purified dendritic cells via multiple chemokine coreceptors. *J. Exp. Med.* **184**, 2433–2438.

Gravell, M., London, W. T., Hamilton, R. S., Stone, G., and Monzon, M. (1989). Infection of macaque monkeys with simian immunodeficiency virus from African green monkeys: Virulence and activation of latent infection. *J. Med. Primatol.* **18**, 247–254.

Grody, W. W., Fligiel, S., and Naeim, F. (1985). Thymus involution in the acquired immunodeficiency syndrome. *Am. J. Clin. Pathol.* **84**, 85–95.

Guillon, C., van der Ende, M. E., Boers, P. H., Gruters, R. A., Schutten, M., and Osterhaus, A. D. (1998). Coreceptor usage of human immunodeficiency virus type 2 primary isolates and biological clones is broad and does not correlate with their syncytium-inducing capacities. *J. Virol.* **72**, 6260–6263.

Guyader, M., Emerman, M., Sonigo, P., Clavel, F., Montagnier, L., and Alizon, M. (1987). Genome organization and transactivation of the human immunodeficiency virus type 2. *Nature* **326**, 662–669.

Harrington, R. D., and Geballe, A. P. (1993). Cofactor requirement for human immunodeficiency virus type 1 entry into a CD4-expressing human cell line. *J. Virol.* **67**, 5939–5947.

Hattori, N., Michaels, F., Fargnoli, K., Marcon, L., Gallo, R. C., and Franchini, G. (1990). The human immunodeficiency virus type 2 vpr gene is essential for productive infection of human macrophages. *Proc. Natl. Acad. Sci. USA* **87**, 8080–8084.

He, J., and Landau, N. R. (1995). Use of a novel human immunodeficiency virus type 1 reporter virus expressing human placental alkaline phosphatase to detect an alternative viral receptor. *J. Virol.* **69**, 4587–4592.

He, J., Chen, Y., Farzan, M., Choe, H., Ohagen, A., Gartner, S., Busciglio, J., Yang, X., Hofmann, W., Newman, W., Mackay, C. R., Sodroski, J., and Gabuzda, D. (1997). CCR3 and CCR5 are co-receptors for HIV-1 infection of microglia. *Nature* **385**, 645–649.

He, J., Choe, S., Walker, R., Di Marzio, P., Morgan, D. O., and Landau, N. R. (1995). Human immunodeficiency virus type 1 viral protein R (Vpr) arrests cells in the G2 phase of the cell cycle by inhibiting p34cdc2 activity. *J. Virol.* **69**, 6705–6711.

Heiber, M., Marchese, A., Nguyen, T., Heng, H. H., George, S. R., and O'Dowd, B. F. (1996). A novel human gene encoding a G-protein-coupled receptor (GPR15) is located on chromosome 3. *Genomics* **32**, 462–465.

Helseth, E., Kowalski, M., Gabuzda, D., Olshevsky, U., Haseltine, W., and Sodroski, J. (1990). Rapid complementation assays measuring replicative potential of human immunodeficiency virus type 1 envelope glycoprotein mutants. *J. Virol.* **64**, 2416–2420.

Heneine, W., Yamamoto, S., Switzer, W. M., Spira, T. J., and Folks, T. M. (1995). Detection of reverse transcriptase by a highly sensitive assay in sera from persons infected with human immunodeficiency virus type 1. *J. Infect. Dis.* **171**, 1210–1216.

Herbein, G., Van Lint, C., Lovett, J. L., and Verdin, E. (1998). Distinct mechanisms trigger apoptosis in human immunodeficiency virus type 1-infected and in uninfected bystander T lymphocytes. *J. Virol.* **72**, 660–670.

Herzog, H., Hort, Y. J., Shine, J., and Selbie, L. A. (1993). Molecular cloning, characterization, and localization of the human homolog to the reported bovine NPY Y3 receptor: Lack of NPY binding and activation. *DNA Cell Biol.* **12**, 465–471.

Hill, C. M., Deng, H., Unutmaz, D., Kewalramani, V. N., Bastiani, L., Gorny, M. K., Zolla-Pazner, S., and Littman, D. R. (1997). Envelope glycoproteins from human immunodeficiency virus types 1 and 2 and simian immunodeficiency virus can use human CCR5 as a coreceptor for viral entry and make direct CD4-dependent interactions with this chemokine receptor. *J. Virol.* **71**, 6296–6304.

Hirsch, V. M., Campbell, B. J., Bailes, E., Goeken, R., Brown, C., Elkins, W. R., Axthelm, M., Murphey-Corb, M., and Sharp, P. M. (1999). Characterization of a novel simian immunodeficiency virus (SIV) from L'Hoest monkeys (*Cercopithecus l'hoesti*): Implications for the origins of SIVmnd and other primate lentiviruses [In Process Citation]. *J. Virol.* **73**, 1036–1045.

Hirsch, V. M., Dapolito, G., Johnson, P. R., Elkins, W. R., London, W. T., Montali, R. J., Goldstein, S., and Brown, C. (1995). Induction of AIDS by simian immunodeficiency virus from an African green monkey: Species-specific variation in pathogenicity correlates with the extent of in vivo replication. *J. Virol.* **69**, 955–967.

Hirsch, V. M., Dapolito, G. A., Goldstein, S., McClure, H., Emau, P., Fultz, P. N., Isahakia, M., Lenroot, R., Myers, G., and Johnson, P. R. (1993a). A distinct African lentivirus from Sykes' monkeys. *J. Virol.* **67**, 1517–1528.

Hirsch, V. M., McGann, C., Dapolito, G., Goldstein, S., Ogen-Odoi, A., Biryawaho, B., Lakwo, T., and Johnson, P. R. (1993b). Identification of a new subgroup of SIVagm in tantalus monkeys. *Virology* **197**, 426–430.

Hoffman, T. L., Stephens, E. B., Narayan, O., and Doms, R. W. (1998). HIV type 1 envelope determinants for use of the CCR2b, CCR3, STRL33, and APJ coreceptors. *Proc. Natl. Acad. Sci. USA* **95**, 11360–11365.

Horuk, R., Hesselgesser, J., Zhou, Y., Faulds, D., Halks-Miller, M., Harvey, S., Taub, D., Samson, M., Parmentier, M., Rucker, J., Doranz, B. J., and Doms, R. W. (1998). The CC chemokine I-309 inhibits CCR8-dependent infection by diverse HIV-1 strains. *J. Biol. Chem.* **273,** 386–391.

Hoxie, J. A., LaBranche, C. C., Endres, M. J., Turner, J. D., Berson, J. F., Doms, R. W., and Matthews, T. J. (1998). CD4-independent utilization of the CXCR4 chemokine receptor by HIV-1 and HIV-2. *J. Reprod. Immunol.* **41,** 197–211.

Huang, Y., Paxton, W. A., Wolinsky, S. M., Neumann, A. U., Zhang, L., He, T., Kang, S., Ceradini, D., Jin, Z., Yazdanbakhsh, K., Kunstman, K., Erickson, D., Dragon, E., Landau, N. R., Phair, J., Ho, D. D., and Koup, R. A. (1996). The role of a mutant CCR5 allele in HIV-1 transmission and disease progression. *Nat. Med.* **2,** 1240–1243.

Huet, T., Cheynier, R., Meyerhans, A., Roelants, G., and Wain-Hobson, S. (1990). Genetic organization of a chimpanzee lentivirus related to HIV-1. *Nature* **345,** 356–359.

Hwang, S. S., Boyle, T. J., Lyerly, H. K., and Cullen, B. R. (1991). Identification of the envelope V3 loop as the primary determinant of cell tropism in HIV-1. *Science* **253,** 71–74.

Imai, T., Hieshima, K., Haskell, C., Baba, M., Nagira, M., Nishimura, M., Kakizaki, M., Takagi, S., Nomiyama, H., Schall, T. J., and Yoshie, O. (1997). Identification and molecular characterization of fractalkine receptor CX3CR1, which mediates both leukocyte migration and adhesion. *Cell* **91,** 521–530.

Innocenti, P., Ottmann, M., Morand, P., Leclercq, P., and Seigneurin, J. M. (1992). HIV-1 in blood monocytes: Frequency of detection of proviral DNA using PCR and comparison with the total CD4 count. *AIDS Res. Hum. Retroviruses* **8,** 261–268.

Jamieson, B. D., Pang, S., Aldrovandi, G. M., Zha, J., and Zack, J. A. (1995). *In vivo* pathogenic properties of two clonal human immunodeficiency virus type 1 isolates. *J. Virol.* **69,** 6259–6264.

Jazin, E. E., Yoo, H., Blomqvist, A. G., Yee, F., Weng, G., Walker, M. W., Salon, J., Larhammar, D., and Wahlestedt, C. (1993). A proposed bovine neuropeptide Y (NPY) receptor cDNA clone, or its human homologue, confers neither NPY binding sites nor NPY responsiveness on transfected cells. *Regul. Pept.* **47,** 247–258.

Jin, M. J., Hui, H., Robertson, D. L., Muller, M. C., Barre-Sinoussi, F., Hirsch, V. M., Allan, J. S., Shaw, G. M., Sharp, P. M., and Hahn, B. H. (1994). Mosaic genome structure of simian immunodeficiency virus from west African green monkeys. *EMBO J.* **13,** 2935–2947.

Joag, S. V., Li, Z., Foresman, L., Stephens, E. B., Zhao, L. J., Adany, I., Pinson, D. M., McClure, H. M., and Narayan, O. (1996). Chimeric simian/human immunodeficiency virus that causes progressive loss of CD4+T cells and AIDS in pig-tailed macaques. *J. Virol.* **70,** 3189–3197.

Joshi, V. V., and Oleske, J. M. (1985). Pathologic appraisal of the thymus gland in acquired immunodeficiency syndrome in children. A study of four cases and a review of the literature. *Arch. Pathol. Lab. Med.* **109,** 142–146.

Joshi, V. V., Oleske, J. M., Minnefor, A. B., Singh, R., Bokhari, T., and Rapkin, R. H. (1984). Pathology of suspected acquired immune deficiency syndrome in children: A study of eight cases. *Pediatr. Pathol.* **2,** 71–87.

Jourdan, P., Abbal, C., Nora, N., Hori, T., Uchiyama, T., Vendrell, J. P., Bousquet, J., Taylor, N., Pene, J., and Yssel, H. (1998). IL-4 induces functional cell-surface expression of CXCR4 on human T cells. *J. Immunol.* **160,** 4153–4157.

Jurriaans, S., Van Gemen, B., Weverling, G. J., Van Strijp, D., Nara, P., Coutinho, R., Koot, M., Schuitemaker, H., and Goudsmit, J. (1994). The natural history of HIV-1 infection: Virus load and virus phenotype independent determinants of clinical course? *Virology* **204,** 223–233.

Kazazi, F., Mathijs, J. M., Foley, P., and Cunningham, A. L. (1989). Variations in CD4 expression by human monocytes and macrophages and their relationship to infection with the human immunodeficiency virus. *J. Gen. Virol.* **70,** 2661–2672.

Kestler, H., Kodama, T., Ringler, D., Marthas, M., Pedersen, N., Lackner, A., Regier, D., Sehgal, P., Daniel, M., King, N., and Desrosiers, R. (1990). Induction of AIDS in rhesus monkeys by molecularly cloned simian immunodeficiency virus. *Science* **248,** 1109–1112.

Kestler, H. W. III., Li, Y., Naidu, Y. M., Butler, C. V., Ochs, M. F., Jaenel, G., King, N. W., Daniel, M. D., and Desrosiers, R. C. (1988). Comparison of simian immunodeficiency virus isolates. *Nature* **331,** 619–622.

Kestler, H. W. III., Naidu, Y. N., Kodama, T., King, N. W., Daniel, M. D., Li, Y., and Desrosiers, R. C. (1989). Use of infectious molecular clones of simian immunodeficiency virus for pathogenesis studies. *J. Med. Primatol.* **18,** 305–309.

Kim, S., Ikeuchi, K., Byrn, R., Groopman, J., and Baltimore, D. (1989a). Lack of a negative influence on viral growth by the nef gene of human immunodeficiency virus type 1. *Proc. Natl. Acad. Sci. USA* **86,** 9544–9548.

Kim, S. Y., Byrn, R., Groopman, J., and Baltimore, D. (1989b). Temporal aspects of DNA and RNA synthesis during human immunodeficiency virus infection: Evidence for differential gene expression. *J. Virol.* **63,** 3708–3713.

Kimpton, J., and Emerman, M. (1992). Detection of replication-competent and pseudotyped human immunodeficiency virus with a sensitive cell line on the basis of activation of an integrated beta-galactosidase gene. *J. Virol.* **66,** 2232–2239.

Kirchhoff, F., Pohlmann, S., Hamacher, M., Means, R. E., Kraus, T., Uberla, K., and Di Marzio, P. (1997). Simian immunodeficiency virus variants with differential T-cell and macrophage tropism use CCR5 and an unidentified cofactor expressed in CEMx174 cells for efficient entry. *J. Virol.* **71,** 6509–6516.

Kitchen, S. G., and Zack, J. A. (1997). CXCR4 expression during lymphopoiesis: Implications for human immunodeficiency virus type 1 infection of the thymus. *J. Virol.* **71,** 6928–6934.

Klatzmann, D., Champagne, E., Chamaret, S., Gruest, J., Guetard, D., Hercend, T., Gluckman, J. C., and Montagnier, L. (1984). T-lymphocyte T4 molecule behaves as the receptor for human retrovirus LAV. *Nature* **312,** 767–768.

Kostrikis, L. G., Huang, Y., Moore, J. P., Wolinsky, S. M., Zhang, L., Guo, Y., Deutsch, L., Phair, J., Neumann, A. U., and Ho, D. D. (1998). A chemokine receptor CCR2 allele delays HIV-1 disease progression and is associated with a CCR5 promoter mutation. *Nat. Med.* **4,** 350–353.

Kourtis, A. P., Ibegbu, C., Nahmias, A. J., Lee, F. K., Clark, W. S., Sawyer, M. K., and Nesheim, S. (1996). Early progression of disease in HIV-infected infants with thymus dysfunction. *N. Engl. J. Med.* **335,** 1431–1436.

Koyanagi, Y., Miles, S., Mitsuyasu, R. T., Merrill, J. E., Vinters, H. V., and Chen, I. S. Y. (1987). Dual infection of the central nervous system by AIDS viruses with distinct cellular tropisms. *Science* **236,** 819–822.

Kozak, S. L., Platt, E. J., Madani, N., Ferro, F. E., Jr., Peden, K., and Kabat, D. (1997). CD4, CXCR-4, and CCR-5 dependencies for infections by primary patient and laboratory-adapted isolates of human immunodeficiency virus type 1. *J. Virol.* **71,** 873–882.

Lapham, C. K., Ouyang, J., Chandrasekhar, B., Nguyen, N. Y., Dimitrov, D. S., and Golding, H. (1996). Evidence for cell-surface association between fusin and the CD4-gp120 complex in human cell lines. *Science* **274,** 602–605.

Lapham, C. K., Zaitseva, M. B., Lee, S., Romanstseva, T., and Golding, H. (1999). Fusion of monocytes and macrophages with HIV-1 correlates with biochemical properties of CXCR4 and CCR5. *Nat. Med.* **5,** 303–308.

Lavi, E., Kolson, D. L., Ulrich, A. M., Fu, L., and Gonzalez-Scarano, F. (1998). Chemokine receptors in the human brain and their relationship to HIV infection. *J. Neurovirol.* **4,** 301–311.

Lavi, E., Strizki, J. M., Ulrich, A. M., Zhang, W., Fu, L., Wang, Q., O'Connor, M., Hoxie, J. A., and Gonzalez-Scarano, F. (1997). CXCR-4 (Fusin), a co-receptor for the type 1 human immunodeficiency virus (HIV-1), is expressed in the human brain in a variety of cell types, including microglia and neurons. *Am. J. Pathol.* **151,** 1035–1042.

Lazaro, I., Naniche, D., Signoret, N., Bernard, A. M., Marguet, D., Klatzmann, D., Dragic, T., Alizon, M., and Sattentau, Q. (1994). Factors involved in entry of the human immunodeficiency virus type 1 into permissive cells: Lack of evidence of a role for CD26. *J. Virol.* **68,** 6535–6546.

Lee, A. H., Han, J. M., and Sung, Y. C. (1997). Generation of the replication-competent human immunodeficiency virus type 1 which expresses a jellyfish green fluorescent protein. *Biochem. Biophys. Res. Commun.* **233,** 288–292.

Lee, B., Doranz, B. J., Ratajczak, M. Z., and Doms, R. W. (1998). An intricate Web: Chemokine receptors, HIV-1 and hematopoiesis. *Stem Cells* **16,** 79–88.

Lee, B., Sharron, M., Montaner, L. J., Weissman, D., and Doms, R. W. (1999). Quantification of CD4, CCR5, and CXCR4 levels on lymphocyte subsets, dendritic cells, and differentially conditioned monocyte-derived macrophages. *Proc. Natl. Acad. Sci. USA* **96,** 5215–5220.

Lee, M. K., Heaton, J., and Cho, M. W. (1999). Identification of determinants of interaction between CXCR4 and gp120 of a dual-tropic HIV-1DH12 isolate. *Virology* **257,** 290–296.

Levy, J. A., Hoffman, A. D., Kramer, S. M., Landis, J. A., Shimabukuro, J. M., and Oshiro, L. S. (1984). Isolation of lymphocytopathic retroviruses from San Francisco patients with AIDS. *Science* **225,** 840–842.

Li, J., Lord, C. I., Haseltine, W., Letvin, N. L., and Sodroski, J. (1992). Infection of cynomolgus monkeys with a chimeric HIV-1/SIVmac virus that expresses the HIV-1 envelope glycoproteins. *J. Acquir. Immune Defic. Syndr.* **5,** 639–646.

Liao, F., Alkhatib, G., Peden, K. W. C., Sharma, G., Berger, E. A., and Farber, J. M. (1997). STRL33, A novel Chemokine receptor-like protein, functions as a fusion cofactor for both macrophage-tropic and T cell line-tropic HIV-1 *J. Exp. Med.* **185,** 2015–2023.

Libert, F., Cochaux, P., Beckman, G., Samson, M., Aksenova, M., Cao, A., Czeizel, A., Claustres, M., de la Rua, C., Ferrari, M., Ferrec, C., Glover, G., Grinde, B., Guran, S., Kucinskas, V., Lavinha, J., Mercier, B., Ogur, G., Peltonen, L., Rosatelli, C., Schwartz, M., Spitsyn, V., Timar, L., Beckman, L., Parmenntier, M., and Vassart, G. (1998). The deltaccr5 mutation conferring protection against HIV-1 in Caucasian populations has a single and recent origin in Northeastern Europe. *Hum. Mol. Genet.* **7,** 399–406.

Liu, R., Paxton, W. A., Choe, S., Ceradini, D., Martin, S. R., Horuk, R., MacDonald, M. E., Stuhlmann, H., Koup, R. A., and Landau, N. R. (1996). Homozygous defect in HIV-1 coreceptor accounts for resistance of some multiply-exposed individuals to HIV-1 infection. *Cell* **86,** 367–377.

Liu, Z. Q., Wood, C., Levy, J. A., and Cheng-Mayer, C. (1990). The viral envelope gene is involved in macrophage tropism of a human immunodeficiency virus type 1 strain isolated from brain tissue. *J. Virol.* **64,** 6148–6153.

Loetscher, M., Amara, A., Oberlin, E., Brass, N., Legler, D., Loetscher, P., D'Apuzzo, M., Meese, E., Rousset, D., Virelizier, J. L., Baggiolini, M., Arenzana-Seisdedos, F., and Moser, B. (1997). TYMSTR, a putative chemokine receptor selectively expressed in activated T cells, exhibits HIV-1 coreceptor function. *Curr. Biol.* **7,** 652–660.

Loetscher, M., Geiser, T., O'Reilly, T., Zwahlen, R., Baggiolini, M., and Moser, B. (1994). Cloning of a human seven-transmembrane domain receptor, LESTR, that is highly expressed in leukocytes. *J. Biol. Chem.* **269,** 232–237.

Loetscher, P., Uguccioni, M., Bordoli, L., Baggiolini, M., Moser, B., Chizzolini, C., and Dayer, J. M. (1998). CCR5 is characteristic of Th1 lymphocytes. *Nature* **391,** 344–345.

Luciw, P. A., Cheng-Mayer, C., and Levy, J. A. (1987). Mutational analysis of the human immunodeficiency virus: The orf-B region down-regulates virus replication. *Proc. Natl. Acad. Sci. USA* **84,** 1434–1438.

Luciw, P. A., Shaw, K. E., Unger, R. E., Planelles, V., Stout, M. W., Lackner, J. E., Pratt-Lowe, E., Leung, N. J., Banapour, B., and Marthas, M. L. (1992). Genetic and biological comparisons of pathogenic and nonpathogenic molecular clones of simian immunodeficiency virus (SIVmac). *AIDS Res. Hum. Retroviruses* **8,** 395–402.

Maddon, P. J., Dalgleish, A. G., McDougal, J. S., Clapham, P. R., Weiss, R. A., and Axel, R. (1986). The T4 gene encodes the AIDS virus receptor and is expressed in the immune system and the brain. *Cell* **47,** 333–348.

Marcon, L., Choe, H., Martin, K. A., Farzan, M., Ponath, P. D., Wu, L., Newman, W., Gerard, N., Gerard, C., and Sodroski, J. (1997). Utilization of C-C chemokine receptor 5 by the envelope glycoproteins of a pathogenic simian immunodeficiency virus, SIVmac239. *J. Virol.* **71,** 2522–2527.

Mariani, R., Wong, S., Mulder, L. C., Wilkinson, D. A., Reinhart, A. L., LaRosa, G., Nibbs, R., O'Brien, T. R., Michael, N. L., Connor, R. I., Macdonald, M., Busch, M., Koup, R. A., and Landau, N. R. (1999). CCR2-64I polymorphism is not associated with altered CCR5 expression or coreceptor function. *J. Virol.* **73,** 2450–2459.

Marthas, M. L., Banapour, B., Sutjipto, S., Siegel, M. E., Marx, P. A., Gardner, M. B., Pedersen, N. C., and Luciw, P. A. (1989). Rhesus macaques inoculated with molecularly cloned simian immunodeficiency virus. *J. Med. Primatol.* **18,** 311–319.

Martin, M. P., Dean, M., Smith, M. W., Winkler, C., Gerrard, B., Michael, N. L., Lee, B., Doms, R. W., Margolick, J., Buchbinder, S., Goedert, J. J., O'Brien, T. R., Hilgartner, M. W., Vlahov, D., O'Brien, S. J., and Carrington, M. (1998). Genetic acceleration of AIDS progression by a promoter variant of CCR5. *Science* **282,** 1907–1911.

Martinson, J. J., Chapman, N. H., Rees, D. C., Liu, Y. T., and Clegg, J. B. (1997). Global distribution of the CCR5 gene 32-basepair deletion. *Nat. Genet.* **16,** 100–103.

Maudru, T., and Peden, K. (1997). Elimination of background signals in a modified polymerase chain reaction-based reverse transcriptase assay. *J. Virological Methods* **66,** 247–261.

McDermott, D. H., Zimmerman, P. A., Guignard, F., Kleeberger, C. A., Leitman, S. F., and Murphy, P. M. (1998). CCR5 promoter polymorphism and HIV-1 disease progression: Multicenter AIDS Cohort Study (MACS). *Lancet* **352,** 866–870.

McElrath, M. J., Pruett, J. E., and Cohn, Z. A. (1989). Mononuclear phagocytes of blood and bone marrow: Comparative roles as viral reservoirs in human immunodeficiency virus type 1 infections. *Proc. Natl. Acad. Sci. USA* **86,** 675–679.

McElrath, M. J., Steinman, R. M., and Cohn, Z. A. (1991). Latent HIV-1 infection in enriched populations of blood monocytes and T cells from seropositive patients. *J. Clin. Invest.* **87,** 27–30.

McIlroy, D., Autran, B., Cheynier, R., Clauvel, J. P., Oksenhendler, E., Debre, P., and Hosmalin, A. (1996). Low infection frequency of macrophages in the spleens of HIV+ patients. *Res. Virol.* **147,** 115–121.

McKnight, A., Dittmar, M. T., Moniz-Periera, J., Ariyoshi, K., Reeves, J. D., Hibbitts, S., Whitby, D., Aarons, E., Proudfoot, A. E., Whittle, H., and Clapham, P. R. (1998). A broad range of chemokine receptors are used by primary isolates of human immunodeficiency virus type 2 as coreceptors with CD4. *J. Virol.* **72,** 4065–4071.

McKnight, A., Wilkinson, D., Simmons, G., Talbot, S., Picard, L., Ahuja, M., Marsh, M., Hoxie, J. A., and Clapham, P. R. (1997). Inhibition of human immunodeficiency virus fusion by a monoclonal antibody to a coreceptor (CXCR4) is both cell type and virus strain dependent. *J. Virol.* **71,** 1692–1696.

Michael, N. L., Chang, G., Louie, L. G., Mascola, J. R., Dondero, D., Birx, D. L., and Sheppard, H. W. (1997). The role of viral phenotype and CCR-5 gene defects in HIV-1 transmission and disease progression. *Nat. Med.* **3,** 338–340.

Michael, N. L., Nelson, J. A., KewalRamani, V. N., Chang, G., O'Brien, S. J., Mascola, J. R., Volsky, B., Louder, M., White, G. C., 2nd, Littman, D. R., Swanstrom, R., and O'Brien, T. R. (1998). Exclusive and persistent use of the entry coreceptor CXCR4 by human immunodeficiency virus type 1 from a subject homozygous for CCR5 delta32. *J. Virol.* **72,** 6040–6047.

Mikovits, J. A., Lohrey, N. C., Schulof, R., Courtless, J., and Ruscetti, F. W. (1992). Activation of infectious virus from latent human immunodeficiency virus infection of monocytes *in vivo*. *J. Clin. Invest.* **90,** 1486–1491.

Miller, M. D., Warmerdam, M. T., Gaston, I., Greene, W. C., and Feinberg, M. B. (1994). The human immunodeficiency virus-1 nef gene product: A positive factor for viral infection and replication in primary lymphocytes and macrophages. *J. Exp. Med.* **179**, 101–113.

Moore, J. P., Trkola, A., and Dragic, T. (1997). Co-receptors for HIV-1 entry. *Curr. Opin. Immunol.* **9**, 551–562.

Mori, K., Ringler, D. J., and Desrosiers, R. C. (1993). Restricted replication of simian immunodeficiency virus strain 239 in macrophages is determined by env but is not due to restricted entry. *J. Virol.* **67**, 2807–2814.

Moriuchi, M., Moriuchi, H., Turner, W., and Fauci, A. S. (1998). Exposure to bacterial products renders macrophages highly susceptible to T-tropic HIV-1. *J. Clin. Invest.* **102**, 1540–1550.

Mörner, A., Björndal, A., Albert, J., Kewalramani, V. N., Littman, D. R., Inoue, R., Thorstensson, R., Fenyö, E. M., and Björling, E. (1999). Primary human immunodeficiency virus type 2 (HIV-2) isolates, like HIV-1 isolates, frequently use CCR5 but show promiscuity in coreceptor usage. *J. Virol.* **73**, 2343–2349.

Mosier, D. E., Picchio, G. R., Gulizia, R. J., Sabbe, R., Poignard, P., Picard, L., Offord, R. E., Thompson, D. A., and Wilken, J. (1999). Highly potent RANTES analogues either prevent CCR5-using human immunodeficiency virus type 1 infection *in vivo* or rapidly select for CXCR4-using variants. *J. Virol.* **73**, 3544–3550.

Mummidi, S., Ahuja, S. S., Gonzalez, E., Anderson, S. A., Santiago, E. N., Stephan, K. T., Craig, F. E., O'Connell, P., Tryon, V., Clark, R. A., Dolan, M. J., and Ahuja, S. K. (1998). Genealogy of the CCR5 locus and chemokine system gene variants associated with altered rates of HIV-1 disease progression [In Process Citation]. *Nat. Med.* **4**, 786–793.

Mummidi, S., Ahuja, S. S., McDaniel, B. L., and Ahuja, S. K. (1997). The human CC chemokine receptor 5 (CCR5) gene: Multiple transcripts with 5′-end heterogeneity, dual promoter usage, and evidence for polymorphisms within the regulatory regions and noncoding exons. *J. Biol. Chem.* **272**, 30662–30671.

Murphy, P. M. (1994). Molecular piracy of chemokine receptors by herpesviruses. *Infect. Agents Dis.* **3**, 137–154.

Naidu, Y. M., Kestler, H. W. III., Li, Y., Butler, C. V., Silva, D. P., Schmidt, D. K., Troup, C. D., Sehgal, P. K., Sonigo, P., Daniel, M. D., and Desrosiers, R. C. (1988). Characterization of infectious molecular clones of simian immunodeficiency virus (SIVmac) and human immunodeficiency virus type 2: Persistent infection of rhesus monkeys with molecularly cloned SIVmac. *J. Virol.* **62**, 4691–4696.

Naif, H. M., Li, S., Alali, M., Sloane, A., Wu, L., Kelly, M., Lynch, G., Lloyd, A., and Cunningham, A. L. (1998). CCR5 expression correlates with susceptibility of maturing monocytes to human immunodeficiency virus type 1 infection. *J. Virol.* **72**(1), 830–836.

Neote, K., DiGregorio, D., Mak, J. Y., Horuk, R., and Schall, T. J. (1993). Molecular cloning, functional expression, and signaling characteristic of a C-C chemokine receptor. *Cell* **72**, 415–425.

Nishiyori, A., Minami, M., Ohtani, Y., Takami, S., Yamamoto, J., Kawaguchi, N., Kume, T., Akaike, A., and Satoh, M. (1998). Localization of fractalkine and CX3CR1 mRNAs in rat brain: Does fractalkine play a role in signaling from neuron to microglia? *FEBS Lett.* **429**, 167–172.

Nomura, H., Nielsen, B. W., and Matsushima, K. (1993). Molecular cloning of cDNAs encoding a LD78 receptor and putative leukocyte chemotactic peptide receptors. *Int. Immunol.* **5**, 1239–1249.

Novembre, F. J., Saucier, M., Anderson, D. C., Klumpp, S. A., O'Neil, S. P., Brown, C. R., 2nd, Hart, C. E., Guenthner, P. C., enson, R. B., and McClure, H. M. (1997). Development of AIDS in a chimpanzee infected with human immunodeficiency virus type 1. *J. Virol.* **71**, 4086–4091.

Nussbaum, O., Broder, C. C., and Berger, E. A. (1994). Fusogenic mechanisms of enveloped-virus glycoproteins analyzed by a novel recombinant vaccinia virus-based assay quantitating cell fusion-dependent reporter gene activation. *J. Virol.* **68**, 5411–5422.

O'Brien, T. R., Winkler, C., Dean, M., Nelson, J. A., Carrington, M., Michael, N. L., and White, G. C. N. (1997). HIV-1 infection in a man homozygous for CCR5 delta 32 [letter]. *Lancet* **349,** 1219.

O'Brien, W. A., Koyanagi, Y., Namazie, A., Zhao, J. Q., Diagne, A., Idler, K., Zack, J. A., and Chen, I. S. (1990). HIV-1 tropism for mononuclear phagocytes can be determined by regions of gp120 outside the CD4-binding domain. *Nature* **348,** 69–73.

O'Dowd, B. F., Heiber, M., Chan, A., Heng, H. H., Tsui, L. C., Kennedy, J. L., Shi, X., Petronis, A., George, S. R., and Nguyen, T. (1993). A human gene that shows identity with the gene encoding the angiotensin receptor is located on chromosome 11. *Gene* **136,** 355–360.

Oberlin, E., Amara, A., Bachelerie, F., Bessia, C., Virelizier, J. L., Arenzana-Seisdedos, F., Schwartz, O., Heard, J. M., Clark-Lewis, I., Legler, D. F., Loetscher, M., Baggiolini, M., and Moser, B. (1996). The CXC chemokine SDF-1 is the ligand for LESTR/fusin and prevents infection by T-cell-line-adapted HIV-1. *Nature* **382,** 833–835.

Ohagen, A., Ghosh, S., He, J., Huang, K., Chen, Y., Yuan, M., Osathanondh, R., Gartner, S., Shi, B., Shaw, G., and Gabuzda, D. (1999). Apoptosis induced by infection of primary brain cultures with diverse human immunodeficiency virus type 1 isolates: Evidence for a role of the envelope. *J. Virol.* **73,** 897–906.

Ohta, Y., Masuda, T., Tsujimoto, H., Ishikawa, K., Kodama, T., Morikawa, S., Nakai, M., Honjo, S., and Hayami, M. (1988). Isolation of simian immunodeficiency virus from African green monkeys and seroepidemiologic survey of the virus in various non-human primates. *Int. J. Cancer* **41,** 115–122.

Owman, C., Garzino-Demo, A., Cocchi, F., Popovic, M., Sabirsh, A., and Gallo, R. C. (1998). The leukotriene B4 receptor functions as a novel type of coreceptor mediating entry of primary HIV-1 isolates into CD4-positive cells. *Proc. Natl. Acad. Sci. USA* **95**(16), 9530–9534.

Owman, C., Nilsson, C., and Lolait, S. J. (1996). Cloning of cDNA encoding a putative chemoattractant receptor. *Genomics* **37,** 187–194.

Owman, C., Sabirsh, A., Boketoft, A., and Olde, B. (1997). Leukotriene B4 is the functional ligand binding to and activating the cloned chemoattractant receptor, CMKRL1. *Biochem. Biophys. Res. Commun.* **240,** 162–166.

Page, K. A., Liegler, T., and Feinberg, M. B. (1997). Use of a green fluorescent protein as a marker for human immunodeficiency virus type 1 infection. *AIDS Res. Hum. Retroviruses* **13,** 1077–1081.

Papiernik, M., Brossard, Y., Mulliez, N., Roume, J., Brechot, C., Barin, F., Goudeau, A., Bach, J. F., Griscelli, C., Henrion, R. *et al.* (1992). Thymic abnormalities in fetuses aborted from human immunodeficiency virus type 1 seropositive women. *Pediatrics* **89,** 297–301.

Patience, C., McKnight, A., Clapham, P. R., Boyd, M. T., Weiss, R. A., and Schulz, T. F. (1994). CD26 antigen and HIV fusion? *Science* **264,** 1159–1160. [discussion 1162–1165]

Peden, K., Emerman, M., and Montagnier, L. (1991). Changes in growth properties on passage in tissue culture of viruses derived from infectious molecular clones of HIV-1LAI, HIV-1MAL, and HIV-1ELI. *Virology* **185,** 661–672.

Peden, K. W. C., and Martin, M. A. (1995). Virological and molecular genetic techniques for studies of established HIV isolates. *In* "HIV: A Practical Approach: Virology and Immunology" (J. Karn, Ed.), Vol. 1, pp. 21–45. IRL Press, Oxford, UK.

Pedroza-Martins, L., Gurney, K. B., Torbett, B. E., and Uittenbogaart, C. H. (1998). Differential tropism and replication kinetics of human immunodeficiency virus type 1 isolates in thymocytes: coreceptor expression allows viral entry, but productive infection of distinct subsets is determined at the postentry level. *J. Virol.* **72,** 9441–9452.

Pleskoff, O., Treboute, C., Brelot, A., Heveker, N., Seman, M., and Alizon, M. (1997). Identification of a chemokine receptor encoded by human cytomegalovirus as a cofactor for HIV-1 entry. *Science* **276,** 1874–1878.

Pöhlmann, S., Krumbiegel, M., and Kirchhoff, F. (1999). Coreceptor usage of BOB/GPR15 and Bonzo/STRL33 by primary isolates of human immunodeficiency virus type 1. *J. Gen. Virol.* **80,** 1241–1251.

Ponten, J., and Macintyre, E. H. (1968). Long term culture of normal and neoplastic human glia. *Acta Pathol. Microbiol. Scand.* **74,** 465–486.

Pope, M., Betjes, M. G., Romani, N., Hirmand, H., Cameron, P. U., Hoffman, L., Gezelter, S., Schuler, G., and Steinman, R. M. (1994). Conjugates of dendritic cells and memory T lymphocytes from skin facilitate productive infection with HIV-1. *Cell* **78,** 389–398.

Pope, M., Gezelter, S., Gallo, N., Hoffman, L., and Steinman, R. M. (1995). Low levels of HIV-1 infection in cutaneous dendritic cells promote extensive viral replication upon binding to memory CD4+ T cells. *J. Exp. Med.* **182,** 2045–2056.

Pyra, H., Böni, J., and Schüpbach, J. (1994). Ultrasensitive retrovirus detection by a reverse transcriptase assay based on product enhancement. *Proc. Natl. Acad. Sci. USA* **91,** 1544–1548.

Qin, S., LaRosa, G., Campbell, J. J., Smith-Heath, H., Kassam, N., Shi, X., Zeng, L., Buthcher, E. C., and Mackay, C. R. (1996). Expression of monocyte chemoattractant protein-1 and interleukin-8 receptors on subsets of T cells: Correlation with transendothelial chemotactic potential. *Eur. J. Immunol.* **26,** 640–647.

Rabin, R. L., Park, M. K., Liao, F., Swofford, R., Stephany, D., and Farber, J. M. (1999). Chemokine receptor responses on T cells are achieved through regulation of both receptor expression and signaling. *J. Immunol.* **162,** 3840–3850.

Ramarli, D., Cambiaggi, C., De Giuli Morghen, C., Tripputi, P., Ortolani, R., Bolzanelli, M., Tridente, G., and Accolla, R. S. (1993). Susceptibility of human-mouse T cell hybrids to HIV-productive infection. *AIDS Res. Hum. Retroviruses* **9,** 1269–1275.

Rana, S., Besson, G., Cook, D. G., Rucker, J., Smyth, R. J., Yi, Y., Turner, J. D., Guo, H. H., Du, J. G., Peiper, S. C., Lavi, E., Samson, M., Libert, F., Liesnard, C., Vassart, G., Doms, R. W., Parmentier, M., and Collman, R. G. (1997). Role of CCR5 in infection of primary macrophages and lymphocytes by macrophage-tropic strains of human immunodeficiency virus: Resistance to patient-derived and prototype isolates resulting from the delta ccr5 mutation. *J. Virol.* **71,** 3219–3227.

Reeves, J. D., McKnight, A., Potempa, S., Simmons, G., Gray, P. W., Power, C. A., Wells, T., Weiss, R. A., and Talbot, S. J. (1997). CD4-independent infection by HIV-2 (ROD/B): Use of the 7-transmembrane receptors CXCR-4, CCR-3, and V28 for entry. *Virology* **231,** 130–134.

Regier, D. A., and Desrosiers, R. C. (1990). The complete nucleotide sequence of a pathogenic molecular clone of simian immunodeficiency virus. *AIDS Res. Hum. Retroviruses* **6,** 1221–1231.

Reichert, C. M., O'Leary, T. J., Levens, D. L., Simrell, C. R., and Macher, A. M. (1983). Autopsy pathology in the acquired immune deficiency syndrome. *Am. J. Pathol.* **112,** 357–382.

Rich, E. A., Chen, I. S., Zack, J. A., Leonard, M. L., and O'Brien, W. A. (1992). Increased susceptibility of differentiated mononuclear phagocytes to productive infection with human immunodeficiency virus-1 (HIV-1). *J. Clin. Invest.* **89,** 176–183.

Rocancourt, D., Bonnerot, C., Jouin, H., Emerman, M., and Nicolas, J. F. (1990). Activation of a beta-galactosidase recombinant provirus: Application to titration of human immunodeficiency virus (HIV) and HIV-infected cells. *J. Virol.* **64,** 2660–2668.

Roos, M. T., Lange, J. M., de Goede, R. E., Coutinho, R. A., Schellekens, P. T., Miedema, F., and Tersmette, M. (1992). Viral phenotype and immune response in primary human immunodeficiency virus type 1 infection. *J. Infect. Dis.* **165,** 427–432.

Rosenzweig, M., Bunting, E. M., and Gaulton, G. N. (1994). Neonatal HIV-1 thymic infection. *Leukemia* 8(Suppl. 1), S163–S165.

Rosenzweig, M., Clark, D. P., and Gaulton, G. N. (1993). Selective thymocyte depletion in neonatal HIV-1 thymic infection. *AIDS* 7, 1601–1605.

Rubbert, A., Combadière, C., Ostrowski, M., Arthos, J., Dybul, M., Machado, E., Cohn, M. A., Hoxie, J. A., Murphy, P. M., Fauci, A. S., and Weissman, D. (1998). Dendritic cells express multiple chemokine receptors used as coreceptors for HIV entry. *J. Immunol.* **160,** 3933–3941.

Rucker, J., Edinger, A. L., Sharron, M., Samson, M., Lee, B., Berson, J. F., Yi, Y., Margulies, B., Collman, R. G., Doranz, B. J., Parmentier, M., and Doms, R. W. (1997). Utilization of chemokine receptors, orphan receptors, and herpesvirus-encoded receptors by diverse human and simian immunodeficiency viruses. *J. Virol.* **71,** 8999–9007.

Ryan-Graham, M. A., and Peden, K. W. C. (1995). Both virus and host components are important for the manifestation of a Nef- phenotype in HIV-1 and HIV-2. *Virology* **213,** 158–168.

Sakuragi, J., Sakai, H., Sakuragi, S., Shibata, R., Wain-Hobson, S., Hayami, M., and Adachi, A. (1992). Functional classification of simian immunodeficiency virus isolated from a chimpanzee by transactivators. *Virology* **189,** 354–358.

Sallusto, F., Mackay, C. R., and Lanzavecchia, A. (1997). Selective expression of the eotaxin receptor CCR3 by human T helper 2 cells. *Science* **277,** 2005–2007.

Samson, M., Edinger, A. L., Stordeur, P., Rucker, J., Verhasselt, V., Sharron, M., Govaerts, C., Mollereau, C., Vassart, G., Doms, R. W., and Parmentier, M. (1998). ChemR23, a putative chemoattractant receptor, is expressed in monocyte-derived dendritic cells and macrophages and is a coreceptor for SIV and some primary HIV-1 strains. *Eur. J. Immunol.* **28,** 1689–1700.

Samson, M., Libert, F., Doranz, B. J., Rucker, J., Liesnard, C., Farber, C. M., Saragosti, S., Lapoumeroulie, C., Cognaux, J., Forceille, C., Muyldermans, G., Verhofstede, C., Burtonboy, G., Georges, M., Imai, T., Rana, S., Yi, Y., Smyth, R. J., Collman, R. G., Doms, R. W., Vassart, G., and Parmentier, M. (1996a). Resistance to HIV-1 infection in caucasian individuals bearing mutant alleles of the CCR-5 chemokine receptor gene. *Nature* **382,** 722–725.

Samson, M., Soularue, P., Vassart, G., and Parmentier, M. (1996b). The genes encoding the human CC-chemokine receptors CC-CKR1 to CC-CKR5 (CMKBR1-CMKBR5) are clustered in the p21.3-p24 region of chromosome 3. *Genomics* **36,** 522–526.

Sanders, V. J., Pittman, C. A., White, M. G., Wang, G., Wiley, C. A., and Achim, C. L. (1998). Chemokines and receptors in HIV encephalitis. *AIDS* **12,** 1021–1026.

Scarlatti, G., Tresoldi, E., Björndal, A., Fredriksson, R., Colognesi, C., Deng, H. K., Malnati, M. S., Plebani, A., Siccardi, A. G., Littman, D. R., Fenyö, E. M., and Lusso, P. (1997). *In vivo* evolution of HIV-1 co-receptor usage and sensitivity to chemokine-mediated suppression. *Nat. Med.* **3,** 1259–1265.

Schmidtmayerova, H., Alfano, M., Nuovo, G., and Bukrinsky, M. (1998). Human immunodeficiency virus type 1 T-lymphotropic strains enter macrophages via a CD4- and CXCR4-mediated pathway: Replication is restricted at a postentry level. *J. Virol.* **72,** 4633–4642.

Schols, D., and De Clercq, E. (1998). The simian immunodeficiency virus mnd(GB-1) strain uses CXCR4, not CCR5, as coreceptor for entry in human cells. *J. Gen. Virol.* **79,** 2203–2205.

Schols, D., Este, J. A., Cabrera, C., and De Clercq, E. (1998). T-cell-line-tropic human immunodeficiency virus type 1 that is made resistant to stromal cell-derived factor 1 alpha contains mutations in the envelope gp120 but does not show a switch in coreceptor use. *J. Virol.* **72,** 4032–4037.

Schols, D., Este, J. A., Henson, G., and De Clercq, E. (1997a). Bicyclams, a class of potent anti-HIV agents, are targeted at the HIV coreceptor fusin/CXCR-4. *Antiviral Res.* **35,** 147–156.

Schols, D., Struyf, S., Van Damme, J., Este, J. A., Henson, G., and De Clercq, E. (1997b). Inhibition of T-tropic HIV strains by selective antagonization of the chemokine receptor CXCR4. *J. Exp. Med.* **186,** 1383–1388.

Schuitemaker, H., Koot, M., Kootstra, N. A., Dercksen, M. W., de Goede, R. E., van Steenwijk, R. P., Lange, J. M., Schattenkerk, J. K., Miedema, F., and Tersmette, M. (1992a). Biological phenotype of human immunodeficiency virus type 1 clones at different stages of infection: Progression of disease is associated with a shift from monocytotropic to T-cell-tropic virus population. *J. Virol.* **66,** 1354–1360.

Schuitemaker, H., Kootstra, N. A., Koppelman; M. H., Bruisten, S. M., Huisman, H. G., Tersmette, M., and Miedema, F. (1992b). Proliferation-dependent HIV-1 infection of monocytes occurs during differentiation into macrophages. *J. Clin. Invest.* **89,** 1154–1160.

Seemayer, T. A., Laroche, A. C., Russo, P., Malebranche, R., Arnoux, E., Guerin, J. M., Pierre, G., Dupuy, J. M., Gartner, J. G., Lapp, W. S. *et al.* (1984). Precocious thymic involution manifest by epithelial injury in the acquired immune deficiency syndrome. *Hum. Pathol.* **15,** 469–474.

Shapiro, S. Z., Maudru, T., and Peden, K. W. C. (1999). Detection of human immunodeficiency virus type 1 after infection of unstimulated peripheral blood mononuclear cells. *J. Gen. Virol.* **80,** 857–861.

Shearer, W. T., Langston, C., Lewis, D. E., Pham, E. L., Hammill, H. H., Kozinetz, C. A., Kline, M. W., Hanson, I. C., and Popek, E. J. (1997). Early spontaneous abortions and fetal thymic abnormalities in maternal-to-fetal HIV infection. *Acta Paediatr. Suppl.* **421,** 60–64.

Shibata, R., Hoggan, M. D., Broscius, C., Englund, G., Theodore, T. S., Buckler-White, A., Arthur, L. O., Israel, Z., Schultz, A., Lane, H. C., and Martin, M. A. (1995). Isolation and characterization of a syncytium-inducing, macrophage/T-cell line-tropic human immunodeficiency virus type 1 isolate that readily infects chimpanzee cells *in vitro* and *in vivo*. *J. Virol.* **69,** 4453–4462.

Shieh, J. T., Albright, A. V., Sharron, M., Gartner, S., Strizki, J., Doms, R. W., and Gonzalez-Scarano, F. (1998). Chemokine receptor utilization by human immunodeficiency virus type 1 isolates that replicates in microglia. *J. Virol.* **72,** 4243–4249.

Shimizu, N., Soda, Y., Kanbe, K., Liu, H. Y., Jinno, A., Kitamura, T., and Hoshino, H. (1999). An orphan G protein-coupled receptor, GPR1, acts as a coreceptor to allow replication of human immunodeficiency virus type 1 and 2 in brain-derived cells. *J. Virol.* **73,** 5231–5239.

Shioda, T., Levy, J. A., and Cheng-Mayer, C. (1991). Macrophage and T cell-line tropisms of HIV-1 are determined by specific regions of the envelope gp120 gene. *Nature* **349,** 167–169.

Silver, J., Maudru, T., Fujita, K., and Repaske, R. (1993). An RT-PCR assay for the enzyme activity of reverse transcriptase capable of detecting single virions. *Nucleic Acids Res.* **21,** 3593–3594.

Simmons, G., Clapham, P. R., Picard, L., Offord, R. E., Rosenkilde, M. M., Schwartz, T. W., Buser, R., Wells, T. N. C., and Proudfoot, A. E. (1997). Potent inhibition of HIV-1 infectivity in macrophages and lymphocytes by a novel CCR5 antagonist. *Science* **276,** 276–279.

Simmons, G., Reeves, J. D., McKnight, A., Dejucq, N., Hibbitts, S., Power, C. A., Aarons, E., Schols, D., De Clercq, E., Proudfoot, A. E., and Clapham, P. R. (1998). CXCR4 as a functional coreceptor for human immunodeficiency virus type 1 infection of primary macrophages. *J. Virol.* **72,** 8453–8457.

Simon, J. H., Miller, D. L., Fouchier, R. A., Soares, M. A., Peden, K. W., and Malim, M. H. (1998). The regulation of primate immunodeficiency virus infectivity by Vif is cell species restricted: A role for Vif in determining virus host range and cross-species transmission. *EMBO J.* **17,** 1259–1267.

Singh, A., Besson, G., Mobasher, A., and Collman, R. G. (1999). Patterns of chemokine receptor fusion cofactor utilization by Human Immunodeficiency Virus Type 1 variants from the lungs and blood. *J. Virol.* **73,** 6680–6690.

Siveke, J. T., and Hamann, A. (1998). T. helper 1 and T helper 2 cells respond differentially to chemokines. *J. Immunol.* **160,** 550–554.

Smith, M. W., Dean, M., Carrington, M., Winkler, C., Huttley, G. A., Lomb, D. A., Goedert, J. J., O'Brien, T. R., Jacobson, L. P., Kaslow, R., Buchbinder, S., Vittinghoff, E., Vlahov,

D., Hoots, K., Hilgartner, M. W., and O'Brien, S. J. (1997). Contrasting genetic influence of CCR2 and CCR5 variants on HIV-1 infection and disease progression: Hemophilia Growth and Development Study (HGDS), Multicenter AIDS Cohort Study (MACS), Multicenter Hemophilia Cohort Study (MHCS), San Francisco City Cohort (SFCC), ALIVE Study. *Science* **277**, 959–965.

Soares, M. A., Robertson, D. L., Hui, H., Allan, J. S., Shaw, G. M., and Hahn, B. H. (1997). A full-length and replication-competent proviral clone of SIVAGM from tantalus monkeys. *Virology* **228**, 394–399.

Sol, N., Ferchal, F., Braun, J., Pleskoff, O., Treboute, C., Ansart, I., and Alizon, M. (1997). Usage of the coreceptors CCR-5, CCR-3, and CXCR-4 by primary and cell line-adapted human immunodeficiency virus type 2. *J. Virol.* **71**, 8237–8244.

Sonza, S., Maerz, A., Deacon, N., Meanger, J., Mills, J., and Crowe, S. (1996). Human immunodeficiency virus type 1 replication is blocked prior to reverse transcription and integration in freshly isolated peripheral blood monocytes. *J. Virol.* **70**, 3863–3869.

Sonza, S., Maerz, A., Uren, S., Violo, A., Hunter, S., Boyle, W., and Crowe, S. (1995). Susceptibility of human monocytes of HIV type 1 infection *in vitro* is not dependent on their level of CD4 expression. *AIDS Res. Hum. Retroviruses* **11**, 769–776.

Sozzani, S., Allavena, P., D'Amico, G., Luini, W., Bianchi, G., Kataura, M., Imai, T., Yoshie, O., Bonecchi, R., and Mantovani, A. (1998). Differential regulation of chemokine receptors during dendritic cell maturation: A model for their trafficking properties. *J. Immunol.* **161**, 1083–1086.

Spina, C. A., Guatelli, J. C., and Richman, D. D. (1995). Establishment of a stable, inducible form of human immunodeficiency virus type 1 DNA in quiescent CD4 lymphocytes *in vitro*. *J. Virol.* **69**, 2977–2988.

Spina, C. A., Kwoh, T. J., Chowers, M. Y., Guatelli, J. C., and Richman, D. D. (1994). The importance of nef in the induction of human immunodeficiency virus type 1 replication from primary quiescent CD4 lymphocytes. *J. Exp. Med.* **179**, 115–123.

Spira, A. I., Marx, P. A., Patterson, B. K., Mahoney, J., Koup, R. A., Wolinsky, S. M., and Ho, D. D. (1996). Cellular targets of infection and route of viral dissemination after an intravaginal inoculation of simian immunodeficiency virus into rhesus macaques. *J. Exp. Med.* **183**, 215–225.

Stephens, E. B., Joag, S. V., Sheffer, D., Liu, Z. Q., Zhao, L., Mukherjee, S., Foresman, L., Adany, I., Li, Z., Pinson, D., and Narayan, O. (1996). Initial characterization of viral sequences from a SHIV-inoculated pig-tailed macaque that developed AIDS. *J. Med. Primatol.* **25**, 175–185.

Stephens, E. B., Mukherjee, S., Sahni, M., Zhuge, W., Raghavan, R., Singh, D. K., Leung, K., Atkinson, B., Li, Z., Joag, S. V., Liu, Z. Q., and Narayan, O. (1997). A cell-free stock of simian-human immunodeficiency virus that causes AIDS in pig-tailed macaques has a limited number of amino acid substitutions in both SIVmac and HIV-1 regions of the genome and has offered cytotropism. *Virology* **231**, 313–321.

Stephens, J. C., Reich, D. E., Goldstein, D. B., Shin, H. D., Smith, M. W., Carrington, M., Winkler, C., Huttley, G. A., Allikmets, R., Schriml, L., Gerrard, B., Malasky, M., Ramos, M. D., Morlot, S., Tzetis, M., Oddoux, C., di Giovine, F. S., Nasioulas, G., Chandler, D., Aseev, M., Hanson, M., Kalaydjieva, L., Glavac, D., Gasparini, P., Dean, M. *et al.* (1998). Dating the origin of the CCR5-Delta32 AIDS-resistance allele by the coalescence of haplotypes. *Am. J. Hum. Genet.* **62**, 1507–1515.

Stevenson, M., Stanwick, T. L., Dempsey, M. P., and Lamonica, C. A. (1990). HIV-1 replication is controlled at the level of T cell activation and proviral integration. *EMBO J.* **9**, 1551–1560.

Stivahtis, G. L., Soares, M. A., Vodicka, M. A., Hahn, B. H., and Emerman, M. (1997). Conservation and host specificity of Vpr-mediated cell cycle arrest suggest a fundamental role in primate lentivirus evolution and biology. *J. Virol.* **71**, 4331–4338.

Tatemoto, K., Hosoya, M., Habata, Y., Fujii, R., Kakegawa, T., Zou, M. X., Kawamata, Y., Fukusumi, S., Hinuma, S., Kitada, C., Kurokawa, T., Onda, H., and Fujino, M. (1998). Isolation and characterization of a novel endogenous peptide ligand for the human APJ receptor. *Biochem. Biophys. Res. Commun.* **251,** 471–476.

Tersmette, M., de Goede, R. E., Al, B. J., Winkel, I. N., Gruters, R. A., Cuypers, H. T., Huisman, H. G., and Miedema, F. (1988). Differential syncytium-inducing capacity of human immunodeficiency virus isolates: frequent detection of syncytium-inducing isolates in patients with acquired immunodeficiency syndrome (AIDS) and AIDS-related complex. *J. Virol.* **62,** 2026–2032.

Tersmette, M., Gruters, R. A., de Wolf, F., de Goede, R. E., Lange, J. M., Schellekens, P. T., Goudsmit, J., Huisman, H. G., and Miedema, F. (1989a). Evidence for a role of virulent human immunodeficiency virus (HIV) variants in the pathogenesis of acquired immunodeficiency syndrome: Studies on sequential HIV isolates. *J. Virol.* **63,** 2118–2125.

Tersmette, M., van Dongen, J. J., Clapham, P. R., de Goede, R. E., Wolvers-Tettero, I. L., Geurts van Kessel, A., Huisman, J. G., Weiss, R. A., and Miedema, F. (1989b). Human immunodeficiency virus infection studied in CD4-expressing human-murine T-cell hybrids. *Virology* **168,** 267–273.

Terwilliger, E., Sodroski, J. G., Rosen, C. A., and Haseltine, W. A. (1986). Effects of mutations within the 3&pirme; orf open reading frame region of human T-cell lymphotropic virus type III (HTLV-III/LAV) on replication and @cytopathogenicity. *J. Virol.* **60,** 754–760.

Theodore, T. S., Englund, G., Buckler-White, A., Buckler, C. E., Martin, M. A., and Peden, K. W. C. (1996). Construction and characterization of a stable full-length macrophage-tropic HIV type 1 molecular clone that directs the production of high titers of progeny virions. *AIDS R. Hum. Retroviruses* **12,** 191–194.

Theodorou, I., Meyer, L., Magierowska, M., Katlama, C., and Rouzioux, C. (1997). HIV-1 infection in an individual homozygous for CCR5 delta 32: Seroco Study Group. *Lancet* **349,** 1219–1220.

Tiffany, H. L., Lautens, L. L., Gao, J. L., Pease, J., Locati, M., Combadière, C., Modi, W., Bonner, T. I., and Murphy, P. M. (1997). Identification of CCR8: A human monocyte and thymus receptor for the CC chemokine I-309. *J. Exp. Med.* **186,** 165–170.

Trkola, A., Ketas, T., KewalRamani, V. N., Endorf, F., Binley, J. M., Katinger, H., Robinson, J., Littman, D. R., and Moore, J. P. (1998). Neutralization sensitivity of human immunodeficiency virus type 1 primary isolates to antibodies and CD4-based reagents is independent of coreceptor usage. *J. Virol.* **72,** 1876–1885.

Tsujimoto, H., Cooper, R. W., Kodama, T., Fukasawa, M., Miura, T., Ohta, Y., Ishikawa, K.-I., Nakai, M., Frost, E., Roelants, G. E., Roffi, J., and Hayami, M. (1988). Isolation and characterization of simian immunodeficiency virus from mandrills in Africa and its relationship to other human and simian immunodeficiency viruses. *J. Virol.* **62,** 4044–4050.

Tsujimoto, H., Hasegawa, A., Maki, N., Fukasawa, M., Miura, T., Speidel, S., Cooper, R. W., Moriyama, E. N., Gojobori, T., and Hayami, M. (1989). Sequence of a novel simian immunodeficiency virus from a wild-caught African mandrill. *Nature* **341,** 539–541.

Tuttle, D. L., Harrison, J. K., Anders, C., Sleasman, J. W., and Goodenow, M. M. (1998). Expression of CCR5 increases during monocyte differentiation and directly mediates macrophage susceptibility to infection by human immunodeficiency virus type 1. *J. Virol.* **72,** 4962–4969.

Valentin, A., Lu, W., Rosati, M., Schneider, R., Albert, J., Karlsson, A., and Pavlakis, G. N. (1998). Dual effect of interleukin 4 on HIV-1 expression: Implications for viral phenotypic switch and disease progression, *Proc. Natl. Acad. Sci. USA* **95,** 8886–8891.

van't Wout, A. B., Kootstra, N. A., Mulder-Kampinga, G. A., Albrecht-van Lent, N., Scherpbier, H. J., Veenstra, J., Boer, K., Coutinho, R. A., Miedema, F., and Schuitemaker, H. (1994). Macrophage-tropic variants initiate human immunodeficiency virus type 1 infection after sexual, parenteral, and vertical transmission. *J. Clin. Invest.* **94,** 2060–2067.

Vazeux, R. (1991). AIDS encephalopathy and tropism of HIV for brain monocytes/macrophages and microglial cells. *Pathobiology* **59,** 214–218.

Verani, A., Pesenti, E., Polo, S., Tresoldi, E., Scarlatti, G., Lusso, P., Siccardi, A. G., and Vercelli, D. (1998). CXCR4 is a functional coreceptor for infection of human macrophages by CXCR4-dependent primary HIV-1 isolates. *J. Immunol.* **161,** 2084–2088.

Vodicka, M. A., Goh, W. C., Wu, L. I., Rogel, M. E., Bartz, S. R., Schweickart, V. L., Raport, C. J., and Emerman, M. (1997). Indicator cell lines for detection of primary strains of human and simian immunodeficiency viruses. *Virology* **233,** 193–198.

Wain-Hobson, S., Sonigo, P., Danos, O., Cole, S., and Alizon, M. (1985). Nucleotide sequence of the AIDS virus, LAV. *Cell* **40,** 9–17.

Wain-Hobson, S., Vartanian, J. P., Henry, M., Chenciner, N., Cheynier, R., Delassus, S., Martins, L. P., Sala, M., Nugeyre, M. T., Guétard, D., and Montagnier, L. (1991). LAV revisited: Origins of the early HIV-1 isolates from Institut Pasteur. *Science* **252,** 961–965.

Warren, M. K., Rose, W. L., Cone, J. L., Rice, W. G., and Turpin, J. A. (1997). Differential infection of CD34+ cell-derived dendritic cells and monocytes with lymphocyte-tropic and monocyte-tropic HIV-1 strains, *J. Immunol.* **158,** 5035–5042.

Watkins, B. A., Crowley, R. W., Davis, A. E., Louie, A. T., and Reitz, M. S., Jr. (1996). Expression of CD26 does not correlate with the replication of macrophage-tropic strains of HIV-1 in T-cell lines. *Virology* **224,** 276–280.

Weiner, D. B., Hubner, K., Williams, W. V., and Greene, M. I. (1990). Species tropism of HIV-1 infectivity of interspecific cell hybridomas implies non-CD4 structures are required for cell entry. *Cancer Detect. Prev.* **14,** 317–320.

Weiner, D. B., Huebner, K., Williams, W. V., and Greene, M. I. (1991). Human genes other than CD4 facilitate HIV-1 infection of murine cells. *Pathobiology* **59,** 361–371.

Weiss, R. A., and Wrangham, R. W. (1999). From Pan to pandemic. *Nature* **397,** 385–386.

Weissman, D., Rabin, R. L., Arthos, J., Rubbert, A., Dybul, M., Swofford, R., Venkatesan, S., Farber, J. M., and Fauci, A. S. (1997). Macrophage-tropic HIV and SIV envelope proteins induce a signal through the CCR5 chemokine receptor. *Nature* **389,** 981–985.

Westervelt, P., Gendelman, H. E., and Ratner, L. (1991). Identification of a determinant within the human immunodeficiency virus 1 surface envelope glycoprotein critical for productive infection of primary monocytes. *Proc. Natl. Acad. Sci. USA* **88,** 3097–3101.

Westervelt, P., Trowbridge, D. B., Epstein, L. G., Blumberg, B. M., Li, Y., Hahn, B. H., Shaw, G. M., Price, R. W., and Ratner, L. (1992). Macrophage tropism determinants of human immunodeficiency virus type 1 *in vivo*. *J. Virol.* **66,** 2577–2582.

Willey, R. L., Martin, M. A., and Peden, K. W. C. (1994). Increase in soluble CD4 binding to and CD4-induced dissociation of gp120 from virions correlates with infectivity of human immunodeficiency virus type 1. *J. Virol.* **68,** 1029–1039.

Winkler, C., Modi, W., Smith, M. W., Nelson, G. W., Wu, X., Carrington, M., Dean, M., Honjo, T., Tashiro, K., Yabe, D., Buchbinder, S., Vittinghoff, E., Goedert, J. J., O'Brien, T. R., Jacobson, L. P., Detels, R., Donfield, S., Willoughby, A., Gomperts, E., Vlahov, D., Phair, J., and O'Brien, S. J. (1998). Genetic restriction of AIDS pathogenesis by an SDF-1 chemokine gene variant: ALIVE Study, Hemophilia Growth and Development Study (HGDS), Multicenter AIDS Cohort Study (MACS), Multicenter Hemophilia Cohort Study (MHCS), San Francisco City Cohort (SFCC). *Science* **279,** 389–393.

Xiao, L., Rudolph, D. L., Owen, S. M., Spira, T. J., and Lal, R. B. (1998). Adaptation to promiscuous usage of CC and CXC-chemokine coreceptors *in vivo* correlates with HIV-1 disease progression. *AIDS* **12,** F137–F143.

Xiao, X., Wu, L., Stantchev, T. S., Feng, Y. R., Ugolini, S., Chen, H., Shen, Z., Riley, J. L., Broder, C. C., Sattentau, Q. J., and Dimitrov, D. S. (1999). Constitutive cell surface association between CD4 and CCR5. *Proc. Natl. Acad. Sci. USA* **96,** 7496–7501.

Yokomizo, T., Izumi, T., Chang, K., Takuwa, Y., and Shimizu, T. (1997). A G-protein-coupled receptor for leukotriene B4 that mediates chemotaxis. *Nature* **387,** 620–624.

York-Higgins, D., Cheng-Mayer, C., Bauer, D., Levy, J. A., and Dina, D. (1990). Human immunodeficiency virus type 1 cellular host range, replication, and cytopathicity are linked to the envelope region of the viral genome. *J. Virol.* **64,** 4016–4020.

Zack, J. A., Arrigo, S. J., Weitsman, S. R., Go, A. S., Haislip, A., and Chen, I. S. (1990). HIV-1 entry into quiescent primary lymphocytes: molecular analysis reveals a labile, latent viral structure. *Cell* **61,** 213–222.

Zack, J. A., Haislip, A. M., Krogstad, P., and Chen, I. S. (1992). Incompletely reverse-transcribed human immunodeficiency virus type 1 genomes in quiescent cells can function as intermediates in the retroviral life cycle. *J. Virol.* **66,** 1717–1725.

Zaitseva, M., Blauvelt, A., Lee, S., Lapham, C. K., Klaus-Kovtun, V., Mostowski, H., Manischewitz, J., and Golding, H. (1997). Expression and function of CCR5 and CXCR4 on human Langerhans cells and macrophages: implications for HIV primary infection. *Nat. Med.* **3,** 1369–1375.

Zaitseva, M. B., Lee, S., Rabin, R. L., Tiffany, H. L., Farber, J. M., Peden, K. W. C., Murphy, P. M., and Golding, H. (1998). CXCR4 and CCR5 on human thymocytes: Biological function and role in HIV-1 infection. *J. Immunol.* **161,** 3103–3113.

Zhang, L., He, T., Huang, Y., Chen, Z., Guo, Y., Wu, S., Kunstman, K. J., Brown, R. C., Phair, J. P., Neumann, A. U., Ho, D. D., and Wolinsky, S. M. (1998a). Chemokine coreceptor usage by diverse primary isolates of human immunodeficiency virus type 1. *J. Virol.* **72,** 9307–9312.

Zhang, L. Q., MacKenzie, P., Cleland, A., Holmes, E. C., Brown, A. J., and Simmonds, P. (1993). Selection for specific sequences in the external envelope protein of human immunodeficiency virus type 1 upon primary infection. *J. Virol.* **67,** 3345–3356.

Zhang, Y., and Moore, J. P. (1999). Will multiple coreceptors need to be targeted by inhibitors of human immunodeficiency virus type 1 entry? *J. Virol.* **73,** 3443–3448.

Zhang, Y. J., Dragic, T., Cao, Y., Kostrikis, L., Kwon, D. S., Littman, D. R., KewalRamani, V. N., and Moore, J. P. (1998b). Use of coreceptors other than CCR5 by non-syncytium-inducing adult and pediatric isolates of human immunodeficiency virus type 1 is rare *in vitro*. *J. Virol.* **72,** 9337–9344.

Zhu, T., Mo, H., Wang, N., Nam, D. S., Cao, Y., Koup, R. A., and Ho, D. D. (1993). Genotypic and phenotypic characterization of HIV-1 patients with primary infection. *Science* **261,** 1179–1181.

Zhu, T., Wang, N., Carr, A., Nam, D. S., Moor-Jankowski, R., Cooper, D. A., and Ho, D. D. (1996). Genetic characterization of human immunodeficiency virus type 1 in blood and genital secretions: Evidence for viral compartmentalization and selection during sexual transmission. *J. Virol.* **70,** 3098–3107.

Zimmerman, P. A., Buckler-White, A., Alkhatib, G., Spalding, T., Kubofcik, J., Combadiere, C., Weissman, D., Cohen, O., Rubbert, A., Lam, G., Vaccarezza, M., Kennedy, P. E., Kumaraswami, V., Giorgi, J. V., Detels, R., Hunter, J., Chopek, M., Berger, E. A., Fauci, A. S., Nutman, T. B., and Murphy, P. M. (1997). Inherited resistance to HIV-1 conferred by an inactivating mutation in CC chemokine receptor 5: Studies in populations with contrasting clinical phenotypes, defined racial background, and quantified risk. *Mol. Med.* **3,** 23–36.

Zoeteweij, J. P., and Blauvelt, A. (1998). HIV-Dendritic cell interactions promote efficient viral infection of T cells. *J. Biomed. Sci.* **5,** 253–259.

Index

Contents of Previous Volumes

Volume 37

Volume 38

Volume 39

Volume 40

Volume 41

Volume 42

Volume 43

Volume 44

Volume 45

Volume 46